RISEHOLME LRC
WITHDRAWN

Large Animal
Clinical Nutrition

Large Animal Clinical Nutrition

JONATHAN M. NAYLOR, B.Sc., M.R.C.V.S., Ph.D.
Diplomate, A.C.V.I.M. and A.C.V.N.
Professor, Department of Veterinary Internal Medicine
University of Saskatchewan
Western College of Veterinary Medicine
Saskatoon, Saskatchewan, Canada

SARAH L. RALSTON, M.S., V.M.D., Ph.D.
Diplomate, A.C.V.N.
Department of Animal Sciences, Cook College
Rutgers University
New Brunswick, New Jersey

Mosby Year Book

St. Louis Baltimore Boston Chicago London Philadelphia Sydney Toronto

Mosby Year Book
Dedicated to Publishing Excellence

Editor: Robert W. Reinhardt
Developmental Editor: Elaine Steinborn
Assistant Editor: Susie Howell Baxter
Project Manager: Linda J. Daly
Designer: Liz Fett

Copyright © 1991 by Mosby–Year Book, Inc.

A Mosby imprint of Mosby–Year Book, Inc.

All rights reserved. No part of the publication may be reproduced, stored in a retrieval system, or transmitted, in any form or by any means, electronic, mechanical, photocopying, recording, or otherwise, without prior written permission from the publisher.

Printed in the United States of America

Mosby–Year Book, Inc.
11830 Westline Industrial Drive, St. Louis, Missouri 63146

Library of Congress Cataloging-in-Publication Data

Large animal clinical nutrition / edited by Jonathan M. Naylor and
   Sarah L. Ralston.
       p.   cm.
     Includes bibliographical references and index.
     ISBN 0-8016-2902-0
     1. Animal nutrition. 2. Feeds. 3. Animal feeding.
   4. Nutritionally induced diseases in animals.   I. Naylor, Jonathan
   M.   II. Ralston, Sarah L.
     [DNLM: 1. Animal Feed. 2. Animal Nutrition. 3. Horses.
   4. Ruminants. 5. Swine.   SF 95 L322]
   SF95.L265   1991
   636.08'52—dc20
   DNLM/DLC
   for Library of Congress   91-10566
                                                                 CIP

92 93 94 95 96 GW/MY/MY 9 8 7 6 5 4 3 2 1

Contributors

FRANK X. AHERNE, M.Sc., Ph.D.
Professor, Department of Animal Science
University of Alberta
Edmonton, Alberta, Canada

THOMAS E. BESSER, D.V.M., Ph.D.
Assistant Professor, Department of Veterinary Microbiology and Pathology
College of Veterinary Medicine
Washington State University
Pullman, Washington

KATHERINE BRETZLAFF, D.V.M., Ph.D.
Assistant Professor, Department of Large Animal Medicine
Texas A & M University
College Station, Texas

MILLARD CALHOUN, Ph.D.
Professor of Animal Science, Agricultural Research and Extension Center
Texas A & M University
San Angelo, Texas

WILLIAM CHALUPA, Ph.D.
Professor of Nutrition, School of Veterinary Medicine
University of Pennsylvania
Kennett Square, Pennsylvania

R.D.H. COHEN, B.Rur.Sc., Ph.D.
Professor, Department of Animal and Poultry Science
University of Saskatchewan
Saskatoon, Saskatchewan, Canada

LARRY R. CORAH, Ph.D.
Professor, Extension State Leader
Department of Animal Science
Kansas State University
Manhattan, Kansas

N.F. CYMBALUK, D.V.M., M.Sc.
Professional Research Associate, Department of Animal Science
University of Saskatchewan
Saskatoon, Saskatchewan, Canada

JAMES D. FERGUSON, V.M.D., M.S.
Assistant Professor of Nutrition, University of Pennsylvania
Kennett Square, Pennsylvania

DAVID T. GALLIGAN, V.M.D., M.B.A.
Assistant Professor of Animal Health and Economics
University of Pennsylvania
Kennett Square, Pennsylvania

CLIVE C. GAY, D.V.M., M.V.Sc., F.A.C.V.Sc.
Professor, College of Veterinary Medicine
Washington State University
Pullman, Washington

WALTER M. GUTERBOCK, D.V.M., M.S.
Production Medicine Clinician, Veterinary Medical Teaching and Research Center
University of California
Tulare, California

GEORGE HAENLEIN, Ph.D., D.Sci.
Professor, College of Animal Science and Agricultural Biochemistry
University of Delaware
Newark, Delaware

H.F. HINTZ, B.Sc., Ph.D.
Professor of Animal Nutrition, Department of Animal Science
Cornell University
Ithaca, New York

R. HIRONAKA, B.Sc., M.Sc., Ph.D.
Animal Nutritionist, Agriculture Canada Research Station
Lethbridge, Alberta, Canada

LYNN R. HOVDA, D.V.M., M.S.
1389 West Hoyt
Falcon Heights, Minnesota

J.T. HUBER, M.S., Ph.D.
Professor of Animal Nutrition, Department of Animal Sciences
University of Arizona
Tucson, Arizona

ED HUSTON, B.S., M.S., Ph.D.
Professor of Range and Animal Science, Texas Agricultural Experiment Station
Texas A & M University
San Angelo, Texas

THOMAS R. KASARI, D.V.M., M.V.Sc.
Diplomate, A.C.V.I.M.
Assistant Professor, Food Animal Internal Medicine
Texas A & M University
College Station, Texas

SHEILA M. McGUIRK, D.V.M., M.S., Ph.D.
Diplomate, A.C.V.I.M.
Associate Professor, College of Veterinary Medicine
University of Wisconsin
Madison, Wisconsin

JAMES G. MORRIS, M.Agr.Sc., Ph.D.
Professor, Department of Physiological Sciences
School of Veterinary Medicine
University of California
Davis, California

JONATHAN M. NAYLOR, B.Sc., M.R.C.V.S., Ph.D.
Diplomate, A.C.V.I.M. and A.C.V.N.
Professor, Department of Veterinary Internal Medicine
University of Saskatchewan
Western College of Veterinary Medicine
Saskatoon, Saskatchewan, Canada

JAMES W. OLTJEN, Ph.D.
Associate Professor, Department of Animal Science
University of California
Davis, California

JOHN F. PATIENCE, B.Sc.(Agr.), M.Sc., Ph.D.
Director, Prairie Swine Center;
Associate Professor, Department of Animal and Poultry Science
University of Saskatchewan
Saskatoon, Saskatchewan, Canada

SARAH L. RALSTON, M.S., V.M.D., Ph.D.
Diplomate, A.C.V.N.
Department of Animal Sciences, Cook College
Rutgers University
New Brunswick, New Jersey

Contributors

SUSAN SHAFTOE, V.M.D.
Diplomate, A.C.V.I.M.
Clinical Assistant Professor, School of
  Veterinary Medicine
University of Wisconsin
Madison, Wisconsin

JOHN C. SIMONS, D.V.M.
Director, Science Division
Eastern Wyoming College
AVMA, WYVMA, Society for Theriogenology,
  ACT (honorary)
Torrington, Wyoming

M.E. SMART, D.V.M., Ph.D.
Professor, Department of Veterinary Internal
  Medicine
University of Saskatchewan
Western College of Veterinary Medicine
Saskatoon, Saskatchewan, Canada

RONALD L. TERRA, B.S., M.S., D.V.M.
Lander Veterinary Clinic
Turlock, California

P.A. THACKER, Ph.D.
Associate Professor, Department of Animal
  and Poultry Science
University of Saskatchewan
Saskatoon, Saskatchewan, Canada

H.H. VAN HORN, B.S., M.S., Ph.D.
Professor Dairy Nutrition and Management
University of Florida
Gainesville, Florida

JAMES N. WILTBANK, Ph.D.
Professor, Department of Animal Science
Brigham Young University
Provo, Utah

RICHARD A. ZINN, B.S., M.S., Ph.D.
Associate Professor, University of California
Imperial Valley Agricultural Center
El Centro, California

Dedication

This book would not have been possible without the support of parents, family and the beneficial influence of many educators and researchers who formed our lives. A special thanks is extended to the authors, many of whom put great effort into their chapters.

Preface

This book is written for the veterinarian and veterinary student. Its objective is to provide a knowledge base for nutritional advice on optimal performance and production. Computerized ration evaluation and ration balancing packages have removed much of the tedium from ration formulation. With the help of the basic knowledge found in this book the veterinarian should be able to gather appropriate information about feed availability and animal requirements on the farm and use it to formulate an appropriate feeding policy. Guidelines are given for checking the results by observing production and changes in animal condition. This approach provides a basis for incorporating nutrition into herd health programs.

Vitamins, minerals, and the principles of ration formulation are included in Unit I, whereas Units II and III are species oriented. Unit II includes sections on beef cattle, dairy cattle, goats, llamas, and sheep. Unit III covers the nonruminants, horses and swine. Each section discusses the organization of the industry, feedstuffs and their use, feeding management for optimal production, nutritional diseases and nutrition of the sick animal. Chapters stress recognition and diagnosis of nutritional diseases and nutrition-related problems. Nutritional management as supportive therapy for individual sick animals is covered where appropriate. The chapters are written by people with practical knowledge of nutritional management. Nutritional requirements for each of the species, as well as compositions of common feedstuffs, are provided in quick reference tables in the Appendix.

The ability to give informed nutritional advice is very important to veterinarians—irrespective of whether one works with individual animals or with herds. A major portion of the costs of keeping animals is associated with the cost of feeds. A number of diseases are the result of poor feeding practices, and others are influenced by nutrition.

In today's world people are increasingly concerned about the role of animals in our society. Although only indirectly involved, veterinarians will have to grapple with the appropriate place for animal products in their own, and in the nation's, diets. Veterinarians should be informed about health and welfare arguments. Vegetarianism means less animals and a less diverse agricultural ecology. Parts of this book deal with the contribution of meat and milk to human nutrition. Some of the complexities in interpreting the data relating serum cholesterol and human diet are discussed. It is important that nutritional policies that may have a major effect on the economy, animal production, and human health be clearly thought out.

JONATHAN M. NAYLOR
SARAH L. RALSTON

Contents

UNIT I Basic Principles of Nutrition

Section A Nutrients

1. Bioenergetics: Basic Concepts and Ration Formulation, 1
   JOHN C. SIMONS and JONATHAN M. NAYLOR
2. Protein Nutrition and Nonprotein Nitrogen, 22
   H.H. VAN HORN
3. The Major Minerals (Macrominerals), 35
   JONATHAN M. NAYLOR
4. Trace Minerals, 55
   M.E. SMART and N.F. CYMBALUK
5. Vitamins, 68
   JONATHAN M. NAYLOR
6. Water: Requirements and Problems, 90
   JONATHAN M. NAYLOR

Section B Feeds and Feeding

7. The Digestive Systems of Horses, Ruminants, and Swine, 97
   JONATHAN M. NAYLOR
8. Control of Feed Intake, 114
   SARAH L. RALSTON
9. Feeds for Livestock, 120
   H.F. HINTZ
10. Principles of Ration Analysis, 131
    SARAH L. RALSTON
11. Effects of Disease on Nutritional Needs, 138
    SARAH L. RALSTON and JONATHAN M. NAYLOR

UNIT II Ruminants

Section A Beef Cattle

12. The Beef Industry, 147
    R.D.H. COHEN
13. Feeding the Beef Cow for Optimal Production, 157
    LARRY R. CORAH
14. Body Condition Scoring in Beef Cattle, 169
    JAMES N. WILTBANK
15. Pasture Management and Forage-Related Health Problems in Beef Cattle, 179
    THOMAS R. KASARI
16. Feeding Feedlot Cattle for Optimal Production, 190
    RICHARD A. ZINN and JAMES W. OLTJEN
17. Nutritional Problems in Cow/Calf Practice, 214
    JONATHAN M. NAYLOR
18. Nutritional Problems in Management and Feedlot Practice, 224
    R. HIRONAKA

Section B Dairy Cattle

19. The Dairy Industry in North America, 231
    RONALD L. TERRA
20. The Role of the Veterinarian as a Nutritional Advisor in Dairy Practice, 239
    DAVID T. GALLIGAN
21. Colostrum and Feeding Management of the Dairy Calf during the First Two Days of Life, 242
    CLIVE C. GAY and THOMAS E. BESSER

22 Evaluating Dietary Management of Hand-Reared Calves: Milk, Preserved Colostrum and Milk Replacers, 248
    JONATHAN M. NAYLOR
23 Nutrition of Dairy Replacement Heifers, 261
    WALTER M. GUTERBOCK
24 Feeding the Dairy Cow for Optimal Production, 274
    J.T. HUBER
25 The Role of Dietary Fat in the Productivity and Health of Dairy Cows, 304
    WILLIAM CHALUPA
26 Body Condition Scoring: Use and Application, 316
    RONALD L. TERRA
27 Nutritional Problems Encountered in Dairy Practice, 323
    JAMES D. FERGUSON
28 Choosing a Computerized Ration-Balancing Program, 332
    DAVID T. GALLIGAN

Section C Goats
29 The Goat Industry: Feeding for Optimal Production, 339
    KATHERINE BRETZLAFF, GEORGE HAENLEIN, and ED HUSTON
30 Common Nutritional Problems: Feeding the Sick Goat, 351
    KATHERINE BRETZLAFF, GEORGE HAENLEIN, and ED HUSTON

Section D Llamas
31 The Llama Industry: Feeding Systems and Special Feeding Requirements, 357
    LYNN R. HOVDA and SUSAN SHAFTOE
32 Common Nutritional Problems in Llamas, 363
    LYNN R. HOVDA and SUSAN SHAFTOE

Section E Sheep
33 Feeding Sheep for Optimal Production, 367
    MILLARD CALHOUN
34 Nutritional Diseases in Sheep, 384
    JAMES G. MORRIS

Section F Alimentation of Clinically Ill Ruminants
35 Nutritional Support for Sick Cattle and Calves, 393
    JONATHAN M. NAYLOR
36 Alimentation of the Clinically Ill Neonatal Ruminant, 397
    LYNN R. HOVDA and SHEILA M. McGUIRK

UNIT III Nonruminants
Section A Horse
37 The Horse Industry and Feeding Horses for Optimal Reproduction and Growth, 407
    H.F. HINTZ and SARAH L. RALSTON
38 Performance Horse Nutrition, 417
    SARAH L. RALSTON
39 Nutritional and Feed-Induced Diseases in Horses, 423
    SARAH L. RALSTON
40 Feeding Sick Horses, 432
    SARAH L. RALSTON and JONATHAN M. NAYLOR

Section B Swine
41 The Swine Industry, 447
    JOHN F. PATIENCE
42 Feeding the Gilt and Sow for Optimal Production, 455
    FRANK X. AHERNE
43 Feeding the Weaned Pig for Optimal Productivity, 474
    P.A. THACKER
44 Feeding the Growing-Finishing Pig for Optimal Production, 480
    P.A. THACKER

APPENDIX, 497

Basic Principles of Nutrition

UNIT I

SECTION A Nutrients

1

Bioenergetics
Basic Concepts and Ration Formulation

JOHN C. SIMONS, JONATHAN M. NAYLOR

Nutrition involves the acquisition and utilization of the chemical substances that are necessary to establish, maintain, and propagate life. These chemicals are called *nutrients* and include water, carbohydrates, fats, proteins, minerals, and vitamins. Energy is supplied by carbohydrates, fats, and proteins.

Bioenergetics is a branch of energetics concerned with energy transformations in living organisms.[8] Energy exists in two states:
1. Kinetic energy is the energy of motion. This energy is actively engaged in doing work (lifting, pushing, driving).
2. Potential energy is stored. This has the potential for doing work. The chemical energy in nutrient molecules is an example of potential energy.[6]

Without energy, life stops. Animals obtain energy from the food they eat. If an animal is deprived of food, it begins to lose weight. The energy stored in the body is eventually used up, and the animal dies. If animals are to survive and reproduce, they must have a constant supply of food that contains energy. Domestic animals require additional energy for the production of marketable products such as food and work.

HISTORY OF BIOENERGETICS

Lavoisier and Laplace and Adair Crawford were the first to measure animal heat using calorimeters in the early 1780s.[3] Crawford constructed his first calorimeter at the suggestion of Joseph Priestley.[18] Lavoisier and Laplace constructed a more sophisticated model that had a better insulating system[18] (Fig. 1-1).

Antoine Lavoisier is usually credited as being "the father of animal nutrition." He laid the foundations for the first law of thermodynamics, which states that energy cannot be created or destroyed but can only undergo conversion from one form to another. This law was formally conceptualized by Robert Mayer, a ship's doctor. Mayer solved a problem that had puzzled Galileo—what happens to the kinetic energy of a pendulum when it is stopped in midswing? Mayer decided that the energy of motion must become heat. At first this concept was treated with skepticism. Mayer convinced his peers by demonstrating that the temperature of water could be increased by shaking.[18] The first law was substantiated in 1843 by James Prescott Joule, a brewer by trade, who experimentally determined the mechanical equivalent of heat.

The second law of thermodynamics states that disorder in the universe constantly increases. It is older than the first and was formulated in 1824 by S.N.L. Carnot, a 28-year-old French army engineer. The concept has been enriched by the mid–nineteenth-century contributions of Clausius, Gibbs, Helmholtz, Kelvin, van't Hoff, and others. This law implies that

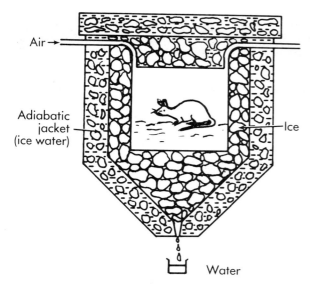

Fig. 1-1 Lavoisier's calorimeter. The outer jacket is filled with a mixture of water and ice. This mixture maintains a constant temperature of 0° C at both the inside and outside walls. No heat flows through it, as heat can flow only from a higher to a lower temperature, and the jacket is referred to as an *adiabatic jacket*. The ice pack around the animal chamber is also at 0° C, and all the heat given off by the animal melts ice. The water thus formed is collected and weighed to measure heat production (80 kcal of heat melt 1 kg of ice).
Redrawn from Kleiber M: The fire of life, an introduction to animal energetics, Melbourne, Fla, 1961, RE Kreiger Publishing Company, Inc. Reproduced with permission.

energy spontaneously converts to less organized forms. As energy is transferred from one molecule to another, some always leaks away and increases the random motion of surrounding molecules. This addition is expressed as heat. Heat is the energy of random molecular motion, and with every transfer of energy, some potential energy is dissipated as heat.[7] Thus there is a progression toward equalization of energy levels. The total energy of a system is comprised of two entities:
1. Energy unavailable for work (entropy). This is the "bound" energy needed to maintain the energy concentration at a level similar to the surroundings.
2. Free energy available for work.[7]

The second law also implies that only a part of an energy source is convertible to work. Work is orderly, relatively large-scale motion, as opposed to heat, which is random, uncontrolled, disorderly molecular motion. It is not possible to transfer energy without loss of some kinetic energy as heat.[7]

The efforts of Samuel Brody of the University of Missouri and Max Kleiber of the University of California further advanced our understanding of animal bioenergetics.[11,18] They showed that maintenance energy requirements are directly proportional to body weight to the ¾ power.[19]

The concept that heat production is a power of body weight was first formulated by Rameaux and Sarrus (1838-1839) and is called the *surface law*. Rameaux and Sarrus stated that:
1. At a constant body temperature, there must be equality between heat loss and heat production.
2. Heat loss is proportional to surface area.
3. Heat production is proportional to oxygen consumption.
4. Thus oxygen consumption and heat production are also proportional to surface area.[19]

They said that body surface area, heat loss, heat production, and oxygen consumption are all proportional to the body volume$^{2/3}$ or to body weight$^{2/3}$.

Brody pointed out that there were problems with this concept. The surface area of a living animal is not constant and is difficult to measure in a consistent fashion. The surface tends to roll up in cold weather and spread out in warm weather. Surface area varies with the body weight$^{2/3}$ only in geometrically similar animals that have comparable specific gravities. Furthermore, representing metabolism as a function of external surface area implies that surface area is the causal factor. In reality, the control mechanisms reside mostly in the neuroendocrine system.[9]

Because of these problems, Brody decided to empirically derive the relationship between metabolic rate and body weight. He measured metabolic rate during basal (resting) metabolism of

Bioenergetics

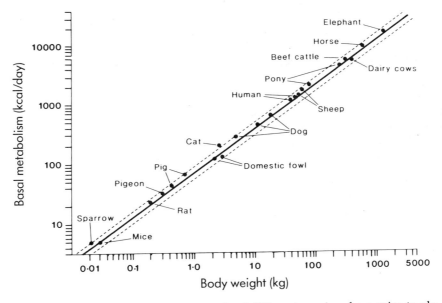

Fig. 1-2 Basal metabolism of mature animals of different species, from mice to elephants, plotted against body weight on a double-logarithmic scale. Basal metabolism of these species is described by the equation kcal of heat produced per day = 70 × body weight, kg.$^{0.75}$
Modified from Kleiber M: The fire of life, an introduction to animal energetics, Melbourne, Fla, 1961, RE Kreiger Publishing Company, Inc.

mature animals for 25 different types of animals including mice, canaries, and elephants[10] (Fig. 1-2). From these data Brody[9,10] proposed that the metabolic energy requirement for all animals is 70.5 kcal/kg of body wt.$^{0.734}$ This was later amended and simplified by Klieber to basal metabolic energy, kcal = 70 × body wt to the 0.75 (¾) power (kg$^{0.75}$). This formula became widely accepted, partly because it was easier to calculate a ¾ power and partly because a ¾ power better reflects the precision of the measurements.

In 1916 DuBois and DuBois published a formula for reliably calculating the body surface area of humans based on body weight and length. It has found little application in farm animals,[19] except in studies of environmental effects.

The energy requirements for most species of domestic animals are often expressed as kilocalories or megacalories per kilogram of body weight to the ¾ power.

BASIC CONCEPTS
Systems of energy measurement

Energy is the most critical consideration in animal nutrition. It is measured as heat expressed as kilocalories (kcal), megacalories (Mcal), or megajoules (MJ). The energy in foods is stored as potential energy, which is released by chemical reactions. The kinetic energy released may be used to make muscles contract, to induce glands to secrete, or to supply energy for storage.

Nutritionists are concerned with the release of food energy and its conversion into forms that are used by the animal. Some of the potential energy that is stored in feeds is lost during digestion and metabolism in fecal material, combustible gases, and urine. The kinetic energy released from nutrients by chemical reactions includes free energy available for work and energy that is dissipated as heat. Energy contents are determined by the amount of heat that is released on complete oxidation (combustion) of the substance. A number of terms are used to describe energy at various stages of transition from food to animal energy (Fig. 1-3):

1. Gross energy (GE) is obtained from the whole feed.
2. Fecal energy (FE) is for the fecal matter produced by the digestion of feed. The fecal material contains undigested feed, enteric microbes and their products, and excretions from the gastrointestinal tract.
3. Combustible gas energy (CGE) is for the nonabsorbable gases (mostly methane) that are produced during digestion. This is most important in ruminants.
4. Digestible energy (DE) equals GE minus FE.
5. Urinary energy (UE) is the energy content of the urine produced by the metabolism of the DE of a specific feed sample.
6. Metabolizable energy (ME) is DE minus (UE plus CGE).
7. Net energy (NE) is ME minus the heat increment (Hi). The heat increment is sometimes called the *specific dynamic activity* (SDA). The heat increment results from energy lost in the production of digestive secretions and in the absorption, circulation, and metabolic conversion of nutrients into end products.

Net energy is divided into two categories:

a. Net energy for maintenance (NEm) includes the energy required for basal metabolism (Hb) and the energy required for normal activity (Ha).

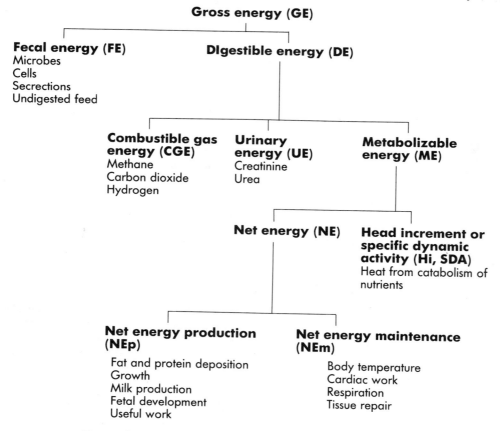

Fig. 1-3 Interrelationships between various forms of energy.

b. Net energy for production (NEp) includes energy deposited in growth and fat storage (NEg), energy deposited in the products of gestation (NEr), energy deposited in milk (NEl), and energy used to perform work (NEw).[7]

For most species, the energy requirements and concentrations in feeds are measured as digestible energy (DE) or metabolizable energy (ME). The mathematical relationships between DE and ME vary according to species and type of diet. For swine, sheep, goats, and beef cattle, the National Research Council (NRC) calculates ME and DE as direct linear functions of each other. In these species either of these categories may be used without loss of accuracy. In dairy cattle, conversion of DE to ME is more complicated, and ME should be the more accurate measurement. See box, compiled from data from the National Research Council.[23-29]

For herbivorous animals, metabolizable energy or digestible energy is not the best energy measurement, especially if the diet contains a large roughage component (p. 10). The heat increment (Hi) that results from metabolism of roughages is significantly greater than for concentrates. Thus the use of digestible energy or metabolizable energy systems results in over-rating roughages as an energy source. For example, it is possible to get tissue retention of 2.8 Mcal of NEg if either 21 Mcal ME of an alfalfa diet or 14 Mcal ME of a concentrate diet are fed.[20]

For this reason, Lofgreen and Garrett developed a net energy system for ruminants. The California Net Energy System is the most accurate method of measuring energy requirements in ruminants; it has been the primary NRC method for measuring energy requirements in cattle since 1972.[20]

Tabulation of net energy requirements requires the measurement of heat production. The pioneers in the field used respiration calorimeters, but only three of these were ever built for large animals, and prior to the late 1960s, little information was available about net energy.

Lofgreen and Garrett used a different system based on the use of simple algebra.[7,20]

$$ME = NE + Hi$$

NE = NE for maintenance (NEm) + NE for production (NEp)

Therefore,

$$NEm + NEp = ME - Hi$$

and

$$NEm + Hi = ME - NEp$$

If ME and the energy deposited in products are measured, the total heat production from a unit of feed can be estimated without using a respiration calorimeter. The energy retained (NEg) in an animal was measured by the "com-

---

Energy Interconversions

1 Mcal = 1,000 kcal = 4.184 megajoules
1 kcal = 3.968 British thermal units

Ruminants, horses

1 kg TDN = 4.409 Mcal of DE
1 lb TDN = 2.000 Mcal of DE
1 kg starch equivalent = 5.082 Mcal of DE
1 lb starch equivalent = 2.305 Mcal of DE

Beef cattle, sheep, goats

ME = 0.82 × DE (Underestimates the ME content of high-grain rations)

Swine

ME = 0.95 × DE

Or more precisely

ME = DE × (0.96 − [0.00202 × crude protein, %])

Goats

NE = DE × 0.464

For ruminant feedstuffs

NEm (Mcal/kg DM) = 1.37 ME − 0.138 ME^2 + 0.0105 ME^3 − 1.12
NEg (Mcal/kg DM) = 1.42 ME − 0.174 ME^2 + 0.122 ME^3 − 1.65

For dairy cattle feedstuffs

ME (Mcal/kg) = 1.01 DE (Mcal/kg) − 0.45

DM = Dry matter; TDN = total digestible nutrient

parative slaughter" technique. Animals were slaughtered at the beginning and end of the feeding trial, and carcass composition was measured. Specific gravity was used to estimate the amount of energy in the carcass because it reflects the relative proportions of fat (high energy, low specific gravity) and muscle.[15,16] This system was used to determine heat production in a large number of animals at various levels of intake, and the net energy requirements for maintenance were calculated.

The Europeans take a related approach. They express feed energy contents in terms of ME and then assign different efficiencies of utilization (k values) according to the energy density of the feed (i.e., the relative proportions of roughage and concentrate) and the purpose for which the energy is being used. These efficiency factors can be regarded as converting ME to NE values.

DE and ME values of feeds are relatively constant. However, NE values and k values depend on the use to which the animal puts the feed.

Chemical composition of feeds

The energy-containing components of food are mainly proteins, lipids, and carbohydrates.

Proteins. Proteins are composed of amino acids, which consist of a short hydrocarbon chain with an amino (NH_2) and a carboxy acid (COOH) group. After deamination (removal of nitrogen), the hydrocarbon chain can be used as an energy substrate. Proteins are discussed in Chapter 2.

Lipids. *Lipid* is the generic term given to fats, which are solid at room temperature, and oils, which are liquid. Oils have shorter carbon chain lengths and tend to be less saturated but their caloric value is similar to fats. Lipid has approximately double the gross energy content of carbohydrates and proteins. Fat is used as an energy store in animals, and oil as an energy store in some seeds. The main role of fats in animal nutrition is as a concentrated energy source. They are particularly useful in situations where it is difficult to meet energy requirements within the constraints of dry matter intake. Additional benefits of adding fat to diets often include an increase in palatability, a decrease in dustiness, and reduced wear on food processing and grinding machinery.

The unsaturated fatty acids—arachidonic, linoleic, and α-linolenic acid—are termed *essential fatty acids* and usually cannot be synthesized in adequate amounts by mammals. However, ruminal microflora synthesize sufficient essential fatty acids to meet the animal's requirements. Among other things, the essential fatty acids are required for prostaglandin and leukotriene synthesis. In their absence the skin becomes flaky and growth rate is reduced. Requirements are small, and deficiency is rarely seen in large animals.

Neonatal calves, lambs, and kids are not functional ruminants and require a dietary source of fats. A level of 10% fat (dry matter basis) is sufficient to supply the essential fatty acids, carry the fat-soluble vitamins, and supply adequate energy to achieve normal weight gains under thermoneutral conditions. Milk replacers that contain more than 10% fat do not enhance muscle and bone development, but do increase fat deposition in tissues, desirable in production systems such as that of veal. Additional dietary fat may be necessary to achieve normal weight gains when the environmental temperature is below the thermoneutral zone.

Monogastrics such as pigs can digest higher levels of dietary fat than ruminants. However, fat can play a role in ruminant nutrition. In dairy cows negative energy balance is a major problem in early lactation. Fat supplementation may be beneficial in reducing weight loss and improving productivity in early lactation (Chapter 25).[12,24]

Carbohydrates (sugars, starches, fiber). The majority of plant and animal carbohydrates occur as sugars and sugar polymers. Carbohydrates are divided into two categories: storage and structural carbohydrates.

Storage carbohydrates are compact polysaccharides used as an energy store by plants and animals. The α linkages that join the sugars can be degraded by mammalian digestive enzymes. Starches are glucose polymers (glucosoglycans). Amylose and amylopectin are the two forms found in plants. Seeds are particularly rich in starch. Glycogen is the animal equivalent of plant starch; it is a rapidly mobilizable store of glucose. Nutritionally, storage carbohydrates are highly digestible and are a concentrated source of energy.

Structural carbohydrates form dietary fiber. The structural carbohydrates are mainly found in the cell wall of plants and tend to be insoluble and provide strength. In structural carbohydrates the sugar subunits are joined in chains by β linkages. These cannot be degraded by mammalian digestive enzymes. Cellulose, a glucose polysaccharide, is the most abundant organic substance in the biosphere. Hemicelluloses are pentose sugar polymers that may also contain hexose subunits. Lignin is a polymer of aromatic alcohols.

Herbivores have solved the problem of fiber utilization by developing a symbiotic relationship with microorganisms. Gut microbes secrete enzymes that degrade cellulose and hemicellulose bonds; in return they are fed by the host (Chapter 7).

The usefulness of various types of fiber as energy sources depends on the degree of lignification; lignin energy is unavailable, even to herbivores. Fiber moderates digestion through a number of effects. It slows digestion because it takes time for microbes to break down fibrous cell walls and thereby release the more rapidly degradable nutrients inside. Soluble fiber physically slows mixing of nutrients and digestive enzymes. In some cases, the digestibility of non-fiber dietary components is reduced. The superior reputation of oats over other grains as a horse feed is probably because the high fiber content of oats moderates starch digestion and reduces the risk of digestive upsets.

Fiber content of feeds is classified according to solubility:
1. Crude fiber is the fiber fraction that is resistant to degradation in acid and alkali.
2. Neutral detergent fiber (NDF) includes hemicellulose, cellulose, lignin, acid detergent insoluble nitrogen, and acid insoluble ash.
3. Acid detergent fiber (ADF) is a subdivision of neutral detergent fiber that excludes hemicellulose.

It is generally accepted that neutral detergent fiber and acid detergent fiber are better indices of fiber digestibility than crude fiber. Determination of the percentage of neutral detergent fiber in a food sample is probably the best single method of assessment, but no single analysis can accurately predict fiber quality and energy values for all feeds.[24] The physical form of fiber is important. Many feeds are ground or rolled to speed digestion by breaking down fibrous structures and to increase the area available for digestion.

Fiber plays a positive role in animal nutrition. It promotes gastrointestinal motility and tends to retain water in the digesta. Low fiber diets are associated with hard, dry feces. High fiber diets are beneficial to normal passage of digesta and the regular production of soft feces.

Fiber is particularly important in herbivores. A minimum amount of fiber of proper quality and physical form is necessary in the diets of ruminants to maintain optimal health and productivity. Ruminants require some form of long-stem, high-fiber food (roughage, e.g., grass, hay, straw) in their diet. Fiber maintains proper ruminal and gastrointestinal motility and stimulates chewing of food and thus saliva production (p. 110). The main adverse effects of inadequate fiber are derangements of ruminal digestion (p. 112) and inadequate butterfat content in the milk of dairy cows.[24,37]

Energy content of feeds

The gross energy content of feedstuffs was originally measured directly using a bomb calorimeter (Fig. 1-4). However, this is time-consuming, and today energy content is usually estimated from the food's chemical composition and from known information about the energy content and digestibility of these constituents. Feed analysis typically involves a number of simple chemical procedures that give an approximate indication of the type of nutrients present in the feed. In the Weende system of proximate analysis, dry matter content is measured directly by drying. Protein content is usually estimated by determining nitrogen content and multiplying by 6.25 to give the crude protein content. This includes both true protein and nonprotein nitrogen sources such as ammonia, which contain no available energy. Fat content is determined by ether extraction, which also removes waxes. Crude fiber is that part of the sample resistant to dilute acid and alkali. The nitrogen-free extract is the remainder and consists mostly of starches and simple sugars.

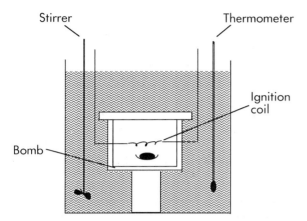

Fig. 1-4 Cross section of a bomb calorimeter. A sample of the material is weighed into a little dish of nickel or platinum and placed inside the bomb. The top of the bomb is then screwed on, and the bomb is filled with oxygen gas at a pressure of 25 to 30 atmospheres. The bomb is immersed in a weighed amount of water in the calorimeter bucket, which is placed inside an insulated jacket. An electric current ignites the sample, and the heat produced by combustion of the sample is measured from changes in temperature of the calorimeter water. Modified from Kleiber M: The fire of life, an introduction to animal energetics, Melbourne, Fla, 1961, Kreiger Publishing Company, Inc.

The gross energy content of simple components is approximately 4 Mcal/kg (1.8 Mcal/lb) for fiber, starches, and sugars; 5.6 Mcal/kg (2.5 Mcal/lb) for true protein; and 9.4 Mcal/kg (4.3 Mcal/lb) for fat. From the chemical composition and known energy contents of the constituents, the gross energy content of a given feedstuff can be calculated. Gross energy contents of many feeds are similar. Differences in nutritive value are mainly the result of differences in digestibility. The energy content can be expressed as DE, ME, or NE, using empirically derived data on the digestibility and metabolism of different feedstuffs.

The fiber content of a diet is inversely related to its digestibility and net energy content. Roughages are generally low in fat. Some feed laboratories measure only fiber content and then use predictive equations to calculate the likely energy content of a feed.

Total digestible nutrients. An old system of estimating the energy content of feeds is the total digestible nutrient (TDN) system. This is the summation of the digestible crude protein, digestible crude fiber, digestible nitrogen-free extract, and digestible ether extract and is expressed as a percentage of the total amount of feed. The digestible ether extract term was multiplied by 2.25 before summation to account for the fact that fat has 2.25 times the energy content of carbohydrates. As a predictor of DE, the protein content should be corrected for its higher energy content than carbohydrates, but unfortunately the system was instituted before the energy content of protein had been accurately determined. TDN is not really a measure of digestible energy. It is better related to metabolizable energy because when protein is catabolized for energy, the energy lost in the deamination process leaves a skeleton of energy content similar to carbohydrates.

Concentrates and roughages. In general, oilseeds have the highest energy contents and availabilities owing to the high caloric value of fat. Starch-containing seeds are next in digestible energy content. These high-energy feeds are referred to as concentrates.

Leaves and stem material are high in fiber. These feeds are referred to as roughages and are less concentrated sources of digestible energy. The gross energy content of roughages is similar to concentrates; the difference is in digestibility. This is because fiber cannot be directly digested by mammals. In general, roughages contain more than 20% crude fiber, concentrates less than 15% crude fiber.

Signs of energy deficiency and toxicity

Energy deficiency and toxicity (excess) produce similar signs in all farm animals. Lack of energy causes loss of weight, which is most visible in the rump and lumbar region, in adults. Fat cover over the ribs is lost late in starvation. Animals die when they have lost about 25% to 30% of normal body weight. Death is thought to be due to protein depletion and loss of respiratory and cardiac muscle function. In young animals, however, energy stores are lower, and death can be due to a terminal hypoglycemia.

In growing animals, energy deprivation initially slows growth, with cessation of growth or weight loss occurring in severe deprivation. Even if food intake is severely restricted, young animals can maintain a constant weight for long periods of time. Body proportions also tend to be maintained in the correct ratio for an animal of that weight. However, brain and skeletal growth continues, and eventually the head becomes disproportionately large. Surprisingly, growth restriction can have few long-term adverse effects. Fusion of growth plates is delayed, and when adequate food becomes available growth resumes and the animal will reach a near normal adult weight. The earlier in life nutritional restriction occurs, the more likely it is that long-term adverse effects will occur.

Energy restriction delays puberty, and in adults it reduces estrus behavior, ovulation rates, and libido. Conception and birth rates fall. In sheep, energy restriction reduces wool growth and the diameter of wool fibers. In severe cases wool growth stops altogether, creating a break (weak spot) in the staple of wool. The wool will then detach, particularly during handling. Energy deficiency also reduces immune function, and the animal becomes more susceptible to infectious and parasitic diseases. Wound healing is adversely affected by malnutrition. Poor body condition prior to an operation increases the risks of postoperative problems and death.[23-27,30]

Energy excess results in obesity. This may exacerbate chronic lameness and reduces fitness for work. There is reduced reproductive efficiency in both males and females, and dystocia is more likely. Obesity predisposes to ketosis and fatty liver in heavily pregnant sheep and recently calved cattle.[26]

Maintenance energy

Basal energy requirement is measured with the animal at rest. It represents the amount of energy used to maintain normal cellular function. Energy requirements for maintenance are about 25% to 100% higher depending on energy expenditures for movement in the natural environment. Maintenance requirements are even higher if the environment is adverse, for example, because of heat or cold stress or because of the need for extensive foraging in search of food.

Cold and heat stress. Maintenance energy requirements are affected by the ambient temperature, thermal radiation, humidity, wind, and rainfall. A breeze may reduce the NE requirement if the weather is abnormally hot or increase it when the ambient temperature is cool.[26] The thermoneutral zone is the zone of climatic conditions in which an animal can live without expending additional energy to maintain body temperature by heat generation (e.g., shivering) or by cooling (e.g., panting). The lower critical temperature is the temperature below which energy has to be expended to maintain body temperature.

The ability to tolerate cool climatic conditions is affected by level of production. High-producing or hardworking animals eat more food and generate heat as a result of the heat increment (specific dynamic activity). Under cool conditions this heat helps maintain body temperature. Thus high-producing dairy cows are unlikely to need additional feed in cold weather. Eventually this heat has to be dissipated to the environment. This is more difficult if the environment is warm. In consequence, lactating dairy cows are particularly susceptible to heat stress.

Adaptation affects heat and cold tolerance. Zebu and Brahman cattle are genetically better adapted to hot climates than cattle of European breeds. As animals go into winter, they develop thick hair coats that greatly increase their ability to withstand cold weather. For example, winter adaptation can reduce the lower critical temperature for mature European types of beef cattle to $-20°$ C ($-4°$ F). Body condition is also important; well-conditioned animals have a better cover of insulating fat and are less susceptible to cold stress.

Below the lower critical temperature, energy expenditure is inversely proportional to environmental temperature in adapted animals.[23] Wind velocity is also important because it reduces the effect of insulation. A wind velocity of 1 km/hr equals a temperature drop of $0.35°$ C (1 mph = $1°$ F).[17] Thus if the environmental temperature is $10°$ C and the wind velocity is 15 km/hr, the adjusted temperature is $10 - (15 \times 0.35) = 4.75°$ C.

Acute heat stress that affects the respiratory pattern increases the NEm requirement. When there is rapid, shallow breathing in cattle, the NEm requirement increases about 7%. If animals are showing open-mouthed panting, the NEm requirement increases 11% to 25%. The problem of acute heat stress is compounded by a concomitant decrease in appetite (30% decreases may be seen in hot, humid weather).[35] In these cases production losses can be curbed only by cooling the animals.

Wool and hair are said to be effective insulators against both heat and cold. The shorter the fleece or hair coat, the higher the lower critical temperature.[1,2,5]

Season. A seasonal shift in the maintenance energy requirements of sheep has been reported in thermoneutral environments. This effect is probably modulated by the photoperiod. The NEm requirement increases in summer (up to 14% in sheep) and decreases similarly in winter. Parallel fluctuations in appetite and voluntary feed intake also occur.[4]

Growth pattern. Liberal amounts of feed and high rates of growth increase maintenance energy requirements. In lambs liberal feeding can increase maintenance energy requirements 35% above that in animals maintained at constant weight. This high basal metabolism is apparently related to high digestive tissue (liver and small intestine) mass. Conversely, low metabolic rates and low digestive tissue mass contribute to survival during periods of feed shortage.[26]

Foraging. Keeping animals under range conditions increases energy requirements. This is related to a greater sensitivity to the weather and to increased activity. The requirement is even greater if the terrain is rolling or steep rather than flat, if the density of grazeable forage is low, and/or if the distance to water is far.[26] Maintenance energy requirements of grazing sheep are 60% to 70% greater than those of pen-fed sheep.

Efficiency of food production

Food production involves the conversion of plant energy into animal energy. The efficiency of this process is affected by the efficiency of growth and by the proportion of total energy that is spent in maintaining the breeding herd.

In species with high reproductive rates, such as poultry and pigs, only a small proportion of total energy consumed by the breed goes to the parent population. In broiler poultry the division of total energy expenditure is 4% to the parents and 96% to the progeny (which are slaughtered and eaten). In pigs, sows consume 20% of energy and the progeny 80%. In contrast, ruminants have much lower birth rates and so a large proportion of energy is diverted to the parents; the dams consume about half the energy in the single suckle beef industry and 70% of the energy in lowland sheep rearing.[39] This situation is tolerated because ruminants (and particularly the parents) consume a large proportion of their energy as fiber. Ruminants convert dietary fiber in grasses, legumes, and forbs (bushes, shrubs) to meat and milk energy useful to people. Without ruminants, energy trapped in fiber would be largely unavailable to people.

In farm animals, body weight increases in a curvilinear fashion (Fig. 1-5). Efficiency of growth depends on three factors. The first is the intake of ME in excess of maintenance requirements. This represents the total energy available for the growth processes. In immature animals given free-choice access to a high-quality feed, voluntary intake of ME exceeds maintenance re-

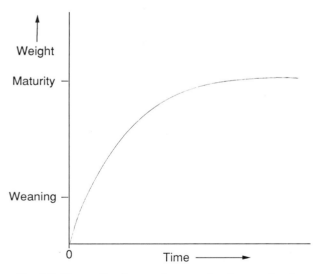

Fig. 1-5 Generalized growth curve for farm animals given free-choice food.
Modified from Brody S: Bioenergetics and growth, 1945, Reinhold Publishing Company.

quirements. The surplus is used for growth. As maturity approaches, ME intake and maintenance requirements converge until at the mature weight the two are in balance. In consequence, daily weight gain declines, and the efficiency of energy conversion into gain (ME retained divided by ME ingested) decreases because only a small amount of surplus ME is available for growth[38,39] (Figs. 1-6 and 1-7).

To some extent the benefits of high feed intakes are counterbalanced by reductions in digestibility secondary to more rapid passage of digesta through the gut. This is particularly important in the case of fiber digestion.[40] In cattle optimal feed conversion occurs if feed intake is somewhat less than maximal. For example, beef cattle grown to slaughter weight over 24 months have better feed conversion ratios than cattle grown to slaughter weight over 18 months.[38]

A third factor that determines the efficiency of weight gain is the proportion of energy laid down as protein versus fat. Protein synthesis is favored in growing animals; fat synthesis becomes dominant as maturity approaches. Meat-producing animals are slaughtered as they approach maturity at a stage when the carcass contains enough fat to make it most palatable to the consumer. More feed energy is required to produce liveweight gain as adipose tissue, mainly because adipose tissue contains little water. Although fat has a higher gross energy content than protein, the ME in food is used much more efficiently for fat than for protein synthesis (75% and 45%, respectively). In consequence the amount of ME required for the synthesis of 1 kg of fat or 1 kg of protein is almost identical. However, adipose tissue is mostly fat whereas other tissues contain a lot of water. For most conditions, the energy content (and thus NEg) of liveweight gain varies between 1.2 kcal/g (550 kcal/lb) (energy content of fat-free body) and 8 kcal/g (3600 kcal/lb) (energy content of adipose tissue).[23]

One approach to improving energetic efficiency in meat-producing animals has been to select for larger animals. Larger individuals take longer to reach maturity. In traditional agricultural systems where animals are slaughtered at fixed weights or fixed ages, this means that these animals will be less mature at slaughter. Daily weight gain is greater in immature animals than in mature animals. In consequence, the efficiency of growth is increased (greater ME intake

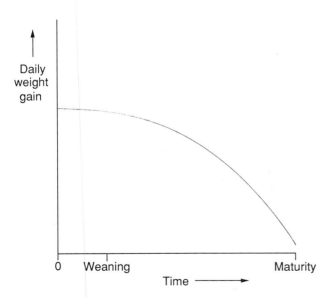

Fig. 1-6 Relationship between daily weight gain and time in animals.

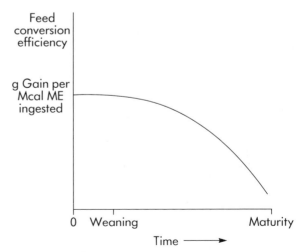

Fig. 1-7 Food conversion efficiency in growing animals.
Modified from Webster AJF: Livestock Prod Sci 32:123, 1973.

excess above maintenance requirement), and the carcasses are leaner. One drawback to the selection of larger individuals is that the costs of maintaining the parent population are increased. In pigs this is of small concern because the parents use only a small proportion of the total energy. In ruminants this effect can be avoided in part by breeding small dam breeds to large sire breeds. This approach combines the benefits of low-maintenance energy expenditures on the dams with large, rapidly growing offspring.[39] Use of excessively large sires may, however, predispose to dystocia.

Male animals are more efficient converters of feed into liveweight gain than females. Males grow to a larger mature size; thus they eat more, have greater daily weight gains, and contain less fat at a given weight than females. In consequence, energy required for a unit of liveweight gain is less than in females. Because males grow to a larger size, they have to be finished at heavier weights than females to ensure a similar fat content.

The feed-conversion ratio is the ratio of feed ingested (in kg or Mcal) to liveweight gain. Low ratios of feed fed per unit of gain indicate efficient production. Feed conversion efficiency and feed conversion rates decline with age (Fig. 1-6).

Compensatory gain

Compensatory gain is increased growth rate and thus increased efficiency of food conversion into gain in animals that were previously restricted in growth rate. When someone speaks of compensatory gain, they are comparing the performance of growth-restricted animals to that of well-grown animals of the same age.

Compensatory gain occurs because growth is determined by nutrient availability and the degree of immaturity of the animal. If nutrients are restricted, then the animal is held back at a point on the growth curve where the potential for growth is greater. When the animal is placed on an unrestricted diet, it eats more and puts on weight more rapidly than a well-grown animal of the same age. This is because the well-grown animal is closer to its mature weight and thus grows more slowly.

Compensatory gain following growth restriction can be seen in a variety of circumstances, for example, when calves kept on poor-quality pasture are moved into a feedlot. Another example is the growth of twin calves compared with single calves following weaning. Diseases that run a transient course can also be followed by compensatory gain.[31]

Dry matter intake

Dry matter intake is controlled by the animal's requirement for energy intake and the physical capacity of the gut. When a high-quality diet is fed and requirements are low, food intake is regulated in accordance with requirements. Gastrointestinal fill is the major limiting factor when requirements are high and ration nutrient density is low (forage based).

Pregnant animals have reduced capacity for food intake compared with nonpregnant animals because the room taken up by the pregnant uterus limits the space available in the abdominal cavity for gastrointestinal fill. This is particularly important in late pregnancy (the majority of fetal growth occurs in the last third of gestation) and in animals that give birth to multiple fetuses that are well developed and thus comparatively large. The restriction of gut fill is particularly important in heavily pregnant, twin- or triplet-bearing ewes on forage-based diets.

Lactating animals can eat more than pregnant animals, but heavily producing animals such as the dairy cow require even more energy than pregnant ewes. In the first weeks of lactation, dairy cattle cannot fulfill energy requirements within the limits of appetite. Milk production builds up more rapidly than appetite. Production peaks 4 to 8 weeks postpartum and appetite at 10 to 14 weeks. The result is a negative energy balance and weight loss in early lactation. As dry matter intake increases and milk production stabilizes or begins to decrease, weight stabilizes. In late lactation weight is usually gained.

Dry matter intake is difficult to predict accurately. An example of the changes in dry matter intake that can occur in response to changes in energy requirement is the increase in dry matter intake from about 2% of body weight in dry cows to about 4% in heavily lactating cows.

RATION FORMULATION

Nutritional advice frequently involves recommendations on the comparative value of different types of feed and the evaluation and formulation of rations for animals.

Feed selection

Information on the nutritional quality of a feed can be obtained by looking up average values in tables (see Appendix). These values may be modified according to one's judgement of the quality of the particular feed sample. Alternatively, a representative sample of the feed can be collected and analyzed, using methods described in Chapter 10.

Economic value of a feed. When choosing feed energy sources for a ration, the feeds (concentrates and roughages) that are commonly grown in the area will generally be least costly. This principle may be affected by unfavorable weather in a local area or an abundance of production in another area. In estimating comparative feed values, consider:

1. The economic substitution value of the feed
2. Analysis of feed cost trends for predicting the most favorable times for purchasing feeds that can be stored
3. Managing the formulation of rations to ensure that producers can benefit from feed price changes.[14]

The economic substitution value of a feed is based on the assumption that its value is equal to the sum of its individual nutrient values. This approach has some shortcomings. Feed intake limits productivity in high-producing dairy cow rations and in rapidly growing animals. In these situations, high-energy-density feeds carry a premium because they allow greater energy intake and more efficient production. The problem of feed density is partly overcome by comparing concentrate feeds with concentrates and roughages with roughages. However, within concentrates, variations in ration density still exist. Another problem is in evaluating the protein component because all proteins are not equivalent. In monogastrics there are variations in the biological value of proteins. High-producing dairy cows may gain special benefits from the provision of bypass (rumen undegradable) proteins. Some feeds contain a much higher portion of bypass protein than others. For these reasons evaluating feeds solely on the basis of the unit price of crude protein can be misleading. Overall, however, comparisons based on unit cost are a useful first step in comparing the economic value and acceptable purchase prices of feeds.

Unit costs are calculated as follows:

1. Choose one concentrate grain and one natural protein supplement. For our example, we will use Number 2 corn and solvent-extracted soybean meal destined for a feedlot beef ration. We will use as-fed concentrations of total protein and energy, converted from the dry matter concentrations listed in the NRC *Nutrient Requirements of Beef Cattle*, 1984. We obtain average energy concentrations by adding the listed NEm and NEg concentrations together and dividing by 2. The as-fed total protein and average energy concentrations are:

Feed	Average NE as fed (Mcal/kg)	Total protein % as fed
Corn	1.68	8.9
Soybean meal	1.54	44.4

For our example the market values of these feeds are:
   Corn = $100/tonne
   Soybean meal = $200/tonne

2. Formulate an equation for each feed that shows the value of the feed to be a function of the protein and energy content.

   Let X = the market value of a unit (1 tonne) of protein and let Y = the market value of 1000 Mcal of average energy as follows:

   a. For 1 tonne of corn:
      $0.089X + 1.68Y = \$100$
   b. For 1 tonne soybean meal:
      $0.444X + 1.54Y = \$200$

3. Use the simultaneous equation method to solve for the values of X and Y as follows:
   a. Divide the coefficient of X in the soybean meal equation by the coefficient of X in the corn equation.

$$\frac{0.444}{0.089} = 4.99X$$

b. Multiply the total corn equation by the quotient that results in part a above.

4.99 (0.089X + 1.68Y) = $100 × 4.99
= 0.444X + 8.38Y = $499

c. Subtract the soybean meal equation from the modified corn equation and solve for Y.

$$\begin{array}{r} 0.444X + 8.38Y = \$499 \\ -\underline{0.444X + 1.54Y = \$200} \\ 6.84Y = \$299 \end{array}$$

Y (value of 1000 Mcal average energy) = $299/6.84 = $43.71

X (value of 1 tonne of total protein) =
$$\frac{\$100 - (1.64 \times \$43.71)}{.089} = \$318.15$$

4. Use the calculated values of X and Y to establish the price that we can pay for a unit of an alternative feed. For example, if a sample of as-fed barley contains 11.9% total protein and 1.54 Mcal average net energy per kg, we can afford to pay (0.119 × $318.15) + (1.54 × $43.71) = $105.17 per tonne for barley.
5. The relative values of alternative concentrates, roughages, and protein supplements can be established by this method.[14]

Economic analysis reveals that in certain periods of the year purchasing feeds for storage can be economically advantageous. Generally speaking, feed prices will be depressed in the harvest season. Prudent producers develop storage areas to enable them to take advantage of favorable markets. The study of fluctuating markets is called *trend analysis* and is often reflected in futures markets.

Variations in the market value of feeds may be such that alternative feed sources should be used in certain seasons or years. Economic efficiency may dictate flexibility in ration formulation as a part of nutrition management. An example is the use of animal fats as energy sources in feedlot rations. The caloric density of fats is very high, and when feed grain prices are high fats may provide a more economical source of ration energy.[14]

Ration evaluation

Ration evaluation is the determination of the suitability of a particular diet for the production goals of the farm. Ration formulation is the design of a ration to meet a particular production goal. Both techniques require an accurate estimation of the animal's nutritional requirements. In ration evaluation the requirements are compared with the nutrient intake from the diet. In ration formulation a diet is formulated to meet the nutritional requirements. These procedures are demonstrated in the following beef cattle example.

Energy requirements. From Appendix Table A-1, it can be calculated that at thermoneutral temperatures a 450 kg (1000 lb) cow requires $0.077 \times 450^{0.75} = 7.5$ Mcal of NEm per day.[23]

However, it may be necessary to modify maintenance energy values to take into account local climatic conditions. Outside the thermoneutral zone (15° to 25° C [60° to 80° F]) the coefficient (0.077 in the thermoneutral range) changes by 0.0007 for each 1° C change from 20° C. For example, for cattle at temperatures of 10° C, the coefficient becomes:

Coefficient = 0.077 + (0.0007 × (20 − 10)) = 0.084

Thus a 450-kg cow kept at 10° C requires $0.084 \times 450^{0.75} = 8.2$ Mcal of NEm.

Three basic questions are presented in feedlot energy nutrition:
1. What are the most economical feed sources of energy in any specific situation?
2. What is the expected production (liveweight gain) when a specific ration is fed to a specific class and weight of cattle?
3. How do I mix the least costly ingredients to achieve the desired objective in the most practical and economical way?

The prediction of expected gain. The following steps can be used to predict the level of production that can be obtained from a given ration.
1. Establish the class and weight of animal to be fed. For our example we will use medium-frame steer calves that have a shrunk weight of 400 kg (880 lb).

2. Define the type of diet being fed. For our example we will use a diet of high-moisture corn, alfalfa hay (sun-cured mid-bloom), and complete supplement.
3. Using this feeds analysis, we can calculate the contribution of each feed to the NE content of 1 kg of diet (Table 1-1).
4. From Appendix Table A-2, the predicted dry matter intake (DMI) is

Bodyweight$^{0.75}$ × [0.1493 NEm − (0.046 × NEm^2) − 0.0196] = 400$^{0.75}$ × [(0.1493 × 2.18) − (0.046 × 2.18^2) − 0.0196] = 7.81 kg per day

Many feedlots weigh the feed fed to the cattle, and this predicted value can be compared with the actual intake. Reasons for discrepancies between the two values are discussed in Chapter 16.

5. From Appendix Table A-1, the NEm requirement of the individual animal is

0.077 × Bodyweight$^{0.75}$ = 0.077 × 400$^{0.75}$ = 6.89 Mcal NEm per day

The DMI required for maintenance is the cow's NEm requirement divided by the ration NEm concentration:

6.89 ÷ 2.18 = 3.16 kg

The NEg available from the ration for gain Mcal is:

(Total DMI − DMI required for maintenance) × (NEg/kg Ration) = (7.81 − 3.16) × 1.49 = 6.93

This is used to calculate the predicted shrunk liveweight gain in kg per day as follows.

From Appendix Table A-1, NEg required for a given liveweight gain in medium-frame steer calves:

NEg, Mcal = (0.0557 × Bodyweight$^{0.75}$) × (Liveweight gain$^{1.097}$)
= (0.0557 × 400$^{0.75}$) × (Liveweight gain$^{1.097}$)
= 4.98 × Liveweight gain$^{1.097}$

Our calculations predict that the ration provides 6.93 Mcal of energy for gain, so:

6.93 = 4.98 × Liveweight gain$^{1.097}$(kg)

Thus, predicted liveweight gain is:

(6.93 ÷ 4.98)$^{0.9116}$ = 1.35 kg/day

(Note that 0.9116 is the reciprocal of 1.097)

If the cattle are being weighed on a regular basis these predictions can be compared to the actual performance.

Table 1-1 *Calculation of ration composition*

A Feed	B Fraction of ration DM	C NEm* (Mcal/kg)	D Contribution to ration NEm	E NEg* (Mcal/kg)	F Contribution to ration NEg
High-moisture corn	0.88	2.33	2.05	1.62	1.43
Alfalfa hay	0.08	1.24	0.10	0.68	0.05
Complete supplement	0.04	0.80	0.03	0.35	0.01
Totals	1.00		2.18		1.49

*NE values are expressed per kg of diet dry matter
Column B contains the proportions the feeds are mixed into the ration. Columns C and E are from laboratory analyses of individual feeds.
The NEm from each feed in 1 kg of complete ration, column D, is calculated as: NEm concentration in feed × fraction of diet DM contributed by feed = Column C × Column B.
The sum of the values in column D is 2.18; this is ration NEm, Mcals per kg of diet dry matter.
The NEg from each feed in 1 kg of complete ration, Column F, is calculated as: NEg concentration in feed × fraction of diet DM contributed by feed = Column E × Column B
The sum of the values in column F is 1.49; this is the total ration NEg, Mcals per kg of diet dry matter.

Calculations for ration formulations

Once the basic ingredients are chosen, a ration must be formulated to meet the production goals. In the feedlot, the production objective is maximum gain. Limiting factors include the dry matter feed intake (Appendix Table A-2). A certain amount of roughage (fiber) also seems to be necessary if feeding periods exceed 60 to 90 days.

High-concentrate rations commonly used to finish cattle generally require supplementation with calcium, phosphorus, sodium and chlorine (salt), trace minerals, and vitamin A. Commercial vitamin-mineral supplements or supplements formulated by the feedlot are commonly used as an integral part of feed rations. A common practice is to feed a complete supplement at 3% to 5% of the ration dry matter. The energy supplied by the vitamin-mineral supplement represents a small amount of the total energy in the ration. In formulating rations it can be calculated with the concentrate energy or even be ignored without significant loss of accuracy.

There are situations where liveweight gains need to be limited. It is not advantageous for prospective female replacement cattle to be fat. A healthy growing condition that enables these heifers to attain early puberty and maximum fertility is generally desired. Most heifers are initially bred as yearlings (13 to 15 months old). For maximum breeding efficiency, the heifers need to attain about 60% of their adult weight by the beginning of the breeding season. To do this the heifers need to gain between 0.6 and 0.7 kg (1.25 to 1.5 lb) per day.

For example, we will formulate a ration for 300 kg large-frame heifer calves being fed to gain 0.6 kg (1.3 lb) per day. The ambient temperature is 10° C (50° F). From Appendix Table A-1, energy requirements are (see p. 16 for details of correction for cold ambient temperatures):

$$\text{NEm Mcal} = 0.084 \times (300)^{0.75} = 6.06 \text{ Mcal}$$

$$\text{NEg Mcal} = 0.0608 \times (300^{0.75}) \times (0.6)^{1.119} = 2.47 \text{ Mcal}$$

We will use a ration in which the desired NEm concentration equals 1.5 Mcal/kg dry matter and the NEg concentration equals 0.9 Mcal/kg. The maximum dry matter intake can be estimated from the formula in Appendix Table A-2:

$$(300)^{0.75} \times [(.1493 \times 1.5) - (0.046 \times 1.5^2) - 0.0196] = 7.3 \text{ kg}$$

To meet our NEm requirement we need
$6.06 \div 1.5 = 4.04$ kg feed.

To meet our NEg requirement we need
$2.47 \div 0.9 = 2.75$ kg feed.

Total DMI required for energy $= 6.79$ kg feed.

It appears that a ration containing these energy concentrations will meet our liveweight gain goals within the constraints of the maximum dry matter intake of the cattle. If the animal could not eat sufficient dry matter, a more energy-dense ration would have to be formulated. If the cows can eat much more dry matter than is required, it may be economically worthwhile to reformulate to a lower energy density so that cheaper feeds can be used.

Once the required energy density of the ration has been determined, we can turn our attention to the mix of ingredients needed. In our example we will assume that the least costly feed sources include crested wheatgrass hay and triticale that have the following dry matter (DM) compositions:

Feed	% DM	NEm Mcal/kg DM	NEg Mcal/kg DM	Crude protein, % DM
Wheatgrass hay	90	1.07	0.52	12.4
Triticale	90	2.06	1.4	17.6

The Pearson square is a commonly used method to mix two feeds that have different concentrations of a specific nutrient to give a desired concentration of that nutrient. The computational steps are:

1. A square is drawn and the desired concentration for the nutrient in question is placed in the middle of the square.
2. One of the feeds and the specific nutrient concentration in that feed are listed at the upper left corner of the square.
3. The other feed and the specific nutrient concentration for that feed are listed at the lower left corner of the square.
4. Subtractions are made across each diagonal. The smaller number is subtracted from the larger, and the differences are listed at the right-hand corners.
5. The differences are added together, and

the sum is placed below each difference as the denominator of a fraction.
6. The fractions are converted to percentages. These percentages represent the proportion of each feed in the mixture.

We will use a Pearson's square to compute the percentages of ingredients (on a dry matter basis) to be used in the ration. We will use NEm concentrations as a base as follows:

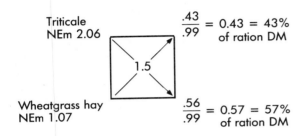

From this information, Table 1-2 can be constructed.

The total protein requirement for these animals is about 630 g/day, which is less than the intake, so no supplementation is necessary.

Once the energy and protein components of the ration have been determined, the mineral and vitamin intake must be determined and an appropriate vitamin-mineral supplement added to meet requirements.

Specific energy considerations

Dairy cows have the largest increases in requirements for lactation of all the farm animal species. Because of this, they are particularly prone to ketosis (a metabolic disease characterized by disordered energy metabolism) in early lactation. Sheep and goats have the greatest requirements of energy for pregnancy (expressed as a percentage of maintenance requirements) and are particularly prone to pregnancy toxemia, a metabolic disease related to ketosis.

The energy requirements for farm animals are best initially approximated by information distilled in the National Research Council publications on nutrition.[23-29] They describe average energy requirements using formulas derived from many different experiments. The formulas are also used to generate charts of energy requirements for many of the more common types and weights of cattle (see Appendix).

Net energy is the most accurate measure of energy requirement in ruminants. However, ME has been measured more frequently than NE in sheep, and ME is often used to calculate nutrient requirements. In goats there is a paucity of direct measurements of energy content and DE, ME, TDN, and NE are calculated as a simple proportion of one another. All are equally accurate; ME is usually used.

In swine the relationship between ME and DE

Table 1-2 *Calculation of wet matter and protein intakes*

A Feed fraction	B DM content of diet	C Fraction of diet DM	D DM intake (kg)	E As-fed intake (kg)	F Crude protein	G Protein intake (kg)
Wheatgrass	0.9	0.43	2.92	3.24	0.124	0.40
Triticale	0.9	0.57	3.87	4.30	0.176	0.76
Totals		1.00	6.79	7.54		1.16

Columns B and F contain the dry matter contents (expressed as a fraction) and crude protein (expressed as a fraction of feed dry matter) contents from the feed analysis. Column C is the fraction of each feed in the total ration (dry matter basis) from the Pearson square.
Dry matter intake of each feed, column D, is calculated as: dry matter intake of cattle (p. 18) × fraction of ration DM contributed by feed = 6.79 × column C.
The as-fed intake of each feed, column E, is calculated as: dry matter intake of feed ÷ dry matter concentration in feed = column D ÷ column B.
Crude protein intake, column G, is calculated as: dry matter intake of feed × fraction of feed that is crude protein = column D × column F. The sum of the values in column G is the total crude protein intake for each animal.

in feedstuffs is fairly constant. Because swine are monogastrics, protein metabolism is not confused by effects of rumen microbes (p. 25). It is possible to take protein content of feeds into account when calculating ME from DE. If excess protein is fed, the amino acids not used for protein synthesis are catabolized and used as an energy source. The excess nitrogen is excreted as urea, carrying energy and reducing ME. ME values are sometimes corrected for nitrogen gained or lost from the body because the energy that is deposited as retained protein cannot be fully recovered if the amino acids are subsequently degraded for energy utilization. Nitrogen retention is usual in growing animals, and the correction is less likely to be necessary.[21,22,27] Net energy is the most accurate measure of energy utilization. However, NE is not ordinarily used in the formulation of swine rations because it is difficult to measure. Furthermore, swine rations are largely composed of concentrate feeds, so difficulties associated with energy utilization of roughages are not a big problem. In conventional swine diets the ratio of NE to ME ranges from 0.66:1 to 0.72:1.[13,32-34,36]

In horses, maintenance requirements are expressed as a simple function of body weight because there is no advantage of weight$^{0.75}$ within this species. As one would expect, requirements for pregnancy and lactation are small in comparison with those of food-producing animals. The major requirement for additional energy is for work.

SUMMARY

The gross energy content of a feed can be found by measuring the heat produced on complete combustion. However, it is a poor index of the usefulness of the feed due to energy losses in digestion (gases and fecal energy), urine, and metabolism (heat increment or specific dynamic activity losses). Net energy reflects the energy available for tissue metabolism and is the most accurate measurement of useful food energy. Net energy is particularly useful in ruminants where there are considerable differences in the energy value of roughages and concentrates. In pigs, metabolizable energy and net energy are closely related, and metabolizable energy is used because measurements of net energy values for feeds are time-consuming.

The major energy-containing nutrients are lipids, carbohydrates, and proteins. Lipids contain more than twice the energy content of carbohydrates. Proteins have amounts of metabolizable energy similar to carbohydrates. Carbohydrates occur as storage carbohydrates and structural carbohydrates. Storage carbohydrates (starches) are compact; the sugar subunits are joined by alpha linkages. They can be directly digested by mammals and have a high digestible energy content. In structural carbohydrates (fiber) the sugar units are joined by beta linkages; these can be digested only by microbes and have a lower digestible energy content. Structural carbohydrates predominate in forages such as grasses and legumes. Storage carbohydrates or lipids predominate in seeds such as grains or oilseeds.

Ration formulation should satisfy nutritional requirements and optimize economic returns. The tedious calculations are now performed mostly by computer.

REFERENCES

1. Ames DR: Normal response of sheep to acute thermal stress, doctoral dissertation, East Lansing, 1969, Michigan State University.
2. Blaxter KL: The energy metabolism of ruminants, Springfield, Ill, 1966, Charles C Thomas.
3. Blaxter KL: Proc Nutr Soc 37:1, 1978.
4. Blaxter KL and Boyne AW: J Agricul Sci 99:611, 1982.
5. Brink DR and Ames DR: J Anim Sci 41:264, 1975.
6. Brody S: Bioenergetics and growth, New York, Reinhold Publishing Company, 1945. Reprint Hafner Press, 1974, pp 1-11.
7. Brody S: Bioenergetics and growth, New York, Reinhold Publishing Company, 1945. Reprint Hafner Press, 1974, pp 12-36.
8. Brody S: Bioenergetics and growth, New York, Reinhold Publishing Company, 1945. Reprint Hafner Press, 1974, pp 59-75.
9. Brody S: Bioenergetics and growth, New York, Reinhold Publishing Company, 1945. Reprint Hafner Press, 1974, pp 352-384.
10. Brody S: Bioenergetics and growth, New York, Reinhold Publishing Company, 1945. Reprint Hafner Press, 1974, pp 470-483.
11. Brody S: Bioenergetics and growth, New York, Reinhold Publishing Company, 1945. Reprint Hafner Press, 1974, pp 307-351.
12. Chalupa W and Ferguson JD: The role of dietary fat in productivity and health of dairy cows. Paper presented at the application of nutrition in dairy practice proceedings of a symposium, The Dairy Production Medicine

Continuing Education Group annual meeting, College of Veterinary Medicine, North Carolina State University, American Cyanamid Company, 1988.
13. Ewan RC: Distill Feed Res Counc Conf Proc 31:16, 1976.
14. Galligan DT: Economic aspects of nutritional monitoring. Paper presented at the application of nutrition in dairy practice proceedings of a symposium, The Dairy Production Medicine Continuing Education Group annual meeting, College of Veterinary Medicine, North Carolina State University, American Cyanamid Company, 1988.
15. Garrett WN: J Anim Sci 51:1434, 1980.
16. Garrett WN and Hinman N: J Anim Sci 28:1, 1969.
17. Johnson DE and Crownover JC: Energy requirements of feedlot cattle in north central Colorado: variation in maintenance requirements by the month. Research highlights of the Animal Science Department, Colorado State University Exp Stn General Series 948, 1975.
18. Kleiber M: The fire of life, an introduction to animal energetics, Melbourne, Fla, 1975, RE Krieger Publishing Company, pp 104-130.
19. Kleiber M: The fire of life, an introduction to animal energetics, Melbourne, Fla, 1975, RE Krieger Publishing Company, pp 179-222.
20. Lofgreen GP and Garrett WN: J Anim Sci 27:793, 1968.
21. Morgan DJ, Cole JA, and Lewis D: J Agricul Sci (Camb) 84:7, 1975.
22. Morgan DJ, Cole JA, and Lewis D: J Agricul Sci 84:19, 1975.
23. National Research Council: Nutrient requirements of beef cattle, ed 6, Washington, DC, 1984, National Academy Press.
24. National Research Council: Nutrient requirements of dairy cattle, ed 6, Washington, DC, 1989, National Academy Press.
25. National Research Council: Nutrient requirements for goats, No 15, Washington, DC, 1981, National Academy Press.
26. National Research Council: Nutrient requirements for sheep, ed 6, Washington, DC, 1985, National Academy Press.
27. National Research Council: Nutrient requirements for swine, ed 9, Washington, DC, 1988, National Academy Press.
28. National Research Council: Predicting feed intake of food producing animals, Washington, DC, 1986, National Academy Press.
29. National Research Council: Nutrient requirements of horses, ed 5, Washington, DC, 1989, National Academy Press.
30. Naylor JM: Current therapy in equine medicine, Philadelphia, 1984, WB Saunders, pp 26-46.
31. Ostrowski SR and others: J Am Vet Med Assoc 195:481, 1989.
32. Pals DA and Ewan RC: J Anim Sci 46:402, 1978.
33. Phillips BC and Ewan RD: J Anim Sci 44:990, 1977.
34. Robles A and Ewan RC: J Anim Sci 55:572, 1987.
35. Sniffen CJ: Balancing rations for carbohydrates for dairy cattle: the application of nutrition in dairy practice proceedings of a symposium, The Dairy Production Medicine Continuing Education Group annual meeting, College of Veterinary Medicine, North Carolina State University, American Cyanamid Company, 1988.
36. Thorbeck G: Studies on energy metabolism in growing pigs. II. Protein and fat gain in growing pigs fed different feed compounds, efficiency of metabolizable energy for growth, Research Report No 424, Copenhagen, 1975, Beret Statens Husdyrbrugsforog.
37. Van Soest PJ and McQueen RW: Proc Nutr Soc 32:123, 1973.
38. Webster AJF: Livestock Prod Sci 7:243, 1980.
39. Webster AJF: Anim Prod 48:249, 1989.
40. Zinn RA and Owens FN: J Anim Sci 56:471, 1983.

Protein Nutrition and Nonprotein Nitrogen

H. H. VAN HORN

The term *protein* embraces an enormous group of closely related but physiologically distinct compounds. The distinguishing features of the various proteins are their amino acid makeup, sequence, and structure. Protein is required in animal nutrition in larger quantity than any other nutrient, with the exception of the total quantity of carbohydrates, soluble fibers, and fats that are required for energy. Carbohydrates, soluble fibers, and fats contain many chemically different entities, but they are all metabolized in the body to carbon dioxide and water; any surpluses are stored as adipose tissue. With protein, however, a different situation exists. The primary units, amino acids, primarily are used for the synthesis of body or secretory proteins. However, amino acids also can provide energy if they are catabolized to carbon dioxide, water, and nitrogenous residues that must be excreted. Thus, protein metabolism is more complex than that of either carbohydrate, fiber, or fat, all of which are concerned mostly with energy transfer.

Proteins are large molecules of connected amino acids with molecular weights ranging from 35,000 to several hundred thousand. Because protein is the principal constituent of the organs and soft structures of the animal body, a liberal and continuous dietary supply is needed for growth and repair of tissues throughout life. Transformation of food protein into body or product protein is a very important part of digestion and metabolism.

In order to be absorbed and utilized by an animal, dietary protein must be broken down by digestion into its constituent amino acids. Proteins from various sources have different combinations of amino acids. The generalized chemical structure of an amino acid is:

$$\begin{array}{c} HH \\ \diagdown\diagup \\ NO \\ |\| \\ R-C-C-O-H \\ | \\ H \end{array}$$

R represents different carbon chains with varying lengths and configurations. For more than a century, it has been known that proteins differ in nutritional value. The variation in nutritional value of proteins for nonruminants has been shown to be due to differences in amino acid content. In ruminants, nutritional value of a dietary source of protein is related to a combination of amino acid content and the ability of the protein to escape degradation within the rumen.

Twenty-five different amino acids are found in animal and plant proteins. Some amino acids can be synthesized within the body from other compounds, but others can be made only by plants and bacteria. Amino acids that cannot be manufactured by the animal's metabolism in adequate amounts are termed *essential amino acids*. Simple-stomached animals require dietary sources of these amino acids. Research has identified dietary requirements for most essential or limiting amino acids for poultry and swine.

Those amino acids that are essential to the animal and are needed in the diet because the animal's body cannot synthesize them fast enough to meet its requirements are:

Phenylalanine Tryptophan   Histidine Glycine (for
Valine   Isoleucine   Arginine   poultry)
Threonine   Methionine   Leucine
                            Lysine

A helpful acronym to remember the 10 essential amino acids is

        PVT   TIM   HALL

A ruminant like the cow has billions of microorganisms in its rumen that are capable of synthesizing proteins from amino acids and nonprotein nitrogen (NPN) derived from the cow's diet. These microbial proteins subsequently are digested and absorbed by the cow and give her a source of all essential amino acids, even though her initial diet may not have contained adequate amounts.

Therefore, protein quality is not of as much importance to the ruminant as it is for simple-stomached animals. Ruminants can live without any preformed protein in their rations so long as nitrogen and an adequate source of carbohydrates and other nutrients are available to ruminal microorganisms for protein synthesis. Ruminant researchers are concerned with whether the amino acids delivered from the rumen to the lower intestine for final digestion and absorption satisfy optimum performance requirements. "Bypass" protein escapes ruminal degradation. A major research area in ruminant nutrition is the design of bypass protein.

The meaning of "an essential dietary amino acid" should not be misunderstood. Physiologically, all amino acids found in animal tissues are essential. Nonessential amino acids can be synthesized within metabolic pathways of animal cells. However, if optimal animal performance is limited because the animal had to degrade essential amino acids to obtain the amino radicals necessary to synthesize "nonessential" amino acids, then even the nonessential amino acids become "essential" when formulating production diets.

Protein (amino acid) deficiencies do not become apparent as rapidly as energy deficiencies. Hair coat quality, immune competence, and wound healing all are adversely affected, however. Over time, growth and milk or egg production will be subnormal in animals fed protein-deficient diets. Severe protein deficiencies reduce growth rate in both the fetus (resulting in small animals at birth) and the young animal. Deficiencies of even a single essential amino acid in the diet will result in subnormal growth, even though there may be an excess of other amino acids available. Restricted growth or limited body stores of protein in young animals also may adversely affect production later in life.

ANALYSIS OF FEEDS FOR PROTEIN

In common with fats and carbohydrates, proteins contain carbon, hydrogen, and oxygen. In addition, they contain a large and fairly constant percentage of nitrogen. Most proteins also contain sulfur, and a few contain phosphorus and iron. Because proteins are unique in containing a large amount of nitrogen while other major nutrients contain no nitrogen, "crude protein" is generally calculated by multiplying the analyzed nitrogen content by 6.25. This figure is derived from the assumed 16% nitrogen content of most amino acids ($^{100}/_{16}$ = 6.25). This procedure is based on an assumption that is only approximately true. Proteins vary in nitrogen content between 15% and 18%. If precise estimates of protein content are to be made from nitrogen analysis, it is more accurate to multiply by a factor specific for the type of protein if it is known to deviate from 16%. Proteins of nonlegume plant origin show the greatest deviation, averaging about 17.5% nitrogen. Milk proteins, however, average only 15.5% nitrogen. Meat, egg, and fish proteins and those of legume plant seeds average 16% nitrogen. Most modern tables of human foods give protein values calculated by using the specific nitrogen content for their protein. Tables for animal feeds normally assume 16% nitrogen in the proteins. Therefore protein content of nonlegume feeds such as grass hay may be overestimated.

Another factor to be considered in protein calculations is that not all nitrogen is a part of protein. Some feeds, particularly green roughages, contain one third or more of their nitrogen as nonprotein nitrogen substances, such as amides, ammonium salts, alkaloids, and other nitrogenous compounds. Because of nonprotein nitrogen, multiplication of total nitrogen content by 6.25 does not give a true protein content but is considered to be *crude protein*, which represents a combination of true protein and nonprotein nitrogen. For ruminant animals that can make use of some nonprotein nitrogen, crude protein is often as good a measure for protein allowances as true protein. Although most nonruminants cannot use nonprotein nitrogen, ingredients with appreciable amounts of nonprotein nitrogen are not usually part of their diet. Thus, estimating protein content of dietary ingredients from analyzed nitrogen content is common for animals with simple stomaches, too.

DIGESTION AND ABSORPTION OF PROTEINS IN NONRUMINANTS

The biological availability of dietary proteins is affected by the ability of an animal to digest them to their constituent amino acids, absorb these amino acids, and utilize them in the body to synthesize new proteins. Proteins must be hydrolyzed in the gut into the constituent amino acids before they can be absorbed (Chapter 7).

After absorption, which is usually sporadic because of the interval between meals, amino acids are carried by the bloodstream to the liver. The liver synthesizes proteins, supplies amino acids to the circulation when needed, and processes nitrogen for excretion. Many chemical reactions take place in protein metabolism within the liver and other tissue cells. Intracellular transmission pathways permit the exchange of ammonia from an amino acid to the keto moeity of an α-keto acid for synthesis of nonessential amino acids. Deamination reactions are similar to transamination, except ammonia is released and shunted through the urea cycle en route to excretion as urea via the urine. In avian species, uric acid is formed instead of urea. After ammonia is removed from excess amino acids (essential or nonessential), the remaining keto acid can be metabolized via the tricarboxylic acid cycle for energy.

Free amino acids, whether from dietary sources or transamination, are carried in the blood to various cells dedicated to the synthesis of specific proteins (e.g., immunoglobulins, muscle, milk protein, and egg protein). The protein mass of the body is in a continuous state of flux, with tissues constantly being catabolized and resynthesized. As amino acids are released, they become available to the general amino acid pool and can be either reused for protein synthesis or utilized as a source of energy.

Aside from total dietary amino acid content, digestibility of dietary protein is the most important variable that affects the quantity of amino acids absorbed. Research by animal nutritionists on protein digestibility has been extensive. *Digestible protein* is usually defined as the amount of crude protein consumed minus the crude protein excreted in the feces. In actuality, this should be called *apparent* digestible protein rather than true digestible protein because part of the nitrogen in feces is metabolic fecal nitrogen. Metabolic fecal nitrogen is not affected by the type of protein consumed; it arises from metabolic functions in the gut, such as residues of bile and digestive enzymes, sloughed epithelial cells, and undigested bacterial residues. The amount tends to vary with dry matter intake rather than protein intake.

Because metabolic fecal nitrogen is independent of the amount of protein consumed, it contributes a larger proportion of the total fecal nitrogen when low-protein feeds are fed than when high-protein diets are fed. Thus, failure to account for metabolic fecal nitrogen artificially reduces apparent digestibility coefficients for low-protein feeds. Because it is very difficult to separate metabolic fecal nitrogen from truly undigested feed nitrogen and because both represent losses to the animal, digestible protein often is used synonymously for apparent digestible protein. Some researchers have taken an alternative approach and convert apparent digestibility to estimated true digestibility based on the assumption that metabolic fecal nitrogen is proportional to dry matter intake (i.e., about 2 g of metabolic fecal nitrogen produced per kg of dry matter intake). In avian species, it is easier to obtain estimates of metabolizable protein (total nitrogen consumed minus nitrogen in feces and urine) than of digestible protein because kidney excretions containing uric acid and fecal contents are excreted as one intermixed product.

Although the digestibility of proteins from different sources is variable, most proteins are 75% to 80% digestible unless some inhibitory factor exists that protects the protein from attack by enzymes or bacteria. Common inhibitory factors include protease inhibitors and heat damage to the protein. For example, soybeans contain at least four proteins that inhibit trypsin or chymotrypsin activity (trypsin inhibitors and hemoglutinins). The presence of antitrypsin factors reduces protein digestibility in nonruminants and results in a reduction in the available energy content of the feed.[6] The hemagglutinins agglutinate erythrocytes in vitro, reduce amylase activity, and also cause a severe growth depression. Most other legume beans contain similar antinutritional factors. Ruminants are not affected by these proteins because rumen fermentation inactivates the factors. Heat treatment also inactivates these factors. Therefore in commercial production of soybean and other oilseed meals, heat treatment is routine. However, if overheating occurs, amino acids and protein moeities form complexes with carbohydrates by means of the Maillard reaction. The complexes are resistant to digestion, dramatically reducing protein availability. Heat damage may also need to be assessed in nonoilseed by-products such as distiller's dried grains, brewer's dried grains, corn gluten feed, and whey. Haylage (high dry matter silage) sometimes gets so hot during the ensiling fermentation that forage protein becomes heat damaged and digestibility is reduced.

Biological value (BV) is a historic method of measuring protein quality, particularly for monogastric animals. BV is defined as the percentage of digested and absorbed protein that is retained in the body for productive functions. It may also be defined as the percentage of absorbed true protein that is utilized for maintenance and production.

$$BV = \frac{\text{Dietary N} - (\text{Urinary N} + \text{Fecal N})}{\text{Dietary N} - \text{Fecal N}} \times 100$$

It is an indirect means of evaluating both digestibility and how well a specific protein delivers the correct ratio of essential amino acids. Whole egg protein has a BV of about 100; meat proteins, 75 to 79; cereal proteins, 50 to 65; and gelatin, 15. If the correct balance of amino acids is not available to animal tissues for maintenance and protein synthesis, the use of absorbed protein is limited by the most limiting amino acid. For example, if absorbed lysine is in short supply, then tissue synthesis of a protein that contains lysine will be limited. Other essential amino acids present in excess of the amount that can be used for protein synthesis will then be used as an energy source and the waste nitrogen excreted in the urine. This situation is wasteful and leads to poor performance and low feed efficiency.

In practice an animal normally consumes proteins from several sources in any given meal. Thus, feeds containing poor-quality protein are apt to be balanced by other sources of higher-quality protein. In ruminants, degradation of dietary protein by rumen microorganisms and synthesis of microbial protein also compensate for poor-quality dietary protein. In monogastrics supplementation of diets with highly digestible protein supplements and proper quantities of the limiting amino acid(s) optimizes the efficiency of protein utilization. Since the biological value of different protein sources easily can be increased if appropriate amino acids are supplemented, the primary criterion in diet formulation for monogastric animals is to supply the required amino acids in adequate amounts without undue concern regarding the biological value of the component feeds.

Commonly supplemented amino acids include lysine and methionine. Lysine is the amino acid required in the greatest amounts in growing pigs and horses and is most likely to be present in limited quantities in conventional diets. Lysine is also important in the diet of breeding stock. Methionine is frequently deficient in conventional diets and is essential for growth and cartilagenous tissue integrity.

DIGESTION AND ABSORPTION OF PROTEINS IN RUMINANTS

In ruminants, digestion of protein and absorption of the digested amino acids are not as closely interrelated as they are in monogastric animals. The combined effects of degradation of major quantities of true protein by enzymes of rumen microbial origin and the utilization of nonprotein nitrogen by rumen microorganism in synthesis of microbial protein are the major differences in protein metabolism between ruminants and nonruminants. Changes due to degradation of dietary protein and gain of protein synthesized from nonprotein nitrogen by rumen microorganisms must be considered. This has a tremendous effect on net quantities of amino acids delivered to the small intestine. Furthermore, the amount of a limiting amino acid in the diet does not necessarily reflect the amount of that amino acid that is absorbed. However, postruminal protein metabolism is essentially the same as for nonruminants. An abbreviated schematic of crude protein metabolism (true protein plus nonprotein nitrogen) is shown in Fig. 2-1.

Another significant difference in protein metabolism between ruminants and nonruminants is the ability of ruminants to conserve nitrogen during protein-deficient periods. Ruminants can recycle urea as a result of secretion into the saliva instead of urine, thereby allowing delivery back to the rumen mixed with masticated feedstuffs. In the rumen, urease releases ammonia from urea, and this can be used by microorganisms in protein synthesis. This extremely important

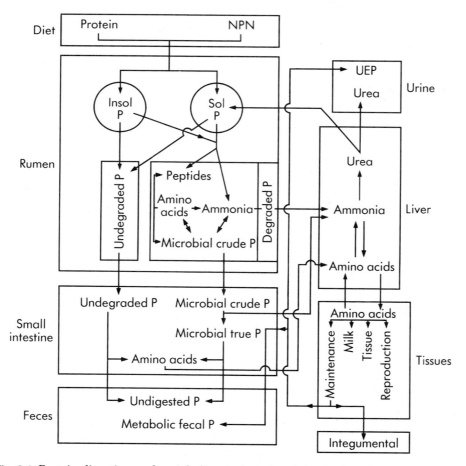

Fig. 2-1 Protein digestion and metabolism in lactating dairy cattle. *NPN* = nonprotein nitrogen; *P* = protein; *Insol* = insoluble; *Sol* = soluble; *UEP* = urinary endogenous protein.
From Chalupa W: J Dairy Sci 67:1134, 1984.

mechanism facilitates survival when dietary protein is deficient.

Basing dietary protein recommendations for ruminants on crude protein or digestible protein estimates of individual ingredients often inadequately describes the total diet. Predictive models that improve diet formulation are being developed. Important features of these models (systems) include consideration of the degradation of dietary protein by rumen microbes and synthesis of protein by rumen microbes. The amount of bacterial protein synthesized is usually considered to be directly related to dietary energy level. In addition, the models estimate metabolic fecal nitrogen losses and digestibility of dietary and microbial proteins.

Iowa State University researchers[2,3] were among the first to propose a quantitative model that estimates protein requirements for ruminants in terms of "metabolizable protein," which is the amount of protein (amino acids) absorbed from the gastrointestinal tract. Metabolizable protein represents protein leaving the rumen (dietary or microbial) that is digested and absorbed. It is the same as true digestible protein for nonruminants. Metabolizable protein can be quite different from crude or true protein in the diet owing to the effect of rumen microorganisms.

Protease activity is quite high in rumen microorganisms. Many proteins are rapidly degraded to amino acids, which are then deaminated. Often about 60% of true dietary protein is degraded in the rumen.[4] However, the proportion of dietary protein that escapes rumen degradation varies tremendously with different dietary ingredients. A major influence is the solubility of protein in the rumen. Heat-denatured proteins, which are relatively insoluble, resist hydrolysis in the mildly acidic rumen but are still readily hydrolyzed by strong acid and pepsin secreted in the abomasum. These proteins have a much higher proportion of their original amino acids absorbed intact from the small intestine than more soluble proteins.

A model that attempts to quantify these factors is illustrated in Table 2-1. The digestibility of undegraded protein postruminally is estimated to be 90%, and the digestibility of ruminally synthesized microbial protein, 80%. Microbial cell growth and protein synthesis are functions of available energy, assuming adequate ammonia is available. It is estimated that 104.4 g of microbial protein is formed per kg of total digestible nutrients in the diet (a dietary energy estimate). The expense of forming fecal metabolic protein is assumed to be directly related to the quantity of dietary dry matter ingested (15 g of protein per kg of dry matter intake).

We can illustrate use of the predictive model with soybean meal: 54% crude protein (100% dry matter basis), 30% of protein estimated to escape degradation in the rumen, and 84% of dry matter as total digestible nutrients. Calculations for this example are included in Table 2-1, A. The model estimates that 162 of 540 g/kg of soybean meal protein are delivered to the small intestine, where 90% is digested and absorbed. Thus, only 145.8 of the original 540 g of soybean meal protein are estimated to be absorbed from the small intestine. An additional quantity of microbial protein is estimated to be generated from energy metabolites derived from soybean meal (87.7 g), from which 15 g are subtracted to offset fecal metabolic protein loss. The microbial protein remaining is estimated to be 80% digested and absorbed (net, 72.7 g × 0.80 = 58.2 g). Therefore, of 540 g of dietary protein/kg of soybean meal dry matter, only 204 net g (20.4% of original dry matter) of protein is estimated to be absorbed, and the amino acid composition of 58.2 g is now that of bacterial protein instead of soybean meal.

Table 2-2 shows estimates of undegraded fractions and metabolizable protein for a number of common dietary ingredients. It is apparent that with high-protein feeds degradation is far greater than net gain from microbial synthesis (also reflected in the urea fermentation potential). A great amount of research is now underway to try to increase the percent of protein that is undegraded without reducing its digestibility in the small intestine.

Solubility of protein within the rumen is a major factor affecting protein degradability within the rumen. It can be readily determined, and it may be a useful index of metabolizable protein.[15] Casein (milk protein) is readily soluble and highly degradable within the rumen. Thus,

Table 2-1 *Use of metabolizable protein computational formulas*

A. Metabolizable protein content of feedstuffs

$$MP(g/kg\ DM) = \{(UP \times 0.90) + [(MCP - 15.0) \times 0.80]\}$$
$$MP\% = [MP(g/kg)/(1000\ g/kg)] \times 100$$

where

- MP = Burroughs' metabolizable protein calculated as g/kg dry matter (DM) or as %.
- UP = Ruminally undegraded protein (g/kg ingredient DM) entering abomasum, calculated as expected fraction of crude protein (CP) of a feed ingredient to bypass rumen degradation × its CP content (g/kg DM)
- 0.90 = Fraction of undegraded protein truly digested postruminally
- MCP = Net amount of ruminally synthesized microbial protein (grams) entering abomasum per kg DM. Estimated synthesis of 0.1044 g MCP/g total digestible nutrients (TDN). It is assumed there is adequate rumen ammonia to permit optimum microbial growth.
- 15.0 = Feed expense (g protein/kg DM) of forming fecal metabolic protein
- 0.80 = Fraction of MCP truly digested postruminally

EXAMPLE: Soybean meal (54% CP [30% UP] and 84% TDN)

$$MP(g/kg) = (540 \times 0.3 \times 0.9) + [(.1044 \times 840 - 15.0) \times 0.80]$$
$$= 204\ g/kg\ DM$$
$$= 20.4\%\ MP$$

B. Calculation of urea fermentation potential:

$$UFP(g/kg\ DM) = (0.1044 \times TDNg - DP) \div 2.81$$

where

- UFP = Urea fermentation potential expressed as grams of urea that can be used for microbial protein synthesis per kg of feed ingredient
- 0.1044 = Same as described above for MCP
- DP = Rumen degradable protein (g/kg DM) whose ammonia contributes to rumen pool
- 2.81 = Factor to convert grams of CP to grams of urea equivalent. Urea at 45% nitrogen has 2.81 times as much nitrogen as an average protein with 16% nitrogen.

Modified from Burroughs W, Nelson DK, and Mertens DR: J Anim Sci 41:933, 1975.

---

although it is a high-quality protein if infused directly into the lower digestive tract, it is similar in value to urea when fed to mature ruminants because of its degradation to amino acids and then to ammonia and volatile fatty acids. Conversely, less soluble proteins, such as zein (corn protein), may largely escape degradation within the rumen and be digested in the lower digestive system. The value of zein is then determined by its amino acid composition and the extent of digestion in the lower digestive tract. If proteins escape degradation in the rumen but are of lower quality than microbial protein, little gain is achieved. Ideally, undegraded protein should be fed to supply the limiting amino acids needed to support optimal production by the animal.

The advantage of the metabolizable model is illustrated in Fig. 2-2, where metabolizable protein intake has a stronger relationship to milk yield than crude protein intake.

Protein Nutrition and Nonprotein Nitrogen

Table 2-2 *Calculated metabolizable and estimated undegraded protein values*

Ingredient	Fractional UP[a]	Dry matter %			UFP[e]
		CP[b]	TDN[c]	Est. MP[d]	
Ground corn	.45	10.0	88	10.2	13.1
Wheat	.30	12.5	92	9.9	3.1
Wheat midds	.25	18.7	84	10.0	−18.8
Barley	.20	13.9	87	8.6	−7.3
Citrus pulp	.25	6.9	80	7.0	11.4
Cane molasses	.10	4.3	80	5.9	16.0
Soybean hulls	.25	12.0	81	8.3	−1.9
Alfalfa hay #1	.25	20.0	64	8.6	−29.7
Alfalfa hay #2	.30	15.0	55	7.4	−16.9
Bermuda grass hay	.25	6.0	48	4.2	1.8
Corn silage	.30	8.0	72	7.0	6.9
Sorghum silage	.40	6.2	59	6.0	8.7
Soybean meal (49%)	.30	54.0	84	20.4	−103.3
Soybean meal (44%)	.30	49.6	84	19.2	−92.7
Peanut meal	.25	54.2	80	17.7	−115.3
Cottonseed meal	.30	44.0	80	17.4	−80.2
Whole cottonseed	.30	24.9	104	15.3	−19.0
Whole soybean	.20	41.7	99	14.6	−82.2
Dist. dried grains	.55	29.8	92	21.2	−13.6
DDGS[f]	.55	29.5	92	21.1	−13.1
Sunflower meal	.25	44.1	72	14.7	−91.3
Urea	0	281.0	0	0	0

[a] *UP*, Estimated fraction of crude protein that is undegraded in the rumen.
[b] *CP*, Crude protein.
[c] *TDN*, Total digestible nutrients.
[d] *Est. MP*, Estimated metabolizable protein.
[e] *UFP*, Urea fermentation potential in g/kg DM of feed ingredient.
[f] *DDGS*, Distiller's dried grains plus solubles.

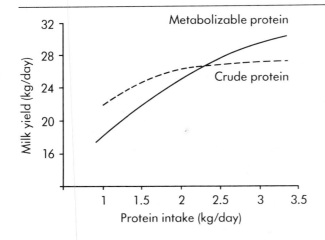

Fig. 2-2 Milk yield response to Burroughs' metabolizable protein intake compared to response to crude protein intake. Increasing the level of metabolizable protein intake increases milk production, whereas there is little yield increase with crude protein intakes above 2 kg/day. Both response curves are adjusted to equal feed intake.
Adapted from Briceno and others: J Dairy Sci 71:1647, 1988.

Importance of amino acids for ruminants

It is difficult to study the quantities of amino acids absorbed postruminally in animals consuming high-performance rations. Also, it is difficult to protect individually supplemented amino acids from degradation. Thus, ruminant nutrition has been slow to identify which amino acids are limiting to production. For lactating dairy cows consuming a diet consisting primarily of corn, corn silage, and limited amounts of alfalfa and grass hay, lysine and methionine appear to be the first limiting amino acids.[11] Probably the main reason that the amino acid content of the diet is not always of critical importance is that the quality of microbial protein for growth and milk production is relatively good.

Table 2-3 estimates how soybean meal and distiller's dried grains compare in their ability to supply lysine and methionine to the intestine. Soybean meal is an excellent source of lysine, containing about 68 g of lysine per kg of protein. If 30% of the protein in soybean meal escapes degradation in the rumen, about 20 g of lysine per kg of soy protein fed would escape to the intestine. Distiller's grains, which are low in lysine but contain a relatively undegradable protein, would be expected to supply 17 g to the animal. Despite the much higher content of "bypass" protein in distiller's grains, it actually supplies less lysine to the intestine than does soybean meal. If lysine is a limiting amino acid, feeding distiller's grains in place of soybean meal may be counterproductive.

In lactation trials, distiller's dried grains have only occasionally shown benefits over soybean meal at equal dietary protein percentages, even though it is relatively certain that with the distiller's grains a higher percent of the total dietary protein escaped ruminal degradation. This may be due to lysine being a limiting amino acid. However, another possible explanation is that heat damage to protein in distiller's grains may have occurred, making the protein unavailable even in the small intestine. Heat damage of distiller's dried grains during the drying process is not uncommon.

With methionine the two supplements rank differently. The protein in distiller's grains has a relatively high methionine content, and this, combined with the low degradability, makes distiller's grains an excellent source of methionine. Corn gluten meal is also resistant to degradation in the rumen and contains relatively large amounts of methionine. For growth in beef cattle, distiller's dried grains and corn gluten meal have supported good growth at lower dietary protein percentages than soybean meal. This implies either that the extra bypass protein was more beneficial than in some lactation trials or that methionine is a more limiting amino acid for growth and maintenance than lysine.

Effect of protein on dry matter intake and digestibility

Often with high-producing or rapidly gaining cattle, the benefits of supplemental protein have been accounted for through increased voluntary dry matter intake and increased dry matter digestibility.[8,12] With animals at a maintenance level of energy intake, increasing crude protein concentrations beyond 10% to 12% of dry matter has little or no effect on dry matter digestibility. However, with high-producing dairy cows consuming more than three times maintenance energy needs, increasing protein from 11% to 13% of dry matter to approximately 16% of dry matter increased digestibility of dry matter by 4 to 8 percentage points.[8] Two important rumen factors that may be influenced by dietary protein and that may affect intake and digestibility are the retention time of feeds in the rumen and the

Table 2-3 *Expected supply of lysine and methionine to the intestine when soybean meal and distiller's dried grains are fed to cattle*

Amino acid	Protein source	Grams AA* per kg protein	Fraction escaping	Grams AA escaping
Lysine	Soybean meal	68.2	.30	20.5
	Distiller's grains	30.3	.55	16.7
Methionine	Soybean meal	11.4	.30	3.4
	Distiller's grains	16.2	.55	8.9

*Amino acid.

rate of microbial growth. If these factors truly are affected by protein intake, then the model of Burroughs[2] will have to be modified to become more dynamic.

Feeding standards for ruminants

Ruminant nutritionists have not yet shown benefits from routine supplementation with specific amino acids. Even if they wished to recommend supplementation, reliable methods are not yet available to deliver a specific quantity of a limiting amino acid to the small intestine in a form ready for absorption. Minimum recommended allowances for crude protein are available for all classes of ruminants with some guidance given on whether supplemental nonprotein nitrogen will be useful. However, consideration of estimated metabolizable protein, or at least of the need for undegraded protein, is now common. For example, the 1988 revision of *Nutrient Requirements of Dairy Cattle*, published by the National Research Council of the U.S. National Academy of Sciences, includes feeding standards for undegraded protein.[7]

NONPROTEIN NITROGEN (NPN) UTILIZATION BY RUMINANTS

NPN is mainly of value in ruminants because proteins and amino acids derived from microbial NPN fermentation are made available in large quantities to the host animal. Protein may be synthesized from NPN by microorganisms in the large intestine (e.g., in the horse), but protein synthesized here is not a major source of absorbed protein except on very low protein diets. It is possible that NPN could be of some benefit for nonruminants if essential amino acids were available at adequate levels but nonessential amino acids were not. Then, with available ammonia and proper carbon structures for transamination, nonessential amino acids might be synthesized in tissues in the absence of microbial intervention.

In ruminants, if inadequate ammonia is available to maintain rumen nitrogen levels of 80 to 100 mg of N/L, not only microbial protein synthesis but also digestion of organic matter will be reduced. Reduction in cellulose digestion is most critical. Associated with reduced digestibility is a reduction in dry matter intake. The combined effects greatly inhibit animal production.

Urea is by far the major feed additive used to supply NPN to rumen microorganisms for use in protein synthesis. The major justification for incorporation of NPN into rations is one of economics. Generally, a reduction in feed costs and increased profits have resulted from inclusion of the maximal amount of NPN in ruminant rations that does not depress growth. In many rations, 3.2 kg (7 lb) of shelled corn and 0.45 kg (1 lb) of urea is equal in net energy and crude protein to 3.6 kg (8 lb) of soybean meal. The amount of savings depends directly on the price of soybean meal, urea, and corn. It is important to note, too, that mineral and vitamin concentrations differ between the two types of rations. The cost of the mineral and vitamin supplements necessary should also be factored into the comparison.

Complete substitution of NPN for protein in diets for ruminants has shown that animals survive, are fertile, and reproduce normally but do not perform as well as when some natural protein is provided; growth, milk production, and feed efficiency are reduced. Attempts at improving performance with amino acid supplementation either singly or in combination have been largely unsuccessful. Ammonia is used by many species of microorganisms; however, some require preformed peptides or amino acids to survive. If these are not provided in the diet, then some microorganisms may disappear from the rumen and change the balance of species. The total quantity of protein synthesized may thus be altered. Ruminants fed diets free of protein have depressed concentrations of branched-chain volatile fatty acids. They also have depressed free blood plasma concentration of essential amino acids and increased concentrations of glycine and serine. However, cattle raised from 84 days of age to 4 years of age on purified diets with all nitrogen in the forms of urea and ammonium salts have normal ruminal vitamin B synthesis. Also, moderate milk production (up to 4500 kg [10,000 lb] per cow yearly) has been obtained.[13] Benefit from urea supplementation in diets containing more than 12% crude protein is unlikely—ammonia accumulates and there is no further increase in microbial protein synthesis.[10]

Figure 2-3 shows an experiment in dairy cows in which dietary nitrogen content was varied

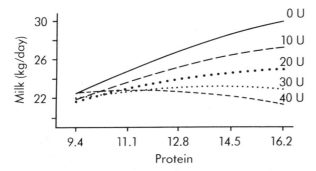

Fig. 2-3 Milk response to dietary protein and urea. *0U* = 0 urea; *10U, 20U, 30U, 40U* = 10%, 20%, 30%, or 40% of diet N from urea.
From Polan and others: J Dairy Sci 59:1910, 1976.

using soybean meal, urea, or combinations of the two. Urea had a detrimental effect on milk production. This exemplifies metabolizable protein principles discussed earlier. The metabolizable protein model presented in Table 2-1, *B* can be used to predict when there is a positive urea fermentation potential, indicating a potential benefit to urea supplementation.[2,3] If the estimated synthesis of microbial protein exceeds the estimated amount of degraded protein, then bacteria in the rumen will not have enough ammonia available from protein breakdown to grow optimally. Thus, there is a benefit to adding some urea to the ration (positive urea fermentation potential). The amount depends on the ingredients. For corn (88% total digestible nutrients, 10% crude protein [55% rumen degradable and 45% undegradable protein]; DM = dry matter):

Microbial protein synthesis = 0.1044 × 880
 = 91.9 g/kg DM

Rumen protein degradation = 100 × .55
 = 55.0 g degraded/kg DM

Net g protein from rumen = 91.9 − 55.0 = 36.9 g/kg DM

Net g N from rumen = 36.9 × .16
 = 5.90 g N/kg DM

UFP = g urea at 45% N to supply 5.9 g N
 = 5.90 ÷ .45
 = 13.1 g/kg DM

So corn, being high in energy and low in rumen degraded protein, is a good ingredient to which to add urea. A mixture of 1 kg of ground corn and 13.1 g of urea would give 13.7% crude protein (on a dry matter basis). With most mixed diets for ruminants containing more than 11% or 12% crude protein (100% dry basis) there rarely is a positive benefit from adding urea to diets. For example, the estimated UFP for 15% crude protein alfalfa hay (55% TDN, 30% undegradable protein) is −16.9 g urea/kg DM. Combining 1 kg corn with 1 kg of hay gives −3.8 g UFP for the mixture (13.1 g UFP + [−16.9 g UFP] = −3.8 g). Thus, there is more than enough rumen ammonia available from the crude protein in alfalfa hay to maximize microbial protein synthesis in the rumen. Urea's main potential dietary benefit is with fattening and some growing diets for ruminants in which relatively low percentages of dietary protein are required.

Palatability

Concentrates containing more than 2% urea result in depressed feed intakes for cattle, even in animals adapted to urea diets. Lower urea levels (1% to 1.5%) can cause feed intake problems when it is part of high-moisture (greater than 14% water) concentrates. This is probably because of partial hydrolysis of urea to ammonia, which can be a serious problem for urea feeds held in storage during warm, humid weather. Adding molasses to diets with high levels of urea improves intake in some cases but not others. Addition of urea to corn silage does not cause palatability problems, probably because ammonia released from urea during ensiling is converted to ammonium salts and also because silage acids mask the taste of urea. Urea should not give palatability problems if it is incorporated at 1% of total ration dry matter, 1.5% of the concentrate portion, or less.

Reproductive performance

For many years, some veterinarians and cattlemen have suggested that feeding rations containing NPN lowers conception rates of cattle. However, in a Michigan survey of more than 600 dairy herds involving 85,281 individual cow lactations and 3157 year-herd observations, no reduction in calving interval was found in herds that were fed urea.[9] Urea was fed about half the time, with an average intake per cow of 80 g (0.18 lb) of urea per day. This and many other studies show that urea per se is not detrimental to reproduction. Feeding high levels of protein has nearly the same effect on blood urea nitrogen levels as feeding urea. Several recent experiments suggest that feeding protein above 18% of total diet dry matter interferes with reproductive performance. Excess ammonia from either protein or urea at critical tissue sites may interfere with conception. If this is the case, then total dietary nitrogen level (crude protein), and particularly total rumen-degradable crude protein, will better predict the possibility of tissue ammonia excess. A review of a number of experiments by Ferguson and Chalupa[4] concluded that a high level of dietary protein, particularly highly degradable protein, contributes to poor reproductive performance.

The detrimental effects of protein on reproduction have been seen with diets in which protein was approximately 20% of total diet dry matter. The studies are not conclusive. Two of the studies with the greatest cow numbers did not find that high protein was detrimental to reproduction; milk yield was increased with high protein in both studies (e.g., Howard and others[5]).

One theory explaining the possible negative impact of high protein is that high intakes may cause excessive ammonia or urea nitrogen concentrations in the reproductive tract leading to a less than optimal environment for conception. Alternatively, high systemic or local urea nitrogen concentrations may reduce luteinizing hormone binding to ovarian receptors on luteal cells and lead to a decrease in progesterone concentration and fertility.[14]

It is good judgment to try to manage feed offerings in a way to keep blood levels of urea nitrogen from peaking higher than necessary without compromising the feeding program for optimum milk production. Thus, with good management of the sequence and frequency of the desired ingredients, feeding for optimal performance still seems compatible with the feeding program necessary for good reproduction.

Relationship to possible nitrate toxicity

Neither logic nor experiment supports the notion that urea and nitrate are additive in their toxic effects. Although both are simple nitrogenous compounds, they are quite different in their physiological actions. Nitrate accumulates in plants grown on soils high in nitrogen, mainly in hot weather, following frosts, or during drought. The nitrogen content of nitrates is relatively small, and only a very small proportion is reduced to ammonia. Conversely, oxidation of ammonia to nitrate, which would aggravate a nitrate toxicity, is highly unlikely because the rumen is a reducing environment.

Ammonia toxicity

It is well known that dietary urea, if consumed in large quantities in a short time, can be toxic. In a Kansas experiment, cattle were dosed with 0.5 g urea per 1 kg body weight, which was added through a rumen fistula.[1] In approximately half of the cases, signs of toxicity were seen. Most of these occurred within 60 minutes of dosing (average time 52.8 minutes). In toxic cases, blood ammonia (29 mg/L blood) and rumen pH (7.41) were elevated 60 minutes after dosing.

Signs of toxicity include uneasiness, dullness, muscle and skin tremors, excessive salivation, frequent urination and defecation, rapid respiration, incoordination, stiffness of the front legs, prostration, tetany, and death. Muscle tremors followed by muscle spasms were the most frequently observed symptoms (all toxic cases). Bloat was never observed. As tetany became severe, most animals lost muscular coordination and became prostrate. When tetany became severe, any loud noise, such as talking, slamming doors, or dropping metal objects, caused a violent involuntary contraction of the muscles (convulsions), characterized by stiffening of the fore and hind legs.

A common recommendation for treatment has been to inject acetic acid intraruminally. Cattle given approximately 1 mol acetic acid (5% vol/vol) per mol nitrogen administered to induce toxicity usually showed improvement if it was administered prior to tetany. Emptying the rumen contents of animals in severe tetany brought about survival in all animals treated this way. Following rumenotomy, the cattle were left in a quiet place, and signs of tetany usually ceased after about 30 minutes. Respiration and heart rate also returned to normal. After about

60 minutes, the animals stood and appeared normal. For severe field cases, emergency rumenotomy should be performed and rumen contents evacuated rapidly. In their studies, Bartley and associates[1] added about 4 L (1 gal) warm water to the rumen when the animal stood and appeared normal. Recovery was dramatic and rumen fermentation was good 48 hours after the experiment.

Ammonia toxicity is not a major concern in relation to utilization of NPN if standard recommendations are followed. Toxicity is associated only with consumption of high levels of urea (more than 44 g [0.11 lb] per 100 kg [220 lb] body weight) in a short period by unadapted cattle. Mistakes such as animals breaking into urea supplies or areas where feed was inadvertently spilled and miscalculation of the intended feeding levels are sometimes found as causes of toxicity.

Bovine bonkers syndrome

Ammoniation of forages is sometimes recommended to help increase dry matter digestibility and reduce molding. However, ammoniation of high-quality forages such as alfalfa, sorghum-sudan hybrids, rye, wheat, oats, brome, or fescue hays may produce 4-methyl imidazole and other toxic compounds. Problems are particularly common when ammonia is injected into bales unevenly (most likely if only a single probe is used) and when the weight of the bales is overestimated and excessive ammonia applied. Clinical signs from toxicosis from feeding hays containing 4-methyl imidazole (bovine bonkers syndrome) include hyperexcitability, wild running, circling, convulsive seizures, and death. The toxins are excreted in the milk and can cause clinical signs in calves while the adults are unaffected. The same syndrome has also been seen when feeding large amounts of ammoniated molasses.

CONCLUSION

Proteins are polymers of amino acids. Essential amino acids cannot be synthesized by mammalian cells. Monogastrics require a dietary supply of essential amino acids. Ruminal microorganisms interconvert nonessential and essential amino acids. In consequence, the amino acid balance in the diet of ruminants is relatively unimportant. Ruminal microflora can also synthesize amino acids from inorganic nitrogen sources; ammonia and urea are the two most commonly used sources. Ruminants can maintain themselves and reproduce successfully on diets in which all nitrogen is supplied as urea. However, such diets are not recommended. In high-producing dairy cows, delivery of certain amino acids to the small intestine may be inadequate, and provision of protein that is resistant to ruminal, but not to intestinal, degradation may be beneficial.

REFERENCES

1. Bartley EE and others: J Anim Sci 43:835, 1976.
2. Burroughs W, Nelson DK, and Mertens DR: J Anim Sci 41:933, 1975.
3. Burroughs W, Trenkle AH, and Vetter RL: Leaflet R767, Ames, 1972, Iowa State University Cooperative Extension Service and Agricultural and Home Economics Experimental Station.
4. Ferguson JD and Chalupa W: J Dairy Sci 72:746, 1989.
5. Howard HJ and others: J Dairy Sci 70:1563, 1987.
6. Maynard LA and others: Animal Nutrition, ed 7, New York, 1979, McGraw-Hill.
7. National Research Council: Nutrient requirements of domestic animals, no 3, Nutrient requirements of dairy cattle, ed 6, Washington, DC, 1988, National Academy of Science.
8. Oldham JD: J Dairy Sci 67:1090, 1984.
9. Ryder WL, Hillman D, and Huber JT: J Dairy Sci 55:1290, 1972.
10. Satter LD and Roffler RE: J Dairy Sci 58:1219, 1975.
11. Schwab CG, Satter LD, and Clay AB: J Dairy Sci 59:1254, 1976.
12. Van Horn HH and others: J Dairy Sci 62:1086, 1979.
13. Virtanen AI: Science 153:1603, 1966.
14. Visek WJ: J Dairy Sci 67:481, 1984.
15. Wohlt JE, Sniffen CJ, and Hoover WH: J Dairy Sci 56:1052, 1973.

3

The Major Minerals (Macrominerals)

JONATHAN M. NAYLOR

Calcium, phosphorus, magnesium, sodium, potassium, and chloride are sometimes referred to as major minerals or *macrominerals* because they comprise a large proportion of body minerals and are required in much larger amounts in the diet than trace minerals (typically 0.1% to 1.0% of the dry matter for macrominerals). Because of the large contribution of bone to body content, calcium and phosphorus are the most abundant minerals in the body (Table 3-1).

Potassium, sodium, and chlorine are highly electrogenic and usually exist in ionized form in the body. Sodium and chloride are the major ions of extracellular fluid. Potassium is mainly found within cells associated with organic anions and chloride. Potassium is the third most abundant mineral in the body, but sodium and chloride are close runners-up (Table 3-1).

Table 3-1 *Mineral composition of the mature bovine carcass*

Mineral	Quantity*
Calcium	16.0
Phosphorus	6.5
Potassium	2.0
Sodium	1.5
Chlorine	1.2
Magnesium	0.4

*All values are g/kg and are on a dry matter basis. Modified from data from technical review by the Agricultural Research Council Working Party, Slough, 1980, Commonwealth Agricultural Bureau.

The macrominerals are absorbed primarily in the small intestine.[48,69,120] An exception is magnesium absorption in ruminants, which occurs mainly in the fore stomachs.[23,120] Sodium and chloride are absorbed from the large intestine in all species,[78,120] and this is also a major site of phosphorus absorption in horses.[107] The absorption (true digestibility, Chapter 7) of sodium, potassium, and chloride is high, partly because these highly electrogenic elements form soluble salts. Absorption of magnesium, phosphorus, and, to a lesser extent, calcium is less efficient.

POTASSIUM
Role

The main functions of potassium are to help maintain the intracellular osmotic pressure and to contribute to the membrane potential by virtue of the Nernst effect. Potassium is also important in a number of electrically activated phenomena,[59] particularly contraction of cardiac and skeletal muscle.

Pathophysiology

Herbivores eating forage typically ingest large quantities of potassium. Potassium salts are water soluble, and much of the ingested potassium is absorbed. Surplus potassium is excreted via the kidneys. Excretion is stimulated in part by aldosterone; increased plasma potassium concentrations are one stimulus to the secretion of this hormone.

Most of the potassium in the body is concentrated inside cells as a result of the activity of the Na-K ATPase, which pumps potassium into cells and excretes sodium into the extracellular

Table 3-2 *Major mineral compositions of some common feedstuffs in relation to nutritional requirements*

	Potassium	Sodium	Chloride	Calcium	Phosphorus	Magnesium
Feed compositions						
Alfalfa hay	2.0	0.15	0.40	1.3	0.2	0.3
Timothy hay	1.6	0.18	0.60	0.4	0.2	0.16
Corn silage	1.1	0.01		0.3	0.2	0.28
Corn grain	0.4	0.01	0.05	0.03	0.3	0.13
Barley grain	0.5	0.03	0.20	0.05	0.4	0.15
Wheat bran	1.4	0.07	0.07	0.1	1.3	0.58
Soybean meal	2.2	0.3	0.03	0.4	0.8	0.30
Requirements						
Growing swine (10-20 kg)	0.3	0.1	0.1	0.7	0.6	0.04
Nonlactating dairy cow	0.8	0.1	0.1	0.4	0.3	0.16
Lactating dairy cow	0.8	0.2	0.4	0.6	0.4	0.20

All values are % of dry matter in the feed; most values are approximate.
Based on data from Nutrient Requirements of Dairy Cattle,[82] Nutrient Requirements of Beef Cattle,[81] and Fettman and others.[33]

fluid. Potassium also enters cells in exchange for hydrogen ions and together with glucose.

Requirements

Potassium requirements for herbivores are on the order of 0.5% to 0.8% of the dietary dry matter.[69,81-83,100,113] The higher requirements are for heavily lactating animals, as milk contains a high concentration of potassium. Deficiency is not usually seen on high-roughage diets because of the relatively high potassium content of grasses and legumes (Table 3-2). Cane and beet molasses also are particularly high in potassium, averaging 5%. Grains contain only 0.3% to 0.5% potassium, and deficiency is possible when diets are relatively high in concentrate and low in forage and plant protein. Pigs have low potassium requirements, about 0.25% of the diet dry matter, and deficiency is unlikely on standard diets.[25,56,74,79] In sick animals potassium deficiency is more common because of the combination of low intake due to inappetence and increased losses due to diarrhea, gastric reflux, and so on. Potassium supplementation has also been reported to reduce morbidity and mortality following transportation in feedlot calves, perhaps because food intake is minimal during

---

Mineral Deficiencies That Can Lead to Pica

Sodium
Potassium
Chloride
Phosphorus

---

transport. There also may be increased losses due to stress-induced diarrhea.

Deficiency

The first sign of potassium deficiency is decreased food intake.[29] Reduced food intake forces a larger percentage of ingested energy to be used for maintenance. As a result, growth, weight gain, milk production, and feed efficiency are reduced.[96,113] These often are the only signs seen with a mild deficiency. Depraved appetite, or pica, may develop in potassium and many other macromineral deficiencies (see box). Moderately potassium-deficient cattle lick at wood and concrete and increase their licking of the hair coat of other animals.[96] With chronic deficiency the hair coat becomes rough, and emaciation develops.[56,96] Weakness and cardiac abnormalities (e.g., prolonged QRS intervals, T wave changes) are seen with severe deficiency.[56] Potassium deficiency can lead to excessive urination (polyuria) and increased water intake (polydipsia).[75] Serum potassium may be reduced in potassium deficiency, but most of the potassium is intracellular and serum levels are unreliable measures of potassium adequacy. At postmortem examination, potassium-deficient animals show either no abnormal findings or histological signs of swelling of the cells of the renal cortex and degeneration of muscle fibers.[56,113]

Toxicity

The maximal tolerable levels of dietary potassium are unknown, but growth is depressed in some species at levels of 3% or greater. High levels of potassium interfere with the absorption of magnesium and predispose animals to hypomagnesemia in magnesium-deficient areas.[14,23,34,41,120]

Supplementation

It is rarely necessary to take specific precautions to prevent dietary potassium deficiency. In high-risk situations, potassium deficiency can be prevented by increasing the amount of roughage in the diet, by mixing potassium chloride with the grain portion of the diet, or by spraying solutions of potassium chloride onto hay.

SODIUM
Role

Sodium is important in controlling the osmotic pressure of extracellular fluid, in the generation of the membrane potential, and in the conduction of electrical impulses in nerve and muscle.[9,59]

Pathophysiology

Sodium is the only mineral for which a clearly defined appetite exists.[5,82] Sodium intake is controlled by the hypothalamus. The sodium ion is one of the few substances that can be directly detected by taste buds in the tongue.[4,6,7] In times of sodium deficiency, animals actively seek out sources of salt and increase their salt consumption. At the same time, increased release of aldosterone from the adrenal gland reduces sodium excretion in urine and saliva and stimulates intestinal sodium absorption. Together these factors maintain serum sodium concentrations by increasing intake and reducing excretion. In times of sodium excess, blood volume increases; this stimulates release of atrial natriuretic factor, which increases sodium diuresis.[20,72,115,121] These homeostatic mechanisms are remarkably efficient, as evidenced by the ability of animals to compensate for increased losses, such as through sweat or a salivary fistula. Some sick animals given access to a choice of salty and fresh water can even increase their sodium intake to match losses through diarrhea or renal disease. Another example of the priority of the sodium drive is seen in the behavior of wild herbivores, which may travel long distances to ingest sodium at special licks where salt is concentrated in the soil. Normal homeostatic control of sodium intake is usually successful providing salt is available in the environment and the animal has not become compromised to the point where control mechanisms fail to function.

Requirements

Sodium is of great nutritional importance because grains and some forages are deficient in this element[82] and there are no readily mobilizable stores within the body (Table 3-2). Sodium requirements are estimated to be about 0.1% of the dietary dry matter for most nonlactating farm animals, and these are close to the levels found in grasses and legumes. However, corn silage and grains have lower levels.[74,76,79,81,82] Heavily lactating or working animals require up to about 0.2% sodium, and these levels are too high to be reliably met by unsupplemented diets.

Deficiency

Sodium deficiency is more common in hot environments and in heavily lactating or strenuously exercising animals (sodium is lost in milk and sweat).[117] The initial sign of deficiency is usually increased salt appetite; animals seek out salt sources and may appear to have pica.[82,117] For example, one foal that was sodium deficient as a result of renal disease repeatedly licked the wood of the stall wall—in the area immediately underneath an empty salt block holder. In cattle a craving for salt usually appears after 1 to 3 weeks on a deficient diet. As sodium depletion progresses, feed intake, growth, body condition, and milk production are reduced. These signs may not appear until months have passed on a deficient diet.[82] Polyuria and polydipsia have also been reported in sodium-deficient cattle; it is thought that lack of sodium interferes with the countercurrent exchange mechanism in the renal medulla.[117] The hypertonic environment surrounding the renal tubules cannot be maintained in the absence of sodium, and there is an insufficient osmotic gradient for water reabsorption from the tubules.

Serum sodium concentrations are of limited use in the diagnosis of sodium deficiency.[76,77,117] In animals with excessive sodium losses (due to diarrhea, for example), water and salt are usually lost together, so the animal becomes dehydrated rather than hyponatremic. In animals with deficient salt intake, the body defends serum osmotic pressure by reducing the release of antidiuretic hormone. This in turn increases renal water loss, and dehydration rather than hyponatremia develops. Salivary sodium and sodium-potassium ratios are much more useful in diagnosis in ruminants. A sample of saliva can be obtained by using tweezers to place a sponge in the mouth between the tongue and the molars in the area of the parotid salivary duct. Deficient ruminants have Na-K ratios that are always less than 10:1 and sometimes as low as 1:1.[76,77,117] Potassium contamination from feed may affect this ratio. Therefore, measurement of salivary sodium concentration may be more reliable than calculation of the ratio to potassium. Normal salivary sodium concentrations in cattle and sheep are in the range of 120 to 140 mmol/L (120 to 140 mEq/L); deficient animals have values less than 100 mmol/L (100 mEq/L).[77,117] In horses normal salivary sodium concentrations may be too variable to be diagnostically useful.[1,110]

Toxicity

Sodium toxicity is influenced by the availability of water low in dissolved salts. The provision of unlimited, good-quality water protects against sodium toxicity.[30,92,108] In some areas where the soil is very saline, natural feedstuff can be high in sodium. When water intake is restricted, diets with more than 2% salt can be toxic. Another problem with high-salt diets is that the excess is excreted in urine; urine-contaminated manure is therefore salty, which may occasionally be a disadvantage to crops upon which it is spread as fertilizer.[69]

There are at least two syndromes produced by salt (sodium ion) intoxication—diarrhea and nervous signs. Occasionally, the ingestion of very large quantities of salt by animals with access to unlimited good-quality water produces diarrhea, polyuria, and polydipsia.[108]

Drinking excessively salty water can reduce productivity and produce diarrhea. It is difficult to be sure at what levels salinity (sodium chloride in water) causes problems; an approximate guide is probably a maximum of 1.3% salt (13,000 ppm), which is isotonic with plasma. A hot environment and high levels of milk production make animals more sensitive to salinity. In high-producing dairy cows 0.25% salt (2500 ppm) decreases milk production.[55] Many natural water sources that are contaminated with salt contain other cations (magnesium, for example) and anions (sulfates and bicarbonate are examples) that can cause a variety of problems at much lower concentrations than sodium chloride.

The more severe form of salt intoxication is seen when animals are exposed to salty foods and have only limited or intermittent access to water. This is referred to as salt intoxication, water deprivation, or chronic salt poisoning because it usually takes at least 2 days before signs of intoxication develop. Chronic salt poisoning is most commonly seen in swine, in part because of the feeding of salty swill and the reliance on nipple drinkers, which are difficult to check for proper water flow.[30,67,88,108] Hot weather also predisposes to toxicity, because water requirements are increased. It is thought that salt loading as a result of ingestion of salty feeds initially produces hypernatremia and a hypertonic plasma. The cells adapt to the high plasma osmotic pressure. When the animal drinks, plasma osmotic pressure falls rapidly, water enters the cells, and cellular edema develops.[108] This is particularly deleterious in the brain, where there is only limited room for expansion; intracranial pressure increases and limits blood flow. The neurons of the cerebral cortex, with their high metabolic rate, are particularly susceptible to damage.[108] Signs of salt poisoning occur during the periods of water restriction or following the reintroduction of water on an ad lib basis. Signs of salt poisoning are similar to those produced by diseases such as polioencephalomalacia (thiamine deficiency), lead poisoning, and cerebral anoxia, which also adversely affect the functioning of the cerebral cortical neurons. In pigs an initial period of pruritus and constipation is occasionally observed.[108] In all species the main signs are caused by dysfunction of the central nervous system. Blindness is an early sign; as the syndrome progresses, head pressing, convulsions, coma, recumbency, and death are seen. Convulsions can start as jerky movements of the head and neck, which progress to paddling and running movements.[30,67,92] Opisthotonos and placing the head against the chest may also be observed. In some pigs the convulsions repeat at a characteristic 7-minute interval.[108] The most useful diagnostic tool is measurement of plasma sodium concentrations. However, there is some overlap in sodium concentrations between normal, sick, and salt-intoxicated pigs. In general, unintoxicated swine have serum sodium concentrations from 135 to 160 mmol/L (135 to 160 mEq/L), whereas sodium-intoxicated swine have concentrations from 150 to 210 mmol/L (210 mEq/L).[30,88,92,108] If poisoning is due to sodium chloride ingestion, then chloride ion concentrations are also elevated. Occasionally, however, other sodium salts are responsible for toxicosis.[30,108] Necropsy findings are mainly cerebral cortical edema and neuronal degeneration.[30,88,104,108] Pigs are unusual in that in the later stages of salt intoxication an eosinophilic meningoencephalitis is seen.[30,88,104,108]

Supplementation

Sodium and chloride requirements are usually met by offering free-choice salt or by incorporating 0.5% of salt into the whole ration (calculated on a dry matter basis). This overall level can also be met by incorporating higher levels into an individual fraction of the feed if this is more convenient; for example, salt is often included as 1% of the concentrate portion of the diet of dairy cows.[69,71,83,101] Sometimes salt is fed at higher levels. Its incorporation at the level of 1% to 2% of the total diet increases water consumption and can decrease the incidence of urolithiasis (stones [calculi] in the urinary tract) in cattle and sheep.[69,95] Even higher levels are sometimes used to reduce consumption of the salted feed under range conditions.

Salt can be offered free choice as a solid block or as loose salt; typical intakes are shown in Table 3-3. Many feed manufacturers make supplements containing a mixture of minerals that are designed to be fed free choice. Because sodium is the only mineral that is ingested in a reliable fashion, it is important that these mixes contain salt (usually at least 25% of the mix) and that the other minerals are present in the correct ratio to sodium so that ingestion to satisfy salt appetite also meets the needs for the other minerals. Provision of different types of salt-mineral mixes on a free-choice basis should be avoided because there may be a preferential or uneven ingestion of one mix. Non–sodium-containing mineral mixes should be mixed with concentrate or other palatable feedstuff that is being ingested in known amount by the animals to ensure an even intake.

Table 3-3 *Free-choice salt consumption of animals*

Animal type	Intake of loose salt (g/day)
Mature ewes	10-30
Feedlot lambs	5-10
Feedlot beef cattle	10-30
Lactating dairy cows	30-80*

*Consumption is halved if block salt is provided.

CHLORINE
Role

Chlorine mainly exists in the body as the chloride ion. It is the principal anion of the extracellular fluid and an important intracellular ion.[81] Chloride ions are important in the maintenance of osmotic pressure and in electrical phenomena.

Pathophysiology
There is no direct hormonal control of serum or total body chloride. Levels are affected indirectly as a result of the control of osmotic pressure and pH.

Requirements
Feed levels are satisfactory to meet requirements under many circumstances. Heavily lactating animals, however, have large needs that may not be met by unsupplemented diets (Table 3-2). With most unsupplemented diets, sodium deficiency is observed before chloride deficiency sets in. Supplementation of diets with salt at generally recommended levels provides more than enough chloride to meet requirements.

Deficiency
In large animals chloride deficiency has been observed only in recent years. Heavily lactating dairy cows appear to be particularly susceptible because of the losses of chloride in milk (three times that of sodium) and because sodium bicarbonate is often added to buffer the feed. These buffered diets may contain sufficient sodium, but if they are not also balanced for chloride a chloride deficiency may result.[33] Signs of deficiency include loss of weight, decreased food intake, and reduced milk production.[8,33] Polyuria and polydipsia can be early signs, which may be related to the need for chloride in the countercurrent exchange mechanism of the renal tubules.[8,17] In advanced deficiency, signs of dehydration and pica are seen. Affected animals may drink urine, lick at pipes, and chew wood.[33] Salivary chloride levels are a poor indication of chloride deficiency.[33] Serum chloride falls progressively with chloride deficiency. This is one of the earliest changes and is diagnostically useful.[33] At the same time a metabolic alkalosis develops as the homeostatic mechanisms defend extracellular fluid volume. Sodium reabsorption in the kidneys is coupled to bicarbonate reabsorption because of the lack of chloride.[33] In consequence, bicarbonate accumulates in plasma and alkalosis develops.

Toxicity
Free chlorine can be poisonous but only chloride ion and organically bound chloride are found in feedstuffs. Toxicity of chloride ion is low. Cations present in foods containing large amounts of chloride ion usually cause problems before signs of chloride toxicity are seen.

Supplementation
Chloride deficiency can be prevented by ensuring that high-producing dairy cows fed sodium buffers also have access to sufficient chloride ion. There is no evidence of a specific drive for chloride intake, so this is best done by incorporating sodium chloride into the feed at levels of 0.5% into the total ration or 1% into the grain portion of the ration.

CALCIUM
Role
Calcium accounts for about 1.5% of body weight[112] and, precipitated with phosphate and water, it forms hydroxyapatite, the major mineral of bone and teeth. The vast majority of calcium in the body (99%) occurs in the skeleton and teeth.[107,116] However, the small proportion of calcium that is found outside the skeleton plays a major role in metabolic functions. Calcium is an important enzyme cofactor and also binds to the cellular regulatory protein calmodulin.[22] Many metabolic processes are controlled by changes in the level of intracellular calcium. In skeletal muscle cells, for example, calcium is stored in compartments of the sarcoplasmic reticulum.[54,90] When an action potential passes over the cell, calcium channels open, and calcium is released into the sarcoplasm of the cell, activating muscle contraction.[40] Other examples of calcium-dependent cellular processes include cardiac and smooth muscle contraction,[9,12,58,89] the secretion of a variety of storage vesicles, and the control of enzymatic processes. Calcium ions plays a role in release of neurotransmitter from some nerves, including the release of acetylcholine at neuromuscular junctions, and in electrical phenomena in some brain and nerve cells.[59]

Pathophysiology
Calcium-dependent processes, particularly those in muscle and nerve tissue, are sensitive

to changes in the amount of ionized calcium present in the extracellular fluid. In general, mammals have total plasma calcium concentrations of around 2.5 mmol/L (10 mg/dl), although slightly higher values are found in horses. About 50% of calcium is bound, and 50% is in the ionized form. Much of the bound calcium is complexed to albumin. A normal ionized calcium level can be maintained at a lower total plasma calcium concentration if the animal is hypoalbuminemic. Acidosis also favors calcium ionization and reduces total plasma calcium concentrations.[116] One example of the effects that changes in ionized calcium can have is illustrated by the effects of rapid intravenous infusion of oxytetracycline into cattle. The oxytetracycline binds calcium and reduces ionized calcium without affecting total calcium.[18] If infusion of oxytetracycline is too rapid, the decrease in ionized calcium results in weakness, and the patient collapses.

Calcium homeostasis is maintained by three hormones: parathormone, calcitonin, and the vitamin D–derived steroid hormone, 1,25 dihydroxy vitamin D_3 (Fig. 3-1). For ease, and because three hydroxylations are involved in its formation, 1,25 dihydroxy vitamin D_3 is sometimes referred to by an alternative name, *calcitriol*.

The main action of calcitriol is to promote small intestinal absorption of calcium.[62,116] Like all steroid hormones, it interacts with the nucleus of the cell. It stimulates intestinal epithelial cells to produce calcium-binding proteins.[116] In addition, calcitriol activates bone osteoclasts to cause bone lysis and release of calcium; this action is most marked in the presence of parathormone.[28,62,116]

Parathormone is released from the parathyroid gland in response to low plasma-ionized calcium concentrations.[62] The calcium in bone is separated from the extracellular fluid of the rest of the body by a lining of osteoblasts and osteocytes. Although much of the calcium in bone is firmly bound as hydroxyapatite, there is also a small but significant pool of more readily available calcium.[116] Parathormone acts on the osteoblasts and osteocytes to stimulate pumping of calcium into the extracellular fluid, which restores plasma calcium levels. In the longer

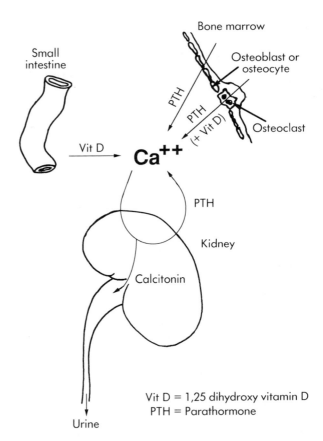

Fig. 3-1 Hormones which affect serum calcium levels. Light arrows represent flows of calcium in and out of the serum pool. The names of the hormones that promote calcium flow are written alongside the appropriate arrows.

term, parathormone also increases the number of osteoclasts and stimulates them to break down bone, releasing considerably more calcium.[38,44,62,116] Parathormone also stimulates renal calcium conservation.[116]

Calcitonin protects against hypercalcemia. It is released from the thyroid gland, promotes renal calcium excretion, and inhibits bone mobilization of calcium.[62]

Requirements

Calcium requirements are particularly high in growing and heavily lactating animals. Dietary concentrations of calcium are highest in legumes such as alfalfa and clover. These feedstuffs contain enough calcium to meet requirements for maintenance, growth, and lactation. Next in order of content are other roughages, such as grasses and hays, which are usually adequate for maintenance calcium needs but not growth or lactation. Oilseed meals may need calcium supplementation to meet maintenance requirements and would have to be supplemented for growing and lactating animals. Oilseed meals contain more phosphorus than calcium and should not be fed as a sole source of calcium. The grains are very low in calcium and by themselves are an inadequate source of calcium, even for maintenance.

The only animal products that are rich in calcium are milk (dairy products are the main source of calcium for people and are important in the prevention of osteoporosis in people[57,103]) and bonemeal or bone flour.

A number of substances such as phosphorus, aluminum, zinc, and organic acids can interfere with calcium absorption. The best-documented example is oxalate (e.g., in spinach and certain tropical grasses), which binds calcium and makes it unavailable.

Deficiency

Hypocalcemia means a reduction in blood calcium and does not necessarily imply the presence of signs of disease. We will refer to that group of diseases brought on by hypocalcemia as *hypocalcemic disease* (incomplete paralysis, or paresis, is the major symptom in ruminants; signs of nerve irritation are more common in horses). In dairy cows hypocalcemic disease is associated with parturition, and it is often referred to as *parturient paresis* or *milk fever*. Hypocalcemic disease in ewes, goats, and beef cattle is seen during late pregnancy and early lactation; sometimes these diseases are referred to as hypocalcemia or *nonparturient paresis*. The hypocalcemic diseases are caused by a deficiency of circulating ionized calcium. They are acute in onset, and total body calcium need not be abnormal. Hypocalcemic diseases are often the result of inadequate homeostasis, not dietary calcium deficiency.

Major strains are placed on calcium metabolism at the initiation of lactation and in late pregnancy, particularly in milking cows and ewes with rapidly growing fetuses, respectively. In these situations there are rapid increases in calcium requirements for either milk production or fetal bone formation. These requirements are too large and rapid in onset to be readily met by dietary input, particularly if feed intake is reduced, as often happens at parturition or during stress.

In dairy cows, parturition is accompanied by a sudden increase in calcium sequestration for the production of milk. The calcium requirement rises to two to five times that of the dry period.[97] The daily calcium outflow into milk is many times the amount of calcium available in the extracellular fluid (Fig. 3-2). This places a major strain on the homeostatic mechanisms. Plasma calcium levels fall, and there are compensatory increases in plasma parathormone and calcitriol, but it takes time for hormones to exert their full effects and particularly for parathormone to

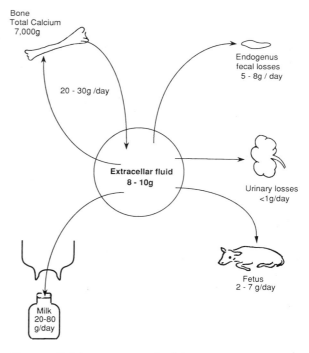

Fig. 3-2 Calcium turnover in dairy cows. Modified from data in Georgievskii VI, Annenkov BN, and Samokhin VT[37] and Reinhardt TA, Horst RL, and Goff JP.[99]

stimulate osteoclastic activity.[52,62] The efficiency of dietary absorption takes about a day to improve, whereas an increase in bone calcium mobilization is not apparent for a week.[97] Plasma calcitonin levels decrease at parturition, reducing renal calcium loss, but this makes only a small difference.[52,62] Nearly all cows develop some degree of hypocalcemia at parturition; in some, hypocalcemia is severe, and clinical signs of milk fever develop.[63] Milk fever is particularly likely to occur in those cows with high calcium drains because of high milk production (e.g., high-yielding herds, cows rather than heifers) and in older cows,[32] which have less efficient gut calcium absorption[43] and slower bone mobilization (Fig. 3-3). Milk fever is an important disease of dairy cows that affects around 8% of cows.[99] It is difficult for modern high-producing dairy cows to meet calcium intakes by dietary calcium absorption in early lactation, so there is a cycle of bone calcium mobilization in early lactation and replenishment in late lactation.

In ruminants, hypocalcemia produces an initial period of tetany and stiffness followed by signs of weakness. Signs of tetany may be the result of the effects of calcium on nerve conduction. The major signs of hypocalcemia are weakness, ruminal atony, and diminished strength of the cardiac contractions. The strength of ruminal contractions decreases progressively with the decline in plasma calcium[53] and may be directly related to an inadequate supply of calcium for smooth muscle contraction. The abomasum is more resistant to hypocalcemia, but motility suddenly ceases at about the same time clinical signs become apparent.[66] Weakness is related to two phenomena: hypocalcemia decreases the strength of cardiac contractions, and cardiac output and blood pressure fall markedly.[27] In addition, there is interference with the release of acetylcholine from the neuromuscular junctions.[15] The hypophosphatemia that usually accompanies hypocalcemia enhances flaccidity and occasionally cows need to be treated with both calcium and phosphorus to effect a recovery.[84] Untreated cows and sheep often die, and the disease should be treated as an emergency situation.

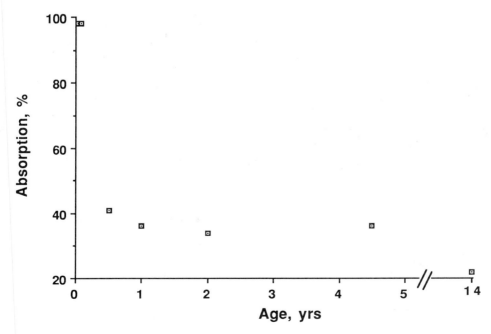

Fig. 3-3 Effect of age on the ability to absorb dietary calcium.
Modified from Hansard SL, Comar CL, and Plumlee MP: J Anim Sci 13:25, 1954.

Signs of hypocalcemic paresis resemble those seen in animals with endotoxemic shock, and it can be difficult to differentiate hypocalcemia from toxic mastitis or metritis as a cause of recumbency in cows. This is further complicated by the fact that endotoxemia can induce a moderate hypocalcemia in some animals.[64,80] The decrease in total calcium in endotoxemia is partly due to decreased albumin binding of calcium, but ionized calcium also decreases as the result of decreased bone mobilization.[122] Intravenous calcium should always be given carefully, with close monitoring of cardiac function, particularly to endotoxemic animals, which are very susceptible to calcium toxicity and may be best treated with subcutaneous calcium infusions.

Cows with milk fever are more susceptible to dystocia (difficulties at parturition) and retained fetal placental membranes,[26] probably because hypocalcemia interferes with uterine and abdominal contractions. Secondary muscle degeneration in milk fever is the result of pressure damage during recumbency, compounded by the effects of low blood flow and the inability of the cow to shift its weight to spread the burden on its muscles.

A dietary factor that tends to increase the incidence of milk fever in dairy cows is inadequate magnesium. Among other things, hypomagnesemia interferes with parathormone release. Low dietary phosphorus contents in the dry cow ration are thought to be protective—possibly because it increases intestinal responsiveness to calcitriol[52] and because high serum phosphorus levels inhibit production of 1,25 dihydroxy vitamin D_3.[99] However, not all studies show an effect of phosphorus intake on the incidence of milk fever.[3] The incidence of milk fever tends to increase with the level of calcium in the dry cow diet. Incidence is lowest when the diets contain less calcium than is required to meet the dry cow's daily needs (approximately 35 g [1 oz] calcium per cow per day). These diets cause an initial drop in serum calcium that stimulates parathormone release and bone calcium mobilization.[39] The calcium-mobilizing system is thus geared up for the demands of lactation. Acidifying the dry cow diet with ammonium salts also reduces the incidence of milk fever.[85]

Chronic dietary deficiency of calcium, phosphorus, or vitamin D results in total body calcium depletion and similar clinical signs (Table 3-4). Although serum calcium concentrations may be reduced, homeostatic mechanisms mobilize bone calcium and often keep plasma calcium above the level at which signs of hypocalcemia are clinically evident. Chronic deficiency results in inadequate skeletal calcium deposition. This is called *rickets* in growing animals and *osteomalacia* in adults. Affected animals are often on high-grain diets with little hay and no mineral supplement. Total confinement housing and the winter feeding period are also often associated with rickets and osteomalacia because lack of sunlight reduces metabolism of vitamin D to the active hormone and because animals are more dependent on conserved feeds, which can be low in calcium (e.g., grain) or phosphorus under these conditions. The earliest signs of calcium deficiency may be reduced feed intake and poor growth.[94] It takes several months of an inadequate diet before signs of bone disease develop, but growing animals and lactating animals may be more rapidly affected because of their high calcium requirements. As skeletal hydroxyapatite deficiency becomes severe in rickets and os-

Table 3-4 *Common signs of calcium, phosphorus, and vitamin D deficiency*

Nutrient	Reduced feed intake	Reduced weight gain	Rickets or osteomalacia	Decreased serum		
				Ca	P	Mg
Calcium	+++	+++	+++	+		
Phosphorus	+++	+++	+++		+++	
Vitamin D	+++	+++	+++	+	++	+

+++ = very common, ++ = common, + = sometimes seen.

teomalacia, lameness becomes a prominent sign, and there may be a hunched stance. In the final stages, pathological fractures of bones occur, primarily in the long bones and vertebrae, resulting in severe lameness and reluctance to move or recumbency (Fig. 3-4). On close examination, the transverse processes of the lumbar vertebrae may be soft and flexible and the teeth loose in their sockets[13,114] (Fig. 3-5).

Rickets is a disease of young animals. Because of the presence of active growth plates, there are additional signs to those seen in osteomalacia. Osteoid laid down at active growth plates fails to calcify and accumulates. In long bones this produces swollen diaphyses that give the impression of swollen joints. Enlarged growth plates at the costochondral junctions of the ribs can sometimes be palpated as a "rachitic rosary" (Fig. 3-6). The enlargement and decreased bone mineralization can sometimes be seen radio-

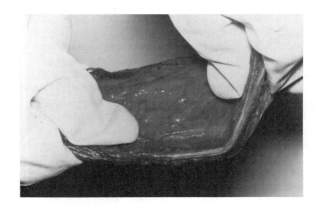

Fig. 3-5 Ribs from a calf with rickets; they are easily bent due to the poor mineralization.

Fig. 3-4 Calves with rickets secondary to feeding on a diet deficient in calcium. Some calves would stand and walk with short, painful steps. Others spent all their time down. At necropsy the calves had long-bone fractures and collapse of some lumbar vertebrae.

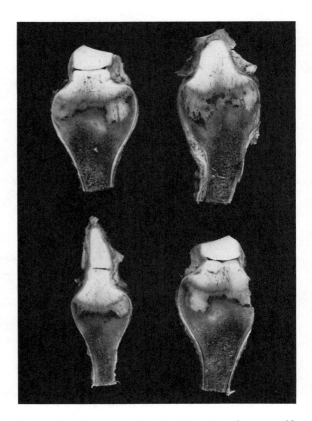

Fig. 3-6 Swollen costochondral junction from a calf with rickets. There is overgrowth of nonmineralized cartilage and hemorrhage.

graphically. In addition, there is generalized skeletal weakening, and the long bones may bend and become curved under the strain of weight bearing.

Osteomalacia (osteoporosis) is a disease of adult animals; dairy cows are particularly susceptible to osteomalacia, which is also called *milk lameness*. Interestingly, the development of spontaneous vertebral fractures in racehorses in Hong Kong may have been reduced by calcium supplementation; this may be an example of a mild form of osteomalacia that produced problems only in horses undergoing hard exercise.[68]

Nutritional secondary hyperparathyroidism (parathyroid hyperplasia secondary to an inadequate diet) is another term for rickets or osteomalacia; it develops when the diet is low in calcium or vitamin D, which promotes absorption of calcium. The parathyroid glands hypertrophy as parathormone output is increased to maintain plasma calcium concentrations. Fibrous osteodystrophy is a variant of rickets and osteomalacia in which fibrous tissue accumulates in the incompletely calcified bone. It is most prominent in horses fed diets low in calcium and high in phosphorus. There is marked accumulation of fibroosseous tissue, which enlarges and softens the bones, particularly those of the mandible and maxilla.[105] This gives rise to bulging facial bones, and the disease is sometimes referred to as *bighead*. *Bran disease* and *miller's disease* are other terms for nutritional secondary hyperparathyroidism. They are named for the high incidence of this disease in horses fed diets composed mainly of unsupplemented bran and grain.

Osteopetrosis is excessive calcification of bone. It is rarely seen although it has been described in bulls receiving high-calcium diets for prolonged periods of time.

Diagnosis of hypocalcemic paresis is best carried out by analysis of serum or plasma calcium in samples obtained before treatment is initiated if possible. Diagnosis of hypocalcemia as a cause of death is more difficult because there are no pathognomonic signs and blood analysis is not valid in dead animals. Concentrations in the vitreous humor of the eye may prove useful in dead cattle. Normal vitreous humor calcium concentrations are lower than serum.[73] Vitreous humor concentrations in hypocalcemia have not been established.

Diagnosis of nutritional lack of calcium (chronic calcium deficiency, osteomalacia) is best carried out by feed and bone analysis. Necropsy findings can be used to diagnose nutritional bone disease, the bones are weak and samples of bone can be collected and ashed to determine their calcium and phosphorus concentration (Table 3-5). The vertebrae are the most labile bones in most species, but the mandible is useful in horses. Although there is some overlap between normal and diseased animals, the amounts of calcium and phosphorus are usually diminished by a tenth to a third in animals with osteomalacia.[37] The ratio of N to Ca in bone ash is the most sensitive index of calcification. The ratio of calcium to phosphorus in bone is con-

Table 3-5 *Minimum mineral content of normal, dry, defatted bone*

| Species | Content (g/kg) | | | | Ca:P | Ca:N | Bone |
	Ash	Ca	P	Mg			
Bovine	490	195	90	4.4	2.1		Rib
Pig	480	170	90	4.1	2.2	7.3	Humerus
Equine	550	210	100		2.2		Rib

Fresh bone contains 33% to 35% ash. The content of calcium (37% to 38%) and phosphorus (18% to 19%) in bone ash is fairly constant; the mineral contents decrease in equal proportion in calcium, phosphorus, and vitamin D deficiencies. Overt rickets is usually associated with bone ash contents between 280 and 450 g/kg and calcium contents between 100 and 160 g/kg. Reductions in bone tensile strength may occur before changes in mineral composition are detectable. Bones that bear only a minor proportion of body weight are more sensitive to dietary changes than weight-bearing long bones. Differences in mineral content exist between bones.

stant, even in disease, and low bone calcium and phosphorus may be caused by calcium, phosphorus, or vitamin D deficiency. Although serum levels of calcium may be reduced on deficient diets, homeostatic mechanisms minimize this change. Serum alkaline phosphatase levels are usually increased because of excessive bone osteoclast activity; in low-calcium-intake rickets the levels are elevated twofold to fourfold.[37] However, many other factors also elevate serum alkaline phosphatase concentrations. Excess dietary calcium is excreted in the urine, particularly in horses, and the fractional excretion of calcium and phosphorus in urine has been used as an index of dietary adequacy in horses.[21,68] Urinalysis has the advantage of giving results somewhat more rapidly than dietary analysis, but the results are affected by hormonal affects and feeding frequency. Results should always be confirmed by dietary analysis.

Toxicity

Calcium has a low toxicity. High concentrations in hay interfere with the absorption of other minerals.

Supplementation

High-grain diets should always be supplemented with calcium (Table 3-2). Calcium supplements may also be needed for animals fed diets based on grass or corn silage, particularly if the animal is growing or lactating. Rations should be routinely checked for calcium content as part of diet formulation. The most commonly used supplements are ground limestone (calcium carbonate), dicalcium phosphate, and bonemeal (Table 3-6). If the animals are fed daily, these supplements can be mixed with the ration when the feed is prepared. Animals at pasture are usually supplemented by mixing calcium supplements with salt. Mineral mixes that only contain calcium and phosphorus are ingested erratically unless they are mixed into the ration.

PHOSPHORUS
Role

Besides its very important skeletal role, phosphorus is used throughout the body as a source of rapidly available chemical energy in the form of high-energy phosphate links in creatine phosphates, ATP, and other highly phosphorylated nucleotides. In addition, phosphates are present in nucleic acids and in a wide variety of phosphoproteins and phospholipids.

Pathophysiology

Phosphorus absorption occurs mainly in the small intestine in most species. Stimulation of calcium absorption by vitamin D also leads to a secondary increase in phosphorus absorption.[28,49] The extracellular pool of phosphorus is small in relation to requirements for lactation (Fig. 3-7). Plasma phosphorus concentrations are not as closely regulated as calcium levels and tend to vary with dietary intake. Although phosphorus is released when parathormone stimulates bone resorption, it is rapidly excreted in the urine because parathormone also acts on the kidneys to increase phosphorus loss.[62] In contrast, calcitriol mobilizes bone calcium and phosphorus without affecting renal excretion, and plasma phosphorus levels often rise in vitamin D intoxication.

Requirements

Phosphate levels tend to be high in oilseeds and oilseed meals. Grains also contain reasonable amounts of phosphorus, particularly in the outer bran layers. Phosphorus contents of grasses and legumes are below NRC require-

Table 3-6 *Approximate calcium content and availability in some common supplements*

Item	Calcium* (%)	Digestibility† (%)
Limestone‡	34	45
Dolomitic limestone§	22	
Oyster shell	38	
Dicalcium phosphate, defluorinated	22	64
Bone flour	32	68

*Values are on a % dry matter basis.
†Values are true digestibilities in young steers.
‡Also contains about 2% magnesium.
§Also contains about 10% magnesium.
Modified from Nutrient Requirements of Beef Cattle[81] and Peeler HT.[93]

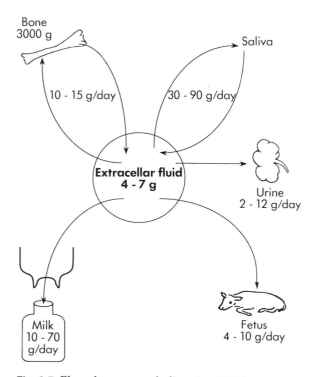

Fig. 3-7 Phosphorus metabolism in a 500-kg dairy cow.
From data in Georgievskii VI, Annenkov BN, and Samokhin VT[37] and Reinhardt TA, Horst RL, and Goff JP.[99]

ments for many species, and phosphorus deficiency is a common problem in forage-fed animals, particularly in the Midwest and prairie regions. Recently it has been suggested that the National Research Council (NRC) overstates the phosphorus requirements for beef cows, and the daily requirement for a 450-kg cow was estimated at 12 g/day instead of the NRC recommendation of 17.5 g/day.[19] If this is the case, the need for phosphorus supplementation may not be as critical in beef cows as it once was suspected to be.

A particular concern to nutritionists has been the digestibility of phytate phosphorus. In phytates, phosphorus is bound to the sugar alcohol inositol; they are found in plant material. Phytates are destroyed by rumen microbes, and the phosphorus is reasonably readily available in ruminants. In monogastrics, phytate phosphorus is poorly available, although horses are less affected due to their colonic fermentative capacity and hind gut absorption of phosphorus. In horses the digestibility of phytate phosphorus is half that of sodium phosphate.[49,50]

Deficiency

Deficiency signs are mainly related to phosphorus's role in the skeleton and in energy metabolism (Table 3-4). It usually takes several months on a deficient ration before clinical signs are evident, although rapidly growing animals are more quickly affected. The skeletal effects of phosphorus deficiency are rickets[114] and osteomalacia;[19] the signs of these diseases have already been described. Not all cases of phosphorus deficiency manifest as skeletal disease. Reduced food intake, milk production, growth rate, and weight loss can be prominent early signs.[19] Phosphorus-deficient animals may exhibit pica and eat bones in an attempt to increase their phosphorus intake. This can result in the development of botulism. In beef and dairy cows, phosphorus deficiency has been linked to problems with fertility as a result of a failure to cycle.[16] However, in experimental studies in beef cows, fertility problems developed very late, when many other signs of disease were present.[19] This leads one to suspect that infertility in cows with low serum phosphates is due either to other coexisting nutrient deficiencies (e.g., protein) or to multiple nutrient deficiencies.[16]

Postparturient hemoglobinuria (red cell hemolysis with loss of hemoglobin in the urine) has been linked to phosphorus deficiency in North America. This may represent a more acute form of phosphorus deficiency in which serum phosphate is rapidly depleted with little drain of bone phosphorus. In this condition, herd serum phosphorus concentrations fall and some multiparous, heavily producing cows develop a regenerative, hemolytic anemia and hemoglobinuria.[31,65] Pale mucosa, inappetence, and weakness are seen. This typically occurs 10 to 40 days postpartum and is thought to be due to a fall in high-energy phosphate levels in erythrocytes that results in decreased glycolysis,[86] inability to maintain normal cellular ionic gradients, and increased osmotic fragility.

Acute phosphorus deficiency can be gauged by analysis of serum phosphate levels; this is best carried out on a group basis. Normal serum values are 2 mmol/L (5 mg/dl), and signs of acute deficiency are not usually seen until concentrations have fallen to half this level. Serum phosphate levels are influenced by a variety of nondietary factors and tend to be higher in young, growing animals and in animals that are off feed. Dietary analysis offers the best means for gauging intake, but there can be problems in the range situation in obtaining a representative sample. Analysis of bone samples at necropsy can assist in assessing the long-term availability of phosphorus in the diet (Table 3-5).

Toxicity

High-phosphorus diets may suppress the absorption of calcium and can predispose to urolithiasis.[51]

Supplementation

Diets are routinely balanced for phosphorus as part of ration formulation. Phosphorus can be incorporated into salt-mineral mixes for range and pastured animals. There are several mineral sources of phosphorus that can be used as supplements; they have different phosphorus contents and availabilities (Table 3-7). Dicalcium phosphate is commonly used. Rock phosphates and products made from untreated rock phosphates are contaminated with fluorine in their natural state and have to be defluorinated before they can be safely fed to animals. Sometimes the high-heat treatments used in the defluorination of rock phosphate result in the formation of pyrophosphates and metaphosphates. This reduces the phosphorus availability in the supplement; the phosphorus in pyrophosphates has only 75% of the digestibility of that in orthophosphates; metaphosphate digestibility is from 0% to 50%.[93] The phosphate levels in rock preparations are variable. Superphosphates are particularly rich in phosphate; however, they are not usually defluorinated and are mainly used as fertilizer. Colloidal phosphate or soft rock phosphates are mixtures of fine rock phosphate and clay and contain less phosphate; untreated soft phosphates are high in fluorine and low in digestibility. Curaçao phosphate is a form of naturally occurring rock phosphate that is low in fluorine; it has a similar composition to defluorinated rock phosphate, and the phosphate is highly available.

Table 3-7 *Approximate phosphorus contents of some common supplements*

Item	Phosphorus (%)	Digestibility* (%)
Dicalcium phosphate, defluorinated	19	58
Rock phosphate, defluorinated	18	
Monosodium phosphate	23	58
Bone flour	14	58

All values are on a % dry matter basis.
*True digestibility in horses.
Modified from Nutrient Requirements of Beef Cattle[81] and Hintz HF and Schryver HF.[49]

CALCIUM-PHOSPHORUS RATIO

The major nutritional requirement is for the diet to contain adequate amounts of calcium and phosphorus. However, very high levels of one mineral relative to the other may interfere with absorption and metabolism. In general, the calcium-phosphorus ratio should be 1:1 to 2:1, although animals often perform satisfactorily over a much wider range of ratios.[50]

Mineral supplements vary in their calcium and phosphorus contents, depending on whether they are designed to be fed to horses or beef cattle grazing grass at pasture or eating high-grain diets in confinement operations. Legumes are high in calcium, a straight phosphorus supplement or a 0.5:1 (calcium to phosphorus, weight basis) mineral supplement will meet requirements. Grasses are low in both calcium and phosphorus and are usually supplemented with 1:1 mineral supplements. Grains are much lower in calcium and are supplemented with 2:1 mineral mixes.

MAGNESIUM
Role

Magnesium is a minor component of bone (about one part magnesium for every 50 parts calcium), but about 70% of total body magnesium is found in the bone. It is the second most abundant intracellular cation, occurring in concentrations of about 15 mmol/L (35 mg/dl) in most tissues.[119] Plasma magnesium concentrations are about 1 mmol/L (2.5 mg/dl) so hemolysis of red cells secondary to prolonged storage or faulty collection techniques increases serum magnesium (and potassium).

Magnesium is involved in membrane electrical phenomena and muscle contraction, where it tends to antagonize the effects of calcium. It is also involved in energy metabolism as a cofactor for several enzymes.[62]

Magnesium plays a role in calcium metabolism.[60] It facilitates absorption of dietary calcium[50] and release of parathormone from the parathyroid glands.

Pathophysiology

Plasma levels of magnesium are not under close hormonal control, although they are influenced by parathormone (reduces renal excretion and mobilizes bone magnesium), vitamin D (mobilizes bone magnesium and aids intestinal absorption), and adrenal gland hormones (inhibits absorption).[106,119] They tend to be maintained around 1 mmol/L (2.5 mg/dl) in part because this is just above the renal threshold for magnesium excretion.[62] Thus, on a sufficient diet, excess magnesium is constantly being excreted by the kidneys. When a deficient diet is fed, renal magnesium excretion ceases and plasma magnesium levels fall. In general, symptoms may be seen when plasma magnesium concentrations fall below 0.5 mmol/L (1.2 mg/dl), and severe signs are often associated with concentrations below 0.25 mmol/L (0.6 mg/dl).[10] There may be a considerable lag between the development of hypomagnesemia and the appearance of clinical signs. Cerebrospinal fluid and cellular magnesium concentrations are much better protected than plasma magnesium. Ventricular cerebrospinal fluid magnesium concentrations are more closely associated with symptoms.[2] Cattle become symptomatic at cerebrospinal fluid magnesium concentrations around 0.5 mmol/L (1.2 mg/dl).[2]

Clinical hypomagnesemia is often complicated by hypocalcemia. This is because hypomagnesemic animals are often lactating and thus lose calcium in milk. Decreased intestinal calcium absorption[50] and reduced mobilization of calcium from bone also occur in hypomagnesemia.[60] Compromised parathormone release is partly responsible for these changes.[98,102] Onset of clinical symptoms of hypomagnesemia, and particularly signs of tetany, often coincides with the superimposition of acute hypocalcemia on a preexisting hypomagnesemia.[47]

Hypomagnesemia can be seen in any ruminant, but in adults it is most common in older, lactating, or heavily pregnant animals. Lactation, advanced pregnancy, and rapid growth drain the serum magnesium pool as a result of magnesium requirements for milk and bone synthesis. In high-producing dairy cows, the daily loss of magnesium in milk is equal to that found in the extracellular fluid. Reduced feed intake predisposes to hypomagnesemia, and fasting produces hypomagnesemia in lactating cattle. Reduction of feed intake is probably the reason for the association of bad weather and stress with episodes of hypomagnesemia. There is proportionally more magnesium in milk than in bone, so mobilization of bone to meet lactational calcium requirements cannot satisfy magnesium requirements.[62] Hypomagnesemic tetany can also be seen in 2- to 4-month-old calves on all-milk diets.[10,42] The efficiency of intestinal magnesium absorption declines as calves age, and by 2 to 4 months absorption from unsupplemented milk can no longer keep up with requirement.[34] The efficiency of magnesium absorption is higher in horses than in ruminants,[48] and hypomagnesemia is rare in healthy horses on traditional diets. There are considerable individual variations in magnesium absorption.[46]

Requirements

Magnesium requirements are greatest in heavily lactating animals. Most feeds contain sufficient magnesium to meet needs, but grains are lower in total magnesium (Table 3-2).

The availability of magnesium is lower in forages than in grains.[93] Mineral interactions affecting magnesium availability begin in the herbage. High soil acidity, potassium, and ammonia inhibit uptake of magnesium by plants; nitrates enhance uptake.[14,118] Magnesium digestion has been studied most in ruminants, in part because these animals are most susceptible to hypomagnesemia. Potassium, ammonium, and calcium ions inhibit ruminal magnesium absorption; sodium enhances absorption.* The inhibitory influence of ammonium ions may explain why diets high in nonprotein nitrogen and low in energy predispose to hypomagnesemia. Both these conditions are likely to result in high ruminal ammonium ion concentrations. Diets high in crude protein increase the renal excretion of magnesium. However, high crude protein intake does not appear to exacerbate the effects of high potassium intakes.[36]

Deficiency

If hypomagnesemia is sufficiently severe, it produces the disease hypomagnesemic tetany. Hypomagnesemic tetany has been divided into acute and chronic forms. In the acute form, there is a fairly rapid decline in serum magnesium. When high-producing cows are put on lush pasture with low concentrations of available magnesium, the fall in serum magnesium can be rapid. Classically, the feedstuffs that are deficient in magnesium are young, rapidly growing spring grasses and green cereal feeds. These feeds give rise to alternate names for the disease: *grass staggers* and *wheat pasture poisoning*. Concomitant hypocalcemia is often more pronounced in wheat pasture poisoning than in grass staggers. It is possible that some cases of wheat pasture poisoning represent hypocalcemic paresis and not hypomagnesemic tetany because serum magnesium concentrations are often normal, whereas hypocalcemia is a common finding.[11,109] The chronic form of hypomagnesemic tetany develops over months (usually the winter feeding period) in response to a diet that is marginally deficient in nutrients. It is seen in both adult ruminants and in calves.

Hypomagnesemia can reduce feed intake and impair metabolism, producing a reduction in growth rate or body condition without other obvious signs.[35] Hypomagnesemia increases nerve excitability, an effect that can be exacerbated by hypocalcemia, and this is responsible for the nervous signs of hyperexcitability and tetany, seen in all species.[10,44] Initially calves show signs of twitching of the ears, kicking at the belly, and a wide palpebral fissure. After a few days, convulsions appear; these can be precipitated by stress and may end in death.[10,91] Sudden death of stressed animals and finding previously "normal" animals dead in the field are well-recognized results of hypomagnesemia. Like most nutritional diseases, the whole herd is involved, although only a few animals may show obvious clinical signs at any one time.

Absence of urinary magnesium excretion, low serum and cerebrospinal fluid magnesium concentrations, and low serum calcium concentrations are all diagnostically useful.[35] Magnesium concentrations are stable in vitreous humor for 48 hours at 23° C (73° F) and may be used for the diagnosis of hypomagnesemia in dead animals.[61,73] The magnesium content of bone ash is not affected by acute hypomagnesemia,[35] although it decreases in the chronic form[10,91] (Table 3-8).

Toxicity

Magnesium can be toxic at the very high oral intakes associated with therapeutic use. For example, cattle treated with large doses (0.5 kg) (1 lb) of magnesium oxide as an antacid or purgative develop metabolic alkalosis.[87] This is often asymptomatic but could be detrimental if the patient was already alkalotic or if repeated treatments are given. There is also the possibility of magnesium intoxication if the patient has diminished renal function. Intravenous magnesium infusion may elevate serum magnesium concentrations to the point at which magnesium toxicity is seen. High serum magnesium concentrations depress neuromuscular transmis-

Table 3-8 *Magnesium content in calf bones (ribs and tibia)*

	Bone ash mineral content		
Health status	Ca (%)	Mg (%)	Ca: Mg ratio
Normal	38	0.6	<65
Hypomagnesemic	38	0.4-0.5	>80

Ribs appear to be more sensitive to changes in magnesium intake than long bones.
Modified from data in Blaxter KL and Sharman GAM[10] and Parr WH.[91]

---

*References 23, 24, 34, 36, 41, 119, 120.

sion and cardiac contractility. If the levels are sufficiently high, death from anoxia occurs, often preceded by a period of excitement.[45] Dietary excess of magnesium has been associated with an increased risk of developing urinary calcium phosphate calculi. This may be the result of magnesium promoting renal calcium excretion.[95]

Supplementation

Feeds are often supplemented by mixing with magnesium-containing minerals as part of ration formulation. The magnesium in magnesium oxide, sulfate, and carbonate is well used, but the magnesium in dolomitic limestone is less digestible.[93]

Hypomagnesemia is of major economic importance in sheep and dairy and beef cattle kept at grass, and several regimens have been developed to deliver supplementary magnesium to high-risk grazing animals. Although magnesium can be incorporated into salt-mineral licks, these often provide insufficient intake to prevent problems during high-risk periods unless they contain high levels of magnesium. Mixing molasses with the minerals may improve palatability and intake. Cattle can be fed supplementary grain or cottonseed meal to which magnesium oxide has been added. There may be problems with these approaches because high-magnesium mineral supplements can be unpalatable, and young pasture can be preferred to supplementary feed. Slow-release magnesium "bullets" that gradually release magnesium into the rumen fluid are available, but the rate of release of magnesium is slow, and a single bullet is unlikely to make a significant difference to ruminants.[46,111] Fertilization of pastures to increase magnesium within the herbage is a long-term process that requires heavy magnesium applications.[70] Instead, pastures can be top-dressed with magnesium sulfate solutions or magnesium oxide (30 kg of Mg per hectare or 25 lb/acre). These compounds stick to the herbage and increase the magnesium intake of grazing animals.[70,109] This approach is especially useful when combined with strip grazing to limit the cattle to recently treated grass.

SUMMARY

Most chronic mineral deficiencies can produce decreased food intake, reduced weight gain, or weight loss. With the exception of calcium and magnesium, deficiency of macrominerals can produce pica.

Calcium and phosphorus are the most abundant minerals in the body and together form the structural crystal hydroxyapatite, which is used to build bones and teeth. Calcium is also important in muscle contraction and nerve conduction. Phosphorus is important in energy metabolism, reproductive function, and the prevention of hemolysis of red cells. Magnesium has important roles in energy metabolism and nerve conduction.

Sodium is the major mineral of the extracellular fluid; it is important in the control of extracellular fluid volume and osmotic pressure. Many feeds are low in sodium, and a well-developed sodium appetite exists; sodium is the only mineral for which free-choice intake of mineral supplements is satisfactory in meeting nutritional needs. Excessive supplementation of food with salt, combined with water deprivation, leads to convulsions and blindness. Potassium is the major ion of the intracellular fluid. Chloride is an important anion in extracellular and intracellular fluid.

REFERENCES

1. Alexander F and Hickson JCD: In Phillipson AT, editor: Physiology of digestion and metabolism in the ruminant, Newcastle-upon-Tyne, Oriel Press, 1970, pp 375-389.
2. Allsop TF and Pauli JV: Res Vet Sci 38:61, 1985.
3. Barton BA, Jorgensen NA, and DeLuca HF: J Dairy Sci 70:1186, 1987.
4. Bell FR: Proc of the First International Symposium on Olfaction and Taste, Oxford, 1963, Pergammon Press, 1:299-307.
5. Bell FR: Proc R Soc Med 65:27, 1972.
6. Bell FR and Kitchell RL: J Physiol 183:145, 1966.
7. Bell FR and Williams HL: J Physiol 151:42, 1969.
8. Blackmon DM and others: Am J Vet Res 45:1638, 1984.
9. Blaustein MP: J Cardiovasc Pharmacol 12 (suppl 5):S56, 1988.
10. Blaxter KL and Sharman GAM: Vet Rec 67:108, 1955.
11. Bohman VR and others: J Anim Sci 57:1352, 1983.
12. Bolton TB and others: J Cardiovasc Pharmacol 12 (suppl 5):S96, 1988.
13. Bonniwell MA and others: Vet Rec 122:386, 1988.
14. Bould C: Vet Rec 76:1377, 1964.
15. Bowen JM, Blackmon DM, and Heavner JE: Am J Vet Res 31:831, 1970.
16. Brooks HV and others: New Zealand Vet J 32:174, 1984.

17. Burkhalter DL and others: J Dairy Sci 62:1895, 1979.
18. Button C and Mulders MSG: Am J Vet Res 45:1658, 1984.
19. Call JW and others: Am J Vet Res 47:475, 1986.
20. Campbell HT, Lightfoot BO, and Sklar AH: Proc Soc Exp Biol Med 189:317, 1988.
21. Caple IW, Doake PA, and Ellis PG: Aust Vet J 58:125, 1982.
22. Carafoli E: J Cardiovasc Pharmacol 12 (supp 3):S77, 1988.
23. Care AD: Br Vet J 144:3, 1988.
24. Care AD and others: Q J Exp Physiol 69:577, 1984.
25. Cox JL, Becker DE, and Jensen AH: J Anim Sci 25:203, 1966.
26. Curtis CR and others: J Am Vet Med Assoc 183:559, 1983.
27. Daniel RCW and Moodie EW: Res Vet Sci 24:380, 1978.
28. DeLuca HF: Proc Soc Exp Biol Med 191:211, 1989.
29. Dennis RJ, Hemken RW, and Jacobson DR: J Dairy Sci 59:324, 1976.
30. Dow C, Lawson GHK, and Todd JR: Vet Rec 75:1052, 1963.
31. Ellison RS, Young BJ, and Read DH: New Zealand Vet J 34:7, 1986.
32. Erb HN and Grohn YT: J Dairy Sci 71:2557, 1988.
33. Fettman MJ and others: J Dairy Sci 67:2321, 1984.
34. Field AC and Suttle NF: J Comp Pathol 89:431, 1979.
35. Fisher DD and others: Am J Vet Res 46:1777, 1985.
36. Fontenot JP, Wise MB, and Webb KE Jr: Fed Proc 32:1925, 1973.
37. Georgievskii VI, Annenkov BN, and Samokhin VT: Studies in the agricultural and food sciences: mineral nutrition of animals, London, 1981, Butterworth.
38. Goff JP, Kehrli Jr ME, and Horst RL: J Dairy Sci 72:1182, 1989.
39. Goings RL and others: J Dairy Sci 57:1184, 1974.
40. Grabarek Z and Gergely J: Biochim Acta 21:S297, 1989.
41. Greene LW, Webb KE Jr, and Fontenot JP: J Anim Sci 56:1214, 1983.
42. Haggard DL, Whitechair CK, and Langham RF: J Am Vet Med Assoc 172:495, 1978.
43. Hansard SL, Comar CL, and Plumlee MP: J Anim Sci 13:25, 1954.
44. Harrington DD: Am J Vet Res 35:503, 1974.
45. Hatch RC: In Booth NH and McDonald LE, editors: Veterinary pharmacology and therapeutics, ed 6, Ames, 1988, Iowa State University Press, p 1147.
46. Hemingway RG and others: J Agricul Sci 60:307, 1963.
47. Hemingway RG and Ritchie NS: Proc Nutr Soc 24:54, 1965.
48. Hintz HF and Schryver HF: J Anim Sci 35:755, 1972.
49. Hintz HF and Schryver HF: J Am Vet Med Assoc 168:39, 1976.
50. Hintz HF and others: J Anim Sci 36:522, 1973.
51. Hoar DW, Emerick RJ, and Embry LB: J Anim Sci 30:597, 1970.
52. Horst RL and Reinhardt TA: J Dairy Sci 66:661, 1983.
53. Huber TL and others: Am J Vet Res 42:1488, 1981.
54. Inesi G: Braz J Med Biol Res 21:1241, 1988.
55. Jaster EH, Schuh JD, and Wegner TN: J Dairy Sci 61:66, 1978.
56. Jensen AH, Terrill SW, and Becker DE: J Anim Sci 20:464, 1961.
57. Johnston CC Jr: Proc Soc Exp Biol Med 191:258, 1989.
58. Joshua IG, Fleming JT, and Dowe JP: Proc Soc Exp Biol Med 189:344, 1988.
59. Kerkut GA: Comp Biochem Physiol (A) 93:9, 1989.
60. Kronfeld DS and Ramberg CF: J Dairy Sci 54:794, 1971.
61. Lincoln SD and Lane VM: Am J Vet Res 46:160, 1985.
62. Littledike ET and Goff J: J Anim Sci 65:1727, 1987.
63. Littledike ET, Young JW, and Beitz DC: J Dairy Sci 64:1465, 1981.
64. Lohius JACM and others: Vet Rec 124:305, 1989.
65. MacWilliams PS, Searcy GP, and Bellamy JEC: Can Vet J 23:309, 1982.
66. Madison JB and Troutt HF: Res Vet Sci 44:264, 1988.
67. Marks SL and Carr J: Vet Rec 125:460, 1989.
68. Mason DK, Watkins KL, and McNie JT: Equine Practice 10:10, 1988.
69. Matsushima JK: Advanced series in agricultural sciences 7, Berlin, 1979, Springer-Verlag.
70. McConaghy S and others: J Agricul Sci 60:313, 1963.
71. McDonald P, Edwards RA, and Greenhalgh JFD: Animal nutrition, Harlow, Essex, 1988, Longman Scientific & Technical.
72. McGowan JA and others: Proc Soc Exp Biol Med 185:62, 1987.
73. McLaughlin BG and McLaughlin PS: Can J Vet Res 52:476, 1988.
74. Mills CF: In Cuthbertson D, editor: Nutritional animals of agricultural importance, Oxford, 1969, Pergamon Press.
75. Milne MD, Muehrcke RC, and Heard BE: Br Med Bull 13:15, 1957.
76. Morris JG: J Anim Sci 50:145, 1980.
77. Murphy GM and Gartner RJW: Aust Vet J 50:280, 1974.
78. Mylrea PJ: Res Vet Sci 7:394, 1966.
79. National Academic Press, Nutrient Requirements of Swine, Washington, DC, 1988.
80. Naylor JM and Kronfeld DS: Can J Vet Res 50:402, 1986.
81. Nutrient Requirements of Beef Cattle, ed 9, Washington, DC, 1984, National Academy of Sciences.
82. Nutrient Requirements of Dairy Cattle, ed 5, Washington, DC, 1978, National Academy of Sciences.
83. Nutrient Requirements of Sheep, ed 4, Washington, DC, 1985, National Academy of Sciences.
84. Oetzel GR: Vet Clin North Am (Food Anim Pract) 4:351, 1988.
85. Oetzel GR and others: J Dairy Sci 71:3302, 1988.
86. Ogawa E and others: Am J Vet Res 48:1300, 1987.
87. Ogilvie TH and others: Can J Comp Med 47:108, 1983.
88. Osweiler GD and Hurd JW: J Am Vet Med Assoc 165:165, 1974.

89. Ozaki H and others: J Biol Chem 263:14074, 1988.
90. Palade P and others: J Bioenerg Biomembr 21:295, 1989.
91. Parr WH: Vet Rec 69:71, 1957.
92. Pearson EG and Kallfelz FA: Cornell Vet 72:142, 1982.
93. Peeler HT: J Anim Sci 35:695, 1972.
94. Peet RL and others: Aust Vet J 61:195, 1984.
95. Petersson KH and others: J Dairy Sci 71:3369, 1988.
96. Pradhan K and Hemken RW: J Dairy Sci 51:1377, 1968.
97. Ramberg CF Jr and others: Am J Physiol 246:R698, 1984.
98. Rayssiguier Y and others: Ann Rech Vet 8:267, 1977.
99. Reinhardt TA, Horst RL, and Goff JP: Vet Clin North Am (Food Anim Pract) 4:331, 1988.
100. Roberts WK and St Omer VVE: J Anim Sci 24:902, 1976.
101. Roy JH: In Cuthbertson D, editor: Nutrition of animals of agricultural importance, part 2: assessment of and factors affecting requirements of farm livestock, Oxford, 1969, Pergamon Press, pp 645-716.
102. Rude RK and others: J Clin Endocrinol Metab 47:800, 1978.
103. Rudy DR: Postgrad Med 86:151, 1989.
104. Scarratt WK, Collins TJ, and Sponenberg DP: J Am Vet Med Assoc 186:977, 1985.
105. Schmidt H: J Am Vet Med Assoc 96:441, 1940.
106. Schneider KM and others: Aust Vet J 62:82, 1985.
107. Schryver HF, Hintz HF, and Lowe JE: Cornell Vet 64:493, 1974.
108. Smith DLT: Am J Vet Res 18:825, 1957.
109. Smith RA and Edwards WC: Vet Clin North Am (Food Anim Pract) 4:365, 1988.
110. Stick JA, Robinson NE, and Krehbiel JD: Am J Vet Res 42:733, 1981.
111. Stuedemann JA, Wilkinson SR, and Lowrey RS: Am J Vet Res 45:698, 1984.
112. Technical review by the Agricultural Research Council Working Party, Slough, 1980, Commonwealth Agricultural Bureau.
113. Telle PP and others: J Anim Sci 23:59, 1964.
114. Thompson KG and Cook TG: New Zealand Vet J 35:11, 1987.
115. Torres M, Barbella Y, and Israel A: Proc Soc Exp Biol Med 190:18, 1989.
116. West JB: Hormonal regulation of mineral metabolism. In JB West, editor: Best and Taylor's physiological basis of medical practice, ed 11, Baltimore, 1985, Williams & Wilkins, pp 893-901.
117. Whitlock RH, Kessler MJ, and Tasker JB: Cornell Vet 65:512, 1975.
118. Wilcox GE and Hoff JE: J Dairy Sci 57:1085, 1974.
119. Wilson AA: Vet Rec 76:1383, 1964.
120. Wylie MJ, Fontenot JP, and Greene LW: J Anim Sci 61:1219, 1985.
121. Yusufi ANK and others: Proc Soc Exp Biol Med 190:87, 1989.
122. Zaloga GP and others: Circ Shock 24:143, 1988.

4

Trace Minerals

M. E. SMART, N. F. CYMBALUK

Awareness of the importance of trace minerals has increased as sophisticated analytical techniques have been developed. As more information becomes available, the veterinarian must become cognizant of the effects of crop and soil management, climatic conditions, animal genetic selection, animal management, and feeding practices on trace mineral requirements and availability for large animals. Overt clinical signs of trace mineral deficiencies or toxicities produce obvious losses in animal health and productivity, but subclinical deficiencies or toxicities can produce equally serious losses in production.

NORMAL FUNCTION

A trace mineral comprises less than 1% of total body ash and is required in the diet at less than 0.1%, or 1 g/kg, feed dry matter. Fifteen micromineral elements are essential. Iron, iodine, zinc, copper, manganese, selenium, cobalt, molybdenum, and chromium are the principal trace elements needed in livestock diets. In controlled nutritional trials, growth responses occur with dietary supplementation of tin, vanadium, fluorine, silicon, nickel, and arsenic. Dietary aluminum may be important because it can interact adversely with macrominerals and microminerals. Excesses of any of the trace minerals also may be detrimental.

Trace minerals act primarily as catalysts or activators in enzyme or hormone systems or as integral parts of the structure and function of metalloenzymes. Clinical symptoms of deficiency occur as the activities of specific biochemical functions are obliterated or reduced. Toxicities are frequently evidenced as induced deficiencies of other trace minerals. Some metalloenzymes and their functions are given in Table 4-1. The clinical sign or production criterion affected by a trace mineral deficiency depends on the specific physiological function of the metalloenzyme. For example, graying of the hair coat is related to the absence of copper in the enzyme tyrosinase, which is needed for melanogenesis.

DEFINING A DEFICIENT STATE

Many attempts have been made to clearly define the existence of a deficient state. Most are based on clinical, pathological, and biochemical criteria of animals made deficient using purified diets in controlled studies. The criteria from these induced deficiencies are then compared with those obtained from animals under field conditions. In the field, a veterinarian can also identify deficiencies by obtaining a positive production response to trace mineral supplementation. Dietary and tissue trace mineral levels also give a good indication that deficiency exists.

A trace mineral deficiency is considered primary when there is an absolute deficiency of the element in the diet. Secondary deficiencies occur from interactions with other microminerals and with other dietary constituents. More than 25 trace mineral–nutrient interactions are recognized. These interactions can be synergistic or antagonistic to trace mineral status. Trace mineral–nutrient interactions occur within the intestinal lumen, at the absorptive site, or at the metabolic site. An example of preabsorptive antagonism is the impairment of copper absorption by excessive dietary zinc or iron. An example of postabsorptive antagonism is the effect of thiomolybdate on copper utilization by the ruminant.

Table 4-1 *Some of the metalloenzymes of the main trace minerals and their functions*

Trace mineral	Metalloenzyme or protein	Function	Effect of deficiency*
Chromium (Cr)	"Glucose tolerance factor"	Glucose metabolism	Reduced glucose tolerance
Cobalt (Co)	Cyanocobalamin (vitamin B_{12})	Methyl transfer	Lowered use of propionic acid (ruminants) Anemia
Copper (Cu)	Metallothionein Ceruloplasmin	Storage, transport, oxidation (especially Fe)	Decreased blood and tissue copper Microcytic anemia
	Dopamine B monooxygenase	Catecholamine production	Ataxia
	Tyrosinase	Melanin formation	Graying of hair coat (achromotrichia)
	Superoxide dismutase	Cytosolic antioxidant	Reduced phagocytosis, tissue O_2 damage
	Lysyl oxidase	Cross-linkages in collagen and elastin	Abnormal bone and collagen; stiffness of gait, lameness, joint enlargement, aneurysm
	Cytochrome C oxidase	Oxidative phosphorylation	Reduced cellular respiration efficiency
Iodine (I)	Thyroxine and triiodothyronine	Regulate metabolic rate	Goiter Failure of tendons to attach to insertions
Iron (Fe)	Hemoglobin/Myoglobin	Oxygen storage and transport	Hypochromic, microcytic anemia Easily fatigued
	Cytochrome C oxidase	Oxidative phosphorylation	Reduced cellular respiration efficiency
	Cytochromes A Xanthine oxidase	Oxidative phosphorylation Pyrimidine, purine metabolism	Anorexia
Manganese (Mn)	Glycosyl transferase	Polysaccharide, glycoprotein synthesis	Abnormal chondrogenesis; shortened bones; abnormal otolith development, ataxia
	Farnesyl pyrophosphate synthetase	Cholesterol synthesis	Impaired steroid hormone metabolism?
	Pyruvate carboxylase	Gluconeogenesis	Impaired carbohydrate metabolism
Molybdenum (Mo)	Xanthine oxidase Sulfite oxidase	Pyrimidine, purine metabolism Sulphite metabolism	Impaired metabolism of S-amino acids and nucleotides
	Nitrate reductase	Reduction of NO_3 to NO_2	

*Reduced growth rate is a nonspecific effect of all trace mineral deficiencies.

Table 4-1 *Some of the metalloenzymes of the main trace minerals and their functions—cont'd*

Trace mineral	Metalloenzyme or protein	Function	Effect of deficiency*
Nickel (Ni)	Urease	Rumen urea metabolism	Reduced rumen microbial function
	Alanine transaminase		Gluconeogenesis impaired
Selenium (Se)	Glutathione peroxidase	Reduction of tissue and RBC hydroperoxides	Tissue oxidative damage, e.g., hepatic and muscle necrosis
Zinc (Zn)	RNA & DNA polymerase	Nucleotide synthesis	Anorexia, skin lesions, decreased apoenzyme synthesis, decreased cell replication, decreased insulin sensitivity, abnormal protein polymer synthesis
	Thymidine kinase	Protein digestion	
	Carboxypeptidase A-C	CO_2 transport	
	Carbonic anhydrase		
	Alkaline phosphatase		
	Dipeptidase		
	Fructose 1-6 diphosphatase	Gluconeogenesis	
	Aminolevulinic dehydratase	Heme synthesis	

*Reduced growth rate is a nonspecific effect of all trace mineral deficiencies.

The nature and severity of the clinical signs for deficiency of a specific trace mineral depend on the severity of the deficiency and the existence of other dietary deficiencies or excesses. Initial signs of deficiency may be minor, yet chronic or severe deficiencies can cause profound metabolic disruption leading to extensive pathology. Production losses often precede overt clinical signs. The site of absorption and some of the factors that influence the availability of the principal trace minerals are outlined in Table 4-2.

FACTORS THAT INFLUENCE TRACE MINERAL STATUS

Traditionally, dietary deficiency has been considered the primary vehicle of clinical trace mineral deficiencies. However, numerous factors contribute to and exacerbate dietary effects. Animal, plant, soil, environmental, and management factors can enhance development of a clinical deficiency syndrome.

Animal factors

Animal factors that can alter trace mineral requirements include age, breed, sex, species, physiological condition (maintenance, gestation, lactation, growth, and athletic performance), and genetic selection. The trace mineral requirements of ruminal microorganisms and interactions occurring in the rumen can alter trace mineral status and requirements of the ruminant. Modern management practices that optimize feed efficiency and production can also change trace mineral requirements.

Plant factors

Genetic selection and bioengineering of plant species can modify plant growth and the ability of food crops to absorb trace minerals from the soil. This alters the trace mineral composition of the plant.

Soil composition

Soil composition, moisture, pH, organic matter, depletion or addition of trace minerals, and loss of topsoil modify trace mineral composition of local and regional plant crops. For example, irrigation has been related to a depletion of selenium in the topsoil and, consequently, reduction of selenium in the crops grown on these soils. The use of purified fertilizers has de-

Table 4-2 *The sites of absorption and some factors affecting availability of the main trace minerals*

Trace minerals	Site of absorption	Factors affecting availability		Other factors
		Facilitatory	Inhibitory	
Copper (Cu)	Small intestine	L-amino acids Vitamin A Bile Lactose Acid soils	High Zn, Fe High Cd, Se Vitamin C Molybdenum/Sulfur (ruminants) Alkaline soils	Age: plasma low, liver high in newborn Species: ruminants higher liver Cu than nonruminants Lush growth has lower Cu content Legumes have more Cu than grasses
Cobalt (Co)	Small intestine	?	?	Higher in plants grown on poorly drained soils Higher in legumes than grasses
Iodine (I)	Small intestine	Iodate > iodide	Thiocyanate Perchlorate Nitrate Methylthiouracil Soybean meal	Seasonal effects (highest in winter) Breed effect (Jersey > Holstein) Highest in early lactation Many brassicas contain goitrogens
Iron (Fe)	Small intestine	Vitamin C Citrate Histidine Cysteine Bile	Phytates High Ca, P, Zn High Cd, Mn High tannins Tetracycline Cimetidine	Low intestinal pH increases absorption Deficient state increases absorption Young animals absorb more than older Higher in legumes than grasses Health status: parasites decrease Fe absorption
Manganese (Mn)	Small intestine Abomasum	Lactose	High Ca, P, Fe^{3+} Isolated soy Phytate	Lower in legumes than grasses Low in straw-based diets
Molybdenum (Mo)	Small intestine	Copper	Sulfate Tungsten	Legumes and cereals more Mo than grasses Higher in plants grown on poorly drained alkaline soils Hexavalent > absorption than tetravalent

Table 4-2 *The sites of absorption and some factors affecting availability of the main trace minerals—cont'd*

Trace minerals	Site of absorption	Factors affecting availability		Other factors
		Facilitatory	Inhibitory	
Selenium (Se)	Small intestine	Selenomethionine > selenite Histidine Vitamin A Vitamin E	Vitamin C Sulfate Cu, Ca, Ag, As Nitrates Unsaturated fatty acids	Plasma content lower in young than adult Low in plants grown on acid, igneous, rock-derived soils Low in areas of high rainfall High in cruciferas, legumes > grasses > cereals
Zinc (Zn)	Small intestine	Histidine Vitamin C Lactose EDTA Citrate	Phytate Oxalate High Ca, Cd High Fe, Sn High orthophosphate	Higher in legumes than grasses Zn in cereal grains less available

creased the amount of trace mineral added to the soil.

Environmental factors

Differences in annual, seasonal, and regional precipitation and ambient temperatures can alter a plant's growth rate and its ability to absorb trace minerals. Trace minerals can be leached from the soil in areas of high rainfall. High rainfall also promotes the lushness of plant growth, which can dilute the trace mineral concentration in the plant.

Past geological events have preordained the trace mineral content of soils regionally. Geological modification contributes to the iodine deficiency of soils of the Great Lakes region, high concentrations of selenium in certain glacial deposits, selenium deficiency in gray-wooded soils, and high molybdenum content in areas of shale rock.

Changes secondary to inflammatory disease

The factors that influence trace mineral status in disease are demonstrated by changes that occur during inflammation. Following an insult by an infectious or inflammatory agent, the leukocytes liberate interleukin-1 (leukocyte endogenous mediator). Interleukin-1 stimulates hepatic synthesis of metallothionein, which binds zinc. The subsequent depression in serum zinc is related to the degree of toxic insult and is accompanied by enhancement of neutrophil function. These events are followed by synthesis of acute phase reactants, including fibrinogen, haptoglobulin, and alpha-2 globulins (ceruloplasmin). Increased synthesis of ceruloplasmin accompanies the movement of hepatic copper to peripheral tissues, which produces an elevation of circulating copper concentration. In addition to changes in plasma copper and zinc, plasma iron concentration decreases because granulocytes release apolactoferrin, which binds iron. The iron-lactoferrin then binds to hepatocytic receptors and is made unavailable for microbial growth and toxin production.

Deficiencies of several trace minerals may reduce interleukin-1 release from monocytes. Copper deficiency may reduce interleukin-1 production directly, whereas zinc deficiency reduces antibody formation.

CLINICAL PROBLEMS

Trace mineral status is not static but fluctuates throughout the animal's life. Figure 4-1 illustrates the changes that occur in plasma and tissue trace mineral concentration and biochemical function as an animal progresses to a clinically deficient state. It is not until the deficiency results in biochemical changes that disrupt cellular function that clinical signs appear. The signs exhibited depend on the functions impaired and on the sequence of impairment; thus primary and secondary deficiencies can cause different clinical signs. In calves, primary copper deficiencies induced by a deficient diet produce skeletal abnormality, but secondary copper deficiencies produced by an iron interaction do not. Although declining plasma and tissue trace mineral concentrations can indicate a declining trace mineral status, this reduction in circulating trace mineral often occurs well before clinical signs become apparent. Multiple deficiencies are fairly common and can confound the uncomplicated diagnosis of a single trace mineral deficiency. These features of a trace mineral deficiency make it difficult for a practitioner or a nutritionist to make a diagnosis based on clinical signs alone. Table 4-3 outlines some of the common clinical signs associated with trace mineral deficiencies.

DIAGNOSING A DEFICIENCY

Several criteria can be used to evaluate the trace mineral status of single animals or herds. These include the following:
1. A detailed history
2. A detailed physical examination
3. A total dietary evaluation
4. Tissue evaluation for
    a. The specific trace mineral
    b. Activity of metalloenzymes that involve the specific trace minerals, end products, or metabolites in which the trace mineral or its metalloenzyme is involved

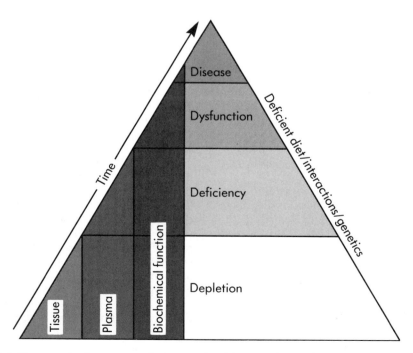

Fig. 4-1 The events that occur when an animal is on a diet deficient in trace minerals before clinical signs occur (disease).

Trace Minerals

Table 4-3 *Some clinical signs associated with trace mineral deficiencies**

	Cattle	Horses	Sheep	Swine
Copper (Cu)	Heart failure Depressed growth rate Lameness, enlarged joints Faded hair coat Anemia Scours Increased susceptibility to disease Infertility Decreased milk production	Osteochondrosis-like lesions? Rupture of uterine artery in aged mares Depigmentation of skin around eyes?	Neonatal ataxia Impaired wool growth	Heart and aortic degeneration Posterior ataxia
Iodine (I)	Weak or hairless young Goiter Abortion, stillbirth Irregular or suppressed estrus Fetal reabsorption Subnormal birth weights Poor growth Infertility Reduced milk yield Decline in male libido	Weak foals unable to nurse Goiter Stillbirths Abnormal estrus Forelimb contracture	Hairless lambs Goiter Reduced quantity and quality of wool Impaired development of wool-producing follicles	Birth of hairless, thick, pulpy-skinned piglets
Manganese (Mn)	Shortened long bones Joint pain Contracted tendons Poor balance Depressed or delayed estrus Poor conception rates	Shortened, bowed limbs Enlarged joints Retarded skeletal growth	Similar to cattle	Crooked and shortened legs Enlarged hock joints Lameness
Selenium (Se)	Nutritional muscular dystrophy Premature or weak calves "Ill thrift" Retained placenta Increased susceptibility to mastitis in early lactation	Nutritional muscular dystrophy Acute heart failure Retained placenta Reduced fertility Delayed estrus Difficulty in suckling and swallowing in foals Predisposes horses to "tying up"	Nutritional muscular dystrophy "Ill thrift" Embryonic death	Muscular dystrophy Mulberry heart disease Low first conception in gilts Hepatosis diatetica Myopathy following iron injections

*Many of the signs are based on clinical rather than scientific data. Those ascribed to only one species does not necessarily preclude the potential appearance in others.

Continued.

Table 4-3 *Some clinical signs associated with trace mineral deficiencies*—cont'd*

	Cattle	Horses	Sheep	Swine
Zinc (Zn)	Parakeratosis and dermatitis Poor growth Decreased feed utilization Decreased milk production Impaired appetite Impaired utilization of vitamin A Swelling and stiffness of joints Foot rot Impaired immune response Impaired gonadal maturation	Parakeratosis (peripheral limb) Poor growth Inappetence Hair loss Poor wound healing Pigmentation loss around eyes?	Posthitis and vulvitis Loss of wool crimp Wool easily shed Changes in hoof structure	Parakeratosis and dermatitis Subnormal growth Depressed appetite and feed efficiency

*Many of the signs are based on clinical rather than scientific data. Those ascribed to only one species does not necessarily preclude the potential appearance in others.

History, clinical examination, and diet evaluation

A thorough history and clinical examination are critical for diagnosis of a suspected trace mineral deficiency. The diet must be evaluated not only for the suspected trace mineral but also for potential nutrient and trace mineral interactions. Water must also be evaluated. Feed analysis is the most important tool in the diagnosis of trace mineral deficiency as low dietary concentrations indicate a potential for development of a deficiency. However, the normality of all nutrients must be appraised. For example, correction of a copper deficiency diagnosed as a cause of bone disorders in livestock would be ineffective if a concurrent calcium deficiency existed.

Tissue and fluid evaluation

Important aids to diagnosis of trace mineral imbalances include analysis of tissue, blood, serum, and plasma. Selection of the appropriate tissue is based on the normal distribution of a trace mineral within tissues. Table 4-4 outlines the most common diagnostic tests and the tissues used to identify specific trace mineral deficiencies.

Liver, bone, and hair have been used most commonly for diagnostic purposes. Hepatic trace mineral is a reliable indicator of mineral stores for those trace minerals preferentially stored in the liver. Thus, copper status can be reliably predicted from liver copper concentrations, but interpretations must be species specific. Liver biopsy is a safe procedure in cattle older than 3 months of age. This technique has also been used (infrequently, owing to risk) in horses. Prior to performing the biopsies, the veterinarian should check with the laboratory to see what size of sample is required and how it should be prepared and stored. There is some debate over whether small biopsy samples are representative of the trace mineral concentration in the total organ. A significant correlation has been found between the concentration of the biopsy sample and the total liver copper concentration when deficiency exists, but not when the total liver copper concentration is high. The trace mineral status or the effectiveness of sup-

Table 4-4 *Some criteria used to diagnose trace mineral deficiencies*

Trace mineral	Diagnostic criteria	Comments
Cobalt (Co)	Serum and urinary methylmalonic acid in cattle	Increased
Copper (Cu)	Serum and plasma Cu	Decreased (serum is lower than plasma in cattle)
	Liver Cu	Decreased
	Ceruloplasmin activity	Decreased
	Leukocyte cytochrome C oxidase	Decreased
	RBC superoxide dismutase	Decreased
	Hair Cu	Questionable value
	RIA,* metallothionein	Recent development that may have future use
Iodine (I)	Serum and plasma thyroxine	Most commonly used—decreased
	Serum protein-bound I	PBI techniques not often available
	Serum, plasma, and milk I	Serum and plasma not often available
Iron (Fe)	Total iron-binding capacity	Normal or increased
	Serum and plasma Fe	Decreased, affected by inflammation
	Ferritin	Decreased
	Hemoglobin and mean corpuscular volume	Decreased
Manganese (Mn)	Plasma and serum Mn	Too low to detect by common methods
	Liver Mn	Low
	Mn superoxide dismutase	Decreased
	Pyruvate carboxylase	Decreased
	Hair Mn	Questionable value
Molybdenum (Mo)	Liver Mo	Decreased
	Milk Mo	Decreased
	Serum or plasma Mo	Decreased
Selenium (Se)	Liver and milk Se	Decreased
	Serum or plasma Se	Decreased
	RBC glutathione peroxidase	Decreased
Zinc (Zn)	Liver, plasma, serum, bone, and hair Zn	Variable, affected by diet, infection, malnutrition
	Serum alkaline phosphatase activity	Difficult to correlate

*Radio immunoassay.

plementation of a herd of meat animals can be evaluated from liver samples harvested at slaughter. Biopsies of bone from the coccygeal vertebra or rib at slaughter have also been used for mineral analysis.

Trace mineral concentrations of hair reflect only the free ionic or loosely bound serum or plasma concentrations at the time of hair formation. Thus, hair clippings usually represent the trace mineral status of the animal 2 to 3 months previously. Hair trace mineral content depends on hair color, site of collection, season, and contamination from sebaceous secretions and the external environment. Sebaceous secretions contain copper, manganese, and zinc. Dust contains high concentrations of zinc, which can contaminate the hair of housed livestock.

Blood, serum, and plasma are the easiest body fluids to sample but may not always be the most useful tissues. These fluids are used to measure the concentration of the trace mineral directly or indirectly by evaluation of the concentration or the activity of enzymes or hormones altered by a deficiency.

A critical factor in accurate measurement of trace minerals in blood is the use of appropriate collection tubes and anticoagulants. For example, butyl rubber contains zinc salts. Butyl rubber in the stoppers of certain blood collection tubes and syringes would therefore contaminate blood for zinc analysis. Many early researchers used these types of collection vessels in early reports. This negates the value of these published data for diagnostic purposes. Collection tubes designed specifically for blood collection for trace mineral analysis are available.

Use of serum rather than plasma can influence the quantification of certain trace minerals. For example, serum copper values for cattle may be lower than plasma copper values because of the precipitation of ceruloplasmin copper in the clot. Hemolysis can also influence the absolute quantity of trace minerals in serum and plasma. Erythrocytes contain relatively high zinc concentrations, and hemolysis markedly elevates plasma or serum zinc content.

Sample storage, sample preparation, and the analytical techniques employed by the laboratory also can alter absolute values obtained. These factors are often overlooked by the veterinarian. Beyond the contamination in sample collection, many possibilities exist for further contamination in sample preparation. Contamination can occur from processing vials, storage bags, or reagents used in analysis. Polystyrene tubes contain high concentrations of zinc, whereas brown paper bags are moderately contaminated with copper and zinc. Diagnostic techniques used among laboratories often differ. For example, zinc analysis by atomic absorption spectroscopy is 100 times more sensitive than chemical colorimetric methods. Normal values should be established within each diagnostic laboratory, and these values should be used by the veterinarian to ascertain whether a deficiency exists.

Evaluation of other factors

When all diagnostic methods are completed, the results must be interpreted in light of physiological factors including age, breed, species, and reproductive condition. These factors can affect tissue trace mineral content independent of the dietary intake. The neonate under 4 days of age normally has plasma copper concentrations one half that of the adult, but plasma zinc concentrations that are twice that of the dam. However, trace mineral concentrations in fetal and neonatal tissues tend to be higher than in the adult. Disease states and stress can also influence blood values independent of trace mineral status. As described previously, infectious diseases are often accompanied by elevated plasma copper content but depressed zinc and iron concentrations. Exercise and fitness can induce changes in plasma trace mineral concentrations similar to those caused by inflammation. Seasonal fluctuations have also been described for some trace minerals.

SUPPLEMENTATION

Before recommending trace mineral supplementation, the veterinarian must determine the amount that is needed based on the trace mineral concentrations of the diet, the interactions that may exist with other nutrients, and the group to be supplemented. An important decision in the choice of supplementation is appraisal of the ability of the owner to implement the recommendations. Table 4-5 outlines the recommended upper tolerance and lower limits for supplementation of the principal trace minerals.

Trace mineral supplementation can be given orally or parenterally.

Oral supplements

The oral route is perhaps the easiest means of supplementation. Oral trace mineral preparations can be incorporated into the feed, into mineral licks, or into water or can be given as

Table 4-5 *Recommended upper and lower limits of mineral elements for livestock in mg/kg (ppm) of dry matter in the total ration**

Mineral	Beef Lower	Beef Upper	Dairy Lower	Dairy Upper	Sheep Lower	Sheep Upper	Horse† Lower	Horse† Upper	Swine* Lower	Swine* Upper
Iron (I)	30	400	100	400	30	300	40	100	40-150	750
Manganese (Mn)	25	850	25	850	40	850	40	1000	10	200
Copper (Cu)	10	100	10	100	5	10	10	800	6	125
Zinc (Zn)	50	500	50	500	50	600	40	500	100	500
Iodine (I)	0.6	20	0.6	20	0.6	20	0.1	5	0.2	10
Selenium (Se)	0.1	5	0.1	5	0.1	5	0.1	2	0.1	0.3 (added)

*Canadian feeds regulations (1982-83).
†1989 NRC for horses.

drenches. Trace mineral products can be fed free choice as licks or "force-fed" by incorporating them into a prepared diet (e.g., pelleted feeds). The simplest method of oral supplementation is feeding fortified mineral or salt in loose or block form. Small ruminants should be offered supplements in a loose form. Intake is, however, governed by the salt (NaCl) content of the diet.

The main disadvantage to oral supplementation is that individual intake may not be adequate to meet requirements. Intake depends on the palatability of the supplement, the total dissolved solids in the water, and the type of diet. Physical characteristics of the mineral feeder, competition, and individuality can adversely affect mineral intake by horses. Salt intake of similar-sized horses can range from 10 to 100 g (0.35 to 3.5 oz) daily. Thus, if salt is used as the carrier for trace mineral supplementation, intakes will be tenfold greater for one animal than for its companion. When free-choice intake of supplement by individuals is determined, levels of trace mineral fortification may have to be altered to ensure adequate intake by the animal consuming the least amount of mineral.

In addition, with oral supplementation, interactions present within the intestinal lumen may not be superseded unless the level of fortification is increased. This may necessitate disproportionate fortification of specific trace minerals in the supplement. The benefits of this increase must be monitored by improvement in production parameters or trace mineral status.

Amino acid chelates of some trace minerals have been developed to overcome rumen interactions. Stable amino acid chelates are not ionized in the intestinal lumen but are "smuggled" in as dipeptides during absorption. Once the amino acid is absorbed, the mineral, if liberated, can participate in reactions.

Various inorganic salts (oxides, phosphates, carbonates, sulfates) are used in trace mineral supplements. These salts vary in solubility and thus availability. Chelates other than the amino acid dipeptides have been developed. The degree of chelate polymerization and the stability constant influence the availability of these mineral chelates. As polymerization increases, efficiency of absorption decreases. A stability constant of under 13 or over 17 reduces availability.

Soluble glass rumen bullets impregnated with trace minerals that slowly release trace minerals over several months are available for use when other forms of supplementation are impractical in ruminants. However, "bullets" can degrade erratically in the rumen and thus do not supply mineral constantly. Copper oxide needles in gelatin capsules are also useful for supplementation of ruminants. The needles are trapped in the abomasum and bypass negative interactions in the rumen. Copper is slowly released over 2 to 3 months.

Parenteral supplements

Various parenteral forms of trace mineral supplementation are available or are in development. These preparations are intended to provide depots of slowly released mineral with minimal local or systemic tissue reactions. The minerals are injected as inorganic or organic chelates. The type of chelate influences the release of the mineral and the extent of the local tissue reaction. Parenteral trace mineral products are valuable if the veterinarian wants to ensure that all animals receive supplementation at a specific rate. However, the veterinarian must be familiar with the potential side effects of each product. For example, extensive tissue reaction and anaphylaxis have been reported in horses treated with parenteral copper and iron compounds.

Toxicities

Whenever a trace mineral supplement is used, there is the potential to create toxic tissue concentrations. This is especially important with copper, iodine, iron, and selenium. High concentrations of certain tissue and milk trace minerals in meat animals have significant public health implications. An example of this is iodine. There is a tendency to increase supplementation when no improvement is seen rather than re-evaluating the problem. Chronic toxicities can occur when an investigator fails to consider all sources of minerals and supplements being supplied to the livestock.

CONCLUSION

Awareness of the importance of trace minerals in the diets of domestic livestock has increased with the development of sophisticated analytical techniques. Improved plant cultivars, crop and soil management, climatic conditions, animal genetic selection, and animal management and feeding practices can have a profound effect on trace mineral requirements and availability for large animals. Fifteen microminerals are considered essential in the domestic animal's diet. Trace minerals are important in maintaining normal metabolic function; as animals gradually become deficient, normal function is slowly disrupted. Overt clinical signs of trace mineral deficiencies produce obvious losses in animal health and production. Subclinical deficiencies are also important and can produce subtle but serious losses in production.

SUGGESTED READINGS

1. Allen VG, Horn FP, and Fontenot JP: J Anim Sci 62:1396, 1985.
2. Ammerman CB: J Dairy Sci 53:1097, 1969.
3. Ammerman CB and Miller SM: J Anim Sci 35:681, 1972.
4. Ashmead D and Christy H: Anim Health Nutr 40:10, 1985.
5. Auer DE, Ng JC, and Seawright AA: Aust Vet J 65:317, 1988.
6. Bohman VR and others: Vet Hum Toxicol 29:307, 1987.
7. Boyne R: Res Vet Sci 24:134, 1978.
8. Burch RE and Sullivan JF, editors: Med Clin North Am 60, 1976.
9. Cameron HJ and others: J Anim Sci 67:252, 1989.
10. Chesters JK: World Rev Nutr Diet 32:135, 1978.
11. Combs DK: J Anim Sci 65:1753, 1987.
12. Cymbaluk NF, Bristol FM, and Christensen DA: Am J Vet Res 47:192, 1986.
13. Cymbaluk NF and Christensen DA: Can Vet J 27:206, 1986.
14. Delpelchin BO and others: Vet Rec 116:519, 1985.
15. Egan AR: Trace elements in soil plant animal systems, New York, 1975, Academic Press, p 371.
16. Fisher GL: Sci Total Environ 4:373, 1975.
17. Fitzgerald JA, Everett GA, and Apgar J: Can J Anim Sci 66:643, 1986.
18. Frank A, Pehrson B, and Petersson LR: J Vet Med A 33:422, 1986.
19. Georgievskii VI, Annenkov BN, and Samokhin VT, editors: Mineral nutrition of animals, London, 1982, Butterworth.
20. Givens DI and Hopkins JR: J Agricul Sci 91:13, 1978.
21. Gleed PT and others: Vet Rec 113:388, 1983.
22. Hansard SL: Nutr Abs Rev—Series B 53:1, 1983.
23. Heffner JE and Repine JE: Am Rev Respir Dis 140:531, 1989.
24. Hidiroglou M: Can J Anim Sci 60:579, 1980.
25. Hill CH: In Prasad AS and Oberleas D, editors: Trace elements in human health and disease, vol 2, New York, 1979, Academic Press.
26. Humphries WR and others: Br J Nutr 49:77, 1983.
27. Hurley LS and Baly DL: Clinical, biochemical and nutritional aspects of trace elements, New York, 1982, Alan R Liss, pp 145-159.
28. Hurley WL and Doane RM: J Dairy Sci 72:784, 1989.
29. Ingraham RH and others: J Dairy Sci 70:167, 1987.
30. Irwin MR and others: J Am Vet Med Assoc 174:590, 1979.
31. Kincaid RL: Nutr Rep Inter 493, 1981.
32. Kincaid RL, Blauwiekel RM, and Cronrath JD: J Dairy Sci 69:160, 1986.
33. Kincaid RL, Gay CC, and Krieger RI: Am J Vet Res 47:1157, 1986.
34. Kincaid RL and Hodgson AS: J Daily Sci 72:259, 1989.

35. Klasing KC: J Nutr 118:1436, 1988.
36. Koopman JJ and Wijbenga A: XI Int Congress of Diseases of Cattle, Tel Aviv, 1980, p 1292.
37. Langlands JP and others: Aust J Agric Res 37:179, 1986.
38. Maas JP: Comp Cont Educ 5:S393, 1983.
39. Mills CF: J Anim Sci 65:1702, 1987.
40. Mills CF, Dalgarno AC, Wenham G: Br J Nutr 35:309, 1976.
41. Puls R: Veterinary trace mineral deficiency and toxicity information, Publication 5139, Ottawa, 1981, Agriculture Canada.
42. Rajagopalan KV: Nutr Rev 45:321, 1987.
43. Sanders DE and Koestner A: J Am Vet Med Assoc 176:728, 1980.
44. Schryver HF and Hintz HF: Comp Cont Educ 4:S534, 1982.
45. Schryver HF and others: Cornell Vet 77:122, 1987.
46. Smart ME: Factors affecting plasma and liver copper and zinc concentrations in beef cattle, doctoral dissertation, Saskatoon, 1984, University of Saskatchewan.
47. Smith BP and others: J Am Vet Med Assoc 166:682, 1975.
48. Spears JW: J Anim Sci 67:835, 1989.
49. Spears JW, Harvey RW, and Samsell LJ: J Nutr 116:1873, 1986.
50. Suttle NF: Vet Rec 119:519, 1986.
51. Suttle NF: Vet Rec 119:148, 1986.
52. Suttle NF and Angus KW: J Comp Pathol 86:595, 1976.
53. Towers NR, Young PW, and Wright DE: New Zealand Vet J 29:113, 1979.
54. Ullrey DE: J Anim Sci 44:475, 1977.
55. Ullrey DE: J Anim Sci 65:1712, 1987.
56. Underwood EJ: Commonwealth Agricultural Bureau, Norwich, 1981, Page Brothers.
57. Wiener G: Livestock Production Science 6:223, 1979.

5

Vitamins

JONATHAN M. NAYLOR

The concept of vitamins entered nutrition between 1900 and 1920. Vitamins are essential, organic nutrients that cannot be synthesized by the body. They must be supplied either in food or by microbial synthesis within the gastrointestinal tract. Vitamins are *micronutrients* because they are required in small quantities in the digesta (usually mg to pg amounts per kg of food). In general, vitamins function metabolically either as coenzymes or as hormones. Requirements are small because most vitamins are recycled within the body. Absorption is usually limited to the small intestine.

Most vertebrates share similar requirements for the same vitamins. There are exceptions, however. Vitamin C is an essential vitamin in humans, but farm animals synthesize this vitamin within their own bodies and therefore it is doubtful that it is essential in farm animals. Ruminants and horses are unusual in that the microbial flora of the rumen and large colon produce large amounts of B vitamins. Usually microbial synthesis is sufficient to meet requirements in ruminants and horses without provision of dietary B vitamins.

Vitamin deficiency impairs normal biochemical function within the animal's body. Deficiency progresses through a number of stages. Initially there is depletion of tissue stores, but by itself this has little deleterious effect. This is followed by a decline in serum concentrations of the vitamin and interference with function. Initially the only outward sign may be suboptimal performance. Growth retardation or weight loss develops, and the animal may be predisposed to infectious or traumatic disease. Severe deficiency leads to classic vitamin deficiency disease. These diseases have manifestations that are characteristic of the vitamin involved.

Vitamin deficiency states are most likely to be observed in young, rapidly growing animals. This is partly because of requirements for vitamins in newly synthesized tissue and partly because vitamin-dependent processes may be more active in the young animal. Vitamin requirements may also be increased in geriatric patients because of decreased absorption and increased tissue needs, and in sick animals because of increased requirements for tissue repair and increases in metabolic rate. This chapter is mainly concerned with optimal vitamin nutrition for production and maintenance of health in growing and adult animals. These requirements are reasonably well documented. Recommendations for geriatric and diseased patients are detailed in the species-specific chapters, particularly in the equine section. Our knowledge of requirements in diseased animals is incomplete. To some extent recommendations are based on our interpretation of shifts in vitamin metabolism in disease. These have not usually been validated by controlled trials showing tangible benefits from supplementation at the recommended level.

Vitamins are divided into two main categories, the fat-soluble and water-soluble vitamins. The fat-soluble group consists of vitamins A, D, E, and K. Vitamins A and D function as hormones, whereas E is an antioxidant and K is an enzymatic cofactor. The water-soluble vitamins are the B group vitamins (including biotin), choline, and vitamin C.

FAT-SOLUBLE VITAMINS
Vitamin A

Forms. There are two types of compounds with vitamin A activity. Retinol (vitamin A alcohol) derivatives are the biologically active form of vitamin A in mammals. Plants contain no vitamin A but are rich in carotenoids, which are also known as provitamin A. Carotenoids are yellow, but they are associated with chlorophyll in plants and are therefore most abundant in green leafy foods. Carrots and tomatoes are also rich in carotene. The interconversion of different measures of vitamin A activity is shown in Table 5-1.

There are a large number of carotenoids but β-carotene, which resembles a retinol dimer, is the most important.

Metabolism. During digestion carotene is split into two retinol molecules; these are esterified in the intestinal mucosa and transported in the chylomicrons and chylomicron remnants to the liver.[118,123] The liver stores vitamin A and regulates plasma retinol concentrations. About half the body's vitamin A content is found in the liver, where it is stored as the palmitate ester.[117] Hepatic parenchymal cells release retinol into the plasma complexed to a binding protein.[118,142] This circulates until specific target cells with receptors for the binding protein take up the retinol.

Role. In the retina of the eye, retinol is converted to the aldehyde and used for synthesis of visual pigments. In most other tissues, retinoic acid is the biologically active form. Vitamin A acts as a hormone; retinoic acid is transported into cell nuclei, where it regulates gene expression.[52,118] A major role of vitamin A is maintenance of the functional integrity of simple epithelial surfaces lining the spinal canal, gut, and respiratory and urogenital tracts. In its absence, simple epithelium undergoes squamous metaplasia, a change that is first apparent in the parotid duct.[45,63,114,115,123] In addition, vitamin A is required for bone remodeling. In the horse and cow some β-carotene is absorbed intact across the gut. This imparts the characteristic yellowish hue to horse serum. It is stored in fat and colors adipose tissue and milk. There is some controversy as to whether dietary carotene is necessary for optimal reproduction in cattle or whether synthesis of carotene and vitamin A in the bovine corpus luteum protects against vitamin A deficiency.[7,21,108,120,139]

Sources. The best sources of vitamin A are the carotenoids in green pasture. Typically during the summer grazing season herbivores accumulate large stores of vitamin A that are gradually depleted over the winter. Rapid loss of carotenoids occur when grass is cut for hay making. Losses occur mainly during the wilting stage because of the combined effects of oxygen, moisture, and sunlight. Once hay is dry and baled, loss of vitamin A activity is slow. Cool, dry temperatures favor preservation of carotene. Carotenoids are present in the largest quantities in the leaf; leafiness and color are an approximate guide to carotene content. Leafy, green hays are reasonable sources of carotene. Badly bleached, stemmy hays contain little carotene. Artificially dried grass is rich in carotenoids because rapid drying and the absence of light reduce the opportunity for degradation.[156] Carotenoids are reasonably conserved in the anaerobic condition found in silages. Yellow corn grain contains some carotene. Most grains contain very little carotene (Table 5-2).

Deficiency. Vitamin A deficiency is most likely to occur in rapidly growing animals fed preserved feeds. High-grain diets fed with straw as the primary source of roughage and poor-quality, badly bleached roughages are particularly apt to cause deficiency. Neonatal calves and growing swine require about 4 to 8 weeks on a vitamin A–deficient diet before liver stores become depleted.[21,64] Deficiency may develop even

Table 5-1 *Interconversion of measures of vitamin A activity*

1 IU of vitamin A = 1 USP unit = 0.3 retinol equivalents
1 IU = 0.300 μg of retinol (vitamin A alcohol)
= 0.344 μg of retinyl acetate
= 0.550 μg of retinyl (vitamin A) palmitate

In humans 1 mg of β-carotene = 557 IU of vitamin A

In pigs 1 mg of β-carotene = 80 to 360 IU of vitamin A

In cattle and horses 1 mg of β-carotene = 400 IU of vitamin A

Other provitamins A are approximately half as useful as β-carotene.

more rapidly if the young failed to obtain colostrum (which is rich in vitamin A) or if the colostrum was deficient in vitamin A because the dam had poor vitamin A reserves (as may happen at the end of winter). Yearling calves that have spent a season at grass require 3 to 4 months for reserves to become depleted[82]; clinical signs are not usually seen until 6 months on a deficient diet. Mature animals are quite resistant to vitamin A deficiency. It usually takes at least a year before the reserves of mature adult animals become depleted.[72,107,158]

The determination of deficiency depends on the criteria used to assess vitamin A adequacy. When an animal is initially switched to a vitamin A–deficient diet, hepatic vitamin A stores are mobilized to maintain plasma vitamin A levels. Plasma retinol declines as hepatic stores become exhausted. Reduced growth rate is an early sign of vitamin A deficiency. The earliest change in growing pigs, lambs, and calves is increased cerebrospinal fluid pressure. This is followed by the development of clinical signs of night blindness and edema of the optic nerve (papilledema). Papilledema is the result of impaired bone remodeling in the growing skull. Bone is added to the optic canal dorsally, but resorption from the ventral aspect is retarded. The canal becomes stenotic and constricts the optic nerve. Papilledema has a characteristic ophthalmological appearance. The normal bovine optic disc is small, and the vessels are straight. In papilledema the disc is large and shaped like an inverted heart, and the blood vessels appear tortuous as they bend over the raised rim of the disc. Tapetal hemorrhages and white areas may also be visible on ophthalmoscopic exam[10,18,120,145,155] (Fig. 5-1). Total blindness then develops; this is often the first sign noticed by the farmer. Sometimes blindness is missed because animals can find their way around a familiar environment by memory. The day blindness produced by vitamin A deficiency can be readily localized to a peripheral nerve or retinal problem because of the absence of the menace and pupillary light responses. The pupils are

Table 5-2 *Carotene content of various feeds*

Feed	Carotene, mg/kg of dry matter
Timothy, fresh	230
Timothy, hay	50
Clover, fresh	380
Clover, hay	190
Alfalfa, hay	50
Alfalfa, dehydrated	100
Corn, silage	40
Corn, grain	6
Oats, grain	0.1
Wheat, growing, early vegetative stage	520
Wheat, hay	80
Wheat, straw	2

Modified from data in National Research Council: Nutrient requirements of horses, ed 5, Washington DC, 1989, National Academy Press.

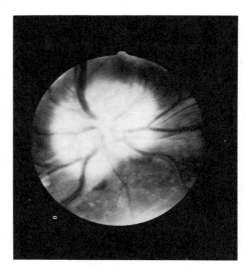

Fig. 5-1 Optic disc of a steer with vitamin A deficiency. The disc is greatly enlarged and heart-shaped instead of round. The border of the disc is indistinct, and the blood vessels curve as they bend over the raised lip of the swollen disc. There are occasional hemorrhages in the outer areas of the disc. Peripheral to the disc, light, mottled areas of damaged, depigmented retina are visible.
From Divers TJ and others: J Am Vet Med Assoc 189:1579, 1986.

fixed and dilated. In contrast, blindness from cerebral cortical dysfunction is characterized by pupils that are responsive to light. Finally ataxia and convulsions develop. These changes are related to compression of spinal nerves. This is due to constriction of the intervertebral foramens by lack of bone remodeling and to degeneration of central nervous system tissue secondary to elevated cerebrospinal fluid pressures. Convulsions are often short-lived, and the animal usually gets up after a few minutes of recumbency and returns to its preconvulsive state.* Vitamin A–deficient animals also tend to be prone to infections at mucosal surfaces, particularly in the enteric and respiratory tracts.[27,145] This is due partly to loss of cilia and goblet cells and partly to compromised macrophage and neutrophil function.[62]

Additional signs that may be observed in cattle are hydrocephalus in neonatal calves,[154] anasarca (subcutaneous edema), heat intolerance, and exophthalmos (protrusion of the eyes).[98,120,145,154] At necropsy a high incidence of pituitary cysts has been observed.[145] Descriptions of vitamin A deficiency in horses are rare but are similar to those in other animals with the exception that day blindness has not been noticed.[61,72] *Xerophthalmia* (thickening of the conjunctiva and drying and opacification of the cornea) is commonly associated with vitamin A deficiency in people but is rare in domestic animals.

Adult animals are quite resistant to vitamin A deficiency. The main effects appear to be confined to the gestating fetus and reproductive failure in adults. The fetus is often affected before signs are noted in the dam or sire. There is a shortened gestation and an increased incidence of stillbirths. Young that are born alive may show blindness or ataxia and fail to thrive. Adult females may show anestrus, and males can have impaired spermatogenesis.

Diagnosis of vitamin A deficiency depends on the presence of clinical signs and the demonstration of vitamin A depletion (Table 5-3).

Toxicity. Toxicity occurs when excessive vitamin A intake saturates retinol-binding protein and free vitamin A circulates in plasma.[142] In consequence nontarget tissues are affected by vitamin A. Most reports of toxicity concern vitamin A intakes about 100 times current recommended levels, but in humans chronic intake of 10 to 20 times the recommended daily allow-

---

*References 9, 27, 37, 40, 42, 64, 98, 108, 114, 145, 155, 158.

Table 5-3 *Serum and liver vitamin A concentrations*

Species	Normal values	Severe deficiency	Units
Plasma			
Bovine	250-550	<100	μg vitamin A/L
	25-55	<10	μg vitamin A/dl
	75-170	<30	IU vitamin A/dl
	1-2	<0.5	μmol vitamin A/L
Porcine	200-300	<50	μg vitamin A/L
	20-30	<5	μg vitamin A/dl
Equine	100-500	<50	μg vitamin A/L
	10-50	<5	μg vitamin A/dl
Liver			
Bovine	>10	<1	μg vitamin A/g
Porcine	>20	<5	μg vitamin A/g

Limited evidence suggests toxic levels for plasma vitamin A are greater than 1 mg/L (100 μg/dl).
Samples for vitamin A analysis should be collected in tubes wrapped in foil to exclude light and then stored in the dark.

ance is toxic.[42,52,107,163] The higher the dose, the shorter the period of time before toxicity signs are observed. The conversion of β-carotene to retinol is reduced at high β-carotene intakes.[137] β-carotene is much less toxic than retinol derivatives. Signs of toxicity include reduced feed consumption, reduced growth, skin erythema in pigs, and rough hair coats in horses. Pigs show stunting, lameness, and *hyperesthesia* (increased reactivity to external stimuli).[163] Horses show depression and ataxia, and the long bones have cortical thinning.[42] Recumbency and death are the final signs in both species.[41,107] Retinol is also teratogenic and causes microphthalmia (small, deformed eyes), crooked legs, and cleft palates in all species studied.

Supplementation. Routine vitamin supplementation of the grain portion of the diet of animals on high-grain, poor-quality roughage is common. Several vitamin premixes are designed to stabilize vitamin A by mixing it with antioxidants and by coating the vitamins to exclude air. Gelatin is used as a coating agent, but it dissolves in humid conditions. Fats may also be used. Destruction of vitamins in moist or liquid feeds is rapid. In dry feeds half the stabilized vitamin A is lost after 6 months of storage.[50,106] An alternative approach is to inject vitamin A. This practice is mainly confined to cattle. Adult and yearling cattle can be supplemented by injection about once every 3 months. Young calves are often given an injection of vitamin A after birth if there is doubt about the amount of colostrum ingested or if the vitamin A content of the colostrum is low because of a poor diet.[41,108]

Vitamin D

Vitamin D was discovered by the efforts of Sir Edward Mellany and of McCollum and Sherman in the early 1920s. These researchers produced rickets (weak, deformed bones) by keeping dogs indoors and feeding them a diet based on oatmeal. The condition could be cured by supplementation with cod liver oil or by exposing the dogs or the diet to ultraviolet irradiation.[38]

Forms. Vitamin D exists in two forms. Vitamin D_2, or ergocalciferol, is formed in plants by ultraviolet irradiation of ergosterol.[159] Vitamin D_3, cholecalciferol, is produced in farm animals exposed to sunlight through the action of ultraviolet irradiation on endogenous 7-dehydrocholesterol.

Vitamin D activity is measured in IU or USP units:

1 IU = 1 USP = Activity of 0.025 μg cholecalciferol[14]

Metabolism. Both forms of vitamin D are processed by the same enzymes located in the liver and kidney to produce the biologically active 1,25 dihydroxyvitamin D_2 or D_3 (Fig. 5-2). Farm animals can make good use of both forms of vitamin D. However, the evidence suggests that vitamin D_3 is more potent. Orally administered vitamin D_3 is preferentially absorbed in cattle and horses,[59,70,143] and pigs[69] hydroxylate vitamin D_3 more effectively than vitamin D_2.

In the kidney hydroxylation produces either the 1,25 dihydroxyvitamin D or a less potent form, 24,25 dihydroxyvitamin D. The production of active 1,25 dihydroxyvitamin D is favored by parathormone (released in response to low serum calcium) and low serum phosphorus concentrations. Although 24,25 dihydroxyvitamin D is thought of as inactive, it may have some effects on bone.[159]

Role. Also known as *calcitriol,* 1,25 dihydroxyvitamin D is a steroid hormone. It circulates in the plasma bound to a binding protein, diffuses into target cells, and is transported to the nucleus, where it effects gene expression.[47,127,159] Its major effect is to stimulate intestinal calcium absorption; it also improves phosphorus and magnesium absorption.[38,68,86,101] Calcitriol increases renal calcium conservation. In the presence of adequate dietary calcium and phosphorus, calcitriol stimulates production of mineralized bone. When dietary calcium is insufficient, active vitamin D stimulates osteoclastic activity, bone resorption, and calcium mobilization from the skeleton. Bone resorption also is favored by high parathormone concentrations.*

A recently recognized function of calcitriol is immune regulation. Calcitriol enhances nonspecific phagocytic defense mechanisms. It stimulates differentiation of myeloid stem cells into macrophages. Defective inflammatory and phagocytic function occur in animals with rick-

---

*References 38, 91, 124, 126, 127, 159.

Vitamins

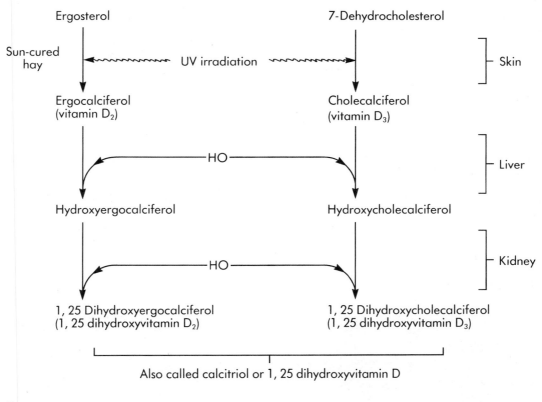

Fig. 5-2 Vitamin D metabolism.

ets. At present it is uncertain whether calcitriol acts locally as a paracrine hormone within the immune system or whether it has systemic effects on macrophage activity.[97,127,150]

Sources. Most feedstuffs are low in vitamin D. However, much vitamin D_2 is formed when plant material is dried (cured) in the sun; in consequence, hay is a good source. Vitamin D deficiency is unlikely to be seen if animals are outside for part of the day or if they are fed hay in reasonable quantities (Table 5-4).

Deficiency. The principal sign of vitamin D deficiency is rickets or osteomalacia (diseases characterized by poorly mineralized bone). Clinical chemistry tests may show hypocalcemia or

Table 5-4 *Vitamin D contents of various feeds*

Item	Vitamin D	
	IU/kg dry matter	IU/lb dry matter
Alfalfa, fresh	220	100
Corn silage	440	200
Wheat straw	660	300
Alfalfa, hay	2,000	900
Brome, hay	2,000	900

Modified from data in National Research Council: Nutrient requirements of horses, ed 5, Washington, DC, 1989, National Academy Press.

hypophosphatemia, but they are not symptomatic. On diets that are marginal in magnesium, signs of hypomagnesemic tetany can be seen (see Chapter 3 for details). The rapidity with which signs develop depends on the stage of growth and severity of deficiency. Young piglets can show signs after a few weeks on a deficient diet. It usually takes several months of deficiency before signs are seen in cattle and horses.* Vitamin D may aid reproductive function, and there are reports of dairy cows on conventional diets benefiting from vitamin D supplementation.[56,108]

Toxicity. Vitamin D toxicity is probably the most common vitamin intoxication. Toxicity is seen following long-term administration of dosages that are above 10 times NRC recommendations. Doses that are 1000 times recommended levels can produce signs within days of administration. Vitamin D_3 is more toxic than vitamin D_2 in farm animals. Intoxication can be seen following inadvertent oversupplementation of rations. Certain plants, notably *Cestrum diurnum*, *Solanum malacoxylon*, and *Trisetum flavescens*, contain 1,25 dihydroxyvitamin D_3 glycosides.[16,75,91,157] These plants are not native to North America, but *Cestrum diurnum* (day-blooming jessamine, day cestrum, wild jasmine) has colonized parts of Florida and Texas and has caused toxicity in animals.[85] The overzealous use of injectable vitamin D as a preventative for milk fever can lead to toxicity syndromes. Vitamin D administration for the prevention of milk fever (hypocalcemia) should be restricted to cows that have a history of previous milk fever. Ten million units of vitamin D_3 can be given by intramuscular injection 2 to 8 days before calving and

---

*References 17, 46, 73, 91, 101, 107, 146.

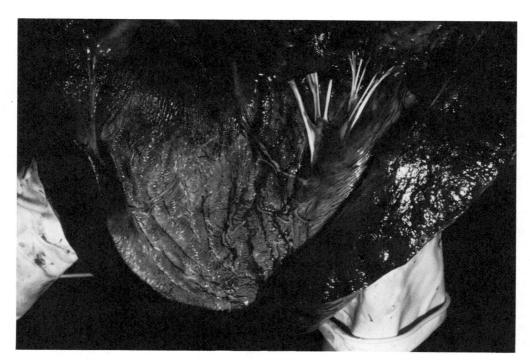

Fig. 5-3 Calcification of the wall of the heart of a cow due to excessive administration of vitamin D. During the dry period, the cow was overtreated with injectable vitamin D to prevent milk fever. Following calving, the cow became weak, spent a lot of time recumbent, and stood with an arched back. She had a poor appetite and lost weight.

repeated a maximum of one time if the cow does not calve.[70,79]

Vitamin D intoxication produces hypercalcemia, hyperphosphatemia, and soft tissue mineralization (Fig. 5-3). Weight loss is the earliest clinical sign in chronic toxicity in horses.[85] Weakness, reluctance to move, and an increased amount of time spent recumbent are common signs.[58,59] Horses show signs of sensitivity and pain when they are palpated,[59,85] and cattle can give a mild positive grunt test. Affected animals can have an elevated heart rate, and the clinical signs can be misdiagnosed as severe infection with shock.[93] Soft tissue calcification is seen at necropsy or on radiography. Calcification is particularly common in the walls of the heart, the large blood vessels, and the kidneys.[59,68,93] Renal dysfunction is fairly common in vitamin D intoxication. Vitamin D also has a direct toxic effect on the gut. In pigs high doses of vitamin D can result in feed refusal and vomiting; a hemorrhagic gastritis is one of the necropsy findings.[93]

Supplementation. Animals that spend most of their time indoors without access to good-quality sun-cured hay should be supplemented with vitamin D. Stabilized vitamin D is usually added to the concentrate part of the diet as part of a multivitamin supplement.

Vitamin E

Vitamin E is a biological antioxidant. It is found in lipids and protects adipose tissue and cell membranes from damage.

Forms. Vitamin E belongs to the tocol group of compounds; d-α-tocopherol is the most common form of vitamin E found in farm animals and in plants. Tocotrienols are also found in plant material, but only α-tocotrienol has significant biological activity. Synthetic forms of α-tocopherol are often racemic mixtures of the d and l forms, but only the d-isomer is biologically active (see box).[4,14,125]

Role. The major role of vitamin E is as an antioxidant. Its protective effects greatly overlap those of the glutathione peroxidase system, a selenium-containing enzyme that inactivates oxidizing agents. Because of this overlap, symptoms of vitamin E and selenium deficiencies are often similar.

Oxidative metabolism generates damaging

Measurement and Relative Potencies of Various Forms of Vitamin E

1 mg dl-α-tocopheryl acetate = 1 IU
1 mg dl-α-tocopherol = 1.1 IU
1 mg d-α-tocopheryl acetate = 1.36 IU
1 mg d-α-tocopherol = 1.49 IU

Relative potencies of other forms of vitamin E in comparison to α-tocopherol:
β-tocopherol 40%
γ-tocopherol 10%
δ-tocopherol 1%
α-tocotrienol 25%

d-α-tocopherol is also known as RRR-α-tocopherol
dl-α-tocopherol is also known as all rac-α-tocopherol.

Modified from Bieri JG and McKenna MC: Am J Clin Nutr 34:289, 1981.

---

by-products,[140] hydrogen peroxide, and the highly reactive free radicals, hydroxyl and superoxide anion, O_2^-. A number of antioxidant systems exist in the aqueous phase to inactivate these compounds. The glutathione peroxidase system, vitamin C, and β-carotene all function as antioxidants.[140,161] These systems spare vitamin E because inactivation of peroxides in the cytosol saves cell membrane lipid from attack (Fig. 5-4). From a nutritional viewpoint, glutathione peroxidase is the most important of these systems.[125] Adequate selenium nutrition almost completely protects against the development of signs of vitamin E deficiency. There are some exceptions, the best example being adipose tissue, which is preferentially protected by vitamin E.[48,105,165] Nutritional requirements for vitamin E are increased when the diet is high in polyunsaturated fat, which is susceptible to damage by oxidizing agents. Exercise increases free radical formation and cell membrane damage[34] as well as susceptibility to vitamin E deficiency.

Sources. The major dietary source of vitamin E is fresh green feed and sprouted grain. Grasses and legumes contain 50 to 200 mg of vitamin E per kg (20 to 90 mg/lb). Hays, silages, grains, and oilseeds are all poor sources, containing 3

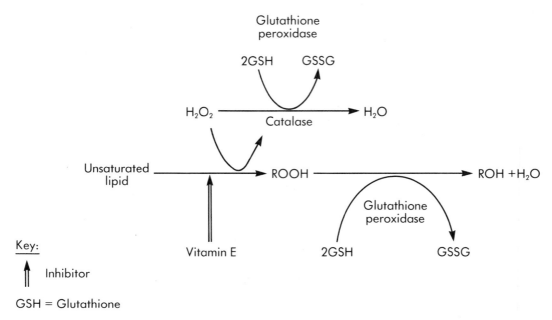

Fig. 5-4 Interrelationships between antioxidant systems.

to 20 mg of vitamin E per kg (1.5 to 9 mg/lb).[32,109,125] There is little transplacental transfer of vitamin E,[154] but colostrum is a rich source, particularly when the dam is on a diet high in vitamin E.[160]

Vitamin E is fairly uniformly distributed within the body. However, levels of vitamin E in fat are several times those in other tissues. There may be some storage in the liver, resulting in higher levels than those found in muscle, when diets containing large amounts of vitamin E are fed.[77,129]

Deficiency. Vitamin E deficiency is most likely to be seen in growing herbivores during the winter feeding period or immediately after they are turned out to fresh pasture in spring. At the end of the winter, vitamin E stores are depleted, and the exercise of spring turnout triggers disease. In pigs, deficiency is associated with indoor rearing and diets high in polyunsaturated fats; *microangiopathy* (damage, blockage, and rupture of the small blood vessels) is a common finding. In all species, deficiency causes myodegeneration (muscle degeneration) and *steatitis (yellow fat disease)*, which is characterized by inflammation of the fatty tissues and results in hard, discolored fat. The onset of clinical signs is often sudden, and rapidly growing animals tend to be preferentially affected.

Myodegeneration may affect the heart, particularly in young animals (under 4 months of age). It usually results in sudden death. In some cases, arrhythmias, tachycardia, cyanosis, or signs of congestive heart failure with peripheral edema may be seen.[48,81,162]

Degeneration of skeletal muscle *(nutritional muscular dystrophy)* appears to be triggered by exercise[6] and is seen in lambs, calves, foals, and possibly horses, which are more active than pigs.[119,138,160,162] In cattle, clinical signs of muscular dystrophy are usually restricted to growing animals under 2 years of age.[119] Signs are often seen following movement or turnout to pasture. At first only a few animals in the group are likely to show clinical signs. Affected animals may be weak, stiff, or reluctant to move. Muscle fasciculations may be seen. Severely affected individuals are recumbent and unable to rise. Serum levels of the muscle-derived enzymes, creatine phosphokinase (CPK) and aspartate serum transferase (AST), are very high. Normal values for CPK are less than 500 IU/L (< 50 IU/dl) in all species; in clinical cases of nutritional muscular dystrophy, values are often greater than

5000 IU/L. Young animals recumbent from other causes typically have some degree of secondary muscle damage with values less than 2000 IU/L. Serum AST values are usually greater than 500 IU/L in nutritional muscular dystrophy, compared with normal values of less than 200 IU/L.[119] The rise in muscle enzyme activity may precede clinical signs, and some clinically normal animals from an affected group also show a rise in serum CPK and AST activity. Myodegeneration can affect the muscles of swallowing and produce a secondary aspiration pneumonia. This is particularly likely in younger animals. An alternative cause of respiratory dysfunction is damage to the intercostal and diaphragmatic muscles resulting in rapid, shallow breathing. Some affected animals have a red-tinged urine,[119] presumably the result of myoglobin release from damaged muscle. There is some controversy as to whether *exertional myopathy* (skeletal muscle damage precipitated by exercise) in horses is the result of vitamin E and selenium deficiency.[60] Some horses have less frequent attacks following vitamin E and selenium supplementation, but other causes are likely.

In pigs, microangiopathy in vitamin E and selenium deficiency is usually widespread. Skin lesions appear as small red hemorrhages about 1 cm in diameter that gradually turn blue.[77] Changes in the liver produce multifocal hemorrhagic hepatic necrosis *(hepatosis dietetica)*.[48,71] Degeneration of cardiac muscle, either as the result of microthrombosis or as direct myodegeneration, is called mulberry heart disease. It is named for the shape and dark red color of the hemorrhages that are found on the epicardium at necropsy in affected swine. Although vitamin E and selenium deficiency produces mulberry heart disease, the finding of normal tissue levels of vitamin E and selenium in some groups of affected pigs indicates that insufficient absorption of these nutrients may not be the only cause.[116] The most common presenting sign in pigs with vitamin E and selenium deficiency is sudden death. Skin hemorrhages, pale mucosae, peripheral edema, and poor growth may be observed in some pigs.[48] At necropsy, signs of microangiopathy, with mulberry heart disease or hepatosis dietetica, are observed. In addition, there may be histological evidence of skeletal or cardiac muscle myodegeneration. Steatitis with yellow discoloration of the fat (reported in swine that have been fed fishmeal) may also be observed.[36,48,107]

Overt vitamin E deficiency may produce nutritional muscular dystrophy in adult animals. Some reports suggest that deficiency may predispose to retention of the placenta in cattle, but other causes also exist.

Vitamin E supplementation can boost humoral immunity, but levels above current NRC recommendations may be required.[8,43,67] Long-term deficiency renders erythrocytes more susceptible to hemolysis.[34,102,147,149] Deficiency can cause nervous tissue degeneration.[112] Vitamin E deficiency may contribute to equine degenerative myeloencephalopathy, but the evidence is not conclusive.[39,67] Very high doses of vitamin E help protect the muscles of stress-susceptible pigs from damage, an effect that appears to be independent of selenium.[44]

Large animals fed diets rich in vitamin E usually have serum and liver vitamin E concentrations greater than 2 mg/L. Deficient animals have concentrations less than 1.5 mg/L (<0.15 mg/dl) in these tissues. Selenium supplementation has little effect on vitamin E concentrations.*

Toxicity. Vitamin E is essentially nontoxic in animals.[12,107] However, supplementation with pharmacological doses of vitamin E in people has led to problems of unexplained death and increased risk of sepsis in premature young.[95,102]

Supplementation. High-grain diets are usually supplemented with vitamin E or selenium to prevent vitamin E and selenium deficiency. Selenium supplementation is usually the cheaper of the two alternatives, although selenium is more toxic than vitamin E. Barley- or corn-based diets contain adequate vitamin E to meet NRC recommendations for swine. Vitamin E supplementation is essential when diets high in polyunsaturated fats (e.g., fish oil) are fed. Vitamin E is reasonably stable in dry feeds but gradually deteriorates in moist feeds. Some stabilized vitamin E supplements are protected in feeds for at least 2 months, even in the presence of moisture.[164]

---

*References 39, 77, 92, 96, 129, 149, 154, 165.

Vitamin K

Vitamin K is the only fat-soluble vitamin that is synthesized by intestinal bacteria. It is also found in green plants and fish meal.

Forms. Vitamin K exists in three forms. Vitamin K_1 is found in plants; it is the safest and most active form. Chemically, vitamin K_1 is called *transphylloquinone* and is a naphthoquinone derivative; the cis isomer of phylloquinone is inactive. Vitamin K_2 exists in the form of a variety of menaquinones and is synthesized by bacteria. Vitamin K_3 is a synthetic naphthoquinone called *menadione*.[14,107,151]

Role. Vitamin K acts as a cofactor in the post-translational modification of blood clotting factors and proteins involved in the control of calcification in bone.[19,104,124] Reduced vitamin K_1 acts as a hydrogen donor in the hepatic conversion of serine proteases to active enzymes of the blood clotting cascade. The hydroxycoumarin type of anticoagulants, warfarin and dicoumarol, block the conversion of oxidized vitamin K_1 back to the active form (Fig. 5-5). Factors II (prothrombin), V, VII, IX, and X are activated by vitamin K, and the major effects of vitamin K deficiency or hydroxycoumarin poisoning are impaired blood clotting manifested as a bleeding tendency.[151] Both intrinsic and extrinsic pathways of blood clotting are compromised. Protein C is also activated by vitamin K; this protein functions as an antithrombotic agent. Thrombosis and skin necrosis are rare manifestations of vitamin K deficiency in people.[151]

Deficiency. Deficiency of vitamin K is uncommon. The vitamin is present in many green feeds. Ruminants synthesize vitamin K in the rumen and absorb it in the small intestine. It is uncertain whether or not coprophagy (eating feces) is necessary for absorption of vitamin K synthesized in the large colon of horses and pigs.

Herbivores show symptoms of vitamin K deficiency when they are fed sweet clover hays contaminated by dicoumarol. Sweet clover is a highly nutritious, drought-resistant plant that can grow as high as 2 m (6 feet). It contains coumarin, which is nontoxic. When sweet clover is preserved, oxygen-dependent fungi may convert the coumarin to toxic dicoumarol. Spoilage is most likely to occur if sweet clover is preserved as hay, particularly if the bales are small and the sweet clover is baled moist. Silage is much less toxic; spoilage in the surface layer can be prevented by placing a top layer of a different feed over the sweet clover and by tightly sealing the trench or silo.[13,54]

Vitamin K_1 deficiency has also been reported in growing swine fed a diet containing antibiotics and high levels of fat. This is a very rare occurrence. It may be related to inhibition of vitamin K absorption by fat. Experimental evidence does not provide strong support for the notion that enteric synthesis of vitamin K is inhibited by oral antibiotics.[135,151]

Although vitamin K is synthesized in the large intestine in pigs, it may be absorbed only

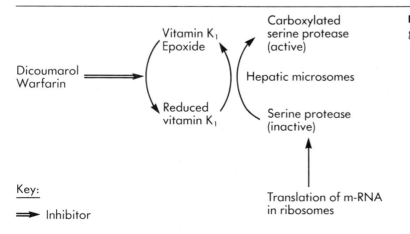

Fig. 5-5 Role of vitamin K as a hydrogen donor.

in the small intestine. Fecal vitamin K reaches the absorptive site as the result of coprophagy or fecal contamination of food. It is unclear to what extent interference with coprophagy contributes to vitamin K deficiency in pigs.[133,134,136]

The main symptom of vitamin K deficiency in pigs or dicoumarol intoxication in cattle is a bleeding tendency. This is usually manifested as hematoma formation and persistent hemorrhage following castration and tail docking. Hematomas tend to form at trauma points—over the widest part of the ribs, the brisket, and hips. Blood loss anemia can be sufficiently severe to result in pale mucous membranes, weakness, and death. Sometimes blood loss is internal. Hemorrhage in the brain can give rise to neurologic signs, lung hemorrhage to breathing difficulties, and joint hemorrhage to lameness. Calves seem more susceptible than cows, and signs of dicoumarol poisoning may be noticed only in perinatal calves born to apparently unaffected dams.[49,134,136]

Presumptive diagnosis of dicoumarol poisoning is usually based on a history of recent feeding of sweet clover, signs of a bleeding tendency, and clinicopathological evidence of increased blood clotting times (whole blood clotting time, prothrombin time, or activated partial thromboplastin time). The feed can be analyzed for dicoumarol at specialist laboratories, but care must be taken to submit a representative sample. The distribution of the poison can be spotty; it is usually found in the highest concentrations in the outer parts of hay bales.

Treatment of vitamin K deficiency in pigs seems to be equally effective whether vitamin K_1, K_2, or K_3 is used. In cattle, however, only vitamin K_1 has been shown to be effective. Large dosages are needed—about 1 to 3 mg/kg (0.5 to 1.5 mg/lb) IM or orally once a day for 3 to 5 days. For poorly understood reasons, vitamin K_3 (menadione) is ineffective in cattle.[3] Contaminated feed should be removed during treatment. Obviously molded feed should be discarded. For economic reasons farmers often wish to use up the contaminated feed. When the cattle have recovered, the remaining feed can be fed on an intermittent schedule. Typically feeding the suspect hay for 2 weeks followed by 3 weeks of other feeds is recommended. Alternatively the spoiled sweet clover can be diluted with other feeds so that it comprises a maximum of 25% of the diet.[54] With severely spoiled feeds, these precautions may be inadequate. All sweet clover feeding should be stopped for at least 3 weeks before and during the calving season. Similar precautions are needed for at least 2 weeks prior to elective surgical procedures. In problem areas always check if sweet clover is fed before castration or dehorning is performed.

Toxicity. Vitamin K_1 is the least toxic of the different forms of vitamin K.[14,103] Horses appear to be particularly sensitive to vitamin K_3 toxicity. Acute renal failure, renal colic, and death were produced by the administration of a commercial vitamin K_3 supplement to horses at the manufacturer's recommended dosage.[55]

Supplementation. Supplementation with dietary vitamin K is not normally required in farm animals. The combination of vitamin K_1 from green plants and enteric synthesis of vitamin K_2 meets requirements.

WATER-SOLUBLE VITAMINS
B vitamins

The B vitamins function as coenzymes and are water soluble. Many are conserved in the body through efficient recycling mechanisms. The nutritional requirements of farm animals for these vitamins vary. Functional ruminants rarely require supplementation because microbes synthesize large quantities of all the B vitamins in the rumen. These are then passed on to the small intestine, where they are absorbed. There is little doubt that horses and pigs have considerable large-intestinal synthesis of all the B vitamins. However, absorption of these vitamins may depend on coprophagy or fecal contamination of food. Pig diets should be fortified with a variety of B vitamins, because intake of conventional diets does not meet requirements (Table 5-5). Experimental studies in horses indicate that large-intestinal synthesis and absorption are inadequate to maintain tissue levels of most B vitamins except pyridoxine[25,26,121,122] (Table 5-6). However, there is little evidence that horses kept under traditional management and feeding systems benefit from B vitamin supplementation.

Brewer's yeast is a good source of all B vitamins except B_{12} and biotin.[109] Rice bran and milk

Thiamin or vitamin B_1 (aneurine, vitamin F)

Role. In cells, thiamin is phosphorylated to the coenzyme thiamin pyrophosphate (TPP).[15] This is required for the decarboxylation of the glycolytic end product pyruvate to acetyl-CoA. This is then used for energy generation in the tricarboxylic acid cycle or for fat synthesis. Thiamin is also required for transketolase reactions in the pentose cycle.

Sources. Most common feeds (grains, oilseeds, grasses, legumes, milk) contain 2 to 5 mg of thiamin per kg (1 to 2 mg/lb) dry matter. Sugar beet pulp and brewer's grains are poor sources with less than 1 mg/kg (0.5 mg/lb). The dietary requirement for adult pigs is only 1.1 mg/kg (0.5 mg/lb) of diet dry matter, so deficiency is rare. Cattle and horses synthesize thiamin in the gut,[26,57,90] but horses subjected to severe stress or exercise may need supplemental dietary sources.[109]

1 IU of thiamin = 1 USP unit = 3 μg thiamin hydrochloride

Deficiency. Clinical signs of thiamin deficiency are observed only when dietary thiamin is destroyed by heat (cooking) or thiaminases in the feed. In ruminants, bacterial thiaminases or inorganic sulfates destroy thiamin in the rumen. Horses suffer from thiamin deficiency when they eat thiaminase-containing plants such as bracken fern (*Pteridium aquilinum*) and horsetails (*Equisetum* spp.). Overt deficiency has not been recognized in horses on standard diets, although growth rate and possibly performance may be improved by thiamin supplementation in some circumstances. Thiamin is one of the first vitamins that becomes deficient when adult horses are fed purified diets; it takes about 4 months before deficiency signs are observed. Oats are rich in thiamin, and supplementation

Table 5-5 *Relationship between dietary intake of B vitamins and nutritional requirements in growing swine (25 kg or 55 lb) fed an unsupplemented corn-soybean meal diet*

Vitamin	Intake as a % of requirements
Thiamin	410
Riboflavin	56
Pantothenic acid	89
Niacin	54
Pyridoxine	600
Biotin	240
Folic acid	116

From data in National Research Council: Nutrient requirements of swine, ed 9, Washington, DC, 1988, National Academy Press.

Table 5-6 *Intestinal B vitamin synthesis in horses*

Vitamin	B vitamin concentrations			Control horses (%) Skeletal muscle
	Dry matter (mg/kg)			
	Feed	Cecum	Rectum	
Thiamin	0.4	4.4	8.5	
Riboflavin*	<0.4	6.6	9.3	50
Pantothenate*	<0.2	<0.1	8.0	10
Niacin*	<0.2	51	15.6	25
Pyridoxine*	<0.01	2.7	10.1	95
Biotin*	<0.01	0.03	0.12	3
Folic Acid*	<0.01	5.3	1.4	30
Inositol*	368	415	425	90

*Values are for a mature horse fed a vitamin-deficient diet for 32 weeks.
Modified from Carroll FD[25] and Linerode PA.[90]

is unlikely to benefit horses on high-oat diets. Pigs may suffer from deficiency if heat treatment of the ration destroys thiamin or if they are fed raw fish containing thiaminases.[25,53,57,107] Very high oral doses of the coccidiostat amprolium (e.g., 400 mg amprolium per kg body weight) result in systemic amprolium absorption. This inhibits the actions of thiamin within the body and can produce signs of thiamin deficiency.[33] The main signs of thiamin deficiency are changes in neurologic and cardiac function.

In ruminants, destruction of thiamin is most common in young animals (cattle younger than 2 years, sheep younger than 6 months) fed diets high in rapidly fermentable carbohydrate (high grain or liquid-molasses based diets). Lush pasture can also be rich in soluble carbohydrate. These diets predispose to ruminal acidosis, which favors thiaminase activity. Two types of bacterial thiaminase exist in the rumen. Type I activity is inhibited by thiamin, and type II activity is not.[57] Thiaminases destroy thiamin and may also produce thiamin analogues that competitively inhibit the action of thiamin within the animal. Thiamin destruction also occurs when the feed or water is rich in sulfate. In the rumen this is converted to sulfite, which destroys thiamin.[57]

In ruminants, neurologic dysfunction is the principal sign of a tissue deficiency of thiamin. Pathological changes seen in the deeper layers of the cerebral cortex give rise to the alternative names for thiamin deficiency of cerebral cortical necrosis (CCN) and polioencephalomalacia. The major pathological findings are destruction of the neurons of the cerebral cortex, cerebral edema, and mild cerebral hemorrhage. These changes are very similar to those produced by salt poisoning and water intoxication. In extreme cases cerebral edema may be so severe that the swollen cortices push caudally through the tentorium cerebelli and the medulla is forced into the foramen magnum.

In ruminants, blindness is an early clinical sign. This is a cortical blindness caused by dysfunction in the cerebral cortices. Pupillary light reflexes are usually intact. This helps differentiate the signs from those caused by vitamin A deficiency, in which the reflexes are absent. If cerebral swelling stretches nerves, strabismus (deviation of the eyes) and constricted pupils may be seen. Common signs of cerebral edema include head pressing and teeth grinding. Lead poisoning causes signs similar to those produced by thiamin deficiency. However, rumen motility is often absent in lead poisoning whereas it is present in early cases of thiamin deficiency. Recumbency and convulsions can also be seen in thiamin deficiency. *Opisthotonos* (dorsiflexion of the head and neck) is an early sign of thiamin deficiency that is particularly common in small ruminants. One should always suspect thiamin deficiency as the likely etiology when examining sheep or goats with opisthotonos. Subclinical deficiency of thiamin is common in herds with clinically affected members; the main herd effect is reduced weight gain.

In swine and horses, reduced appetite, growth, and heart rate are the main signs. Ataxia may be seen in swine and is common in horses. Pigs may die suddenly.[15,26,33,107]

The erythrocyte transketolase reaction is usually used to diagnose thiamin deficiency. The efficiency of this reaction is measured in vitro with erythrocytes incubated with and without supplemental thiamin. When the rate of appearance of sedoheptulose-7-phosphate is used as the marker for transketolase activity, increases of 30% to 50% with thiamin supplementation indicate subclinical deficiency and increases of 80% to 100% generally indicate clinical deficiency. If the rate of disappearance of ribose-5-phosphate is measured, then values greater than 25% indicate deficiency and 15% to 25% subclinical deficiency.[33,57]

Toxicity. Thiamin is virtually nontoxic in animals. High doses have been claimed to have a sedative effect in horses, but controlled, double-blind studies showed no effect from thiamin injection.[76] Occasionally horses injected with high doses show transient signs of excitement.

Usually very high doses, 10 mg of thiamin per kg (5 mg/lb) body weight, are used to treat thiamin-deficiency diseases. Early treatment is important, and thiamin is best given intravenously initially. If possible, four treatments should be given in the first 24 hours, followed by once-a-day treatment. Treatment should be continued until 1 to 2 days after improvement of clinical signs ceases. Oral treatment may be beneficial because type I thiaminases are inhibited by thiamin. Animals that are treated early may recover completely. In others, irreversible changes such as necrosis of neurons may have occurred, and little improvement is seen.

Supplementation. Routine supplementation of diets with thiamin is not usually practiced. However, group supplementation is likely to be worthwhile when some members have shown clinical signs. Routine supplementation may also be indicated in areas with high-sulfate water.

Thiamin, B_2, B_6, and B_{12}, nicotinamide, and pantothenic acid are taken by human athletes.[132] Supplementation with B_1, B_6, and B_{12} may improve fine motor skills. Whether performance horses would show subtle benefits from supplementation is unknown.

Riboflavin or vitamin B_2 (vitamin G)

Role. Riboflavin is a component of the flavin nucleotides, FAD and FMN. Flavin nucleotides are coenzymes for the flavoprotein group of enzymes. They also form part of the electron transport chain in mitochondria. Deficiency impairs cellular respiration.

Sources. Green foods such as grass, legumes, and hays are rich in riboflavin (10 to 20 mg riboflavin per kg [4 to 9 mg/lb] dry matter). Grains are poor sources (less than 2 mg/kg, or 1 mg/lb), and oilseed meals are only slightly better.[15,78,109] For comparison the requirements for adult swine are 4.3 mg/kg (2 mg/lb) of diet dry matter. Neonates receive adequate amounts of this vitamin in milk.

Deficiency. Practical swine diets tend to contain marginal levels of riboflavin, and supplementation improves feed consumption and growth rate.[84,100,107] Riboflavin deficiency has been produced with experimental diets in swine, preruminant calves, and lambs. Severe deficiency produces epithelial lesions, including ulcerative colitis and diarrhea. The hair coat is rough, and a dry, scaly dermatitis develops. Cataracts form in the eyes and may lead to blindness. The gait becomes stiff.[15,83,88,107,153] In early gestation, sows secrete riboflavin into their uterus to support the developing blastocyst. Deficiency of riboflavin leads to anestrus and early embryonic death.[15,107]

Supplementation. Deficiency is prevented by the routine supplementation of grain-based swine diets. Neonates should not need supplementation if fed milk or milk-based products. Unsupplemented soy-based milk replacers are deficient in this vitamin.

Pantothenic acid (vitamin B_3)

Role. Pantothenic acid is a component of coenzyme A, which is required for tricarboxylic acid cycle function and fat metabolism.

Sources. Pantothenic acid is found in high levels in alfalfa (30 mg/kg or 14 mg/lb dry matter). Most other feeds contain pantothenic acid, but the levels are generally somewhat lower than the dietary requirement for adult swine of 14 mg/kg (6.4 mg/lb) dry matter. Corn grain contains about 6 mg pantothenic acid per kg dry matter (3 mg/lb).[107,109]

Deficiency. True deficiency syndromes have not been described in ruminants or horses. In swine, mild B_3 deficiency slows growth. Requirements are influenced by the levels of vitamin B_{12} in the diet and by the presence of antibiotics. Both of these factors reduce the need for pantothenic acid.[28] A cold environment increases requirements. In growing pigs, severe deficiency produces gastrointestinal ulceration with vomiting and diarrhea. There is a rough hair coat and dry skin. In advanced cases, characteristic nervous signs of a goose-stepping hind limb gait and standing on one leg and kicking rhythmically with the other leg are seen.[15,74,94,107] In adults, poor reproductive performance is the main effect of deficiency. Poor immune function can also result from pantothenic acid deficiency.

Supplementation. Most swine diets are routinely supplemented with pantothenic acid to optimize growth rates and reproductive performance. This is particularly important when corn-based diets are fed. Animals whose diets are based on small grains (e.g., barley, wheat) are less likely to benefit from supplementation.[20] Calcium dl pantothenate is usually used as a supplement because it is more stable than the acid form of the vitamin. Only the d isomer is biologically active. Calcium dl pantothenate has 46% of the activity of d pantothenic acid.[107]

Niacin (vitamin PP)

Role. Niacin and nicotinic acid are biologically active as nicotinamide, a component of the coenzymes NAD and NADP. These play a vital role in the tricarboxylic acid cycle and in glycolysis.

Sources. Niacin is found in most common feedstuffs at levels of 10 to 200 mg/kg (5 to 90 mg/lb) dry matter. Canola meal contains high levels, and brewer's yeast contains nearly 500 mg/kg (200 mg/lb).[107] Ruminants synthesize niacin in their rumen, and much of this is absorbed in the small intestine. In addition most mammals (but not cats) can synthesize niacin directly from dietary tryptophan. This option, however, is available only if the diet contains an excess of tryptophan over that needed for protein synthesis. Corn is deficient in both niacin and tryptophan.

Deficiency. Normally, dietary levels exceed the requirement for adult pigs of 11 mg niacin per kg of dry matter (5 mg/lb). However, deficiency may be observed on practical diets because the niacin in cereal grains is bound and has a poor availability. Some studies show benefits of niacin supplementation in swine. There are also suggestions that high-producing dairy cows and rapidly growing beef cattle may benefit from niacin supplementation.

The classic sign of niacin deficiency is pellagra, a disease that was common in people in the southern United States until the 1930s, when the cause was discovered. In swine, niacin deficiency produces slow growth, anorexia, dermatitis, vomiting and diarrhea, rough hair coat, stiffness, paralysis of the hind quarters, irritability, and anemia.[15,74,107]

Supplementation. High-producing cattle may benefit from niacin supplementation. In high-yielding dairy cows, niacin supplementation results in small, usually insignificant increases in milk production.[57,108] Niacin also may reduce fat mobilization and decrease the incidence of ketosis and fatty liver.[57,108] Therefore, there may be benefits to supplementation in early lactation or late gestation if milk yield is high or the cows are obese. In finishing beef cattle, niacin supplementation tends to produce small improvements in growth rate and feed efficiency (0% to 10%). Benefits are most apparent in the early finishing period and are lost later in the finishing phase.[57]

Swine diets usually are supplemented with niacin, although this may not always be necessary.[15,107] Cattle rations are not routinely supplemented with niacin. In cattle, daily intake of 6 g of niacin per head has been recommended.[51] Milk replacers may require niacin fortification.[108]

Pyridoxine (vitamin B_6)

Role. Pyridoxine is required for the coenzyme pyridoxal phosphate. The production of active pyridoxal phosphate within the body depends on NAD (from niacin) and FAD (from riboflavin). Thus niacin or riboflavin deficiency can produce signs of pyridoxine deficiency.[15]

Deficiency. Pyridoxine deficiency in pigs results in poor growth, discharge around the eyes, ataxia, periodic convulsions, vomiting, diarrhea, and anemia. The anemia is due to an inability to synthesize heme and is *microcytic* (small red cells) and *hypochromic* (low hemoglobin content) with elevated serum iron concentrations.[15,87]

Supplementation. There is no known requirement for pyridoxine supplementation in farm animals on traditional diets. Most feeds contain 2 to 10 mg of pyridoxine per kg dry matter (1 to 5 mg/lb), which is greater than the dietary requirement for adult swine of 1.1 mg/kg (0.5 mg/lb) of dry diet.[107,109] Deficiency can be produced in swine only by using experimental purified diets in housing that is kept scrupulously clean.[74] Environmental pyridoxine from fecal material, bacteria, and fungi is more than sufficient to meet requirements in swine[74] and presumably other species.

Biotin (vitamin H)

Role. Biotin is a coenzyme that takes part in biological carboxylation and decarboxylation reactions. It is involved in the tricarboxylic acid cycle, gluconeogenesis, and fat synthesis.[15,107]

Sources. Excellent levels of biotin are found in canola meal, safflower meal, young green grasses, growing green cereals, and legumes (greater than 0.4 mg biotin per kg [0.2 mg/lb] dry matter, highly available). Adequate levels are found in soybean meal and oats (0.3 mg/kg, or 0.14 mg/lb); most grains contain low levels of biotin (less than 0.2 mg/kg or 0.1 mg/lb), much of which is bound and poorly available.[5,32,109,135] Oral antibiotics reduce the intestinal synthesis of biotin. The requirement for adult pigs is 0.2 mg/kg of diet (0.1 mg/lb).[107] A dietary requirement has not been established for ruminants or horses.

Deficiency. Traditional diets contain enough biotin to support optimal growth of pigs. Deficiency affects the integument, particularly the hooves. Both growing and adult pigs fed biotin-deficient rations have soft, friable hooves and a higher incidence of foot lesions. However, this does not always translate into an increased incidence of lameness.* More severe deficiency can produce noticeable coat changes with a reduction in the amount of hair and a dry, scaly skin that is more sensitive to abrasions.[15] In adult sows, biotin deficiency reduces reproductive performance, increases the interval from weaning to first estrus, and reduces litter size.[15,23,107]

There is some evidence that horses may suffer biotin deficiency. Biotin is one of the vitamins that shows the greatest amount of tissue depletion on deficient diets (Table 5-7). Horses with soft, crumbly hooves or hooves that are misshapen and horizontally ridged with areas of broken hoof wall may respond to biotin supplementation. Daily dosages of 10 to 30 mg of biotin per horse have been used. Because growth of new hoof is slow, it usually takes at least 3 months of supplementation before improvement is seen[31]; 9 months may be needed before the cure is complete.

Biotin deficiency has been produced experimentally in neonatal calves. Signs of paralysis and incoordination are seen.[83]

Supplementation. The need for biotin supplementation in the diets of growing pigs depends on the perceived benefits of reducing hoof defects and lameness in the herd. Adult swine should probably be supplemented to ensure optimal hoof condition and reproductive performance. Horses may benefit from supplementation if they are afflicted with weak, crumbly hooves.

Folacin or folate (vitamin B_C, vitamin M)

Role. Folate is active in the body as tetrahydrofolate. Its main function is in the transfer of the single carbon units that are required for synthesis of purines and pyrimidines, which are components of DNA. Folate can also transfer methyl groups to homocysteine to form methionine.[15,29] There are complex interrelationships between folate and the other B vitamins. Folate and vitamin B_{12} cooperate in the synthesis of methionine. Adequate dietary methionine partially overcomes the effects of folate deficiency.[29] Pyridoxine and vitamin C deficiencies exacerbate folate deficiency.[15]

Sources. Green feeds (e.g., grasses, legumes, green hay) are excellent sources of folate (greater than 1.0 mg folate per kg or 0.5 mg/lb dry matter). Barley and soybean meal are good sources (0.6 to 1.0 mg/kg, or 0.3 to 0.5 mg/lb). The other grains contain levels that are usually slightly above the minimum recommended intake for adult pigs of 0.35 mg folate per kg (0.16 mg/lb).[2,107,109]

Deficiency. Folate deficiency is not usually seen in animals fed traditional diets. Occasional studies show growth rate benefits of supplementation in young pigs and litter size benefits in supplemented sows.[89,107] The classic signs of deficiency are poor growth, fading hair coat, hypersegmented neutrophils, large platelets, and macrocytic anemia.[15,29,65,107]

Supplementation. The combination of dietary folate and synthesis in the gut usually supplies sufficient folate to farm animals without the need for further supplementation. Some reports have suggested that horses stabled long-term on poor-quality hay or under severe stress may benefit from supplementation, although conclusive evidence is lacking. Doses needed are not well established.

Vitamin B_{12} (cobalamins)

Forms. Cobalamins consist of a nucleoside side chain and a *corrinoid ring* (which resembles heme) containing a central cobalt molecule. There is a wide variety of corrinoids in nature, some of which are detected by standard bacterial assays for vitamin B_{12}. However, only cobalamins are biologically active in animal tissue.[66] A variety of biologically active cobalamins exist, including cyanocobalamin, which contains a cyanide grouping.

Role. Like folate, vitamin B_{12} functions as a coenzyme in transmethylation reactions, and deficiency impairs DNA synthesis. Vitamin B_{12} is also required for tricarboxylic acid cycle function. This is particularly important in ruminants because propionate, a major end product of ruminal fermentation of carbohydrates, can neither enter the tricarboxylic acid cycle nor be used

---

*References 15, 22, 24, 80, 107, 141.

for gluconeogenesis in the absence of B_{12}. Propionate is an important substrate in ruminants, and these effects of vitamin B_{12} deficiency on energy metabolism help explain the signs of poor growth and ketosis seen in cobalt-deficient ruminants.

Metabolism. The metabolism of vitamin B_{12} is well understood in people,[66] and the mechanisms involved may also be applicable to monogastric animals.[11] Vitamin B_{12} in food is associated with protein. In the stomach, acid hydrolysis frees corrinoids and cobalamines, which attach to salivary R factor, a corrinoid-binding protein that is digested in the small intestine. *Intrinsic factor,* a mucoprotein synthesized in the stomach, then specifically binds vitamin B_{12}. Ileal receptors recognize intrinsic factor, and the complex is absorbed by endocytosis. There is also very efficient enterohepatic recirculation of B_{12}. Vitamin B_{12} is excreted in the bile complexed to R factor, digested, and reclaimed in the small intestine. Because of this recycling, dietary requirements are less than 1 μg/day in people. Indeed, it takes about 20 years before adult humans consuming deficient diets manifest deficiency symptoms (anemia, pancytopenia, neuropathy). However, if absorption is impaired by lack of intrinsic factor or gut disease, then deficiency signs occur much more rapidly because endogenous B_{12} is lost. Colonic bacteria synthesize large amounts of vitamin B_{12}, but it is not absorbed. Therefore, strict vegetarians (vegans) who eat food free from fecal contamination develop signs of vitamin B_{12} deficiency, which can be cured with a water-based supplement made from their own fecal material.[66]

Sources. Cobalamins are synthesized by enteric bacteria and by certain bacteria associated with root nodules of legumes. Fermented soy products contain corrinoids, but these are not active in human nutrition. Dietary intake of B_{12} in herbivores arises from contamination of food with fecal matter or root nodule material. Ruminants directly absorb vitamin B_{12} synthesized by rumen microbes. Animals accumulate vitamin B_{12} in their tissues; therefore, meat, milk, and eggs are rich in vitamin B_{12}.

Pigs synthesize vitamin B_{12} in the large colon and probably absorb it as the result of ingestion of fecal material.[15,107] Vitamin B_{12} deficiency can be produced experimentally in baby pigs.[113] However, naturally occurring cases of deficiency have not been observed. Vast amounts of vitamin B_{12} are synthesized in the equine large intestine (about 2000 μg/day).[36,148] An enterohepatic recirculation occurs in horses,[131] but it functions at the level of the small intestine. When horses are placed on deficient diets, blood levels rapidly fall and then stabilize at a low level despite large-scale intestinal synthesis.[148] This suggests that little colonic B_{12} is directly absorbed. Because fecal production is several hundred times the likely requirement, only a small amount of food has to be contaminated with fecal vitamin to maintain adequacy. Studies clearly demonstrate that horses on natural diets have adequate vitamin B_{12} status. If a horse is injected with vitamin B_{12}, all the vitamin is promptly excreted in feces and urine.[1]

However, synthesis of vitamin B_{12} in the gut is dependent on an adequate supply of cobalt. Ruminants are particularly susceptible to cobalt deficiency. Several months are required before animals on cobalt-deficient pastures manifest signs of deficiency. The major signs are a gradual loss of appetite, poor growth, and emaciation. In lactating cattle, cobalt deficiency produces ketosis. Anemia may develop with accumulation of iron as hemosiderin because production of new red cells is blocked.

Deficiency. Deficiency of cobalt in sheep is associated with plasma vitamin B_{12} levels under 250 ng/L.[128,152] This cutoff point is reasonably reliable as long as the sheep are not off feed. Prolonged feed withdrawal increases serum B_{12} levels twofold to threefold.[99] Methylmalonic acid accumulates in the serum of vitamin B_{12}–deficient animals; values above 5 μmol/L (60 μg/dl) indicate deficiency.[128] Work in cattle suggests that serum vitamin B_{12} less than 200 ng/L and methylmalonic acid concentrations greater than 4 μmol/L are associated with deficiency. Methylmalonic acid concentrations between 2 and 4 μmol/L are associated with subclinical deficiency and also with impaired neutrophil function.

Supplementation. Dietary supplementation with cobalamins is not required in monogastric large animals on conventional diets. Ruminants grazing cobalt-deficient pasture can be supplemented with vitamin B_{12}, but it is usually cheaper to supplement with cobalt.

Choline

Role. Choline is not a true B vitamin because it does not function as a coenzyme. It acts as a source of methyl groups and as a component of phospholipids and acetylcholine.[15,107]

Sources. Requirements are high but so is dietary intake. Methionine has a sparing effect on choline. Most common feeds for swine contain over 1 g of choline per kg (0.5 g/lb), but corn grain contains less (about 0.6 g/kg or 0.3 g/lb).

Deficiency. Severe experimental deficiency of choline in growing pigs produces poor growth, anemia, and ataxia.[107] Deficiency has not been reported in ruminants or horses.

Supplementation. Conventional swine diets contain more than enough choline to support adequate growth. In growing and finishing swine, supplementation is detrimental; growth rates are reduced at high levels of supplementation. Pregnant sows on corn- and soy-based diets benefit from choline supplementation as evidenced by improved conception rates and increased number of live pigs born. The NRC recommends an intake of 1.4 g/kg (0.6 g/lb) of dietary dry matter.[107] There are no lactation benefits to supplementation of conventional diets.*

Ascorbic acid (vitamin C)

Role. Ascorbic acid functions as a water-soluble biological antioxidant.[140] It also is required for cross-linking of collagen.[15] Poor collagen cross-linking leads to the symptoms of scurvy in humans.

Sources. Ascorbic acid is an essential dietary vitamin only in primates and a few unusual species such as the fruit-eating bat and the red vented bulbul bird. Farm animals synthesize vitamin C within their tissues.[83] Most fresh forages and leafy plants contain vitamin C. Citrus fruits are particularly good sources of the vitamin.

Supplementation. There is no convincing evidence of the need for supplemental vitamin C in the diets of farm animals. Ruminants destroy dietary vitamin C in the rumen. Reviews indicate that overall small growth responses (about 2%) can occur with vitamin C supplementation of growing and finishing swine. However, this observation may be biased because trials that

---

*References 15, 107, 110, 111, 130, 144.

show no benefit of vitamin C supplementation are less likely to be reported.[30] Studies of horses supplemented with vitamin C do not yield any convincing evidence of a need for supplementation.[109,148] However, plasma levels of vitamin C are decreased in horses following prolonged transportation or severe illness.

OTHER VITAMINS
Inositol

Inositol is found in plant phytates. High levels are present in a variety of feeds. No nutritional requirement for this substance has been demonstrated in large animals.

Vitamin P complex

The vitamin P complex substances are a group of bioflavonoids obtained from citrus pulp. They may have a role in maintaining capillary permeability, but no dietary requirement has been demonstrated in large animals.

CONCLUSION

Vitamins are essential nutrients required in small quantities in the diet. The fat-soluble vitamins are A, D, E, and K. Vitamins A and D function as steroid hormones. Vitamin A is necessary for the integrity of epithelial surfaces, vision, and bone remodeling. Vitamin D promotes intestinal calcium absorption and bone calcium mobilization. Vitamin E functions as an antioxidant; it is particularly important in free-radical reduction in adipose tissue. Vitamin K is concerned with the synthesis of the blood clotting proteins. It is the only fat-soluble vitamin that can be synthesized by rumen bacteria.

The B and C vitamins are water soluble. The B vitamins function as enzyme cofactors. Vitamin C is synthesized in the tissues of farm animals and is not usually required in the diet. Ruminants and horses do not usually require B vitamins in their diets because of rumen and colonic microbial synthesis of these vitamins. Pigs require dietary supplementation with many B vitamins. Thiamin is involved in energy metabolism; deficiency results in neurologic signs and may arise from the presence of thiaminases. Vitamin B_{12} occupies a unique position in ruminant diets because a dietary supply of cobalt is necessary for its synthesis. Deficiency leads to weight loss, ketosis, and anemia.

REFERENCES

1. Alexander F and Davies ME: Br Vet J 125:169, 1969.
2. Allen BV: Vet Rec 103:257, 1978.
3. Alstad AD and others: J Am Vet Med Assoc 187:729, 1985.
4. Ames SR: J Nutr 109:2198, 1979.
5. Anderson PA, Baker DH, and Mistry SP: J Anim Sci 47:654, 1978.
6. Arthur JR: J Nutr 118:747, 1988.
7. Austern BM and Gawienowski AM: Lipids 4:227, 1969.
8. Baalsrud KJ and Overnes G: Equine Vet J 18:472, 1986.
9. Barlow R: In Practice 11:64, 1989.
10. Barnett KC and others: Br Vet J 126:561, 1970.
11. Batt RM and Horadagoda NU: Am J Physiol 257:G344, 1989.
12. Bendich A and Machlin LJ: Am J Clin Nutr 48:612, 1988.
13. Benson ME, Casper HH, and Johnson LJ: Am J Vet Res 42:2014, 1981.
14. Bieri JG and McKenna MC: Am J Clin Nutr 34:289, 1981.
15. Blair R and Newsome F: J Anim Sci 60:1508, 1985.
16. Boland RL: Biomed Environ Sci 1:414, 1988.
17. Bonniwell MA and others: Vet Rec 122:386, 1988.
18. Booth A, Reid M, and Clark T: J Am Vet Med Assoc 190:1305, 1987.
19. Bovill EG and Mann KG: Adv Exp Med Biol 214:17, 1986.
20. Bowland JP and Owen BD: J Anim Sci 11:757, 1952.
21. Brief S and Chew BP: J Anim Sci 60:998, 1985.
22. Bryant KL and others: J Anim Sci 60:136, 1985.
23. Bryant KL and others: J Anim Sci 60:145, 1985.
24. Bryant KL and others: J Anim Sci 60:154, 1985.
25. Carroll FD: J Anim Sci 9:137, 1950.
26. Carroll FD, Goss H, and Howell CE: J Anim Sci 8:290, 1949.
27. Carrigan MJ, Glastonbury JR, and Evers JV: Aust Vet J 65:158, 1988.
28. Catron DV and others: J Anim Sci 12:51, 1953.
29. Chanarin I: J Clin Pathol 40:978, 1987.
30. Chiang SH and others: Nutr Rep Inter 31:573, 1985.
31. Comben N, Clark RJ, and Sutherland DJB: Vet Rec 115:642, 1984.
32. Cuddeford D: In Practice 11:211, 1989.
33. Cymbaluk NF, Fretz PB, and Loew FM: Am J Vet Res 39:255, 1978.
34. Davies KJA and others: Biochem Biophys Res Commun 107:1198, 1982.
35. Davies ME: Br Vet J 127:34, 1971.
36. Davis CL and Gorham JR: Am J Vet Res 15:55, 1954.
37. Davis TE, Krook L, and Warner RG: Cornell Vet 60:90, 1970.
38. DeLuca HF: Proc Soc Exp Biol Med 191:211, 1989.
39. Dill SG and others: Am J Vet Res 50:166, 1989.
40. Divers TJ and others: J Am Vet Med Assoc 189:1579, 1986.
41. Donoghue S: Int J Vitam Nutr Res 58:3, 1988.
42. Donoghue S and others: J Nutr 111:365, 1981.
43. Droke EA and Loerch SC: J Anim Sci 67:1350, 1989.
44. Duthie GG and others: Res Vet Sci 46:226, 1989.
45. Eaton HD and others: J Dairy Sci 53:1775, 1970.
46. El Shorafa WM and others: J Anim Sci 48:882, 1979.
47. Ena JM and others: Biochem Int 19:1, 1989.
48. Ewan RC and others: J Anim Sci 29:912, 1969.
49. Fraser CM and Nelson J: J Am Vet Med Assoc 135:283, 1959.
50. Fritz JC and others: Fed Proc 15:551, 1956.
51. Gerloff BJ: Vet Clin North Am (Food Anim Pract) 4:379, 1988.
52. Goodman DS: N Engl J Med 310:1023, 1984.
53. Gooneratne SR and others: Can Vet J 30:139, 1989.
54. Goplen BP: Can Vet J 21:149, 1980.
55. Green EM and Green SL: Mod Vet Prac: 625, 1986.
56. Halloran BP: Proc Soc Exp Biol Med 191:227, 1989.
57. Harmeyer J and Kollenkirchen U: Nutr Res Rev 2:201, 1989.
58. Harrington DD: J Am Vet Med Assoc 180:867, 1982.
59. Harrington DD and Page EH: J Am Vet Med Assoc 182:1358, 1983.
60. Harris P: In Practice 11:3, 1989.
61. Hart GH, Goss H, and Guilbert HR: Am J Vet Res 4:162, 1943.
62. Hatchigian EA and others: Proc Soc Exp Biol Med 191:47, 1989.
63. Hayes KC: Nutr Rev 29:3, 1971.
64. Hentges JF Jr and others: J Am Vet Med Assoc 120:213, 1952.
65. Herbert V: Am J Clin Nutr 46:387, 1987.
66. Herbert V: Am J Clin Nutr 48:852, 1988.
67. Hintz HF: Equine Pract 9:6, 1987.
68. Hintz HF and others: J Anim Sci 36:522, 1973.
69. Horst RL, Napoli JL, and Littledike ET: Biochem J 204:185, 1982.
70. Horst RL and Reinhardt TA: J Dairy Sci 66:661, 1983.
71. Hove EL and Seibold HR: J Nutr 56:173, 1955.
72. Howell CE, Hart GH, and Ittner NR: Am J Vet Res 2:60, 1941.
73. Huffman CF, Duncan CW, and Lightfoot CC: J Dairy Sci 18:511, 1935.
74. Hughes EH and Squibb RL: J Anim Sci 1:320, 1942.
75. Hughes MR and others: Nature 268:347, 1977.
76. Irvine CHG and Prentice NG: N Z Vet J 10:86, 1962.
77. Jensen M and others: J Anim Sci 66:3101, 1988.
78. Jones TC, Maurer FD, and Roby TO: Am J Vet Res 19:67, 1945.
79. Julien WE and others: J Dairy Sci 60:431, 1977.
80. Kempson SA, Currie RJW, and Johnston AM: Vet Rec 124:37, 1989.
81. Kennedy S and Rice DA: Am J Pathol 130:315, 1988.
82. Kohlmeier RH and Burroughs W: J Anim Sci 30:1012, 1970.
83. Kon SK and Porter JWG: Nutr Abs Rev 17:31, 1947.
84. Krider JL, Terrill SW, and Van Pouche RF: J Anim Sci 8:121, 1949.

85. Krook L and others: Cornell Vet 65:26, 1975.
86. Lachenmaier Currle U and Harmeyer J: Biol Neonate 53:327, 1988.
87. Lehrer WP Jr and others: J Anim Sci 10:65, 1951.
88. Lehrer WP Jr and Wiese AC: J Anim Sci 11:244, 1952.
89. Lindemann MD and Kornegay ET: J Anim Sci 63:35, 1986.
90. Linerode PA: Am Assoc Equine Pract 13:283, 1967.
91. Littledike ET and Goff J: J Anim Sci 65:1727, 1987.
92. Logan EF and others: Vet Rec 126:163, 1990.
93. Long GG: J Am Vet Med Assoc 184:164, 1984.
94. Luecke RW, McMillen WN, and Thorp F Jr: J Anim Sci 9:78, 1950.
95. Machlin LJ: Int J Vitam Nutr Res (suppl) 30:6, 1989.
96. Maenpaa PH, Pirhonen A, and Koskinen E: J Anim Sci 66:1424, 1988.
97. Manolagas SC, Hustmyer FG, and Xiao-Peng Yu: Proc Soc Exp Biol Med 191:238, 1989.
98. Markusfeld O: J Am Vet Med Assoc 195:1123, 1989.
99. Millar KR, Albyt AT, and Bond GC: N Z Vet J 32:65, 1984.
100. Miller CO and Ellis NR: J Anim Sci 10:806, 1951.
101. Miller ER and others: J Nutr 83:140, 1964.
102. Mino M: Int J Vitam Nutr Res Suppl 30:69, 1989.
103. Molitor H and Robinson HJ: Proc Soc Exp Biol Med 43:125, 1940.
104. Mount ME, Feldman BF, and Buffington T: J Am Vet Med Assoc 180:1354, 1982.
105. Muth OH and others: Science 128:1090, 1958.
106. Myburgh SJ: Onderstepoort J Vet Res 29:269, 1962.
107. National Research Council: Nutrient requirements of swine, ed 9, Washington, DC, 1988, National Academic Press, p 34.
108. National Research Council: Nutrient requirements of dairy cattle, ed 6, Washington, DC, 1989, National Academic Press, pp 42-44.
109. National Research Council: Nutrient requirements of horses, ed 5, Washington, DC, 1989, National Academic Press.
110. NCR-42 Committee on Swine Nutrition: J Anim Sci 42:1211, 1976.
111. NCR-42 Committee on Swine Nutrition: J Anim Sci 50:99, 1980.
112. Nelson JE: Biology of vitamin E (Ciba Foundation Symposium 101), London, 1983, Pitman Books, pp 92-105.
113. Neumann AL and others: J Anim Sci 9:83, 1950.
114. Nielsen SW and others: Res Vet Sci 7:143, 1966.
115. Nielsen SW and others: Am J Vet Res 27:223, 1966.
116. Nielsen TK and others: Vet Rec 124:535, 1989.
117. Nutr Rev 44:311, 1986.
118. Nutr Rev 45:221, 1987.
119. Oksanen HE: Acta Vet Scand 6(suppl 2):1, 1965.
120. Paulsen ME and others: J Am Vet Med Assoc 194:933, 1989.
121. Pearson PB and Schmidt H: J Anim Sci 7:78, 1948.
122. Pearson PB, Sheybani MK, and Schmidt H: J Anim Sci 3:166, 1944.
123. Phillips RW: In Booth NH and McDonald LE, editors: Veterinary pharmacology and therapeutics, ed 6, Ames, 1988, Iowa State University Press, pp 688-697.
124. Price PA: Adv Exp Med Biol 214:55, 1986.
125. Putnam ME and Comben N: Vet Rec 121:541, 1987.
126. Reinhardt TA, Horst RL, and Goff JP: Vet Clin North Am (Food Anim Pract) 4:331, 1988.
127. Reinhardt TA and Hustmyer FG: J Dairy Sci 70:952, 1987.
128. Rice DA and others: Vet Rec 121:472, 1987.
129. Roneus BO and others: Equine Vet J 18:50, 1986.
130. Russet JC and others: J Anim Sci 49:708, 1979.
131. Salminen K: Acta Vet Scand 16:84, 1975.
132. Saris WH and others: Int J Vitam Nutr Res Suppl 30:205, 1989.
133. Sasaki Y and others: Jpn J Vet Sci 44:933, 1982.
134. Sasaki Y and others: Jpn J Vet Sci 47:435, 1985.
135. Scheiner J and DeRitter E: J Agricul Food Chem 23:1157, 1975.
136. Schendel HE and Johnson BC: J Nutrition 76:124, 1962.
137. Schoene F and others: Arch Tierernahr 38:193, 1988.
138. Schougaard H and others: Nord Vet Med 24:67, 1972.
139. Schweigert FJ and Zucker H: J Reprod Fertil 82:575, 1988.
140. Sies H: Int J Vitam Nutr Res Suppl 30:215, 1989.
141. Simmins PH and Brooks PH: Vet Rec 122:431, 1988.
142. Smith JE and Goodman DS: Fed Proc 38:2504, 1979.
143. Sommerfeldt JL and others: J Dairy Sci 64(suppl 1):157, 1981.
144. Southern LL and others: J Anim Sci 62:992, 1986.
145. Spratling FR and others: Vet Rec 77:1532, 1965.
146. Spratling FR and others: Br Vet J 126:316, 1970.
147. Stevenson LM and Jones DG: J Comp Pathol 100:359, 1989.
148. Stillions MC, Teeter SM, and Nelson WE: J Anim Sci 32:252, 1971.
149. Stowe HD: Am J Clin Nutr 21:135, 1968.
150. Suda T: Proc Soc Exp Biol Med 191:214, 1989.
151. Suttie JW: Hepatology 7:367, 1987.
152. Suttle NF and others: Vet Rec 126:192, 1990.
153. Terrill SW and others: J Anim Sci 14:593, 1955.
154. van der Lugt JJ and Prozesky L: Onderstepoort J Vet Res 56:99, 1989.
155. Van Donkersgoed J and Clark EG: Can Vet J 29:925, 1988.
156. Waite R and Sastry KNS: J Agricul Sci 39:174, 1949.
157. Wasserman RH and others: Science 194:853, 1976.
158. Webb KE Jr, Mitchell GE Jr, and Little CO: J Anim Sci 30:941, 1970.
159. West JB: Hormonal regulations of mineral metabolism. In JB West, editor: Best and Taylor's physiological basis of medical practice, ed 11, Baltimore, 1985, Williams & Wilkins, pp 893-901.

160. Whiting F, Willman JP, and Loosli JK: J Anim Sci 8:234, 1949.
161. Willson RL: Biology of vitamin E (Ciba Foundation Symposium 101), London, 1983, Pitman Books, pp 19-44.
162. Wilson TM and others: J Am Vet Med Assoc 169:213, 1976.
163. Wolke RE, Nielsen SW, and Rousseau JE Jr: Am J Vet Res 29:1009, 1968.
164. Young LG and others: J Anim Sci 40:495, 1975.
165. Young LG and others: J Anim Sci 47:639, 1978.

6

Water: Requirements and Problems

JONATHAN M. NAYLOR

Water is a major component of the planet. It is essential to life; with the exception of oxygen, deficiency of water produces death more rapidly than deficiency of any other nutrient.

Water accounts for approximately 70% of total body mass. Animals' requirements for water are proportional to energy consumption. In a thermoneutral environment, a liberal estimate of water requirement is approximately 1 ml of water for every kcal of metabolizable energy used.[1] This implies that maintenance water requirements, like energy requirements, are proportional to weight to the 0.75 power. Three-quarter powers increase more slowly than the first power of body weight, so water requirements per kg of body weight decrease as animals get larger. Assuming maintenance ME requirements are 140 kcal per $kg^{0.75}$ (twice the basal ME requirement), a 50-kg ($50^{0.75}$ = 18.8) animal requires about 50 ml water per kg of body weight, whereas a 500-kg ($500^{0.75}$ = 105.7) animal requires about 30 ml/kg. Cattle are exceptional because they produce fairly liquid feces. They require about 1.5 to 2.0 ml of water per kcal. Water requirements are also increased above this general level by hot environments, which increase water losses through sweating and panting. Beef cattle weighing 450 kg (1000 lb) may drink 28, 41, and 66 L (7, 10, and 17 gal) of water at 4, 21, and 32° C (40, 70, and 90° F), respectively.[1] Lactation and high levels of minerals in water or feed also tend to increase water intake. Lactating cattle may drink more than 100 L of water per day. Guidelines for water intake of groups of animals are shown in Table 6-1; however, much variation occurs.[20]

Drinking has a higher priority than eating. Animals that voluntarily eat normal amounts of feed almost always have satisfactory water intakes. Water deprivation causes a reduction in food intake and production, although it may take 24 hours before the effects are obvious. Weight loss in water deprivation is rapid because of water loss from tissues and gut contents.[19] Gut water tends to act as a reservoir for the animal, and the feces become harder as the digesta gives up its water to maintain body hydration.

It is estimated that there is a total of 273 L of water for every cm^2 of the earth's surface. The majority is in the sea (268.4 L/cm^2 of earth surface) but a significant portion is also tied up in the continental ice sheets (4.5 L). This leaves 100 ml of fresh water for every cm^2 of earth.[1] Fresh water can come from different sources. Surface water occurs in streams and rivers, which have rapid water turnover. Turnover is slower in

Table 6-1 *Expected mean water intakes in temperate climates*

Animal	L/day	Gal/day
Beef cattle	26-66	6.5-17
Dairy cattle	38-110	9.5-28
Horses	30-45	7-12
Swine	10-30	2.5-8
Sheep and goats	4-15	1-4

Modified from a report of the subcommittee on nutrient and toxic elements in water[1] and Madec F, Cariolet R, and Dantzer R.[20]

ponds and lakes. Open waters flow, whereas closed waters have no outflow and lose water only by evaporation; the Great Salt Lake is an example. Groundwater is the major source of water for livestock. Groundwater bodies form when water percolating into the earth's crust is trapped by a layer of impermeable rock or clay. The porous layers above this then become saturated with water. This water is tapped using wells. Groundwater accounts for about 60% of water used for livestock in the United States.[1]

Water per se is almost nontoxic, but problems with water arise from contamination with microbes, parasites, minerals, and various poisons.

MICROBIAL CONTAMINANTS

A wide variety of microbes may contaminate water; they are probably the major hazard in drinking water.[17]

Enteric organisms

Enteric organisms are of particular importance as water contaminants. *Salmonella* species may be the most important bacterial contaminant. These organisms can survive for long periods in drinking water[25] and have a fairly wide host range. Viruses should not be overlooked because they also survive in water for long periods.[13,28] Enteric protozoa can also be a problem, with giardiasis the most common cause of water-related illness in people,[17] partly because this parasite can survive chlorination.[8,24] Although disease caused by *Giardia* is rare in farm animals, it can produce diarrhea in young animals.

Enteric organisms may enter water as a result of direct contamination of water with fecal material. For example, animals may defecate directly into water sources. Alternatively, runoff may wash fecal organisms into pools or streams. If the animal that produced the feces had an enteric infection or was a carrier excreting organisms in its feces, then animals subsequently drinking the water would be at risk. This is one method by which disease can spread within a group of animals.

Sewage outfalls, runoff from slurry lagoons, and leakage from cracks in sewage pipes may allow fecal microbes to contaminate water sources. Surveys of both untreated and treated sewage show that it is often contaminated by *Salmonella*.[5,15,23] In general, surface water is most likely to be contaminated by microbes; spring and well waters tend to be cleaner.[18] Contamination is most likely following heavy rainfalls[18,29] and near sewage outfalls.[23] Piped waters may become contaminated if there are cracks in the pipes (particularly if nearby sewage pipes are also cracked). Well waters may be contaminated by percolation.

The degree of fecal contamination of surface waters is usually assessed on the basis of coliform counts. In North America, most samples of chlorinated potable water destined for human consumption contain no coliforms per 100 ml of water. In other countries, higher levels are accepted (e.g., 50 coliforms per 100 ml).[7,18] Samples with no coliforms may still contain viruses.[27]

Other infectious agents

Contaminated water is one method by which *Leptospira* spread.[2] Snails associated with surface water may harbor the intermediate stages of liver flukes.

Toxic blue-green algae

Blue-green algae (cyanobacteria) are normal inhabitants of ponds and lakes. They cause a problem when they multiply in large numbers in shallow water rich in organic nutrients (eutrophic water bodies). This is likely to happen in periods of warm weather, particularly in summer and fall. Because organic nutrients favor algal growth, runoff of nitrogen and phosphate from slurry lagoons or from fertilizer applied to fields may promote problems. Steady, gentle prevailing winds contribute to toxicity by concentrating the algae at one end of the body of water. Sometimes the algae are visible as a surface scum; in other cases the algae may be mixed with the water.

Blue-green algae produce disease as the result of toxin production. This toxin may be released into the water following death of the algal bloom. The toxic species are *Oscillatoria agardhii*, *Nodularia spumingena*, *Aphanizomenom flos-aquae*, *Coelosphaerium kutzingianum*, *Gloeotrichia echinulata*, *Anabaena spiroides*, *Anabaena flos-aquae*, *Anabaena circinalis*, and *Microcystis aeruginosa*. The last species causes most outbreaks of severe disease.*

---

*References 1, 4, 6, 9, 10, 16, 21, 22.

The signs of blue-green algae intoxication are variable. Some animals die rapidly without any signs. Others may show tremors, weakness, or recumbency. Other neurological signs, including convulsions and opisthotonos, may be seen. Bloody diarrhea is common in some types of algal poisoning. Some animals become dehydrated and may show signs of thirst; pigs may vomit. At necropsy acute hepatic necrosis is often seen. Liver damage is presumably responsible for the photosensitization that may develop in some chronic cases. Other postmortem findings may include hemorrhages on the serosal surfaces and excess transudate in the body cavities.*

Blue-green algal poisoning is difficult to diagnose. Examination of ponds and lakes may reveal heavy growths of the organisms. However, changes in the weather may disrupt algal blooms by the time an investigation is mounted. In cases of sudden death, clumps of organisms may be visible in the intestinal contents.[16]

The disease can be controlled by fencing suspect water sources or by removing animals from fields with affected water. Copper sulfate added to a final concentration of 0.0001% in pond water has been used successfully to kill algal blooms.[16] However, this treatment is likely to have harmful effects on other types of aquatic life.

MINERAL CONTAMINANTS

Ten substances account for 99% of dissolved minerals in water: sodium, potassium, calcium, magnesium, hydrogen, carbonate, bicarbonate, chloride, sulfate, and silicate ions.[1] Water can be classified according to the predominant minerals. Of the cations, calcium and magnesium dominate in 87% of water samples in the United States, and sodium and potassium predominate in the remaining 13%. The major anions are carbonate and bicarbonate, which most commonly occur in association with calcium and magnesium. Sulfates and chlorides tend to be found with sodium and potassium.[1]

Solubilization of minerals in water

Minerals dissolve in water as it runs over the soil or percolates through the soil, sediment, and rock layers. Important factors that affect dissolved mineral content are the contact time; the amount of mineral present in soil, sediment, and rock; and the solubility of the minerals. Surface waters have limited contact with soil; water runs off land into streams and lakes fairly quickly, and dissolved mineral content is usually low. Runoff is most rapid following heavy rain or a spring thaw, and mineral content is lowest at these times. However, heavy runoff may allow solids to be suspended in the water. Closed surface waters tend to be very variable in mineral content; they may have high levels because although water runoff carries small amounts of minerals into closed lakes and ponds, these are then concentrated as water is lost by evaporation.[1] For example, prairie potholes (slews) tend to have the lowest dissolved mineral content in the spring as a result of ice melt and spring rainfall; concentrations rise markedly in late summer because of evaporation of water.[1] Groundwater percolates through soil and rock for long periods of time and usually has a higher mineral level than surface water.[1] Deep goundwater often has a long percolation time and an even higher mineral content.

The amount of mineral available for solubilization in water depends on the extent to which the mineral has been leeched out by previous contact with water. In high-rainfall areas, much of the soluble mineral already has been removed, and water has a low mineral content. In contrast, in more arid areas (for example, the central plains), soil and rock mineral content is high, and water tends to contain more dissolved mineral.

Different types of rocks and sediments have different solubilities. Limestones and other carbonate-containing rocks are readily soluble.[1] Salt and potash deposits from old seabeds are also highly soluble. High solubilities help account for the predominance of calcium-magnesium-carbonate-bicarbonate and sodium-potassium-chloride-sulfate waters.

Nutritive value

Dissolved minerals in water are not necessarily harmful. Better production may be obtained with well or pond water than with distilled water.[1] Although water is usually ignored as a source of minerals, it does make a contribution to mineral intake (Table 6-2). This can become a

---
*References 4, 6, 9, 10, 16, 21, 22.

Table 6-2 *Mean percentages of daily requirements of minerals supplied by drinking water*

Mineral	Cattle	Sheep	Swine	Horses
Salt (NaCl)	20-40	6-7	7-28	6-8
Calcium (Ca)	7-28	5-8	2-3	10-16
Magnesium (Mg)	6-10	5-6	8-11	4-5
Sulfur (S)	20-45	10-11		18-23
Iodine (Io)	81-173	23-28		369

About 10% of cobalt requirements may be supplied by water.
Less than 5% of manganese requirements are supplied by water.
Usually 1% or less of potassium, phosphorus, iron, zinc, selenium, and copper requirements are supplied by water.
From data in a report of the subcommittee on nutrient and toxic elements in water, Washington, DC, 1974, National Research Council.

problem when water contains high levels of minerals, which alone can supply more than the required amounts of sodium chloride, magnesium, sulfur, manganese, and iodine. Problems are particularly likely if high levels of minerals are also present in the feed (e.g., high levels of salt may be added to the feed to reduce intake). Long-term feeding of high levels of minerals to animals results in excessive mineral secretion in feces and urine. This may leach into surface water and further increase its mineral load.

Water quality

Water quality can be measured in a variety of fashions. The term *total dissolved solids* refers to the total amount of mineral present in water. It is a measure of the degree of *salinity* and is probably the single most useful measure of water quality. The term *hardness* refers to the tendency of water to form insoluble precipitates when mixed with soap or boiled; it is mainly due to dissolved calcium and magnesium salts. There is little relationship between salinity and hardness (i.e., hard water is not necessarily highly saline, and vice versa). The terms *acidity* and *alkalinity* refer to the pH. Early settlers often referred to waters high in total dissolved solids as alkali waters. However, problems with these waters are mainly associated with salinity, not with alkalinity.[1]

The most common problems associated with poor-quality water are salt poisoning, refusal to drink, diarrhea, and reduced production. The ability to tolerate high-salt waters is affected by the period of exposure, climatic factors, and level of production.

Table 6-3 *Guidelines for interpretation of total dissolved solids content in waters*

Total dissolved solids, mg/L (ppm)	Interpretation
<1000	Suitable for all classes of livestock
1000-1999	Satisfactory for all classes of livestock; may produce transient diarrhea in animals not accustomed to them
2000-4999	Temporary water refusal and diarrhea may be seen when animals are introduced to them May reduce productivity in dairy cattle
5000-6999	Likely to reduce productivity in dairy cattle May reduce growth rates May result in water refusal and diarrhea Avoid if possible
7000-10,000	Unfit for swine, very risky in all other species Animals may subsist on them Avoid
>10,000	Dangerous, avoid

Waters containing more than 10,000 mg/L (ppm) of total dissolved solids are totally unsuitable for animals (Table 6-3), although some animals can subsist without apparent problems on these waters. Water refusal, poor production, scouring, and, at very high levels (e.g., 20,000 ppm), overt salt poisoning may be seen.[1]

Table 6-4 *Recommended limits of concentration of some potentially toxic substances in water*

Item	Safe upper limit, mg/L (ppm)
Arsenic (As)	0.2
Cadmium (Cd)	0.05
Chromium (Cr)	1.0
Cobalt (Co)	1.0
Copper (Cu)	0.5
Fluoride (F)	2.0
Lead (Pb)	0.1
Mercury (Hg)	0.01
Nickel (Ni)	1.0
Nitrate-N (NO_3^-)	100.0
Nitrite-N (NO_2^-)	10.0
Vanadium (V)	0.1
Zinc (Zn)	25.0

Modified from a report of the subcommittee on nutrient and toxic elements in water, Washington, DC, 1974, National Research Council.

Waters containing between 5000 and 10,000 ppm total dissolved solids are unsuitable for high-producing livestock, particularly for lactating dairy cows.[3,14,26] Temporary water refusal and diarrhea may be seen when animals are introduced to these waters; growth rates and milk yield may be reduced. These waters may be satisfactory for subsistence but should be avoided if at all possible.[1,3,14,26]

Waters containing between 2000 and 5000 ppm total dissolved solids may reduce production in high-producing cows. Temporary water refusal and diarrhea may be seen when animals are introduced to these waters.[1]

Waters containing less than 1000 ppm total dissolved solids should be satisfactory for all classes of livestock. Waters containing between 1000 and 2000 ppm may cause a mild diarrhea when animals are suddenly introduced to them.[1]

A variety of minerals may accumulate at toxic levels in water. Guidelines for maximum acceptable concentrations are shown in Table 6-4.

Sulfates may cause problems independent of their association with salinity. Sodium sulfate is not as well tolerated as sodium chloride in water.[31] Sulfate can be present in high concentrations in some waters (e.g., those from some prairie potholes, and some deep well waters in areas with sodium-potassium types of waters). In ruminants, sulfate may reduce copper availability by combining with copper in the rumen to form insoluble sulfides and thiomolybdates. Sulfate may also destroy thiamin in the rumen. High-sulfate waters have been associated with outbreaks of polioencephalomalacia in cattle.[11,12] Precise guidelines for sulfate contents are not presently available. However, sulfate concentrations greater than 1500 ppm are aversive to cattle and may reduce feed consumption and growth rates.[30,31]

PESTICIDES AND HERBICIDES

Runoff of pesticides and herbicides into surface waters is of concern because of potential deleterious effects on aquatic life and long-term ecological damage.[18] Reports of farm animals' being poisoned by consuming contaminated water appear to be rare. However, by drinking contaminated water animals may accumulate tissue residues that render them unfit for human consumption.

CONCLUSION

Water is one of the most important essential nutrients. Without good-quality water, production rapidly declines. Open (flowing) surface waters are prone to microbial contamination; well waters are prone to mineral contamination. Closed surface waters may suffer from poisonous algal blooms. In arid areas, mineral contamination can also be a problem.

The main indexes of water quality are coliform counts and total dissolved solids. Waters with less than 2000 ppm total dissolved solids should present few problems; more than 10,000 ppm renders water totally unsuitable for animal consumption.

REFERENCES

1. A report of the subcommittee on nutrient and toxic elements in water, Washington, DC, 1974, National Research Council.
2. Carroll AG and Campbell RS: Aust Vet J 64:1, 1987.
3. Challis DJ, Zeinstra JS, and Anderson MJ: Vet Rec 120:12, 1987.
4. Chengappa MM, Pace LW, and McLaughlin BG: J Am Vet Med Assoc 194:1724, 1989.
5. Clegg FG and others: J Hyg Lond 97:237, 1986.
6. Codd GA: Vet Rec 113:223, 1983.

7. Coghlan JD: In Cruickshank R and others, editors: Medical microbiology, ed 12, vol 2, Edinburgh, 1975, Churchill Livingstone, p 278.
8. deRegnier DP and others: Appl Environ Microbiol 55:1223, 1989.
9. Falconer IR and others: J Toxicol Environ Health 24:291, 1988.
10. Galey FD and others: Am J Vet Res 48:1415, 1987.
11. Gooneratne SR and others: Can Vet J 30:139, 1989.
12. Harries WN: Can Vet J 28:717, 1987.
13. Hurst CJ and Goyke T: Can J Microbiol 32:649, 1986.
14. Jaster EH, Schuh JD, and Wegner TN: J Dairy Sci 61:66-77, 1978.
15. Johnston WS and others: Vet Rec 119:201, 1986.
16. Kerr LA, McCoy CP, and Eaves D: J Am Vet Med Assoc 191:829, 1987.
17. Levine WC, Stephenson WT, and Craun GF: MMWR CDC Surveill Summ 39:1, 1990.
18. Lindskog RU and Lindskog PA: J Trop Med Hyg 91:1, 1988.
19. Little W and others: Res Vet Sci 37:283, 1984.
20. Madec F, Cariolet R, and Dantzer R: Ann Rech Vet 17:177-184, 1986.
21. Main DC and others: Aust Vet J 53:578, 1977.
22. McBarron EJ and others: Aust Vet J 51:587, 1975.
23. Menon AS: Can J Microbiol 31:598-603, 1985.
24. Payment P and others: Can J Microbiol 35:932, 1989.
25. Pokorny J: J Hyg Epidemiol Microbiol Immunol 32:361, 1988.
26. Ray DE: J Anim Sci 67:357-363, 1989.
27. Schwartzbrod L and others: Zentralbl Bakteriol Mikrobiol Hyg (B) 181:383-389, 1985.
28. Smith EM and Gerba CP: Survival and detection of rotaviruses in the environment, Proceedings of the 3rd international symposium on neonatal diarrhea, University of Saskatchewan 67, 1980.
29. Tunnicliff B and Brickler SK: Appl Environ Microbiol 48:909, 1984.
30. Weeth HJ and Capps DL: J Anim Sci 33:211, 1971.
31. Weeth HJ and Hunter JE: J Anim Sci 32:277, 1971.

SECTION B Feeds and Feeding

7

The Digestive Systems of Horses, Ruminants, and Swine

JONATHAN M. NAYLOR

Foals, piglets, calves, lambs, and kids all rely on milk for their early nutrition. During the neonatal period, digestion is based on the typical monogastric pattern and is very similar across the species. As the young develop, their gastrointestinal tracts differentiate to accommodate the adult diet.

The major source of energy for farm animals is from dietary carbohydrate (Chapter 1). Storage carbohydrates are sugar polymers joined by α linkages and can be digested by mammalian enzymes. Starches such as amylose and amylopectin, which are composed of glucose subunits, are a good example. The structural carbohydrates or fiber, such as cellulose, are joined by β linkages that are resistant to attack by mammalian enzymes (Fig. 7-1).

Pigs are unique among the farm animals in that they are omnivores. In the wild, however, plant material is a major food source for swine. Wild pigs root for food and ingest a diet rich in roots and tubers, which are rich in starches. In consequence, the adult porcine digestive system emphasizes small intestinal enzymatic digestion of carbohydrates (Fig. 7-2).[15]

Adult horses and ruminants are herbivores. Herbivore digestion is fundamentally different from that of omnivores. Their diet is composed mainly of plants that have a high fiber content. Fiber cannot be digested by mammalian enzymes, which is a major disadvantage because fiber contains a considerable amount of the photosynthetic energy trapped by plants. Herbivores overcome this problem through a symbiosis with bacteria, fungi, and protozoa.[5] The animal provides a carefully controlled microenvironment within its gut suitable for the microbial population. The host also takes on the job of seeking out food and grinding it to increase the surface areas for microbial attack. In return, microbes digest fiber by microbial fermentation and produce short chain acids (mainly volatile fatty acids) that are absorbed by the host and used for energy production. Microbes also form amino acids and vitamins; depending on the digestive approach used by the herbivore, these may also be available to the animal.

The advantages of herbivorous digestion are the ability to use a wider variety of plant feeds and less reliance on the diet for essential amino acids, essential fatty acids, and vitamins. The cost is reduced digestive efficiency because of the energy cost in feeding microbes.[15]

The two main approaches to herbivorous digestion are exemplified by horses and ruminants (Fig. 7-2). Horses digest fiber by microbial

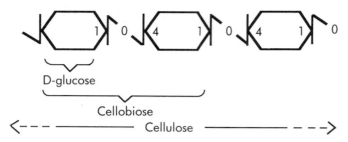

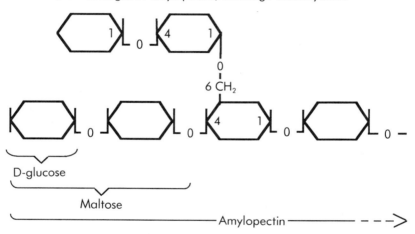

Fig. 7-1 Glucose linkages in structural and storage carbohydrates.

fermentation in a greatly expanded large colon. Volatile fatty acids are absorbed, but the availability of microbial amino acids and vitamins to the horse is limited. Horses depend on an adequate dietary supply of amino acids and of many vitamins. The advantage of this type of digestion is that on high-quality, low-fiber diets there is less energy cost in feeding the microbes because nutrients are absorbed in the small intestine before reaching the fermentation area.

Most food-producing herbivores are ruminants. Ruminants have a fermentation chamber at the beginning of the gastrointestinal tract. In consequence, microbial products are available for digestion and absorption. Volatile fatty acids are absorbed across the rumen wall. Microbial amino acids and vitamins are absorbed in the small intestine. Dietary protein quality, essential fatty acid content, and vitamin intake are often of no concern to ruminants because of microbial synthesis and absorption in the small intestine. The disadvantage is that microbes feed first, reducing availability of storage carbohydrates when compared with nonruminant digestion.

The domestic herbivores are adapted to different ecological niches. Horses can move rapidly and seek out the better-quality forages suited to their digestive system. Ruminants have

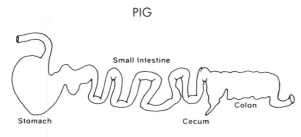

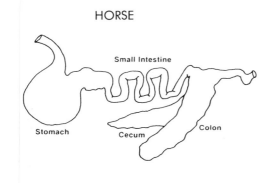

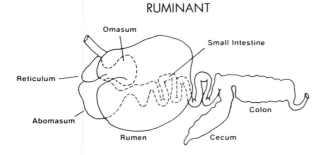

Fig. 7-2 Digestive tracts of farm animals.
From Moran ET Jr: Comparative nutrition of fowl and swine: the gastrointestinal systems, 1982. Used with permission.

bulkier digestive tracts. They move more slowly but digest poor-quality feeds efficiently. Sheep are adapted to grazing herbage extremely close to the ground, whereas goats browse forbs (shrubs, bushes) that cattle tend to avoid.

Omnivores such as pigs possess limited fiber-digestion capabilities owing to colonic fermentation. In addition, a small amount of bacterial fermentation occurs on the wall of the stomach above the liquid phase. The amounts and sites of volatile fatty acid production from microbial fiber digestion in the various species are shown in Fig. 7-3. Table 7-1 shows the relative volumes (percent of total tract volume) of different parts of the digestive tract in the three types of digestive systems.

OVERVIEW
Ingestion and mastication

The process of digestion begins in the mouth. Feed is ground into small particles, moistened, and buffered. These changes increase the surface area and provide a milieu for enzymatic and microbial attack. During this process amylase, lipase, or both are mixed with the feed in some species.

Esophageal transport

The esophagus transports feed from the mouth to the stomach or, in the case of ruminants, the rumen. The ingesta is moved by peristaltic waves of coordinated muscle activity. The mucous membrane of the upper parts of the alimentary tract, where physical grinding and transport of undigested feed take place, are protected by a lining of stratified squamous epithelium; this may be cornified in high-wear areas.[15] Mucous glands provide a protective and lubricative coating to the mucosa.

Gastric digestion

In monogastrics, food enters directly into the true stomach (the abomasum is the equivalent in the ruminant), where protein and fat digestion commence. The stomach of a monogastric is divided into an esophageal area lined by stratified squamous epithelium and a glandular portion covered by a single layer of epithelium (simple epithelium).[15] Gastric glands contain parietal (oxyntic) cells, which secrete hydrochloric acid, and chief (peptic, zymogen) cells, which secrete the protease precursors, prochymosin and pepsinogen. In addition, mucous cells protect the epithelium.[22]

Pepsin is the protease found in adults. It is formed from pepsinogen, is active at acidic pHs, and cleaves feed proteins into peptides. At the same time lipases from the stomach or saliva

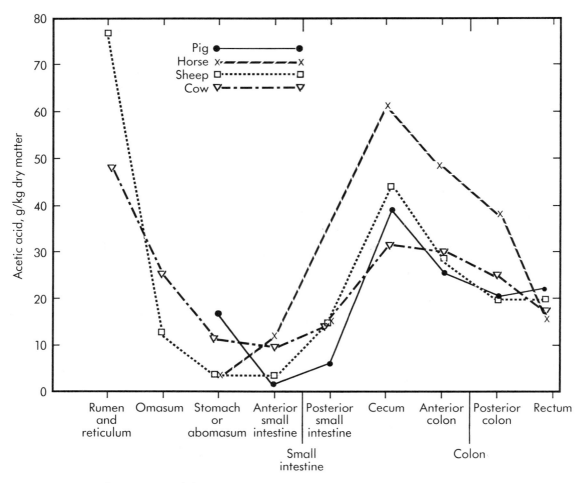

Fig. 7-3 Comparison of the sites and rates of volatile fatty acid production from microbial fiber digestion in the various species.
Modified from Elsden SR and others: Volatile fatty acids in the digestion of ruminants and other animals, J Exp Biol 22:191-202, 1946.

Table 7-1 *Relative volume of different parts of the gastrointestinal*

Species	Type of digestion	Forestomachs	Stomach	Small intestine	Large intestine
Pig	Monogastric		29	34	37
Horse	Large gut fermentor		8	14	78
Sheep	Ruminant	52	6	28	14

Values are % of total gastrointestinal volume.

begin the process of lipid digestion. In animals that secrete amylase (ptyalin) in their saliva, starch digestion occurs in the center of the food bolus. The stomach gradually releases nutrients through the pylorus and helps turn the intermittent ingestion of nutrients in meal-eating animals into a more gradual process of small intestinal digestion and absorption. Gastric acid also acts as a barrier to the entry of pathogens into the intestines.

Pancreatic and biliary secretions

Pancreatic juices and bile are added to the digesta at the head of the small intestine. Pancreatic fluid and the secretions produced by intestinal epithelial cells and glands in the upper part of the small intestine are bicarbonate rich and partially neutralize stomach acidity.[15]

Small intestinal digestion

The small intestinal lumen is covered by a simple epithelium and has a vast surface area as a result of its length, folds of villi and ridges, and of the microvilli that cover the surface of many of the epithelial cells. These adaptations increase the surface area 600 times over that of a simple tube. The microvilli form the brush border of the epithelial cells of the villi. The vascular system in the villi is arranged in a countercurrent exchange mechanism, similar to that found in the kidney. This creates a hypertonic plasma at the villous tip and facilitates water absorption. Mixing contractions, known as *segmental contractions*, bring the digesta into close contact with the cells and digestive juices.

The intestinal lining is protected from abrasion and digestion by a number of mechanisms. Mucus-producing goblet cells protect the epithelium and lubricate the passage of digesta. They tend to be more numerous in the distal areas, where the digesta becomes drier and more fibrous. Attached to the microvilli of the intestinal epithelial cells (enterocytes) is a glycocalyx lining formed of polysaccharides. This prevents large food particles from coming into contact with the enterocytes. Small nutrient molecules, however, can penetrate to reach the digestive enzymes and absorptive surfaces.

The initial part of the small intestine is the duodenum, a short tube in which the gastric outflow is mixed with pancreatic and biliary secretions. Pancreatic proteases and peptidases cleave peptides into oligopeptides and dipeptides in the gut lumen. Digestion is completed by aminopeptidases located on the brush border of the epithelial cells of the small intestine, where amino acids and dipeptides are absorbed. Protein digestion and amino acid absorption occur throughout the length of the small intestine.

Proteases are divided into endopeptidases (pancreatic trypsin and chymotrypsin, which cleave bonds in the interior of peptide molecules) and exopeptidases, which attack the free ends of the peptide chains (examples are pancreatic carboxypeptidases and intestinal aminopeptidases). To prevent autodigestion of secretory cells, the endopeptidases are secreted as precursors (proenzymes or zymogens) and are activated only within the intestinal lumen. Gastric proteases are activated by hydrochloric acid, which converts inactive pepsinogen into active pepsin. Pancreatic trypsinogen is cleaved to trypsin by the intestinal epithelial brush border enzyme enteropeptidase (enterokinase). Trypsin then activates pancreatic exopeptidases and the lipases colipase and phospholipase.[7]

Digestion of starches is initiated by salivary amylase in species such as pigs that secrete this enzyme in their saliva. Pancreatic amylase continues the process. This happens mainly in the intestinal lumen, although amylase may be absorbed onto the mucosal lining, thus increasing the chances that its products will be taken up by epithelial cells and not intestinal bacteria. Intestinal dextrinases cleave branched oligosaccharides produced by amylase. Disaccharidases on the intestinal brush border complete digestion, and simple sugars are absorbed. The most important disaccharidase in the adult is maltase, which cleaves maltose, formed by amylase digestion, into glucose (Fig. 7-4). In some texts the term *maltase* is used to describe all intestinal wall enzymes that cleave α linkages between sugars; the specific enzyme that cleaves glucose disaccharides is called *isomaltase*. Lactase is the most important disaccharidase in the neonate. It cleaves lactose, which is found naturally only in milk, into glucose and galactose. Some of the nonglucose sugars are converted to glucose by the epithelial cells, so there is more glucose in

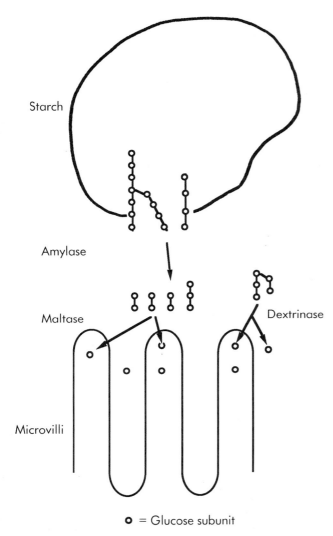

Fig. 7-4 Steps in starch digestion in the small intestine.

actively pumped out of cells by the sodium-potassium pump at the basolateral cell membrane, thus maintaining a low intracellular sodium concentration. This drives the entry of sodium from the lumen down the concentration gradient.

The majority of dietary lipids occur as esters of glycerol and fatty acids and are known as *acyl esters*. Triglycerides are the most common form of acyl ester, but phospholipids and other glycerides also occur. Acyl esters have to be hydrolyzed to free fatty acids before they can be absorbed across the cell membrane of the intestinal enterocytes. This hydrolysis is initiated by salivary and gastric lipases.

In the small intestine, fat digestion is assisted by bile that helps solubilize fat. The bile salts are initially synthesized in the liver from cholesterol; in pigs and cattle the gallbladder stores bile between meals or when the animal is fasting. Bile is excreted into the duodenum through the biliary system. Horses lack a gallbladder, and bile is secreted continuously into the gut. The majority of bile salts are reabsorbed in the terminal small intestine and then are reclaimed from portal blood as it passes through the liver. This *enterohepatic circulation* recycles bile acids. If the bile duct is blocked or recycling of bile acids fails owing to malabsorption, then fat maldigestion occurs.

Pancreatic lipases complete the degradation of triglycerides and other fatty acid esters to fatty acids, which are absorbed. Absorption of long-chain fatty acids into enterocytes occurs mainly in the upper small intestine and is enhanced by bile salts. Fatty acids are converted back into triglycerides within the intestinal epithelial cells and enter lymphatics as a component of chylomicrons. Chylomicrons are micelles of cholesterol, proteins, and phospholipids. They transport triglyceride to the peripheral tissues, where triglyceride is removed for metabolic use or fat synthesis. Medium- and short-chain fatty acids (fewer than 10 to 12 carbon units) travel through the enterocytes and enter the portal blood system directly.[8]

Nucleic acids are broken down by pancreatic ribonuclease and deoxyribonuclease. The pyrimidine bases are absorbed in the small intestine directly into the portal system.

portal blood than is absorbed from the gut. Digestion of starches and sugars tends to occur in the proximal small intestine.

In the absorptive phase, simple sugars and amino acids are carried by specific transport proteins on the enterocyte cell membrane from the gut lumen into the epithelial cells. The absorption of glucose, galactose, and neutral amino acids is an active process because it is linked to cotransport of sodium into the cells. Sodium is

The small intestine reclaims the major portion of the sodium excreted into the gut, and water follows. It is the sole site for calcium, magnesium, and phosphorus absorption in monogastrics. Vitamins are almost exclusively absorbed in the small intestine.

Large intestinal digestion

Digesta are moved progressively toward the large intestine by peristaltic waves in the small intestinal wall. The large intestine does not secrete any digestive enzymes and in most species lacks sugar and amino acid transport mechanisms. However, its importance should not be overlooked. Horses may be able to absorb a limited amount of amino acids in the cecum and large colon. Digesta spend a large proportion of time in the large intestine. The large intestine completes the process of water and sodium reclamation. It is also an area of fiber fermentation; microbes produce volatile fatty acids, which are absorbed and enter the portal blood. In the rectum the digestive remains are stored until evacuation.

Fluid and electrolyte fluxes

Digestion is associated with major fluid and electrolyte fluxes throughout the gastrointestinal tract. To facilitate enzymatic attack and diffusion of nutrients, the digesta are moistened to about 20% dry matter in the stomach; dry matter content declines further to about 10% in the small intestine.[12] The total volume of fluid that passes through the gut in one day is large. The secretions of the major lobulated glands of the gastrointestinal tract add a daily fluid volume greater than the extracellular fluid volume (Table 7-2). This is only part of the fluid flux; glands in the wall of the gut and individual epithelial cells add additional fluid. Overall total daily fluid fluxes are greater than the total body water volume. Along with this fluid travel large quantities of sodium and potassium ions. In consequence, digestive disturbances that interfere with fluid absorption can rapidly lead to dehydration and electrolyte imbalances.

Acid and alkali are pumped in and out of the gastrointestinal tract. The stomach has a resting pH of 2 to 3 that rises following feeding. Much of the gastric acid is neutralized as the digesta leave the stomach, and the pH then slowly rises toward neutrality as the digesta pass through the small intestine. Bacterial fermentation in the large intestine produces volatile fatty acids and a fall in pH in the cecum. As the acids are absorbed, pH slowly rises (Fig. 7-5). Problems with loss of fluid from the stomach or obstructions to the upper small intestine frequently lead to the development of a systemic alkalosis due to loss of acid from the body. Loss of fluid from the intestine in diarrhea leads to loss of pancreatic and intestinal bicarbonate and a systemic acidosis.

Table 7-2 *Liters of fluid secreted into the gut daily and relation to total body water*

Secretion	Pig*	Pony*	Cow†
Saliva	15	18	120
Gastric or abomasal fluid			30
Pancreatic fluid	8	8	13
Bile	4	4	
Total	27	30	163

*Body weight 100 kg; extracellular fluid volume 23 L.
†Body weight 500 kg; extracellular fluid volume 115 L.

Liver

The liver is the largest organ in the body. About 20% of cardiac output travels through the liver, mainly by the portal venous circulation. The liver occupies a unique position at the head of the portal system that drains the stomach and intestines. Hepatic tissues, therefore, are the first to be exposed to nutrients absorbed from the gut. The endocrine pancreatic hormones (insulin, glucagon, somatostatin, and pancreatic polypeptide) are released directly into the portal system. The liver is a major site for insulin and glucagon action, in part because it is exposed to much higher concentrations of these hormones than are other tissues of the body.

The liver performs several important roles. Digestively the liver stores absorbed nutrients against times of need and converts nutrients into forms that are utilizable by other tissues of the body. Following a meal in monogastrics, the

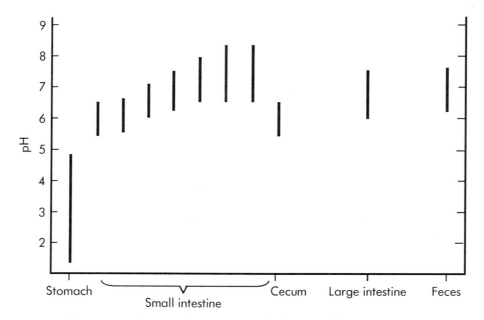

Fig. 7-5 Changes in pH along the gastrointestinal tract of the weaned pig. Modified from Kidder DE and Manners MJ: Digestion in the pig, Bristol, England, 1975, Scientechnica.

liver converts absorbed glucose into glycogen. Glycogen is converted to glucose and released back into blood when the supply from the intestines is inadequate. In herbivores very little glucose is absorbed. Propionate from microbial digestion is taken up by the liver. It enters the Krebs cycle and is converted to oxaloacetate, where its major fate is glucose synthesis. The liver releases glucose for use by peripheral tissues in a similar fashion to monogastric metabolism.

The liver is important in the storage and regulation of several vitamins, including A and D, and some trace minerals (e.g., copper).

Hepatic macrophages help to remove any live bacteria that may have inadvertently traversed the gut mucosal barrier. The liver also detoxifies toxins, including endotoxin, that may have been absorbed.

Metabolism

The main energy substrates for cells are glucose in the fed animal and fatty acids and ketone bodies (acetoacetate, β hydroxybutyrate) in fasted animals. Following a meal, glucose is initially stored in the liver as glycogen. Long-term energy storage is accomplished in adipose tissue by converting glucose to fatty acids; these are esterified with glycerol to form triglycerides. Fatty acids absorbed from the gut can be used immediately as an energy source. Alternatively they may be converted to milk fat or be stored as triglycerides in adipose tissue.

Ruminants ferment most of the carbohydrates in feed to volatile fatty acids in the rumen. Little glucose is absorbed from the small intestine. In consequence, ruminants have unique metabolic adaptations that are discussed later in this chapter.

DIGESTIBILITY

Digestibility is used as a measure of the availability of nutrients in a feed. It usually is measured after animals have been fed the feed for several weeks, which allows gut microbes and tissue time to adapt to the feed. Then the amount of feed offered, the amount of feed refused (known as *orts*), and the amount of feces produced are collected and weighed. Usually collections are made over a 5- to 7-day period. The animals are fed below ad lib intake to minimize refusals. Metabolism stalls or special harnesses are used to separate and contain feces and urine. Representative samples of the feed and feces are analyzed and digestibility is calculated as follows:

$$\text{Digestibility (\%)} = \frac{\text{Amount of nutrient ingested} - \text{Amount of nutrient in feces}}{\text{Amount of nutrient ingested}} \times 100\%$$

This calculation gives apparent digestibility of the nutrient in the feed. Some of the fecal nutrients come from endogenous gut secretions and not from the diet. Endogenous production of nutrients can be measured by radiotracer methods. Alternatively, the amount of nutrients in the feces at various levels of feed intake can be compared. Then by extrapolation the amount of nutrient in the feces at a theoretical zero feed intake can be found; this represents losses from endogenous secretions. True digestibilities are found by subtracting the amount of nutrient in endogenous secretions from the amount in the feces before performing the preceding calculation. Thus true digestibilities are larger than apparent digestibilities.

Digestibilities tend to be high when the nutrients are water soluble. Digestibility is increased by grinding, chopping, or rolling feeds, which increases surface area and breaks down protective fibrous casings. High levels of feed intake speed the passage of digesta through the gut and reduce digestibility.

NEONATES
Milk composition

Milk is composed of protein (principally casein), the sugar lactose, and fat (Table 7-3). Fat is present as triglycerides, mainly of long-chain fatty acids, but there are also some short-chain (C 4) acids. In addition it is rich in calcium and phosphorus.

Gastric digestion

Milk digestion begins in the stomach. In ruminants milk tends to bypass the rumen because the esophageal groove closes, passing fluid directly from the esophagus into the abomasum. Closure of the groove is promoted by sucking. When large quantities of milk are ingested, some spills into the rumen. This is particularly likely to happen if milk is drunk from a bucket rather than sucked from a nipple or teat. Ruminal contractions rapidly pass milk collected in the rumen into the abomasum.

Table 7-3 *Milk composition as percent of dry matter*

Solid	Mare	Cow	Goat	Ewe	Sow
Fat	15	29	34	38	42
Protein	23	27	25	32	29
Sugar	59	38	31	25	24
Fiber	0	0	0	0	0
Total solids, as fed (%)	12	12	13	19	20

From data in Alexander F and Hickson JCD[1] and Naylor JM and Bell R.[16]

In the stomach or abomasum, milk casein is rapidly clotted by the action of acid and chymosin. Chymosin (rennin) is a protease that is secreted as an inactive precursor and is activated by gastric acid. Unlike pepsin, the gastric protease of adults, it is active under less acidic conditions. It specifically cleaves casein chains in such a way that the resultant peptides can polymerize together to form a clot.[6,22] It also spares immunoglobulins so that colostral antibody passes unharmed into the small intestine, where it can be absorbed intact by a special type of pinocytotic absorption that usually is present only on the first day of life (Chapter 21).

The casein clot traps the micelles of milk fat, but whey, which contains lactose and whey proteins, is squeezed out by gastric and chemical contraction. In calves 85% of the whey passes into the small intestine within 6 hours of feeding.[22] In contrast, the clot takes 12 to 18 hours to be digested.[9] The old clot frequently forms a nidus for the clot from the next feed. The milk clot is digested by the actions of acid, which denatures casein, and chymosin. This process is aided by pepsin, which also is secreted in neonates. Peptic digestion becomes more important with age.

Lipases added to saliva by glands in the mouth (calves) or secreted by gastric cells (pigs) break triglycerides into fatty acids.[17] In calves salivary lipase accounts for 30% of milk fat digestion.[9,22]

Small intestinal digestion

Lactose, protein and fat digestion are completed in the small intestine in young farm animals. Neonatal intestine is rich in lactase, which splits lactose into glucose and galactose molecules. Both sugars are transported into epithelial cells on the same carrier, but glucose is absorbed first, in the upper small intestine, because it has a higher affinity for the transport.[20,22] Sodium is cotransported with glucose and normally drives the process. However, high concentrations of glucose in the lumen create a favorable concentration gradient for glucose entry, and this facilitates absorption of both glucose and sodium. This mechanism is used to advantage in oral fluid therapy in neonates, where oral electrolyte solutions often contain glucose and neutral amino acids such as glycine. These drive sodium absorption through the glucose and neutral amino acid cotransports. Sodium absorption draws in water, and together these help restore hydration.

Maturation of digestive system

The neonate has little pancreatic amylase, intestinal maltase, or sucrase activity.[15,19,20,22] Thus neonates digest starches and sucrose very poorly. As the animal matures, lactase activity declines, and amylase, maltase, and dextrinase activities increase (Figs. 7-6 and 7-7). The decline in lactase appears to be partially diet dependent, but age maturation is probably the major factor.[12,15,19,21,22] Pigs and probably horses develop more pancreatic amylase activity than cattle. Sucrase is found in the intestinal tract of the adult pig and horse but is not present in the bovine.[5,15,20] In the stomach, acid and trypsin secretion increases with age, whereas chymosin secretion declines.[15,22] The large intestine—and the rumen in ruminants—also develops with

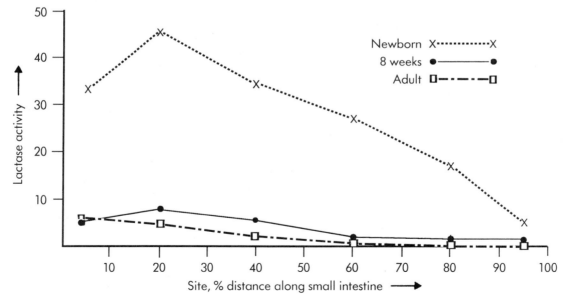

Fig. 7-6 Distribution of lactase activity along the small intestine of pigs and the effect of age. The position along the small intestine is given as the percentage of the total small intestine length.
Drawn from data in Kidder DE and Manners MJ: Digestion in the pig, Bristol, England, 1975, Scientechnica.

age, and the animal becomes capable of fiber digestion.

PIG

Ingestion and mastication

The adult pig finds its feed mainly by smell, and a well-developed sense of taste aids in food discrimination. The importance of olfaction in feed location is illustrated by the use of pigs for locating truffles, an edible fungus that grows 5 to 30 cm beneath the ground. Pigs use their snouts and forward-pointing lower incisors to shovel feed into their mouths. The upper and lower incisors cut large feed particles into lengths. The premolars perform the initial grinding, and the molars the final fine processing.[15]

Gastric digestion

In the stomach, feed and gastric liquid tend to settle into the more ventral glandular portion. Dietary fiber protects the esophageal portion of the stomach from attack by gastric acid. Dissolved fiber makes the gastric fluid more porridgelike and reduces its tendency to slop onto the nonglandular mucosa. Pigs fed high-concentrate, finely ground feed (e.g., feed ground through a 1.5-mm screen) lose the protective benefits of fiber and are particularly prone to gastric ulceration.[13,15]

The most acid-soluble and the smallest particles of food are released from the stomach first. They are propelled down the small intestine by abrupt waves of peristalsis.[13]

Pancreatic and small intestinal digestion

The porcine pancrease secretes large amounts of amylase and lipase and makes a major contribution to starch and lipid digestion. Intestinal maltase digests maltose produced by the action of amylase; sucrase splits sucrose to glucose and fructose. Starch and protein digestion is very efficient; 70% to 90% of dietary protein is ab-

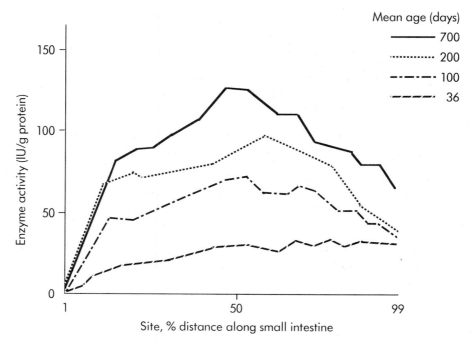

Fig. 7-7 Distribution and development of dextrinase activity in the pig. The position along the small intestine is given as the percentage of the total small intestine length. Modified from Kidder DE and Manners MJ: Digestion in the pig, Bristol, England, 1975, Scientechnica.

sorbed by the time the digesta reach the ileum.[19,25] In pigs on high-grain diets about 65% of dry matter is digested in the small intestine.

Large intestinal digestion

Some undigested peptides and storage carbohydrates reach the large intestine, together with the majority of the ingested fiber. The large intestine accounts for about 40% of gastrointestinal volume.[15] Transit through the large colon takes 35 hours and accounts for 80% of the time feed particles spend in the gut. In comparison an average period of 4 hours is spent in the stomach and 3.5 hours in the small intestine.[15] Transit of large particles is slower than that of small particles. The large intestine accounts for about 15% of dry matter digestion in pigs fed a high-grain diet. Passage of feed down the small intestine is speeded by dietary fiber, but, although it reduces the time for intestinal digestion, only energy yield appears to be affected.[13,14,15]

Pigs digest neutral detergent fiber better than acid detergent fiber. About 80% of fiber digestion occurs in the large intestine. Volatile fatty acids formed by microbial digestion are absorbed; they may contribute up to about 20% of maintenance energy requirements on high-fiber diets.[14,15] Undigested amino acids are utilized by large intestinal microbes, and some are deaminated to ammonia and amines, which are absorbed. This absorbed ammonia is not available to the pig for protein synthesis. The disappearance of dietary nitrogen in the large intestine falsely inflates protein digestibility figures when rectal collections of digesta are used. For this reason, the most meaningful measurements of protein and amino acid digestibilities are made with cannulated pigs where digesta can be collected directly from the terminal ileum. Only the protein that disappears in the small intestine is absorbed as amino acids and is available to the pig.[25]

HORSES
Gastric, pancreatic, and small intestinal digestion

Digestion of feedstuffs before they reach the cecum is similar to the basic pattern of digestion described earlier. The horse eats frequently, and the stomach has a relatively small capacity. There is no gallbladder for bile storage, and bile is excreted continuously.

The small intestine and pancreas contain the full complement of enzymes for digesting starches, sugars, and proteins. On high-grain (80% of the ration dry matter) diets, about 70% of starches and sugars are digested in the small intestine. However, digestion of crude protein and dry matter is less efficient than in the pig; only about 45% of crude protein and 35% of dry matter are absorbed.[9] On high-roughage diets, small intestinal digestion of starches and dry matter is greatly reduced, and a larger proportion enters the large colon.[9]

Large intestinal digestion

The large colon and cecum are greatly expanded in the horse and function as fermentation vats. They comprise about 80% of the total volume of the intestinal tract,[2] and, although not essential to life, make an important contribution to energy digestion.[3] The large intestine is effectively divided into four compartments—the cecum, ventral colons, dorsal colons, and small colon—by barriers at the junctions between these areas.[2] These barriers are in part physical, due to constriction of the gut and the arrangement of valve-like mucosal folds, and in part physiological, due to the pattern of intestinal motility. To-and-fro mixing contractions occur in the large colon. In addition, a pacemaker site at the pelvic flexure initiates propulsive contractions. From this site, waves pass orally and aborally along the digestive tract. The waves that travel proximally push larger feed particles back up the intestinal tract for continued digestion. The distal waves move small particles down the tract. The pelvic flexure acts as a major barrier to gut flow and is a common site for impactions on diets high in poor-quality roughage.[23]

The fluid of the large intestine is buffered by phosphate salts, which are absorbed in the terminal intestine. Within the gut is a large range of flora similar to that found in the rumen. The cecum and ventral colon are the major sites of microbial fermentation. The major volatile fatty acids produced are acetic, butyric, and propionic. The proportion of propionic acid increases on high-grain diets.[10] The volatile fatty acids are absorbed and pass through the portal vein to the liver and peripheral tissues.[2] On high-grain diets, about 80% of nutrients are digested, about half of this in the large intestine. On high-fiber diets, just over half the dry matter ingested is absorbed from the gut. About two thirds of the total dry matter, three quarters of the structural carbohydrates, half of the sugars and starches, and one third of the crude protein are digested and absorbed in the large intestine.[10] The equine large intestine also may possess limited ability for vitamin absorption.

RUMINANTS
Ingestion and mastication
Cattle pull a clump of herbage into the mouth with the tongue and then shear it off by the action of the lower incisors against the dental pad of the upper jaw. Cattle have fairly rigid lips and backward-pointing papillae on their tongue and are nonselective grazers. Sheep and goats have more mobile lips and use these in food selection. Once the feed is in the mouth, well-developed molars grind it. During grazing, the initial grinding is fairly short. Residence in the mouth is only long enough to ensure the feed is well lubricated with saliva before it is swallowed. This practice is thought to be the result of an adaptation by wild ruminants to reduce the vulnerable time period at pasture. Once the beast has finished grazing, it can find some sheltered area to hide. Ruminants then regurgitate their feed, which is thoroughly ground into smaller particles and reswallowed. This process is known as *rumination*. Ruminants produce a copious amount of saliva; it is rich in bicarbonate and functions to buffer the rumen fluid. There is no amylase in ruminant saliva.[5]

Digestion in the rumen
The rumen is a specially modified forestomach. It contains a volume of fluid equal to the extracellular fluid volume and 60% of the gastrointestinal dry matter. Over half of the organic matter present in feed is digested and absorbed in the rumen.[5]

The rumen sac is a carefully controlled fermentation chamber. Temperature is regulated by the animal's homeothermic mechanisms, with the acids produced by ruminal fermentation buffered by salivary bicarbonate. The sealed structure ensures an anaerobic environment. The products of fermentation are continuously removed, so end-product inhibition of enzymatic digestion is minimized.

Ruminal bacteria secrete a wide variety of enzymes that cleave the α linkages of storage carbohydrates and the β linkages of structural carbohydrates. No single microorganism is capable of completely digesting food by itself; instead there is a cooperative and synergistic attack. Enzymes from one microbe attack one type of food molecule to reveal other components that can be attacked by enzymes from different microbes. Bacteria are the main ruminal microbes. Fungi degrade the bonds between lignin and cellulose. Protozoa also are found in low numbers; they do not appear to be essential to the digestive process.[5] However, the presence of protozoa is a good indication of a healthy ruminal flora.

Simple sugars produced in the rumen by digestion are further modified by microbial metabolism. The main products are the volatile fatty acids—acetic (two-carbon chain, C 2), propionic (C 3), and butyric (C 4) acids (Fig. 7-8). Acetic acid predominates on high-roughage diets and propionic acid on high-grain diets. Some poorly preserved silages are rich in butyric acid, which increases the proportion of this acid in the rumen.

The rumen is lined by a stratified epithelium that is lightly cornified. It has both protective and absorptive roles. The epithelium is folded into papillae that increase the surface area for absorption.[5] Volatile fatty acids are absorbed directly across the rumen epithelium. Some butyric acid is converted to β-hydroxybutyrate and acetoacetate as it passes through the epithelial cells.[24] Propionic acid passes through the portal blood circulation to the liver, where it is either metabolized in the Krebs cycle to provide energy or is converted to glucose. Acetate is used by a variety of tissues as an energy source. Unlike monogastrics, ruminants lack the enzymes to use glucose for fat synthesis. Acetate is the major substrate for adipose tissue and milk fat synthesis in ruminants.

Volatile fatty acids tend to acidify the rumen fluid. Rumen pH is particularly acidic when large amounts of grain are fed. Grains are more

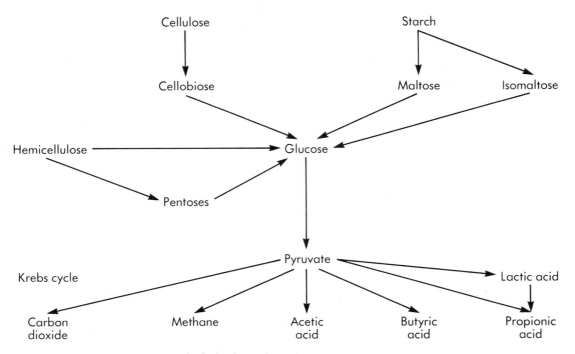

Fig. 7-8 Carbohydrate degradation within the rumen.

rapidly degraded than fiber, and so acid production is more rapid. Because of their rapid degradation, grains leave the rumen quickly and more can be eaten. High-grain diets are low in fiber and so require less chewing, which means less saliva is secreted. On high-grain diets, rumen pH can be as low as 5.[5] Low rumen pHs have a number of adverse effects. The microbial population changes, altering the efficiency of digestion. Severely acidic conditions are directly caustic to the rumen epithelium. This problem is sometimes called *ruminal acidosis*. Reduced feed intake may reduce milk production. Ruminal fermentation shifts away from acetate production, which is used by the mammary gland for butterfat synthesis, and the fat percentage in milk is reduced. Grain overload, which is described later in this chapter, is a more severe variant that produces clinical illness. To compensate for problems of ruminal acidosis, buffer often is added to the diets of dairy cows eating large amounts of grain. This improves feed intake and butterfat percentage.

There are several unique aspects to lipid digestion in ruminants. Salivary lipase works best on triglycerides containing 4-carbon-length fatty acids. Thus butyric acid containing triglycerides are preferentially hydrolyzed, and the butyric acid absorbed across the rumen wall. Ruminant microbes tolerate only low levels of dietary fat. They secrete lipases and these hydrolyze most of the dietary lipids to fatty acids. Unsaturated fatty acids are converted to more saturated fatty acids in the reducing environment of the rumen. Rumen bacteria synthesize essential fatty acids so there are no dietary requirements for fat. As a result of microbial fat synthesis and because long-chain fatty acids are not absorbed in the rumen, the amount of fat entering the duodenum is often greater than the amount of fat ingested.

Much of the protein that enters the rumen is degraded into amino acids or ammonia and is then resynthesized into microbial protein. Energy availability often limits the rate of bacterial synthesis. Microbial proteins leave the rumen via the omasum to be digested in the abomasum and small intestine. One unique feature of ruminant digestion is the ability to recycle urea. Most of the urea produced in the ruminant's body from protein catabolism is excreted in the urine. Some, however, passes into the saliva and reaches the rumen. Rumen microbes convert urea and other nonprotein nitrogen sources into amino acids by using energy from carbohydrate digestion. These amino acids are then available to the animal when the microbes are digested further down the gut. Urea recycling reduces the need for dietary protein on low-protein diets. Under many circumstances protein quality is of minor concern to the ruminant because much is degraded to nonprotein nitrogen and resynthesized into essential and nonessential amino acids of bacterial proteins. It is the microbial protein, which has a fairly constant composition, that is digested by the host. Recently, however, there has been interest in bypass of this process. Milk production in high-producing cows is limited by amino acid availability[4]; if additional amino acid can be supplied to the small intestine, then milk production increases. One method of doing this is to feed proteins that are resistant to microbial attack in the rumen but that can be digested in the small intestine. Poorer-quality protein and nonprotein nitrogen can be used to satisfy rumen microbes.

The time food spends in the rumen increases with particle size and fiber content and decreases with increased feed intake. There is a limit to the rate of emptying, and physical fill of the rumen can limit feed intake, particularly on high-fiber diets. Feeding high levels of concentrate-based diets reduces ruminal retention and ruminal degradation, particularly of protein and fiber.[26] Grinding feed increases the surface area for microbial attack and improves digestibility. However, if no long-stem roughage is fed, rumination is reduced, which in turn reduces salivation and buffering of the rumen. Mixing contractions also may be reduced. In consequence, digestion is impaired and ruminal acidosis may develop.

By-products of rumen fermentation are methane and carbon dioxide, which are removed by eructation. In this process, ruminal contraction forces gas toward the esophageal opening, which is cleared of liquid and relaxes to allow passage of gas. Cows produce 150 L of gas a day, and so it is easy to understand why they can develop severe bloat (ruminal distension) if esophageal blockage, ruminal atony (loss of the ability of the rumen to contract), or foam formation in the rumen prevent eructation.[5]

Abomasal digestion
The abomasum corresponds to the stomach of monogastrics and begins the process of peptic digestion of proteins.

Small intestinal digestion
Digestion of proteins, of bacterial storage carbohydrates and lipids, and of storage carbohydrates and lipids that escape ruminal degradation is performed in the small intestine in a similar fashion to that in monogastrics. Because much of the carbohydrate has been removed in the rumen, the intensity of carbohydrate digestion is lower than in monogastrics.

Large intestinal digestion
The cecum and colon function as a secondary fermentation chamber as well as a site for fluid and electrolyte resorption. Some protein, carbohydrate, and fiber escape both ruminal and small intestinal digestion and are fermented in the large intestine. Cornstarch is somewhat resistant to ruminal degradation and is particularly likely to result in cecal fermentation.[24] Volatile fatty acids formed from microbial fermentation are absorbed and used as energy sources.

ADAPTATION
Rapid dietary changes can lead to a variety of problems, including decreased feed intake and diarrhea. This is because of time required for both digestive enzymes and microbial flora to adapt to the new environment. Production and secretion of many of the digestive enzymes are regulated in accordance with the amount of substrate present in the gut lumen.[19] Microbial populations can take 2 to 3 weeks to adapt completely to new diets.[5] The physical structure of the gut also is affected by diet. The size of the rumen and the development of ruminal papillae are enhanced by high-roughage diets. Intestinal villi shrink during starvation when no food is present in the lumen. The size of the liver also varies considerably in proportion to dietary intake.

To avoid problems, dietary changes should be made gradually in all species. This is particularly important in herbivores, when changing to a higher plane of nutrition or to energy-dense, rapidly fermentable diets. Complete adaptation to a new diet may take more than a month. Diets should be gradually changed over a minimum of 5 to 10 days.

FERMENTATION RELATED PROBLEMS (GRAIN OVERLOAD)

If herbivores gorge on large amounts of starch-containing feed (particularly finely ground feed), microbial fermentation can get out of hand. The pH of the gut contents drops, and the intestinal flora changes. Large amounts of osmotically active lactic acid are produced, which can give rise to systemic acidosis and dehydration. Gut motility is reduced, and the acid may corrode and destroy the epithelial cells of the gut wall. The animal goes off feed and may become weak or recumbent or may even die. These problems are due to dehydration, acidosis, and absorption of toxins across the damaged gut wall. For reasons that are not completely understood, laminitis may develop. This disease primarily affects cattle and horses.

CONCLUSION

Mammals secrete enzymes into the gastrointestinal tract, which digests storage carbohydrates, proteins, fats, and nucleic acids. The main site for digestion and absorption is the small intestine, which is also the primary site for water, mineral, and vitamin absorption.

A major problem faced by mammals is that they cannot digest the β linkages found between sugars of the structural carbohydrates. These account for the major portion of photosynthetic energy trapped by plants. Fiber digestion is accomplished by a symbiosis with microbes. This reduces the efficiency of digestion in herbivores owing to the energy cost of feeding the microbes. The advantage is the ability to use a wider variety of feedstuffs.

Pigs are omnivores and have well-developed small intestines and glandular structures for digestion of dietary protein and storage carbohydrates. Fiber digestion is limited to the large intestine and contributes only to energy metabolism. Horses occupy an intermediate role between pigs and ruminants; they have gastric and small-intestinal digestion equivalent to other monogastrics but greatly enhanced colonic fermentation. Ruminants ferment feed at the beginning of the tract. Digestive efficiency is reduced, but ruminants use microbial products to supply energy, essential amino acids, essential fatty acids, and vitamins.

Herbivores allow humans to harness the energy in plant fiber for their own nutritional purposes and for work. Ruminants also improve the biological value of feeds through harnessing the microbial synthesis of essential amino acids, fats, and vitamins. Pigs compete with humans for nutrients and are used to convert surplus vegetable feed into protein and fat when the cost of grain is low and the cost of meat is high. They also are useful in the recycling of waste feed left over from food processing.

REFERENCES

1. Alexander F and Hickson JCD: In Phillipson AT, editor: Physiology of digestion and metabolism in the ruminant, Newcastle-upon-Tyne, Oriel Press, 1970, pp 375-389.
2. Argenzio RA: Cornell Vet 65:301, 1975.
3. Bertone AL, van Soest PJ, and Stashak TS: Am J Vet Res 50:253, 1989.
4. Chalupa W: J Dairy Sci 67:1134, 1984. Published by DC Church, Corvallis, Oregon.
5. Church DC: Digestive physiology and nutrition of ruminants, vol 1, Digestive physiology, Corvallis, Ore, DC Church, 1975.
6. Foltmann B: In Just A, Jorgensen H, and Fernandez JA, editors: Proceedings of the 3rd international seminar on digestive physiology in the pig, Norwalk, Appleton and Lange, 1985, pp 120-123.
7. Ganong WF: Review of medical physiology, ed 14, Norwalk, Appleton and Lange, 1989, p 408.
8. Ganong WF: Review of medical physiology, ed 14, Norwalk, Appleton and Lange, 1989, p 398.
9. Hill KJ, Noakes DE, and Lowe RA: In Phillipson AT, editor: Physiology of digestion and metabolism in the ruminant, Newcastle-upon-Tyne, Oriel Press, 1970, pp 166-179.
10. Hintz HF and others: J Anim Sci 32:245, 1971.
11. Hintz HF, Argenzio RA, and Schryver HF: J Anim Sci 33:992, 1971.
12. Kidder DE and Manners MJ: Digestion in the pig, Bristol, 1975, Scientechnica.
13. Lawrence TLJ: In Haresign W and Coles DJA, editors: Recent advances in animal nutrition—1984, London, 1984, Butterworth, pp 97-109.

14. Low AG: In Just A, Jorgensen H, and Fernandez JA, editors: Proceedings of the 3rd international seminar on digestive physiology in the pig, 1985, pp 157-179.
15. Moran ET Jr: Comparative nutrition of fowl and swine: the gastrointestinal systems, Guelf, Ontario, ET Moran, Jr, 1982.
16. Naylor JM and Bell R: Vet Clin North Am (Equine Pract) 1:169, 1985.
17. Newport MJ and Howarth GL: In Just A, Jorgensen H, and Fernandez JA, editors: Proceedings of the 3rd international seminar on digestive physiology in the pig, 1985, pp 143-148.
18. Ozimek L, Sauer WC, and Ozimek G: In Just A, Jorgensen H, and Fernandez JA, editors: Proceedings of the 3rd international seminar on digestive physiology in the pig, 1985, pp 146-148.
19. Rerat A: In Batt RM and Lawrence TLJ, editors: Function and dysfunction of the small intestine: proceedings of the 2nd George Durrant memorial symposium, Liverpool, Liverpool University Press, 1984, pp 21-38.
20. Roberts MC, Hill FWG, and Kidder DE: Res Vet Sci 17:42, 1974.
21. Roberts MC: Res Vet Sci 18:64, 1975.
22. Roy JHB: In Batt RM and Lawrence TLJ, editors: Function and dysfunction of the small intestine: proceedings of the 2nd George Durrant memorial symposium, Liverpool, Liverpool University Press, 1984, pp 95-132.
23. Sellers AF and Lowe JE: Equine Vet J 18:261, 1986.
24. Sutton JD: J Dairy Sci 68:3376, 1985.
25. Tanksley TD Jr and Knabe DA: In Haresign W and Cole DJA, editors: Recent advances in animal nutrition—1984, London, 1984, Butterworth, pp 75-95.
26. Zinn RA and Owens FN: J Anim Sci 56:471, 1983.

8

Control of Feed Intake

SARAH L. RALSTON

GENERAL FEEDING BEHAVIOR
Regulation of body weight

It is important to understand the factors that regulate feed intake on both a short- and long-term basis in order to optimize ration intake. Animals adapted to ad libitum access to a single ration usually consume amounts sufficient to maintain a constant "set point" body weight. However, intake varies according to individual physiological status and environmental conditions.[3,23] Animals that evolved in seasonally variable climates (horses, sheep, cattle) eat amounts sufficient not only to maintain lean body mass but also to store body fat for the "lean winter months."[27] This trait is desirable when fattening is the production goal. However, it may make production of lean meat more difficult. Dairy cattle and goats have such high nutrient demands for production that feed intake must be maximized in order to maintain adequate body condition. Reduced intake during late pregnancy is a common problem, particularly in sheep carrying two or more fetuses.[23] Highly stressed performance horses are often reported to be "picky eaters" or hard keepers that are difficult to maintain in adequate condition for competition. It doesn't matter how well balanced a ration is if the animals will not eat it.

Feed intake is controlled by a complex interplay of metabolic, gastrointestinal, and sensory cues. Internal and external cues are integrated in the central nervous system. The animal's current energy status, motivational state, and, in some cases, previous experience with the feed presented all affect feed intake.[3,5,26,27] The size and frequency of individual meals are under short-term control but are influenced over time by homeostatic mechanisms that regulate body weight. A single factor such as palatability may override all other inputs. As a rule, however, a multitude of factors regulates the amount of feed an animal consumes in a day.[3,27,28]

It is important to recognize the upper limits of dry matter intake when balancing rations (see Chapter 1). Animals may be unable to consume adequate amounts of a poor-quality forage to meet their nutrient requirements. Ruminants and horses appear to be able to compensate for sparsely available forage by increasing the rate and duration of eating activity. However, they are unable to adjust to feeds of very low energy density.[9,27] Complex equations that take into account stage of life, production goals, and type of ration have been developed to predict feed intake of food-producing animals and horses.[21-23] Table 8-1 gives rough estimates of the upper level of dry matter intake in the various species.

Nutritional wisdom

Domestic animals tend to lack "nutritional wisdom," which is the ability to select from a variety of feeds a diet that is balanced for all nutrients.[37] Archer[2] reported that the species of grasses and forbs selected by horses did not correlate with their nutritional needs. Horses, cattle and sheep are apparently unable to balance their mineral needs if offered a variety of mineral mixes.

Specific appetites. Specific appetite is the drive to seek out and consume specific substances based on nutritional deficits. Specific appetites exist for energy, salt, water, and perhaps phosphorus in all species. If salt deficient, most animals actively seek out and consume sources

Table 8-1 *Percent body weight of dry matter intake possible per day in mature animals**

Species	Percent body weight
Horses†	3.0
Cattle, beef†	2.1
Cattle, dairy†	2.5
Goats†	3.2‡
Sheep†	2.2
Swine	5.0

*These estimated maximums assume normal body condition. Thin animals consume more on a percent body weight basis; obese animals consume less. Young, growing animals are generally assumed to be able to consume 1.5 to 2 times as much dry matter on a per body weight basis.
†Assumes at least 50% roughage in ration.
‡Can go as high as 5.7% on 80% concentrate diet.

of sodium chloride.[12,19,33] Thirsty animals drink to alleviate dehydration. Phosphorus-deficient cattle chew on bones, rocks, and dirt.[12,19] The specificity of these drives may result in the ingestion of other nutrients in excess of need and may not reflect true "wisdom" on the part of the animal. For example, if a salt-deficient animal is offered a trace mineral mix, it will consume it in amounts sufficient to correct the sodium deficit, regardless of the other minerals present. Most animals routinely consume more salt by free choice than their calculated requirements.[33] Voluntary selection of minerals may be contrary to needs. Lambs and pony yearlings selected high-phosphorus mineral salts when fed calcium-deficient diets.[7,32] Habit and learned taste preference may control mineral intake to a greater degree than actual need.[26,33] It is better to force-feed minerals in a concentrate than to rely on the ability of the animal to voluntarily ingest a balanced diet.

Learned aversions. Another form of nutritional wisdom is *learned aversion*, which is avoidance of a feed previously associated with malaise. The ability to develop learned aversions is not as strong in herbivores as it is in omnivorous species (rats, swine, dogs).[37] It is difficult to make a horse or lamb avoid a familiar feed.[8,37] Cattle apparently are not able to extrapolate aversion from a dried form of toxic weed to the same plant in the vegetative form.[24] Aversions, however, can be created with lithium chloride or other drugs that cause gastrointestinal malaise when either incorporated into a feed or injected into an animal just after a ration has been consumed.[24,26]

Novelty appears to be a major factor in determining learned aversion, especially in older animals.[8,26,37] Young animals are much more likely to sample new feeds than are older individuals. Foraging behavior may be learned in part from older animals in a herd.[26]

Grazing and eating behavior

Foraging behavior consists of both the pattern of meals and the selection of plant types. It influences dietary intake and determines the ability of an animal to survive under harsh or novel conditions.[1,11,14,25,26] Llamas, for example, are able to subsist on coarser forages in dry regions of the Andes than are alpacas or sheep. This may result from longer periods of grazing and greater selection of plants by llamas than by sheep.[25] Voluntary intake is determined not only by the duration of grazing but also by bite size and rate. Bite size[14] has the greatest effect on intake, although it is affected by sward height and type. Rate of bites and time spent eating are more variable and are influenced by the degree of hunger and environmental conditions.[9,26] Computer models are being utilized to predict productivity and supplemental requirements of animals under range conditions.[14]

Horses, cattle, sheep, and pigs voluntarily consume 8 to 14 discrete meals a day, rarely fasting for more than 3 or 4 hours.[5,13,31] Eating activity is usually most intense during the relatively cool hours of dawn and dusk. Smaller, shorter meals are consumed at frequent intervals throughout the day and night.[3,13,31] Temperature has a greater influence on eating activity than light cycle.[13] Extremes of either hot or cold weather reduce feeding activity regardless of time of day.[1,13] The pattern of eating also may be disrupted by crowding,[38] social stimuli,[16,26] and other distractions such as rain and biting flies.[6]

Selection of feeds appears to be influenced strongly by early experience and social learning, at least in sheep.[26] It has been suggested that young animals should be exposed to a wide variety of feeds and forages, ideally in the presence of older, more experienced animals. This helps to maximize grazing efficiency through acceptance of new feeds and range conditions.[26] Livestock on unfamiliar ranges are more likely to select toxic plants than are animals raised on the terrain.[26] Introduction of young animals to a new range should be done with caution, ideally in the presence of older animals familiar with the region.[26]

CONTROL OF APPETITE

Appetite is the desire to seek out and consume feeds; it is not necessarily based on nutritional requirements. It is determined by a variety of exogenous and endogenous cues.

Oropharyngeal cues

The taste, texture, and odor of a feed help to determine its acceptability to a hungry animal. Flavor preferences differ between species and may be modified by previous experience.[3,27] Most domestic livestock prefer sweet and salty flavors, although excess salt (greater than 3% to 5%) limits intake in most species.[20,33] Ponies generally prefer unsalted to heavily salted (2%) rations.[33] More work is needed on the taste preferences of domestic livestock.

Texture of feeds influences the amounts consumed by an animal. Ruminants and horses generally consume more grain or pelleted feed in a single meal than if fed loose hay.[23,35] This is thought to result from (1) more rapid consumption due to reduced chewing time, and (2) faster rate of emptying of grain than hay particles from the rumen or stomach.[10,23,29] However, the increased rate of passage reduces overall digestibility in cattle, thereby reducing the benefit of greater intake.[23]

Although it is generally accepted that odor influences the acceptability of a feed, little work has been done on odor preferences of large animals. Ruminants, horses, and swine all sniff their feed before ingesting it, and horses may reject feeds on the basis of odor alone. If feeds that previously resulted in gastrointestinal distress are identifiable by odor, they may be rejected.[26] Swine are thought to be particularly sensitive to odors but preferences have not been determined.

The overall importance of oropharyngeal stimuli in the determination of feed intake has been investigated in ponies.[28] Results suggest that in ponies oropharyngeal stimuli alone are sufficient to generate physiological and psychological sensations of satiety during a given meal. If extremely hungry after a 24-hour fast, a pony's appetite may exceed available gastrointestinal capacity, which results in limitation of meal size.[28,30] Frequency of meals and degree of hunger at the time of initiation of a meal are regulated strongly by gastrointestinal and metabolic stimuli. However, the size and duration of a meal consumed by ponies, once initiated, is determined most strongly by oropharyngeal cues alone.[28] Such data are lacking in ruminants and swine.

Gastrointestinal cues

The rate of gastric emptying of feeds appears to be a major limiting factor in the short-term regulation of feed intake.[17] Nutrients in the gastrointestinal tract stimulate nerves and the release of peptide hormones such as insulin, bombesin, cholecystokinin, and gastrin. These have been strongly implicated in the control of food intake in sheep and other animals.[4,5] Most amino acids (especially tryptophan) stimulate the release of cholecystokinin, which inhibits appetite. Arginine is a strong stimulus for insulin release, which may enhance subsequent food intake. Fats slow gastric emptying to a much greater extent than other nutrients.

Metabolic cues

Postabsorptive stimuli generated by nutrients and nutrient status influence subsequent feeding activity. Peripheral blood glucose concentration per se does not regulate feeding activity but tends to reflect the degree of hunger a horse or pig is experiencing under normal conditions.[27,31,36] The availability of glucose for metabolism in the cells is the key factor in determination of the degree of hunger or satiety an animal experiences. For example, phloridzin blocks peripheral utilization of glucose. Ponies treated with this drug tend to be hyperglycemic and yet have increased appetites presumably because of decreased availability of glucose for cellular metabolism.[30] Infusion of glucose, resulting in hyperglycemia, reduces appetite only if the animal is capable of utilizing the increased level of carbohydrate.[36] Metabolic cues that regulate the motivation to eat have not been well defined. In general, anabolic agents such as steroids tend to stimulate appetite, whereas strong catabolic stimuli such as interleukin 1 (cachexin) reduce motivation to eat.[18] Interleukin 1 is released from mononuclear cells during autoimmune or infectious processes and may be a factor in anorexia associated with severe sepsis or hemolytic crises.[18] Antiprostaglandin drugs such as aspirin or phenylbutazone may stimulate appetite by blocking the actions of interleukins.

Central nervous system cues

All cues that influence feeding are integrated in the brain via hormonal and neural pathways that converge on the hypothalamus. The hypothalamus is considered to be the major integrative center for appetite control. The lateral hypothalamus generates sensations of hunger, and the ventromedial hypothalamus regulates satiety. Other loci, however, also influence feeding activity (Table 8-2).

Several neurotransmitters and substances influence food intake (Table 8-3). The systems are not mutually exclusive. Adrenalectomized rats will not eat in response to centrally administered opioids or α-adrenergic agonists unless an exogenous source of corticosterone is provided. The effect of a given system also is not constant. For example, mixed adrenergic agents (amphetamine) reduce feeding activity, whereas α-adrenergic agonists (norepinephrine, clonidine) stimulate eating in rats. It is thought that feeding activity observed after mild stimulation such as petting or social interactions is due to α-adrenergic stimulation. Serotonin presynaptic agonists usually reduce appetite.[15] Peptide hormones such as cholecystokinin and bombesin, previously considered to be active only in the periphery, have been demonstrated to play a physiological role in the central regulation of feeding activity of sheep.[4,5]

Anorexia

Anorexia, the lack of motivation to eat despite nutritional deficits, may be caused by gastrointestinal malaise. Pain, generalized malaise, and stress cause release of catecholamines and serotonin, which also result in reduced appetite

Table 8-2 *Brain loci implicated in the control of food intake*

Locus	Effect of stimulation	Effect of ablation	Pathways or neurotransmitters affecting response
Hypothalamus			
Lateral	Hunger	Anorexia	Norepinephrine Dopamine
Ventromedial	Satiety	Obesity	Norepinephrine Opioids
Caudate nucleus	Anorexia	?	GABA
Paraventricular nucleus (requires presence of corticosterone)	Hunger	Anorexia	Opioid Norepinephrine Neuropeptide Y
Nucleus of the solitary tract	Feeding activity	Lack of glucoprivic feeding response	Taste afferents Metabolic cues
Area postrema	Learned aversion; nausea, vomition	Failure to regulate body weight; no learned aversion	Vagal afferents Metabolic cues

Modified from Macy D and Ralston SL: In Bonagura JD, editor: Current therapy in veterinary medicine X, Philadelphia, 1989, WB Saunders, p 302.

Table 8-3 *Neural pathways and substances that influence central control of feeding activity*

Substance	Effect on food intake
Serotonin agonists	
5-Hydroxytryptamine	Inhibits
Serotonin antagonists	
Cyproheptadine (central action)	Stimulates
Chlorimipramine	Stimulates
Opioids	
Naloxone	Inhibits
Endorphins (β and γ) d-ala-enkephalin, α-neoendorphin	Stimulate
Morphine (low dose)	Stimulates
Catecholamines	
α-Adrenergic agonists	
Norepinephrine	
Low dose	Stimulates
High dose	Inhibits
Amphetamines	
Low dose	Stimulates
High dose	Inhibits
Dopaminergic/GABA	Stimulate
Benzodiazepines (low dose)	Stimulate

Modified from Macy D and Ralston SL: In Bonagura JD, editor: Current therapy in veterinary medicine X, Philadelphia, 1989, WB Saunders, p 302.

(Table 8-3). Fever-induced anorexia is thought to be mediated by interleukin 1.[18,27]

Stimulation of appetite. Reduction of fever, pain, and stress usually results in greatly improved intakes. Feeding small amounts frequently (three or more times a day) can also increase daily intake. Ponies, for example, eat small amounts of a novel feed presented in a new container even when they have just voluntarily completed a meal.[30] Palatability of feeds is also important, although preferences may be altered by disease. For example, horses and cattle normally prefer concentrates or silages but reject them in favor of straw or coarse hay when they are ill.

Chemical or nutritional stimulation of appetite in farm animals has not received a great deal of scientific attention (although it is commonly attempted in the field). One of the first signs of B vitamin or protein deficiency is anorexia. Protein and B vitamin supplementation may improve appetite in clinically ill or debilitated animals.[34] Chemical agents reported to increase feed intake in domestic animals include anabolic steroids and benzodiazepine derivatives; their use is controversial, however, and they should be used with caution.[27]

CONCLUSION

Feed intake is governed by a wide variety of endogenous stimuli and environmental factors. It is necessary to consider the relative importance and contribution of each in order to maximize feeding efficiency and productivity in large animals.

REFERENCES

1. Adams DC and others: J Anim Sci 62:1240, 1986.
2. Archer M: J Br Grasslands Soc 28:123, 1973.
3. Baile CA and Forbes JM: Physiol Rev 54:160, 1974.
4. Baile CA and McLaughlin C: J Anim Sci 64:915, 1987.
5. Baile CA and others: Fed Proc 46:173, 1987.
6. Brindley EL: Appl Anim Behav 24:31, 1989.
7. Burghardi SR and others: J Anim Sci 54:410, 1982.
8. Burrit EA and Provenza FD: J Anim Sci 67:1732, 1989.
9. Demment MW and Greenwood GB: J Anim Sci 66:2380, 1989.
10. Deswysen AG and Ellis WC: J Anim Sci 66:2678, 1988.
11. Dougherty CT and others: Grass and Forage Sci 44:335, 1989.
12. Ensminger ME and Olentine CG: Feeds and nutrition—complete, Clovis, Calif, 1982, Ensminger Publishing, p 102.
13. Feddes JJR and others: Appl Anim Behav Sci 23:215, 1989.
14. Forbes TDA: J Anim Sci 66:2369, 1989.
15. Garratini JD and others: Appetite 7:15, 1986.
16. Jenkins TG and Leymaster KA: J Anim Sci 65:422, 1987.
17. Koopmans HS: Am J Clin Nutr 42:1044, 1985.
18. McCauley CA: Am J Clin Nutr 42:1139, 1985.
19. Maynard LA and others: Animal nutrition, ed 7, NY, 1979, McGraw Hill.
20. Muirhead S: Feedstuffs 12, June 1986.
21. NRC: Nutrient requirements of horses, Washington, DC, 1989, National Academy Press.
22. NRC: Nutrient requirements of goats, Washington, DC, 1981, National Academy Press.
23. NRC: Predicting feed intake of food-producing animals, Washington, DC, 1987, National Academy Press.
24. Olsen JD and others: J Anim Sci 67:1980, 1989.
25. Pfister JA and others: Appl Anim Behav Sci 23:237, 1989.
26. Provenza FD and Balph DF: J Anim Sci 66:2356, 1989.

27. Ralston SL: Comp Cont Educ 6:S628, 1984.
28. Ralston SL: J Anim Sci 59:1354, 1984.
29. Ralston SL and Baile CA: J Anim Sci 56:302, 1983.
30. Ralston SL and Baile CA: Unpublished data.
31. Ralston SL and others: J Anim Sci 49:838, 1979.
32. Schryver HF and others: J Equine Med Surg 2:337, 1978.
33. Schryver HF and others: Cornell Vet 77:122, 1987.
34. Stokes SR and others: J Anim Sci 66:1778, 1988.
35. Willard JG and others: J Anim Sci 45:87, 1977.
36. Woods SC and others: Am J Clin Nutr 42:1063, 1985.
37. Zahorik DM and Houpt KA: In Kamil AC and Sargent TD, editors: Foraging behavior, New York, 1981, Garland, pp 289-310.
38. Zinn RA: J Anim Sci 67:853, 1989.

Feeds for Livestock

H. F. HINTZ

A great volume of literature is available on the subject of feeds for livestock. The classic reference, Morrison's *Feeds and Feeding*,[9] still remains a very useful source of information, even though the last edition (the twenty-second) was published in 1957. *Feeds and Nutrition* by Ensminger and associates[4] is a more recent comprehensive reference.

The National Agricultural Library in Beltsville, Maryland, maintains a data bank on feed composition, in which each feed is given a 5-digit International Feed Number. The use of these numbers is required by some scientific journals in order to facilitate comparisons among reports. The feed number system uses the following classes: (1) dry forages and roughages; (2) pasture, range plants, and forage fed green; (3) silages; (4) energy feeds; (5) protein supplements; and (6) mineral supplements.

In this discussion, feeds in the first three of these categories are combined. For the most part, the stage of the plant at harvest and the method of harvest determine in which category it belongs rather than the species of plant.

PRESERVED FEEDS

Dry forages and roughages usually contain 85% to 90% dry matter and a relatively high fiber content. The usual types of materials in this category are hay and straw. *Straw* is the residue after harvest of seeds such as oats, wheat, and barley. Straw usually contains a much higher indigestible fiber content than is found in hay. It is not commonly used as a feed but usually is used for bedding.

Silage is material that is produced by controlled fermentation of high-moisture herbage. It may contain as little as 25% dry matter. When silage is stored under anaerobic conditions, lactic, acetic, and butyric acids are produced; thus pH drops. The acid production prevents the growth of harmful microorganisms, and the material is preserved. Silage has several advantages. It is an excellent way to use the entire plant. In the case of corn, silage yields the greatest amount of energy per acre. Weather losses, which can be considerable in hay making, are avoided. Silage making provides flexibility, extends harvest time, and results in a product that can be readily handled mechanically and easily used in a complete mixed ration. The disadvantages of silage are that it must be made properly or else losses can be high, it must be fed soon after removal from storage, and storage costs can be high because a great amount of water is stored.

FORAGES AND ROUGHAGES (MODERATE ENERGY FEEDS)

Forages can be divided into two general categories: legumes and grasses. Legumes have a symbiotic relationship with bacteria that enables the fixation of atmospheric nitrogen. The bacteria are usually contained in nodules on the roots. Thus legumes store an abundant amount of protein, particularly in the leaves. Legumes also contain a higher concentration of calcium in their leaves than is found in grasses. Properly cured legumes are also excellent sources of vitamins. Under optimal conditions, legumes yield more nutrients per unit area than do grasses.

Grasses, however, have a wider geographical range than do legumes. They can tolerate humid weather, severe cold weather, and poor soil better than legumes and can persist and maintain production with a lower management level. The appearance of grass can vary greatly among the species. For example, compare the appearance of bluegrass, corn, and the largest grass of all, bamboo.

Evaluation of forages

The most important factor influencing nutritive value is stage of maturity at harvest. As the plant matures, the relative amount of stem increases, and the relative amount of leaf decreases. Stems contain a higher percentage of fiber and a lower percentage of protein, soluble carbohydrate, vitamins, and minerals than do leaves.

The stage of maturity is indicated by the heading and flowering of grasses and the presence of flower buds or flowers in legumes. Examples of the effect of stage of maturity on nutritional value are shown in Table 9-1. Thus the earlier the plant is harvested, the greater the quality; however, if the plant is cut too young, the total yield is decreased. Estimates of stage of maturity at harvest to obtain maximal benefit are shown in Table 9-2.

Optimal cutting time is often delayed because of weather conditions. Weather is of concern because hay that is mowed and then rained on before harvest may lose 40% to 50% or more of the nutritional value. For example, when hay has to be reraked in order to facilitate drying, many leaves are lost. Immediately after cutting forage may contain 75% to 85% water. Hay normally should contain not more than 20% to 22% water to prevent spoilage during storage. Wet hay may mold, which generates heat that causes the carbohydrate to bind to protein (browning or Maillard reaction) and therefore decreases energy and protein digestibility. The worst consequence is that excessively wet hay can undergo spontaneous combustion. Many barns have burned down because baled hay that was too wet was stored in them. A decrease in drying time would decrease chances of weather damage. Drying time can be decreased by using mechanical conditioners that crush the stems at the time of mowing. The application of certain chemicals such as potassium carbonate or potassium hydroxide can also decrease drying time. Hong and associates[7] reported that the time for alfalfa to reach 80% dry matter was 109 hours for the control but only 46 hours for alfalfa treated with potassium carbonate. Another method of reducing potential weather damage is to harvest the hay at higher than normal moisture content and then treat with preservatives such as propionic acid.[17]

Hay can be dried in the barn by the use of artificial drying systems. Such systems, however, rely on fossil fuels for energy and are relatively expensive to install and use.

Other factors that influence nutritive value of forages include level of fertilization and amount of foreign material, such as weeds and dust or mold. Roberts and associates[13] concluded that near infrared reflectance spectroscopy (NIRS) could accurately quantify mold in alfalfa. Color can be an indication of quality. Well-cured hay should be green on the inside of the bale. Rain damage, mold, or late harvest results in brown or discolored hay.

The most reliable method of evaluating hay

Table 9-1 *Total digestible nutrients (TDN) and digestible protein concentrations in dry hay*

	Grasses		Legumes	
Harvest date	TDN (%)	Protein (%)	TDN (%)	Protein (%)
June 1	63	12.2	63	16.8
June 15	57	9.0	57	13.1
July 1	50	6.0	50	9.2
July 15	44	3.3	44	5.9

From Slack ST, Stone JB, and Merrill WG: Feeding the dairy cow for maximum returns. Cornell Ext Bulletin 1156.

Table 9-2 *When to harvest grasses and legumes*

Crop	Stage of maturity
Alfalfa	Full bud
Red clover	Quarter to half bloom
Bird's-foot trefoil	Quarter bloom
Smooth bromegrass	Early to medium head
Timothy	Early head
Orchard grass	Boot to early head

From mimeograph by R Seaney, Cornell University.

is proper forage analyses. There are many forage testing laboratories that provide instruction and materials for forage sampling at a modest cost (see Chapter 10). Proper sampling technique is essential.

There are about 1500 species of grass and 400 species of legumes in the United States.[6] A brief discussion of some of the more economically important ones follows.

Legumes

Alfalfa. Alfalfa *(Medicago sativa)* is called "the Queen of the Forages." It is one of the oldest known forage legumes and remains the most popular. Alfalfa can be used for hay, silage, green chop, pasture, or a dehydrated product. It is an excellent source of protein, calcium, carotene, tocopherol, and water-soluble vitamins. Alfalfa is highly palatable to most species of herbivores and swine. A deep-rooted crop, it does best on well-drained soils with a pH of 6.5 or higher. More than 50% of all hay harvested in the United States is alfalfa or an alfalfa mixture.[20] Properly managed alfalfa can maintain a stand for several years and provide several cuttings per year. Most varieties of alfalfa do not produce well in the southern United States because of a high susceptibility to insects and diseases in the South, but considerable efforts are being made to develop more productive varieties that will do well under hot, humid conditions.

Alfalfa can be contaminated with blister beetles that are toxic to horses (see Chapter 39).

Alfalfa, particularly alfalfa leaf meal, is an excellent source of protein. In fact, alfalfa may produce more protein per unit area than soybeans. The development of alfalfa protein for human consumption is being studied.

Bird's-foot trefoil. Bird's-foot trefoil *(Lotus corniculatus)* is a yellow-flowered, fine-stemmed, warm-season perennial (Fig. 9-1). The name is derived from the seedpods, which resemble toes of a bird (Fig. 9-2). It can be grown as a permanent pasture or for hay, depending on the variety. Bird's-foot trefoil can tolerate more acidic soils that are less well drained than can alfalfa. However, alfalfa greatly outproduces bird's-foot when the conditions are good for alfalfa. Cox reported that a major advantage of bird's-foot is that it does not cause bloat in ruminants: "No case [of bloat] has ever been recorded in animals grazing on trefoil."[2]

Fig. 9-1 Bird's-foot trefoil.

Fig. 9-2 Seed pods of bird's-foot trefoil.

Clovers. Clovers of many types are used for pasture, hay, and silage. Alsike *(Trifolium hybridum)* grows best in a cool, moist climate and prefers heavy silt or clay. It has low palatability to horses and may cause photosensitization if horses graze on it exclusively. It can be used for hay or pasture. The flowers vary from almost white to rose.

Crimson clover *(Trifolium incarnatum)* has a large, vivid blossom and is one of the most attractive of all the clovers. It is an annual winter legume that is widely used for temporary winter grazing in the southern United States. If harvested properly, it can also be used for hay. It is used as a green manure crop in pecan and other orchards.[6] Because it will grow under a wide range of soil conditions and has an attractive flower, it has been used extensively for roadside stabilization and beautification throughout the southeastern United States.[6]

Red clover *(Trifolium pratense)* is a thick-stemmed legume that can be highly productive. It is probably the most popular of all the clovers. It has been plagued, however, with short life because of low resistance to disease such as northern anthracnose and damage from pests such as clover root borer and root carculio. Because of the pests, clover stands often die out within 3 years.[2] Black patch *(Rhizoctonia leguminicola)*, a mold that grows on red clover, produces slaframine, an alkaloid that causes excessive salivation in livestock. Heath and associates[6] suggested that black patch problems can be reduced by timely and frequent harvests, diluting the affected hay with nonmoldy forages, or storing the hay for one year.

Sweet clover is a rapidly growing plant with a large, coarse stem that makes drying for hay difficult (Fig. 9-3). Sweet clover contains coumarin that is converted by mold to dicoumarol, a potent anti–vitamin K compound. Vitamin K is necessary for normal blood clotting. Blood of animals fed moldy sweet clover does not clot readily; thus the animals may bleed to death, either internally or externally, from the slightest injury. In spite of these handicaps, sweet clover is still widely used in certain areas. Goplen[5] reported that sweet clover produces more forage per unit area than any other legume grown in western Canada because it is drought resistant and well adapted to the climate and soil. He also reported that sweet clover was valuable in soil improvement, good for silage or hay production when properly harvested, and a prized crop for the honey producer. Furthermore, low-coumarin cultivars of sweet clover have been developed. There are two common types: white-flowered *(Melilotus alba)* and yellow-flowered *(M. officinalis)*. Goplen[5] stated that the farmers and ranchers in Canada prefer the latter.

White clover *(Trifolium repens)* is widely distributed but grows best during cool, moist seasons on well-drained soil. White clover of the low- and intermediate-growing varieties is often found in pasture and lawns (Fig. 9-4). The terms

Fig. 9-3 Sweet clover.

Fig. 9-4 White clover.

Dutch clover or *common clover* are sometimes used for these varieties.

The large white clover most frequently used is Ladino. Ladino clover can be highly nutritious and can be used for hay or pasture. It has a shallow root system, however, and therefore cannot tolerate long dry spells. Furthermore, Ladino clover pasture is one of the most likely legumes to produce bloat in cattle.

Crown vetch. Crown vetch *(Coronilla varia)* is a perennial legume that is frequently used as ground cover for soil conservation. It can be used for hay and pasture, but it is difficult to make crown vetch hay because of its long stems.[2] It can be used as pasture, but grazing reduces persistence.[6]

Lespedeza. Cullison and Lowry[3] reported that lespedeza hay can almost equal alfalfa hay in feeding value if of good quality. Lespedeza was once widely grown as a hay crop in southern states, but its popularity waned because it is usually good for only one cutting per season, is relatively low yielding, and often has a high weed content; other, more desirable, hay crops such as coastal Bermuda grass have been developed for the South.

Grasses

Bluegrass. Kentucky bluegrass (*Poa pratensis*) is perhaps the most famous member of this group. The grass is not blue. The name might have come from the bluish anthers of the flowers or perhaps because it was considered to be royalty among grasses.[6] Kentucky bluegrass can be found in all of the states of the United States but is best adapted to the northcentral and northeastern sections. It is a long-lived perennial that forms a dense sod. It tolerates close grazing better than most grasses but has low midsummer production.

Bahia grass. Bahia grass (*Paspalpum notatum*) is grown over much of the southern coastal plains. It will not tolerate harsh winters and is not as productive as coastal Bermuda grass. It is grown primarily for pasture but can make good hay if properly harvested.

Bermuda grass. There are many varieties of Bermuda grass including common Tifton 44, Tifton 78, and coastal. Coastal Bermuda grass (*Cynodon dactylon*) is, according to Cullison and Lowry,[3] the most extensively produced hay crop over much of the southern United States. Reasons suggested for the popularity are that it is high yielding, it maintains its stand indefinitely when properly fertilized, and it provides three or more cuttings per year. Disadvantages are that it can suffer from winterkill and it loses much of its nutritive value when overmature.

Bromegrass. Smooth bromegrass (*Bromus inermis*) does well for hay, pasture, haylage, and silage, particularly when seeded with alfalfa. If mixed with alfalfa and several cuttings are made, however, the persistency of bromegrass may decrease. Bromegrass is usually more winter hardy and more productive than fescue, reed Canary grass, or orchard grass.

Fescue. Tall fescue (*Festuca arundinacea*) is one of several types of fescue. It is an important forage in the southern central United States because it is more productive during hot weather than other grasses such as bluegrass, it does well in heavy soils, it is deep rooted and hardy, and it tolerates trampling well.

Tall fescue, however, does have some serious problems. Cattle grazing tall fescue may exhibit fescue foot, which is characterized by lameness and dry gangrene in the extremities, or summer syndrome, which is characterized by rough hair coat, reduced rate of gain and milk production, rapid breathing, and an increased body temperature. Mares grazing fescue during late gestation may have prolonged gestation, thickened placentas, abortions, weak foals, or agalactia.

Not all tall fescue has problems. The preceding problems are caused by tall fescue infested with an endophyte, *Acremonium coenophialum*. There are numerous studies underway to eliminate the endophyte problem. For example, endophyte-free varieties have been developed, but tend to be less hardy than the endophyte-infested fescue.

Orchard grass. Orchard grass (*Dactylis glomerata*) is one of the earliest maturing of the common grasses (Fig. 9-5). It tolerates shade and hence is often found in orchards. It is not a dense sod former but instead grows in clumps. Orchard grass is especially well adapted to the cli-

Fig. 9-5 Orchard grass.

mate and soils of Maryland, Pennsylvania, West Virginia, Virginia, Kentucky, and Tennessee.[6]

Reed Canary grass. Reed Canary grass *(Phalaris arundenacea)* is a tall, coarse, vigorous perennial. It is often used in waterways, ditch banks, and other areas with wet soil, but it is also one of the most drought-resistant forage grasses.[2] Palatability is low, particularly late in the growing season, because of a high alkaloid content. Plant breeders, however, are developing low-alkaloid strains that have an improved palatability. Some of the alkaloids present in reed Canary grass can cause disqualifications of performance horses in some states if detected in the urine (see Chapter 39).

Ryegrass. Ryegrass *(Lolium spp.)* is thought to be one of the most widely used of all pasture grasses (Fig. 9-6). It is an important cool-season grass throughout the world. Both perennial and annual ryegrass (often called *Italian ryegrass*) are used. Ryegrass can be highly competitive with legumes and tends to be finer stemmed than most forage grasses.[2]

Sudan grass. Sudan grass and sorghum-Sudan hybrids are annual grasses that are often used for green chop or silage. These plants may contain prussic (hydrocyanic) acid. The young growing plant and regrowth has the highest concentration of prussic acid. To decrease chances of poisoning, Sudan grass should be grazed when it is 18 inches (45 cm) or taller and sorghum or sorghum-Sudan hybrids grazed at 24 inches (60 cm) or taller. Regrowth after a frost or period of dry weather should not be grazed.[2] Sudan grass and sorghum-Sudan hybrids regardless of the stage of maturity cause cystitis in horses (see Chapter 39).

Timothy. Timothy *(Phleum pratense)* is one of the most popular grasses used for hay (Fig. 9-7). Cox[2] reported that when cut in early or mid-June in New York, timothy surpasses bromegrass and orchard grass cut at the same time in nutrient yield. Timothy is adapted to cool, humid climates but does not withstand drought.

Other forages

Cereal forages are used to produce hay and winter forage, particularly for cow-calf operations. Wheat and oats are used much more commonly than barley or rye. All cereal forages are annuals.

Fig. 9-6 Ryegrass.

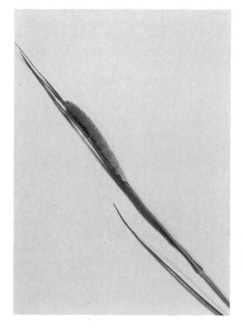

Fig. 9-7 Timothy.

The variety and climate determine if they are planted in fall or spring. Cereal forage may contain high levels of nitrate.[4] Nitrate can be converted by the rumen bacteria to nitrite. Nitrite oxidizes hemoglobin to methemoglobin, which does not carry oxygen. In severe cases, the blood becomes almost chocolate brown, and nonpigmented areas of the skin and mucous membranes have a brownish discoloration. Death may result because of anoxia.

Turnips, rape, and kale (members of the *Brassica* family) can be used as a fall or winter forage in warmer climates or as summer forage. For example, turnips may be sown into wheat stubble or on early potato fields in August in the Columbia Basin for autumn pasture for beef cattle or sheep.[6] Excessive intake may result in thyroid enlargement, as most members of the *Brassica* family contain goiterogens that interfere with thyroid hormone synthesis.[19]

There is considerable interest in the utilization of chemically treated high-fiber roughages such as oat and wheat straw. Ammonia, sodium hydroxide, and urea have been used to improve nutritive value. For example, Sundstol and associates[16] reported that the ammoniation of low-quality forages may increase the energy value of cereal straws by as much as 80% above that of untreated material. Further information on several chemical and physical treatments is provided by Sundstol.[15]

Feeding high-protein forages that have been ammoniated can cause neurological signs including hyperexcitability, circling, and convulsions in ruminants.[21] The condition commonly called "crazy cow disease" or "bovine bonkers" has been suggested to be caused by an imidazole produced during ammoniation[8] (Chapter 2).

HIGH-ENERGY FEEDS
Grains

Grains contain a high concentration of soluble carbohydrate (starch) and thus have a high digestible energy content. They contain almost no calcium but some phosphorus. The availability of the latter depends on the phytic acid content, particularly for nonruminants. Grains are not good sources of most vitamins (Fig. 9-8).

Barley. Barley (*Hordeum vulgare*) is an excellent livestock feed. It contains more fiber than

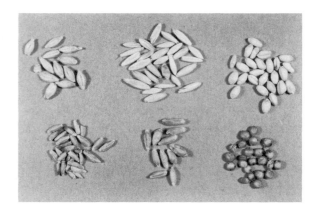

Fig. 9-8 Common grains. *Top row,* barley, oats, wheat; *bottom row,* rye, triticale, milo.

corn and thus is often used in coarse-textured feeds for ruminants. Barley can be grown in northern areas that are not suitable for corn because corn requires a longer growing season. Barley can also tolerate limited rainfall in the summer. It is usually considered to be less palatable than corn for poultry. Barley has a hard kernel and thus should be ground or otherwise processed.

Buckwheat. Buckwheat (*Fagopyrum esculentum*) is grown primarily for flour for human consumption. It has a short growing season and does better on poor, acidic soils than most other grains. It contains a high fiber content and is not very palatable to livestock.

Corn. Corn (*Zea mays*) is the number one grain of North America (Table 9-3). It is very productive where the growing season is long enough and moisture is adequate. Corn has a high digestible energy and a low fiber content. Yellow corn provides some carotene. It is one of the most palatable grains for livestock. The limiting amino acids are lysine and tryptophan. There has been considerable effort to develop varieties of corn that contain increased levels of these amino acids. For example opaque-2 was developed by scientists at Purdue University; however, the yield was lower and the resistance to fungi and disease was less than that of conventional corn. Scientists at the International Center for the Improvement of Maize and Wheat near Mexico City reported that a recently de-

Table 9-3 *Grain production in 1987 in the United States*

			Production	
	Hectares harvested (1000)	Acres harvested (1000)	1000 tonnes	1000 tons
Corn	23,945	59,167	179,436	197,792
Wheat	22,635	55,930	57,295	63,156
Soybeans	22,840	56,437	51,838	57,141
Sorghum grain	4291	10,604	18,819	20,744
Barley	4470	11,046	11,474	12,648
Rice	276	2330	5793	6386
Oats	80	683	501	552

From USDA: Agricultural statistics, Washington, DC, 1989.

veloped variety—quality protein maize (QPM)—has about the same amount of protein as common corn but twice the usable protein because the quality and biological value of the protein is so much higher.[12] Worldwide studies on QPM are needed.

Millet. Several different kinds of millet such as pearl, foxtail, proso, finger, and Japanese are grown in various parts of the world. The fiber content may be almost as high as that of oats, but some varieties have a digestible energy content similar to that of corn.

Rice. Rice *(Oryza sativa)* is the staple food for much of the human population. Little rice is used in animal feeds because of economics, but two by-products of rice that are often used in animal feed are rice bran (pericarp or bran layer and germ of the rice) and rice polishings (obtained in the milling operation of brushing the grain to polish the kernel). Both contain 11% to 14% ether extract (fat).

Rye. Rye *(Secale cereale)* is a winter-hardy cereal with a composition similar to wheat. It does not produce as well as other grains on good soils but can grow on land such as poor, sandy soil that is not suited for other grains. Rye is considered relatively unpalatable to livestock. Thus it is usually recommended that grain mixtures not contain more than 33% rye. Most of the rye production goes for human consumption as bread or whiskey. Rye may be contaminated with ergot, a mixture of alkaloids that can cause abortion. Antoniou and Marquart[1] reported that rye should not be fed to chicks during the first week of life because the pentosans in rye interfered with nutrient absorption and caused retention of water in the intestine.

Sorghum grain. Sorghum grain *(Sorghum vulgare)* has a composition similar to corn. It is more drought resistant than corn because the leaves and stems have a waxy covering that reduces water losses. Thus sorghum outproduces corn when water is limiting, but corn is much more productive than sorghum when water is available. There are many hybrids from which to choose. Some varieties have a high content of tannin that can decrease palatability.

Spelt. Spelt *(Triticum spelta)* and emmer *(Triticum dicoccum)* are relatives of wheat, but the hulls are not usually removed during harvesting; thus they are high-fiber grains. They can be used to replace oats.

Triticale. Triticale is a cross between wheat and rye. The cross was designed to combine the productivity and palatability of wheat with the hardiness of rye. The name came from the genus of wheat *Triticum* and the genus of rye *Secale*. It is expected that triticale production will increase, but now it is mostly used on poor soils where water is limited. Some of the problems that limit production are late maturity, excessive head breaking, and seed shatter.[18]

Wheat. Wheat supplies about 20% of all calories consumed by humans and more than 50% of the world grain trade. It is harvested nearly every month somewhere in the world.[11] Because of economics, it is not extensively used in animal feeds. However, it is a high-energy feed that can be used effectively. Because the protein gluten has a high degree of elasticity, wheat is often considered to have a "pasty nature"; some recommend that the grain ration not contain more than 50% wheat. Wheat should be processed (e.g., flaked or rolled) for all classes of livestock because of the small, hard kernel that must be broken for maximum digestibility.

Other sources of energy

Fats and oils. The use of fats and oils in livestock feed has increased in recent years.[10] The high energy content (2.25 times that of carbohydrate or protein) has been used to great advantage in dairy cattle, swine, and horse rations. A rule of thumb is that when fat is less than 2.5 times the price of corn per pound it is an economical source of energy.[14] However, Soule[14] also points out that fats and oils have additional advantages. The addition of 1% fat to alfalfa meal reduces dustiness and facilitates handling. Fat reduces the wear on pellet mills.

Both animal and vegetable fats and oils are used. Calcium soaps of fats are of particular value to ruminants because the fat escapes rumen degradation.

Molasses. Molasses can be an economical energy source, improve palatability, decrease dustiness, provide minerals (depending on the source of molasses), and help in pellet manufacture. Several different kinds of molasses are used in animal feeds.

Cane molasses is a by-product of the manufacture of sugar from sugarcane. Beet molasses is a by-product of the manufacture of sucrose from sugar beets. Citrus molasses is obtained during the manufacture of dried citrus pulp. Molasses can also be obtained during the processing of cornstarch.

By-products. The use of by-products is ecologically sound and often an economical practice. The types and amounts of by-products most commonly used in animal feeds are shown in Table 9-4.

Many by-products such as almond hulls, apple pomace, citrus pulp, cocoa meal, copra meal, and dried bakery product are used in animal feed. The reader is referred to the *Feed Industry Red Book* (published annually by Communications Marketing Inc., Eden Prairie, MN 55344)

Table 9-4 *Use of by-product feeds in animal feeds in 1987*

Type of feed	Amount used (1000 tonnes)
Wheat mill feeds	5610
Corn gluten feed and meal	1515
Distiller's dried grains	816
Rice mill feed	580
Brewer's dried grains	136

From USDA: Agricultural statistics, Washington, DC, 1989.

for further by-product information. Following are a few definitions based on those provided by the Association of American Feed Control Officials.

Beet pulp is the dried residue from sugar beets after the sugar has been extracted. Pulp contains a high content of highly digestible fiber. It can be used in horse rations as replacement for hay but has an energy content greater than the hay. It contains relatively low levels of protein and phosphorus, a high level of calcium, and no vitamin A.

Brewer's dried grains are the dried extracted residue of barley malt alone or in mixture with other cereal grains or grain products resulting from the manufacture of beer. The energy content is similar to that of barley, but the protein content is high (26%).

Corn gluten feed and meal is obtained after the cornstarch and most of the germ has been removed in the manufacture of cornstarch or syrup. It contains a high concentration of insoluble protein. The protein content varies from 20% to 60%, depending on the ratio of meal to feed.

Distiller's dried grains are the residue obtained after alcohol is removed from the yeast fermentation of a grain or grain mixture. The predominating grain should be declared as the first word in the name.

Several different products result from the processing of wheat. Wheat bran is the coarse outer layer that is separated in milling. Wheat germ is the germ plus some bran. Wheat shorts are fine particles of wheat bran, wheat germ, and some flour. Wheat mill run is coarse wheat bran, wheat shorts, and wheat red dog. Wheat red dog is primarily offal from the tail of the mill plus fine particles of wheat bran, wheat germ, and wheat flour. Wheat bran has a low-calcium, high-phosphorus content. Livestock, particularly nonruminants, fed significant amounts of wheat bran without calcium supplementation are in danger of nutritional secondary hyperparathyroidism.

PROTEIN SUPPLEMENTS

The types and amounts of protein supplements used in animal feeds are shown in Table 9-5. Of course, as mentioned earlier, legume forages such as alfalfa and by-products such as corn gluten meal also supply a significant amount of protein.

Vegetable proteins

Canola meal is the material remaining after the oil is removed from the canola seed. Canola was developed from rape, a *Brassica*. Rape contains erucic acid and glucosinolates that reduce animal performance and can be toxic. Rape, however, can be effectively raised in northern climates whereas soybean meal cannot. Therefore, scientists in Canada developed from rape a plant low in erucic acid and glucosinolate that could tolerate the Canadian climate; it was named *canola*. The use of canola meal has been steadily increasing. The lysine content is usually higher than that for most vegetable proteins except soybean meal.

Cottonseed meal is obtained after the removal of the oil from the cottonseed by a solvent extraction process. Cottonseed meal must not contain less than 36% protein but may contain the toxin gossypol. Swine are more susceptible to gossypol toxicity than are ruminants or horses. Signs include panting, weakness, and anorexia. Lesions include hydrothorax, hydropericardium, hydroperitoneum, and edema of the lungs. Gossypol may cause off-color (green) yolks if fed to laying hens. Degossypolized and glandless low-gossypol forms of cottonseed meal are available. Lysine is the first limiting amino acid.

Linseed meal is obtained after the oil is obtained from flaxseed. At one time it was highly prized for its ability to give a bloom to the coats of cattle and horses. However, the use of solvent extract has apparently decreased the benefit on coat condition because of the more efficient extraction of oil. Linseed meal is low in lysine.

Peanut meal is obtained after the oil is removed from peanuts. It is lower in methionine content than other vegetable proteins but is considered to be an excellent source of niacin and pantothenic acid. Some peanut meal may contain high levels of alfatoxin.

Soybean meal is the most important protein supplement in the United States. The popularity of soybean meal is the result of the relatively low price and good array of amino acids. Soybean meal contains a higher content of lysine and methionine than do most vegetable proteins.

Soybeans contain, as do most legume seeds, several factors that can limit their utilization. For example, trypsin inhibitors decrease protein utilization. Fortunately most of the trypsin inhibitors and other factors such as saponins and hemagglutinins are heat labile. Therefore, soybean meal must be properly heat treated before use as a feed. Excessive heating can cause reduced energy and protein availability because of the browning (Maillard) reaction.

Sunflower meal is the material remaining after removal of the oil from sunflower seeds. If the hulls are left on, the fiber content can be high. Sunflower meal, as for most common vegetable proteins, is inferior to soybean meal in

Table 9-5 *Use of high-protein feeds in 1987*

Meal	Quantity (1000 tonnes)
Plant sources	
Soybean	19,092
Cottonseed	1315
Sunflower	405
Linseed	118
Peanut	109
Animal sources	
Tankage and meat meal	2449
Fish meal	522
Dried milk	364

From USDA: Agricultural statistics, Washington, DC, 1989.

Table 9-6 *Analyses of calcium and phosphorus supplements*

Supplement	Calcium (%)	Phosphorus (%)
Animal bone, steamed, dehydrated	29	14
Dicalcium phosphate	22-26	18-21
Defluorinated phosphates	29-36	12-18
Limestone, ground	34-37	—
Calcium phosphate	17	21
Sodium phosphate	—	22
Diammonium phosphate	—	20
Oyster shell	35-38	—

lysine content but contains a relatively high content of methionine.

Animal proteins

Fish meal can be made from a variety of fish. Menhaden are caught primarily for their oil content. The dried residue after oil removal can provide a high-quality protein supplement with an excellent amino acid content. Anchovy, herring, salmon, whitefish, and sardines are also used to produce fish meal. Fish meal is usually much more expensive per unit of protein than vegetable proteins but has a better protein quality. In recent years the interest in fish meal in ruminant feeding has greatly increased because it is an excellent source of escape protein for ruminants.

Meat-and-bone meal is the rendered product of mammal tissues including bone but excluding blood, hair, hoof, horn, hide trimmings, and contents of the digestive tract. Meat-and-bone meal tankage is similar to meat-and-bone meal except that it can also contain blood or blood meal.

MINERAL SUPPLEMENTS

Most species of livestock utilize inorganic supplements effectively. The composition of several calcium and phosphorus sources is shown in Table 9-6. Raw rock phosphate may contain toxic levels of fluorine. The defluorinated product is efficiently utilized if not heated excessively.

REFERENCES

1. Antoniou T and Marquart R: Poult Sci 60:1898, 1981.
2. Cox WJ: Cornell recommendations for field crops, Ithaca, NY, 1989, Cornell University, p 26-61.
3. Cullison AE and Lowry RS: Feeds and feeding, ed 4, Englewood Cliffs, NJ, 1987, Prentice-Hall.
4. Ensminger ME, Oldfield JE, and Heineman WW: Feeds and nutrition, Clovis, Calif, 1990, Ensminger Publishing.
5. Goplen BP: Can Vet J 21:149, 1980.
6. Heath ME, Barnes RF, and Metcalfe DS: Forages: the science of grassland agriculture, ed 4, Ames, 1985, Iowa State University Press.
7. Hong BJ, Broderick GA, and Walgenbach RP: J Dairy Sci 71:1851, 1988.
8. Morgan SE and Edwards WC: Vet Hum Toxicol 28:16, 1986.
9. Morrison FB: Feeds and feeding, ed 22, Ithaca, NY, 1957, Morrison Publishing Company.
10. Palmquist DL and Jenkins TC: J Dairy Sci 63:1, 1980.
11. Pomeranz Y: Modern cereal science and technology, New York, 1987, VCH Publishing.
12. Raloff J: Science News 134:104, 1988.
13. Roberts CA and others: J Dairy Sci 70:2560, 1987.
14. Soule R: Feed industry red book: reference and buyer's guide, Eden Prairie, Minn, 1987, Communications Marketing.
15. Sundstol F: Livestock Prod Sci 19:137, 1988.
16. Sundstol F, Coxworth F, and Mowat DN: World Anim Rev 26:13, 1978.
17. Thomas JW: J Anim Sci 47:721, 1978.
18. Todorov NA: Livestock Production Science 19:47, 1988.
19. Underwood E: Trace elements in human and animal nutrition, ed 4, New York, 1977, Academic Press.
20. USDA: Agricultural statistics, Washington DC, 1989.
21. Weiss WP and others: J Anim Sci 63:525, 1986.

10

Principles of Ration Analysis

SARAH L. RALSTON

It is necessary to know the nutrient content of the ingredients before rations can be evaluated or formulated (Chapter 1). *Visual evaluation alone of a feed cannot accurately predict its feeding value.* Published values for feeds may be used to estimate general nutrient content of rations. Average nutrient concentrations in commonly used feeds are available in the National Research Council's publications for each species and the *U.S.-Canadian Tables of Feed Composition.*[6] These are fairly reliable for concentrates, although trace mineral concentrations may vary from region to region. Regional averages based on common soil types and growing season may be available from local extension agents.

When balancing rations for high-producing, growing, or stressed animals, obtain an analysis of forages to be used rather than relying on published averages. The concentrations of protein, soluble fiber, and minerals in forages vary with climate, maturity, soil type, and fertilization (see Chapter 9). Hay cut from the same field at different times of year may have dramatically different nutrient contents between cuts (Table 10-1). Repeated analyses may not be practical if forages are purchased in small amounts. However, in systems where forages are purchased at intervals greater than 2 months and production goals are high, each load should be subjected to analysis. Other situations in which feed analyses are recommended include: (1) use of nontraditional or commercial mixed feeds for which complete analyses are not available and (2) when nutritional problems are suspected on the basis of clinical signs.

The critical concerns in feed analysis are:

1. Obtaining a truly representative sample for analysis.
2. Submission of appropriate requests for analysis. Not all laboratories do all analyses. There also are various techniques for analysis that vary in accuracy and reliability.
3. Interpretation of the results of an assay. Which units were used? Are the results presented as dry matter or as fed concentrations? Are the results believable?

Table 10-1 *Nutrient analysis (dry matter basis) of two cuttings of clover-grass mix hay cut from the same field in a single growing season in New Jersey*

Nutrient	First cut	Third cut
Dry matter	91.5	88.0
Mcal DE/kg*	1.9	2.4
Crude protein %	10.5	19.5
Acid detergent fiber %	40.9	31.0
Neutral detergent fiber %	64.0	39.1
Calcium %	.25	.93
Phosphorus %	.10	.10
Magnesium %	.84	.24
Potassium %	1.9	2.5
Iron ppm	56	78
Copper ppm	4	7
Zinc ppm	40	50
Manganese ppm	4	4
Selenium ppm	.02	.01

*Horse digestible energy.

SAMPLING OF FEEDS

Getting a sample that truly represents the overall feed is not always easy. Techniques differ according to the type of feed sampled and the equipment available.

Dry or ensiled forages

Hay samples should be collected from 20 small bales or 20 sites on large, round bales of hay.[4] Ideally a hay sampler (see Table 10-2 for vendors) should be used to drill cores. The tubular sampler may also be used to sample silages in pits, where cores should be drilled from 20 sites[2,4,7] on the freshly exposed face of the silage. At least 500 g (1 lb) of the forage to be tested should be taken.[4] The samples should be thoroughly mixed in a bucket or bag, and a representative subsample (20% of total) submitted for analysis. If a forage sampler is not available, "grab samples" may be taken by hand.[2,7] Grab samples are less accurate and harder to mix for subsampling than core samples.[4,7] However, loose hays or silages may be difficult to core and require the "grab sample" technique. If long-stem forages are sampled by grab sample, the samples should be chopped with a scissors to facilitate mixing. Mixing is even easier if the sample is dried before cutting or grinding; however, the moisture content should be determined first if a dried sample is submitted (see following discussion).

Silage or haylage being loaded into a silo or pit is easily sampled by catching 10 to 12 handfuls at various intervals during loading (at least 500 g total).[2,4] The samples should be combined and ground before taking the subsample (at least 100 g, or 4 oz) for analysis. Some advocate drying before grinding to determine moisture content and facilitate mixing.

Certain precautions are necessary. When submitting silage for analysis, it is important that the sample be placed in an air-tight bag. If sampling is done by hand, latex gloves should be worn, especially if trace minerals are of concern. *Contamination from trace minerals on the skin may alter the analyses.*

Grains or concentrates

Samples (50 to 100 g, or 2 to 4 oz, each) should be taken from at least 20 sites and at a variety of depths in binned or bulk feeds. If sacks or bags are used to store the feeds, samples should come from at least two sites from 10 bags. Mixed grains with supplemental protein or mineral in powder form should be sampled only after thorough mixing or from both top and bottom of the bags to avoid bias due to settling of fines. The samples should be mixed and a representative subsample submitted (at least 100 g) for analysis. Latex gloves should be worn, as mentioned previously, to take samples by hand.

Table 10-2 *Suppliers of forage core samplers*

Company	Address	Telephone number
Forageurs Corp.	8500 210th Street West Lakeville, MN 55044	(612) 469-2596
Hadge Products, Inc.	P.O. Box 1326 El Cajon, CA 92022	(619) 444-3147
NASCO	901 Janesville Avenue Fort Atkinson, WI 53538	(800) 558-9595
Oakfield Apparatus, Inc.	P.O. Box 65 Oakfield, WI 53065	(414) 583-4114
Techni-Serve, Inc.	P.O. Box 848 Madras, OR 97741	(503) 475-2209
Utah Hay Sampler	P.O. Box 1141 Delta, UT 84624	(801) 864-5380

Pastures

To obtain representative samples from pastures, 10 sites should be selected in areas utilized by the animals. *Samples should not be taken from overgrown areas that are obviously not being grazed.* Although easier to sample, these areas would not represent what the animals are consuming. If the entire pasture is being utilized in a uniform fashion, marking sampling sites in an X pattern is often recommended to avoid bias.[2] At each site a 0.3-m (1-ft) square should be marked out and all forage within the square clipped to 2-cm (1-inch) height. However, plants known to be avoided by the animals should not be taken. For example, horses usually avoid spurge (*Euphorbia* species) unless they are starving or adequate forage is lacking.[3] Samples from all sites should be mixed (some advocate drying and grinding as with dried or ensiled forages) and a subsample of 20% of the total submitted for analysis.

METHODS OF ANALYSIS

The most critical nutrients in balancing rations are water, energy, fiber, protein, calcium, and phosphorus. Virtually all feed analysis laboratories are equipped to provide estimates of these in one form or another. Electrolytes (Na, Mg, Cl, K) and common trace minerals (Cu, Fe, Zn, Mn, S, Co) are also commonly assayed. Certain trace mineral analyses (Se, I, Mo) require special analytical methods and are not performed in many laboratories. It is important to ascertain if the laboratory to which you are submitting samples is equipped to do the requested analyses. It is also necessary to specify which analyses are to be performed, rather than submitting a feed sample with the simple request of "analyze the nutrient content." Without specific instructions, laboratories may give more or less information than anticipated. Some forage analysis laboratories are listed in Table 10-3. Local extension

Table 10-3 *Forage analysis laboratories in selected regions of North America*

Laboratory	Address	Telephone number
Agricultural Testing Lab	Hills Building University of Veterinary Medicine Burlington, VT 05405	(802) 656-3030
Agri-Food Laboratories	Unit 1 1-503 Imperial Road Guelph, Ontario, Canada N1H 6T9	(519) 837-1600
Animal Disease Lab	State of Illinois Shattuc Road Centralia, IL 62801	(618) 532-6701
Chandler Analytical Lab	283 North Arizona Avenue Chandler, AZ 85224	(602) 963-2495
Chemical Service Laboratory, Inc.	Suite 360 3600 Chamberlain Lane Louisville, KY 40241	(502) 429-5238
Cooperative Federée de Quebec	1055 Marche Central Montreal, Quebec, Canada H4N 1K3	(514) 384-6450
DANR Diagnostic Lab	LAWR: Hoagland Annex University of California Davis, CA 95616	(916) 752-0147
Florida Dairy Farmers Association	1301 West Main Street Avon Park, FL 33825	(813) 452-0433

Continued.

Table 10-3 *Forage analysis laboratories in selected regions of North America—cont'd*

Laboratory	Address	Telephone number
Holmes Laboratory, Inc.	3559 US 62-Star Route Millersburg, OH 44654	(216) 893-2933
Litchfield Analytical Services	535 Marshall Street P.O. Box 457 Litchfield, MI 49252	(517) 542-2915
Minnesota DHIA	134 Lake Boulevard Buffalo, MN 55313	(612) 625-5277
Northeast DHIA	730 Warren Road Ithaca, NY 14850	(607) 257-1272
North West Labs, Inc.	901 North Lincoln Jerome, ID 83660	(208) 324-7511
Norwest Labs	9938 67th Avenue Edmonton, Alberta, Canada T6E 0P5	(519) 837-1600
Rock River Laboratory, Inc.	Route 3 Box 291 Watertown, WI 53094	(414) 261-0446
Servi-Tech, Inc.	1602 Parkhost P.O. Box 169 Hastings, NE 68902	(402) 463-3522
Servi-Tech, Inc.	1816 East Wyatt Earp P.O. Box 1397 Dodge City, KS 67801	(316) 227-7123
State of Wisconsin Soil and Forage Analysis Lab	8396 Yellowstone Drive Marshfield, WI 54449	(715) 758-2178
Triple S Laboratories, Inc.	Box 678 Loveland, CO 80539	(303) 667-5671
University of Alaska Fairbanks	AFES Palmer Research Center 533 East Firewood Palmer, AK 99645	(907) 745-3278

This is not a comprehensive list. Local extension agents or feed stores can be contacted for local laboratories. Complete listings of certified laboratories are available from National Forage Testing Association, PO Box 371115, Omaha, NE 68137; (402) 333-7485.

agents or feed stores can supply the names of nearby laboratories. Large feed companies often perform forage analyses for their clients.

The type of analysis used in the laboratory is important. The several techniques currently used to determine nutritional content of feeds differ in their accuracy and therefore their usefulness. Many laboratories are equipped to do more than one type of analysis; what is employed on a given sample affects cost and accuracy. Common systems used in commercial laboratories to determine nutrient content are discussed in ascending order of accuracy and cost of obtaining results.

Near infrared reflectance spectroscopy

Near infrared reflectance spectroscopy (NIRS) is based on the assumption that the spectrum of radiation absorbed and emitted by the organic components of feed is similar between feeds of the same chemical and biochemical composition.[5,7,12] It is calibrated on the basis of complex mathematical equations and computer prediction models that compare NIRS spectra of feeds to the chemical analyses (see following discussion).[7,12] The larger the data base on which a laboratory can calibrate its NIRS, the more accurate the interpretations.[5,7] The accuracy of this technique for feeds that fall outside the standard range or that are from different regions of the country is questionable. Although NIRS is extremely useful in the initial stages of evaluation and formulation of rations, it is of limited value when exotic or nontraditional feeds are used. The technique is 25 times more rapid than more traditional assays and relatively inexpensive.[7] NIRS reports usually include dry matter, crude protein, ADF, NDF, TDN, NEI, NEm, NSC, Ca, P, Mg, and K. *It is important to recognize that NIRS is based on predictive equations only, and nutrients are not measured directly.*

Wet chemical analyses

These laboratory assays involve a variety of chemical and biochemical techniques to accurately measure mineral and biological materials.[7,9-11] They are considered to be the most accurate (and traditional) methods of analysis.[2,7] However, there are problems even with these standard techniques.

Proximate (Weende) analysis. In a series of chemical extractions, the crude fiber, crude protein (nitrogen content × 6.25), crude fat (ether extract), and ash (total mineral) content of feed dry matter is determined in proximate analysis.[2] From these values the nitrogen-free extract, theoretically reflecting the soluble carbohydrate content of the feed, is calculated. Total digestible nutrient (TDN) content of feeds is based on proximate analysis (see Chapter 1). Until recently it was considered to be the standard form of feed analysis, and most laboratories are equipped to perform it.

Despite its widespread use, the many inaccuracies inherent in proximate analysis limit its usefulness in modern ration-balancing programs.[8] Ether extraction removes not only fats but also waxes and other fat-soluble materials, which may result in erroneously high estimates of fat content. Use of Kjeldahl analysis of feed nitrogen to derive the estimated total crude protein is based on several assumptions that are more or less true for all feeds.[7,8] These assumptions are:

1. All proteins contain 16% nitrogen.
2. All nitrogen in a feed is in the form of protein.
3. The protein in the feed is totally digestible.

The Weende method of fiber determination used in proximate analysis is considered to be inadequate for herbivorous animals in that it does not distinguish between the fermentable (cellulose and hemicellulose) and nonfermentable (lignin) components of fiber. Calculation of nitrogen-free extract compounds the inherent errors to an even greater degree.[8] Digestibility of protein and energy sources is not estimated by these techniques. Proximate analysis does not measure individual minerals or vitamins.

Proximate analysis is still useful to obtain a rough estimate of a feed's value.[7] However, crude fiber and protein analyses are being rapidly replaced by the detergent system and other assays.

Detergent system analysis. This system is a modification of proximate analysis that uses improved methods for estimating the value of fiber and protein developed by Van Soest.[8] It recently has been accepted as the standard technique for analysis of forages for food animals. It uses a series of neutral and acid detergent extractions for analysis of fiber quality. Used in conjunction with Kjeldahl analysis of nitrogen and a variety of other chemical treatments, detergent system analysis provides an estimate of the digestible versus indigestible portions of both fiber and protein, in addition to soluble carbohydrates and ash.[8] It is important to note that the "soluble carbohydrates" in this system include fibrous materials—pectins and gums—that are highly available only to herbivorous animals, in addition to the universally available sugars and starches. Ether extraction is still used to give an estimate of fat content.

Detergent analysis is most useful in evaluation of forages for ruminants and horses, which rely heavily on fermentable fiber as an energy source. It is of limited use in feeds such as grains that have high soluble carbohydrate but low fiber content. This system overestimates the caloric value of feeds for omnivorous animals such as swine.

Results of detergent system analysis usually include values for dry matter, soluble carbohydrates, neutral detergent (NDF) and acid detergent fiber (ADF), digestible protein, and ash. This system does not estimate individual minerals or vitamins.

Mineral chemistries. Laboratories conducting chemical determination of minerals most commonly use atomic absorption, colorimetric, or spectrophotometric techniques.[1,7,9,10] Some techniques are more accurate for a given mineral than others, but in general most wet chemistries for the macrominerals are fairly accurate. Atomic absorption is considered to be the most accurate technique but frequently is not available. Problems arise in trace mineral analyses, which are extremely susceptible to errors due to contamination and require microanalysis technology. Each year, however, assay techniques are improved and refined. As long as the sample provided is truly representative, free of contaminants, and submitted to a reputable, reliable laboratory, wet chemistry results usually closely approximate the actual concentration of minerals in the feed.

Availability of the minerals to the animal, however, can be accurately determined only by digestion trials. For example, a ration suspected to be deficient in selenium, based on clinical signs exhibited by animals consuming it, is submitted for selenium analysis. The result of the analysis is a low normal concentration of selenium. This does not necessarily mean that the animals are not suffering selenium deficiency. The ration should also be examined for substances that reduce the absorption or enhance the excretion of selenium, such as arsenic, copper, and molybdenum. Unfortunately, the interactions of minerals are quite complex and not entirely understood at this time.

Vitamin assays. Vitamin assays are much more difficult and less available than are mineral assays. The only vitamin routinely measured at commercial laboratories is vitamin A. Usually the assays for vitamin A measure total carotenoids, which overestimate the actual amount of vitamin A activity but at least give an idea of the vitamin status of the feed. With increasing use of high-performance liquid chromatography,[11] assays for vitamin C, B vitamins, and vitamin E may become more available. Previously biological assays, using either animal or bacterial response to an extract of the feed or substance in question, were used to determine concentrations of most vitamins. These assays are tedious, lengthy, and expensive. Vitamins that serve as coenzymes in specific biochemical reactions (i.e., niacin, riboflavin, and thiamin) may be determined by measuring by-products of the reaction the vitamin mediated by either colorimetric reactions or high-performance liquid chromatography (HPLC). These assays, however, are cumbersome and require a level of sophistication usually found only in research laboratories.

Forage moisture

The quality of hay, haylage, and silage is determined in large part by the moisture at which it was baled or ensiled. All commercial laboratories do determinations of moisture content. It is, however, sometimes advantageous to be able to rapidly determine the percent moisture in a forage before submitting it for nutrient analysis. Freeze-drying is the most accurate method and results in minimal alteration in the nutrient composition of the feed. If freeze-drying equipment is not available, moisture determination can be performed in a microwave oven, a Koster Crop Tester* or other commercially available moisture probes used directly on the bales or in the silos (Table 10-4).

Determining moisture content of a forage, haylage, or silage sample in a microwave oven requires minimal equipment: a microwave oven, a gram weighing scale, paper plates, and a glass of water. A separate paper plate must be weighed for each sample to be analyzed and the weight written on the plate. Exactly 100 g of chopped forage is weighed onto the plate. The material is spread evenly over the plate and placed in the microwave with a half-full glass of water set near the plate to reduce the chance of burning the sample. Silage and haylage samples

---

*Koster Crop Tester, Inc., 21079 Westwood Drive, Strongsville, Ohio 44136; (216) 572-5615.

Table 10-4 *Forage moisture testers*

Instrument	Manufacturer	Address and telephone number
PreAgro Moisture Tester	Omni Mark, Inc.	1631 East 79th Street Suite 134 Bloomington, MN 55420 (612) 854-2161
Delmhorst Hay Tester	Delmhorst Instrument Co.	51 Indian Lane East Towaco, NJ 07062 (800) 222-0638

assumed to contain more than 50% moisture should be heated at medium power for 4 minutes initially. Hay samples (usually less than 30% moisture) should be heated for only 3 minutes initially. The sample is then removed from the oven and weighed. The forage on the plate is stirred and returned to the oven for 1 minute. Then it is weighed, stirred, and reheated for 30 seconds. The drying and weighing are continued until the weight becomes constant. The final weight subtracted from the original weight gives the percent water in the feed. If the sample gets charred, the analysis must be redone on a fresh sample with shorter drying times.

CONCLUSION

When formulating or evaluating rations, it is necessary to know the nutrient content of the ingredients. Although published average values for concentrates may be used, it is strongly recommended that, when economically feasible, forages be submitted for chemical analysis. To sample a feed, at least 20 individual samples should be taken from a variety of sites in the lot and thoroughly mixed, and the subsample (20% of total sample) submitted for nutrient analysis. To ensure getting a truly representative sample of hay, haylage, or silage, it is strongly recommended that a hay corer be used rather than grab samples. If grab samples are taken or samples are mixed by hand, latex gloves should be worn, especially if trace mineral content is a concern. It is important that the type of analyses desired be specified. NIRS is the most rapid and inexpensive analysis; however, it may be inaccurate if forages from outside the region or nontraditional feeds are submitted. Proximate analysis provides a reasonable estimate of forage quality, but it is not as accurate as the newer detergent system of analysis. It is strongly recommended that minerals be assessed by wet chemistry techniques rather than estimated by NIRS. Vitamin A is the only vitamin for which commercial laboratories commonly have assays.

REFERENCES

1. Calcium and phosphorus in animal nutrition, International Minerals and Chemical Corp, 1982.
2. Ensminger ME and Olentine CG: Feeds and nutrition—complete, Clovis, Calif, 1978, Ensminger Publishing.
3. Holland C and Kezar W: Pioneer forage manual: a nutritional guide, Des Moines, Iowa, 1990, Pioneer Hi-Bred International, Inc.
4. Kingsbury JM: Information bulletin 104, Ithaca, NY, 1980, Cooperative Extension, New York State College of Agriculture and Life Sciences, Cornell University.
5. Marten GH: Invitational ECOP Workshop NIRS, Madison, Wisc, 1985.
6. National Research Council: United States–Canadian tables of feed composition, Washington, DC, 1982, National Academy Press.
7. Quade Z, Holland C, and Kezar W, editors: Pioneer forage manual, Des Moines, Iowa, 1990, Pioneer brand products.
8. Van Soest PJ: Nutritional ecology of the ruminant, Corvallis, Ore, 1982, O & B Books.
9. Varma A: Handbook of atomic absorption analysis, vol 1, Boca Raton, Fla, 1984, CRC Press.
10. Werner M: CRC handbook of clinical chemistry, vol 2, Boca Raton, Fla, 1984, CRC Press.
11. Werner M: CRC handbook of clinical chemistry, vol 3, Boca Raton, Fla, 1985, CRC Press.
12. Williams P and Norris K, editors: Near infrared technology and the food industry, St Paul, Minn, 1987, American Association of Cereal Chemists.

11

Effects of Disease on Nutritional Needs

SARAH L. RALSTON, JOHATHAN M. NAYLOR

It will come as no surprise that sick animals perform poorly compared with their healthy counterparts. In addition to overt losses from clinical illness, there also are considerable economic losses from subclinical disease. In animals with no overt signs of illness, production is less severely reduced than in clinical disease. However, economic losses can be substantial. This is particularly the case if a large proportion of the animals in a group is affected and the disease is long lasting. The classic examples of chronic, economically important diseases are gastrointestinal parasitism and chronic pneumonia. Clinically ill animals often require special diets. Both situations are addressed in this chapter.

DISEASE AND ANIMAL PRODUCTION

Growth rates and feed conversion ratios often are reduced in animals with subclinical infections.[14,15,25] One reason for these changes is reduced feed consumption.[39] As a result proportionally more of the ingested energy is required for maintenance and less is available for production, including growth. Reduced feed intake inevitably means that feed conversion ratios are increased because the proportion of energy available for production is a smaller fraction of ingested energy (ratio of diet net energy [NE]: NE used for production).[42] In animals kept for meat production, the time to reach the desired finished weight is increased.[40] The final weight of the animal is not usually reduced because animals are usually slaughtered only when they have attained a weight consistent with satisfactory carcass quality.[27] However, the amount of feed and time required to attain this goal are increased.

In some circumstances feed conversion rates may be reduced despite an increased feed intake and a normal growth rate.[11] This may be because illness increases metabolic requirements as a result of fever.[4] In addition nutrients are diverted to the inflammatory response, tissue repair, and defense mechanisms. As part of this response, muscle protein is catabolized to fuel the synthesis of acute phase reactant proteins in the liver.[40]

Many of the diseases that adversely affect production are particularly important in growing animals. There are a number of reasons for this predilection for the young. Immunity to infectious agents is not well developed in young stock. The spread of infectious agents is facilitated by mixing as animals move from farms where they were born to growing units. Furthermore, intensive rearing methods allow close contact between infected and healthy animals.

Parasitism has a major adverse effect on the economics of production in many areas,[10,35] particularly where the weather is warm and wet. Parasites reduce the efficiency of the utilization of nutrients through a number of mechanisms. To some extent they compete for nutrients within the gut. Much more importantly, some invade the gut wall. When this happens there is an inflammatory reaction. In the stomach or abomasum, inflammation destroys glandular tissue and reduces the secretion of acid and digestive enzymes; in the small intestine, destruction of the villi reduces the number of cells available to digest and absorb nutrients.[38] This loss of gastrointestinal function may reduce the digestibility of nutrients. In addition, inflammation and ulceration of the gut wall allow the escape of body constituents. Intestinal losses of serum proteins, and sometimes of red cells, can be considerable in advanced parasitism.[1] These changes can be partially compensated for if dietary protein levels are increased.[1] Food intake may also be reduced by parasitism. This can be the critical factor in the host's ability to withstand the deleterious effects of parasitism.[1]

Routine deworming in animals exposed to high level of parasitism not only prevents clinical disease but also improves growth and food conversion rates.[9,23,35] *Deworming is particularly important in sheep and growing cattle kept on pasture.* It may also benefit lactating cows, even though clinical parasitic disease is uncommon in this class of livestock.[1,19]

Gastrointestinal infections include a number of bacterial diseases that can run a chronic course and adversely affect nutrient utilization. Classic examples include Johne's disease of cattle and swine dysentery. Johne's disease causes extensive damage to the small intestine with interference of nutrient digestion and loss of serum proteins. Feed intake is often normal or increased, but negative energy balance and loss of serum protein lead to emaciation and edema. In pigs, swine dysentery reduces food intake and may interfere with digestion and absorption. Intestinal damage allows serum proteins and blood to escape from the body. In farms with this problem, long-term feed medication with antibiotics improves growth and food conversion.[25]

NUTRITIONAL SUPPORT

Clinical nutrition can be defined as the dietary management of clinically ill animals. The primary goal of clinical nutrition is to optimize nutrient intake, minimize catabolism and utilization of body nutrient stores, and maximize wound healing and immune competence in a sick animal.[12,16,43] Nutritional support has a valuable role in treatment of chronically ill or severely compromised patients. It replenishes tissue stores, improves immune function and wound healing, and prevents death from cachexia.[6,7,20,30]

Most patients make rapid recoveries (2 to 7 days) following an infectious or traumatic insult. In some cases, however, the clinical condition persists for more than a week. In this circumstance nutritional status can be severely compromised by the combined effects of reduced feed intake and increased metabolic demands secondary to fever and tissue catabolism. Prolonged anorexia (longer than 7 to 10 days) or inappropriate diets compromise immune competency and enhance or alter clinical signs of disease.[31,41,43] The duration of anorexia that can be tolerated before nutritional support becomes critical is not known. Nutritional support becomes important after 3 to 7 days of severe illness (less in neonates).

Recommendations for dietary management of clinically ill large animals are based primarily on data from other species because there are few studies of nutritional needs of injured or ill farm animals. All recommendations in this chapter are therefore guidelines only and should be modified according to the clinical response of the individual.

Alterations in nutritional needs

Although starvation alone usually decreases metabolic rate, nutrient requirements increase following trauma, sepsis, or stress, even in anorexic animals.[16,24] The degree of alteration in metabolic rate is dictated by the severity of the insult to the organism. Certain nutrients are of special concern in clinically ill animals.

Glucose. Tissue utilization of glucose is increased by inflammation. In acute sepsis, inflammatory cells, particularly macrophages, release a compound with insulinlike activity. In addition, interleukin-1 stimulates release of insulin from the pancreas. These hormones stimulate glucose entry into cells and enhance glucose utilization.[13,26,36] If the cardiovascular system is severely compromised, glucose uptake is further stimulated. This is because inadequate delivery of oxygen to the tissues forces energy metabolism to increase use of anaerobic glycolysis.[2] Another factor that may contribute to increased glucose use in inflammation is the activation of neutrophils. Glucose is essential for the generation of bactericidal radicals within neutrophils. Glucose administration therefore is beneficial in the acute phase of sepsis.

Most tissues, however, utilize fatty acids as well as glucose as energy sources during the adaptive phase after injury or sepsis (1 to 5 days after acute response).[16,24] In this phase relative insulin resistance and hyperglycemia are also common. Prolonged infusion of 5% dextrose may therefore be counterproductive because it further stimulates hyperinsulinemia without a significant increase in available calories (less than 200 kcal/L).[16] Lipolysis is suppressed by the increased insulin.[32] The reduction of available energy from fatty acids enhances gluconeogenesis from protein catabolism.

Fat. Severely injured or septic ruminants and horses may utilize volatile fatty acids from fiber fermentation more efficiently than simple carbohydrates from grain. Vegetable or seed oils are excellent sources of energy. Horses can tolerate up to 10% total diet dry matter as edible oils or fat.[34] Lower levels (less than 5%) are recommended for ruminants.

Protein. Protein metabolism is greatly altered by inflammation. Activated macrophages release interleukin-1, which stimulates the catabolism of skeletal muscle protein. Amino acids released are transported to the liver, where they have two important roles. In part they may be deaminated and used for glucose synthesis. Also, intact amino acids act as building blocks for the synthesis of the acute-phase reactant proteins.[9,39] This diverse group of proteins is important in the localization of tissue damage and in host defense. An example is fibrinogen. It is released from the liver and converted to fibrin in areas of inflammation in order to surround and entrap foci of bacteria.

Minerals. Trace mineral concentrations are altered as part of the inflammatory response. Serum iron and zinc concentrations decrease and copper increases. These changes are mediated by interleukin-1 and are believed to be beneficial. In vitro experiments demonstrate that the combination of high temperature (fever) and low iron availability reduces the growth of some pathogenic bacteria.

Routine supplementation of trace minerals to sick patients is contraindicated. Iron is the mineral most commonly supplemented; however, it is usually not necessary. Synthesis of red cells is decreased in inflammation, and this may lead to anemia. The combination of low serum iron and a mild anemia in a patient with a chronic infection is more likely to be a normal response than evidence for iron deficiency anemia. *Iron should be administered only when there is clear evidence of tissue iron depletion.* Evidence compatible with iron deficiency include a history of blood loss, low levels of hemosiderin in bone marrow, and a hypochromic anemia, the first two parameters being the most reliable indicators. In these instances iron supplementation may be beneficial. About 70% of total body iron resides in red cells. Iron deficits are calculated on the assumption that the degree of anemia parallels total body iron depletion. The average body content of iron is 7 g/100 kg body weight. Thus,

Total body iron depletion (g) =
$$[(PCV_n - PCV_a) \div PCV_n] \times 7 \times (W \div 100)$$

Where PCV_n = Normal packed cell volume
PCV_a = Packed cell volume in diseased individual
W = Body weight, kg.

Iron is absorbed and utilized slowly so there is little point in administering all the estimated requirements of iron at one time, either orally or parenterally. It is important to recognize that other causes of hypochromic anemia such as copper deficit are unrelated to iron deficiency. As mentioned previously, excessive administration of iron to clinically ill animals may adversely affect their ability to resist bacterial infections.

Vitamins. Thiamin, folic acid, and ascorbic acid are utilized and excreted in greater amounts than normal in the hypermetabolic animal.[3,16,43] Thiamin is not stored in the body and is normally obtained from the diet or bacterial synthesis. In stressed, anorexic animals, thiamin may be depleted. Serum folate is reduced in stressed or strenuously exercised horses. In other species, folic acid rapidly becomes depleted during periods of anorexia, disease, or stress.[43] Ascorbic acid is synthesized in the liver and stored in the adrenal glands and ovaries. However, it is released during periods of stress, and stores may be depleted in prolonged illnesses. All of these vitamins are essential for energy metabolism and normal immune competence. Although requirements in diseased an-

Effects of Disease on Nutritional Needs

imals have not been established, safe oral doses are estimated to be 500 μg thiamin, 1 to 10 mg folate, and 30 to 50 g ascorbyl palmitate per 500 kg body weight per day.

Tumors

Tumors may interfere with the host metabolism either by directly draining nutrients for their own purpose or by increasing the amount of futile cycling in the body. Futile cycles are the consumption of energy in the cyclic synthesis and degradation of a molecule. For example, glucose could be degraded by glycolysis and then resynthesized by gluconeogenesis in the liver[5] with no net change in glucose content but an expenditure of ATP to power the interconversions. Factors released from tumors can interfere with the regulation of cellular processes so that both synthesis and degradation run concurrently.[28] Energy is wasted and the efficiency of energy utilization reduced. Reduced efficiency of energy generation and decreased feed intake may also contribute to energy deficit and weight loss in patients with cancer.[28]

Calculating requirements

Oversupplementation of energy to a critically ill animal may be of greater detriment than underfeeding. Despite stress- or disease-induced alterations in metabolic rate, the clinically ill animal is usually less active. When calculating energy requirements of hospitalized animals, it is best to use the estimated requirement for animals confined to stalls and multiply these basal energy requirements by a "fudge factor" that reflects the severity of the animal's disease and stress[34] (Table 11-1). Any changes in diet (either in type of feed or amount) should be made slowly, especially in critically ill animals. Give 25% of the amounts of feed estimated to meet caloric requirements in three or four feedings on the first day of a dietary change, and gradually increase the amounts fed over 3 to 4 days.[29,34]

Modes of alimentation

Oral supplementation. This is the preferred mode because digestion, absorption, and metabolism of nutrients are enhanced if the animal is allowed to taste, chew, and swallow its feed. If recommended intakes exceed the maximum voluntary intake of the animal, more concentrated sources of energy may be added to the ration. Grains are excellent sources of energy but generally should not comprise more than 50% of the total ration. In the severely ill animal, however, a higher percentage may be required. Adaptation to a low-fiber, high-concentrate diet must be done slowly over the course of 7 to 10 days. Seed or vegetable oils are the most concentrated sources of energy, but high levels may reduce palatability of the ration.

Anorexia. Clinically ill animals are frequently anorexic. A variety of feeds should be offered in small amounts to determine preferences. Any

Table 11-1 *"Fudge factors" for estimating caloric needs of clinically ill animals. Multiply the estimated basal maintenance requirements (see text) by the appropriate factor*

Clinical problem*	Severity of problem		
	Mild/minor	Moderate	Severe/extensive
Sepsis	1.15	1.35	1.50
Trauma	1.15	1.35	1.50
Burns	1.20	1.45	1.75
Organ failure†	1.10	1.25	1.30
Diarrhea	1.10	1.30	1.50
Tumors/neoplasia	1.10	1.30	1.50
Head trauma	1.50	1.75	2.00

*If two or more problems are present, use the single highest factor. Use maintenance estimates alone for all other conditions.
†Cardiac, renal, or hepatic.

ruminant or horse that consistently refuses to graze fresh green grass is probably a candidate for extraoral supplementation. Bitter or unpalatable medications should not be given before feeding or in the ration. The mouth and pharynx should be checked for lesions. If painful oral lesions are present, sweetened mashes or "soups" made of complete pelleted feeds or alfalfa pellets may be more readily consumed than roughage. See Chapter 40, for other techniques of appetite stimulation.

Intragastric supplementation. If an animal is unwilling or unable to eat and has fairly normal GI function (e.g., in botulism, pharyngitis, esophagitis, or severe debilitation), supplemental nutrients may be provided by stomach tube. Enteral alimentation, if possible, is always preferable to intravenous routes.[17] Complete pelleted feeds soaked in water may be used. The volume of water necessary to get the particles down a stomach tube, however, often precludes delivery of adequate amounts. Soybean meal, alfalfa meal, vegetable oil, molasses, propylene glycol, and casein may be added to a pellet slurry to increase energy and protein density. Oral electrolyte replacement solutions for calves and foals rarely contain adequate protein or energy to meet an animal's needs[18] (see Chapter 17).

Intravenous alimentation. Intravenous supplementation of energy and protein in adult large animals is possible when absolutely necessary.[21,22,37] Because of the cost of solutions, total nutrition is practical only in neonates or very valuable adult animals. Partial supplementation, however, is often indicated and should be used when other forms of supplementation are not possible or are grossly inadequate. Candidates for IV nutrient supplementation include animals with severe, acute diarrhea or anterior enteritis.[37] Complications associated with intravenous feeding include septicemia, thrombosis, hyperglycemia, and glucosuria. (See Chapter 36.)

CONCLUSION

Clinical and subclinical diseases alter the ability of an animal to digest and metabolize a variety of nutrients. In subclinical diseases and parasitism, the effects are manifest primarily in a reduction of growth rate and feed conversion ratios. In clinically ill animals, supplementation of adequate sources of energy, protein, and vitamins may improve immune function and wound healing and reduce the risk of death if the disease process lasts more than 3 to 7 days. Oversupplementation, however, should be avoided. Trace mineral supplementation is usually not indicated, and, as in the case of iron, may be detrimental. Trace mineral intake should be increased only if clear evidence of a deficiency is apparent. Oral supplementation, ideally through voluntary consumption by the patient, is preferable to intravenous therapy. However, anorexia is frequently a problem. Extraoral or parenteral nutrition are alternatives to feeding animals that do not respond to managerial or chemical stimulants of appetite.

REFERENCES

1. Abbot EM, Parkins JJ, and Holmes PH: Vet Parasitol 20:291, 1986.
2. Astiz M and others: Circ Shock 26:311, 1988.
3. Baker EM: Am J Clin Nutr 20(6):583, 1967.
4. Baracos VE, Whitmore WT, and Gale R: Can J Physiol Pharmacol 65:1248, 1987.
5. Baranyai JM and Blum JJ: Biochem J 258:121, 1989.
6. Brenner MF and others: JPEN 10(5):446, 1986.
7. Brinson RR, Curtis WD, and Singh M: J Am Coll Nutr 6(6):517, 1987.
8. Brown RF, Houpt KA, and Schryver HF: Pharmacol Biochem Behav 5:495, 1976.
9. Caldow GL, Taylor MA, and Hunt K: Vet Rec 124:111, 1989.
10. Catchpole J and Harris TJ: Vet Rec 124:603, 1989.
11. Ciprian A and others: Can J Vet Res 52:434, 1988.
12. Clowes GHA and others: Ann Surg 22:446, 1985.
13. Cornell RP: Circ Shock 28:121, 1989.
14. Dalton PM and Ryan WG: Vet Rec 122:307, 1988.
15. DiFranco E and others Can Vet J 30:241, 1989.
16. Dudrick SJ and Rhoads JE: Metabolism in surgical patients: protein, carbohydrate and fat utilization by oral and parenteral routes. In Sabison CO, editor: Davis-Christopher textbook of surgery: the biological basis of modern surgical practice, ed 12, vol 1, Philadelphia, 1981, WB Saunders, pp 147-171.
17. Feldman EJ and others: Gastroenterology 70:712, 1976.
18. Fettman MJ and others: J Am Vet Med Assoc 188:397, 1986.
19. Foged NT, Nielsen JP, and Jorsal SE: Vet Rec 125:7, 1989.
20. Gleck A: Proc Nutr Soc 39:125, 1980.
21. Hansen TO, White NA, and Kemp D: AAEP Newsletter 1:18, 1985.
22. Hansen TO, White NA, and Kemp DT: Am J Vet Res 49(1):122, 1988.

23. Kennedy MJ and ZolBell DR: Can Vet J 29:566, 1988
24. Kudsk KA, Stone J, and Sheldon GF: Surg Clin North Am 61(3):671, 1981.
25. Kyriakis SC: J Vet Pharmacol Ther 12:296, 1989.
26. Lang CH and others: Circ Shock 23:131, 1987.
27. Malone FE and others: Vet Rec 122:203, 1988.
28. McAndrew PF: Surg Clin North Am 66:10003, 1986.
29. Naylor JM, Freeman DE, and Kronfeld DS: Comp Cont Educ 6(2):593, 1984.
30. Naylor JM and Kenyon SJ: Proc AAEP 24:505, 1978.
31. Newton DJ, Clark RD, and Woods HF: Proc Nutr Soc 39:141, 1980.
32. Rakanic J and others: J Parenter Enter Nutr 11:513, 1987.
33. Ralston SL: Comp Cont Educ 6(11):5628, 1984.
34. Ralston SL: J Equine Vet Sci 5:336, 1985.
35. Rowlands D and others: Vet Rec 125:55, 1989.
36. Shearer JD, Amaral JF, and Caldwell MD: Circ Shock 25:131, 1988.
37. Spurlock SL: Equine Vet Data 9(9):131, 1988.
38. Straw BE and others: J Am Vet Med Assoc 195:1702, 1989.
39. Tian S and Baracos VE: Comp Biochem Physiol (A) 94:323, 1989.
40. Tian S and Baracos VE: Biochem J 263:484, 1989.
41. Tomkins AM: Proc Nutr Soc 45:289, 1986.
42. Webster AJF: Livestock Prod Sci 7:243, 1980.
43. Wilmore DW: The metabolic management of the critically ill, New York, 1977, Plenum.

Ruminants

UNIT II

SECTION A Beef Cattle

12

The Beef Industry

R. D. H. COHEN

Cattle have been valued as a source of food, clothing, and power for as long as people have domesticated animals. In today's societies, beef consumption is still a major indicator of the standard of living in a population.

The beef cattle of the world belong to two species: *Bos taurus* and *Bos indicus*. *Bos taurus* cattle are largely European in origin and are adapted to temperate climates. Breeds include Hereford, Angus, Shorthorn, Simmental, Charolais, Maine-Anjou, and Limousin. *Bos indicus* cattle are Afro-Asian in origin and are adapted to tropical climates. The most important breeds are Brahman, Red Sindhi, and Sahiwal.

Beef cattle were first introduced to North America from Spain by Christopher Columbus on his second voyage in 1493. Although European breeds of cattle were introduced to North America in the early seventeenth century, there was little interest in selection and improvement of cattle until the late eighteenth century. Cattle were initially required to graze the existing native rangelands with no provision for supplementary feeding during winter, drought, and other extreme climatic conditions. There was little or no correction of nutritional deficiencies in the forage. This method of husbandry still exists today in many parts of the world.

The economic value of cattle in developed societies has increased dramatically during the twentieth century. For example, in Canada in 1988, beef production accounted for 55.2% of the farm receipts from livestock and 26.9% of the total farm receipts from all agricultural enterprises.[2] Nevertheless, the beef industry is still the least intensive of all agricultural industries.

Because of low fecundity and the great length of time necessary for progeny to reach slaughter weight, the beef cow produces only about 70% of her body weight as slaughter weight per year. This compares unfavorably with the sow, who produces about eight times her body weight, and the hen, who produces about 100 times her body weight as slaughter weight per year. However, cattle consume only 17% of their feed as concentrates compared with 85% for swine and 94% for poultry and are, therefore, much less competitive with humans for feed resources. Cattle derive their agricultural importance from the effective conversion of energy not usable by people (e.g., grass and crop aftermath) into high-quality protein and other essential nutrients for human consumption.

STRUCTURE OF THE BEEF INDUSTRY

The beef industry is divided into two enterprises: cow-calf and feeder. They may be conducted independently, with transfer of ownership of calves at weaning to independent premises; simultaneously, with retention of calves on the original farm through to slaughter; or as an integrated enterprise, with retention of ownership of the cattle, but transfer of them to some independent premises at some time between weaning and slaughter.

The cow-calf enterprise

The cow-calf enterprise can be subdivided into the production of purebred and commercial cattle. Purebred cattle are sources of germ plasm—most commonly bulls and occasionally heifers—for other purebred and commercial breeders. Purebred cattle are eligible for entry in a breed association registry and represent a product descended from a selected long line of ancestors of similar type and breed. They are produced with the aim of improving efficiency of production, frame size, and carcass quality. Calves that are not kept in the herd for replacement purposes are not usually fed to slaughter but are fed to at least 1 and frequently 2 years of age and sold as breeding stock. The costs associated with feeding are usually greater than those for commercial cattle. Promotion and advertising of the product through the media or show ring are also integral and expensive parts of the purebred enterprise. Although showing beef cattle has declined in importance in the purebred enterprise, it remains a popular way of advertising in the industry. Performance testing of cattle has increased in popularity in the last 20 to 30 years because of its ability to provide proof of superiority. However, there is evidence that performance testing may have been used more as a marketing tool than as a means of improving productivity in the purebred industry.[34]

In the commercial cow-calf enterprise, calves are produced to be sold at weaning, carried through to slaughter, or reared as replacement heifers. Crossbred cows are frequently the basis of the commercial herd, and these cows are usually bred to bulls of a third breed (terminal sire). For example, popular crossbred cows are Hereford × Angus or Hereford × Simmental, and Charolais bulls are popular terminal sires, but combinations of all breeds are in use in the industry.

Most beef calves are born in late winter or early spring to keep winter feeding costs at a minimum and to allow the cow to make maximum use of fresh spring and summer pasture growth for lactation. Nevertheless, fall calving is practiced in some herds to reduce the incidence of neonatal calf diarrhea,[1,5,19,41,57] to take advantage of higher prices for calves sold in spring,[5,41,47] or to increase pasture utilization by calves.[5,33,41] However, both growth rate and weaning weight are reduced in fall-born compared with spring-born calves.[8,39,51] Milk yield is also lower in fall-calving cows than in spring-calving cows; the advantage in favor of spring-calving cows increases if ambient temperatures during winter fall to $-5°$ C (23° F) or below.[33] In addition, fall-calving cows must be bred during winter, but bull fertility is lower during winter than summer[35] and highest in spring.[39] Furthermore, there is an increased requirement for feed in the winter if the calves are born in the fall because the nutritional requirements of the lactating cow are greater in winter than in spring and summer.[14,22,41] The provision of additional feed for fall-born calves is also important to ensure adequate nutrition through winter.[39,43,46,51] All these requirements increase feed costs relative to spring-born calves.

Many cow-calf operators sell their calves at weaning to operators in the feeder enterprise, but retention of ownership through to slaughter is becoming increasingly common. Ownership may be retained either by feeding the calves from weaning to slaughter on the home farm or ranch or by moving the calves at some stage between weaning and slaughter to the premises of a feeder enterprise operator. The latter is usually referred to as *custom feeding*.

Plant breeders have been effective in providing varieties of crops that are better adapted to lower rainfall, local temperatures and day length, and fewer frost-free days and that are more resistant to pests and diseases. This has resulted in a continual encroachment of profit-motivated cash cropping into the traditional rangeland areas grazed by the cow-calf herd. The result has been a declining number of farms and ranches that depend on beef production as their major source of income. For example, Canada has approximately 80,000 cow-calf herds, and the average size is fewer than 50 cows per herd. Fewer than 15% of Canadian cow-calf herds derive their major source of income from beef cattle.[25] In the United States, approximately 74% of farms and ranches have fewer than 50 cows.[9]

A highly profitable cow-calf operation depends on:

1. A high percentage of calf crop at time of weaning
2. A high level of longevity in the cows
3. Heavy weaning weights of calves
4. A high pregnancy rate in the replacement heifers
5. Maximum utilization of pasture and cheap roughage
6. Low investment per cow in land, buildings, equipment, and labor
7. Effective investment in feed supplements and fertilizers

Reproductive performance (the annual cow cycle). On the average, the gestation length of the cow is 283 days, and postpartum anestrus about 40 days. This allows just 42 days for the cow to conceive if she is going to produce a calf every 365 days (Fig. 12-1). The cow must then successfully carry the fetus to term and rear and wean the calf to remain a fully functional unit within the herd.

A high level of reproductive performance is essential to an efficient cow-calf operation. The percentage calf crop is a measure of fertility, mothering ability, and calf survival and has a direct bearing on production costs and profitability (Table 12-1). Although disease and genetics influence reproductive performance, the greatest benefits are achieved from careful consideration of herd nutrition.

The cow must remain as a productive unit in the herd for as long as possible. This means she must breed at 12 to 15 months of age and produce her first calf by the time she is 2 years old. The heifer becomes an economically productive unit only when her first calf is sold. The nutritional management of the replacement heifer herd has a major influence on her reproductive performance[53] (Table 12-2).

Data from Australia indicated that the incidence of estrus in Hereford heifers increased from about 20% to 80% as live weight increased from 220 to 250 kg (485 to 550 lb)[15] (Fig. 12-2). Although the frame sizes of Australian cattle are considerably smaller than those of North American cattle of the same breed, these data do illustrate that relatively small increases in live weight can have a marked effect on heifer fertility.

The beef cow reaches her maximum productive potential when she is 4 to 5 years old and should maintain that level until at least 10 years of age. Failure to conceive at any time before 10 years of age represents a major loss of productive potential.

Heavy weaning weights of calves are directly associated with the nutrition of the calf and nursing cow. However, calf weaning weight is also influenced by age at puberty in the dam and by the length of the postpartum anestrous period. For example, calves conceived in the first and second estrous cycles of the breeding season (i.e., within the first 21 and 42 days, respectively, of exposure of heifers to bulls) grew significantly faster from birth to weaning and weighed more at weaning than calves conceived later in the season[37] (Table 12-3). Furthermore, heifers that calved early at the first calving had higher average annual lifetime calf production than heifers that calved late at the first calving.

It is recommended that more replacement heifers than are required to maintain the herd structure should be exposed to bulls, pregnancy tested, and culled if not pregnant. Nonpregnant heifers can be fed out for slaughter. If after calving some surplus heifers remain, the early calv-

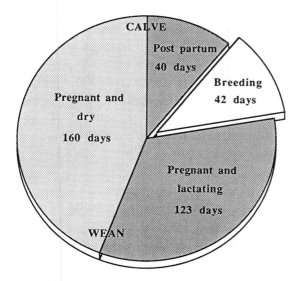

Fig. 12-1 The annual cycle of a cow-calf operation.

Table 12-1 *Economic implications of calf crop percentage (calves weaned per 100 cows exposed to bulls) on cost and net income*

Calf crop (%)	Total kg of calf weaned	Gross income ($)	Cost per calf weaned ($)	Net income per calf weaned ($)
100	21,500	38,700	285	102
90	19,350	34,830	317	70
80	17,200	30,960	356	31
70	15,050	27,090	407	−20
60	12,900	23,320	475	−88

The assumptions made are that cow costs = $285/yr; herd size = 100 cows; calf weaning weight = 215 kg (475 lb); calf price = $1.80/kg ($0.82/lb).

Table 12-2 *The effect of nutrition during the first winter (December to May) on the fertility of replacement heifers*

	Heifers fed to gain (kg/d)		
Attribute	0.28	0.45	0.68
Initial live weight (kg)	146	150	151
Gain in 150 days (kg)	43	68	103
Age at puberty (days)	433	411	388
Weight at puberty (kg)	238	248	259
Pregnancy rate (%)	50	86	87

Data from Short RE and Bellows RA: J Anim Sci 32:127, 1971.

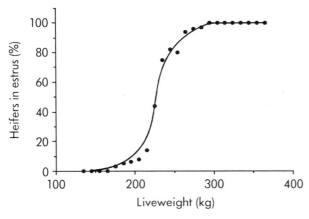

Fig. 12-2 The relationship between live weight and the incidence of estrus in Hereford heifers.

ing heifers should be given some preference in selection.[37] Frequently surplus 2-year-old heifers can wean a calf, be placed on feed for slaughter, and still receive top-quality carcass grades.

Profitability in the cow-calf enterprise is greatly influenced by bull fertility. The only function of a bull in the herd is to impregnate female cattle and pass on his genetic code. He should be genetically superior, be fertile, have high libido, and be free from physical defects. Scrotal circumference in bulls is highly correlated with sperm production, and bulls with large testicles can be successfully bred to more cows than bulls with small testicles.[7] Scrotal circumference is highly heritable and positively correlated with growth rate and live weight. Testicular development is very rapid around pu-

Table 12-3 *The effect of interval between exposure of heifers to bulls and subsequent calving date on calf production at Havre, Montana*

Interval from exposure to calving (days)*	Birth weight (kg)	Survival rate (%)	Weaning weight (kg)	Weaning age (days)	Average daily gain (kg)
<283 (premature)	30	93	198	204	0.78
283-304	35	100	206	202	0.86
305-326	36	99	190	186	0.84
327-348	36	93	172	167	0.83
>348	36	83	152	146	0.78

*Number of days from the first day of the breeding season to the day of calving.
From data in Lesmeister JL, Burfening PJ, and Blackwell RL: J Anim Sci 36:1, 1973.

Table 12-4 *Economic benefits of pregnancy testing*

	Pregnancy rate (%)			
	40-head herd		100-head herd	
	93	80	93	80
Per open cow ($)	21.36*	46.93	41.74	54.06
Per cow ($)	1.50	9.39	2.92	10.81
Per herd ($)	59.82	375.42	292.15	1081.15

*Benefits are additional profit above costs, including veterinary fees.
From Rupert PE: The economic feasibility of pregnancy testing Saskatchewan beef cattle, BSA thesis, University of Saskatchewan, 1983.

berty, and feeding from 6 to 13 months of age is very important, particularly with respect to energy,[18] protein,[45] and trace minerals.[29] Evidence of the importance of bull soundness on herd fertility comes from data from a 4000-cow herd in Australia in which culling of unsound bulls each year increased calving percentage from 70% to 85% and maintained it at the latter level.[6]

A nonpregnant cow is a nonproductive unit in the herd, and it is not an economical proposition to carry open cows for 18 months between weaning one calf and producing another. The economic benefit of selling an open cow in fall compared with feeding her through winter and selling her in spring when she fails to calve are shown in Table 12-4. The data are from a 10-year study in Saskatchewan and show that the benefits of pregnancy testing increase with increasing herd size and decreasing pregnancy rates.[48]

The feeder enterprise

The feeder segment of the beef industry involves the feeding of calves from weaning to slaughter. In most areas of the world, it is done mainly by ensuring high-quality pasture with or without a concentrate (grain, protein, or mineral) supplement. In North America, however, the feedlot has evolved as a sophisticated, efficient segment of the feeder enterprise, largely because of the need to use grain that is surplus to or unsuitable for the human food market by feeding it to cattle to convert into beef. Cattle may enter the feedlot as weaned calves where, depending on circumstances such as breed, availability of feedstuffs, market requirements, and cost-price relationships of feed and cattle, they are fed high-forage, low-grain rations during an initial period designed to stimulate growth of the skeleton and muscle with little fat deposition (backgrounding). This is followed by high-grain, low-forage rations designed to allow deposition of enough fat to give the meat its tenderness and cooking qualities. Alternatively, cattle may be fed a high-grain ration throughout their time in the feedlot to stimulate maximum rate of gain and feed conversion efficiency. As a third option, cattle may be backgrounded in dry lots or on pasture through winter or early summer on the home farm and subsequently moved to the feedlot for a short period of high-grain feeding (Fig. 12-3).

There is a continuing trend toward more specialization in the beef industry and toward larger but fewer feedlots. It is also becoming more common for the cow-calf operator to retain ownership of the cattle in the feedlot, where they are custom fed and the owner is charged for costs associated with feeding the cattle such as feed, bedding, use of facilities, veterinary fees, fuel, electricity, and labor. The feedlot operates on small profit margins per head. Therefore, the operator attempts to maximize rate of gain, efficiency of feed conversion, and the number of animals put through the feedlot each year, as well as minimize the death losses and feed costs. Feed cost-effectiveness is achieved by buying in large lots, contracting, self-supplying, or using opportunity feeds such as surplus fruits and vegetables.

Because the greatest risk of death losses occurs within the first 3 weeks of cattle entering the feedlot[12] as a result of respiratory disease (shipping fever), a program of preconditioning of calves has been encouraged and introduced to most areas of North America. The program is designed to reduce the stress inflicted on an animal once it is weaned and shipped directly to a feedlot. Preconditioning requires that animals be weaned and conditioned to feeding in confinement; individually identified, dehorned, and castrated; immunized against *Clostridium* spp., infectious bovine rhinotracheitis (IBR), parainfluenza (PI_3), and perhaps other infectious agents; and treated for internal and external parasites. Other optional treatments such as implanting with growth-promoting substances may also be included. These treatments must be carried out at least 3 weeks before sale, and the cattle must be accompanied by a certificate signed by the owner and a veterinarian to be eligible for classification as preconditioned. The specific details of the program may vary from one jurisdiction to the next.

The success of a preconditioning program depends upon the holding of special preconditioned cattle sales and the willingness of the buyers to pay a premium to cover the costs to the owner. In Alberta, for example, the premium paid for preconditioned calves in 1980 through 1987 was $0.11 per kg (5 cents/lb) for steers and $0.09 per kg (4 cents/lb) for heifers.[50] If the owner preconditions the calves for 6 to 8 weeks before sale, the costs can also be recovered in the extra weight gained during that time and the

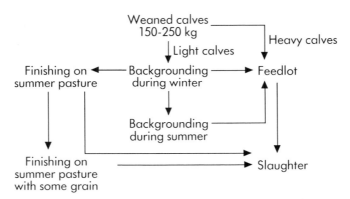

Fig. 12-3 Options in feeding beef calves after weaning.

healthier cattle being offered for sale. For example, data from 6438 calves in the Alberta program from 1982 and 1984 through 1987 indicated that the average cost of preconditioning was $53.75 per calf. The average daily gain during the preconditioning program was 0.84 kg/day (1.85 lb/day), and the estimated additional return net of costs was $56.46 per calf.[50]

The advantages of the preconditioning program are:
1. The cow-calf operator sells a better-quality calf and builds a reputation that helps to attract premium prices.
2. The feedlot operator loses fewer calves from shipping fever and therefore increases the margin of economic return.

For example, in Alberta between 1980 and 1987, 9.1% of preconditioned calves were treated for shipping fever in the feedlot and 0.6% died, compared with 21.4% and 1.6%, respectively, for nonpreconditioned calves.[50] Similar data are available from Ontario.[60] Furthermore, weight gains in preconditioned calves have been shown to be greater than in nonpreconditioned calves.[50,60]

BEEF AND HUMAN HEALTH
Lipids and cholesterol

Beef is a rich source of protein and other essential nutrients for human nutrition. Moreover, the essential amino acids present in beef are well balanced with respect to human requirements. Beef is not a concentrated energy food. A 100-g (3.5 oz) serving of roast beef provides only 8% of human daily caloric requirements, but 57% of the protein, 34% of the vitamin B_{12}, 32% of the zinc, 18% of the niacin, and 12% of the iron requirements and is a good source of other essential minerals such as calcium, phosphorus, and copper.

Despite its obvious food value, the intake of beef throughout most of the developed world has declined markedly in recent years. This decline is due to the notion that because coronary heart disease and atherosclerosis are associated with high blood cholesterol, modification of dietary cholesterol intake may reduce heart disease. This hypothesis has never been demonstrated to be true.[10,11] Factors such as sex, obesity, exercise, cigarette smoking, and alcohol consumption and total fat intake have a greater effect on blood cholesterol than cholesterol intake.[10] Furthermore, when some of these factors were removed by multivariate analysis, the effects of total fat, meat, eggs, and milk became nonsignificant.[3] Brisson concluded that "the intake of dietary cholesterol has no significant effect on the concentration of cholesterol in the blood of healthy persons representative of the general population," and "we synthesize perhaps 5-6 times more cholesterol than we get as food," and furthermore "the sustained fear of dietary cholesterol is not justified based on the scientific evidence presently available."[11]

Although the population believes that chicken meat is healthier than beef, this is probably not true. In response to public pressure, the beef industry has reduced the amount of total lipid and cholesterol in beef by as much as 20% in the last 10 years.[28,32,49,59] Some analyses show that a 100-g (3.5 oz) serving of roast lean beef provides exactly the same amount of cholesterol as a 100-g (3.5 oz) serving of light chicken meat without the skin.[42]

Antibiotics

For the past 40 years, antibiotics have been added to feedlot rations to improve growth rate and feed efficiency as well as to prevent infectious diseases. This practice has raised concerns regarding safety to human health.

An outbreak of salmonella poisoning occurred in 18 people in four midwestern U.S. states. They were infected with a strain of *Salmonella newport* that was resistant to tetracycline, ampicillin, and carbenicillin. Thirteen of the subjects were found to have consumed hamburger that was traced back to a South Dakota feedlot where the cattle had been fed chlorotetracycline at 100 g/t of feed. It was concluded that the antibiotics in the feed caused a proliferation of resistant salmonella, which were passed on to the consumers of the meat.[31] However, the evidence was largely circumstantial, and the meat could have been contaminated during slaughtering or processing.[54] Nine of the 13 infected people had been taking antibiotics for medical reasons other than diarrhea for 24 to 48 hours before the onset of salmonellosis.[31]

If antibiotic resistance has resulted from subtherapeutic use in animal feeds, it might be expected that the population most at risk would be the animals themselves. However, after 40 years of subtherapeutic use of antibiotics in animal feeds, animal performance is still improved by such subtherapeutic use.[27] Furthermore, the use of antibiotics in the human population is approximately twice that in animals, and natural selection for resistance is therefore more likely to have occurred in the human population. In one study, 18% of salmonella isolates from meat showed resistance to tetracycline, whereas 45% of isolates from hospitals showed resistance.[56] Every year there are reported cases of food poisoning in humans related to salmonella, and sometimes these are traced to food animal products. Often the salmonella involved are antibiotic sensitive, and these incidents are the result of poor food preparation or kitchen hygiene rather than the use of antibiotics in feed.[58]

In 1969, the Swan Report in the United Kingdom recommended that restrictions be placed on human medicinal antibiotics that were being used in animal feeds. These restrictions have not reduced disease or infections in the human population due to bacteria of animal origin.[26] They have not improved the efficacy of therapy of human or animal diseases due to a decrease in resistant bacteria, nor has the bacterial contamination of food of animal origin been reduced.[26] The use of antibiotics is 16 times greater in human medicine than in veterinary medicine in the United Kingdom.[58] One might ask, "In what sector is overuse more likely to be occurring?"

The beef industry must continue to be objective and diligent in gathering information related to human health. It has been estimated that a ban on subtherapeutic use of antibiotics in the United States would increase production costs in the feedlot industry by $225.6 million, increase the losses due to liver abscesses by $21.37 million, and add a further $56.6 million in costs associated with increased mortality.[23] Similar figures have been reported from Canada.[52]

Implants

The use of agents such as estradiol-17β, testosterone, progesterone, trenebolone, and zeranol for the promotion of growth in cattle is widespread. The anabolic effects of these substances cannot be disputed. For example, an advantage of almost 12 kg (26 lb) of live weight for each of four implants of Ralgro (zeranol) at 100-day intervals and an increase of 12.3 cm^2 in rib eye area at slaughter have been reported.[17] The Scientific Working Group on Anabolic Agents of the Council of the European Economic Community concluded that none of these compounds or their metabolites produced residues of a significant genotoxic potential and that, provided accepted animal husbandry practices were followed, they did not present a harmful effect to health.[36] Nevertheless, the EEC instigated a ban on these substances in a move that may have been more related to a need to reduce agricultural subsidy payments than to reduce a possible threat to human health. However, the ban has fueled the fires of public speculation that those substances pose a threat to human health. This further challenges the perceived integrity of the beef industry, and diligent and objective public relations efforts are required to correct this attitude.

ANIMAL WELFARE

The welfare of animals kept in confinement is under continual and increasing scrutiny. Although the beef industry is not under such close scrutiny as are the more intensive animal industries such as swine and poultry, it cannot afford to be complacent. Many of the husbandry practices such as branding, dehorning, castration, and confining of animals in feedlots are receiving attention from animal rights activists. The industry must objectively assess the effects of these practices on animal welfare and, where appropriate, seek alternate methods of husbandry. It is not sufficient to challenge the animal rights movement to prove that these husbandry practices are not in the interests of animal welfare. Such a challenge will tend only to provoke an emotional response. The industry must gather objective data based on sound scientific methods and, where necessary, adopt new technologies.

Examples of these changes in technology are given by breeding for polledness in cattle and the introduction of electronic identification systems. Although bulls gain weight faster, produce leaner carcasses, and convert feed more efficiently than steers, and although there is little difference in the quality and palatability of meat from bulls and steers,[4,21] castration of male cattle continues to be a common practice in North and South America, Oceania, and many other countries, although the practice is less common in Europe. This is because antagonistic behavior among feedlot bulls may offset these advantages[30] and because of the greater risk of dark red muscle coloration among bulls than steers or heifers.[13,40] This dark coloration ("dark cutters") is caused by a greater stress susceptibility among bulls at slaughter.[21,40] The darker meat is unacceptable to the consumer and therefore returns a greatly reduced price to the feedlot operator. Tranquilizers delay the onset of antagonistic behavior in bulls but do not decrease the intensity of the behavior.[38] However, antagonistic encounters among bulls that were regrouped (mixed with bulls from other pens) during transport to slaughter were responsible for a much higher incidence of dark cutting meat (73%) than for bulls not regrouped (2%).[44] Thus, if more care was taken during the transport of bulls to and handling of bulls at the slaughterhouse, the feedlot industry may more readily accept the feeding of bulls and thus reduce the need for castration of male cattle. It is important to obtain information on the amount of stress experienced by cattle as a result of these practices.[20,55] Nevertheless, alternate methods of castration such as chemical castration[16] and immunocastration[24] should continue to be investigated.

THE EFFICIENCY OF BEEF PRODUCTION

There are many ways of measuring production efficiency. On one hand, it can be argued that beef production is highly efficient because it uses resources not suited for other forms of agricultural production. On the other hand, the fecundity of beef cattle is low, and the interval from conception to food is longer for beef cattle than for any other livestock. The greatest improvements in beef production are most likely to be made by increasing fecundity in the beef cow.

Approximately 75% of the gross value of each beef calf weaned in the annual beef cow cycle is invested in the maintenance of the cow unit (Table 12-1). Rarely, if ever, are 100 calves weaned per 100 cows exposed to bulls. Frequently this statistic is between 65 and 95 calves per 100 cows exposed to bulls. Such statistics would not be tolerated in any other meat-producing animal.

The world's rangelands are dwindling because of the encroachment of crop production and those that remain are, almost without exception, overgrazed. A cow consumes three to four times as much forage as her calf from parturition to weaning. The present level of beef production could be maintained—and the pressure on the land base eased—by reducing the cow herd size, increasing the rate of twinning in the beef herd, and thus increasing the efficiency of production per cow. The challenge to researchers to increase the rate of twinning either by genetic inheritance or by controlled manipulation of the number of viable ova produced at each ovulation, to select superior milking strains of cows that are capable of supporting more than one calf per lactation, and to determine and meet the nutritional and management requirements of twin-bearing cows and their calves is considerable, but compelling.

The future of the beef industry depends on production cost-effectiveness achieved through research, technology transfer, and adoption; on marketing a product that is seen by the public as safe, healthful, and humanely produced; and on continued global economic development, particularly in the Third World.

REFERENCES

1. Acres SD: Calf scours, Vido views fact sheet no. 1, Saskatoon, Saskatchewan, 1980, Veterinary Infectious Disease Organisation.
2. Agriculture Canada: Market commentary June, 1989, Market Outlook and Analysis Division of the Policy Branch, Ottowa, Agriculture Canada.
3. Armstrong BK and others: J Chron Dis 28:455, 1975.
4. Arthaud VH and others: J Anim Sci 44:53, 1977.
5. Bagley CP and others: J Anim Sci 64:687, 1987.
6. Blockley MAdeB: Proc Aust Soc Anim Prod 13:50, 1980.
7. Blockley MAdeB: Proc Aust Soc Anim Prod 13:52, 1980.
8. Bolton RC and others: J Anim Sci 65:42, 1987.
9. Boykin CC, Gilliam HC, and Gustafson RE: Structural characteristics of beef cattle raising in the United States, AER No. 450, Washington, DC, 1980, USDA.
10. Brisson GJ: Lipids in human nutrition: an appraisal of some dietary concepts, Fort Lee, NJ, 1981, Jack K Burgess Inc.

11. Brisson GJ: Fats and human health, Proc 4th West Nutr Conf, Saskatoon, 1983, pp 302-323.
12. Bristol RF: J Am Vet Med Assoc 150:69, 1967.
13. Buchter L: In Cole DJA and Lawrie RA, editors: Meat, Westport, Conn, 1975, AVI Publishing Co.
14. Christopherson RJ, Hudson RJ, and Christophersen MK: Can J Anim Sci 59:611, 1979.
15. Cohen RDH, Garden DL, and Langlands JP: Anim Prod 31:221, 1980.
16. Cohen RDH, Janzen ED, and Hunter PR: J Anim Sci 61(suppl 1):422 (abstr), 1985.
17. Cohen RDH, Janzen ED, and Nicholson HH: Can J Anim Sci 67:37, 1987.
18. Coulter GH and Kozub GC: J Anim Sci 59:432, 1984.
19. Davidson K: Cattlemen 49:34, 1986.
20. Fell LR and Shutt DA: Can J Anim Sci 66:637, 1986.
21. Field RA: J Anim Sci 32:849, 1971.
22. Fredeen HT and others: J Anim Sci 64:714, 1987.
23. Gilliam HC Jr and Martin JR: J Anim Sci 40:1241, 1975.
24. Goubau S: Active immunization against gonadotropin releasing hormone in sheep and cattle, M.Sc. thesis, Saskatoon, 1987, University of Saskatchewan.
25. Gracey C: Problems and opportunities in the Canadian cattle industry. Paper presented at the International Minerals Conference, St. Petersburg, Fla, 1981.
26. Guest GB: J Anim Sci 42:1052, 1976.
27. Hays VW and Muir WM: Can J Anim Sci 59:447, 1979.
28. Health and Welfare Canada: Nutrient value of some common foods, Ottawa, 1979, Health Services and Promotions Branch and Health Protection Branch.
29. Hidiroglou M: J Dairy Sci 62:1195, 1979.
30. Hinch GN: In Wodzicka-Tomaszewska M, Edey TN, and Lynch JJ, editors: Behaviour, reviews in rural science, series IV, Armidale, Australia, 1980, University of New England Publishing Unit.
31. Holmberg SD and others: N Engl J Med 311:617, 1984.
32. Jones SDM: J Can Diet Assoc 46:40, 1985.
33. Kartchner RJ, Rittenhouse LR, and Raleigh RJ: J Anim Sci 48:425, 1979.
34. Koots KR, Cohen RDH, and Nicholson HH: Can J Anim Sci 68:965, 1988.
35. Kozhemyakin VP: Seasonal effects on reproductive function of Kalmyk bulls, Osnov napravleniya v selektsii skota myas porod Orenburg, USSR, 1983, pp 72-78.
36. Lamming GE and others: Vet Rec 122:389, 1987.
37. Lesmeister JL, Burfening PJ, and Blackwell RL: J Anim Sci 36:1, 1973.
38. MacNeil MD, Gregory KE, and Ford JJ: J Anim Sci 67:858, 1989.
39. Marlowe TJ, Mast CC, and Schalles RR: J Anim Sci 24:494, 1965.
40. Martin AH, Fredeen HT, and Weiss GM: Can J Anim Sci 51:305, 1971.
41. Mueller RG and Harris GA: J Range Manage 20:67, 1967.
42. NCA: Beef cattle research needs and priorities, Beef Business Bulletin, Englewood, Colo, 1982, National Cattlemen's Association.
43. Peterson GA and others: J Anim Sci 64:15, 1987.
44. Price MA and Tennessen T: Can J Anim Sci 61:205, 1981.
45. Rekwot PI and others: Anim Reprod Sci 16:1, 1988.
46. Robison OW, Yusuff MKM, and Dillard EU: J Anim Sci 47:131, 1978.
47. Rosaasen KA, Kulshreshtha SN, and Yu G: A monthly forecasting model of beef cattle-calves sector in Canada, Saskatoon, 1979, Department of Agricultural Economics, University of Saskatchewan.
48. Rupert PE: The economic feasibility of pregnancy testing Saskatchewan beef cattle, BSA thesis, University of Saskatchewan, 1983.
49. Sahasrabudhe MR and Stewart L: Can Inst Food Sci Technol 22:83, 1989.
50. Schipper C, Church T, and Harris B: Can Vet J 30:736, 1989.
51. Sellers HI, Willham RL, and deBaca RC: J Anim Sci 31:5, 1970.
52. Sharby TF: Can J Anim Sci 59:333, 1979.
53. Short RE and Bellows RA: J Anim Sci 32:127, 1971.
54. Sun M: Science 62:74, 1984.
55. Thomas LR and others: J Anim Sci 61 (suppl 1):394 (abstr), 1985.
56. van Houweling CD and Gainer JH: Can J Anim Sci 46:1413, 1978.
57. Walter-Toews D, Martin SW, and Meek AH: Prev Vet Med 4:125, 1986.
58. Walton JR: J Anim Sci 62:74, 1986.
59. Watt BK and Merril AL: Composition of foods, Agriculture Handbook No. 8, Washington, DC, 1975, U.S. Department of Agriculture.
60. Woods GM, Mansfield M, and Webb R: Can J Comp Med 37:249, 1970.

Feeding the Beef Cow for Optimal Production

LARRY R. CORAH

The essence of beef cow-calf production is the economical production of the net calf crop or, more appropriately stated, the weight of calf produced per cow in the herd. The four principal factors influencing the profitability of a cow-calf operation are (1) weaning weights, (2) percentage of cows weaning calves, (3) cost of maintaining the cow per year, and (4) price of calves. Factors controlling the kg of calf weaned per cow are (1) the reproductive efficiency of the cows, (2) calf survival, and (3) genetic growth potential of the calf optimized by the cow and measured as weaning weight.

Reproductive efficiency encompasses numerous parameters, but in a cow herd it means optimum conception rates with the cow producing a live, healthy calf annually with ease at parturition. To maximize conception rates in the most economical manner possible, the beef cow must show behavioral estrus and conceive as early as possible in the breeding season. Relatively uncontrollable factors such as environment, as well as controllable factors such as nutrition and animal health, have a direct effect on overall conception rates and cow productivity. Reproductive efficiency, however, entails more than just conception rate. It includes the ease with which a cow calves, the percentage of cows breeding early, and the achievement of as high a conception rate in as short a breeding season as is practicable.

EFFECT OF NUTRITION ON REPRODUCTION AND CALF GROWTH

Because of its economic impact, the net calf crop at weaning has historically been a concern of cattle owners. Reports in the United States have indicated that calf crops may range from 25% to 100% with, unfortunately, national averages being in the 75% to 85% range.[12,18,20]

Calf crop at weaning is determined by six factors: (1) percentage of cows bred, (2) fertilization rate, (3) embryonic losses, (4) fetal losses, (5) perinatal losses, and (6) calf losses from 2 weeks of age to weaning.

The failure of a cow to show estrus and conceive is an important reason for low calf crop.[8] Reports show first-service conception rates that ranged from 39% to 80%, with ranges of 55% to 60% commonly encountered.[25] Nutrition plays a key role in influencing both conception rates and the percentage of cows cycling during the breeding season.

It has been well documented that the quality and quantity of pasture and range has a dramatic influence on the overall conception rate of the cows.[6,12,22] Studies involving stocking rate in eastern Montana found that the average calving percentage was 92.5% of moderately and lightly stocked range, compared with 77.5% on heavily stocked ranges.[16] There was also a tendency for cows on heavily grazed pasture to calve every other year.

Prepartum nutrition

Numerous trials have documented that nutritional deficiencies affect reproductive performance. The level of prepartum nutrition—both protein and particularly energy—has a major impact on how soon the cows cycle after calving. It also has a major impact on the birth weight, calf survival, subsequent milk production, and growth rate of the calf. Early data clearly documented that low levels of energy prepartum delay the onset of postpartum estrus, reduce the percentage of cows cycling early in the breeding season, and subsequently delay the percentage of cows bred during the first 21 days of the breeding season.[4,26] A few studies have shown that cows that lose weight during gestation but then compensate greatly during the postpartum period actually have higher conception rates.

The effect of nutrition during gestation on calf survival from birth to weaning is well documented. Two separate experiments observed a marked effect of precalving nutrition on weaning percentage.[10,11] High and low planes of nutrition resulted in weaning percentages of 95% and 75%, respectively. The losses were attributed to poor vitality because of a combination of low birth weight and reduced suckling, which resulted in starvation. Supplementation during gestation can come in many forms. Supplementation is often used to enhance the intake of grazed forage and compensate for deficiencies that may be present in the forage or native range. In general, prepartum protein levels do not have nearly the influence on reproductive efficiency that energy has, except in instances in which low-quality forage is being grazed.

Although it has been well documented that energy levels prior to calving have some effect on birth weight, most of the studies have shown that this increased birth weight does not have an adverse effect on calving difficulty.

Generally, the increased birth weight has been associated with improved growth rate and increased weaning weights, which are probably a function of both the cow's production of additional milk and the fact that higher birth weights are associated with faster rates of gain to weaning.[7]

Postpartum nutrition

Postcalving nutrition, particularly energy, has its greatest effect on calf growth and cow conception rates. One study reported that low levels of nutrition after calving reduced weaning weight by 13.6 kg (30 lb) compared with calves of normally fed cows.[11] Calves weaned from cows restricted both before and after calving were 21 kg (46 lb) lighter than calves on fully fed cows. With mature cows, one study showed no effect of postcalving nutrition on conception when cows were fed adequately before calving, but 44% of the cows remained open if restricted both before and after calving.[11] Another study showed an 18% reduction in conception rates in cows on low levels of energy after calving, whereas 80% of the cows restricted both before and after calving were open.[26] The use of increased energy or "flushing" rations during the breeding season has shown inconsistent results. Workers in Louisiana reported that energy supplementation exceeding NRC recommendations (125%) during the breeding season increased conception rates by 6% to 10% and shortened the interval of the first estrus.[15] By contrast, other research reports have shown no effect of flushing during the postpartum period.

Protein levels during the postpartum period have their major impact on level of milk production. In instances in which protein supplementation influences feed intake and thus increases the amount of energy the cows are consuming, marked effects on reproductive performance and the calf crop can result.

Factors influencing nutritional requirements

Effect of stage of production on nutrient requirements. To put together a nutritional program for commercial beef cows, the producer needs to recognize the cow's stage of production. Table 13-1 breaks down the periods into four distinct nutritional periods with the requirements listed. A brief description of the four nutritional periods follows.

Period 1. Period 1 is the 80-day period after calving when the cow is lactating at her highest level and trying to maintain a high level of calf growth. In addition, the cow must undergo uterine involution, start recycling, and rebreed during this period. This is the most critical nutritional period for the beef cow.

Period 2. During this period, the cow should be in the early part of pregnancy while still lactating and maintaining a calf. Spring-calving cows should be gaining weight and laying on some energy reserve as body weight and fat to prepare for the winter months.

Period 3. This period follows the weaning of the calf and is referred to as *midgestation*. Basi-

Table 13-1 *The 365-day beef cow year by periods (500-kg cow with average milk production)*

Nutritional needs	Period 1 80 days (postcalving)	Period 2 125 days (pregnant and lactating)	Period 3 110 days (midgestation)	Period 4 50 days (precalving)
TDN (kg/day)	6.6	5.20	4.30	5.10
NEm (Mcal/day)	14.9	12.2	9.2	10.3
Protein (kg/day)	1.04	0.86	0.64	0.72
Calcium (g/day)	33	27	17	25
Phosphorus (g/day)	25	22	17	20
Vitamin A (IU/day)	39,000	36,000	25,000	27,000

From the National Research Council.[17]

cally, during this time the beef cow must primarily maintain her developing fetus. During this period, the beef cow's nutrition needs are at the lowest level of any stage of the year.

Period 4. This period is the second most important period during the beef cow year and again is a period when many producers fail to feed the cows as well as is necessary. During this period, 70% to 80% of the total fetal growth occurs. In addition, the cow is preparing for lactation.

Cow condition. Monitoring the effectiveness of a nutrition program is often one of the difficult parts of cow-calf management. Recently, research reports have evaluated the use of cow condition scores as an indicator of the nutritional status of cows.[24] Studies have shown that cow condition at calving time can be a key factor in determining the productivity of that female.[24] Cows in thin condition at calving time have longer postpartum intervals and conceive late in the breeding season. Thus, evaluating cow condition at least 80 to 100 days prior to calving is important in cow-calf operations. The effect of prepartum weight change is of less importance for cows in good body condition.[7] For cows in moderate body condition, it is important to ensure weight gain prior to calving; net weight gain is critical for cows in thin condition. Greater detail on the role of body condition is presented in Chapter 14.

Cow size and milk production. As producers strive for more productive cows, emphasis on growth and milk production have been foremost in the selection emphasis of many cow-calf producers. This has tended to increase cow size and level of milk production. It can be seen that a 2.3-kg (5-lb) increase in milk production per cow per day increases the energy requirements by 10% and the crude protein requirement by about 15% (Table 13-2).

A common question asked by today's beef producer is, "Can we maintain reproductive efficiency in higher-producing cows?" The question is really not whether we can maintain reproductive efficiency in higher-producing cows, but rather *whether commercial cattle producers will adjust their management programs and nutritional philosophies to accommodate the added nutrient demands of higher-producing cows.* Ample research indicates that normal reproductive performance can be maintained in more productive cows, provided the added nutrient needs are met. The real dilemma facing the commercial cow-calf producer is the fact that the nutritional needs will be increased and thus some change in managerial philosophy must occur to accommodate the more productive cow. This is often more easily intended than accomplished.

In making the decision to have a more productive cow, the producer needs to consider the resources available. If there is an ample supply of high-quality feed, a heavier, larger milking cow can often be maintained. If the feed supply is limited or if environmental conditions that reduce reproductive rates (e.g., drought) frequently occur, then maintaining a slightly smaller, somewhat lower-producing cow may be the most functional for that environment.

Age of the cattle. A good management prac-

Table 13-2 *Relationship between cow size and milk production**

| Cow size (kg) | Milking Level | Milk/cow/ day (kg) | Requirements ||| Crude protein (kg) |
| | | | Energy ||| |
			TDN (kg)	NEm (Mcal)	ME (Mcal)	
450	Average	5.0	5.3	11	18.8	.91
	Above average	6.8	5.7	12.7	20.8	1.04
	Superior	10.0	6.4	14.4	22.7	1.19
500	Average	5.0	5.6	11.5	19.9	.96
	Above average	6.8	6.0	13.2	21.9	1.04
	Superior	10.0	6.8	14.9	23.8	1.25
550	Average	5.0	5.9	12.1	21.0	1.00
	Above average	6.8	6.3	13.8	22.9	1.09
	Superior	10.0	7.1	15.5	24.9	1.30

From National Research Council: Nutrient requirements of beef cattle, Washington, DC, 1984, National Academy of Sciences, National Academy Press.

Table 13-3 *Increased maintenance energy costs for cattle per °C of coldness*

| | Cow weight (kg) |||||
| Coat description | 400 | 450 | 500 | 550 | 600 |
	Percentage increase per °C of coldness				
Summer coat or wet coat	3.6	3.6	3.6	3.4	3.4
Fall coat	2.5	2.5	2.3	2.3	2.3
Winter coat	2.0	2.0	1.8	1.8	1.8
Heavy winter coat	1.3	1.3	1.3	1.1	1.1

Modified from Ames DR: Proc Range Beef Cow Symp IX:11-16, 1985.

tice used by many cattle producers is to sort cattle by age. The nutritional requirements are different for young heifers than for mature cows. When animals are in a growth stage, it is important to have adequate energy and protein present in the ration to maintain growth. In contrast, with mature cows, particularly those that enter the fall in "good" condition, some weight loss can occur during the winter with no adverse effect on productivity.

Environmental effects. In areas of North America where environmental conditions are severe during the winter, the environment may play a major role in influencing the nutritional requirements of the cow.

The degree to which the environment influences cow productivity and reproductive performance is not well documented; however, among cattlemen it is commonly known that the environmental influence has a major impact on productivity. The degree to which the environment affects a cow is influenced by a number of factors including the hair coat and degree of fleshiness of the cow (Table 13-3). It appears the critical temperature for most cows with a normal hair coat and normal body condition is somewhere in the vicinity of 0° C (30° to 35° F).[13] As temperatures drop below this level, the main nutrient affected is energy requirement; there is approximately a 1% increase in energy requirement per 0.5° C (1° F) drop below the critical temperature.

Nutritional development of replacement heifers

The replacement heifer is a mixed blessing for most cow-calf operators. She represents the future profitability and genetic improvement of the cow herd. Thus, her selection and development are of paramount importance to the continued success of the operation.

The replacement heifer is also an inconvenience. Her smaller size and higher nutritional requirements dictate that she be raised and managed separately from the rest of the herd. Yet, the fact that she is essentially nonproductive for the first two years of life makes her easy prey for mismanagement. The growth and development of the replacement female from birth until she produces her first calf are of critical importance in order for her to become a highly productive part of the cow herd.

The selection and development of replacement heifers can be divided into three phases: (1) preweaning, (2) weaning to breeding, and (3) breeding to calving.

Preweaning. During this phase, we largely depend on the dam to nurture and care for the replacement heifer. However, the influence of a few management practices should be mentioned. Producers are encouraged to individually identify all calves at birth so replacement heifer birth dates are known. Working with the local vet, a sound vaccination program must be developed.

Because puberty depends on both age and weight, selecting early-born replacement heifers that weigh at least 180 to 270 kg (400 to 600 lb) at weaning—(depending upon breed, frame size, and feed supply)—is the first important step. However, it is important that this weight is the result of true skeletal and muscle growth without a substantial amount of fat. Research at several locations[8,13] has shown that feeding a high-energy creep feed to suckling heifers of British breeding (relatively small frames) hinders their subsequent milking ability because of fat deposition in the developing udder.[13] However, a summary of studies of large-framed heifers of European breeding showed no effect of creep feeding on subsequent maternal performance.[8]

Thus, the creep feeding of replacement heifers, when economically feasible, should depend on the breeding and growth potential of the calves. No creep or limit-fed creep rations should be used on smaller-framed heifers.

Implanting suckling calves is a highly profitable practice used by cow-calf operators to increase weaning weights. The implants currently cleared for use on nursing heifers are zearanol (Ralgro) and progesterone with estradiol (Synovex-C). Research on the effect of implanting heifers subsequently selected for replacements has been somewhat inconsistent. Several trials[3,21] have shown little or no effect of implanting heifers once with zearanol, at 1 to 4 months of age, on subsequent conception rate and milk production of these females. However, recent trials have shown that implanting within a few days after birth, at weaning time, or twice during the suckling period will likely reduce the reproductive performance of heifers.[9,23] These studies have shown that implanted heifers have larger pelvic areas at 1 year of age, but one study in which the heifers were followed to calving noted no reduction in calving difficulty. Because producers typically do not know which heifers will be kept for replacements at implanting time, an easy way to avoid possible problems is to implant only those heifer calves born during the last half of the calving season and to select replacements from the older group of nonimplanted calves.

In selecting replacement heifers, it is normally advisable to keep 20% to 50% more heifers than are needed as replacements so that further culling can be practiced on yearlings and after breeding. Keeping only those heifers that conceive in a short 45- to 60-day breeding season is an ideal way to enhance cow herd fertility and shorten the calving season.

Weaning to breeding. Once the replacement heifer is selected at weaning, growth and proper development prior to breeding have a profound impact on her subsequent productivity. Replacement heifers need to weigh about 65% to 70% of their mature weight in order to breed consistently as yearlings. Thus, a good nutrition program is essential. Generally, heifers should gain about 0.5 kg/day (1.1 lb/day) from weaning to breeding, depending upon their weaning weight and the length of the feeding period prior to breeding. Usually, this means the average British-breed heifer needs to gain about 110 kg (240 lbs) in order to weigh the 275 to 300 kg (600 to 660 lb) necessary to reach puberty. With the larger-framed European breeds and crosses, a target breeding weight of 320 to 365 kg (700 to 800 lb) is usually necessary.

Puberty in heifers is a function of breed, age, and weight. Recent research illustrates that the degree of development from weaning to breeding influences not only how soon heifers cycle (reach puberty) as yearlings but also their subsequent productivity and rebreeding rate after they calve as two-year-olds. Research at Kansas State[19] and at Purdue[14] indicates the impact of inadequate growth and development during this phase on subsequent calving ease, rebreeding, and calf growth (Tables 13-4 and 13-5).

It should be emphasized that replacement heifers need to be fed separately from the rest of the herd. Because of their size and age, as well as higher nutritional demands, they simply cannot compete with the rest of the cow herd, nor can they be expected to utilize poorer-quality forages efficiently and still breed as yearlings.

Cow-calf operators are encouraged to have their replacement heifers cycling early enough to be bred 3 to 4 weeks before the rest of the cow herd. The stress of calving is greater on heifers than on older cows and more likely to be accompanied by calving difficulty. Thus, breeding replacement heifers essentially one heat cycle earlier than mature cows allows the producer more time to watch the heifers at calving time and gives the heifer the extra time she needs to start recycling and breed back "in sync" with the rest of the herd. In addition, the wean-

Table 13-4 *Effect of first winter nutrition on subsequent performance of heifers*

	Kilograms of grain per head daily (plus low-quality fescue hay)		
	0	1.4	2.7
No. heifers	112	113	112
Fall weight (kg)	225.5	228	224
Daily gain in winter (kg)	.03	.23	.36
Breeding weight (kg)	230	262	279
% Conceiving as yearlings	69.2	73.9	83.5
Subsequent production:			
% Rebreeding after first calf	67.3	75.4	87.1
Weaning weight of first calf, kg	184	197	201

Modified from Lemenager RP and others: J Anim Sci 51:837, 1980.

Table 13-5 *Effect of heifer nutritional development on subsequent performance*

% of mature weight at breeding as yearling	No.	Preweaning weight (kg)	Calving weight (kg)	Calf birth weight (kg)	Calving difficulty (%)	Calf death loss (%)	Fall pregnancy rate (%)
55	60	272	379	32.2	52	6.2	85
65	61	310	408	33.3	29	4.5	94

Modified from Patterson DJ and others: Kansas Agri Expt Stat Report of Prog 514:60, 1987.

Feeding the Beef Cow for Optimal Production

ing weights of calves from replacement heifers that are bred 3 to 4 weeks early are usually increased 13 to 20 kg (28 to 45 lb) because the calves are older.

If it is impractical for cow-calf producers to breed replacement heifers prior to the rest of the cow herd, they should then stress a very short breeding season with their heifers. The place to start emphasizing reproductive efficiency in a cow herd is with the replacement heifers. Utilizing a short breeding season (34 to 45 days) ensures fertile replacements that conceive promptly. This also forces the heifers into a short calving season so that producers can give them more attention.

Breeding until calving. The final step in the profitable management of the replacement heifer is to ensure her adequate growth and development from breeding until she calves as a 2-year-old at about 85% to 90% of her mature weight. During this time, the bred heifer should gain about 0.4 kg (1 lb) per day, or a total of 110 to 160 kg (240 to 350 lb). British-breed heifers and crossbred heifers of British breeding should go into the calving season weighing 380 to 430 kg (836 to 946 lb), and the larger-framed breeds and crosses should weigh 430 to 450 kg (946 to 990 lb). It is preferable to have the heifer growing continuously throughout this phase. For spring-calving herds, summer pasture is usually adequate for the first half of this period. However, it is important to recognize that most of the fetal growth occurs during the last 50 to 60 days prior to calving. Adequate nutrition, especially adequate energy and feeding separately from the mature cows, is essential for proper development of the fetus and to prepare the heifer for calving and lactation.

Research at several stations[2,5] has consistently shown that poor heifer development prior to calving results in lighter, weaker calves at birth with no decrease in calving difficulty, greater calf sickness and mortality, lower milk production, slower return to estrus, and poorer overall reproductive performance. Thus, "shorting" the heifer nutritionally prior to calving is an invitation to disaster.

The major nutritional requirements of replacement heifers from weaning through calving and rebreeding are shown in Table 13-6. This information can be used as a guide to feeding these females. However, genetics and environment can substantially influence the heifer's needs. Thus, body condition and weight should be used to modify the feeding program.

Management practices to control cow maintenance costs

Optimum use of range and pasture resource. Cattle owners must keep in mind that they are really grass managers and strive to maximize production without damaging the grass resource. Several key requirements to attain this goal include:

1. *Utilization of proper stocking rates.* When a moderate stocking rate is maintained, approximately half of the forage produced each year is consumed, with the remaining forage required to sustain next year's forage production.
2. *Maintenance of good grazing distribution.* In too many pastures, grass adjacent to water and minerals is overgrazed, and grass far removed from these areas is undergrazed. The net result is a reduction in the effective size of a pasture. Water, minerals, salt, and fencing must be placed to obtain good grazing distribution.

Maximum use of low-quality forages. A careful analysis indicates that of the $260 to $350 (1985 to 1990) required annually to maintain a cow, approximately half is spent on feed, with a fair portion of this spent on winter feed and supplement. To help cut winter feed costs, low-quality roughages can be utilized to a greater extent than is currently practiced. Specific examples of underutilized forages include:

1. *Wheat straw.* Despite its abundance, wheat straw is not fully utilized in some North American cow rations. Moreover, it can be used very effectively as part of the diet at certain times of the year without reducing production. For example, during the period from the time calves are weaned to 60 days prior to calving, wheat straw can be effectively fed, but it should be blended with other roughages or protein and energy supplements. In periods of drought, it can be used at low levels throughout the year to reduce feed costs. To utilize wheat

Table 13-6 *Major nutritional requirements of replacement and first-calf heifers*

Body weight (kg)*	Daily gain (kg)†	Minimum daily DM intake (kg)‡	Crude protein		TDN§		DE Mcal	Calcium (% of DM)	Phosphorus (% of DM)	Vitamin A IU
			% of DM	kg/day	% of DM	kg				
Replacement heifers										
180	.7	4.6	11.8	.6	68	3.1	13.7	.45	.24	10,000
230	.7	5.5	10.7	.6	68	3.7	16.3	.38	.22	12,000
275	.6	6.2	9.6	.6	65	4.0	17.6	.30	.21	14,000
320	.6	7.0	9.2	.6	65	4.5	19.9	.27	.20	16,000
Bred yearling heifers: last third of pregnancy										
320	.5	7.0	8.5	.6	56	3.9	17.2	.27	.20	19,000
360	.5	7.6	8.4	.6	55	4.2	18.5	.27	.20	21,000
410	.5	8.3	8.3	.7	54	4.5	19.9	.26	.20	23,000
Two-year-old heifers: milking 4.5 kg (10 lb)/day: calving through rebreeding										
340	.25	7.6	11.0	.8	64	4.9	21.6	.34	.24	30,000
385	.25	8.4	10.6	.9	63	5.3	23.4	.33	.23	33,000
430	.25	9.1	10.2	.9	62	5.7	25.2	.31	.23	35,000

Trace mineralized salt should be provided either free choice in a mineral supplement or mixed into the ration at 0.3% of the dry matter to all cattle.
*Average body weight during feeding period.
†About 0.4 kg gain per day during the last third of pregnancy is made up of fetal growth.
‡Minimum dry matter intake required to provide needed amounts (kg per day) of protein and TDN based on the dietary concentrations (% of DM) reported in the table. If intake is above or below this level, nutrient concentrations can be adjusted accordingly.
§The energy (TDN) levels reported are sufficient in relatively mild climates. As a general rule, increase the amount of TDN by 1% for each 0.5° C (1° F) decrease in the windchill temperature below 0° C (30° F) for cattle with dry, winter hair coats, or below 13° C (55° F) for wet or summer hair coats.
Requirements are on a daily basis and are adapted from National Research Council: Nutrient requirements of beef cattle, Washington, DC, 1984, National Academy of Sciences, National Academy Press.

straw effectively, it must be harvested immediately after grain harvest while it is still bright and unweathered. This increases the consumption and nutritional value of the straw. Another management strategy that improves the energy and protein value of wheat straw is ammoniation.
2. *Corn and milo stover*. Although these residues are widely grazed, full and proper use can greatly help reduce the cow feed bill. In some cases, this may require harvesting the residues for use after weather prohibits grazing. As with wheat straw, they are most useful when fed to dry, pregnant cows but can be utilized effectively at other times if properly supplemented.
3. *Other low-quality forages*. In addition, other low-quality forages, such as volunteer wheat, alfalfa aftermath, and possibly low levels (<50% of ration, feed to low-producing cattle) of fireweed *(Kochia),* offer opportunities to significantly reduce forage costs.

Complementary forages. Research has indicated that summer annuals such as sudan-sorghum hybrids and wheat grown on land adjacent to pasture lands can be used as supplemental grazing or even as creep grazing for calves. The use of complementary forage might be an economical way of cutting grass costs because it reduces the area of pasture or range required to carry a cow.

Feeding the Beef Cow for Optimal Production

Table 13-7 *Costs of protein for common protein supplements in Kansas*

Supplement	Cost $/tonne	Cost $/ton	Crude protein (%)	Cost per kg of crude protein (cents)
Soybean meal	165	150	44	37.5
Soybean meal	220	200	44	50.0
Soybean meal	276	250	44	62.5
Alfalfa hay	44	40	15	29.4
Alfalfa hay	66	60	15	44.1
Alfalfa hay	88	80	15	36.4
Alfalfa hay	66	60	18	36.8

Efficient use of supplemental protein and minerals. Cattlemen spend many dollars providing supplemental protein and minerals each year. Unfortunately, in many cases, a high-priced source of protein is utilized, yet the supplement may not be meeting the supplemental nutritional needs of the cow. These practices should be followed:

1. *Compare the price of supplements.* Protein and mineral supplements can be priced on the basis of cost per unit of protein or mineral supplied. For example, Table 13-7 illustrates the cost of crude protein for several potential supplements. As can be seen, a considerable price range exists per unit of protein supplied, indicating the need to compare costs. This comparison assumes that the protein in these feedstuffs is of equal value.

2. *Assess the value of protein in supplements.* Some protein supplements include high levels of nonprotein nitrogen (NPN) in the form of urea, biuret, or ammonium phosphate. Although these products are utilized efficiently by cattle on high-energy rations (e.g., feedlot cattle), they are often of limited value to cows on low-quality roughages. Thus, as a general recommendation, natural protein sources should be used for cows on milo stover, corn stover, wheat straw, low-quality hay, or other low-energy feedstuffs; otherwise, much of the cost of supplemental protein is wasted.

3. *Utilize protein and mineral supplements only when needed.* Many cattle owners feed costly supplements when they are not required. For example, dry pregnant cows on milo stover in the early fall probably do not require a protein supplement, whereas cows in late pregnancy on this same feedstuff in the winter do require supplementation. Furthermore, some cattle owners feed straight alfalfa when it could be mixed with low-quality, cheaper forages to reduce ration costs. The nutritional requirements of the cow herd must be evaluated to avoid overfeeding costly feedstuffs.

Use of proven products and practices. During hard economic times, it is imperative that cattlemen stick with proven products and practices.

Practical application of nutrient needs

Crop residue fields. One way of reducing the annual costs of running a cow herd is to graze crop residue fields. Obviously, the extent that this can be done is greatly influenced by weather.

There are a few things to consider in grazing crop aftermath. One of these is what type of supplementation is needed. Part of this will be influenced by the condition score of the cows. If cows enter the fall grazing period in thin condition, it may be important to feed protein and energy supplements. In contrast, if the cows are in fairly good "flesh" in the fall, supplementation may not be necessary. It is important to monitor the cows' weight and condition change while they are grazing stalk fields. Fields should

Table 13-8 *Nutrient intake of cows grazing cornstalks*

Nutrient	Required for 500-kg (1100-lb) cow during midgestation	Supplied by cow eating 9.1 kg (20 lb) of cornstalks*
TDN	4.30 kg (9.5 lb)	4.10 kg (9.0 lb)
Crude protein	0.64 kg (1.4 lb)	0.45 kg (1.0 lb)

*Cornstalks contain 45% TDN and 5% protein on a dry matter basis.

Table 13-9 *Nutrient intake of cows grazing dry grass*

Nutrient	Required for 500-kg (1100-lb) cow during midgestation	Supplied by cow eating 7.7 kg (17 lb) dry grass
TDN	4.30 kg (9.5 lb)	3.30 kg (7.3 lb)
Crude protein	0.64 kg (1.4 lb)	0.31 kg (0.7 lb)

be utilized to the fullest extent but not to the point that the stalk fields are overgrazed and the cows lose considerable weight.

A cow consuming 1.8% of her body weight in corn stalks (dry matter basis) comes fairly close to meeting her energy needs but is deficient in protein (Table 13-8).

If during the early part of the foraging season cows can do some selective feeding including consuming unharvested grain, then the corn protein content of the residue will be well above the 5% figure. Thus, during early grazing, cows will gain weight, and only mineral supplementation is needed. In contrast, after selectively grazing the higher-quality material, cows still on stalk fields during December and January will require supplementation with other feeds. Unfortunately, many producers graze cows on stalk fields too long; as a result, cows lose weight and enter the calving season too thin.

Dry grass. A common wintering program in much of the North American continent is maintaining cows on dry native or cool-season grasses such as brome or fescue. In this situation, cows consume from 7 to 8.5 kg (15 to 19 lb) of dry grass per day. This dry grass varies in quality, depending on weather and growing conditions. The nutrient intake of cows consuming grass with 43% TDN and approximately 4% crude protein is shown in Table 13-9.

Cows grazing dry grass often are in more critical need of supplementation than cows on some of the other feeding systems used. The cows can often go unsupplemented during the early part of the year (October, November, and December) without severe weight loss. Part of this is because of the cows' ability to selectively graze higher-quality forage. However, cows that are maintained on dry grass during the critical months of January, February, March, and early April need supplementation. During this period, feeding harvested forages such as hay, grain, or silage is necessary.

Feeding hay. A common feeding program during the winter months is feeding hay. The following example illustrates feeding 9.1 kg (20 lb) of brome hay and assumes a 50% TDN and 7% protein content (Table 13-10).

In this type of feeding system, only mineral supplementation is needed. The exception would be during severe cold weather, when the cows need extra energy. Extra energy can be provided by increased hay intake or 1 kg (2 lb) or so of grain.

Mineral needs

The key mineral that needs to be considered in formulating a cow-calf nutrition program is

Table 13-10 *Nutrient intake of cows supplemented with hay*

Nutrient	Required for 500-kg (1100-lb) cow during midgestation	Supplied by feeding 9.1 kg (20 lb) brome hay
TDN	4.30 kg (9.5 lb)	4.50 kg (9.9 lb)
Crude protein	0.64 kg (1.4 lb)	0.63 kg (1.4 lb)

phosphorus. In many situations in which cows are grazing dry winter grass or crop residue fields, there may well be a deficiency of phosphorus in the diet.

There are many ways to meet the phosphorus need. One common method is to buy one of the commercial mixes marketed. Some producers choose to mix their own. A mix that works very well is a blend of half salt and half a phosphorus source like dicalcium phosphate, which contains 18.5% phosphorus. Mixing these equally gives a mixture of 50% salt, 11.5% calcium, and 9.3% phosphorus. A pregnant cow needs about 18 to 22 g of phosphorus per day; a lactating cow requires from 22 to 26 g of phosphorus per day.

How much phosphorus is supplied by the forage? An example is a cow eating 10 kg (22 lb) of brome hay that contains 0.15% phosphorus on an as-fed basis (an analysis can be helpful in determining exactly what the phosphorus content of the feed is because regional variations occur). To convert to g consumed per cow per day, multiply the 10 kg (22 lb) of feed $\times$ 0.15% phosphorus; the result is 15 g (0.5 oz) phosphorus. That means pregnant cows should have at least 3 to 7 g (0.1 to 0.23 oz) of additional phosphorus. Eating 100 g (3.5 oz) of a half salt and half dicalcium phosphate mix would supply 9 g (0.3 oz) of phosphorus.

A common question that comes up in the cattle industry relates to the value of trace minerals. There is considerable discrepancy in the research data on the importance of trace minerals. Much of the early research showed very little response. However, new information indicates isolated areas of trace mineral deficiencies, particularly copper. With the higher-producing cow, it is logical to expect that these cattle would have a higher need for specific trace minerals. To be on the safe side, many individuals formulating their own mineral mix add a well-fortified trace mineral premix to protect against trace mineral deficiencies. Most commercial mineral mixes are generally well fortified with various trace minerals.

Vitamin A

The most important vitamin in a cow-calf nutrition program is vitamin A. Vitamin A can be supplied to the cow in many ways. Many producers use high-quality alfalfa hay, which is an excellent source of vitamin A. Many commercial mineral mixes are fortified with vitamin A, and a producer mixing minerals can also add a small amount of vitamin A to ensure that vitamin A requirements are met.

Injectable vitamin A given once or twice during the winter feeding period is another very effective method used by some producers to supply vitamin A.

The exact value of vitamin A in a cow-calf nutrition program is often hard to determine. However, the expense of supplementation to ensure that the vitamin A requirements of cows is met is small.

COMPLETE NUTRIENT REQUIREMENTS OF COWS AND HEIFERS

Please refer to the Appendix for a complete list of nutrient requirements for cows and heifers.

REFERENCES

1. Ames DR: Proceedings of range beef cow symposium IX, 1985, pp 11-16.
2. Bellows RA and others: J Anim Sci 55:18, 1982.
3. Bolze RP: Doctoral thesis, 1985.
4. Corah LR, TG Dunn, and CC Kaltenbach: J Anim Sci 41:819, 1978.
5. Corah LR and others: J Anim Sci 41:819, 1975.
6. Donaldson LE: Aust Vet J 4:493, 1968.

7. Dunn TG: Paper read at National Beef Symposium and Oklahoma Cattle Feeders Seminar, 1982.
8. Friedrich RL and others: Proc Western Section ASAS 26:14, 1975.
9. Goehring TB and others: Kansas Agricul Exp Sta Report of Prog 494:62, 1986.
10. Hight GK: N Z J Agricul Res 9:479, 1966.
11. Hight GK: N Z J Agricul Res 11:71, 1968.
12. Hilts WH: Nev Agricul Ext Cir 57:3, 1925.
13. Hixon DL and others: Illinois beef cattle day report, 1980, p 12.
14. Lemenager RP and others: J Anim Sci 51:837, 1980.
15. Loyacano A and others: LSU 11th Annual Livestock Producers' Day, 1971, p 104.
16. Marsh H, Swingel KF, and Woodward RR: Mont Agricul Exp Sta Bull 549, 1959.
17. National Research Council: Nutrient requirements of domestic animals, No. 4. Nutrient requirements of beef cattle, Washington, DC, 1984, National Academy Press.
18. Patterson DJ: Master's thesis, Bozeman, 1979, Montana State University.
19. Patterson DJ and others: Kansas Agricul Exp Sta Report of Prog 514:60, 1987.
20. Rice FJ and others: Mont Agricul Exp Sta Bull 461, 1961.
21. Sprott LR and others: Kansas Agricul Exp Sta Report of Prog 350:19, 1979.
22. Sprott LR and others: KSU cattlemen's day, 1981, p 37.
23. Staigmiller RB and others: J Anim Sci 57:527, 1983.
24. Whitman RW: Doctoral thesis, Fort Collins, 1975, Colorado State University.
25. Wiltbank JN and others: J Anim Sci 20:409, 1961.
26. Wiltbank JN and others: J Anim Sci 21:219, 1962.

Body Condition Scoring in Beef Cattle

JAMES N. WILTBANK

Body condition scoring is a subjective measure of the amount of fat in an animal's body. Body condition influences production in a beef cow herd in three ways:
1. Growth rate of a calf suckling a thin cow is decreased.
2. Pregnancy rate in thin cows is lowered.
3. Thin cows become pregnant later in the breeding season and thus calve later the following year.

The main factors determining the kg of calf weaned in a beef cow herd are growth rate of the calf, the proportion of cows pregnant, and the calving pattern of the cow herd. Consequently, each person working with beef cows should (1) know how to estimate body condition, (2) recognize the influence body condition has on production, (3) know how and when to change body condition, and (4) be able to calculate the costs and returns for changing body condition.

The purpose of this chapter is to make it possible for a person to perform those procedures and thus make a beef herd more profitable.

DETERMINING BODY CONDITION

Several studies have shown that body condition score is highly correlated with the amount of fat present in the body.[2,4] Consequently, it is a reliable way to estimate body fat. The two methods currently used to determine body condition are (1) measurement of covering over the ribs, shoulder, and back (Nebraska System); and (2) measurement of covering over the short rib region and tailhead (Scottish System). Both methods appear to be accurate, and the system used is largely a matter of preference. In the Nebraska System cows are scored from 1 (thin) through 9 (fat), and in the Scottish System from 1 (thin) to 5 (fat). Table 14-1 outlines the two systems. In this chapter the Nebraska System is used.

PREGNANCY RATE AND POSTPARTUM ESTRUS

Numerous studies have shown a decrease in pregnancy rates in thin cows. A group of cows in Florida is used here as an example.[6] These cows were suckling calves and were scored for body condition between birth of the calf and the start of the breeding season. The cows were in two groups, and each body condition was represented in each group. The two groups were grazed on similar pastures during the 120-day breeding season. The proportion pregnant after 120 days of breeding varied from 23% in cows scoring 2 to 95% in cows scoring 7 (Table 14-2).

The proportion of cows pregnant after 60 days of breeding was low in cows scoring 4 or less. For each increase in body condition score, an increase in pregnancy rate was noted.

A study in South Carolina reported similar results.[3] In these two studies pregnancy rate was lower in thin cows at all stages of the breeding season. Also, the calving pattern was altered, with fewer thin cows becoming pregnant early in the breeding season.

The pregnancy rate is lower in thin cows because the onset of estrus following calving is delayed in thin cows. The results from several groups of cows in Nebraska summarized by Whitman are used to illustrate this phenome-

Table 14-1 *Description of two systems for body condition scoring*

Group	Nebraska	Scottish	Description
Emaciated	1	0.5	Poor—no palpable fat cover along backbone or ribs
	2	1	Very thin—Some fat present along backbone but no fat cover over ribs
Thin condition	3	1.5	Thin—Fat along backbone and slight amount of fat cover over ribs
Borderline condition	4	2	Borderline—Some fat cover over ribs
	5	2.5	Moderate—Fat cover over ribs feels spongy
Optimum moderate condition	6	3	Moderate to Good—Spongy fat cover over ribs, and fat beginning to be palpable around tailhead
	7	3.5	Good—Spongy fat cover over ribs and fat around tailhead
Fat condition	8	4	Fat—Large fat deposits over ribs, around tailhead, and below vulva
	9	5	Extremely fat—Cow extremely overconditioned

Modified from Richards MW, Spitzer JC, and Warner MB: J Anim Sci 62:300, 1986.

Table 14-2 *Body condition and pregnancy rate in a group of cows in Florida*

	Body condition near calving				
	2 (Very thin)	3 (Thin)	4 (Borderline)	5 and 6 (Moderate)	7 (Good)
Number of cows	115	545	564	344	234
Pregnancy rate (%)					
60 days after breeding	5*	15	19	40	56
120 days after breeding	23*	51	73	86	95

*Values are cumulative percent pregnant.

Table 14-3 *Effect of body condition on the proportion of cows showing estrus at different times after calving*

	Body condition		
	2-3 (Thin)	4-6 (Moderate)	7-8 (Good)
Number of cows	272	364	50
Days postcalving			
30	3*	7	13
40	19	21	31
50	34	45	42
60	46	61	91
70	55	79	96
80	62	88	98
90	66	92	100

*% of cows showing estrus by this time.

non.[5] Most of the cows in good body condition had shown heat by 60 days postcalving. Comparable levels were attained by cows in moderate body condition by 80 days postcalving, whereas thin cows never attained comparable levels of estrus (Table 14-3).

The onset of estrus after calving is more rapid in cows in moderate or good condition than in thin cows. Interestingly, Whitman also reported that in cows in moderate or good condition estrus tended to be expressed more rapidly in cows that were losing condition postpartum. Body condition at calving has more influence on the onset of estrus than the postcalving plane of nutrition (Table 14-4).

Similar results were reported in a South Carolina study.[3] The number of cows bred in the first 20 days of the breeding season was not influenced by the plane of nutrition after calving

Table 14-4 *Effect of body condition and plane of nutrition on the expression of estrus postpartum*

| | Body condition | | | |
| | Thin (2-3) | | Moderate or good (4-8) | |
	Losing weight	Gaining weight	Losing weight	Gaining weight
Number of cows	13	259	120	294
Days postcalving				
40	31*	8	36	20
50	46	24	61	43
60	54	45	71	63
70	54	64	87	78
80	54	78	91	89
90	54	84	92	94

*% of cows showing estrus by this time.

Table 14-5 *Effect of body condition score (BCS) and postpartum feed level on reproductive performance in the first 20 days of the breeding season*

| Feed level postcalving | Breeding rates | | Pregnancy rates | |
| | BCS at calving | | BCS at calving | |
	4 or less	5 or more	4 or less	5 or more
High	84*	82	46	48
Moderate	83	93	42	51
Low	67	89	30	55
Low-flush	71	92	41	52

*% of cows.
From data in Richards MW, Spitzer JC, and Warner MB: J Anim Sci 62:300, 1986.

if the cows were in good condition (score greater than 5) at calving (Table 14-5).

Both sets of data emphasize that body condition at calving is important in determining pregnancy rate. Cows in moderate or good body condition start cycling early, whereas the onset of estrus and pregnancy is delayed in cows calving in body condition of 4 or less.

GROWTH RATE OF THE SUCKLING CALF AND MILK PRODUCTION OF THE COW

In a study with 2-year-old heifers in Nebraska, milk consumption and calf growth were lower in heifers receiving 8.0 Mcal digestible energy precalving (approximately half current NRC recommendations) compared with those receiving 16 Mcal digestible energy. The effect of level of feeding after calving was dependent on the level of feeding before calving. Heifers fed poorly before calving showed only a small response to an increase in their plane of nutrition after calving. Heifers fed well before calving responded to an increase in their plane of nutrition to 26 Mcal digestible energy after calving by increasing their milk production and the growth of their calves; further increases in plane of nutrition had little effect. This indicates that cows that calve thin are adversely affected for the whole lactation and cannot respond properly to additional feeding after calving (Table 14-6).

Table 14-6 *Effect of feeding different levels of energy to heifers before and after calving on their milk production and growth of their calves*

	Number of calves	Age of calf (days)					
		53		81		109	
Precalving energy level (Mcal DE/day)							
Low (8)	83	59.9*	(5.4)†	84.4	(5.2)	111.2	(3.6)
High (16)	79	65.8	(7.3)	91.2	(6.9)	121.6	(4.6)
Difference		5.9	(1.9)	6.8	(1.7)	10.4	(1.0)
Low precalving energy level							
Level postcalving (Mcal DE/day)							
26	41	58.6	(5.9)	81.3	(5.1)	111.7	(3.6)
44	42	61.3	(7.9)	87.2	(5.4)	111.2	(3.5)
Difference		2.7	(2.0)	5.9	(0.6)	0.5	(0.1)
High precalving energy level							
Level postcalving (Mcal DE/day)							
14	42	58.6	(5.5)	84.4	(5.2)	113.0	(2.9)
26	37	65.4	(6.3)	91.2	(5.4)	124.4	(3.2)
44	42	66.3	(8.8)	90.8	(8.0)	118.9	(6.0)

*Calf weight in kg.
†24-hour milk consumption at this stage of lactation in kg.
NRC recommends about 16 Mcal of DE precalving and 26 Mcal of DE postcalving.

Table 14-7 *Weight changes needed to change various initial body condition scores (BCS) to a final score of 5 at different stages of the cow's reproductive cycle*

Reproductive stage	Initial BCS score	Increase in weight needed (kg)			ADG needed to make change (kg)			
		Fat and muscle	Calf, fluid, and membranes*	Total	70 days	100 days	150 days	200 days
Dry, pregnant	2	109	45	154	2.2	1.5	1.0	0.8
Lactating	2	109	0	109	1.5	1.1	0.7	0.5
Dry, pregnant	3	73	45	118	1.7	1.2	0.8	0.6
Lactating	3	73	0	73	1.0	0.7	0.5	0.4
Dry, pregnant	4	36	45	81	1.2	0.8	0.5	0.4
Lactating	4	36	0	36	0.5	0.4	0.2	0.2
Dry, pregnant	5	0	45	45	0.6	0.4	0.3	0.2
Lactating	5	0	0	0	0	0	0	0

*Last 100 days of pregnancy.

Table 14-8 *Energy levels required to achieve body condition of 5 (for cows weighing 500 kg with frame score 5)*

	Dry				Open			
	Pregnant cow				Cow suckling calf			
	Body score 100 days before calving				Body condition score at calving			
	2	3	4	5	2	3	4	5
ADG (kg/day)*	1.5	1.2	0.8	0.4	1.5	1.0	0.5	0
Energy requirements:								
For maintenance (Mcal DE/d)*	14	14	14	14	14	14	14	14
For muscle & fat (Mcal DE/d)	12	8	4	0	12	6	3	0
For fetus (Mcal DE/d)*	4	4	4	4	—	—	—	—
For milk (Mcal DE/d)†	—	—	—	—	18	18	18	18
Total (Mcal DE/d)	30	26	22	18	44	38	35	32
Feed requirements: alfalfa hay to meet requirements (kg/d)	13	11	10	8	22	19	16	14
Maximum free choice alfalfa hay intake (kg/d)	13	13	13	13	15	15	15	15

*Digestible energy per day. Last 100 days of pregnancy for a dry cow; first 70 days of lactation for a wet cow.
†First 70 days of lactation.

CHANGING BODY CONDITION

Body condition scores predict total body fat and total body energy ($r = 0.71$ to 0.85). There is also a decrease in muscle and, consequently, protein in thin cows.[2,4] Therefore, the amounts of fat and protein in a cow increase as body condition improves. The increase in body weight needed to change body condition one score depends upon the reproductive status of the cow. Changing body condition from 4 to 5 requires a 36 kg (80 lb) increase in fat and muscle. The pregnant cow must also gain the weight of the calf, fluids, and membranes. The amount of weight gained by the calf varies according to breed and stage of gestation. An estimate of the increase in weight due to calf, fluid, and membrane is 45 kg (100 lb) in a 500 kg (1100 lb) crossbred cow in the last 100 days of pregnancy. Therefore, to change body condition from 4 to 5 in a pregnant cow in the last 100 days of pregnancy requires an increase in weight of 81 kg (180 lb) (Table 14-7), whereas the same change in body condition in the nonpregnant, nursing cow requires a weight change of 36 kg (80 lb). Note that a pregnant cow scoring 5 must gain 45 kg (100 lb) in the last 100 days of pregnancy to maintain a body condition score of 5.

To change body condition from 2 to 5 in a pregnant cow requires a weight change of 153 kg (340 lb). Although this represents a large increase in body weight, the average daily gain (ADG) required to make the change is relatively small (0.77 kg or 1.7 lb) if a 200-day period is available. Weight gain in pregnant cows can be facilitated by early weaning. As an example, to change a cow from 2 to 5 in 100 days requires an ADG of 1.54 kg (3.4 lb); if 200 days are available, the ADG needed would be 0.77 kg (1.7 lb). Once the cow has calved, very little time is available to change body condition before the cow is bred. In most cows only 60 to 80 days is available from calving to the start of breeding. To change a wet cow from a body condition of 2 to a score of 5 in 70 days requires an ADG of 1.54 kg (3.4 lb). Careful examination of Table 14-7 shows that reasonable changes in body condition require time and planning.

As an example of the energy required to make weight changes, consider the problem of changing a dry, pregnant cow from a body condition score of 2 to 5. The cow must make a weight gain of 153 kg (340 lb) or 1.5 kg/day (3.4 lb/day) if she is 100 days from calving. The amount of digestible energy required to make the change is 30 Mcal/day. This can be provided by 13 kg (28 lb) of alfalfa hay or equivalent in grass or other feeds (Table 14-8). It is estimated that a 500-kg (1100 lb) cow can eat 13 kg (28 lb) of hay. Therefore, a cow scoring 2 could be changed to condition score 5 by full-feeding alfalfa hay or its equivalent for 100 days.

The energy levels needed for maintenance and fetal growth are similar, whether a cow scores 2 or 5. The level needed for increase in muscle and fat varies from 12 Mcal/day for cows scoring 2 to 0 Mcal for cows scoring 5.

The energy level needed for adding muscle and fat in the 70 days after calving varies from 0 Mcal in cows scoring 5 to 12 Mcal in cows scoring 2. The alfalfa hay needed to meet the energy requirement varies from 14 kg (30 lb) in a cow scoring 5 to 22 kg (48 lb) in a cow scoring 2. It is estimated that the wet cow will eat 15 kg (33 lb) of alfalfa hay. The energy requirements of a cow scoring 5 can be met by full-feeding alfalfa hay or its equivalent. Changing body condition for cows scoring 2 or 3 to 5 in 70 days by feeding hay or an equivalent is impossible. The cow cannot eat enough. Even changing body condition from 4 to 5 requires some concentrate. To change a cow from 3 to 5 in 70 days requires a ration containing 55% to 60% concentrate. To meet this requirement, a cow must eat 8 to 9 kg (18 to 20 lb) of either corn or barley. This indicates that body condition in cows calving in thin body condition is difficult and expensive to change. By contrast, body condition can be changed rather inexpensively in the pregnant cow, particularly if the change is begun at least 100 days prior to calving.

IMPROVING PRODUCTION BY CHANGING BODY CONDITION

Improvement in body condition improves the weight of weaned calves by improving calf growth rates and cow conception rates and by promoting early calving (Table 14-9).

The total economic impact of a thin cow is difficult to assess completely. Cows calving late in the next calving season tend to be late calvers the rest of their lives, and open cows have to be either replaced or carried open for 1 year. However, the economic impact over a 2-year period is relatively easy to assess (Table 14-9).

The calf suckling a moderate-condition cow is worth $36 more than the calf from a thin cow because the calf from the thin cow weighs 18 kg (40 lb) less. Next year fewer thin cows will calve and more early calves will be born to cows in moderate body condition (Table 14-10). Consequently, the kg of calf weaned per cow bred will be only 147 kg (324 lb) in thin cows as compared to 199 kg (439 lb) in moderate-condition cows. This means the moderate-condition cow will return $104 more next year. This assumes cows in both groups are in moderate condition at the start of calving in the second year. In the 2-year period the thin cow will return $140 less than the cow in moderate body condition.

Each day 18 Mcal of digestible energy or 7.7 kg (17 lb) of alfalfa hay are required to maintain the cow scoring a 3 (Table 14-11). To change the

Table 14-9 *Economic impact of body condition in cows (calves at $2 per kg)*

	Return from calf this year		Return from calf next year		Total Two years	Return
Body condition	Weaning weight (kg)	Return ($)	Weaning weight per cow bred (kg)	Return ($)	Weight weaned per cow (kg)	Return ($)
Moderate	211	422	199	398	410	820
Thin	193	386	147	294	340	680
Difference	18	36	52	104	70	140

Table 14-10 *Improvement in performance of moderate-condition cows (calving period 80 days)*

Body condition (score)	Weaning weight (kg)	Pregnant 80 days (%)	Calving next year (%)	
			First 20 days	First 40 days
Moderate (5)	211	94	44	72
Thin (3)	193	70	30	50
Difference	18	24	14	22

Assumptions: 40% calved by 20 days, 80% by 40 days, 92% by 60 days, 100% by 80 days. Calf average daily gain 0.9 kg/day in moderate-condition cows and 36-kg birth weight.

Table 14-11 *Economic impact of changing body condition from an initial score of 3 at 100 days precalving to a score of 5 at calving*

Body condition score at calving	Energy needed (Mcal DE per day)	Alfalfa hay per day (kg)	Alfalfa needed (kg)	Cost of alfalfa* ($)	Two-year return† ($)
5	26	11.3	1130	100	820
3	18	7.7	770	68	680
Difference	8	3.6	360	32	140

*Cost per 1000 kg is $88.
†Return $2/kg of calf.

score from 3 to 5 requires 26 Mcal energy per day or 11.3 kg (25 lb) of hay per day. In a 100-day period, the cow scoring 5 at calving eats 360 kg (800 lb) more hay. If the hay costs $88 per 1000 kg (2200 lb), then it will cost $32 to change body condition from 3 to 5. The $32 investment will be repaid by an increase in calf weaning weight the first year. The $104 return the second year will be a bonus.

For an investment of $32, the body condition of a cow can be changed from 3 to 5. The return on this investment is $140; the investment is quadrupled.

PREDICTING REPRODUCTIVE PERFORMANCE USING BODY CONDITION

Body condition scores and fetal age can be used to predict pregnancy rate and time of pregnancy the next breeding season. Using these figures, the time of calving and the kg of calf weaned the next year can be predicted. The veterinarian can then help the producer look at alternative strategies for increasing net income as shown in the following example. The information presented in Table 14-12 was collected on a herd of cows at the time cows were checked for pregnancy.

Using the information obtained at pregnancy check and the data presented in Table 14-3, an estimate can be made of the number of cows in heat at various times in the breeding season. The average calving date is estimated, and days from calving to the end of the first 20 days of breeding next spring is calculated for the six groups at pregnancy check time. Then, using data in Table 14-3, an estimate of cows that will be in heat is made. An average conception rate of 60% is used to estimate the number of cows pregnant (Table 14-13).

Table 14-14, the "Cow Worksheet," is a form to help make these calculations.

The expected pregnancy rates vary from 93% in moderate-condition cows calving from 3-15 to 4-4 to 56% in thin cows calving from 4-25 to

Table 14-12 *Results obtained at time of pregnancy check (October 9, 1989)*

Estimated fetal age (days)	Body condition (score)	Number of cows	Expected calving date* 1990	Period
	Thin (2 or 3)			
100-120		100	3-15 to 4-4	1
80-100		100	4-5 to 4-24	2
60-80		100	4-25 to 5-14	3
	Moderate to good (5 or 6)			
100-120		100	3-15 to 4-4	1
80-100		100	4-5 to 4-24	2
60-80		100	4-25 to 5-14	3

*Month-day.

Table 14-13 *Expected pregnancy rates according to body condition and breeding date*

Body condition	Breeding date 1990	Expected calving date (and period) in 1990		
		3-15 to 4-4 (1)	4-5 to 4-24 (2)	4-25 to 5-14 (3)
Thin				
	6-6 to 6-26*	40†	33	20
	6-27 to 7-16	58	53	41
	7-17 to 8-5	65	63	56
Moderate				
	6-6 to 6-26	55	47	27
	6-27 to 7-16	82	74	58
	7-17 to 8-5	93	90	78

*Month-day.
†Percent expected to be pregnant in this 20-day period.
The values in each column are cumulative for cows of similar body condition.

Table 14-14 *Cow worksheet*

Body condition	Calving period	Average calving date*	First 20 days of breeding season (157-177*)						Second 20 days of breeding season (178-197*)			
			Days P P†	No. of cows	In heat		Conception rate, %	No. pregnant.‡	Days P P†	No. of cows	In heat	
					%	No.					%	No.
Thin	1	84	93	100	66	66	60	40	113	100	70	70
Thin	2	104	73	100	55	55	60	33	93	100	66	66
Thin	3	124	53	100	33	33	60	20	73	100	55	55
Moderate	1	84	93	100	92	92	60	55	113	100	100	100
Moderate	2	104	73	100	79	79	60	47	93	100	92	92
Moderate	3	124	53	100	45	45	60	27	73	100	79	79

*Days from beginning of year
†Days PP = Days post partum
‡No. pregnant = Number of cows in heat × % Conception rate ÷ 100
§Number of cows in heat in this period = (No. of cows × % Heat ÷ 100) − No. already pregnant

5-14. Using this information, the kg of calf weaned can be calculated. The ADG made by calves in previous years on the ranch is used in the calculation. In this herd calves have averaged 0.9 kg (2 lb) ADG from birth to weaning. Calves suckling thin cows gain 0.1 kg/day (0.22 lb/day) less than calves suckling cows in moderate body condition. Consequently, calves suckling cows in moderate body condition and weaned at 200 days of age weigh 20 kg (44 lb) more at weaning than calves suckling thin cows. The expected weaning weight of calves born in the first 20 days of the calving season was calculated by determining the number of days (200) from the average birth date (March 25) to weaning (October 11). This was multiplied by ADG (0.9 kg or 2 lb) for calves suckling cows in moderate condition and 0.8 kg (1.8 lb) for calves suckling cows in thin condition to obtain the gain from birth to weaning (180 kg or 397 lb, in moderate-condition cows and 160 kg or 353 lb, in thin cows). The birth weight of 36 kg (80 lb) was then added to obtain the weaning weight (216 kg or 477 lb, for cows in moderate body condition and 196 kg or 433 lb for cows in thin body condition) for calves expected to be born in the first 20 days of the calving season. To obtain weaning weights for calves born in the second 20 days, 18 kg (40 lb) was subtracted for calves suckling cows in moderate body condition and 16 kg (35 lb) for calves suckling cows in thin condition. Additional weight was subtracted to obtain an estimate of the weaning weight of calves born from 40 to 60 days of the calving season.

These estimated weaning weights and expected pregnancy dates were then used to estimate the kg of calf weaned by different groups and gross return per cow (Table 14-15). The gross return on the thin cows was more than $100 less than on moderate-condition cows calving at the same time.

It is noteworthy that the early-calving thin cow returned $67 less than the late-calving cow in moderate body condition.

Determining the economic impact of body condition changes in a beef cow herd is an important management tool. A method for calculating the costs for changing body condition has been shown; consequently, an assessment can be made of whether making the change will pay.

CONCLUSIONS

Body condition of cows can be measured easily and accurately. Body condition has been shown to be related to production efficiency of beef cows. Improving body condition will increase the kg of calf weaned in a beef cow herd by increasing the pregnancy rate, increasing the number of calves born early in the calving season, and increasing the milk production of the cow. Changes that can be made in production can be predicted; costs can be calculated; and, consequently, economic decisions can be made. In many situations, improving body condition has a positive impact on net return.

Table 14-14 *Cow worksheet—cont'd*

| | | | | | | | | Third 20 days of breeding season (198-217*) | | | | |
| | | | | | | In heat | | | | | | |
No. already pregnant	Heat this period§	Con-ception rate, %	No. preg-nant	Days PP	No. of cows	%	No.	No. already pregnant	Heat this period*	Con-ception rate %	No. preg-nant	Total pregnan-cies
40	30	60	18	133	100	70	70	58	12	60	7	65
33	33	60	20	113	100	70	70	53	17	60	10	63
20	36	60	21	93	100	66	66	41	25	60	15	56
55	45	60	27	133	100	100	100	82	18	60	11	93
47	45	60	27	113	100	100	100	74	26	60	16	90
27	52	60	31	93	100	92	92	58	34	60	20	78

*Days from beginning of year
†Days PP = Days post partum
‡No. pregnant = Number of cows in heat × % Conception rate ÷ 100
§Number of cows in heat in this period = (No. of cows × % Heat ÷ 100) − No. already pregnant

Table 14-15 *Expected kg of calf weaned, 1990*

Calved 1990	Expected calving dates, 1991			Total	Dollars per cow ($2/kg)
	3-15 to 4-4	4-5 to 4-24	4-25 to 5-14		
Thin body condition					
3-15 to 4-4					
No calves	40	18	7	65	
Ave wt (kg)	196	180	164	—	
Total wt (kg)	7840	3240	1148	12,228	244
4-5 to 4-24					
No calves	33	20	10	63	
Ave wt (kg)	196	180	164	—	
Total wt (kg)	6468	3600	1640	11,708	234
4-25 to 5-14					
No calves	20	21	15	56	
Ave wt (kg)	196	180	164	—	
Total wt (kg)	3920	3780	2460	10,160	203
Moderate body condition					
3-15 to 4-4					
No calves	55	27	11	93	
Ave wt (kg)	216	198	180	—	
Total wt (kg)	11880	5346	1980	19,206	384
4-5 to 4-24					
No calves	47	27	16	90	
Ave wt (kg)	216	198	180	—	
Total wt (kg)	10152	5346	2880	18,378	378
4-25 to 5-14					
No calves	27	31	20	78	
Ave wt (kg)	216	198	180	—	
Total wt (kg)	5832	6138	3600	15,570	311

REFERENCES

1. Dunn TG: Master's thesis, Lincoln, 1964, University of Nebraska.
2. Dunn TG and others: Sec Proc ASAS 34:56, 1983.
3. Richards MW, Spitzer JC, and Warner MB: J Anim Sci 62:300, 1986.
4. Swingle RS and others: J Anim Sci 48:913, 1979.
5. Whitman RW: Doctoral thesis, Fort Collins, 1975, Colorado State University.
6. Wiltbank JN: Unpublished data.

Pasture Management and Forage-Related Health Problems in Beef Cattle

THOMAS R. KASARI

Grasses are usually the most important plant category of forages consumed by grazing cattle. A knowledge of the physiology of growth of grasses and the ecology of the various species of grasses growing on dryland or irrigated pastures is a necessary first step toward prudent and efficient pasture management. This applies regardless of whether stocker cattle or a cow-calf unit is used to harvest the forage. Coupled with intensively managed improved pastures are several diseases that have a negative impact on the health and well-being of grazing animals. This chapter provides information on optimization of production of grazing beef cattle through management of forage and prevention of naturally occurring diseases associated with forage.

PASTURE MANAGEMENT
Plant physiology and ecology

Leaves, stems, crowns, roots, and seed heads are basic components of most grasses.[7] Grasses are uniquely suited to grazing because the growing point is protected inside the plant for most of the growing season.[2] Each stem has a growing point that develops either more nodes, leaves, or seed heads. A blade of grass originates from a node. Once the leaf has emerged and unfolded, its growth is complete. In the leaves minerals, water, and carbon are converted to plant sugars through the process of photosynthesis. Photosynthesis takes place in each leaf in the presence of chlorophyll and sunlight, which furnishes the energy to drive the reaction. Anything that negatively affects the process of photosynthesis slows or arrests plant growth. Overgrazing is particularly detrimental to the photosynthetic process because cattle preferentially eat the leaves.

Plant roots serve chiefly as an anchor for the plant. Other important root functions include extraction of minerals and water from the soil and storage of reserve supplies of plant nutrients, including carbohydrates from photosynthetic activity. The stem functions as a pathway for movement of raw nutrients and water from the roots up to the leaves. The stem transports excess leaf-derived plant sugars down to the roots. In addition, the inherent rigidity displayed by a particular plant species is due primarily to stem structure.

Because photosynthesis is a biological process, changes in environmental and soil temperature and the availability of water and minerals have an important impact on the efficiency of this process. Consequently, drought, low soil fertility, and extremes in environmental temperature reduce forage production. Compound these factors with wanton overgrazing, and the end result is a decrease in plant vigor and eventual death of grass species. However, if the rancher gains a greater appreciation of the critical periods in the plant life cycle necessary for sustained growth and reproduction, overgrazing will have minimal adverse effects on the plant community in the pasture.

There are three critical periods in the life cycle of a plant when energy must be available to ensure plant survival.[2] The first period is early spring, when new growth is initiated. Energy for growth is provided by carbohydrate stores in the roots. The next critical period is active reproduction from flower stalk to seed. Energy is derived from photosynthesis in upper leaves. The final period is fall regrowth, during which energy comes from stored carbohydrates in the roots. In addition, plants need energy to replace grazed leaves and to withstand drought.

The amount of new spring growth depends on the level of energy stored the previous season. Roots begin growth before leaves.[2] In fact, this process may take place for several weeks prior to any visible growth of grasses. The general pattern of energy or carbohydrate depletion is fairly similar among grass species, but the pattern of storage is not. As much as 75% of the entire supply of stored energy may be needed for a plant to make as little as 10% of the next season's growth. Consequently, energy and food reserves of perennial plants are at their seasonal lows soon after early spring growth starts. Plants must grow beyond these early stages to restore carbohydrate losses. Most perennial grasses store energy up to the time of seed maturity.

Plants grow in different seasons of the year. Cool-season plants experience their principal growth during the cool weather in the spring or late fall; warm-season plants generally experience their principal growth during the frost-free period and develop seed in the summer or early fall. In general, managed spring grazing and grazing after maturity do not greatly affect energy storage, provided that less than 50% of grass is removed.[2] If grazing is stopped at this point and regrowth of lost foliage is allowed, energy storage continues. Problems arise when moisture is limited or grazing pressure is severe and prolonged. Plants are not able to maintain sufficient leaf for extraction of moisture and nutrients. In consequence, energy stores become depleted, and the size of the root system is reduced. This affects the ability of the plant to achieve normal growth the next year. Low energy reserves also weaken the plant's tolerance to cold winter temperatures. Drought resistance is reduced because a continuously closely grazed plant cannot supply its own needs and is starved from lack of an active recharging mechanism.

Grazing must be adjusted to allow for continued reproduction of the plant. Annual plants must produce seed to survive from year to year,[2] or, alternatively, they must be reseeded into a pasture each growing season. Close grazing combined with poor growing conditions reduces active photosynthetic leaf surface, resulting in slower growth and less viable seed. In contrast, perennial plants live from year to year and produce leaves and stems from the same crown. These plants reproduce by seeds, stems, bulbs, and underground rootstocks. Perennials can better survive adverse grazing patterns because seed production is not mandatory for survival.

By recognizing the major grasses in a particular pasture and how they grow, the producer should be able to set the most advantageous time to graze each area and also to predict the level of use the plants will tolerate and still produce well. Pasture managers should determine whether the forage in their pastures is primarily stemmed grasses, stemless grasses, or a combination of these. Stemmed grasses tend to have a high ratio of reproductive to vegetative stems. They also tend to be more robust and productive.[2] They are more numerous in cool growing conditions. Some species have the ability to grow more or less prostrate and can escape grazing. In contrast, most stemless grasses are less susceptible to grazing because their growth points are at or below ground level for most of the growing season, and their leaves emerge from below ground. In stemmed grasses, the first four or five leaves develop in a manner similar to stemless grasses. Thereafter, the internodes of the stem start to elongate and the growing point is lifted, whether it produces seed or not. If the growing point is removed, there will be no more growth on that stem. This cannot happen with stemless grasses because the growing point is too low to be grazed. For the stemmed group, all new growth comes from inactive buds at the base of the grass. If sufficient soil moisture exists, buds may develop, but in dry climates there is insufficient moisture left for much regrowth. Sometimes this factor can be capitalized on; for example, crested wheatgrass may be grazed heavily enough that its growing point will be removed and the new buds will form leaves, not flowers.

Grazing animals, the pasture, associated microorganisms, soil, and climatic conditions in a given area are all part of the grazing ecosystem.[26] The pasture affects grazing animals through the quantity and quality of feed available. Grazing animals have an adverse effect on this ecosystem through defoliation, treading, and trampling of plants during grazing. The adverse effects of treading on the plant community is greatest under wet conditions and least in dry summer months. A positive effect of cattle on the plant community is the return of nutrients to the soil from manure and urine. The beneficial nutrients in excreta include nitrogen, potassium, and phosphorus compounds. The actual amounts furnished depend upon stocking rate, size and age of animals, and the palatability and chemical composition of herbage. About 60% to 70% of excreted nitrogen and 80% to 90% of potassium are voided as urine and is freely available. The remainder, in manure, is more slowly available, depending on the composition and digestibility of herbage. There is usually little phosphorus in urine, and that in manure is mostly slowly available, although this depends on the phosphorus content of the herbage. During any given year, between 4% and 20% of a pasture will be covered by urine, depending on stocking rate and type of livestock. The nitrogen and potassium in urine stimulate pasture growth for approximately 7 months. However, this herbage is usually rejected for about 2 months. Subsequent pasture quality is reduced, and the proportion of clover and shorter grasses, intolerant of shade, declines. Between 1% and 5% of a pasture is covered by manure pats each year, providing nitrogen, potassium, and phosphorus. The plants directly below manure may be temporarily killed, but surrounding pasture growth is stimulated, particularly in the first 2 to 3 months following deposition. Cattle may not consume forage growing in this manure pat for periods up to 18 months. Harrowing pastures breaks up large manure pats and reduces the amount of underutilized forage. Climate has large effects upon total forage production and the seasonal pattern of production. There are smaller effects upon forage quality and plant composition. Soil factors, especially fertility, have large effects on forage production; there are lesser effects on plant composition, seasonal pattern of production, and forage quality.

Forage management is concerned with forage quality and forage quantity.[26] The quality of a forage is a function of palatability and digestibility of the plant. The quantity of forage available to an animal is a function of stocking rate and yield of the pasture. The greatest challenge in grazing management is to juggle quantity and quality to maximize weight gains of cattle. In summary, environmental factors (i.e., climate and soil fertility) have the greatest effect on pasture production and quality. The efficiency with which this forage is then converted into animal products depends more on animal factors, especially stocking rate.

Grazing systems

Stocking rate and grazing management are the two most important management practices affecting herbage production, pattern of production, herbage quality, and botanical composition.[26] Grazing systems and the intensity of grazing or stocking rate can be evaluated by their effect on individual animals (e.g., weaning weight or average daily gains) or the number of animals per unit area (e.g., hectare or acre) of grazing. Grazing systems that utilize low stocking rates show high gains per animal. Heavy stocking rates result in lower body weight gains per animal, but high overall gains per area of pasture grazed.[27,30] At low stocking rates both animal and pasture growth approach their potential maximum rates. High stocking rates severely reduce both herbage production and intake per animal.[30] There is a linear decline in productivity per animal unit as stocking rate increases; however, productivity per unit area usually increases with stocking rate.

The term *stocking rate* refers to the number of animals per area of pasture grazed. There is no universal formula for determining stocking rate for a given pasture. As a starting point, the manager should seek advice from range scientists or county or state extension forage specialists for specific recommendations for the plant species in a given locale. However, in situations where the supply of animal drinking water is a major concern, stocking rate should be set relative to the carrying capacity of the water source(s) and not the area of pasture.[21] This is particularly true of pastures in the arid areas.

Continuous grazing has been the traditional

method of utilizing beef cattle to harvest pastures.[28] A continuous grazing system allows cattle to graze one pasture the entire grazing season.[26] In this situation spot grazing usually occurs, which is characterized by some patches of grass that are closely cropped while in adjoining areas grass plants mature and seedheads form. Obvious inefficiencies in forage utilization and higher infestations of weeds are typical of continuously grazed pastures. Several alternative grazing systems have been developed over the years to optimize forage harvest by livestock and maximize weight gains without adversely affecting grass growth and survival. These grazing systems include deferred, rotational, deferred-rotational, rest-rotational, controlled, forward, and creep grazing.[21,28] The success or failure of each of these grazing systems in maximizing forage harvest without damage to grass growth and survival is dependent on several factors. These include stocking rate, kind and class of animals, the distribution of water relative to the spatial and temporal distribution of plant communities, and terrain constraints such as slope, rockiness, or gullies.[29]

Deferred grazing systems were developed as a means of delaying grazing until the most important plants have set seed, thereby ensuring continued propagation of the plant community.[28] Nutritional value declines as plants mature, so deferred grazing is not usually an annual event, but rather is practiced every few years.

The rotational grazing system involves the movement of cattle from one pasture to another during the grazing season. The number of pastures utilized is quite variable. An example would be three pastures that receive up to several weeks grazing by animals in rotation. This grazing system was based on the assumptions that large numbers of animals make a more uniform use of forage and that a rest from grazing is beneficial to the plant. However, the pasture must support a greater number of animals during the shorter time when it is grazed.

A controlled grazing system is a more intensively applied form of rotational grazing. It utilizes a very high stocking rate (e.g., 70 to 120 animals per hectare or 30 to 50 animal units per acre) on small areas of pasture (e.g., 20 to 40 paddocks, each 0.4 hectares or 1 acre in size) for short grazing periods (e.g., 2 or 3 days). The goal of this grazing system is to increase total animal gain per unit area through a greater utilization of forage. Individual animal gain may be less than with other grazing systems. Because the density of animals is higher, however, total weight of beef produced is greater per unit of land. Two recent variations in controlled grazing are forward grazing and creep grazing.[21] Forward grazing is quite applicable to cow-calf operations. The principle is to have the class of cattle that will benefit most from highest-quality forage do the initial grazing. An example would be allowing thin cows nursing young calves to perform the first grazing. Following this group, another class of cattle is introduced to graze remaining forage. Creep grazing involves providing an area of a pasture containing high-quality forage for the exclusive use of calves. Because a calf is a very selective grazer, this allows greater forage consumption than grazing calves with the cow herd.

Deferred-rotation grazing involves deferment of grazing on one part of the range or pasture during 1 or more years. By rotation, other areas are successively given the benefit of deferment until all have been deferred.[28] It is desirable under this system to avoid a change in rotational order in less than 2 years. This allows seeds produced in the first year to germinate in the second year. The young plants are given protection from grazing while they are becoming established. The rest-rotation grazing system rests part of the range for an entire year. This system is similar to deferred-rotation grazing, differing mainly in a longer rest period and heavier use of the grazed portion, because, unlike deferred-rotation grazing, the rested portion is not grazed at all. This is a popular grazing system where seasonal grazing is practiced and cool-season grasses (e.g., tall fescue) make up most of the vegetation.

FORAGE-ASSOCIATED NONINFECTIOUS DISEASES IN BEEF CATTLE

Veterinarians should be cognizant of several noninfectious diseases that frequently occur in grazing beef cattle. These conditions include hypomagnesemic tetany, ruminal tympany, nitrate poisoning, fescue toxicity, internal parasitism, and *Brassica* spp.–related disease syndromes, including hemolytic anemia, polioencephalomalacia, and pulmonary emphysema.

Hypomagnesemic tetany

Hypomagnesemic tetany (HT) is a metabolic disease that usually afflicts cattle grazing on improved pastures (see Chapter 3). Magnesium imbalance is unlikely to occur in cattle that graze native, unimproved pastures. Veterinarians should closely monitor the grazing activity of cattle on rapidly growing, cool-season grasses and winter cereal grain pastures (e.g., wheat, oats, or rye) for signs consistent with HT. These pastures predispose cattle to HT and are particularly dangerous when forage contains 0.2% magnesium or less concomitant with a high nitrogen concentration (more than 4% nitrogen or 25% crude protein) and high potassium concentration (more than 3%).[25] Increasing concentration of plant potassium also interferes with calcium absorption. Hypocalcemia often accompanies hypomagnesemia in affected cattle.

The plant community that makes up the forage has a bearing on the magnesium concentration of harvested forage. Legume and mixed pastures have a much higher concentration of magnesium than those without legumes.[3] Magnesium, potassium, nitrogen, and calcium concentrations should be routinely determined on forage samples prior to and during grazing to assess the potential for HT in cattle grazing on high-risk pastures. Although not routinely determined on a forage analysis, high forage organic acid content (above 1%), particularly in cereal grains, also predispose grazing livestock to HT. The ratio of potassium to the sum of calcium and magnesium in grazed forage appears to be useful in predicting the onset of HT. A ratio of 2.2:1 is associated with a marked increase in cases of HT.[8]

Animal factors also play an important role in the development of HT. Older cows, particularly those individuals older than 6 years, that are in advanced pregnancy or within 30 days postpartum are at greatest risk.[3] Some breeds appear to be more resistant to development of hypomagnesemia. Brahman and Brahman-influenced breeds of cattle seem to have the greatest capability for magnesium absorption and retention and, hence, reduced susceptibility to HT.

In cattle displaying clinical signs consistent with HT, determination of the magnesium status of that animal should be made by utilizing blood, cerebrospinal fluid, or urine. Although cerebrospinal fluid magnesium concentration below 0.6 mmol/L (1.45 mg/dl) is a more reliable indicator of HT, plasma or serum magnesium is more often used because of ease of sampling. Normal laboratory values for magnesium in either serum or plasma of cattle have been reported as 0.5 to 1.44 mmol/L (1.2 to 3.5 mg/dl)[3] and 0.7 to 0.9 mmol/L (1.8 to 2.3 mg/dl).[14] Urine magnesium concentrations less than 0.4 mmol/L (1.0 mg/dl) are presumptive evidence of low magnesium status.

When HT is the suspected cause of death, vitreous humor and cerebrospinal fluid can be collected for determination of magnesium concentration. Cerebrospinal fluid should be collected within 12 hours of death.[3] Vitreous humor can be collected from the eyes for up to 48 hours after death, provided the environmental temperature did not exceed 23° C (74° F) after 24 hours.[16] Urine should be collected within 24 hours of the death.

The magnesium requirement of cows grazing pastures is 0.20% on a dry matter basis.[3] Unfortunately, most rapidly growing spring grasses usually contain 0.10% to 0.12% magnesium. It is prudent to provide supplemental magnesium to cattle during periods of the year when the risk of this metabolic condition is greatest.[3] Supplemental magnesium salts must be consumed daily because cattle are unable to store this macroelement in a readily metabolizable form.

Several methods of delivering supplemental magnesium to cattle are available to the veterinarian. These include free-choice salt-mineral mixes in a loose or block form, inclusion of magnesium salts directly to a supplemented feed source, magnesium fertilization of the soil, dusting of standing forage with magnesium salts, and addition of magnesium to a water supply. The most popular method of magnesium supplementation appears to be use of salt-mineral mixes. These mixes should contain at least 10% magnesium to prevent tetany. However, because of poor palatability of all magnesium salts, intake can be erratic if offered free choice. Daily intake should be 50 g/head/day. If this consumption is not reached, the mixture should be changed to include molasses or other concentrate that increases palatability.[11]

Several magnesium compounds have been utilized as a source of this mineral, including MgO, MgOH, and Mg citrate. The bioavailability is greater with Mg citrate and MgOH. However, MgO is most often utilized as the source of supplemental magnesium because several feed grades exist, whereas MgOH and Mg citrate have to be purchased as reagents. Magnesium salts must be fed daily. Supplementation should begin at least 2 weeks prior to anticipated grazing in an effort to stabilize intake of magnesium. Total dry matter intakes >1.0% magnesium should be avoided because diarrhea and poor performance of cattle are real possibilities.[9] Application of calcined magnesite (50 to 70 kg per hectare or 45 to 65 lb per acre) as foliar dust to pastures prior to grazing is a practical method of preventing hypomagnesemia. A long-term solution is increasing soil magnesium through fertilization.[11]

Ruminal tympany

Ruminal tympany is overdistension of the rumen and reticulum with the gases of fermentation, either in the form of a persistent foam mixed with the rumen contents (frothy bloat) or in the form of free gas separated from the ingesta (free gas bloat). Frothy bloat is referred to as *primary* bloat, is dietary in origin, and occurs most frequently in cattle on legume pasture and in feedlot cattle on high-level grain diets. Free gas bloat is referred to as *secondary* bloat and is usually due to failure of eructation of free gas because of a physical interference with eructation. The discussion here is limited to frothy bloat.

Alfalfa, red clover, white clover, lucerne clover, and subterranean clover pastures as well as hay from these plant species are the most important predisposing causes of frothy bloat.[12] However, frothy bloat is also a common problem among stocker cattle grazing winter wheat pastures in the south-central United States. The principal bloat-causing legumes are rapidly digested by rumen microorganisms, whereas bloat-safe legumes (e.g., bird's-foot trefoil and arrowleaf clover) are digested more slowly.[6,12]

The rapidity of digestion is a function of the number of rumen microorganisms colonizing the leaf surface of these legumes. The number of rumen microorganisms on the leaf surface is dictated, in turn, by the amount of nutrients that leach from the stomata of the plant and the resistance to maceration (i.e., softening and disruption of fresh herbage) during rumen microorganism penetration into plant structures.

A number of other substances in bloatagenic leguminous forage, including saponins, soluble forage proteins, and small feed particles, have been investigated as to their ability to create frothiness of rumen fluid.[12,13] The long-held belief that saponins were the primary initiating factor in this type of bloat has been dispelled by recent research. In contrast, soluble forage proteins, namely Fraction I protein and Fraction II protein, can stabilize foam. Soluble protein is not responsible for an immediate onset of frothiness in rumen fluid. It may contribute to frothiness as part of a complex with other substances such as modified salivary mucoproteins, high-molecular-weight salivary mucin, and lipid.

Small feed particles, specifically chloroplast membrane fragments from legumes, along with adherent rumen microorganisms, are now considered the primary factor responsible for frothiness of rumen fluid.[12] The froth in rumen contents is not a true foam but, rather, a dispersion of gas and feed particles in liquid. There is a liquid lamellae between these gas bubbles and fragments of chloroplast membranes dispersed in the liquid. Coalescence of gas bubbles occurs naturally when these liquid lamellae drain away. Frothiness occurs when a stable dispersion of small feed particles prevents drainage of rumen fluid between bubbles of gas. Alternatively, frothiness may develop when these particles complex with soluble proteins.

Attempts have been made to breed nonbloating forage legumes. Specific plant components that reduce the likelihood of bloat include condensed tannins, epicuticular waxes, leaf structure that increases resistance of cells to rupture, low concentrations of soluble protein, and a low ratio of potassium to sodium.[6] Condensed tannins appear to prevent bloat by acting as protein precipitants or by inhibiting microbial invasion and digestion. Epicuticular waxes are lipids that possess surface-active properties that may function as antifoaming agents or reduce microbial penetration into plants during digestion. Increased cell wall and leaf tissue strength are characteristic of non–bloat-producing plants. Reticular vein structure and the thickness of mesophyll and epidermal cell walls are critical. Thick, hard cell walls are more important in resisting cellulolytic enzymes than dense epicuticular wax. Despite a considerable effort directed toward producing a non-bloating forage legume, there have been no dramatic successes to date.

Nitrogen content and initial rate of digestion are related to day-to-day occurrence of bloat.[12] Maturity of forage is the major plant factor affecting bloat incidence; young, rapidly growing plants are more apt to precipitate an episode of bloat. Environmental variables, principally ambient temperature, may affect bloat occurrence by accelerating or retarding stand maturation.

Cattle vary in their susceptibility to frothy bloat. Susceptibility to bloat has a medium-to-high degree of heritability.[20] Cattle that have a propensity to bloat have a larger rumen volume and a difference in protein composition of saliva compared with bloat-resistant cattle. Various genetic markers have been investigated to differentiate between cattle with high and low levels of susceptibility to bloat. Most promising is band 4 protein of low-molecular-weight proteins of bovine saliva.[13] Monoclonal antibodies have been generated to this protein and a suitable assay for quantitating band 4 protein in bovine saliva has been developed. If further testing confirms the relationship between band 4 protein and bloat susceptibility, then this genetic probe could be utilized to distinguish between more- and less-susceptible cattle.

The approach to control of frothy bloat includes genetic selection of bloat-resistant forages and animals, pasture management, and chemical intervention practices in cattle that are consuming bloat-producing forages. However, the only satisfactory method available for the prevention of bloat in cattle on these pastures is the administration of antifoaming agents. The options available for delivery of these substances to cattle include bloat blocks, pasture spraying, direct injection of the agent into individual cattle, water treatment, and oral drenching.[10] The ultimate choice is dependent upon the cost of the product, the method and ease of administration, and the absence of side effects.

The basic chemical choices of antifoaming agents are either oils and fats or synthetic non-ionic surfactants.[3,10] More recently, polyether antibiotics such as lasalocid and monensin have been shown to reduce the incidence of legume and grain bloat.[1] Vegetable oils, mineral oil, and emulsified tallow are all effective as antifoaming agents (approximately 120 g or 4 oz per head per day). However, the duration of action is limited to only a few hours per day. These substances are best combined in concentrates or as a 2% emulsion in drinking water. The nonionic surfactants have enjoyed a greater utilization than the fats and oils. The best known of these compounds is poloxalene, a polyoxythylene polyoxypropylene block polymer. Daily intakes of 10 to 20 g/head/day are recommended. This should commence several weeks prior to grazing bloat-producing pastures and continue through the grazing period. A single treatment prevents bloat for approximately 12 hours. Poloxalene can be top-dressed onto feed, added to concentrates, mineral blocks, or mixed with the pluronic L64 in drinking water. Because of low palatability, users should be cognizant that animal intake can be erratic. Dry hay or grass pasture should also be available to these cattle when they are grazing legume pastures. It is prudent not to introduce cattle to a legume pasture immediately after a rain or when a heavy dew is on the forage.

Nitrate poisoning

Common plants known to accumulate nitrates to dangerous levels are sweet clover, Johnsongrass, oats, rape, alfalfa, rye, Sudan grass, wheat, and corn.[5] The abnormal accumulation of nitrate in plants is influenced by a number of factors, the most important being content and form of nitrogen in the soil. Soils high in nitrate content or ammonia levels supply nitrate more readily to plants. Soil conditions that favor nitrate uptake include adequate moisture, acid soils, low molybdenum, sulfur deficiency, phosphorus deficiency, low temperature (less than 13° C or 55° F), and soil aeration. Drought conditions, decreased light (light is required to maintain activity of the enzyme nitrate reductase), and herbicide treatment with phenoxyacetic herbicides (the 2, 4-D herbicides are plant hormones that favor increased growth rate and nitrate accumulation in early stages) are also factors in the incidence of nitrate poisoning in grazing cattle.

Nitrates accumulate in vegetative tissue and not in fruits or grain. Levels are most elevated just prior to flowering and drop off rapidly after pollination and setting of grain or fruit. Acute poisoning may be expected when forage nitrates exceed 1.0% nitrate (dry matter basis) or 1500 ppm nitrate in water. The nitrate ion itself is not particularly toxic. However, the nitrite ion, the reduced form of nitrate, is readily absorbed and quite toxic. Ruminants readily reduce nitrate to nitrite. The nitrite ion oxidizes hemoglobin to methemoglobin, which is unable to carry oxygen for transport to tissues.

When nitrate poisoning is suspected in a group of cattle, nitrate and nitrite levels in the rumen or stomach contents, plasma, serum, urine, forage, and water may be examined. Because nitrate levels in body fluids are rather labile, it is probably best to take representative samples from the offending forage. General recommendations are to obtain four to five samples from different areas of the field by clipping the forage down to about the same level that it would ordinarily be grazed.

Treatment of individual animals is aimed at converting methemoglobin (Fe^{3+}) to hemoglobin (Fe^{2+}) by using methylene blue. The suggested dose is 2 mg/kg administered intravenously as a 2% to 4% solution. The response should be immediate, but treatment may need to be repeated because absorption of nitrite from a full rumen can continue. For forage nitrite poisoning, purging with saline cathartics and control of bacterial nitrate reduction with intraruminal antibiotics and 12 to 20 L (3 to 5 gal) of cold water may be beneficial.

The potential for nitrate poisoning is greatest when plants are simultaneously affected by a number of adverse factors. Predisposing factors include long periods of overcast weather with cold temperatures, drought, and damage to plant material, such as occurs with trampling. In situations such as these, provision of additional roughage and concentrate will reduce the incidence of nitrate poisoning. Additional feeds dilute nitrate and assist in the metabolism of nitrate to ammonia and ultimately microbial protein.

Fescue toxicity

Fescue toxicity in cattle includes three distinct entities, fescue foot, fat necrosis, and summer syndrome.[19] These syndromes occur in cattle consuming tall fescue *(Festuca arundinacea)* grass, a high-yield, cool-season perennial, infected with an endophytic fungus *(Acremonium coenophialum)*. Fescue seed, hay, soilage, silage, and pasture are all toxic. The endophyte affects neither growth nor outward appearance of the grass and cannot be detected macroscopically. This fungus is apparently present in more than 95% of tall fescue swards. The endophyte appears to be transmitted only through infected seed. The toxic principle responsible for decreased performance in cattle grazing on infected grass may be due to three ergopeptide alkaloids; ergovaline, ergosine, and ergonine.[23]

Cattle consuming high–endophyte-infected fescue, compared with low–endophyte-infected fescue, have decreases in feed intake (5% to 10%), weight gains (5% to 10%), milk yield, pregnancy rates, and animal production per unit area of land. Individually affected animals exhibit tachypnea, hyperthermia, increased salivation, increased water consumption, increased urine output, harsh hair coat, nervousness, and increased time interval spent in the shade. Hot, humid weather exacerbates these effects. Data suggest that with growing animals, for every 10% increase in plants infected, there is a reduction in daily gain of about 0.05 kg (0.1 lb). Animals removed from offending pastures generally require up to 6 weeks for clinical abnormalities to disappear.

When this disease is suspected as a cause of decreased animal performance, the following procedure is suggested for isolation of the fungus: A representative sample of plants from the pasture must be taken. Fields up to 2.5 hectares (6 acres) should have a minimum of 12 samples/hectare (5 samples/acre).[17] These samples should be collected from sites evenly distributed throughout the field. Larger fields, seeded at one time, should have 30 samples taken for a representative sample of the entire area. Replacing a high-endophyte fescue sward with a new endophyte-free variety can greatly improve animal performance. A field to be replanted should not be allowed to go to seed in the year before reestablishment. Seed head formation can be prevented by heavy grazing, clipping, or application of chemicals. Preventing seed formation ensures that any seed remaining in the soil will be more than 1 year old at the time of the new seeding. The endophyte eventually dies in seed, usually within 1 year. Therefore, most volunteer plants would be endophyte free. On steep land, where crop rotation and a prepared seedbed are not feasible because of erosion hazard, chemical kill of infected stands, followed by direct drilling, is a viable alternative.

Internal parasitism

Cattle managed on intensively grazed permanent pastures and pastures seeded with winter grains can suffer from parasitism. The strongylid group of nematodes, particularly *Haemonchus placei* and *Ostertagia ostertagi*, are the most economically important.[4,18,32] Feeder calves in warm climates are most susceptible to *H. placei*. Expected clinical signs are blood loss anemia, weight loss, lack of thriftiness, and weakness. In adult cattle, *O. ostertagi* is more important because mature cattle are usually resistant to *H. placei*. It is not unusual to encounter a 0.25 kg (0.5 lb) or more reduction in daily weight gain in subclinically affected cattle.

The period of greatest passage of eggs onto pastures varies with the geographic locale in North America. In the southern regions of the United States, large numbers of over-wintered larvae infect cattle during the spring. As temperatures rise and precipitation declines (early summer season), the number of larvae decreases dramatically while egg elimination in the feces increases. At midsummer, the level of pasture contamination with larvae has reached a minimum. As environmental temperatures begin to decline as fall approaches and rainfall begins anew, eggs in feces develop to infective third-stage larvae and reinfect cattle. In northern climates the period of pasture contamination is shifted; larval pasture contamination is maximal during the spring and summer months and minimal during winter.

In stocker cattle (weaned, growing calves or yearlings) operations utilizing annual grasses, the possibility of preexisting parasitic contamination of pasture is reduced. A minimally parasitized pasture is one that has not had grazing cattle for 6 months and has been completely tilled and reseeded with annual grasses. The recommended deworming program is to treat all cattle upon arrival and at least 24 hours prior to turnout with any of the approved commercial anthelmintics. When *O. ostertagi* is of concern, the anthelmintic should have good efficacy against the inhibited larval forms of this nematode. This single anthelmintic treatment should be sufficient for the grazing period. When it is questionable whether pastures are "parasite-safe," all stocker cattle should be dewormed a second time 3 to 4 weeks after the start of the grazing period. When the grazing period persists for 120 days or longer, a third deworming should be instituted.

In farms and ranches that have spring-calving cows, cows and all calves older than 45 days can be dewormed at least 24 hours prior to movement onto new pastures. At midsummer, all cows and calves should be dewormed. In the southern United States, this date is during June and July. At weaning (October or November) all animals should be dewormed again and moved to new pastures. Fall-calving cows are dewormed after calving and moved to fresh autumn pasture. Both cows and calves are dewormed at weaning and moved to separate pastures. Cows and previously weaned fall calves are dewormed again before being moved to pasture in the spring.

Other recommendations for internal parasite control include avoiding overcrowding at weaning time. If possible, a pasture should be grazed for only one period in a calendar year. Following weaned calves with suckling calves on a pasture should be avoided. Harrowing idle pastures during hot, dry periods facilitates desiccation of parasite eggs.

Brassica spp. disease syndromes

Brassica species include the wild and cultivated mustards, charlock, kale, rape, brussels sprouts, cabbage, cauliflower, broccoli, kohlrabi, rutabaga, and turnip.[15] Kale, rape, and, more recently, turnips are used as forage for a variety of classes of cattle. Specific health-related problems encountered include acute pulmonary emphysema, polioencephalomalacia, hemolytic anemia, and gastrointestinal atony. In addition, bloat, nitrate poisoning, infertility, and goiter have been documented. Pulmonary emphysema, hemolytic anemia, and polioencephalomalacia are associated with ingestion of kale, rape, and turnips.[15,28] Complete gastrointestinal atony appears to be a specific disease entity of rape poisoning.[15]

The exact cause of *Brassica* spp.–induced pulmonary emphysema is not precisely known. Onset of respiratory distress usually occurs within 6 to 9 days following introduction of cattle onto pastures.[28] Morbidity is variable but usually under 15%. Case fatality rates are generally high (50% to 80%).[28] The clinical course is short; most fatalities occur within 2 days of onset of signs. It is a common practice to remove cattle from offending pastures at the first signs of illness. However, this practice is of unproved efficacy. Producers have indicated that if groups of cattle experiencing pulmonary emphysema are allowed to continue grazing turnips, no new cases occur 1 week after the initial onset. Because respiratory distress appears to be self-limiting in mild to moderately affected animals, treatment is recommended for severe cases only and consists of 0.4 to 1.0 mg/kg (0.2 to 0.5 mg/lb) furosemide given intravenously or intramuscularly every 12 hours, with restriction of drinking water.[28] Recently, antiprostaglandin therapy (2.2 mg/kg or 1 mg/lb of flunixin meglumine intravenously once a day), initiated at the onset of illness, has been shown to promptly alleviate clinical signs and lung pathology of experimentally induced acute bovine pulmonary emphysema and edema.[24] Whether this drug would be effective in cattle with overt disease is unknown.

A nervous syndrome referred to as rape blindness has been identified in cattle since the early 1940s.[15] Polioencephalomalacia is the suspected cause of this clinical syndrome based upon the lesions in the brains of affected animals. Thiamin deficiency and high-sulfate feed or water are associated with this disease. The disease occurs sporadically, usually during the first 10 days on pasture but sometimes anytime during the grazing period, and it seldom affects more than one or two animals in a given pasture. Although reports from producers and veterinarians indicate that polioencephalomalacia on these pastures is responsive to thiamin treatment, the necessary laboratory tests to confirm thiamin deficiency have not been performed. *Brassica* spp. contain high amounts of sulfur, which are released in considerable quantities as they are digested in the rumen. Sulfur can degrade thiamin in the rumen. Thiamin hydrochloride, administered intravenously at 10 mg/kg (5 mg/lb) every 3 hours for a total of five times, is the treatment of choice.[31] Improvement usually begins within 6 to 12 hours if therapy is successful. Ancillary therapy includes dexamethasone (1 mg/kg, or 0.5 mg/lb) to reduce cerebral edema. The numbers of new cases during an outbreak might be minimized by providing a liquid supplement containing high levels of thiamin.

Hemolytic anemia is evident after 1 to 3 weeks of feeding on the various *Brassica* species. The incidence of clinical anemia appears to be more severe in high-producing or recently calved cows. Clinical signs include icterus, occasional hemoglobinuria, anorexia, and fever. The primary toxic factor in *Brassica* spp. responsible for hemolysis is S-methyl cysteine sulfoxide, which is converted into dimethyl disulfide, a powerful oxidant. Heinz body anemia is produced as a result of this oxidant's effect on hemoglobin. Treatment is purely symptomatic and may include blood transfusions in the severely affected individual. Immediate removal of affected cattle from offending pastures and provision of other feed sources usually result in complete recovery in 4 to 6 weeks.

General preventative measures for *Brassica* spp. poisoning include either dietary management or medical measures. These should be instituted approximately 2 weeks prior to, and continued 2 weeks after, introduction of cattle onto pastures.[31] An abrupt change in diet should be avoided. If cattle have been on a low plane of nutrition, they should be fed appropriate amounts of protein and energy for 2 weeks before they are permitted access to *Brassica* spp. pastures. This approach allows ruminal microorganisms to adapt to normal amounts of energy and protein. It decreases the risk of pulmonary emphysema from moving cattle from a poor feed source to lush pasture. Cattle should always have a good fill just before they enter a *Brassica* field. An adjustment period of limited grazing for only a few hours a day for a week to 10 days is recommended to reduce risk of disease. However, this is often difficult to accomplish. Another approach is to supplement cattle with dry roughage to dilute ingested fresh herbage and help maintain a normal balance of ruminal microorganisms. This is particularly important when grazing turnips. At least 1 kg (2 to 3 lb) per day of grain hay, grass hay, or straw or else access to an adjacent dry pasture is recommended. In addition, isolating portions of a field with an electric fence or other suitable barriers and holding cattle until they have consumed all available forage before moving them to a fresh section can be an alternative. This reduces the risk of disease from consumption of large quantities of *Brassica* herbage.

Although clinical trials have not proven their worthiness on *Brassica* pastures, the use of polyether antibiotics to prevent the onset of pulmonary emphysema may be of value. These antibiotics (lasalocid and monensin) prevent the onset of acute bovine pulmonary emphysema and edema by blocking the conversion of L-tryptophan to 3-methylindole by ruminal bacteria.[22] Polyether antibiotics can be fed in pellets with either an energy or a protein supplement formulated to provide 200 mg of monensin or lasalocid per head per day. Treatment should commence 1 week before and continue 1 week after introduction of cattle to pastures. Inadequate intake of these polyether antibiotics usually occurs when a lick block is utilized. Thiamin administration to cattle fed *Brassica* plants may prevent polioencephalomalacia. The recommended daily intake of this B vitamin is 5 to 10 mg/kg (2-5 mg/lb) of dry feed. A thiamin injection immediately before introduction of animals to *Brassica* plants may also take cattle safely through the 2-week high-risk period of polioencephalomalacia.

REFERENCES

1. Bartley EE and others: J Anim Sci 56:1400, 1983.
2. Bedell TE: Cow-calf management guide, cattleman's library, 1981, pp 510.1-510.4.
3. Blood DC and Radostits OM: Veterinary medicine, ed 7, London, 1989, Bailliere Tindall.
4. Brunson RV: Vet Parasitol 6:185, 1980.
5. Buck WB and others: Clinical, diagnostic veterinary toxicology, Dubuque, Iowa, 1973, Kendall-Hunt Publishing Co, p 55-60.
6. Caradus JR: Breeding non-bloating forage legumes, Proc Dairy Cattle Soc NZ Vet Assoc, 1987, p 31.
7. Dorsett DJ: Beef cattle sci handbook 23:204, 1989.
8. Fontenot JP and others: Fed Proc 32:1925, 1973.
9. Gentry RP and others: J Dairy Sci 61:1750, 1978.
10. Heyes I: Proc Dairy Cattle Soc NZ Vet Assoc, 1987, p 47.
11. Hoffsis GF and others: Comp Cont Educ Pract Vet 11:519, 1989.
12. Howarth RE and others: Control of digestion, metabolism in ruminants, Englewood Cliffs, NJ, 1986, Prentice-Hall, pp 516-527.
13. Jones WT: Proc Dairy Cattle Soc NZ Vet Assoc 1987, p 17.
14. Kaneko JJ: Clinical Biochemistry of Domestic Animals, ed 3, New York, 1980, Academic Press, pp 792-795.
15. Kingsbury JM: Poisonous plants of the United States and Canada, Englewood Cliffs, NJ, 1964, Prentice-Hall, pp 158-171.
16. Lincoln SD and Lane VM: Am J Vet Res 46:160, 1985.
17. Martin T and Edwards WC: Vet Med :1162, 1986.
18. Michel JF: Adv Parasitol 14:355, 1976.
19. Miksch D and Lacefield G: Bovine Pract 22:187, 1987.
20. Morris CA: Proc Dairy Cattle Soc NZ Vet Assoc 1987, p 39.
21. Morrow RE: Beef 23:25, 1987.
22. Nocerini MR and others: J Anim Sci 60:232, 1984.
23. Porter JK and others: J Agricul Food Chem 33:34, 1985.
24. Selman IE and others: Bovine Pract 20:124, 1985.
25. Smith RA and Edwards WC: Vet Clin North Am (Food Anim Pract) 4:365, 1988.
26. Snaydon RW: Grazing animals, Amsterdam, 1981, Elsevier, pp 13-31.
27. Stoddart LA and others: Range management, ed 3, New York, 1975, McGraw-Hill, pp 256-289.
28. Stoddart LA and others: Range management, ed 3, New York, 1975, McGraw-Hill, pp 290-314.
29. Stuth JW: Beef cattle sci handbook 23:225, 1989.
30. Vickery PJ: Grazing animals, Amsterdam, 1981, Elsevier, pp 55-77.
31. Wikse SE and others: Comp Cont Educ Prac Vet 9:F112, 1987.
32. Williams JC: Vet Clin North Am (Food Anim Pract) 5:183, 1980.

16

Feeding Feedlot Cattle for Optimal Production

RICHARD A. ZINN, JAMES W. OLTJEN

THE PROBLEM IS THE PROBLEM

The term *nutrition* refers to the mutual relationships between intake, digestion, and assimilation of nutrients to achieve some productive function. Applied to feedlot cattle, the primary productive functions are growth and fattening, and within the practical limits of the modern industry, the standards for achieving those purposes are clearly detailed.[5] Thus, it would appear that the task for the feedlot nutritionist is straightforward and simple: Drawing from experience on limitations of individual feed ingredients and assisted by one of the many computer programs available, diets are balanced for least-cost gain.

Unfortunately the bigger challenge for the nutritionist is not in such straightforward solutions, but rather in determining the primary concerns of a particular feedlot operation, defining the problems, and then establishing clear-cut production goals. Although it can be said of all feedlot operations that primary objectives are profit driven, the criteria for achieving and appraising a perceived "optimal" may be quite varied. One feedlot may attract customers on the basis of feed conversion (feed/gain), whereas another functions on the basis of feed cost of gain. Because the major source of variation in net returns among pen lots of cattle is usually associated with the buy-sell margin, some feedlots may place greater emphasis on timing and marketing, while others stress animal health and consistent rather than maximal growth performance.

Once growth-performance goals for a feedlot have been established, the next step for the nutritionist is to tailor a feed management program to achieve those goals. There are six key factors in accomplishing this: (1) the ability to assess the growth potential of feeder cattle by type and classification, (2) an appreciation of specific nutrient requirements and constraints, (3) a familiarity with the feeding values of the variety of feedstuffs that may be available for use in diet formulations, (4) the development of feeding strategies, (5) a knowledge of the practical limits of a particular feedlot for feed mixing and delivery, and (6) monitoring the feeding program.

TARGETING FEEDLOT GROWTH AND PERFORMANCE

Growth and performance of feeder cattle can vary tremendously. Intrinsically, the most important factors involved are sex, genotype, and previous environmental influences.[1] The precision of nutrition and management programs and, ultimately, the profitability of a cattle-feeding venture depend on careful assessment of these three factors.

Sex

Although most feeder cattle are either steers or heifers, a number of producers have found market niches for fast-gaining young bulls. In general, bulls exhibit daily gains about 10% to 15% greater than steers, and steers about 10% to 15% greater than heifers. Feed intake is greater (10%) for steers than heifers, resulting in roughly a 5% improvement in efficiency of gain. However, heifers are ready to market at about 20% lighter weights than steers, due to their increased gain of fat. At the other extreme, bulls resist fattening unless grown on high-energy diets; therefore, they must be fed to weights 20% greater than those of steers if similar carcass fat is to be achieved.[7] In practice, however, bulls are not usually allowed to reach that degree of finish. Adverse social interactions among pen lots of bulls can often be overcome if they are placed on feed as calves and marketed for slaughter as yearlings.

Genotype

Most feeder calves are selected primarily on a weight basis. However, weight measures alone can be very misleading because they do not adequately reflect differences among cattle in size and body composition. The USDA has developed a grading system that can be used in conjunction with live weight to assist selection of feeder calves. The system classifies cattle based on frame size and muscle thickness. Frame size (small, medium, and large) is a measure of the weight at which an animal will attain a specified level of fatness (grade low to choice) and is related to mature height. Muscle thickness (graded 1 to 3, with 1 the thickest) measures carcass muscle-to-bone ratio and is related to conformation, particularly of the rear quarters. Most of the variation in growth rate within cattle type is related to frame size.[10] Larger- and smaller-frame cattle gain about 10% faster or slower, respectively, than medium-frame cattle. There is no relationship between muscle thickness and growth rate. In general, steers rated 1 in muscle thickness are those of beef breeds, whereas those rated 3 are predominantly of dairy, longhorn, and Brahman breeding. Irrespective of USDA class, several breeds (Angus, Holstein, Jersey) are known for their ability to deposit intramuscular fat (marbling) and may reach the choice carcass grade at lower levels of total body fatness than other breeds of similar frame size. Others (longhorn, Brahman) may reach choice later. Heterogeneity of breed types and age uncertainties among feeder cattle make implementation of the feeder cattle grading system difficult, particularly in lightweight feeder calves.

Cattle background

Rate of feed intake and body weight gain depends on animal fleshiness (i.e., previous nutrition and rate of gain). In the feedlot, feed intake soon plateaus for most pens of cattle (Fig. 16-1). However, feed intake generally increases with the initial weight at which the cattle were first fed the high-energy ration. Thus, cattle placed on feed as yearlings have greater intakes at a given weight than if placed on feed as calves (Fig. 16-2). Cattle fed high-energy diets after a period of restricted feeding (either by feeding low-energy forage diets or limiting intake of higher-energy diets) exhibit compensatory growth. These animals eat more than those of similar weight grown without a restricted period.[2] Also, less energy is required for maintenance, and efficiency of energy use for weight gain is improved for a short time when compensation occurs. This is due both to carryover effects of decreased metabolic rate and to an initially higher proportion of lean tissue gain (p. 14) in compensating animals.

As mentioned previously, cattle are brought into feedlots and fed high-energy rations at different ages after different feeding management regimes. Usually animals entering the feedlot have been fed some type of forage-based diet, with somewhat restricted weight gains. The

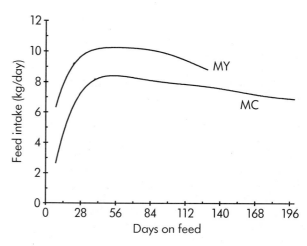

Fig. 16-1 Influence of days on feed on dry matter intake of medium-frame calves (MC) and yearlings (MY).

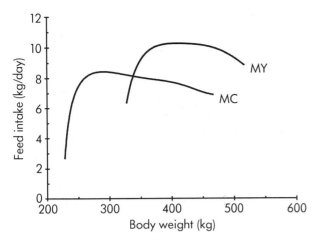

Fig. 16-2 Influence of body weight on dry matter intake of medium-frame calves (MC) and yearlings (MY).

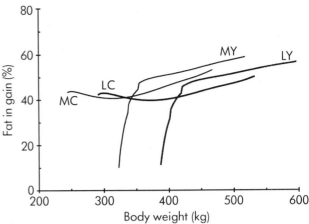

Fig. 16-3 Influence of body weight on percentage fat in live weight gain of medium-frame calves (MC), large-frame calves (LC), medium-frame yearlings (MY), and large-frame yearlings (LY).

compensatory growth seen when cattle are placed in the feedlot is usually of economic value. A notable exception is the feeding of young, recently weaned calves. These animals are generally in good body condition and exhibit little compensatory growth when fed a high-energy diet. Daily weight gain of cattle placed on feed as calves is somewhat less than gains of cattle fed as yearlings. They usually reach "chemical" maturity (28% fat in the carcass) at lighter weights than cattle placed on feed as yearling. Consequently, calves are usually marketed for slaughter at a lighter weight than yearlings.

A question often asked by cattle feeders is whether energy intake affects body composition (fat) of the growing animal. Early research suggested that body composition is a function of body weight (maturity) within cattle of the same breed and sex.[9] However, it is now known that rate of gain also affects composition of gain, with those cattle fed to gain faster (e.g., high-energy feedlot diets) having increased fat to lean ratios at a given weight.[8] Smaller-frame cattle or heifers reach a body composition near low choice (about 28% body fat) at weights that may be less than desirable for the market. Conversely, large-frame cattle or bulls may be too heavy. Nutritional treatment may be used to mollify this situation. For example, restricting energy intake of the smaller-frame calves (or heifers) until they reach physiological maturity (270 to 320 kg, or 600 to 700 lb) allows the cattle to be finished at heavier slaughter weights without sacrificing yield. In contrast, bulls and larger-frame cattle fit best in today's production system if fed to gain at maximum levels from weaning on in order to achieve acceptable carcass fatness before becoming too large. Medium-frame steers are most versatile, being neither too small upon reaching the choice grade if fed high-energy feedlot diets from weaning nor too large if grown on restricted energy levels before being fed for a minimum number of days in the feedlot.

After a period of restricted growth, cattle refed a high-energy diet initially gain a greater proportion of protein (compared to fat) than their continuously feedlot-fed counterparts at equal weights (Fig. 16-3). However, the proportion rapidly changes, and after a month on feed they gain a greater proportion of fat than continuously fed cattle.[8] This allows the restricted, refed cattle, which were initially thinner, to catch up in terms of body composition. To achieve similar carcass fat, however, cattle that have undergone a period of restricted feeding prior to being placed on full feed of a high-energy diet should be heavier at slaughter than those that entered the feedlot as calves.

NUTRIENT CONSIDERATIONS IN DIET FORMULATION

Nutrients considered in formulation of diets for feedlot cattle generally include moisture, net energy (NE), crude protein, fat, calcium, phosphorus, potassium, sulfur, salt, trace minerals, and vitamin A. Fat and salt are discussed later in this chapter on pp. 197-199.

Moisture

Moisture content of the feed per se has not been shown to have consistent effects on energy intake by feedlot cattle. Diets that contain greater than 14% moisture are susceptible to mold growth and thus should be fed fresh. Complete mixed diets containing less than 14% moisture have been stored for longer than 30 days with no detrimental effects on cattle performance.[3]

Net energy

The NE system for expressing the energy requirements of feedlot cattle is based on the convention that there are two partial efficiencies for utilization of dietary energy: one for maintenance (NEm) and one for gain (NEg). The close relationship between the two factors across a broad range of feedstuffs (NEg = 0.877 NEm − 0.41, R^2 = 0.99[5]) indicates that the NE values are additive. Furthermore, it is implied from the equation that as diet NE declines the relative efficiency of NE use for gain also declines. Thus, the greater the energy content of a diet, the greater the efficiency with which the energy is utilized for growth and vice versa.

Feedlot cattle growth-performance is a predictable function of NE intake.[11] The capacity of the digestive tract (particularly the rumen) limits energy intake and thus growth rate when the energy density of the diet (NEm) falls below roughly 1.80 Mcal/kg (dry matter basis[6]). Typical growing-finishing diets contain between 2 and 2.3 Mcal/kg NEm (dry matter basis).

Crude protein and fat

The crude protein (N × 6.25) requirements for feedlot cattle are currently estimated using the factorial approach.[5] General assumptions to this approach are: (1) feed N, irrespective of source, has a net protein utilization factor of 0.594 (biological value of 66%, true digestibility of 90%); (2) protein deposition is related to live weight (W) and rate of weight gain (G); (3) metabolic fecal losses are related to feed intake (FI); and (4) endogenous urinary and scurf losses are directly related to body size. The general equation for predicting protein requirements in grams (PR_g) is as follows:

$$PR_g = (\{G_{kg} [268 - (29.4\{[W_{kg}^{.75}(0.0527G_{kg} + 0.00684G_{kg}^2)] \div G_{kg}\})]\} + 33.4\, FI_{kg} + 2.75\, W_{kg}^{.5} + 0.2 W_{kg}^{.6}) \div 0.594$$

Primary limitations of the factorial equation are the generalized nature of the approach and the uncertain reliability of its components. With respect to the latter, there is considerable controversy over what constitutes metabolic fecal N. As it presently stands, that loss alone represents nearly 50% of the total estimated crude protein requirement of feedlot cattle!

The factorial equation does not allow for distinction among feed N sources in terms of ruminal degradability or quality of amino acid profile. It assumes that protein passage to the small intestine is equivalent to crude protein intake, which is probably valid for diets containing 12% to 14% crude protein. However, the equation will likely underestimate protein passage to the small intestine with diets containing less than 12% crude protein.

The proteolytic capacity of the rumen far exceeds the practical limits of protein intake. For most feedstuffs, primary factors that influence the extent of dietary protein degradation in the rumen are exposure rate to the proteolytic process and ruminal retention time. Both factors are affected by the nature of the complete diet and plane of nutrition. Consequently, in feedlot cattle the extent of ruminal protein degradation for a given protein source generally is higher for receiving and growing diets than for growing-finishing diets.

In addition to meeting the protein needs of the animal, the requirements of the ruminal microbial population must also be considered. The factorial equation does not provide specifically for the needs of the ruminal microbial population. Insufficient ruminal available N may limit microbial growth and the extent of ruminal fermentation. However, this latter limitation can be overcome if care is taken to ensure that diets for feedlot cattle contain approximately 7% ruminal available crude protein.

Growing-finishing diets for feedlot cattle typically contain between 11% and 12.5% crude protein. In most cases, urea as well as other forms of nonprotein N are as effective as natural protein supplements, provided that the total estimated ruminal available crude protein does not exceed 7%. As a general rule, urea should not exceed 1.3% of diet dry matter.

Fat supplementation increases diet caloric density while tending to decrease ruminal fermentation and, consequently, microbial protein synthesis. As a result, additional ruminal escape protein may be needed when supplemental fats are fed. This appears to be more of a factor with wheat- or barley-based diets, in which ruminal escape feed N is expected to be low. Thus, it is recommended that wheat- or barley-based growing-finishing diets are supplemented at the rate of 0.12 g ruminal escape protein per gram supplemental fat.

Calcium

Calcium is the most likely macromineral to be limiting in diets for feedlot cattle. Cereal grains, which generally comprise the bulk of growing-finishing diets, are particularly low in calcium, averaging less than 0.02% dry matter. Thus the majority of the animals' calcium needs are provided for by means of calcium supplements.

Calcium is excreted in feces in association with nondigested feed residues. However, digestibility of feedlot diets or level of fat supplementation does not appear to have an appreciable effect on calcium absorption from the small intestine.[4]

The dietary calcium requirement for optimal feedlot growth-performance is not clear. Current standards are based on estimates for maintenance (1.54 g/100 kg body weight or 0.70 g/100 lb) and protein gain (7.1 g/100 g protein gain or 32 g/lb protein gain).[5] Typical feedlot diets are formulated to contain between 0.5% and 0.8% calcium (dry matter basis). Higher levels (1% to 1.2%) of dietary calcium have been fed. However, in those cases the additional calcium was provided as a buffer to augment digestion. Experimental support for feeding higher levels (greater than 0.8%) of calcium is limited; because adverse effects on feed intake have been occasionally observed, it is not presently recommended.

Phosphorus

Phosphorus is the second most likely macromineral to be limiting in diets for feedlot cattle. Cereal grains are a fairly good source of phosphorus, averaging 0.38%. Forages may be low in phosphorus, typically averaging less than 0.25% (phosphorus content of forages also tends to decline markedly with maturity and is particularly low in crop residues). Thus, diets that are higher in forage, such as receiving and growing diets, may require phosphorus supplementation.

The dietary phosphorus requirement for optimal feedlot growth-performance is not clear. Like calcium, current standards[5] are based on estimates for maintenance (2.8 g/100 kg or 1.27 g/100 lb body weight) and protein gain (3.9 g/100 g or 17.7 g/lb protein gain). Typical feedlot diets are formulated to contain 0.25% to 0.4% phosphorus (dry matter basis), with most formulations ranging between 0.25 and 0.3%. The higher levels (0.4%) are more common in receiving diets for lightweight (115 to 180 kg), rapidly growing feeder calves, such as Holsteins. Higher levels (greater than 0.4%) of dietary phosphorus may precipitate urinary calculi, particularly during winter months when water consumption is low. Increasing both the calcium and potassium level of the diet can reduce the probability of this occurring.

Potassium

There is considerable controversy over what constitutes the potassium requirement for feedlot cattle. Current standards[5] are 0.5% to 0.6% of diet dry matter. Increasing the potassium level of receiving diets to 1% to 1.2% diet dry matter has been found to decrease morbidity and increase the rate of recovery of purchase weight of shipping stressed calves.

Sulfur

Most diets fed to feedlot cattle contain adequate sulfur to meet the animals' needs (0.1% diet dry matter). However, when high concentrate corn or milo-based diets containing nonprotein nitrogen as the primary source of supplemental crude protein are fed, ruminally available sulfur may be insufficient to sustain optimal microbial growth. As a result, feed intake and animal performance may be depressed. Care should be taken to assure that the nitrogen to sulfur ratio of the diet is between 10:1 and 15:1.

Trace minerals

Within the practical limits of feedlot diet formulations, the trace minerals most likely to limit growth-performance are cobalt, copper, iodine, iron, manganese, molybdenum, zinc, and selenium. Requirements for these elements are roughly 0.1, 10, 1, 50 to 100, 40, 2, 30 to 50, and 0.1 to 0.3 mg/kg feed dry matter, respectively. Individual feed ingredients can be quite variable in these elements, the cost of routine feed analysis is high, and the practicality of having a separate trace mineral package for each diet formulation is low. For the present, provision of the trace mineral requirements via a trace mineral supplement seems like good insurance.

Vitamins

The B vitamins and vitamin K are synthesized by ruminal microorganisms and bacteria in the digestive tract in sufficient amounts to meet the animals' theoretical requirements. Cattle can biosynthesize vitamin C. Cattle exposed to sunlight can synthesize adequate vitamin D. Thus, vitamins for which supplementation may be necessary are A and E.

Vitamin A is the vitamin most likely to be deficient in diets for feedlot cattle (p. 68). In terms of feedlot performance, inadequate vitamin A can result in depressed feed intake and rate of weight gain, increased susceptibility to calculi during winter, and reduced tolerance to heat during the summer. Cattle grazing high-quality green forage may accumulate enough vitamin A to sustain them for periods of up to 6 months. Unless the background of the cattle is certain, however, it may not be wise to rely on liver stores for meeting or even supplementing the vitamin A requirement during the feedlot phase. The requirements for vitamin A are currently set at 2200 IU/kg (1000 IU/lb) dry feed,[5] which corresponds to a daily intake of roughly 50 IU/kg (20 IU/lb) live weight. These recommendations of daily requirements have been criticized because they are based on levels of feedlot growth-performance that are considerably below (less than 65%) what is typical in today's industry. Calves (particularly Holstein) placed on feed at light weights and managed for rapid gains throughout the growing-finishing periods may benefit from vitamin A supplementation at levels as high as 6000 IU/kg (2727 IU/lb) dry feed. Dietary carotenoids can be converted by cattle to vitamin A (see p. 69 for interconversions). However, it is customary to provide at least 100% of the estimated vitamin A requirement in the supplement. An alternative to feeding supplemental vitamin A is to administer it via injection. A single intramuscular injection of 1 million IU of vitamin A may be adequate to meet an animal's vitamin A requirements for up to 4 months.

Vitamin E is essential for normal health and performance of feedlot cattle. One of the most important functions of vitamin E is as an antioxidant. Vitamin E has also been found to enhance disease resistance by protecting leukocytes and macrophages during phagocytosis. The more classical symptoms of vitamin E deficiency are related to muscle degeneration (white muscle disease, p. 76).

The requirements for vitamin E are far from clear. Selenium and vitamin E share similar roles, and with adequate dietary selenium, depressed health and growth-performance of cattle may not be manifest, even with very low dietary vitamin E. The NRC recommends that diets for young growing calves contain 15 to 60 IU vitamin E per kg dry feed.[5] Typical feedlot diets should fall within that range and thus there may not usually be a need for supplemental vitamin E. Unlike monogastric nutrition, even very high levels of fat supplementation do not appear to increase the vitamin E requirement of feedlot cattle.

INGREDIENT CONSIDERATIONS IN DIET FORMULATIONS
Grains

Barley. Barley should be stored for a minimum of 30 days following harvest. Feeding "green" barley can result in a high incidence of bloat. Barley can be quite variable in bushel weight. Below densities of 0.64 kg/L (50 lb/bu), the feeding value of barley declines markedly. For bushel weights above 0.64 kg/L the relationship between density and feed value is less apparent. The relative advantage of steam-rolling versus dry-rolling barley are small, compared to that of corn or sorghum. Steam-rolled barley is less prone to produce bloat than dry-rolled barley, and this may account for occasional differences in cattle performance observed. The growth-performance response to barley-based growing-finishing diets is usually improved by the addition of buffers, for example, 0.75% sodium bicarbonate ($NaHCO_3$).

Corn. Corn is the primary grain fed to livestock in the United States. It tends to be less variable in composition than other feed grains. Number 2 grade yellow corn is the predominant feed corn. It is characterized by a density of 0.70 kg/L (54 lb/bu) and 3% broken kernels and foreign material. When greater than 90% of the diet is corn, whole shell corn is utilized with as much efficiency as dry-rolled corn. Adding moisture back to the grain by tempering or soaking prior to rolling increases the NE value of corn by roughly 5% over dry rolling. Steam flaking (final density 0.31 kg/L or 24 lb/bu) increases the NE value of corn by as much as 14%. High-moisture ensiled corn (greater than 30% moisture) is equivalent to steam-flaked corn in feeding value. High-moisture ensiled corn that contains less than 25% moisture may not be different in feeding value from tempered corn. As a general rule, the response to steam-flaked corn is optimal in diets containing low levels (8% to 12%) of forage. High-moisture ensiled and dry-rolled corn give the greatest response in diets containing higher levels (16% to 20%) of forage. Feed intake and daily weight gain response to steam-flaked corn–based growing-finishing diets may be improved by the addition of buffers.

Sorghum. To obtain satisfactory results, sorghum grain should be processed prior to feeding. Dry-rolled sorghum has approximately 88% the feeding value of corn. Steam flaking increases the feeding value of sorghum by roughly 14%. Properly flaked (final density 0.31 to 0.36 kg/L or 24 to 28 lb/bu) sorghum has roughly 92% to 95% the feeding value of steam-flaked corn (dry matter basis). Improvements in the feeding value of sorghum with reconstitution have been inconsistent but are probably intermediate to that of steam flaking.

Wheat. Like barley, the relative advantage of dry-rolled versus steam-rolled wheat is comparatively small. However, dry-rolled wheat is considerably more prone to produce acidosis. If wheat is to be dry rolled, the rolls should be adjusted to allow for roughly 5% whole kernels to pass through the rolls intact. Properly processed wheat has approximately 98% of the value of steam-flaked corn, 105% of the value of steam-flaked sorghum, and 103% to 105% the value of barley (density greater than 0.64 kg/L). There appears to be an associative effect between wheat and corn. Substituting 25% to 50% of the steam-flaked corn with steam-rolled wheat may result in growth-performance equal or superior to that of steam-flaked corn alone. Response to wheat-based growing-finishing diets is improved by the addition of buffers.

Forages

Forages typically comprise 8% to 30% of growing-finishing diets (dry matter basis). Although all-concentrate diets have been successfully fed to feedlot cattle, it is generally considered that the inclusion of a minimum of 8% to 10% forage (dry matter basis) is beneficial for keeping cattle "on-feed." Feedlot performance in terms of daily weight gain is usually not affected by inclusions of from 10% to 20% forage in the diet. When the forage level of the diet is marginal, the tendency for ruminal dysfunctions is increased and feedlot growth-performance is correspondingly depressed.

The choice of types of forage to include in the diet depends on comparative cost, available nutrients, palatability, and anticipated level of incorporation. When the diet dry matter contains 10% or less of forage, the differences among forage types with respect to intrinsic nutritive value is minimized and primary concerns are palatability and cost. At higher levels of incorporation, however, the differences in nutritive value among the various forage types are more distinct.

Disappearance of fiber from the rumen is dependent on particle size reduction. Cellulolytic activity is depressed at ruminal pH below 6.5, and rumination is minimal with high-concentrate diets. As a result, it is important that forages be sufficiently processed to prevent ruminal accumulation and, consequently, depressed feed intake. When the forage level of the diet is 20% or less, legumes such as alfalfa hay should be ground or chopped to pass through screens with a minimum diameter of 5 cm, whereas grass hays and straws or stubbles should be ground or chopped to pass through a 2.5-cm-diameter screen. The basis for the larger screen size for alfalfa hay is to prevent development of excessive fines and loss of roughage characteristics.

Protein supplements

Protein supplements consist of those feeds that contain more than 20% crude protein. In recent years increased attention has been given to differences among protein sources with respect to their ability to resist microbial degradation in the rumen. Because the requirements of ruminal microorganisms for growth can largely be met by less expensive nonprotein nitrogen sources such as urea, microbial degradation of intact feed protein is considered wasteful.

Due to availability and cost, oilseed meals represent the most common source of supplemental protein used for feedlot cattle. They include canola, cottonseed, peanut, soybean, and sunflower meal. As a class, these protein sources may be considered moderately to highly degradable in the rumen (55% to 85%).

Supplemental protein sources that are more resistant to ruminal degradation (17% to 45% rumen degradable) include blood meal, corn gluten meal, distiller's dried grains, feather meal, fish meal, and meat meal. Blends of these more resistant protein sources have resulted in faster and more efficient gains in lightweight feeder calves (110 to 160 kg, or 242 to 352 lb) than oilseed meal controls. Blood meal–corn gluten meal combination (40:60) has been found to be particularly complementary. Blood meal is high in lysine, whereas corn gluten meal is low in lysine. However, these more resistant supplemental protein sources are usually more expensive as well.

The effects of protein supplementation on feedlot performance appear to be directly related to a stimulation of energy intake. Differences in feed efficiency are largely a reflection of differences in daily weight gain rather than improved energetics. In some situations the cost of maximizing daily gain may not equate to optimal cost of gain.

Fat

Commercial feed fats are commonly added to growing-finishing diets. Benefits include increased caloric density of the diet, reduction in fines, improved condition of complete mixed diets, and improved longevity of grinding, mixing, and conveying equipment. Adding as little as 1% fat to hay prior to grinding markedly reduces dust loss.

When fat is consumed by cattle, it is readily hydrolyzed by ruminal microorganisms to form free fatty acids and glycerol. Glycerol is further metabolized to propionic acid and absorbed from the rumen. Free fatty acids are hydrogenated. Because of the saturated nature of fat entering the small intestine, digestibility of fat in ruminants is lower than for monogastric species, averaging roughly 80%. Although fat sources differ in quality, the generalized NE value for maintenance and gain are 6 and 4.85 Mcal/kg (2.7 and 2.2 Mcal/lb), respectively.

The practical constraints or limits for optimal utilization of supplemental fats in growing-finishing diets for feedlot cattle have not been resolved. Of particular concern is how the level of fat supplementation influences its comparative feeding value. Fat is typically supplemented into the diet at levels ranging from 2% to 5% of diet dry matter. Linear improvements in both rate of weight gain and feed conversion have been noted for levels of supplementation up to 8% of diet dry matter. However, in some instances these higher levels of supplementation have resulted in drastic reductions in feed intake and associated poor performance. Intestinal digestibility of fat decreases approximately 3.5% for each percent increase in level of fat supplementation above 4%. The basis for occasional poor performance with higher levels of fat supplementation is generally attributable to improper diet formulation or failure to adapt cattle properly to supplemental fat.

Growth-performance response to supplemental fat is generally better with small-grain–based diets such as wheat and barley. When diets are comprised largely of corn or milo, the level of fat supplementation probably should not exceed 4% (dry matter basis). As mentioned previously, response to supplemental fat may be considerably less than optimal when inadequate ruminal escape protein is fed. Fat supplementation may also increase the requirement for vitamin A.

Fat supplementation per se does not influence carcass quality (marbling score) or measures of external fat thickness. Measures of internal fat (kidney, pelvic, and heart fat) are consistently increased with fat supplementation.

Yellow grease and blended animal-vegetable fats are primary sources of commercially available feed fats used in diet formulations for feedlot cattle.

Yellow grease. The term *yellow grease* is descriptive of its yellowish appearance. It also may be referred to as *restaurant grease* or *kitchen grease*, as it is composed of any combination of waste greases collected from bakeries, restaurants, school cafeterias, and the like and of rendered animal fat. Because of this diversity of source, yellow grease is not uniform in composition from area to area or plant to plant. According to standards set by the American Fats and Oils Association, yellow grease has a maximum of 15% free fatty acids, no refined or bleached color, and a maximum of 2% of moisture, impurities, and unsaponifiables.

Blended fats. Blended animal-vegetable fats are a mixture, in any proportion, of rendered animal fat or grease, yellow grease, hydrolyzed animal fat or vegetable oil, and acidulated vegetable or animal soap stocks. As with yellow grease, blended animal-vegetable fat is not uniform in composition, and thus it may be misleading to generalize or typify its characteristics. Compared with yellow grease, however, it is dark in appearance and usually higher in free fatty acids and unsaponifiable matter. Blended animal-vegetable fats also tend to be higher in iodine value. Typical quality specifications for blended animal-vegetable fat are 90% minimum total fatty acids, 50% maximum free fatty acids, 1.5% maximum moisture, 1% maximum impurities, and 3.5% maximum unsaponifiables.

Conditioners

In addition to proper nutrient balance, another important consideration in diet formulation is the general condition and palatability of the diet. Diets that are too dry may not feed well from the delivery truck, particularly on a windy day. If certain ingredients in the diet are especially unpalatable, there may be sorting by the animals, resulting in a buildup of fines in the feed bunk. Smaller feed particles may segregate or settle out from the rest of the diet during the course of the day. These problems are usually addressed by inclusion of some type of conditioning agent in the formulation. The least expensive option is water. However, water may lose its conditioning properties very quickly, particularly during dry weather. Molasses (cane, beet sugar, or hemicellulose extract) is another popular alternative. In addition to providing excellent conditioning properties, it is also very palatable and may help to mask off odor or taste of less palatable feed ingredients. Molasses is also a good source of energy and certain minerals; however, it is often priced too high to be considered on this basis alone. Molasses has its highest value when added to growing-finishing diets that contain high levels of low-quality forage or when palatability of other dietary ingredients is suspect. In diets that contain wet feeds and in which palatability is not considered a problem, benefits from the inclusion of molasses in the formulation are not likely to be realized. With the development of highly effective suspending agents, molasses has also become a convenient carrier for supplement preparations.

Beet molasses. Beet molasses is a by-product of the manufacture of sugar from sugar beets. On a dry matter basis, it contains 11% ash, slightly lower than cane molasses. However, it is very high in potassium (6%), which contributes to the high laxative characteristics of beet molasses. For this reason it is recommended that beet molasses not exceed 10% of diet dry matter. Beet molasses is a very poor source of phosphorus (0.03%). Consequently, inclusion of higher levels of beet molasses in the diet may require additional phosphorus supplementation. Beet molasses is higher in protein than cane molasses (8% versus 4%). In both cases, however, the quality of the protein is low. Aside from its conditioning properties, the feeding value of beet molasses can be largely attributed to its total sugars content (78%). The fermentability of these sugars in the rumen is virtually 100%. Beet molasses and cane molasses may be diluted prior to delivery to reduce viscosity. This should be taken into consideration when determining their comparative feeding value. Beet molasses that contains 72% dry matter has a feeding value of 69% of the value of corn.

Cane molasses. Cane molasses, also called *blackstrap,* is a by-product of the manufacture of sugar from sugarcane. On a dry matter basis, it is high in ash (13%) and like beet molasses is a particularly good source of potassium (3.8%). It is low in protein (4%) and phosphorus (0.11%). Like beet molasses, the primary component of its feeding value (other than its conditioning properties) is its total sugars content (74%). Cane molasses that contains 72% dry matter has a feeding value of 65% of the value of corn.

Masonex. Hemicellulose extract or *masonex,* as it is called, is a by-product of the hardboard industry. It is low in protein and ash. On a dry matter basis, it contains 60% fermentable carbohydrates. The major sugars in masonex consist of xylose and mannose, as opposed to the sucrose found in beet and cane molasses. Masonex is palatable and has been fed without deleterious effects at levels as high as 10% of diet dry matter. Masonex that contains 72% dry matter has a feeding value of 55% of the value of corn.

Salt

Except perhaps under conditions of heat stress, the requirement for salt is roughly 0.1% of diet dry matter. However, it is customary to add 0.2% to 0.5% salt to diets of feedlot cattle to enhance palatability and stimulate intake.

Buffers

Feeding high levels of readily fermentable carbohydrates predisposes cattle to digestive dysfunctions associated with depressed ruminal pH (persistent ruminal pH below 5.5). Feeding buffers such as $NaHCO_3$ or $KHCO_3$ (0.75% of diet dry matter) can be an effective means of minimizing this risk. Limestone ($CaCO_3$) has also been considered for this purpose. However, it has very low ruminal solubility and thus plays a very minor role in controlling ruminal pH. Magnesium oxide (MgO) has also been used with some success, although it has low palatability. Aside from animal health considerations, the direct effect of adding buffers to feedlot diets on animal performance is increased feed intake and consequently an increased rate of daily weight gain. Improvements in feed efficiency are secondary to increased daily weight gain. Using sodium or potassium buffers in combination with an ionophore (e.g., monensin, lasalocid) may nullify the effect of the ionophore on feed efficiency. If buffers are to be supplemented into the diet, it may not be cost-effective to use an ionophore as well.

By-product feeds

By-product feeds are usually much more variable in composition and quality than the more conventional feed ingredients. However, their inclusion in diet formulations for feedlot cattle can be an effective means of reducing feed cost of gain. Indeed, in many feedlots in the southwestern United States, where conventional feed grains are comparatively expensive, by-product feeds may constitute as much as 40% of the complete mixed diet for growing-finishing cattle.

Almond hulls. Almond hulls are high in fiber and lignin and low in protein (2%). They are very palatable and serve as an excellent roughage source. The feeding value is approximately 90% that of good-quality alfalfa hay in growing-finishing diets for feedlot cattle. Due to its physical characteristics (larger particle size), greater segregation can occur with almond hulls; thus it is recommended that it not exceed 5% of diet dry matter for growing-finishing cattle.

Bakery waste. Bakery waste consists of stale breads and other bakery products. It is low in fiber and high in fat (13%). Because of high solubility, very little of the protein in bakery waste is expected to escape the rumen. The feeding value is similar to that of wheat. Because of physical characteristics and high fermentability, it should not exceed 10% of diet dry matter for feedlot cattle.

Beet pulp. Sugar beet pulp has 85% of the feeding value of barley. It is highly palatable, but because of its bulk and laxative effects it should not exceed 20% of diet dry matter in diets for growing-finishing cattle.

Carrots. Cull carrots are highly palatable. Primary disadvantages are their low dry matter (10% to 15%) content and short seasonal availability. They require no additional processing prior to feeding. It is customary to feed the mill feed and carrots separately by placing the mill feed along one portion of the feed bunk and cull carrots along the other. Carrots are high in carotene. Consequently, if fed too near the completion of the finishing period (less than 60 days), they may impart a yellow color to the fat (a factor discriminated against by most packers). On a dry matter basis, cull carrots have an energy value similar to that of barley. On an as-fed basis their value is roughly 13% to 15% that of barley.

Citrus pulp. Citrus pulp is the residue from manufacture of grapefruit and orange juice (and to some extent lemon juice). The composition can vary somewhat, depending on whether or not the citrus molasses is returned to the pulp prior to drying. Compositionally it resembles beet pulp except that it is slightly higher in energy and lower in protein. It has about 95% of the feeding value of barley. However, palatability can be a problem, and thus it should not exceed 15% of diet dry matter for feedlot cattle.

Cotton gin trash. Cotton gin trash consists of the boll residue, leaves, stems, lint, and dirt aftermath of the ginning process. It is high in fiber and ash and low in nutritive value (less than 90% the value of cottonseed hulls). It may also be contaminated with pesticide residues and as such may not be a suitable feed for livestock. Gin trash should not exceed 10% of dry matter in growing-finishing diets.

Cottonseed hulls. Cottonseed hulls are a high-fiber, low-protein by-product of the cotton industry. The NEg of cottonseed hulls is roughly 0 Mcal/kg. Consequently, it contributes little nutritive value to the diet. Because cottonseed hulls require no additional processing prior to incorporation into the diet, store well, and are palatable and relatively inexpensive, they have become a popular roughage source for growing-finishing diets. As a roughage source, however, cottonseed hulls are less effective than other equally indigestible crop residues, such as straw. Cottonseed hulls should not exceed 15% of diet dry matter for growing-finishing cattle.

Grape pomace. Grape pomace is a by-product of the wine and grape juice industry. It is high in fiber and lignin (in part, because of high stem content). The feeding value is roughly 60% that of alfalfa hay. Grape pomace is palatable and has been fed to feedlot cattle at levels of up to 20% of diet dry matter with no reduction in rate of gain. However, it is not recommended as the sole roughage source.

Hominy. Hominy feed consists of corn bran, corn germ, and a portion of cornstarch. It usually contains greater than 6% fat but is otherwise similar in composition and feeding value to ground corn. Hominy feed is palatable and has replaced up to 100% of the grain in finishing diets with no adverse effects on daily weight gain. From the standpoint of feed efficiency, however, it appears to be better utilized when restricted to not more than 10% of diet dry matter.

Peanut hulls. Peanut hulls are high in fiber and lignin. They contain appreciable amounts of protein (8%), but otherwise the comparative feeding value is similar to cottonseed hulls. Peanut hulls should not exceed 10% of diet dry matter for growing-finishing cattle.

Rice hulls. Rice hulls are low in protein and high in fiber and contain especially high levels of lignin (14%) and silica (22%). The palatability of rice hulls is comparatively low. Furthermore, the high silica content can contribute to urinary calculi. Rice hulls should not exceed 5% of diet dry matter for growing-finishing cattle.

Straw. Straw residues, although differing somewhat depending on species, variety, and the agronomic conditions under which they were produced, contribute very little nutritive value per se to the diets of feedlot cattle. They do, however, have good roughage characteristics. Straw should be ground through a 2.5-cm (1-inch) screen prior to feeding. The contribution of straw to the diet is maximized when the diet contains 90% or more of concentrates. Adding more than 10% straw to the diet dry matter can result in depressed energy intake and gain in growing-finishing cattle. At 10% or less of the diet dry matter, straw has 110% the feeding value of cottonseed hulls.

Tapioca (cassava meal pellets). Tapioca is a product made from cassava starch cakes. It is high in energy (similar to barley) but low in protein (2.5%). Depending on variety, it may contain high levels (0.03%) of hydrocyanic acid, which gives it a bitter taste and may reduce diet palatability.

Tomato pomace. Tomato pomace is the pulp residue from the manufacture of tomato juice and catsup. It is high in protein (20%) and fat (12%). The feeding value of tomato pomace is equivalent to that of good-quality alfalfa hay. It should not exceed 10% of diet dry matter for growing-finishing feedlot cattle.

Wheat middlings. Wheat middlings consist of very fine bran, germ, and starch particles. It is high in protein (18%) and phosphorus (1%). The protein fraction is highly soluble, and consequently very little is likely to escape ruminal degradation. The feeding value of wheat middlings is roughly 75% of the value of barley. Middlings are usually limited to not more than 10% of dry matter in growing-finishing diets.

Wheat mill run. Wheat mill run consists of all the milling residue combined together (bran, germ, shorts, and other matter). It is high in protein (17%) and phosphorus (1.3%) and has a feeding value of roughly 85% that of barley. Mill run is usually limited to not more than 10% of dry matter in growing-finishing diets.

FEEDING MANAGEMENT
Feed mixing

Prior to formulating diets for a given feedlot, the feedmill should be visited and its limitations carefully considered. Safe allowances for error in formulation should be given to feed ingredients that are less palatable or potentially toxic. Some of the greatest obstacles for optimizing a feeding program are associated with feed mixing. It is not enough that a diet is properly balanced to meet the desired growth-performance requirements. Care must also be taken to ensure that the diet can be safely and consistently produced within the given limitations of a particular mixing system.

There are basically two types of feed mixing systems: proportion and batch. It is not unusual to find mills that incorporate both systems. With batch mixing, individual feed ingredients are first weighed and then dispensed into a mixer. The advantage of batch mixing is that it allows for greater precision in combining feed ingredients. Disadvantages are generally associated with challenges of obtaining uniform mix. Because the process is discontinuous, the tendency, in practice, is to mix for too short a period of time in order to increase mill output. With proportion mixing, feed is dispensed and mixed as it passes along the length of an auger or belt. The amounts of ingredients combined in the mix are governed by the rate at which they are dispensed rather than by weight. The advantage of proportion mixing is that it is continuous and generally capable of producing comparatively larger volumes of feed per unit time. Furthermore, the feed is mixed in small proportions, allowing for more uniform distribution of ingredients in the finished feed. The disadvantage, however, is that any change in density, flowability, or viscosity of individual feed ingredients or intrinsic metering fluctuations can result in appreciable variance between formulation and resultant feed.

The order in which feed ingredients are mixed can be important. Grain and forage should be combined first, followed by supplement. If the supplement is dry, it should be allowed to mix thoroughly with the grain and forage prior to addition of liquid ingredients.

Feed allocation

Feedlot cattle are normally allowed ad libitum access to feed (access to feed at all times). This is usually accomplished by providing roughly 10% more feed in a day than the cattle will consume. One perceived advantage of this type of feeding program is that it allows cattle to maximize their voluntary intake and, consequently, to maximize their rate of weight gain. There are, however, some intrinsic limitations to this type of feeding system. Anticipating how much feed cattle will consume on a given day may be more art than science. Providing too much feed may allow cattle to become selective in what they eat, resulting in an accumulation of fines. Furthermore, as excess feed builds up in the bottom of bunks the chances for mold and off flavor development increase. This feed then has to be discarded because forcing cattle to consume it depresses overall feed intake and consequently cattle performance.

Under ad libitum feeding conditions, cattle often exhibit wide interday and intraday fluctuations in feed intake. These effects may be most pronounced during frontal weather systems and are thought to be directly related to rapid falling and rising of barometric pressure. Wide fluctuations in feed intake can cause serious digestive disturbances including acidosis and bloat. One approach for limiting the magnitude of the upswings in intake is to set an upper limit or do-not-exceed value for daily pen feed allotments. This should be done on an individual pen basis.

The consequences of underfeeding are variable. If cattle are fed 90% to 95% of ad libitum, growth rate may be depressed slightly. However, feed efficiency is often improved over that obtained under ad libitum feeding conditions.

The effect of underfeeding is often confused with the hazards of abrupt reduction or cessation in feed intake. During severe weather cattle may be prevented from consuming feed. This could be the result of a number of factors, including stress-related inappetence, impaired access to the feed bunk, feed that is covered with snow or water, and inability to deliver feed to the feed bunk. If the return to full feed is equally abrupt, serious digestive upsets may result. Many problems can be avoided if whenever feedlot cattle go off feed they are readapted or brought back up onto full feed in increments with transition diets.

An alternative to ad libitum feeding is to restrict intake to an amount that permits animals to attain some predetermined daily weight gain. Advantages to this type of feeding include minimized day-to-day variation in feed intakes, improved feed bunk management, and greater control over feed inventories. Challenges of restricted feeding programs include the ability to select an optimum target daily weight gain, the reliability of estimates of net energy value of the diet, the general applicability of the net energy equations for partitioning energy intake into components of gain, the necessity that animals consume their daily feed allotments, and uncertainties with respect to competition among animals for feed. A feed intake generator for use in "programming" cattle performance is shown in Table 16-1.

Time and frequency of feeding

There is considerable controversy over the merits of once-daily versus twice-daily feeding. A primary advantage of feeding twice daily is that it allows for closer adjustment of feed allotments. This may be particularly important in situations where the condition or stability of the diet is in question or where feed bunk space is limited.

Time of feeding may influence the efficiency with which feed is utilized. If cattle are to be fed once daily, there is some indication that it may be preferable to feed in the late afternoon rather than in the early morning. This might result in a more efficient use of the heat increment for maintaining body heat during the winter months and stimulate greater feed intake by allowing for greater dissipation of the heat load during the summer months. With twice-daily feeding, the feed bunk should be fairly clean sometime between the morning and afternoon feeding, and the afternoon feeding should be the heavier feeding.

Feed delivery

It is usually not beneficial to spread feed too thinly along the length of the feed bunk. Roughly 10 to 15 cm (4 to 6 inches) of bunk space per animal is adequate in most situations. If a pen provides more bunk space than this, spreading feed along the length of the manger often increases sorting and the accumulation of fines.

Feed bunk space allotment

The general recommendation has been that 15 cm (roughly 6 inches) linear bunk space is adequate for cattle with ad libitum access to feed and that restricted fed steers require 60 cm (24 inches) linear bunk space. The basis for the greater bunk space allotment for restricted fed cattle is the assumption that is necessary for all cattle to have access to the feed at once. In practice, however, this does not appear to be the case. Within the practical constraints of growing and finishing, a linear bunk space allotment of 10 cm appears to be adequate for both ad libitum

Table 16-1 *Basic language program for generating daily feed allotments for achieving target weight gains in feedlot calves*

```
10 PRINT"                    Animal Types"
20 PRINT"type 1 = Medium-frame steer calves"
30 PRINT"type 2 = Large-frame steer calves"
40 PRINT"           Compensating medium-frame yearling steers"
50 PRINT"           Medium-frame bull calves"
60 PRINT"type 3 = Large-frame bull calves"
70 PRINT"           Compensating large-frame yearling steers"
80 PRINT"type 4 = Medium-frame heifer calves"
90 PRINT"type 5 = Large-frame heifer calves"
100 PRINT"          Compensating yearling heifers"
110 PRINT"type 6 = Mature thin cows"
120 PRINT"type 7 = Holstein steer calves"
130 PRINT " "
140 PRINT " "
150 INPUT "animal type";TYPE
160 INPUT "Do you want to generate a maintenance
adjustment factor (Y/N)";Y$
170 IF Y$="N" THEN INPUT "What is the maintenance coefficient (ie
    .077 for beef types, .085 for Holstein)";MQ:GOTO 410
180 INPUT "How many close-outs will be used to generate
adjustment factor";N1
190 PRINT " "
200 MQ1=0
210 FOR N=1 TO N1
220 INPUT "Initial weight, kg (shrunk)";PW
230 INPUT "Final weight, kg (shrunk)";SW
240 INPUT "Average daily gain, kg/d";G
250 INPUT "Average daily feed, kg/d";FI
260 INPUT "NEm of diet, Mcal/kg";NEM
270 NEG=(.877*NEM)-.41
280 AW=(PW+SW)/2
290 IF TYPE=1 THEN EG=(.0557*(AW^.75)*(G^1.097))
300 IF TYPE=2 THEN EG=(.0493*(AW^.75)*(G^1.097))
310 IF TYPE=3 THEN EG=(.0437*(AW^.75)*(G^1.097))
320 IF TYPE=4 THEN EG=(.0686*(AW^.75)*(G^1.119))
330 IF TYPE=5 THEN EG=(.0608*(AW^.75)*(G^1.119))
340 IF TYPE=6 THEN EG=6.2*G
350 IF TYPE=7 THEN EG=(.0557*(AW^.75)*(G^1.097))
360 MQ=((FI-(EG/NEG))*NEM)/(AW^.75):PRINT FI;G
370 MQ1=MQ1+MQ:NEXT N
380 MQ=MQ1/N1
390 PRINT TAB(20)"adjusted maintenance coefficient";MQ
400 PRINT " "
410 LPRINT TAB(16)"Animal";TAB(29)"Feed Intake, kg/feeding"
420 LPRINT TAB(5)"Week";TAB(15)"Wt, kg";TAB(30)"    AM";TAB(40)"
    PM"
430 LPRINT " "
440 PRINT "FEED INTAKE GENERATOR FOR FEEDLOT CATTLE"
450 INPUT "Pen #";PN
460 INPUT "Cattle/pen";A
470 INPUT "Initial weight, kg (shrunk)";PW
```

Continued.

Table 16-1 *Basic language program for generating daily feed allotments for achieving target weight gains in feedlot calves—cont'd*

```
480 INPUT "Target final weight, kg (shrunk)";SW
490 INPUT "Target daily gain, kg/d";G
500 INPUT "NEm value of the diet, Mcal/kg";NM
510 NG = (.877*NM) − .41:AW = (SW + PW)/2
520 CW = PW + ((G*7)/2)
530 W = 1
540 IF TYPE = 1 THEN
    FA = ((.0557*(AW^.75)*(G^1.097))/NG) + ((MQ*(AW^.75))/NM)
550 IF TYPE = 2 THEN
    FA = ((.0493*(AW^.75)*(G^1.097))/NG) + ((MQ*(AW^.75))/NM)
560 IF TYPE = 3 THEN
    FA = ((.0437*(AW^.75)*(G^1.097))/NG) + ((MQ*(AW^.75))/NM)
570 IF TYPE = 4 THEN
    FA = ((.0686*(AW^.75)*(G^1.119))/NG) + ((MQ*(AW^.75))/NM)
580 IF TYPE = 5 THEN
    FA = ((.0608*(AW^.75)*(G^1.119))/NG) + ((MQ*(AW^.75))/NM)
590 IF TYPE = 6 THEN FA = ((6.2*G)/NG) + ((MQ*(AW^.75))/NM)
600 IF TYPE = 7 THEN
    FA = ((.0557*(AW^.75)*(G^1.097))/NG) + ((MQ*(AW^.75))/NM)
610 FI = (FA*A)/2:PRINT FI;FA;A
620 LPRINT TAB(5)W;TAB(15)CW;TAB(30)FI;TAB(40)FI
630 CW = CW + (G*7):AW = CW
640 IF SW>CW THEN W = W + 1:GOTO 540
650 INPUT "Do you want to change inputs and recalculate analysis
    (Y/N)";Y$
660 IF Y$ = "Y" THEN CLS:GOTO 410
670 END
```

Adapted from National Research Council: Nutrient requirements of beef cattle, ed 6, Washington, DC, 1984, National Academy of Sciences, National Academy Press.

and restricted fed steers. There is, however, some indication that the more aggressive, rapidly growing cattle particularly benefit, in terms of feed intake and growth rate, from the competition associated with more restricted bunk space allotment, whereas less aggressive, slower-growing cattle may benefit from greater space allotments.

FEEDING STRATEGIES
Receiving diet

The objective of the receiving program is to adapt calves to their new environment as quickly as possible. The primary concern during this period is health, as indicated by morbidity, mortality, and rate of recovery of purchase weight. The receiving period is a lag time during which calves are backgrounded or tooled up for the pending growing-finishing period. Approaches vary, depending on the background and weight of calves entering the feedyard.

Holstein calves. Holstein calves are typically removed from the cow 3 days postpartum and transferred to calf-rearing units. They remain in these units for 120 to 130 days. During the first 50 to 70 days, they are fed milk or milk replacer and allowed ad libitum access to a high-energy concentrate mix that is usually supplemented with antibiotic. Following weaning from milk or milk replacer, they are retained on a high-energy concentrate mix. During this period Holstein calves typically gain 0.7 kg/day (1.5 lb/day) and are shipped to the feedlot with an average purchase weight of 120 kg (265 lb). The receiving diet for Holstein calves may be formulated (dry matter basis) as follows: NE, 2 Mcal/kg (0.9 Mcal/lb); crude protein, 16%; calcium, 0.8%; phosphorus, 0.4%; potassium, 1%; trace mineralized salt (NaCl, 93.5%; $CoSO_4$, 0.068%; $CuSO_4$, 1.04%; $FeSO_4$, 3.57%; ZnO, 0.75%; $MnSO_4$, 1.07%; KI, 0.052%), 0.5%; and vitamin A, 6000 IU/kg (2700 IU/lb). The receiving diet should not contain more than 25% roughage, 3% supplemental fat, 8% molasses, and 0.5% urea (dry matter basis). Reasons for the upper limit on roughage include (1) optimum finish at traditional slaughter weights of 445 to 500 kg (980 to 1100 lb) are attained when Holsteins are fed for maximum rate of gain, (2) calves are already well adjusted to consuming a high-energy concentrate diet, and (3) diets with high energy density reduce girth development (excess girth development in Holsteins during the feedlot period has been associated with reduced feed efficiency and dressing percentage). Because it is highly probable that calves were receiving antibiotic supplementation prior to shipment, it may be advisable to continue its incorporation into the diet at therapeutic levels during the first 5 to 10 days following arrival. An ionophore may also be introduced into the diet during the receiving period. The receiving diet should be fed until calves attain an average weight of 160 to 180 kg (350 to 400 lb), at which time they should be switched over to a growing-finishing diet.

Conventional feeder calves. Feeder calves typically enter the feedlot following a series of weaning, marketing, transportation, and handling stresses. Consequently, morbidity is usually high. Furthermore, calves are likely to be unaccustomed to eating from a feed bunk. Most sickness is observed during the first 14 days following arrival. Starting the calves on higher-energy receiving diets during this period accelerates the recovery of purchase weight; however, it also increases morbidity and mortality. A practical approach has been to provide the calves with ad libitum access to good-quality hay, either chopped or loose, during the first 2 weeks, after which a high-energy receiving diet is introduced. Provided the receiving diet has at least 25% forage, there is usually no need to step up the calves to the new diet. For feeder calves weighing less than 180 kg (400 lb), the receiving diet may be as follows (dry matter basis): NE, 1.90 Mcal/kg (0.86 Mcal/lb); crude protein, 14%; calcium, 0.8%; phosphorus, 0.3%; potassium, 1%; trace mineralized salt, 0.5%; and vitamin A, 2200 IU/kg (1000 IU/lb). The receiving diet should not contain more than 3% supplemental fat, 8% molasses, and 0.5% urea (dry matter basis). The receiving diet should be fed for 28 days or until calves attain an average weight of 205 kg (450 lb). For feeder calves weighing more than 180 kg (400 lb), the receiving diet may be similar to that for lighter calves except that the crude protein level is reduced to 12.5%. Urea (or total nonprotein nitrogen) should not exceed 1.3% of diet dry matter. The receiving diet should be fed a minimum of 28 days.

Transition diet

The decision to proceed from the receiving program to the growing-finishing program is based on body weight, general appearance, condition, and average daily consumption. As a group, calves should be in good health and thrifty condition. They should be alert and have the glossy hair coats and abdominal fill indicative of cattle on full feed. Feed intake should be stable for a period of at least 1 week. Once these criteria are met, calves can be transitioned or moved up onto the growing-finishing diet. Because the receiving diets are already comparatively high in energy, only one transition diet may be necessary. This diet may be formulated (dry matter basis) as follows: NE, 2.10 Mcal/kg (0.95 Mcal/lb); crude protein, 12.5%; calcium, 0.8%; phosphorus, 0.28%; potassium, more than 0.6%; trace mineral salt, 0.4%; and vitamin A, 2200 IU/kg (1000 IU/lb). The transition diet should contain 20% forage. Urea or related nonprotein nitrogen equivalents should not exceed 1.3% of diet dry matter. Supplemental fat may be increased to 4% of the diet, if economically feasible. The duration of the transition period should be roughly 2 weeks. Calves should not be placed on the growing-finishing diet until transition diet intake has been stable for at least 3 consecutive days.

Growing-finishing diet

The growing-finishing diet for Holstein calves may be formulated (dry matter basis) as follows:

NE, 2.2 to 2.3 Mcal/kg (1.0 to 1.1 Mcal/lb); crude protein, 12.5%; calcium, 0.8%; phosphorus, 0.4%; sulfur, 0.14%; trace mineral salt, 0.4%; and vitamin A, 2200 IU/kg (1000 IU/lb). The diet should contain 10% roughage (not to exceed 15%) and not more than 8% molasses, 1.3% urea (or nonprotein nitrogen equivalent), and 6% supplemental fat. Both growth rate and feed conversion have been found to increase with level of fat supplementation up to 9% of diet dry matter. However, exceeding 9% may cause cattle to go off feed, and once this occurs they may not fully recover, even when fat is completely removed from the diet. Thus, placing the limit at 6% supplemental fat provides a margin of safety for mixing errors. It is advisable to also include 0.75% $NaHCO_3$. This should result in increases in both rate of gain and feed efficiency. The incidence of liver abscesses is often higher in Holsteins than in conventional feeder calves. In areas where this problem is more prevalent, it may be wise to include an appropriate antibiotic in the diet formulation.

The growing-finishing diet for conventional feeder calves may be formulated (dry matter basis) as follows: NE, 2.1 to 2.3 Mcal/kg (0.95 to 1.05 Mcal/lb); crude protein, 12.5%; calcium, 0.8%; phosphorus, 0.28%; sulfur, 0.14%; trace mineral salt, 0.4%; and vitamin A, 2200 IU/kg (1000 IU/lb). The diet should contain a minimum of 10% forage and not more than 8% molasses, 1.3% urea (or nonprotein nitrogen equivalent), and 6% supplemental fat. As indicated previously, the inclusion of 0.75% $NaHCO_3$ may afford some protection against digestive disturbances and thus enhance weight gain and feed conversion responses. During the winter months when water consumption normally decreases, it may be advisable to increase the potassium in the diet to 1% (KCl may be used for this purpose). This should markedly reduce the incidence of urinary calculi. Alternatively, supplementing the diet with 0.5% ammonium sulfate (dry matter basis) has also been found to be an effective means of reducing the incidence of urinary calculi.

MONITORING THE FEEDING PROGRAM

The most important tool for monitoring the feeding program is regular visual inspection of each pen lot. Periodic check weights along with computer-assisted evaluation of consumption data are useful ancillaries. However, efficient feedlot operation requires a quicker response to needed change. There is no substitute for regular visual inspection of each pen lot. Each pen lot should be considered individually. The evaluation of a pen of cattle should include three factors: days on feed, feed consumption, and the condition of cattle. One approach to conducting an evaluation is as follows:

1. Upon arriving at the pen, take note of general background data: initial weight, estimated current weight, days on feed, diet the cattle are receiving, and feed consumption.
2. Look at the cattle. Consider their age, stature and type, weight, uniformity, and general disposition. Do the cattle show bloom? Are they alert? Do they show signs of self- or mutual grooming? Look for indications of poor health in the cattle and note the consistency of the manure. Pay particular attention to digestive dysfunctions such as bloat or acidosis.
3. Target the current daily weight gain for the cattle based on observations of their age, weight, type, and days on feed. Shown in Table 16-2 are periodic consumption and daily gain values for various types of cattle fed in the desert southwest and high plains feedlots. This table provides historical data that can be useful as benchmarks for evaluating performance.
4. Using the previous 14-day average daily feed consumption, estimate their actual weight gain (in some feedlots this estimate may already appear on the pen record). A simple to install BASIC language program that can be used for accomplishing this is shown in Table 16-3. This program can be easily installed in a hand-held pocket computer.
5. Divide the target daily gain by the estimate based on feed consumption. Ideally the ratio will be 1. Note the variance and record it both for the evaluation of the next lot of cattle and for comparison to the next evaluation of the current pen lot. If estimated daily weight gain varies greatly (either plus

Table 16-2 *Growth-performance attributes of feedlot cattle in desert southwest and high plains regions*

	Days on feed																			
	14	28	42	56	70	84	98	112	126	140	154	168	182	196	210	224	238	252	266
Southwest feedlots																			
Brahman (calf)																			
Weight, kg	152	166	186	206	225	244	264	282	299	316	332	348	364	378	392	404	416	428	439
Gain, kg/d	.09	1.06	1.47	1.29	1.33	1.46	1.35	1.27	1.67	1.18	1.10	1.15	1.14	.94	.95	.76	.78	.91	.72
DM intake, kg/d	1.8	3.8	5.1	5.1	5.5	6.3	6.3	6.4	6.5	6.5	6.6	7.0	7.2	6.6	6.9	6.2	6.4	7.1	6.5
Brahman (light yearling)																			
Weight, kg	227	234	251	271	291	310	330	349	366	381	395	410	424	436	450	461			
Gain, kg/d	0	.85	1.34	1.54	1.49	1.41	1.29	1.34	1.20	.97	1.07	.96	.95	.90	.93	.82			
DM intake, kg/d	.8	4.4	6.1	7.1	7.4	7.5	7.3	7.9	7.6	6.8	7.5	7.2	7.4	7.3	7.6	7.2			
Crossbreds (calf)																			
Weight, kg	155	168	189	208	229	250	270	288	307	325	341	359	375	390	450	420	432	445	457
Gain, kg/d	.40	1.04	1.51	1.40	1.34	1.46	1.58	1.18	1.37	1.30	1.21	1.16	1.17	1.05	1.09	.94	.92	.83	.79
DM intake, kg/d	2.3	3.7	5.1	5.2	5.5	6.3	7.1	6.0	7.0	7.0	7.0	7.1	7.4	7.1	7.5	7.0	7.1	6.8	6.8

Continued.

Table 16-2 *Growth-performance attributes of feedlot cattle in desert southwest and high plains regions—cont'd*

	Days on feed																		
	14	28	42	56	70	84	98	112	126	140	154	168	182	196	210	224	238	252	266
Southwest feedlots—cont'd																			
Crossbreds (light yearling)																			
Weight, kg	256	266	285	304	324	343	363	380	398	415	431	447	462	475					
Gain, kg/d	.56	.90	1.37	1.39	1.35	1.46	1.30	1.20	1.29	1.18	1.26	1.04	1.02	.90					
DM intake, kg/d	3.8	4.9	6.7	7.1	7.3	8.1	7.8	7.6	8.3	8.1	8.7	7.9	8.0	7.6					
Holsteins (calf)																			
Weight, kg	131	148	165	181	200	220	241	264	285	306	326	345	363	380	401	419	440	458	
Gain, kg/d	1.15	1.22	1.14	1.15	1.37	1.44	1.59	1.54	1.48	1.44	1.36	1.38	1.27	1.25	1.34	1.33	1.49	1.34	
DM intake, kg/d	3.3	3.8	3.9	4.3	5.1	5.8	6.6	7.0	7.2	7.4	7.5	7.9	7.8	8.0	8.7	9.0	10.1	9.6	
High Plains feedlots																			
Light yearlings																			
Weight, kg	233	249	270	294	318	340	361	380	398	415	429	443	454	465					
Gain, kg/d	.18	1.11	1.51	1.70	1.70	1.62	1.50	1.38	1.26	1.16	1.05	.95	.86	.79					
DM intake, kg/d	2.3	6.2	7.9	8.4	8.4	8.2	8.0	7.9	7.8	7.7	7.5	7.3	7.1	7.0					
Yearlings																			
Weight, kg	336	359	384	408	431	454	475	495	511										
Gain, kg/d	1.33	1.62	1.76	1.76	1.70	1.61	1.50	1.36	1.22										
DM intake, kg/d	6.0	9.0	10.0	10.2	10.2	10.2	9.9	9.6	9.2										

Table 16-3 *Basic language program for estimating daily gain based on live weight, daily feed intake, and diet NEm*

```
10 PRINT "Daily gain estimator, kg/d"
20 INPUT "Continue";Y$
30 CLS
40 PRINT"                    Animal Types"
50 PRINT"type 1 = Medium-frame steer calves"
60 PRINT"type 2 = Large-frame steer calves"
70 PRINT"             Compensating medium-frame yearling steers"
80 PRINT"             Medium-frame bull calves"
90 PRINT"type 3 = Large-frame bull calves"
100 PRINT"            Compensating large-frame yearling steers"
110 PRINT"type 4 = Medium-frame heifer calves"
120 PRINT"type 5 = Large-frame heifer calves"
130 PRINT"            Compensating yearling heifers"
140 PRINT"type 6 = Mature thin cows"
150 PRINT"type 7 = Holstein steer calves"
160 PRINT " "
170 PRINT " "
180 INPUT "animal type";TYPE
190 INPUT "Live weight, kg (shrunk)";AW
200 INPUT "Feed intake, kg/d (DM basis)";I
210 INPUT "Diet NEm, Mcal/kg (DM basis)";NM
220 G = .5
250 EM = .077*(AW^.75)
260 IF TYPE = 7 THEN EM = .086*(AW^.75)
270 IF TYPE = 1 THEN EG = (.0557*(AW^.75)*(G^1.097))
280 IF TYPE = 2 THEN EG = (.0493*(AW^.75)*(G^1.097))
290 IF TYPE = 3 THEN EG = (.0437*(AW^.75)*(G^1.097))
300 IF TYPE = 4 THEN EG = (.0686*(AW^.75)*(G^1.119))
310 IF TYPE = 5 THEN EG = (.0608*(AW^.75)*(G^1.119))
320 IF TYPE = 6 THEN EG = 6.2*G
330 IF TYPE = 7 THEN EG = (.0557*(AW^.75)*(G^1.097))
340 NG = (NM*.87) - .41
350 IF EG> = (I - (EM/NM))*NG THEN GOTO 370
360 G = G + .01:GOTO 260
370 PRINT TAB(10)"DAILY GAIN, KG";TAB(30)G
380 INPUT "continue (Y/N)";Y$:IF Y$ = "Y" THEN GOTO 190
390 END
```

Adapted in part from National Research Council: Nutrient requirements of beef cattle, ed 6, Washington, DC, 1984, National Academy of Sciences, National Academy Press.

or minus) from the target for this pen lot but not for others, this may indicate that there was an error in recorded feed allocations (not an altogether unlikely situation). However, a consistent negative variance (ratio is less than 1) is more likely an indication of a problem with the diet (formulation, mixing, or feed-processing error). A positive variance (ratio is greater than 1) is a likely indicator of a possible animal problem that increases the energy requirement, such as poor health, bad weather or buller heifers.

6. Look at the feed bunk, particularly for indications of underfeeding or overfeeding. With twice-daily feeding, the feed bunk

should be fairly clean sometime *between* the morning and afternoon feeding, since the afternoon feeding should be the heavier meal. Note the condition of the feed. Is the feed fresh? Are there any hints of mold (heat or odor)? Is the ingredient distribution in the feed uniform?

7. Look at the waterers. Are they clean? Is there adequate space (i.e., 30 cm or 12 inches per 10 calves). Is the water quality good?
8. Look at the general pen conditions. Are they too dry (dusty) or too wet (muddy)? Is there adequate slope away from the feed bunk so that cattle do not have to stand in mud to eat? Is there adequate shade or shelter?

In addition to regular visual inspection of the cattle, the feedmill should also be evaluated. The estimated use of commodities should be checked against inventories. This is particularly important for supplements, liquid feeds (molasses and fat), and drugs (if added separately). Also, meters and scales should be checked for accuracy. Screens, grinding equipment, and grain-processing equipment should be checked. Is the grain being processed according to specifications? The daily monitoring of grain processing should be reviewed, ensuring that regular moisture analysis is being conducted on processed feeds to which moisture is added. All complete mixed diets should be analyzed weekly for moisture, crude protein (CP), fat (EE), acid detergent fiber (ADF), ash, calcium, and phosphorus. Feed samples should be obtained as soon as possible after the feed is delivered to the pens. Feed samples should be obtained from five or six locations along the feed bunk, composited, and then placed in air-tight bags. Laboratory analysis should be conducted as soon as possible following feed sample collections.

Check weights should be obtained on cattle as often as is practical (usually at times of implanting or roughly 56- to 70-day intervals). Both interim (check weight) and final close-out performance for each lot should be evaluated.

Two simple-to-install BASIC language programs that can be used to evaluate check-weight and close-out performance are shown in Tables 16-4 and 16-5. Table 16-4 contains a program that estimates NEm or NEg values of the diet based on animal weight, daily weight gain, and feed intake. This program can be very useful for comparing estimates of diet energy values based on formulation or chemical analysis with those derived from animal performance. This can be particularly helpful when a new feed ingredient is added to the diet for which there is little information or when there is concern about the adequacy of a grain-processing technique. The program also has particular application for comparing treatments when experiments are conducted within a given feedlot site. The estimates of NE are very precise and unlikely to be biased by the confounding effects of feed intake on daily weight gain (a problem with the use of feed per gain as the criterion for comparing treatment differences).

The program shown in Table 16-5 compares observed feed intake with predicted intake based on daily weight gain and the average NE value of the diet consumed during the period under consideration. Necessary inputs include initial weight, final weight (or check weight), daily weight gain, and NEm of the diet. The NEm (Mcal/kg) value can best be obtained based on diet formulation and tabular NE values from tables of feed standards.[5] Alternatively, the NE value can be estimated from feed analysis based on the following equation:

NEm = 0.054 Fat + 0.187 NFE − (0.072ADF/CP) + 0.567

where NFE is 100 − (EE + ADF + CP + ash). The primary limitation of using NE values based on chemical analysis is that the equation does not assess added value due to grain processing.

The computations used in estimating feed intake are based on generalized relationships between energy intake and energy gain. Although the accuracy of the program is normally good, it should be considered as only a tool for evaluating animal performance. Many factors can contribute to variance between observed and predicted performance, including genetics, health, weighing conditions, weather, pen (location, condition, bullers), animal management, and estimates of the NE value of the diet. Much of the variation from animal type, feedlot site, and prevailing weather can be removed by using a subroutine located at the beginning of the pro-

Table 16-4 *Basic language program for estimating diet NEm and NEg based on live weight, daily weight gain, and daily feed intake*

```
10 PRINT "NEm NEg (Mcal/kg) estimator "
20 INPUT "Continue";Y$
30 CLS
40 PRINT"                    Animal Types"
50 PRINT"type 1 = Medium-frame steer calves"
60 PRINT"type 2 = Large-frame steer calves"
70 PRINT"           Compensating medium-frame yearling steers"
80 PRINT"           Medium-frame bull calves"
90 PRINT"type 3 = Large-frame bull calves"
100 PRINT"          Compensating large-frame yearling steers"
110 PRINT"type 4 = Medium-frame heifer calves"
120 PRINT"type 5 = Large-frame heifer calves"
130 PRINT"          Compensating yearling heifers"
140 PRINT"type 6 = Mature thin cows"
150 PRINT"type 7 = Holstein steer calves"
160 PRINT " "
170 PRINT " "
180 INPUT "animal type";TYPE
190 INPUT "initial weight, kg (shrunk)";IW
200 INPUT "final weight, kg (shrunk)";FW
210 INPUT "daily gain, kg/d";G
220 INPUT "intake";I
230 NM = 1.6
240 AW = (IW + FW)/2
250 EM = .077*(AW^.75)
260 IF TYPE = 7 THEN EM = .086*(AW^.75)
270 IF TYPE = 1 THEN EG = (.0557*(AW^.75)*(G^1.097))
280 IF TYPE = 2 THEN EG = (.0493*(AW^.75)*(G^1.097))
290 IF TYPE = 3 THEN EG = (.0437*(AW^.75)*(G^1.097))
300 IF TYPE = 4 THEN EG = (.0686*(AW^.75)*(G^1.119))
310 IF TYPE = 5 THEN EG = (.0608*(AW^.75)*(G^1.119))
320 IF TYPE = 6 THEN EG = 6.2*G
330 IF TYPE = 7 THEN EG = (.0557*(AW^.75)*(G^1.097))
340 NG = (NM*.87) - .41
350 IF EG< = (I - (EM/NM))*NG THEN GOTO 370
360 NM = NM + .01:GOTO 260
370 PRINT TAB(10)NM;TAB(20)NG
380 INPUT "continue (Y/N)";Y$:IF Y$ = "Y" THEN GOTO 190
390 END
```

Adapted in part from National Research Council: Nutrient requirements of beef cattle, ed 6, Washington, DC, 1984, National Academy of Sciences, National Academy Press.

gram. The subroutine utilizes prior close outs that have been selected to typify the cattle type and feeding period to create an adjusted maintenance coefficient, which is then inserted into the equations for predicting feed intake.

Usually, observed feed intake is roughly 10% greater than predicted from weight gain during the initial 56-day period. This decrease in energetic efficiency is most likely a reflection of health-related stress. Throughout the remainder of the feeding period, a variance between observed and predicted intake of greater than 5% can be considered as significant, and efforts should be made to ascertain its cause.

Table 16-5 *Basic language program for evaluating variance in check-weight and close-out performance data*

```
10 PRINT"                   Animal Types"
20 PRINT"type 1 = Medium-frame steer calves"
30 PRINT"type 2 = Large-frame steer calves"
40 PRINT"            Compensating medium-frame yearling steers"
50 PRINT"            Medium-frame bull calves"
60 PRINT"type 3 = Large-frame bull calves"
70 PRINT"            Compensating large-frame yearling steers"
80 PRINT"type 4 = Medium-frame heifer calves"
90 PRINT"type 5 = Large-frame heifer calves"
100 PRINT"            Compensating yearling heifers"
110 PRINT"type 6 = Mature thin cows"
120 PRINT"type 7 = Holstein steer calves"
130 PRINT " "
140 PRINT " "
150 INPUT "animal type";TYPE
160 INPUT "Do you want to generate a maintenance adjustment factor      (Y/N)";Y$
170 IF Y$="N" THEN INPUT "What is the maintenance coefficient (ie .077 for beef types, .085 for Holstein)";MQ:GOTO 410
180 INPUT "How many close-outs will be used to generate adjustment     factor";N1
190 PRINT " "
200 MQ1=0
210 FOR N=1 TO N1
220 INPUT "Initial weight, kg (shrunk)";PW
230 INPUT "Final weight, kg (shrunk)";SW
240 INPUT "Average daily gain, kg/d";G
250 INPUT "Average daily feed, kg/d";FI
260 INPUT "NEm of diet, Mcal/kg";NEM
270 NEG=(.877*NEM)-.41
280 AW=(PW+SW)/2
290 IF TYPE=1 THEN EG=(.0557*(AW^.75)*(G^1.097))
300 IF TYPE=2 THEN EG=(.0493*(AW^.75)*(G^1.097))
310 IF TYPE=3 THEN EG=(.0437*(AW^.75)*(G^1.097))
320 IF TYPE=4 THEN EG=(.0686*(AW^.75)*(G^1.119))
330 IF TYPE=5 THEN EG=(.0608*(AW^.75)*(G^1.119))
340 IF TYPE=6 THEN EG=6.2*G
350 IF TYPE=7 THEN EG=(.0557*(AW^.75)*(G^1.097))
360 MQ=((FI-(EG/NEG))*NEM)/(AW^.75):PRINT FI;G
370 MQ1=MQ1+MQ:NEXT N
380 MQ=MQ1/N1
390 PRINT TAB(20)"adjusted maintenance coefficient";MQ
400 PRINT " "
410 PRINT "Check-weight or close-out analysis for feedlot cattle"
420 INPUT "Owner of cattle";OWN$
430 INPUT "Lot # for cattle";LT$
440 INPUT "Initial weight, kg (shrunk)";PW
450 INPUT "Final weight, kg (shrunk)";SW
460 INPUT "Observed average daily gain, kg/d";G
470 INPUT "Observed feed intake, kg/d";FEED
480 INPUT "Average NEm value of the diet, Mcal/kg";NM
490 NG=(.877*NM)-.41:AW=(SW+PW)/2
500 IF TYPE=1 THEN
    FA=((.0557*(AW^.75)*(G^1.097))/NG)+((MQ*(AW^.75))/NM)
510 IF TYPE=2 THEN
```

Table 16-5 *Basic language program for evaluating variance in check-weight and close-out performance data—cont'd*

```
          FA = ((.0493*(AW^.75)*(G^1.097))/NG) + ((MQ*(AW^.75))/NM)
520 IF TYPE = 3 THEN
          FA = ((.0437*(AW^.75)*(G^1.097))/NG) + ((MQ*(AW^.75))/NM)
530 IF TYPE = 4 THEN
          FA = ((.0686*(AW^.75)*(G^1.119))/NG) + ((MQ*(AW^.75))/NM)
540 IF TYPE = 5 THEN
          FA = ((.0608*(AW^.75)*(G^1.119))/NG) + ((MQ*(AW^.75))/NM)
550 IF TYPE = 6 THEN FA = ((6.2*G)/NG) + ((MQ*(AW^.75))/NM)
560 IF TYPE = 7 THEN
          FA = ((.0557*(AW^.75)*(G^1.097))/NG) + ((MQ*(AW^.75))/NM)
570 FI = ((SW − PW)/G)*FA
580 FCONV = FA/G:ADF = FA
590 FCG = (FA*FC)/G
600 PRINT TAB(5)"Owner: ";OWN$
610 PRINT TAB(5)"Lot #: ";LT$
620 PRINT TAB(5)"Breed type: ";TYPE
630 PRINT " "
640 PRINT " "
650 PRINT TAB(5)"Initial weight, kg (shrunk)";TAB(35)PW
660 PRINT TAB(5)"Final weight, kg (shrunk)";TAB(35)SW
670 PRINT TAB(5)"Observed weight gain, kg/d";TAB(35)G
680 PRINT TAB(5)""
690 PRINT TAB(5)"Observed feed intake, kg/d";TAB(35)FEED
700 PRINT TAB(5)"Expected feed intake, kg/d";TAB(35)ADF
710 PRINT TAB(5)"Expected/observed feed intake";TAB(35)ADF/FEED
720 PRINT TAB(5)""
730 PRINT TAB(5)"Observed feed/gain";TAB(35)FEED/G
740 PRINT TAB(5)"Expected feed/gain";TAB(35)FCONV
750 PRINT TAB(5)"Expected/observed
          feed/gain";TAB(35)FCONV/(FEED/G)
760 PRINT ""
770 PRINT ""
780 PRINT ""
790 INPUT "Do you want to change inputs and recalculate analysis
          (Y/N)";Y$
800 IF Y$ = "Y" THEN CLS:GOTO 410
810 END
```

Adapted from National Research Council: Nutrient requirements of beef cattle, ed 6, Washington, DC, 1984, National Academy of Sciences, National Academy Press.

Fine-tuning the cattle-feeding process to achieve an optimum level of performance requires a great deal of personal attention, hard work, and, most of all, persistent effort.

REFERENCES

1. Fox DG, Sniffen CJ, and O'Connor JD: J Anim Sci 66:1475, 1988.
2. Hicks RB and others: J Anim Sci 68:254, 1990.
3. Lofgreen GP: California Feeders Day Report. 1973, pp 35-42.
4. Nath K and Agarwala ON: J Agricul Sci 100:359, 1983.
5. National Research Council: Nutrient requirement of beef cattle, ed 6, Washington, DC, 1984, National Academy of Sciences.
6. National Research Council: Predicting feed intake of food-producing animals, Washington, DC, 1987, National Academy of Sciences.
7. Oltjen JW and others: J Anim Sci 62:86, 1986.
8. Oltjen JW and Garrett WN: J Anim Sci 66:1732, 1988.
9. Reid JT and White OD: Proc Cornell Nutr Conf 1977, pp 16-27.
10. Tatum JD and others: J Anim Sci 62:121, 1986.
11. Zinn RA: J Anim Sci 66:1755, 1988.

Nutritional Problems in Cow/Calf Practice

JONATHAN M. NAYLOR

Beef cattle are often used to harvest poor-quality forage. Many of the problems in this type of enterprise are associated with nutrient deficiencies.

ENERGY AND PROTEIN

Energy and protein deficiencies often go hand-in-hand in beef cattle because poor-quality pasture is often low in both energy and protein density. Furthermore, inadequate protein intake reduces digestion of feed in the rumen, compounding the energy deficit.

Protein-energy malnutrition (protein-calorie malnutrition, starvation) is usually seen in the fall and winter months, when cattle are grazing on crop residues or are being fed preserved feeds. During periods of drought, malnutrition may also occur at pasture. Sometimes protein-energy malnutrition develops because managers misjudge the holding capacity of a field. When cattle graze crop residues after harvesting, they initially eat leftover grain and seeds, which are high in digestible energy. Later they eat the leaves. As time passes they are forced to eat poorer-quality, fibrous, stem material, which is low in digestible energy.[40] At this time the cattle may lose condition even though casual inspection of the field may suggest adequate forage. However, although stemmy material may be abundant, the more nutritious residues will have been consumed.

Protein-energy malnutrition usually occurs when feed is in short supply. Managers often cut back on the amount of conserved feed offered to cattle or extend grazing periods past the time that pasture or crop residues are capable of supporting the cattle. Managers may overestimate the amount and quality of conserved feed on hand. For example, large hay bales are difficult to weigh, and digestible energy content can decrease by 13% during storage if they are damaged by bad weather. Wastage of 20% to 40% of hay can occur during unrolling if the bales are fed on the ground.[40]

Heavily pregnant cattle are most prone to protein-energy malnutrition. The fetus has a high priority and takes nutrients at the expense of the dam. Young pregnant heifers are at a particular disadvantage because they also need energy for growth; furthermore, mature cattle may push pregnant heifers aside if there is competition for feeding space. Poor dentition, internal or external parasitism, and cold, wet weather also increase the susceptibility to malnutrition. Parasite loads may also be greater in malnourished animals, presumably because host defenses are weakened.

In protein-energy malnutrition cattle progressively lose body condition. Weight loss may be partly masked by increased gut fill if poorly digestible feeds are eaten. Cattle often appear alert and have good appetites until the terminal stages. In the later stages body temperature may be subnormal, particularly if the weather is cold. Cattle may become recumbent, particularly if they are heavily pregnant. As body stores are used up, affected cattle sink into oblivion. Death usually occurs 1 to 2 weeks after they become recumbent.[41] Diarrhea is sometimes seen in cattle suffering from malnutrition; this can be the result of unsuitable feeds (e.g., moldy hay) or the sudden reintroduction of large amounts of grain.

In some circumstances malnourished cows survive to parturition; however, only small amounts of colostrum and milk are produced. The calves then suffer from starvation and debilitation. At first they may show signs of hunger and bawl for food.[42] Infectious diseases can be particularly severe in malnourished calves because lack of colostrum and poor immune function secondary to starvation increase the calves' susceptibility to disease.[17,30,49]

Diagnosis of protein-energy malnutrition is usually based on the body condition of the cattle, feed availability, and appetite. Cattle that are in poor condition secondary to malnutrition usually have good appetites, whereas cattle that are thin as a result of disease often eat poorly. The availability of feed and its quality should be assessed. It is important to quantify the contribution of parasitism and other diseases to poor body condition.[41,42,49] Serous atrophy of fat and absence of disease at necropsy helps confirm the diagnosis of protein-energy malnutrition.

Prevention of protein-energy malnutrition involves recognition of the low nutritional value of poor-quality forages, provision of an appropriate balanced ration, and planning so that the number of cattle over wintered does not exceed the amount of feed available. If additional feed has to be brought in to support cattle, it may be sensible to make the purchases early in the season. As winter progresses, feed gets shorter in supply and becomes more expensive. If inadequate feed is available, cattle should be sold before they get in poor condition. Selling cattle early in the feeding season means they weigh more and bring more money. Furthermore, the amount of stored feed required is reduced. If cattle are culled when the feed runs out, much valuable feed has already been spent in maintaining the culls. At times when feed shortages seem imminent, for example, following a summer drought, it is especially worthwhile for veterinarians and farmers to discuss feeding and management policies. Body scoring the cattle as they enter the feeding period is helpful. Cattle in good condition can use body fat to help tide them over winter. Poor-condition cows need to be fed well at all times and are particularly susceptible to cold. Culling poor cows early reduces demands for feed.

Abomasal impaction

Abomasal impaction usually arises secondary to traumatic reticuloperitonitis or to feeding excessive amounts of fibrous feeds.[4] Traumatic reticuloperitonitis causes impaction by interfering with abomasal motility. The cause is contamination of feed by sharp objects, not nutrition per se.

Dietary abomasal impaction is most commonly seen in beef cattle wintered on straw with readily digestible feed such as grain or hay. Straw is a useful feed for maintaining beef cattle if handled appropriately; problems arise when it is fed at high levels in an imbalanced ration. Diets containing 80% or more straw are most likely to produce impaction.[4,8] Chopping reduces particle size and increases digestibility. However, if a screen smaller than 1.5 cm (0.5 inches) is used in the machine, the risk of abomasal impaction increases because of reduced retention time in the rumen. If straw is not chopped and cattle gorge on it, then ruminal impaction may result. Although some believe that cold weather may increase feed intake and the incidence of abomasal impaction, one study failed to find an association with cold weather.[4] Nutritional considerations are consistent with the latter finding. Cattle fed diets high in straw are likely to be eating as much as possible to support maintenance requirements even at normal temperatures; the large amounts of heat released in the digestion of straw should help protect against cold stress.

Abomasal impaction is characterized by abdominal distension that is particularly pronounced low in the right flank. It is often possible to feel the hard mass of the impacted abomasum by ballottement through the body walls. Appetite and fecal output are reduced; fecal consistency is usually normal or dry, but occasionally is watery. Differentiation of dietary impaction from traumatic reticuloperitonitis depends on dietary, historical, and epidemiological information. Cattle with impaction secondary to traumatic reticuloperitonitis tend to be sick longer than cattle with dietary impactions (4 to 20 days versus 2 to 4 days in one study). Dietary impactions often involve several to many animals in the group, whereas impactions secondary to traumatic reticuloperitonitis usually occur singly. Necropsy, surgical exploration, and sometimes radiology are useful in definitively ruling out traumatic reticuloperitonitis.

Abomasal impaction can be prevented in straw-fed cattle by also feeding grain and a protein source. The protein source can be good-quality hay or an oilseed meal. Alternatively, the straw can be ammoniated. Straw intake should be limited to a maximum of 60% of the dry matter, and the total ration should be balanced. A 2-cm (0.75-inch) screen in the chopping machine is a reasonable size, although there is considerable machine variation in the fineness of the chop. These practices allow cattle to ingest adequate amounts of energy and protein within the confines of dry matter intake. The energy and protein from the more digestible feeds are used by rumen microbes to help degrade straw. In cold climates, heat release associated with straw digestion helps maintain body temperature, thus conserving other nutrients for maintenance and gestational requirements.

MACROMINERALS
Phosphorus
Simple phosphorus deficiency is a common problem in cattle at pasture. Anecdotal evidence suggests that it may be more common in the Great Plains and prairie areas of North America. The major signs of phosphorus deficiency are poor growth and reproductive rates, weight loss, and weakened bones. Postparturient hemoglobinuria may also be seen; this syndrome is discussed in more detail in Chapter 3, pp. 42-44.

Magnesium
Hypomagnesemia is an important problem in cattle at pasture (Chapter 15, p. 183) and in wintered cattle (Chapter 3, p. 51).

TRACE MINERALS
Copper
Copper deficiency can be *primary* (due to inadequate concentration of copper in the diet) or *secondary* (when copper metabolism is inhibited by other minerals, most commonly molybdenum and sulfates). The latter minerals interfere with common metabolism at a number of levels including feed copper availability, digestion of copper, and metabolism of copper within the body. In the rumen, sulfates are reduced and react with molybdenum to form thiomolybdates. They interfere with ruminal absorption and also increase the rate of copper excretion from the body.[47] High levels of calcium, zinc, selenium, and iron also interfere with copper absorption.[6,35,39,47,53,54,60] Secondary deficiency due to molybdenum toxicity is a common problem in certain soil types and is referred to as *teart scours, peat scours,* or *salt sickness.*[8]

Signs of copper deficiency are usually seen in cattle at pasture because copper availability is low in roughages, particularly in fresh grass. Availability is higher in concentrates and *Brassicas* (e.g., canola) fed to housed cattle. Improved pastures are particularly likely to produce copper deficiency because molybdenum or lime (calcium oxide) fertilizers may have been applied to improve plant growth. Hay making and ensiling increase copper availability in the feed by breaking down bonds between thiomolybdates and sulfides. Consequently, copper deficiency is rare in housed cattle, but poisoning is more common.[54]

Calves are particularly susceptible to copper deficiency. Although colostrum is rich in copper, milk is a poor source, particularly in copper-deficient cattle. Molybdenum is concentrated in milk.[8]

The principal sign of copper deficiency in beef cattle is reduced growth rate in calves.[54,60] This may be accompanied by loss of sheen and lightening of hair coat color. Cattle with black coats may develop bleached gray or dull red hair. In some cases light colored hair around the eyes can give a spectacled appearance.[8] Unfortunately, hair coat changes are an unreliable sign because they are only seen in advanced deficiency. Furthermore, in cattle with normal copper status, coat hairs may turn red as they are shed. Some breeds naturally have light hair. Severely deficient calves may show epiphyseal swelling, lameness, angular limb deformities, and reduced amounts of cortical bone. They are more prone to fractures.[36]

Other possible signs of copper deficiency are anestrus and reduced fertility in adult cattle.[23,24,54] Copper deficiency can weaken blood vessel walls and cause cardiac hypertrophy, myodegeneration, and heart failure.[47] Anemia caused by defective erythropoiesis can be seen; the copper-containing protein ceruloplasmin is required to reduce ferrous to ferric ions prior to transport of iron via transferrin to hematopoietic tissue from reticuloendothelial stores or mucosal cells. Copper deficiency has been implicated in outbreaks of postparturient hemoglobinuria in Australia. This is probably because the copper-containing enzyme superoxide dismutase helps to inactivate oxidizing agents that otherwise damage red cell membranes.[32,46,47]

In sheep, in utero copper deficiency interferes with myelination of spinal cord neurones. Clinically affected lambs show signs of posterior ataxia. It may be present at birth or develop within the first 6 months of life. The disease is called *swayback* or *enzootic ataxia*.[47] Enzootic ataxia is extremely rare in cattle.

Copper supplementation of pregnant cattle can have long-term advantages to the calves as a result of improved milk production or improved transfer of copper to the calf via the placenta, colostrum, and milk.[38]

Diarrhea is particularly common in cattle with copper deficiency secondary to molybdenum toxicosis; it is less common in primary copper deficiency.[47,60] Villous atrophy is partly responsible for diarrhea and poor growth rate.[36] Copper deficiency can impair immune function[47,55] and may increase the susceptibility of cattle to enteric infections.

Diagnosis of copper deficiency can be difficult. Definitive evidence is afforded by improvement in health or productivity following copper supplementation. Diarrhea due to molybdenum-induced copper deficiency responds rapidly to copper supplementation. Disadvantages of copper supplementation trials are the need for an untreated control group, the time delay needed to document subtle improvements in growth rate or health, and the danger, especially in sheep, of toxicity if copper was not deficient.

Tissue copper analysis and enzyme function tests are usually used to identify situations where copper supplementation may be beneficial. The two most commonly used measures are plasma and liver copper concentration. These two tests have different limitations and often need to be used together to obtain a more accurate picture of herd copper status. Liver copper reflects the cumulative history of copper balance in the animal. Samples of liver can be obtained at necropsy, at slaughter, or by biopsy. Many laboratories require approximately a gram of tissue for this analysis. This means that special large-bore biopsy instruments with an internal diameter of 3 to 5 mm need to be used to obtain liver samples from live animals. A technique to obtain large biopsies successfully has been described.[50] Low liver copper levels indicate that deficiency is either in progress or imminent; high levels are a warning of potential toxicosis (Table 17-1). Liver copper should be checked when high levels of copper are being fed long-term to cattle so that toxicity can be avoided. Liver concentrations are a poor guide of immediate copper balance. Levels may be low in recently weaned calves as a result of poor copper availability in pasture. However, supplementation may be unnecessary if the calves have just been switched to a diet of conserved feeds for the winter. Copper availability in the preserved feeds may be sufficient to meet requirements without further supplementation.

Plasma and serum copper and ceruloplasmin concentrations are highly correlated.[7] Measurements of plasma copper concentrations are generally used, partly because of the much greater body of knowledge regarding plasma copper concentrations in health and deficiency. Plasma is also preferred over serum because clotting removes variable amounts of copper and ceruloplasmin. In general, plasma copper concentrations are about 15% higher than serum values.[28] The major problem with plasma copper levels is that they are poor indicators of sufficiency. Levels fall only in severe deficiency when liver copper stores are exhausted (Fig. 17-1). Levels are reasonably constant over a wide range of marginal to excessive levels of copper intake. Concentrations rise above the normal range only during toxic crises. A problem with plasma copper estimations is that all the copper may not be biologically available. For example, cattle supplemented with iron or molybdenum can have identical serum copper levels, but the molybdenum-treated animals will show signs of deficiency. Presumably some of the circulating plasma copper is bound to thiomolybdates,

Table 17-1 *Interpretation of serum and liver copper concentrations*

Status	Plasma (per L)	Liver (per kg dry matter)*
Intoxication		
μmol (plasma), mmol (liver)	>50	>6
mg (ppm)	>3	>400
Normal		
μmol (plasma), mmol (liver)	8-24†	1.5-6
mg (ppm)	0.5-1.5†	100-400
Marginal‡ (production response likely)		
μmol (plasma), mmol (liver)	<8	<0.6
mg (ppm)	<0.5	<40
Deficient (deficiency disease signs likely)		
μmol (plasma), mmol (liver)	<3	<0.08
mg (ppm)	<0.19	<5

*Divide values by 3.5 to obtain concentration per kg wet matter of liver.
†Plasma values can be in the normal range even though the liver is deficient in copper.
‡For evaluating plasma copper status, samples should be obtained from a representative proportion of the herd. Occasional cattle with normal copper status have plasma values in the deficient range (less than 5% of plasma samples from cattle with normal copper status are in the marginal range).

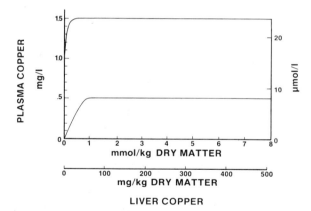

Fig. 17-1 Relationship between plasma and liver copper concentrations. Plasma copper levels fall only when the liver is severely depleted.

which are unavailable to tissues. Inflammation elevates plasma ceruloplasmin and copper concentrations; normal copper levels in animals with inflammatory problems do not mean that dietary status is adequate.[5]

Measurement of the activity of copper-containing enzymes such as superoxide dismutase and tyrosinase is useful in documenting biochemical dysfunction. However, these enzymes may not mirror health and production responses. Furthermore, red cell superoxide dismutase changes only slowly with dietary copper availability because red cells have a long life span in the blood.

Hair copper concentration reflects plasma copper concentrations during the growth phase of the hair. It is thus retrospective and may be useful in confirming copper deficiency in animals showing clinical signs. It is of limited use as a predictor of when to supplement to avoid deficiency. A major limitation is the problem of obtaining a sample of recently formed hair; old hair could reflect nutritional status months in the past.[14,25]

Copper deficiency is usually diagnosed on the combination of pathological or clinical signs of copper deficiency disease and low serum and liver copper concentrations. If plasma copper concentrations are well into the deficient range, liver copper is usually low, and estimation of liver status is not usually necessary. When plasma copper concentrations are normal, liver concentrations may be marginal, and a positive response to copper supplementation is possible. In some situations a controlled copper response trial may be warranted to determine copper adequacy, particularly if dealing with large numbers of cattle.

Nutritional policy should aim at maintaining liver copper levels in the range of 100 to 300 mg/kg (ppm) on a dry matter basis to prevent both deficiency and toxicity. As a general guideline, the diet should contain 10 ppm copper. The exact requirements of copper depend on the levels of other minerals and the type of feed. The ratio of copper to molybdenum should be at least 2:1; ratios of 5:1 may be safer, particularly when fresh grass is the main feed.[22,25,37] Sulfur levels greater than 0.16% of the dry matter exceed requirements and reduce copper availability. Although it is not unusual to supplement cattle diets with 20 mg of copper per kg of dry matter, in some circumstances as little as 4 mg of copper per kg of dry matter may be sufficient.

Cattle are relatively resistant to copper poisoning; however, cases have been seen. It is particularly likely to happen when diets containing more than 100 ppm copper (dry matter) are fed long-term.[6] In some circumstances, production efficiency may be impaired at levels of copper supplementation greater than 40 ppm copper (dry matter basis).[38]

Dietary supplementation with copper is easiest for cattle being fed concentrates on a daily basis. Copper can be added as a soluble salt to the mineral portion of the ration so that the desired copper concentration (usually 10 to 20 ppm copper on a dry matter basis) is obtained. There is some evidence that copper supplementation to pregnant beef cattle over the winter feeding period can subsequently benefit their calves if the summer pasture is deficient in copper.[38]

The group of cattle most likely to benefit from copper supplementation—calves reared on copper-deficient pasture—are the most difficult to supplement. The pasture can be dusted with copper sulfate at the rate of 5 to 7 kg/hectare (4.5 to 6 lb/acre). This treatment usually boosts herbage levels for 2 to 3 years. Salt blocks or licks containing 0.2% to 0.6% copper can be placed on the pasture. A loose salt mix containing 0.4% copper can be made by mixing 16 g of feed-grade copper sulfate ($CuSO_4 \cdot 5H_2O$, 25% copper) on an as-fed basis per kg of salt.[31] Assuming a 200 kg (440 lb) calf ate 10 g of salt and 4 kg of dry matter daily, this salt mix would provide an additional 10 mg of copper per kg of ration.

The alternatives to dietary copper supplementation are injection of slow-release preparations or oral dosing with copper-containing boluses. Copper injections are usually given subcutaneously; copper is released into the bloodstream and stored in the liver.

The two injectable preparations that are in common use worldwide for prevention and treatment of copper deficiency in cattle are copper glycinate (United States) and copper calcium EDTA (United Kingdom). Calcium EDTA releases copper more slowly than the sodium EDTA salt but is still too rapid for the product to be safe in sheep. Copper glycinate, while delivering copper much more slowly than the EDTA compounds, is associated with a higher rate of abscesses and adverse tissue reactions. The copper in these preparations is greater than 90% available.[52] The usual dosages given are about 50 mg of copper to calves less than 100 kg (220 lb) and 100 mg to larger cattle. Severely deficient adult cattle may be given 200 mg (consult manufacturer for specific recommendations). Benefits usually last about 4 months in growing calves but may be as short as 1 month in severely deficient areas.

Copper can be administered orally as part of trace-mineralized glass boluses or as copper oxide needles (fine rods 1 to 10 mm long) in a gelatin capsule. Glass boluses slowly release copper as they degrade in the rumen. Copper oxide needles lodge in the abomasum. They are preferred in areas of severe copper deficiency because they release more copper over longer periods. There is little risk of acute toxicity because of the slow release of copper, although chronic poisoning from oversupplementation is possible. Copper oxide needles provide protection for the entire grazing season (7 months). They are usually given at the rate of 0.1 g copper oxide per kg (2.2 lb) body weight.[15,44,45]

Selenium

Selenium and vitamin E function as antioxidants. Nutritionally they should be considered together because the deficiency of one can be mitigated by supplementation of the other.

The major defects caused by selenium and vitamin E deficiency in beef cattle are degeneration of skeletal and cardiac muscle in young, rapidly growing calves at pasture (white muscle disease).[3,26] Signs of skeletal muscle degeneration are muscle stiffness, soreness, and recumbency.[21] Signs are usually seen in well-grown calves at pasture or following movement to a new pasture. Cardiac muscle degeneration produces sudden death.[26]

A syndrome of poor growth and increased mortality not associated with white muscle disease or myodegeneration has been associated with selenium deficiency.[57] Deficiency has also been implicated as a cause of weakness and increased mortality in neonatal calves[51] and as a cause of retained placenta in cows.

Selenium enhances immune function, antibody production, and neutrophil bactericidal activity.[10,29,43,55,56] Selenium-deficient cattle are more susceptible to mastitis and recover more slowly from metritis. Morbidity in calves is also increased by selenium deficiency.[11,18,59]

Areas most likely to be deficient in selenium are the northwestern United States, the eastern seaboard, and the states bordering on the Great Lakes (Fig. 17-2). High soil sulfur concentrations inhibit uptake of selenium by plants and animals. Fish meals are often high in selenium.

Vitamin E deficiency is mainly seen in animals fed straw, poor-quality hay, or root crops. Vitamin E is destroyed in acid-treated, high-

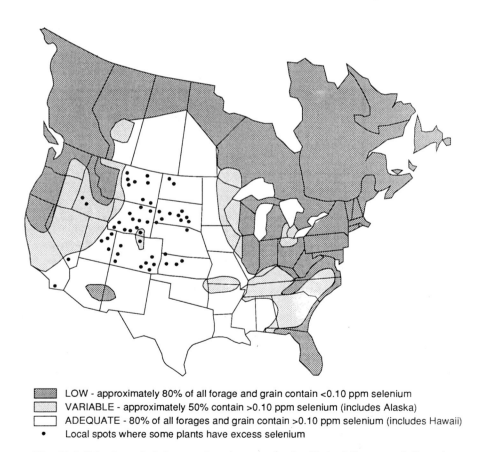

LOW - approximately 80% of all forage and grain contain <0.10 ppm selenium
VARIABLE - approximately 50% contain >0.10 ppm selenium (includes Alaska)
ADEQUATE - 80% of all forages and grain contain >0.10 ppm selenium (includes Hawaii)
• Local spots where some plants have excess selenium

Fig. 17-2 Selenium deficient and toxic areas in the United States and Canada.

Table 17-2 *Guidelines for the interpretation of selenium, glutathione peroxidase, and vitamin E levels in blood and liver*

Status	Blood selenium (ppb or µg/L)	Plasma selenium (ppb or µg/L)	Blood glutathione peroxidase (IU/g Hb)	Liver selenium (ppm or µg/g dry matter)
Normal	>100	>35	>50	0.90-1.75
Marginal	50-70	16-25	15-50	0.60-0.9
Deficient	10-49	2-15	<15	0.07-0.6

Normal plasma α-tocopherol concentrations are greater than 1.5 mg (3.5 µmol) per L in adult cattle. Transfer to the fetus is poor, and calves are born with half these levels. A level of less than 0.7 mg per L of α-tocopherol has been used to indicate vitamin E deficiency.

moisture grains. Vitamin E requirements are increased if diets rich in polyunsaturated fats are fed to calves. In situations where vitamin E is deficient, selenium supplementation may also be beneficial.

Diagnosis of selenium deficiency is complicated by the fact that vitamin E availability and amount of exercise influence whether clinical signs are seen.

In general, serum or plasma selenium status responds most rapidly to changes in dietary selenium intake. Serum selenium concentrations respond rapidly to changes in dietary intake and reach stable levels within 1 to 3 weeks of a change in diet. Erythrocyte glutathione peroxidase activity reflects historical selenium intake because this selenium-containing enzyme is formed when the red cells are synthesized. As a result, erythrocyte glutathione peroxidase activity takes several months to stabilize following a change in diet.[13,20] Whole-blood selenium concentrations are highly correlated with glutathione peroxidase activities.[2] About three quarters of blood selenium is found in the erythrocytes. Blood selenium changes at a rate intermediate between serum and erythrocyte glutathione peroxidase activities (Table 17-2).[48]

Prevention of deficiency usually relies on administration of a selenium supplement. If mixed rations are being fed, selenium can be incorporated into the diet. Cattle at pasture can be supplemented satisfactorily using trace-mineralized salt blocks. Although these usually contain 20 ppm Se, recent work suggests that higher levels—perhaps 40 ppm selenium—may give better results in some circumstances.[12,56]

Injectable selenium preparations are available. They have the disadvantage of a relatively short duration of action and require individual handling of the cattle. They do have the advantage of ensuring that all animals get treated with a consistent dosage. However, selenium-mineralized salt preparations are just as effective and require less handling.

In some parts of the world selenium is available in long-release ruminal boluses. A selenium-iron pellet provides about a 1-year supply of selenium.[20,34,57] Long-acting selenium-mineralized glass boluses are available but do not appear to be as effective.[19] Reaction cements gradually release selenium over a 6-month period.[33]

Selenium is efficiently transferred across the placenta to the fetus.[58]

The margin between beneficial and toxic selenium concentrations is narrow, and care must be taken to avoid oversupplementing with selenium. Selenium toxicity can be a problem in some localities where plants and soil naturally contain very high levels of selenium (Fig. 17-2).

Cobalt

Cobalt is required for the synthesis of vitamin B_{12} by ruminal microbes. The signs of cobalt deficiency are those of vitamin B_{12} deficiency and include poor growth and ill thrift in growing cattle and weight loss and ketosis in adult cattle. In addition, clinical signs of anemia may be present. Cobalt-deficient cattle are more prone to internal parasitism. Cobalt-deficient areas are

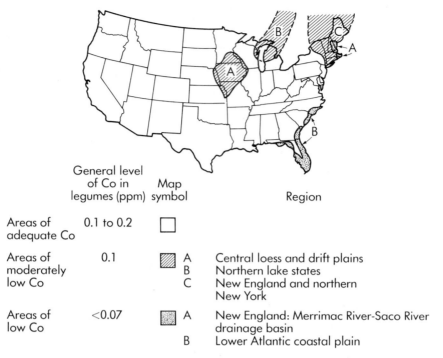

Fig. 17-3 Cobalt-deficient areas of the United States and Canada. Modified from Ammerman CB: J Dairy Sci 53:1097, 1970.

shown in Fig. 17-3. Vitamin B_{12} levels are used to diagnose cobalt deficiency. Further details on the clinical signs of vitamin B_{12} deficiency and the interpretation of vitamin B_{12} levels are given in Chapter 5, p. 85.

REFERENCES

1. Ammerman CB: J Dairy Sci 53:1097, 1970.
2. Anderson PH, Patterson DSP, and Berrett S: Vet Rec 103:145, 1978.
3. Arthur JR: J Nutr 118:747, 1988.
4. Ashcroft RA: Can Vet J 24:375, 1983.
5. Auer DE and others: Vet Rec 124:235, 1989.
6. Banton MI and others: J Am Vet Med Assoc 191:827, 1987.
7. Blakley BR and Hamilton DL: Can J Comp Med 49:405, 1985.
8. Blood DC and others: Veterinary medicine: a textbook of the diseases of cattle, sheep, pigs, goats and horses, ed 7, 1989.
9. Claypool DW and others: J Anim Sci 41:911, 1975.
10. Droke EA and Loerch SC: J Anim Sci 67:1350, 1989.
11. Erskine RJ and others: Am J Vet Res 50:2093, 1989.
12. Eversole DE and others: Cornell Vet 78:75, 1988.
13. Field AC and others: Vet Rec 123:97, 1988.
14. Fisher DD and others: Am J Vet Res 46:2235, 1985.
15. Gallagher J and Cottrill BR: Vet Rec 117:468, 1985.
16. Grasso PJ and others: Am J Vet Res 51:269, 1990.
17. Griebel PJ, Schoonderwoerd M, and Babiuk LA: Can J Vet Res 51:428, 1987.
18. Harrison JH and others: J Dairy Sci 69:1421, 1986.
19. Hidiroglou M and Proulx J: Ann Rech Vet 19:187, 1988.
20. Hidiroglou M, Proulx J, and Jolette J: J Dairy Sci 68:57, 1985.
21. Hoshino Y and others: Nippon Juigaku Zasshi 51:741, 1989.
22. Humphries WR and others: Br J Nutr 49:77, 1983.
23. Ingraham RH and others: J Dairy Sci 70:167, 1987.
24. Kappel LC and others: Am J Vet Res 45:346, 1984.
25. Kellaway RC, Sitorus P, and Leibholz JML: Res Vet Sci 24:352, 1978.
26. Kennedy S and Rice DA: Am J Pathol 130:315, 1988.
27. Khan AA and others: Can J Vet Res 51:174, 1987.
28. Kincaid RL, Gay CC, and Krieger RI: Am J Vet Res 47:1157, 1986.
29. Kiremidjian-Schumacher L and others: Proc Soc Exp Biol Med 193:136, 1990.

30. Logan EF: Br Vet J 133:120, 1977.
31. Maas J and Smith BP: In Smith BP, editor: Large animal internal medicine, pp 829-832, St Louis, 1990, The CV Mosby Co.
32. MacWilliams PS, Searcy GP, and Bellamy JEC: Can Vet J 23:309, 1982.
33. Manston R and Gleed PT: J Vet Pharmacol Ther 8:374, 1985.
34. McClure TJ, Eamens GJ, and Healy PJ: Aust Vet J 63:144, 1986.
35. Miller WJ and others: J Dairy Sci 72:1499, 1989.
36. Mills CF, Dalgarno AC, and Wenham G: Br J Nutr 35:309, 1976.
37. Miltimore JE and Mason JL: Can J Anim Sci 51:193, 1971.
38. Naylor JM and others: Can J Vet Res 53:343, 1989.
39. Nederbragt H, van den Ingh TS, and Wensvoort P: Vet Q 6:179, 1984.
40. Oetzel GR and Berger LL: Comp Cont Educ 7:S672, 1985.
41. Oetzel GR and Berger LL: Comp Cont Educ 8:S16, 1986.
42. Oetzel GR and others: Comp Cont Educ 6:S277, 1984.
43. Reffett JK, Spears JW, and Brown Jr TT: J Nutr 118:229, 1988.
44. Richards DH and others: Vet Rec 116:618, 1985.
45. Rogers PAM and Poole DBR: Vet Rec 123:147, 1988.
46. Sas B: Vet Hum Toxicol 31:29, 1989.
47. Saunders DE: Comp Cont Educ 5:S404, 1983.
48. Scholz RW and Hutchinson LJ: Am J Vet Res 40:245, 1979.
49. Schoonderwoerd M and others: Can Vet J 27:365, 1986.
50. Smart ME and Northcote MJ: Comp Cont Educ 7:S327, 1985.
51. Spears JW, Harvey RW, and Segerson EC: J Anim Sci 63:586, 1986.
52. Suttle NF: Vet Rec 109:304, 1981.
53. Suttle NF: Vet Rec 119:519, 1986.
54. Suttle NF: Vet Rec 119:148, 1986.
55. Suttle NF and Jones DG: J Nutr 119:1055, 1989.
56. Swecker WS Jr and others: Am J Vet Res 50:1760, 1989.
57. Underwood EJ: The mineral nutrition of livestock, ed 2, Slough, England, 1981, Commonwealth Agricultural Bureaux.
58. Van Saun RJ, Herdt TH, and Stowe HD: J Nutr 119:1156, 1989.
59. Waltner-Toews D, Martin SW, and Meek AH: Can J Vet Res 50:347, 1986.
60. Whitelaw A: In Practice 7:98, 1985.

Nutritional Problems in Management and Feedlot Practice

R. HIRONAKA

The nutrition of feedlot cattle requires special attention because the cattle are expected to gain weight rapidly. Cattle entering the feedlot may be deficient in nutrients necessary for high weight gains. Nutrition may also play a role in the animal's susceptibility to infectious diseases and metabolic and digestive disorders. Thus, proper nutrition, management, and health care are each part of the overall feedlot health program. A nutritional program for healthy cattle includes proper quantities and balance of nutrients and appropriate feed processing and feeding practices. It must also be economically viable. Least-cost diets or those that give the fastest gain do not necessarily give the greatest profit to the feedlot owner.

BRINGING CATTLE INTO THE FEEDLOT

Cattle entering a feedlot are often under stress from weaning, handling, shipping, and changes in feed and water, all of which result in weight loss. These stresses may increase the animals' susceptibility to disease. New arrivals should be given special attention, and those that show signs of disease should be removed from the pen and medicated. Stress can be minimized by immediately providing newcomers with adequate clean, dry bedding, fresh water, and a palatable feed such as loose hay. Because incoming cattle are often deficient in vitamin A, phosphorus, and various trace minerals, a starter diet with relatively high levels of these nutrients is recommended. Subclinical deficiency of vitamin A is especially common in some regions and may predispose cattle to respiratory ailments, resulting in higher losses. Commonly, 2 to 3 million IU of vitamin A are injected into all cattle entering a feedlot. This must be done using sterile techniques because bacteria introduced will cause an abscess and make the vitamin A unavailable to the animal. It is also possible to mix vitamin A with water and spray it on the hay just prior to feeding. About 500,000 IU per head should be provided in this manner for 4 consecutive days. This procedure also reduces the stress of putting the cattle through a chute to inject the vitamin A.

Cattle entering the feedlot require special dietary treatment to reduce morbidity and death losses. Appetite or willingness to consume feed is depressed during the first 1 to 3 weeks after arrival at a feedlot (Table 18-1).[21] It may be necessary to supplement the diet during this entire period to reduce the development of mineral and vitamin deficiencies.

Transportation increases weight loss (shrink) more than would be expected from feed and water restriction alone; this suggests that changes in metabolism may be caused by transportation stress, which results in weight loss.[21] Stress from transportation is believed to increase aldosterone secretion, which causes sodium retention and potassium secretion. Although 0.7% to 0.9% potassium in the diet meets the long-term potassium requirements of cattle, apparently an increased loss of potassium occurs dur-

Table 18-1 *Feed intake of newly arrived cattle*

Time after arrival at feed lot (days)	Dry matter intake (% of body weight)
1-7	0.5-1.5
8-14	1.5-2.5
15-28	2.5-3.5

ing stress. This loss of potassium, coupled with the reduced feed intake of stressed cattle, increases the dietary potassium requirement for incoming cattle. Hay diets generally provide about 2% potassium, which is sufficient if the hay is consumed in normal quantities. Less potassium is present in grain and corn silage-based diets. The addition of 1% to 3% potassium chloride to basal diets with 0.7% to 0.9% potassium over the first 2 weeks reduces morbidity and mortality in transported cattle.[22] The addition of 5% potassium chloride is excessive and results in mortality similar to that caused by nonsupplemented diets. Electrolyte supplements containing potassium that can be added to the water are also available and appear to be effective when feed intake is depressed.

PROCESSING FEEDS

Grains must be processed to improve digestibility.[25] Overprocessing, so that there is a large percentage (more than 25%) of fine-particle–size feed, may lead to disorders such as bloat, ruminitis, and acidosis as well as reduced feed intake.[16,17] Kernels need only be cracked to optimize rumen microbial digestion.[4] A coarsely cracked feed that contains some whole kernels is better than fully cracked feed with a high proportion of fine particles. Cattle eat more coarse than fine feed, resulting in increased weight gain and feed efficiency.[16] Feed that is composed of coarse, rather than fine, particles also reduces wind loss, produces less dust, and reduces irritation of the eyes and respiratory tract.

The feedlot operator can prevent fine-particle–size feed and maintain coarse particles by steam rolling or tempering the grain, which increases the moisture content to about 14% before processing.[18] The rollers in steam rolling should be set so that they only crack the grain and do not roll it out into a thin disk. To temper grain, experienced operators have found that adding 3% to 4% cold water to grain the first time that it is moved from storage after harvest is successful if the environmental temperature is above $-10°$ C ($14°$ F). In freezing temperatures, however, if the grain is moved before tempering, the water freezes before it penetrates the grain. Below $-10°$ C, hot water should be added.

Cattle are even less likely to eat a high percentage of fine particles in a particular meal if the grain is mixed with silage. If cattle must eat feed that contains many fine particles, digestive upsets can be reduced by adding 4% salt to the feed to slow the rate of intake, but this should be done only for a few weeks.[5]

High-moisture grains are made by adding water to raise the moisture content to about 30%. The advantage gained from feeding high-moisture grain is attributable to a reduced proportion of fine particles in the rolled grain. A similar effect can be achieved by tempering, without the need for airtight silos or preservatives.

NUTRITIONAL REQUIREMENTS
Use of high-concentrate feeds

Cattle accustomed to a high-roughage diet should be introduced gradually to high-concentrate diets to avoid digestive upsets. Ruminants that are accustomed to forage diets and then suddenly ingest a large quantity of starch develop ruminal acidosis, dehydration, and ruminal epithelial damage.[10] This is known as *grain overload* or *ruminal acidosis*. Starchy feeds and feeds that contain sugars such as fruit, some root crops, and molasses are capable of initiating the syndrome. Animals that are suddenly fed a high level of grain have a marked lowering of rumen pH (less than 6.0) and a high concentration of ruminal ethanol.[1] The ill effects of a sudden change in the quantity of grain are mediated through the rumen microflora. This has been demonstrated by transferring rumen contents from animals that were accustomed to grain to hay-fed animals just prior to feeding wheat and alfalfa pellets. Animals that received a transfer of ruminal contents from animals that were accustomed to grain did not become sick, but those that did not receive ruminal contents exhibited typical symptoms of grain overload.

There are many ways to gradually increase the amount of concentrate intake until the cattle are on full feed. Starter diets can help put the cattle on a full, high-concentrate diet safely and quickly.[13] Cattle can be introduced to a high-concentrate diet with an allowance equal to 1% of the body weight of each animal, provided in two equal feedings per day, and increased at the rate of 0.1% of body weight daily until the cattle are on full feed. Another method is to start with a ratio of 30% grain to 70% hay. The proportion of concentrate is increased by 10% and the proportion of hay decreased by 10% every second day until the cattle are eating the level of concentrate desired.

Protein

Rumen microorganisms convert nonprotein nitrogen such as urea to proteins that the animal can use. Because high levels of urea are toxic, no more than one third of the dietary nitrogen should be provided as urea. Urea should be used if it provides nitrogen at the lowest cost. If a protein supplement that contains urea is used, the amount that the animal consumes at one time must be limited and the urea mixed thoroughly with the feed to avoid toxicity. Pelleting a supplement containing urea appears to mask the taste of urea, which may result in high consumption and urea toxicity. A relatively high level of urea in liquid or salt supplement may be fed without toxic effects because cattle usually reduce intake of the urea-containing feed.[14] To prevent toxicity a lick-wheel mechanism should be used to limit the rate of intake of liquid supplement that contains urea.

Charred grain or overheated feed such as black silage may be deficient in available protein.

Minerals

High-concentrate diets, which are low in calcium and high in phosphorus, are detrimental (see Chapter 3, pp. 44-47). These problems can be avoided or corrected by adding 1% to 1.4% calcium carbonate (limestone or calcium flour) to the concentrate so that the Ca to P ratio is at least 1:1.[15] The addition of 1.8% or more limestone reduces feed intake, rate of gain, and feedutilization efficiency. Legume forages contain fairly high levels of calcium and moderate levels of phosphorus; thus diets with a legume generally do not require a limestone supplement. It may be necessary to add silage or about 2% molasses to prevent separation of the limestone supplement from the grain, especially if the grain is coarse.

A mild zinc deficiency may produce symptoms similar to vitamin A deficiency and predispose cattle to respiratory problems, rough dry hair, and ringworm.[23] Copper deficiency may be encountered when feed or water has a high molybdenum and sulfate content. Symptoms of copper deficiency include fading of the hair color and failing to shed the winter coat in spring. With more prolonged deficiency, cattle become anemic, and their rate of gain declines. With advanced deficiency in lambs, bone deformities occur.

Vitamins

Vitamin A deficiency symptoms include watery eyes, rough coat, a high incidence of respiratory disease, loss of appetite, reduced rate of gain, and susceptibility to ringworm. In cases of severe deficiency, cattle become blind and may even lose weight. Feedlot cattle require about 20,000 IU of vitamin A at a live weight of 200 kg (440 lb) and about 30,000 IU at 400 kg (880 lb). Although this is a daily requirement for each animal, up to a 1-month supply can be given in one dose. Vitamin A may be given orally or by injection, or it can be dissolved in water. If vitamin A is injected intramuscularly, a clean, sterile needle must be used for each animal. Vitamin A often is sold in combination with vitamins D and E.

Zinc is required in the metabolism of vitamin A.[29] Because feeds are commonly low in zinc, animals with apparent vitamin A deficiency often respond to a trace-mineralized salt containing about 0.5% zinc.

Carotene in hay and silage is not efficiently converted to vitamin A by cattle. The vitamin A equivalents of the carotene in the forage must be divided by 4 to determine the effective amount of vitamin A.[12] Vitamin A deficiency has been found in cattle fed silage diets even though chemical analysis indicated sufficient carotene to meet the vitamin A needs of the animal.[28] Some silages appear to increase the vitamin A requirement to a level above that usually recommended. The reason is not known but is thought to be due to a compound produced during fermentation that ties up vitamin A and causes it to be excreted.

Thiamin (vitamin B_1) deficiency is responsible for nutritional polioencephalomalacia, which is characterized by depression, total lack of appetite, blindness, stumbling gait, convulsions, opisthotonus, and finally coma and death.[11] The malady appears to be caused by feeding diets with a high level of readily available carbohydrates or by high-sulfate feed and water.[3] The condition responds to thiamin if caught early.[24]

PROBLEMS IN FEEDLOT PRACTICE
Urinary calculi

Urinary calculi (water belly, urolithiasis) is a common feedlot problem. Stones formed in the kidney or bladder block the ureter. Symptoms of urinary calculi include dribbling of urine or complete lack of urination, uneasiness, twitching of the tail, kicking at the belly, and standing with hind legs extended.[8] If the condition is not corrected, the bladder will rupture and spill the urine into the abdominal cavity, resulting in death from uremia. The usual treatment is surgical removal of the stones from the urethra. Occasionally, injection of a muscle relaxant to relax the urethra so that the stone may be passed is helpful. There are several types of stones, but the majority of stones in feedlot cattle are formed of phosphates or silicates. The latter are believed to be formed from the silica in grass and are in the cattle before they arrive at the feedlot.[2] Although several hypotheses have been raised about the etiology of phosphate stones, it appears that a low calcium to phosphorus ratio is the major cause. This occurs in high-concentrate diets that are not adequately supplemented with a calcium source such as limestone.[19] The diet should meet NRC recommendations for calcium and phosphorus intake. Calcium to phosphorus ratios should be at least 1:1.

Bloat

Bloat, a major problem in feedlots, is caused when gas that is naturally produced by the microflora in the rumen during feed digestion is trapped. It causes distension of the rumen and pressure on the diaphragm; eventually the animal is no longer able to breathe. Cattle normally expel the gases produced in the rumen by belching. The gas may be trapped by viscous rumen fluids and is termed *frothy bloat*. The condition develops rapidly, sometimes in less than 30 minutes from the time that an animal is observed to be normal until it is in critical distress.

Secondary bloat develops when a lack of rumen motility allows the rumen contents to continually block the esophagus so that the gas is trapped in the rumen. Free gas accumulates at the top of the rumen; this is termed *free-gas bloat*. This type of bloat is usually slower to develop than frothy bloat.

Many feedlot operators blame high-energy diets for causing bloat. However, all-concentrate diets have been fed with a very low incidence of bloat,[16] so it is now obvious that a factor other than high-energy diets is the cause. Frothy bloat occurs when there is a combination of readily available carbohydrate and fine particles in the rumen. This condition is created by feeding grain or forage diets that have been finely ground or chopped. Leaves of legumes such as alfalfa and clovers are broken down in the rumen, creating the condition of readily available carbohydrates and fine particles.[20] This combination results in slime secretion by rumen microorganisms and cell lysis that releases cell contents into the rumen fluid.[7] The resultant viscous rumen fluid traps gases produced by the rumen microflora in small bubbles. The entrapped gas is seen as a frothy mixture when released from the rumen. The high viscosity of the rumen fluid develops over a period of about 2 weeks on diets with fine-particle–size feed and readily available energy.[6] Mixtures of grain and legume hays are considered to be especially prone to causing frothy bloat.

Numerous treatments have been suggested for bloated animals, including drenching cattle with compounds such as mineral oils, adding detergent at 0.05% of the diet, using antifoaming agents, releasing the gas with a rubber hose, and cannulating or surgically fistulating the rumen. Because a major cause of feedlot bloat appears to be fine-particle–size feed, prevention by processing grain so that fine particles are avoided rather than treatment of bloat is recommended. A change in diet ingredients or proportions to reduce the amount of fine particles and legumes may eliminate the problem. Grass hay or even some legumes, which are digested more slowly than alfalfa, reduce the rate of digestion, reduce the rate that energy is made available to the rumen microflora, and can decrease the incidence of bloat.

Feed toxins

Ergot. Ergot is a fungal infection of the seed heads of grasses and grains. Rye often contains ergot, which reduces palatability and may cause restriction of blood flow to the extremities. More than 0.1% ergot in the total diet is potentially toxic. Rye, which is lower in energy content and less palatable than barley, should constitute no more than 25% of the concentrates in the diet.

Newly harvested grain. Consumption of newly harvested grain may cause digestive upsets and even mortality in cattle. The cause is unknown, and the occurrence is not predictable. A similar condition in pigs, called *hepatosis diatetica*, occurs in the fall when newly harvested grain is fed.[27] To avoid the problem, grain should be stored for at least 6 weeks after harvest before it is fed.

Frozen grain, smut, and rust. Immature grains frozen before harvest can be fed to cattle without ill effects. However, frozen green flax straw should not be fed to cattle because it can be toxic. Slightly frozen immature grain and unfrozen grain have similar nutritional value. Severely frozen immature grain has a higher fiber content and lower nutritional value than unfrozen grain. Digestible energy decreases roughly in proportion to the increase in fiber content.

Grains infected with smut, rust, and ergot have a higher fiber content and consequently a lower nutritional value than grain free of these diseases.

Sprouted grain. Although energy appears to be lost as heat from sprouting grain, diets containing up to at least 36% sprouted grain do not interfere with the rate of gain or feed efficiency of feedlot cattle.

Mycotoxins. Mycotoxins are toxic chemicals produced by molds which grow in moist feeds. Adult cattle are fairly resistant to some mycotoxins. However, special precautions should be taken to protect handlers of moldy feeds.[30] Calves with access to moldy straw that had high fungal concentrations of *Acrospeira macrosporoides* were afflicted with polioencephalomalacia.[9] This was attributed to a thiaminase produced by the fungi.

Miscellaneous feeds and by-products. Because beet tops are laxative, the amount fed should be limited to about 3 kg (6.5 lb)/day for calves and 12 kg (26 lb)/day for older cattle. Although beet tops contain moderate levels of calcium, they also contain oxalic acid, which combines with calcium and makes it unavailable to the animal. Feedlot cattle fed beet tops have been observed to develop a nervous condition that responded to the addition of about 1% limestone to the wet beet tops. Beet tops also may be toxic because of a high nitrate content. Cattle can choke on beet roots, and the feedlot operator must be prepared to treat choke if feeding beet roots.

Potato sprouts and sunburned (green-colored) potatoes should not be fed in large quantities because they contain glycoalkaloids, which are toxic.[26] The concentration of the toxic compounds is increased with exposure of the sprouts to light and warm, moist conditions. The symptoms of toxicity are dilated pupils and staring appearance, trembling, staggering, weakness, and sometimes convulsions.

Moldy sweet clover hay contains coumarin, which prevents the clotting of blood. Cattle that are to be dehorned, castrated, or operated on in any way should not be fed sweet clover hay for a few weeks before such an operation. To be safe, it is recommended that sweet clover be fed alternately at about 2-week intervals with another type of hay.

REFERENCES

1. Allison MJ, Bucklin JA, and Dougherty RW: J Anim Sci 23:1164, 1964.
2. Bailey CB: Can J Anim Sci 55:187, 1975.
3. Beke GJ and R Hironaka: Sci Total Environ 101:281, 1991.
4. Broadbent PJ: Anim Prod 23:165, 1976.
5. Cheng K-J and others: Can J Anim Sci 59:737, 1979.
6. Cheng K-J and Hironaka R: Can J Anim Sci 53:417, 1973.
7. Cheng K-J and others: Can J Microbiol 22:450, 1976.

8. Church DC: In Digestive physiology and nutrition of ruminants, vol 2, Corvallis, Ore, 1971, DC Church.
9. Davies ET and Pill AH: J Am Vet Med Assoc 154:1200, 1968.
10. Dunlop RH and Hammond PB: NY Acad Sci 119:1109, 1965.
11. Edwin EE and others: J Agricul Sci 87:679, 1976.
12. Edwin EE and Jackman R: Vet Res Commun 5:237, 1982.
13. Ensminger ME and Olentine CG: Feeds and nutrition, Clovis, Calif, 1978, Ensminger Publishing.
14. Hironaka R: Can J Anim Sci 49:181, 1969.
15. Hironaka R: Can J Anim Sci 58:795, 1978.
16. Hironaka R: Can J Anim Sci 68:179, 1988.
17. Hironaka R, Kimura N, and Kozub GC: Can J Anim Sci 59:395, 1979.
18. Hironaka R and others: Can J Anim Sci 53:75, 1973.
19. Hironaka R and Sonntag BH: Agriculture Canada Publication 1591, Ottawa, 1981, Research Branch.
20. Hoar DW, Emerick RJ, and Emery LB: J Anim Sci 30:597, 1970.
21. Howarth RE and Cheng K-J: In Proc 4th West Nutr Conf, Saskatoon, Saskatchewan.
22. Hutcheson DP and Cole NA: J Anim Sci 62:555, 1986.
23. Hutcheson DP, Cole NA, and McClaren JB: J Anim Sci 58:700, 1984.
24. Lamand M: Ir Vet J 38:40, 1984.
25. McDonald JW: Aust Vet J 58:212, 1982.
26. Morgan CA and Campling RC: J Agricul Sci 91:415, 1978.
27. Nicholson JWG: Agriculture Canada Publication 1527, Ottawa, 1974, Research Branch.
28. Oksanen HE: In Muth OH, Oldfield JE, and Weswig PH, editors: Selenium in biomedicine, 1967.
29. Smith GS and others: J Anim Sci 23:625, 1964.
30. Smith JC Jr and others: Science 181:954, 1973.
31. Trenholm HL and others: Agriculture Canada Publication 1827E, Ottawa, 1988, Research Branch.

SECTION B Dairy Cattle

19

The Dairy Industry in North America

RONALD L. TERRA

PRODUCTION AND CONSUMPTION TRENDS

The dairy industry has played an important role in American society since the first colonists arrived in the early seventeenth century. The small numbers of cows that first reached these shores have now multiplied into the 10 million plus that contribute to the nutritional support of many nations. In addition, hundreds of thousands of people make their living from one of the myriad jobs involved with the production and marketing of dairy products. Dairy farmers utilize the services of milkers, feeders, veterinarians, nutritionists, and other consultants.

Fresh milk is transported by refrigerated truck to a processing plant, where some is homogenized and pasteurized for sale as liquid milk. The remainder is processed into cheese, yogurt, ice cream, or powdered and evaporated milk. These products are sold by retailers to the consumer. At each point along this marketing trail, the services of many people are required. In 1987 the fluid milk industry alone accounted for the employment of 73,000 people with an annual payroll of more than $1.5 million.[2]

Because of this large and efficient industry, there is a readily available supply of milk and other dairy products. This was not always the case; only recently has milk been so available. With advances in the fields of processing, packaging, refrigeration, and distribution, a wide variety of dairy products has become available. This is reflected by the large increase in the total value of milk and dairy products produced by the milk industry over the past 20 years. Between 1967 and 1985 the total value increased from $6.6 billion to $24 billion.[2] More than half of these sales were from packaged fluid milk and related products (flavored milk and cream products).

Over the past 30 years dairy sales have increased. There has been a switch to lower-fat products, but milk has become an increasingly important part of our diet (Table 19-1). In the 1980s consumption of dairy products rose by 2% to 3% annually. Since 1950 milk production increased over 25% to 66 billion kg of milk in 1989.

The five largest milk-producing states (Wisconsin, California, New York, Minnesota, and Pennsylvania) produced 37% of the total milk in 1950 and 52% of the total in 1986, an increase in production of 73%.[2] During the same period the other 45 states realized a small decrease in milk production. In 1986 they produced 31 billion kg of milk. The state with the largest increase was California, with a greater than threefold increase; Wisconsin milk production doubled during this time period. At present 40% of the nation's dairy cattle reside on only 10% of the nation's dairy farms.[4] Whereas 96.5% of California's cattle are on dairies with more than 100 cows, the majority of cows in New York, Wisconsin, Minnesota, Michigan, and Pennsylvania

Table 19-1 *Per capita consumption of milk products in the United States*

Product	1960	1986
Whole milk (L)	104	51
Lowfat and skim milk (L)	6	47
Yogurt (kg)	0.1	2
Heavy cream (L)	0.5	0.5
Cheese (kg)	3.5	9.5
Butter (kg)	3.0	1.5

are in herds of 30 to 100 cows.[4] In the 1980s cows on Washington and California dairies increased production at a rate of 2.5% to 2.75% per annum. Cows in Minnesota and Wisconsin had a smaller increase of 1.5% to 2%.[5]

ROLE OF THE DAIRY COW IN HUMAN NUTRITION

In the 1970s and 1980s, North Americans became increasingly aware of the role of nutrition in maintaining health. Because the consciousness is directed toward a leaner diet, one lower in fat and cholesterol, one would think that dairy products would have shown a decrease in consumption. Milk and other dairy products have a high food value, however, and many low-fat products are available. As a result milk products have maintained a significant share of the average American diet. Additionally, the milk industry has constantly tried to improve both the acceptance and the nutritional value of its product. This is evidenced by the recent release of "Extra Light 1% Milk," which has a higher protein content (5.5%) and lower fat content (1%) than whole milk (3.5% fat) and represents a further reduction in fat content on standard 2% fat semiskimmed milk. Dairy products (excluding butter) contributed 10% of the total energy needs in the average American diet in 1986. Additionally, they accounted for 21% of the protein, 76% of the calcium, 36% of the phosphorus, 20% of the magnesium, 35% of the riboflavin, 11% of the fat, and 14% of the cholesterol intakes.[2] Oils and fats including butter accounted for 46% of the fat in the diet.[2]

Milk contains from 3% to 5% protein, depending on the breed of cow and the milk flow. Jerseys are generally accepted as being "the protein breed" and are followed by the Brown Swiss, Guernsey, and Ayrshire breeds. Holsteins usually have a lower percentage of protein in their milk. This could be a reflection of the higher milk production (flow) of the Holstein. Regardless of the percentage, the quality of that protein is consistent across the breeds. Milk protein (casein) has a biological value of 74% that of albumin, the standard protein utilized in comparisons.[6] Because of this high quality, milk can lower the amount of protein required to meet the daily needs of the average adult. When the major protein source is from soy, the requirement is 37 g per day. Supplying the protein as milk protein decreases the requirement to 34 g.[6] One cup of milk supplies the same amount and quality of protein as one large egg.[3] Dairy products enable people to balance their diets readily.

Currently, much attention is directed toward the level of calcium in the diet and its importance in the prevention of osteoporosis. The recommended daily dietary allowance for calcium is 800 mg for adults and young children and 1200 mg for adolescents and nursing mothers.[2] Dairy products contribute 76% of the total dietary calcium daily. The next major source is vegetables, contributing 10%. One serving of milk (1 cup, or approximately 250 ml) contains 291 mg of calcium. So three servings of milk per day would meet the needs of the majority of the population.[3] This also could be accomplished by consuming 90 g of cheese or 450 g of yogurt.[2] Whole milk, skim milk, buttermilk, and yogurt account for 51% of the calcium supplied in dairy products. Cheese accounts for 27%, and ice cream and canned or dry milk provides 21%.[2] It is difficult to ingest sufficient calcium on diets low in dairy products.

Riboflavin, vitamins A and B_{12}, fat, and cholesterol are other major nutrients that are supplied in significant amounts by dairy products. Additionally, most processed fluid milk has vitamin D added and is therefore a major dietary source of this vitamin. Because of the high levels of fat and cholesterol in whole milk, the American public is consuming greater quantities of imitation dairy products.[12] These products are lower in fats and cholesterol than most dairy products; however, they also are lower in overall quality.[12] Lower fat consumption is also the reason for the increased consumption of nonfat and lowfat milk. Current research is directed toward developing high-calcium milk and low-cholesterol milk. However, these are not yet commercially available.

Another aspect of human nutrition that is met by the dairy industry is the production of meat and meat products. With each lactation, a calf is produced. If it is male and not of the genetic quality to justify raising, it is sold. Either these cull calves enter the food chain as veal calves or they are raised as dairy beef and slaughtered at 12 to 18 months of age. The dairy also contributes to the production of beef through the sale of cull cows. On the average dairy approximately 30% of the cows are culled each year, primarily due to reproductive inefficiency or poor milk production. As of 1987, it was estimated that there were 10.3 million dairy cows in the United States.[8] At the average culling rate, this would result in the production of 300,000 to 310,000 head of beef each year. This makes a significant contribution to the processed beef and meat products market.

The use of dairy products makes balancing the diet easier and more economical. Keeping the level of production of these commodities at a high level is dependent on a steady supply of high-quality fluid milk. In order to accomplish this, the dairy industry must operate at its most efficient level. To be efficient, the dairy farm has to incorporate ancillary services into its management system in such a way as to complement each other. Along this line, the ancillary aides (e.g., the veterinarian, nutritionist, and machinery analyst) need to be cognizant of the management system so that their advice benefits that particular system.

MANAGEMENT SYSTEMS

Many factors influence the particular style or system of dairy enterprises in a certain geographical area. Environment plays a key role. The open drylot or free-stall system (Figs. 19-1 and 19-2), common in the southern and western states, would be impractical in the Great Lakes region due to the cold winters. Herd size plays a role in the determination of the management system. Likewise, the management system af-

Fig. 19-1 Large, open drylot. Over the feed bunk is an example of a water-mister cooling system.

Fig. 19-2 Cattle reclining in free stalls, under shade.

fects herd size. Financing also influences the type of dairy system chosen. Finally, the personal preference of the dairy operator is of import. As a nutritional consultant, one needs to be very familiar with the area's management systems.

Housing systems influence the feeding system utilized on the dairy. In the northern regions, enclosed housing is predominant. When it is cold, the cattle are indoors most of the time, perhaps being let out on sunny days if the lots are relatively dry. While inside, they are most likely kept in individual tie stalls that are bedded with straw. Some systems even allow the cow to be milked in these stalls rather than being milked in a separate parlor. Under these conditions, the cows are generally fed separately, usually with weighed amounts of the same mix. The advantage of this system is that each cow is fed a ration that more closely meets her production needs. It is a very labor-intensive system that does not lend itself to feeding large numbers of cattle.

In contrast, large open dairy systems are common in the warmer regions of the United States. Because the climate is milder throughout the year, cattle can be left outside in open lots, free stalls, or combinations thereof. This system lends itself to large numbers of cattle. These animals are penned according to production, reproduction, or whatever other method the owner employs. The feed-delivery system is set up to provide the same ration to an individual pen or group of pens rather than the individual cow. Automation is the key word here. Feedstuffs are moved from storage bin and weighed into the feed wagon by front-end scoop loaders (Figs. 19-3 and 19-4). Weigh cells measure the amount of feed added and displays that weight on a liquid crystal digital display (Fig. 19-5). In this manner one to several feedstuffs are added into the feed wagon and mixed (Fig. 19-6). The feed wagon (or truck) then delivers the complete mixed ration to the cattle by discharging it into

Fig. 19-4 A feeding system in which the feedstuffs are piled outside separately.

Fig. 19-5 An LCD digital display common to most feed wagons.

Fig. 19-3 A commodity storage barn. Individual feedstuffs are placed in each compartment in tonnages that allow feeding out over at least 1 month.

Fig. 19-6 A combination mixer-feeder wagon.

a feed bunk or onto a slab as the truck drives along. This system is more labor efficient than the previously discussed system; however, it tends to feed the cattle more indiscriminately.

There are many different ways of feeding that further differentiate the systems. The roughage, particularly dry hay, can be fed separately from the concentrates (Fig. 19-7). Some dairies also feed silage separately from concentrate and dry hay (Fig. 19-8). When roughages are not mixed in with concentrates (Fig. 19-9), the roughages can be thrown out of the bunk and the more palatable concentrates selected by the cow. Some dairies feed a total mixed ration (TMR) in which the concentrates and the roughages are mixed together and fed as one feed. When the cattle are fed total mixed rations, there is no means to selectively refuse portions of the ration. Feeding a total mixed ration has the drawback of the larger initial expense to obtain the machinery necessary for TMR feeding. In order to incorporate dry hay into the TMR, it needs to be ground up into particles small enough to be handled by the loader and the mixer augers of the feed wagon. Some problems have been noted with the use of grinders. The hay can be ground too small, which results in loss of some of the benefits associated with feeding long-stem hay to ruminants. There is also the potential for a great deal of loss of hay during grinding, especially on windy days. With the advent of newer four-auger wagons, this grinding has been eliminated. These wagons allow for the hay to be added to the mix as whole bales, which are broken apart and shredded by the action of the four augers. As would be expected, these wagons are more expensive; however, they do allow a better ration to be delivered to the cows more efficiently.

Some operations still feed grain in the milking parlor (Figs. 19-10 and 19-11). The most common reason for parlor feeding is that it entices the cows into the stanchions and decreases milking time. However, delivering the grain mix to the

Fig. 19-7 A feedbunk where only hay is fed. Cattle are fed their silage and grain mix in the bunk in the background.

Fig. 19-8 A round feeder used for silage or hay.

Fig. 19-9 Cattle being fed concentrate mix on top of their roughage allowance.

Fig. 19-10 A California walk-through style of flat barn adapted to deliver concentrates to the cows during milking.

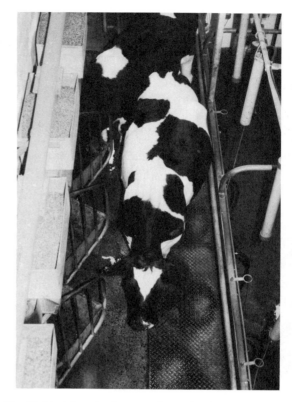

Fig. 19-11 A herringbone-style parlor barn. The grain-delivery hoppers and feeders are to the left of the cows.

cows individually increases labor input and therefore costs. Feed movement can be automated using augers to move grain from a tank into the barn and then into feeders. These systems force the use of grain mixes that contain smaller particles so that the auger does not plug. These grain mixes are often more expensive.

In certain dairy systems, pasture still accounts for a major share of the ration nutrients. Allowing the cow to forage for her own nutrients is probably the most labor-efficient system available. It is probably the least expensive feed-delivery system, although this depends on land values and the cost of maintaining a high-quality pasture. An inherent problem associated with strict pasture feeding is that milk production is lower. The animal obtains most of her dry matter intake from the pasture. Even if the pasture is of the highest quality, the energy and protein density is lower than in most concentrate-based diets. Also, pasture quality varies during the year. Early spring growth is very digestible and contains the highest concentration of nutrients on a dry matter basis. Later in the year the pasture is drier, but fiber content increases and digestibility decreases. Irrigation can help assure a more uniform pasture quality, but this is not practical, physically or economically, in all parts of the country.

Rations are made up of many different components, and these vary across the country depending on the regional availability. Alfalfa and oat hay and corn and oat silage are available throughout most of North America. Utilization of concentrates is dependent upon availability and economics. As an example, in the Midwest soybean meal is the major protein supplement. In the South and Southwest soybean meal is more expensive due to transportation costs. In these states, cotton is grown extensively and whole cottonseed and cottonseed meal are utilized as the major sources of protein. A ration in these regions often consists of beet pulp, whole cottonseed, cottonseed meal, and barley. In the Midwest rations more commonly contain corn, distiller's grains, soybean meal, and wheat mill run.

The use of by-products as feedstuffs is becoming more common. This is primarily due to their low cost. With no other market available, sale of these feeds to the dairy industry is quite attractive. More and more, however, the dairy industry is realizing that some of these feedstuffs have values over their inherent savings in feed costs. For example, fish meal and feather meal are excellent sources of lysine and methionine, the limiting amino acids in milk production. The processing involved in the production of by-products can alter the components of the feed. This may decrease ruminal degradation of nutrients (particularly of protein) and allow bypass of the feed components to the abomasum.

Numerous additives can be found in rations. A vitamin-mineral premix is most common. Buffers, rumen bypass fat, and volatile fatty acids may also be included. Sodium bicarbonate or sesquicarbonate and magnesium oxide are the most common buffers. The ruminal microbial population operates optimally when the pH is between 6 to 6.8.[1] Rumen bypass fat is a recent addition to rations fed to high-producing dairy cattle. The theory is to get as much fat as possible to the small intestine so that the energy can be utilized by the animal rather than being utilized by the rumen microbes. The addition of mineral salts of volatile fatty acids (VFAs) can result in more VFAs being absorbed into the bloodstream.

No one system is consistently better than any other. Environmental factors have the greatest impact on the type of system utilized, which in turn dictates the herd size and the feed-delivery systems.

PRICE-SUPPORT SYSTEMS

Because milk is perishable, the dairy farmer cannot withhold the product from market to increase the price. Without some type of regulatory system in place, the dairy farmer would be at the mercy of processors and handlers. There is also a need to equalize the costs for raw milk incurred by processors across the country. The regulations in place today were developed by a general consensus of producers and processors in an attempt to enable each to realize a fair profit.

In 1989 there were 43 federal milk marketing orders. These establish the price paid for approximately 70% of the fluid milk produced in the United States.[2,11] Thirteen states also have established their own marketing orders separate from the federal orders. All orders base the price paid for fluid milk upon the concentration of components within milk. For example, in California the price is based upon the butterfat, lactose, and protein contents.[7,9] The federal orders are based only on fat test.[11]

Under the federal milk marketing orders laid down by Congress, all fluid milk is class 1, soft and frozen products are class 2, and all other dairy products are class 3.[11] The class 3 price is set by formula and is the average price paid for manufacturing milk according to a monthly survey of 275 plants in the states of Minnesota and Wisconsin.[11] Pricing is adjusted based on the fat percentage. The value of the milk is initially based on a 3.5% test and is adjusted upward or downward by utilizing the butterfat differential determined from the grade A butter price on the Chicago Mercantile Exchange.[11] Once the support price is established, the Commodity Credit Corporation (CCC) purchases surplus cheese, butter, and powdered milk at this price.[11] Because the CCC cannot resell these products at a profit, it realizes a loss. The net losses of the CCC are part of the budget of the U.S. Department of Agriculture. Because of overproduction of milk and the large losses incurred by the federal government, the whole-herd buyout program was instituted in 1985. This program paid dairy farmers to slaughter their cattle and resulted in a decrease in milk production and a decrease in CCC purchases.[2,11]

The California pricing system discriminates based on the use of the milk. The highest price is paid for fluid milk; it is based upon the cost of milk production in California, the average weekly earnings of California workers, and the California price for butter and nonfat dry milk.[7,9] The price for butter and nonfat dry milk is determined by a formula based upon the higher of commercial prices or government support prices for butter and powder, minus the make allowance.[8,9] The make allowance is the difference between the sale value of butter and nonfat dry milk and the cost associated with making these products.[7,9] The federal price systems also affect the price paid for milk through its price-support policy for butter and powdered milk.

Within many states and provinces a quota system attempts to regulate the production of milk. The total production quota is usually based upon the historic usage of fluid milk with an allowance for growth.[7,9] The dairy farmer is paid the highest price for milk that is shipped within the quota limit. Milk shipped over quota receives a lower price, which decreases the overall profit margin. The dairy farmer striving to be more efficient attempts to ship more milk within quota to realize a larger profit.

Producers are selling more and more milk to processors, which pay on an end product pricing system. An example of this is the cheese yield pricing system utilized by certain cheese manufacturers. The dairy farmer is paid based upon a formula that predicts the yield of cheese from the protein and fat contents of milk. As the yield of cheese is influenced more by the protein content of the fluid milk than by the fat content, the higher the protein percent of that milk, the greater the cheese yield. Under this pricing system, dairies that have a large population of cattle of the protein breeds, the Jerseys and Brown Swiss primarily, would benefit more than those with high-milk-yield breeds such as Holsteins.

Nutritional management of dairy cattle is affected by the way a dairy farmer is paid for milk. Those shipping to processors that pay on a cheese yield basis would benefit from rations that attempt to maintain or increase the protein percent of the milk. When they are paid according to the more conventional mode based on butterfat content, nutritional advice should be geared toward supporting fat production. This is only one aspect of the management system that affects the nutritional status of the herd. Environment, feed availability, and management aspects all affect the recommendations that nutritionists give to producers. Knowledge regarding the way these factors interact is invaluable. To succeed in dairy consulting, an attempt should be made to monitor these aspects within the area of practice.

REFERENCES

1. Garry F and Kallfelz FA: Comp Cont Educ 5(3):S159, 1983.
2. Milk facts, Washington, DC, 1987, Milk Industry Foundation.
3. Nutrition: food at work for you, Home and Garden Bulletin No. 1, Washington, DC, 1978, U.S. Department of Agriculture.
4. Outlook, Dairy Today, p 44, Sept 1989.
5. Outlook, Dairy Today, p 32, Feb 1990.
6. Pike RL and Brown ML, editors: Nutrition: an integrated approach, New York, 1975, John Wiley & Sons.
7. State of California, Bureau of Milk Pooling: Pooling plan for market milk, 1983.
8. State of California, Bureau of Milk Pooling: California dairy industry statistics, 1987.
9. State of California, Bureau of Milk Stabilization: Stabilization and marketing plans for market milk, 1984.
10. Today's news: industry briefs, Dairy Today, p 19, Nov-Dec 1989.
11. USDA, Agricultural Marketing Service: The federal milk marketing order program, Marketing Bulletin 27, 1981.
12. Wilkinson M: Bovine Proceedings 20:42, 1988.
13. Where's the milk, Dairyman, p 22, Oct 1989.

20

The Role of the Veterinarian as a Nutritional Advisor in Dairy Practice

DAVID T. GALLIGAN

Effective herd health programs are based on strategies to control disease prevalence so that farm profit is maximized. Historically, dairy farm practice has focused on improving production efficiency primarily in reproduction, mastitis control, and calf management. Although veterinary nutritional consulting only recently has become an integral component of herd health programs, it gradually is becoming recognized that sound management advice in this area may have a greater impact on economic efficiency than other traditional services.

The veterinarian is an ideal provider of nutritional advice for a number of reasons. The practitioner is on the farm routinely, so problems can be detected and solved early. Training in total animal health gives the veterinarian an integrated approach to health and production problem solving. This integrated approach is necessary because of complex production systems in which the economic benefits of solving one problem may not be realized unless other problems also are addressed. Moreover, the veterinarian, unlike feed company nutritionists, usually is not slanted to a particular product and thus can offer an unbiased opinion in management decisions. Feed constitutes approximately 60% of the total cost of milk production;[1] minor improvements in feeding efficiency are directly reflected in higher farm profits (Table 20-1). The operational protocol of a veterinary nutritional herd health service consist of three main components:[2]

1. Routine farm visits for regular surveillance and analysis of the feed management program
2. Ration nutrient and economic analysis
3. Ration formulation and feeding management instructions

FARM VISIT

Farm visits are critical to monitor and continually improve the feed management program. Routine inspection of the condition of the animals and feeds is essential for identifying potential problems and options for improving the operation. Farm visits for nutritional consultation

Table 20-1 *Distribution of predicted feed savings that could accrue from use of a veterinary nutritional advisory service*

Projected feed savings*	
Projected feed savings (% of current feed cost)	Number of farms
0-5	0
6-10	6
11-15	3
16-20	4
21-25	2
Mean (SD) 14.4 (6.3) Total	15

*Farms visited by veterinarians in the Section of Nutrition, New Bolton Center, University of Pennsylvania.

should be established as separate visits or as a component of the routine herd health visits. It is essential that adequate time be allocated for review of the feed management.

The farm manager and other personnel involved in feed management are interviewed to evaluate any feeding changes that might have occurred since the last visit. The integrity of the feed-delivery system is evaluated by checking scales and weighing procedures used in allocating feedstuffs. In analyzing the feed-delivery system, it is important to identify factors that could result in inefficient nutrient delivery to cows for each stage of lactation. These inefficiencies can result in high cost associated with the actual delivery of feedstuffs (e.g., labor and depreciation of feeding equipment). Inefficiencies can also lead to nutrient imbalances and result in impaired production or increased prevalence of metabolic disease. The most desirable feeding program is a system that is flexible enough to meet nutrient requirements and to minimize feed and feeding costs over time.

RATION NUTRIENT AND ECONOMIC EVALUATION

Evaluation of a feeding program is composed of three main steps. The first step is to evaluate the physical quality of feed ingredients. The second step is to access the economic efficiency of the feed ingredient selection. Third, one should determine how efficiently nutrients are distributed within the herd according to requirements.

Physical quality

The physical quality of feed ingredients can be inspected visually by noting the color, degree of maturity, length of chop, moisture, and presence of mold or foreign bodies. Hay bales should be opened and assessed for mold and dust. The quality of hay can be judged best by the stalk thickness rather than by color. Stems should be thin and pliable, not woody. Haylages should be free of mold and have a pleasant, acidic odor and moisture levels above 55%. Silage should be cut at 0.6 to 1 cm (¼ to ⅜ inch) to ensure adequate packing so fermentation can proceed normally. Grains should be examined for foreign bodies, excessive separation, and dust.

Economic evaluation

Feeds can be economically assessed by (1) their nutrient composition and the market price paid for these nutrients and (2) the effect these feed ingredients have on ration cost. The method used to estimate ingredient value is based on the assumption that the economic value of an ingredient is the sum of its nutrient composition multiplied by the cost for each nutrient[3,6] in the ingredient. Energy and protein content are the major nutrients that determine economic value. Costs (economic weight factors) for nutrients can be determined by using feeds that are commonly traded as references. Normally an energy source (shelled corn) and a protein source (soybean meal) are selected as reference feeds. The substitution value of a particular feed ingredient can be calculated (see Chapter 1). However, it is important to realize that this method does not take into account the value of nutrient density in the feed but only the amounts of nutrients. Hence, the combination of reference feeds needed to deliver the same amount of nutrients in one kg of the feed being evaluated may exceed 1 kg in total weight, which is a disadvantage if the cow is already consuming her maximum food intake.

To investigate the economic value of nutrient density in feed ingredients, one can use the reduced cost calculated in most linear programming ration formulation packages.[2] The reduced cost is the cost change required for a feed ingredient currently not in the ration to be included. Reduced cost is calculated under the assumption that all feed prices and constraints remain constant. It is recommended that reduced cost be recalculated with all feed costs and constraints updated on a regular basis.

Feed delivery

The third component of ration evaluation is to calculate the total nutrient intake from all feeds and compare them to required levels (many computer programs exist to facilitate this calculation). Each nutrient is summed and should be displayed as (1) total nutrient weight (or Mcal), (2) nutrient density (nutrient weight/dry matter intake), (3) percent of National Research Council requirement (NRC),[4] and (4) percent of

Table 20-2 *Advantages and disadvantages of feed delivery systems*

Feed system	Advantages	Disadvantages
Stanchion	Each cow is fed based on production and condition score Precise nutritional and cow observation	Pulsatile grain feeding Labor-intensive Poor estimates of forage intakes
Total mixed rations	Efficient use of labor Possible to focus attention on groups of cows in similar stages of lactation Possible to manage larger numbers of cows More constant nutrient delivery Consistent feed source	Labor for regrouping cows Decreases in production with cows changing groups Require several groups for ration balancing Difficult to monitor individual cows Potential over- and underfeeding of individual cows

National Research Council requirement[4] adjusted for dry matter deviations (see Chapter 28 on computer ration formulation programs). Many nutrient imbalances are found only at extremes of production, and thus several production levels should be evaluated. Evaluation of only the average production level often masks potential problems.

RATION REFORMULATION

Ration formulation is started by defining the feed-delivery system on the farm. Feed-delivery systems are a function of the physical setup of the farm, the production level, and the types of feeds being fed. Most feed-delivery systems are a compromise between feeding the individual animal (stanchion barn feeding system) and a group of animals (total mixed-ration system). Each system has its advantages and disadvantages[2] (Table 20-2).

Routine reformulation of rations is driven primarily by changes in forage quality, forage type, market feed prices, feed delivery, and by the relative frequency of cows at certain stages of lactation (early- versus late-lactation animals). Nonroutine reformulation may be required with abnormal production, reproduction, or body condition scores. Responses to suggested changes should be monitored and critically analyzed, and appropriate adjustments made.

REFERENCES

1. Black JR and Hlubik J: J Dairy Sci 63:1366, 1980.
2. Ferguson J and others: Comp Cont Educ 9(5): F192, 1987.
3. Galligan D and others: J Dairy Sci 69:1656, 1986.
4. National Research Council: Nutrient requirements of dairy cattle, Washington, DC, 1989, National Academy of Science.
5. Noordhuizen JPTM and others: Prev Vet Med 3:289, 1985.
6. Peterson WE: J Dairy Sci 15:293, 1932.
7. Weaver LD, Olivas MA, and Gallard JC: J Dairy Sci 71:1104, 1988.

21

Colostrum and Feeding Management of the Dairy Calf during the First Two Days of Life

CLIVE C. GAY, THOMAS E. BESSER

Colostrum, the milk present in the cow's mammary gland at the time of calving, is significantly different from midlactation milk both nutritionally and immunologically (Table 21-1). Nutritionally, first-milking colostrum contains considerably more protein, fat, minerals, and vitamins than does milk. Immunologically, colostrum is an essential source of immunoglobulins to the newborn calf.[3,6]

The primary importance of colostrum rests in its ability to provide passive immune protection to the calf. The absorption of adequate amounts of colostral immunoglobulins is essential to the health of the calf during the first 2 months of life.[1] The calf is born virtually devoid of circulating immunoglobulin and relies on antibody acquired from colostrum to protect it against environmental pathogens. The predominant colostral immunoglobulin is IgG1 because of selective concentration by active, receptor-mediated transfer across the mammary secretory epithelium from the cow's blood.[8] The transfer of IgG1 to colostrum begins approximately 4 to 6 weeks before calving and results in colostral IgG1 concentrations in first-milking colostrum that are 2 to 10 times those of maternal serum. Other immunoglobulin classes (IgG2, IgA, and IgM) are also present in colostrum but at considerably lower concentrations than IgG1. Non-IgG1 immunoglobulins are secreted by nonse-

Table 21-1 *Composition of colostrum and whole milk*

Parameter	Colostrum	Midlactation milk
Specific gravity	1.056	1.032
Fat (%)	6.7	4.0
Total protein (%)	14.0	3.1
Casein (%)	4.8	2.5
Total Ig (%)	6.0	0.09
IgG1 (mg/ml)	46.4	0.58
IgG2 (mg/ml)	2.87	0.06
IgM (mg/ml)	6.77	0.09
IgA (mg/ml)	5.36	0.08
Lactose (%)	2.7	5.0
Ash (%)	1.11	0.74
Ca (%)	0.26	0.13
Mg (%)	0.04	0.01
K (%)	0.14	0.15
Na (%)	0.07	0.04
Vitamins		
A (μg/100 ml)	295	34
D (IU/g fat)	1.5	0.4
E (μg/g fat)	84	15
Thiamin (μg/ml)	0.53	0.38
Riboflavin (μg/ml)	4.83	1.47
Vitamin B_{12} (μg/100 ml)	4.9	0.6
Folic acid (μg/100 ml)	0.8	0.2
Choline (mg/ml)	0.70	0.13

Adapted from Butler JE,[3] and Foley JA and Otterby DE.[6]

lective processes from both circulating and locally synthesized immunoglobulins. Average concentrations of immunoglobulins in first-milking Holstein colostrum are shown in Table 21-1. However, there is substantial cow-to-cow variation in the final concentrations of immunoglobulins in first-milking colostrum, and this variation has significant implications for the volume of first-milking colostrum needed by the calf.

Immunoglobulin concentrations in colostrum fall dramatically after calving: The concentrations in second-milking colostrum are approximately half those in the first milking, and by the fifth milking concentrations approach those found during the remainder of lactation.[9,10]

When colostrum is ingested by a calf during the first few hours after birth, a significant proportion of the colostral immunoglobulins are transferred across the epithelial cells of the calf's small intestine and transported via lymphatics to the blood. Immunoglobulins are then distributed to the extravascular fluids and body secretions.[2,11] Antibodies in absorbed colostral immunoglobulin are highly effective in preventing systemic invasion by microorganisms and septicemic disease. Colostral antibodies also contribute to enteric disease prevention, both by immunoglobulins remaining unabsorbed in the intestine and by absorbed immunoglobulins gradually resecreted back into the gut.

Calves that fail to acquire sufficient circulating immunoglobulin from colostrum are at increased risk for infectious disease during the neonatal period.[1] The concentration of circulating immunoglobulin required to minimize risk of infectious disease depends on the calf's environment and the severity of exposure to potentially pathogenic agents. Thus, calves tethered at pasture or reared in isolated hutches during the first 2 weeks of life face less risk and have a lower immunoglobulin requirement than calves in a veal calf rearing unit. Because IgG1 is the predominant immunoglobulin in colostrum, its concentration is commonly used as a measure for testing the adequacy of passive immunity. Field studies suggest that serum IgG1 concentrations greater than 10 g/L are adequate to minimize the risk of infectious disease in calves in most environments. A goal for colostral feeding practices is for all calves to exceed 10 g/L (1000 mg/dL) serum IgG1.

The two prime determinants for optimum absorption of colostral immunoglobulins by calves are the time after birth of colostrum feeding and the immunoglobulin mass in the colostrum that is fed.[2,7,12] The ability of the small intestine to transfer ingested immunoglobulin from the lumen to the circulation decreases rapidly following birth. It is markedly diminished by 12 hours after birth and essentially absent by 24 hours. In view of this, it is essential that the earliest colostrum feeding contain an adequate mass of immunoglobulin. Subsequent feedings may assist in protection against enteric disease but will have progressively less effect on concentrations of immunoglobulins in circulation.

There is also considerable calf-to-calf variation in the efficiency of absorption of ingested immunoglobulin. Although some calves with high absorptive efficiency may acquire adequate circulating immunoglobulins following the ingestion of a moderate mass of colostral immunoglobulins, others with low efficiencies of absorption attain only low circulating concentrations. In order to promote adequate absorption in calves with low absorptive efficiencies, a minimum mass of 100 g of colostral immunoglobulin at the first feeding is required,[7] with 150 g or more being desirable. The mass of immunoglobulin fed is a function of the concentration of immunoglobulin in colostrum and the volume of colostrum fed. Calves that ingest 2 L of colostrum with 50 mg/ml IgG1 acquire approximately 100 g of IgG1, and so failure of passive transfer should not be a problem. However, approximately 55% of colostrum from American Holstein cows contains less than 100 g IgG1 in a 2-L volume, and 10% contains less than 100 g IgG1 in a 4-L volume (Fig. 21-1). For this reason dairy calves should be fed as large a volume of colostrum as possible at the first and second feedings.

A method to allow selection of colostrum of average or higher immunoglobulin content would be of practical importance. Although there are known influences on immunoglobulin concentration of colostrum, none can effectively eliminate all low-immunoglobulin-concentration colostrums. For example, one can select co-

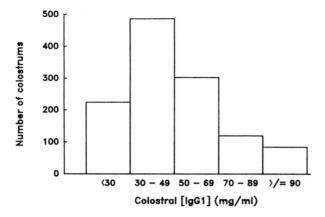

Fig. 21-1 Immunoglobulin concentrations in first-milking colostrum from American Holsteins.

lostrum for calf feeding from older dams because the average immunoglobulin concentration of colostrum increases with parity to reach its maximum at, or beyond, parity 3. However, cow-to-cow variation in colostral immunoglobulin concentration within any parity far exceeds the variation in average colostral immunoglobulin concentration between parities. The immunoglobulin concentration in the first milking after calving is significantly lower in those cows that are milked prior to calving and in those cows that stream milk prior to first milking. Colostrum from these cows should not be used for a first feeding to newborn calves. Colostrum immunoglobulin concentration also is influenced by the length of the dry period. Colostrum from cows with dry periods less than 30 days should not be used for first feeding. In theory, a delay between calving and milking could reduce colostral immunoglobulin concentration. However, this factor is of little practical importance if the delay is less than 12 hours. Very-high-volume colostrum tends to have a lower immunoglobulin concentration and where possible should not be used for first feeding.

There is a linear relationship between the immunoglobulin concentration of colostrum and its specific gravity.[5] The use of whole colostrum with a specific gravity of 1.050 or greater eliminates much of the poorest immunoglobulin content colostrum. Air incorporated into colostrum during machine milking gives false low readings if the specific gravity is measured immediately following milking. The measurement should be made with a hydrometer after the colostrum has been allowed to stand for 1 hour.

Specific gravity estimations also are temperature dependent. Measurement of the specific gravity of colostral whey closely estimates colostral immunoglobulin content but is more difficult to implement in the field.

Severe undernutrition during pregnancy has a deleterious effect on the immunoglobulin content of colostrum and can be a particular problem in beef herds. Although severe undernutrition is unlikely in well-managed dairy herds, colostrum from cows with condition scores of less than 3 at calving may contain less immunoglobulin and should not be used for first feeding.

METHODS OF FEEDING COLOSTRUM
Natural sucking

In temperate zones, dairy calves may be born at pasture and left with their dams for 2 days until the cow is introduced to the milking herd. This practice minimizes the exposure of the calf to environmental pathogen challenge but can result in poor immunoglobulin transfer unless calving and subsequent suckling is closely supervised. In less temperate climates, calves may be left with their dams in calving pens for 2 days following calving, which allows for greater supervision. There is considerable variation in the volume of colostral intake by nursing calves, both within and between different breeds of cattle (Table 21-2). Numerous studies have shown that failure of passive transfer of colostral immunoglobulin is very common in natural suck-

Table 21-2 *Volume of colostrum ingested with natural sucking*

Breed	Volume ingested in kg (observation period)
Ayrshire	3.4 ± 0.54 (12 hours)
Belgium Blue	2.75 ± 0.28 (24 hours)
Friesian	2.5 ± 0.18 (24 hours)
Holstein	2.4 ± 1.5 (24 hours)
German Red/Black	3.1 ± 0.5 (4 hours)

ing situations in dairy herds, and failure rates of 40% or higher have been observed. The colostral intake in natural sucking situations is influenced by the sucking drive and vigor of the calf at birth as well as by the mothering ability of the dam. There can be a considerable delay between birth and the first intake of colostrum; this impairs immunoglobulin absorption due to the decreasing ability of the intestine to transfer immunoglobulin following birth. A low sucking drive is particularly prevalent following a difficult birth, and colostrum should be fed by artificial means in these situations. Pendulous udders and large, bulbous teats also can delay the time of initial ingestion of colostrum by the calf; cows with these characteristics should be identified so that their calves can be fed adequate volumes of colostrum soon after birth by artificial means.

Natural sucking of colostrum is best practiced in small dairy herds where calving is infrequent and calving and calf behavior are subsequently intensively monitored. Calves should be carefully observed and if they have not suckled within 2 hours of birth, the cow should be milked and a minimum of 2 L of colostrum should be given by nipple bottle to the calf.

In beef breeds, natural sucking is the preferred method for the intake of colostrum. The prevalence of failure of passive transfer of colostral immunoglobulin is much lower in beef calves than in dairy calves, due in part to their better sucking drive and in part to the much higher concentrations of immunoglobulin in beef cow colostrum. The volume of colostrum ingested by normal natural sucking results in the intake of an adequate mass of immunoglobulin. Calves that fail to nurse within 2 hours of birth should be assisted to nurse or tube fed colostrum. Blanket intervention with the feeding of colostrum by artificial means is not indicated for the beef breeds.

Artificial feeding

Artificial feeding systems for colostrum ensure a known intake of colostrum at a known time after birth. In all artificial colostrum-feeding systems, it should be emphasized that the first feeding should be given within the first 2 hours after birth. Although this overstates the rapidity of the loss of transfer capacity of the intestine, it emphasizes the importance of early colostrum intake to the feeder. Less stringent recommendations (for example, first feeding during the first 6 to 8 hours of life) imply less urgency and result in a substantial decrease in the efficiency of immunoglobulin absorption from the first feeding of colostrum.

A principal disadvantage of artificial colostrum-feeding systems is that colostrum must be obtained for feeding. If colostrum is obtained from the dam, either by hand or machine milking, either a substantial disruption of routine work procedures or a significant delay in the acquisition of colostrum may result. This problem can be eliminated by the use of stored colostrum. Stored first-milking colostrum from a cow that has previously calved is immediately available for a newly born calf. The use of stored colostrum has the additional advantage that high-quality colostrum can be selected on the basis of specific gravity or other parameters. Colostrum can be stored at 4° C (37° F) for 7 days with little loss of immunoglobulin. In order to meet the colostrum-feeding requirements of peak calving frequencies, a store of colostrum should be frozen at $-20°$ C ($-4°$ F). Colostrum kept in this manner can be stored virtually indefinitely. It is preferable to store frozen colostrum in flat plastic bags rather than gallon jugs to facilitate rapid thawing. Thawing should be conducted in warm water (35° C or 95° F) rather than in hot water. Microwave ovens can be used satisfactorily, but care must be taken not to overheat the colostrum, which would denature the immunoglobulins.

Stored colostrum can be mixed together to form a pool. Colostrum pools have a theoretical superiority over individual colostrums in that they represent the antigenic experience of a number of cows. Pooling colostrum also tends to average the immunoglobulin concentration and avoid the possibility of feeding an individual colostrum of low concentration. However, in practice, colostrum pools almost invariably have low immunoglobulin concentrations due to the diluting effect of colostrum with high volume but low immunoglobulin concentration. Where pools are used for first feeding, the colostrum should be selected on specific gravity (>1.050), and high-volume colostrums should be avoided.

With all artificial feeding systems, first-milking colostrum should be fed for at least the first two feedings. Second-milking colostrums can be used for second-day feedings.

Bucket feeding. Training calves to drink from a bucket requires considerable time and patience by the feeder. Such training at the first feeding of colostrum is not compatible with the early ingestion of adequate amounts of colostrum and should be discouraged.

Nipple bottle feeding. Feeding from a nipple bottle represents a more natural sucking for the calf and is superior to bucket feeding. However, considerable patience is required to ensure an adequate colostrum intake from a nipple bottle. The lack of sucking drive and vigor that causes problems in natural sucking of colostrum is also a major problem with nipple bottle-feeding and can result in inadequate intakes of colostrum. The time required to nipple bottle-feed a newborn calf varies from a few minutes to over an hour, depending upon the vigor of the calf. The target is to feed a colostral volume of 70 ml/kg (7%) of calf body weight at first feeding, with 50 ml/kg (5.0%) at the next three feedings. Not all calves ingest 70 ml/kg (1 oz/lb) at first feeding, and even if this intake is achieved there will be a proportion of calves with failure of passive transfer of immunoglobulin due to the feeding of colostrum with low immunoglobulin concentration. Nevertheless, this system of colostrum feeding can be relatively effective, given sufficient patience and interest by the feeders, and is frequently successful on small family farms where calves are likely to be fed by interested family members. However, on large commercial farms other demands on labor may limit the time available for feeding the newborn calf, and nipple bottle-feeding can be accompanied by high rates of failure of passive transfer of immunoglobulin.

Esophageal tube feeding. Large volumes of colostrum can be administered to the newborn calf in a short period by use of an esophageal feeder. Intakes at a single feeding can be far in excess of those achieved by natural sucking or methods of artificial feeding that rely on sucking. A proportion of colostrum fed by this method enters the immature rumen rather than being bypassed into the abomasum by the esophageal groove.[4] Consequently the efficiency of absorption of immunoglobulin from colostrum fed in this manner is slightly less than that with sucking, and the esophageal feeder should not be used to give small volumes of colostrum. Colostral volumes greater than 10% of the body weight of the calf can be safely administered in a single feeding. In practice, a set large volume (4 L or 1 gallon for Holstein calves) is given for the first feeding of colostrum, as newborn calves cannot be weighed conveniently. The esophageal probe is introduced approximately one third of the way down the esophagus and the colostrum is allowed to flow by gravity. Feeding colostrum by an esophageal feeder has several advantages. It allows the administration of colostrum to calves regardless of their sucking vigor. The volume fed promotes the intake of a large immunoglobulin mass, even with colostrum of low immunoglobulin concentration. It is rapid and minimally disruptive to other labor activities. It is particularly well suited to dairy herds large enough to maintain a supply of fresh stored (4° C or 37° F) colostrum, and significant decreases in the rate of failure of passive transfer of immunoglobulin can be achieved with the adoption of this method of feeding. It is less suited to small herds without a ready supply of stored colostrum. Generally, the esophageal feeder is used for the first feeding, and subsequent feedings are given by nipple bottle. Calves fed these large colostrum volumes at the initial feeding may not willingly suck until 18 to 24 hours later.

CONCLUSION

Monitoring calf serum immunoglobulin concentrations should be done routinely to assess the efficiency of colostral feeding on a farm and can also be used to check the abilities of individual feeders. Zinc sulfate turbidity testing or serum total protein estimations using a refractometer are quite adequate for this purpose and are more practical than radial immunodiffusion. The levels desired vary from farm to farm, depending upon the infection pressure and the management. In general, the aim is to achieve values above 20 ZST units or 55 g/L (5.5 g/dl) for serum proteins.

REFERENCES

1. Besser TE and Gay CC: Vet Clin North Am (Food Anim Pract) 1:445, 1985.
2. Bush LJ and Staley TE: J Dairy Sci 63:672, 1980.
3. Butler JE: Vet Immunol Immunopathol 4:43, 1983.
4. Chapman HW, Butler DG, and Newell M: Can J Vet Res 50:84, 1986.
5. Fleenor WA and Stott GH: J Dairy Sci 63:973, 1980.
6. Foley JA and Otterby DE: J Dairy Sci 61:1033, 1978.
7. Kruse V: Anim Prod 12:661, 1970.
8. Larson BL, Leary HL, and Devery JE: J Dairy Sci 63:665, 1980.
9. Newby TJ, Stokes CR, and Bourne FJ: Vet Immunol Immunopathol 3:67, 1982.
10. Oyeniyi OO and Hunter AG: J Dairy Sci 61:44, 1978.
11. Staley TE and Bush LJ: J Dairy Sci 68:184, 1985.
12. Stott GH and Fellah A: J Dairy Sci 66:1319, 1983.

Evaluating Dietary Management of Hand-Reared Calves
Milk, Preserved Colostrum, and Milk Replacers

JONATHAN M. NAYLOR

Hand-rearing is a procedure used with calves born to dairy cows. To facilitate management, cows usually are separated from their calves within the first few days of birth. Heifer calves often are reared on the farm of origin as future milking stock. Bull calves usually are sold for beef or veal production. Occasionally calves are fostered in groups of four onto dairy cows and are reared at grass. Usually calves born to dairy cows are hand-reared.

MANAGEMENT
Rearing systems

In dairy- and beef-rearing systems, the principal objective of the early period is to wean a healthy calf onto solid food as quickly as possible. Early weaning reduces expenditure on high-cost milk or milk replacers and may reduce susceptibility to diarrhea. The criterion for weaning is that the calf's rumen and digestive enzymes be sufficiently well developed that solid food can be digested. This means that the calf must be eating at least 1 kg (2 lb) of solid food a day and be gaining weight. In practice these criteria are usually fulfilled by about 5 to 9 weeks of age. Occasionally weaning is carried out at 4 weeks of age; however, this demands very high quality solid food and careful management. Late weaning carries no nutrititonal penalty; secretion of the enzymes necessary for milk digestion is maintained if milk feeding is continued.[3] Late-weaned calves grow more rapidly in early life.[6] This is advantageous in systems where rapid maturity is required, for example, in calves being intensively fattened for veal production. If heifers are being grown rapidly for early breeding at 16 to 18 months, weaning toward 9 weeks of age makes it easier to obtain the necessary growth.[52] The advantage of late weaning gradually disappears with time, however, particularly in systems where growth rates are not maximal, because early-weaned calves catch up by eating more concentrate postweaning. Therefore, early weaning also can produce satisfactory results, depending on what is desired. This is illustrated by one study of calves weaned at 6 weeks or 9 weeks of age and fed restricted or ad lib milk replacer. Calves fed milk replacer ad lib performed better in early life (average daily gain 0.8 versus 0.7 kg/day [1.8 versus 1.5 lb/day]). However, by the time of slaughter at 16 to 23 months, calves restricted to 400 g (0.91 lb) of milk replacer powder a day and weaned at 6 weeks of age had similar overall performance to calves fed ad lib milk replacer and weaned at 9 weeks of age.[61]

Ruminal development is stimulated by the ingestion of solid food.[2,6] To facilitate ingestion and digestion of solid food, only high-quality feeds are used. Solid food is first offered at about 7 days of age. Grass or legume hays provide nutrients and stimulate ruminal development;

the hay should be soft and leafy and of high quality. A compounded creep feed or calf starter that typically contains 3.5% fat, around 6.5% crude fiber, and at least 18% crude protein is also offered. Calves can do well on a variety of protein sources in the calf starter.[66] Rolled or flaked grains, oilseed meals, fish meals, skim milk powder, and molasses are ingredients commonly used in formulating calf starters. Coarse mixes may be more attractive than pellets, but both are used.[63] All calves should have free-choice access to water. This is necessary to maintain health and also increases consumption of solid food.[29]

Veal calf raising is favored during periods of low calf and low milk or milk replacer costs. Calves are fed large amounts of milk and either no or limited amounts of solid food. The iron content of the diet may be limited to produce a light-colored meat. Age at slaughter is variable but may be around 3 to 4 months.

Important management practices and their relation to nutrition

Irrespective of the purpose for which calves are reared, all should get an adequate intake of colostrum (Chapter 21). This is probably the single most important factor in the health of hand-reared calves.[9,37] The source of calves is very important. Farmers rearing their own calves are likely to have the least problems. Buying in from a single reliable source is next in desirability,[9] and least favorable is purchasing calves at markets. Market calves come from a variety of sources, are likely to carry various disease agents, and often have low levels of colostral antibody. Unfortunately, markets are often the most convenient and flexible method of buying and selling calves and are widely used. The person in charge of calf rearing is also very important.[18] The type of nutritional regimen is usually of secondary importance. Occasionally poor nutrition is the major cause of calf-rearing problems, and nutritional factors are also more important when enteric disease is rife. In North America, individual housing is usually preferred for rearing calves. It is thought to reduce the opportunity for spread of disease-causing organisms. Studies of dairy calf-rearing enterprises show little health benefit to individual housing.[10,18] This may be because the level of disease challenge in dairy units was relatively low. However, dairy calves reared in individual pens produce more milk when they enter the milking string.[4]

Feeding systems

It is difficult to define a natural feeding system for calves born to dairy cows because dairy cows produce more milk than the calf can consume. Calves can tolerate 16% to 20% of body weight daily as fresh cow's milk without suffering from diarrhea or maldigestion.[36] In the first month of life, suckled beef calves daily drink about 12% of body weight as milk, divided between five feeds.[5,41] Average gain is about 0.75 kg (1.6 lb) daily. Naturally raised calves obtain their milk by sucking a teat, which stimulates closure of the esophageal groove, whereby milk is directed into the abomasum for digestion.

A system commonly used to raise dairy calves is twice daily feeding of milk or milk replacer at a total daily liquid intake of about 8% to 10% of body weight. Once-a-day feeding is sometimes practiced as a labor-saving device. I do not recommend this practice, although it may work satisfactorily on an occasional basis (for example, for Sunday feeding) and in calves older than 3 weeks. Feeding may be from either buckets or nipple pails. Nipple-feeding stimulates closure of the abomasal groove and reduces the speed at which milk is drunk.[65] It probably benefits digestion[30] and may give small improvements in growth and feed-conversion rates. Nipple bottles or nipple pails may require some form of purpose-designed holder in the stall. Some farmers find it easier to allow the calves to drink from open buckets. Calves usually have to be trained to bucket feed, which is usually done by wetting a finger with milk and allowing the calf to suck on the finger while gently directing the calf's mouth (but not the nose) into the milk. Bucket training usually takes a couple of sessions. Recalcitrant calves are allowed to suck from a nipple or are tube fed.

Ad lib feeding is often used for group-reared calves. Feeding is usually from a nipple. It may be attached to an automatic dispenser that mixes—and often warms—milk replacer in small batches on an as-needed basis (Fig. 22-1).

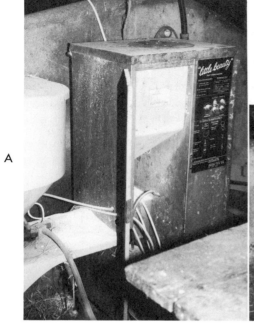

Fig. 22-1 A, Machine for automatic mixing and dispensing of milk replacer. It is connected by tubes to nipples in the stalls, **B.**

Alternatively, it may be connected to a bulk tank of premixed milk replacer via a nonreturn valve. In the latter system, the milk replacer is acidified to prevent bacterial degradation and facilitate storage. These systems have the advantages of allowing the calves to nurse small amounts of milk at frequent intervals and of requiring less labor. Automatic milk replacer dispensers are expensive, however. Feeding acidified milk replacers enables milk to be mixed by hand in batches once a day and avoids the need for expensive automatic dispensers. Another advantage is that frequent feeds and low milk replacer pH may also help preserve the abomasal pH barrier. Abomasal pH in fasting calves is around 2 to 3, which inhibits passage of infectious agents. When a calf drinks a large milk meal, there is a rapid rise in abomasal pH,[16] and microorganisms may be able to cross into the small intestine.[9,15,35] In ad lib feeding systems, milk intake is often limited as weaning approaches to reduce consumption of expensive replacer. This is done by constricting the milk supply tube, removing powder so that only water is fed for part of the day, or turning the machine off for part of the day.

Milk may be fed warm or cold. Excessively cold milk will be refused or drunk slowly.

TYPES OF LIQUID FEEDS FOR CALVES
Cow's milk

Whole cow's milk is the best feed for calves. Until the 1950s milk was the only commonly used feed for calves. Milk contains approximately 3% to 4% fat, 3% to 4% protein, and 4% to 5% lactose, and the total dry matter content is 12% to 14%. Gross energy content is approximately 5.6 Mcal/kg dry matter or 0.7 kcal/ml. Digestibility is 95% so DE content is about 0.67 kcal/ml. Fat is present in micelles. Almost all micelles are smaller than 10 μ, and most are smaller than 1 μ. Casein accounts for 80% of milk protein. Calcium content is about 10 g/kg of dry matter, and phosphorus is 9 g/kg of dry matter.[15,25,63]

Calves grow best and have the least problem with diarrhea when fed whole cow's milk.[19,28,46,59] The only problem likely to occur is when calves are underfed. Although rare, this can occur when poorly informed personnel rear calves for the first time. Calves require 50 kcal DE per kg of weight for maintenance (130 kcal/kg$^{0.75}$) and 3 kcal DE per g body weight gain.[15,47] Thus a 45-kg (100 lb) calf requires 2250 kcal DE or 3.4 L (7.6% of body weight) of milk a day. In practice, calves are fed at 8% to 10% of body weight as milk a day to promote growth. Feeding at 10% of body weight supports 0.25 kg/day (0.55 lb/day) of gain. Creep feed gives additional growth. Milk intake should be adjusted upward in small calves as they grow to help them catch up with their bigger herdmates. Milk intake is held constant as weaning approaches to encourage calves to eat solid food. A problem that occurs rarely is that some cows excrete *Salmonella* in their milk that could infect calves.

There is some evidence that milk-fed calves may benefit from supplemental vitamins. Colostrum is normally rich in vitamin A. Supplemental injection of vitamin A may benefit colostrum-deprived calves and calves born to dams fed preserved feeds low in carotene, for example, poor-quality hay or straw-grain diets. Vitamin E supplementation has been reported to decrease subclinical muscle damage and to improve laboratory measures of immune function. The dosages used to obtain the immune effects were high: 1400 mg of dl-α-tocopherol once a week by injection or 2800 mg dl-α-tocopheryl acetate orally once a week.[48,49] Further work is needed to see if these dosages produce clinical benefits.

Preserved colostrum

Feeding preserved colostrum is used by some dairy farms as a means of utilizing the first six milkings of unsalable milk. The first milking is true colostrum and is highest in immunoglobulin, which is fed to the newborn calf. Subsequent milkings from recently calved cows, *fresh cow's milk,* are intermediate in composition with true milk. Farmers often call colostrum all the milkings collected in the first 72 hours postpartum and refer to feeding fresh cow's milk as *feeding colostrum.* Waste milk from antibiotic-treated cows may also be fed to calves.

In the first six milkings, heifers and cows produce an average of 32 kg and 50 kg of fresh cows milk, respectively. Most farms that feed fresh cow's milk sell their bull calves at a few days of age, and so almost double these quantities are available for rearing heifer calves. Even so, there is only enough available to feed for about the first 3 weeks of life without supplementing with whole milk or milk replacer.

Surplus colostrum, fresh cow's milk, and waste milk must be preserved so that it can be stored to ensure an even supply for calves. This can be done by refrigeration, natural or assisted fermentation, or acidification. Refrigeration obviously involves more expense.

Preservation by direct acidification can be carried out by adding 1% (v/v) of propionic acid to the fresh cow's milk. The final pH should be between 4 and 5. There is little effect on the nutritive value of the milk.[42,46,51]

Preservation by fermentation usually involves allowing the milk to stand in a large plastic bin (usually a purpose-bought trashcan) wherein naturally occurring lactobacilli and other microbes convert lactose to lactic acid. Milk pH drops from 6.66 to 4 or 5. Curd (clotted casein) formation occurs at a pH of 4.7.[68] One system for preserving colostrum involves three bins, and the bins are rotated between filling with colostrum, feeding, and washing. The bins should be stirred each day and before feeding to break up the curd. Fermentation is most rapid at high temperatures. Adding bacterial starters does not affect the final composition or improve feed conversion rates.[7,42] Using milk containing blood may lead to putrefaction, identifiable by its foul smell.[42] Properly fermented milk should have a strong acidic or natural yogurt smell. Antibiotics in waste milk from cows treated for mastitis slow fermentation.[31] Naturally fermented milk contains a wide variety of microorganisms, including large numbers of coliforms.[26] Usually these are nonpathogenic and do not harm calves (providing the calves are older than 2 days; the guts of younger calves may still be permeable to macromolecules and bacteria). However, dairy cows can excrete *Salmonella* in their milk. Although a highly acid pH can kill *Salmonella,* use of fermented colostrum may disseminate this disease through the calves. If enterotoxigenic *E. coli* gains entry to the fermentation vats, problems could also arise. Waste milk contains antibiotics and organisms that cause mastitis; these do not adversely affect short- or long-term health if calves are raised in single pens and are fed waste milk only after 2 days of life.[32] In group-rearing situations, teat sucking might lead to the spread of mastitis-causing organisms.[54]

Fresh cow's milk contains more solids than milk and is usually fed diluted 2:1, or sometimes 1:1 with water. This also dilutes the acid and improves palatability. Limited refusal may occur, particularly in hot weather. The author recommends a minimum daily intake of 3 L of fresh cow's milk diluted with an additional 1.5 L of water, although some farmers may feed as little as 2 L of colostrum. Waste milk should not be diluted and should be fed at levels similar to whole milk. Most studies show that preserved milk gives better weight gains and less diarrhea than milk replacer. However, it is inferior to fresh milk. Direct acidified fresh cow's milk gives better performance than fermented milk.*

Feeding surplus colostrum and fresh cow's milk is best suited to medium-sized dairy herds (50 to 100 cows) with low disease levels. In small herds calving may be too erratic to ensure an even supply of fresh cows. In large herds there is a greater chance of finding an infected cow that could contaminate the fermentation. In situations where pathogens contaminate milk, refrigeration and pasteurization of colostrum and waste milk before feeding can give better results.[55]

In the field the main factors to consider in evaluating colostrum and waste milk feeding are the level of feeding and the presence of enteropathogenic microbes in the feed. Mastitic organisms may be able to cause long-term problems if there is the opportunity for teat sucking between calves receiving mastitic milk. Occasionally fermentations may go bad and putrefy. This is most likely to happen if there is blood contamination of the colostrum or milk.

---

*References 7, 27, 43, 46, 50, 68.

Milk replacers

The ready availability of milk replacer is the result of political decisions to pay farmers a constant price for milk. After it leaves the farm gate, milk used for manufacturing is sold more cheaply than milk drunk as liquid (see Chapter 19). In consequence, milk-based milk replacer powders can be sold back to farmers at a profit.

Milk-based milk replacers are usually made from skim milk, a by-product of butter making. This is defatted cow's milk. It is low in fat-soluble vitamins and rich in casein, lactose, B vitamins, and minerals. Whey also is used. It is the product of cheese making. Most of the fat and casein are absent in whey, leaving lactose, globulins, and albumin.[25,63] Fat is added to increase the energy content. Butterfat has the highest digestibility of the available fat sources. Adding butterfat to skim milk produces reconstituted whole milk. Obviously this is economic only if surplus skim milk powder or butterfat is being sold off at cheap prices (e.g., to get rid of surpluses accumulated for price-support reasons). Usually animal or vegetable fats are used as the fat source. In general, milk replacers are less successful feeds than either whole cow's milk or surplus colostrum.[7,28,43,46] High-fat (20% of dry matter), acidified milk replacers are likely to give the best results.

The economic value of feeding milk replacers as compared to milk can be calculated using the assumption that 170 g (0.4 lb) of *high-quality* milk replacer powder gives a gain equivalent to that obtained with 1 kg (2.2 lb) of whole cow's milk.[52]

EVALUATION OF FEEDS

The evaluation of milk replacers starts with the amount fed. The person feeding the calves is the most reliable source for how much is actually given. If there are discrepancies with the label recommendations, the manager should find out why more or less is fed. Milk replacers are often prepared by using cup measures, which contain variable amounts of powder (Fig. 22-2). It is a good idea to take a plastic bag along and measure out several cups of powder. The bag can then be weighed and the actual amount of powder fed calculated. The amount of powder required depends on the composition of the milk

Fig. 22-2 A variety of sized cups can be used to measure out milk replacer; these cups hold from 120 to 160 g (4.2 to 5.6 oz) of powder. Courtesy of Dr. Matt Schoonderwoerd.

Table 22-1 *Approximate digestible energy contents of milk replacer ingredients*

Constituent	Gross energy (kcal/g)	Digestibility (%)	Digestible energy (kcal/g)
Ash	0		0
Fiber	4	0	0
Lactose	4	95	3.8
Protein	5.7	65-95	4.8
Fat	9.4	88-97	8.7

From data in Hand MS, Hunt E, and Phillips RW: Vet Clin North Am (Food Anim Pract) 1:589, 1985.

Table 22-2 *Calculation of approximate digestible energy content of milk replacer*

Constituent	Percentage	Digestible energy (kcal)
Ash	8*	0
Fiber	1*	0
Protein	25*	25 × 4.8 = 120
Fat	10*	10 × 8.7 = 87
Lactose	48†	48 × 3.8 = 182
Total, kcal/100 g		389
Total, kcal/g		3.89

*From label.
†Lactose content can be calculated as:

$$100 - (\text{Ash} + \text{fiber} + \text{protein} + \text{fat} + \text{moisture})$$

Assume values of 7.5% for moisture and 10% for ash plus fiber if an analysis is not available.

replacer, which determines digestible energy content (Table 22-1). Powder requirements increase as fat contents decrease. The label usually states the amounts of fat, crude protein, ash, and fiber present, so the energy value can be calculated (Table 22-2). A typical 10%-fat milk replacer has a digestible energy concentration of 3.89 kcal/g; if fat content is 20%, then digestible energy content is 4.38 kcal/g. Thus a 45-kg calf will require 578 g and 514 g of powder, respectively, for maintenance.

Sometimes powder is fed at much lower levels. Occasionally this results from poor labeling or mislabeling by the manufacturer. On some milk replacer packets, it is easier to find out how to apply for promotional material than to find the instructions on calf feeding! Some calf rearers intentionally overdilute milk replacer, perhaps because restricted feeding or overdilution of milk replacers is sometimes recommended for diarrheic calves in an attempt to reduce gut overload and maldigestion. The calf rearer may then adopt this as a general prophylactic measure.

Underfed calves usually do well at first but gradually lose condition. Death losses tend to occur at around 2 to 3 weeks of age. Calves that make it past this age usually survive because voluntary intake of solid food in older calves compensates for underfeeding of milk. Calves dying of underfeeding become comatose before death, and blood glucose concentrations are very low (Fig. 22-3). Affected calves are emaciated (Fig. 22-4), and at necropsy there is serous atrophy of fat (Fig. 22-5). The rumen and abomasum may be full of roughage instead of milk (Fig. 22-6). Underfeeding also reduces lymphoid tissue mass and immune function,[12] which in turn may favor attack by infectious agents.

Dilution affects digestibility and the incidence of scours. Both underdilution and overdilution

Fig. 22-3 Calves dying of starvation usually appear healthy until shortly before death, when they collapse and become comatose. This calf was hypothermic and bradycardic and had severe hypoglycemia (plasma glucose concentration below the level of detection). The owner had been hand-rearing it on 2 L of milk a day.

Fig. 22-4 Closeup of calf in Fig. 22-3 showing atrophy of lumbar and gluteal musculature.

are harmful. Reconstitution to 9% to 13% dry matter and feeding the liquid at a rate of 8% to 10% of body weight give optimal health performance and efficiency of feed conversion.[15] The more concentrated solutions and feeding at 10% of body weight are required to meet energy requirements with low-fat milk replacers.

Reading the label of the milk replacer is the next step in evaluation of a calf-feeding program. The list of ingredients is helpful in determining the suitability for the calf's digestive system. An example of a good label is shown in Table 22-3. The ideal protein source is skim milk powder. The method of preparation is rarely given but low-temperature spray-drying is better than high-temperature methods. Whey is used in some milk replacers. Unlike casein, whey proteins do not clot in the abomasum, and there is little abomasal digestion.[69,70] They contain more immunoglobulin than skim milk. Nonmilk protein sources may be included; for example, soybean proteins and fish meal are usually used. Pea protein concentrate and bacterial proteins have been tried experimentally, but none are as digestible as milk protein.* Soy, pea, and fish proteins can provoke an immune reaction that leads to villous atrophy.† The incidence of diarrhea is higher with milk replacers containing these products rather than 100% skim milk.[59,60] Soybean protein sources contain a factor that inhibits the secretion and action of the digestive enzyme trypsin. They may also contain undigestible polysaccharides.[15,33,53] Heat, acid, or alkali treatment destroys the trypsin inhibitor and improves its feeding value; extraction with hot aqueous alcohol under carefully controlled conditions can produce further improvements by denaturing proteins that stimulate allergic reactions. However, the product is still not as good as skim milk. Additional methionine and possibly lysine may improve the digestibility of soy-based milk replacers.[62] Utilization of nonmilk feeds improves with age (Table 22-4).[1,8,15] Milk replacers containing non-

---

*References 1, 8, 13, 14, 33, 39, 52, 57, 58.
†References 8, 15, 38, 40, 59, 60.

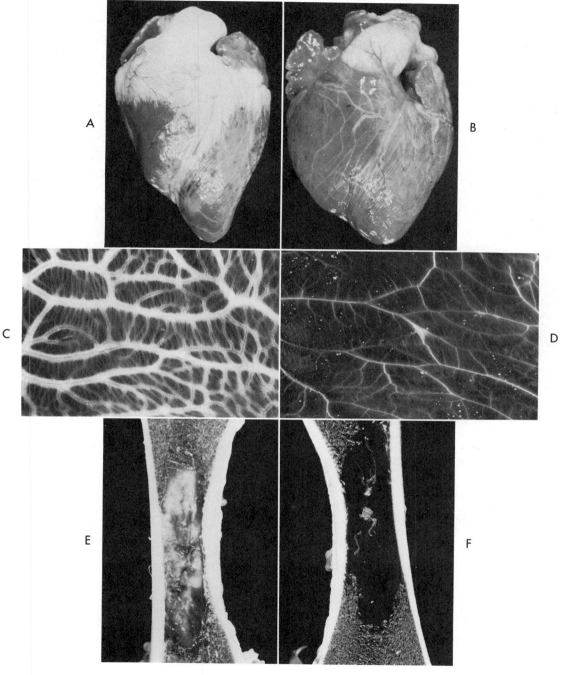

Fig. 22-5 Comparison of necropsy findings in well-fed and starved calves.
Well-fed:
A, Cardiac fat
C, Mesenteric fat
E, Bone marrow
Starved:
B, Cardiac fat
D, Mesenteric fat
F, Bone marrow
Photographs courtesy of Dr. Matt Schoonderwoerd.

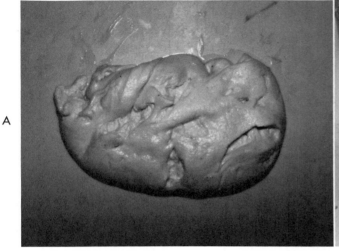
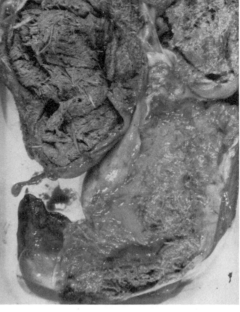

Fig. 22-6 A, Milk clot from abomasum of a calf fed whole cow's milk; **B,** rumen and abomasum full of roughage from a calf receiving inadequate amounts of milk replacer.

Table 22-3 *Composition of a high-quality milk replacer*

Constituent	Amount
Animal or vegetable fat (%)	17-20
Crude soya bean lecithin (%)	1-2
Skim milk solids (%)	78-82
Minerals (mg/kg or ppm dry matter)	
Magnesium	300
Iron	75-100
Manganese	40
Copper	10
Zinc	20
Cobalt (μg/kg)	100
Iodine (μg/kg)	120
Vitamins (amount/kg dry matter)	
Vitamin A (IU)	12,000-30,000
Vitamin D (IU)	1800-3500
Vitamin E (mg)	>20
Vitamin B_{12} (μg)	30

Modified from Roy JHB: The calf, ed 4, London, 1980, Butterworths.

Table 22-4 *Apparent digestibility of crude protein from various sources in calves of different ages*

	Apparent digestibility (%)	
Protein source	10-15 days	30-35 days
Skim milk	85	92
Whey protein		85*
Isolated soy protein		78
Soy protein concentrate†	57	71
Full-fat soy flour†	42	58

*Value calculated by interpolation of results of whey feeding experiments to calves of different ages.
†Plus supplemental methionine.
From data in Dawson DP and others,[8] Akinyele IO and Harshbarger KE,[1] and Khorasani GR and others.[33]

milk proteins perform better if they contain some milk protein, preferably casein; otherwise, they are best reserved for calves over 3 weeks of age.

The amount of fat in milk replacers varies between 10% and 25%. Although products containing 20% to 25% fat may be called *high-fat* replacers, they contain less fat than whole milk, which contains approximately 30% fat on a dry matter basis. A minimum of 20% fat has been recommended when the effective environmental temperature is less than 5° C (40° F).[15,47] All calves grow faster and suffer less diarrhea on high-fat replacers.[64] This is probably because low-fat replacers contain more lactose, which can overload the gut, particularly if it is compromised by enteric infections.[67] Digestibility of fat varies with the source (Table 22-5). Groundnut and corn oil have been associated with an increased incidence of calf diarrhea.[15,23,24]

Adding lecithin or glyceryl monostearate helps emulsify fat and may improve digestibility. Antioxidants, butylated hydroxyanisole, or butylated hydroxytoluene is usually added to prevent oxidation.[52] If fat oxidizes, fatty acids are liberated and it becomes rancid. Free fatty acids have lower digestibilities than triglycerides.[23]

Crude fiber content has been recommended as an index of milk replacer suitability. In general, the more vegetable protein, the more fiber present in the analysis. In practice, however, fiber content is an inadequate reflection of milk-replacer quality. Most crude fiber analyses are insufficiently precise to get excited about, with only tenths of a percentage difference in fiber content. The crude fiber content of vegetable protein supplements is quite variable and may be very low in some preparations. Furthermore, some studies show that the addition of fiber can reduce the incidence of diarrhea.[35]

Calcium and phosphorus are normally adequate if the milk replacer is made from all-milk products. Nonmilk products should be supplemented. Calves growing at 0.3 to 0.4 kg/day (0.7 to 0.9 lb/day) require 6 to 8 g of calcium and 4 g to 5 g of phosphorus daily.[15] Milk and unsupplemented milk-based milk replacers are deficient in magnesium. Calf starter rations are the major source of magnesium for calves. Clinical signs of hypomagnesemia may develop in calves fed large amounts of milk replacer on a long-term basis, although signs of deficiency are rare in calves weaned at 5 weeks. The suggested dietary magnesium requirement for calves is 0.07% of dietary dry matter.

Trace minerals should be included in milk replacers as soluble salts. Supplementation with vitamins A, D, E, and possibly B_{12} is also required (see Table 22-3).

Antibiotics are frequently included in milk replacers at low levels. High levels of antibiotics, particularly of chloramphenicol and neomycin, may interfere with enterocyte growth and cause malabsorption.[17] However, trials by manufacturers indicate that the levels commonly used in milk replacers reduce the incidence of diarrhea and respiratory disease under intensive raising conditions.[15]

Visual inspection of the color of the powder and its solubility is useful in detecting badly burnt skim milk powders and insoluble vegetable constituents. A brown color usually means that the skim milk was badly overheated during processing (Fig. 22-7), which reduces digestibility considerably. Milk replacers should dis-

Table 22-5 *Apparent digestibility of different types of fat*

Type of fat	Mean digestibility (%)	Range (%)
Butterfat	97	95-98
Coconut	95	93-96
Unhydrogenated palm	94	
Bone fat	94	93-95
Unhydrogenated palm	93	90-95
Groundnut	93	
Lard	92	87-96
Tallow, coconut (2:1)	91	
Hydrogenated palm kernel	91	87-95
Tallow, lard (1:1)	90	80-95
Tallow	89	85-93
Corn (maize)	88	88-90
Unhydrogenated palm kernel	88	

From data in Jenkins KJ, Kramer JK, and Emmons DB[23] and Roy JHB.[52]

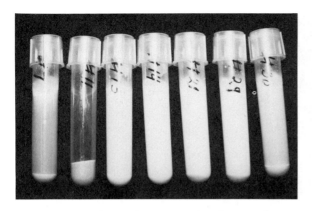

Fig. 22-7 The color and amount of sediment after milk replacer solutions are centrifuged is a useful guide to quality. The solution at the extreme left contains burnt skim milk. The milk replacer second from the left is of very poor quality; there is a lot of insoluble material and the solution is brown. Photograph courtesy of Dr. Matt Schoonderwoerd.

solve readily. An insoluble residue means either that milk was overheated during preparation or that vegetable constituents are present. These may be plant starch grains, which cannot be digested by young calves, or vegetable proteins, which are inferior to milk proteins.

Milk pH influences its keeping and feeding properties. Fresh milk has a pH of 6.6 and sweet (unacidified) milk replacers have a pH higher than 6. Acidified milk replacers are a new addition to the market. These are produced as mild acid (pH about 5.7) and high acid (pH about 4.2) forms. Mild acid replacers keep for at least 24 hours following dilution and can be made from skimmed milk. Incidence of diarrhea is lower on mild acid milk replacers than on sweet milk replacers. High acid replacers cannot use casein, which would cause a clot to form. Whey is the best protein source in this type of replacer. High acid replacers have extended keeping qualities but the absence of casein means they are best suited for calves older than 3 weeks.[63]

Microscopic examination can be helpful in identifying starch grains in the sediment from milk replacer solutions. It may also allow the fat globules to be sized. Globules over 4 μ are less efficiently digested.

Chemical analysis can confirm that the concentrations of major nutrients given on the label are correct. Both too much and too little supplementation with minerals and vitamins A and D can lead to problems. Toxic concentrations of some trace minerals are shown in Table 22-6. Detailed analysis is useful in detecting abnormal concentrations of trace minerals, vitamins, heavy metals, and poisons. The main problems with milk replacers are due to the source of proteins. This information cannot be obtained by routine chemical analysis. Recently, electrophoresis has been used to identify the presence of pea protein in milk replacers.[56]

Milk clotting tests also have been advocated as a measure of milk quality. They work on the principle that rennet (see Chapter 7) should clot casein in milk replacer.[34] The test is performed by adding a small amount of concentrated rennet solution to the reconstituted sample and incubating at 37° C. Clotting can be measured by a number of methods. One method is to place a drop of the mixture on a microscope slide and examine it for evidence of flocculation. Whole cow's milk is usually run as a control. Milk replacers that are based on whey, nonmilk proteins, or heat-denatured skim milk do not clot. Alkalizing agents, low levels of calcium, and other factors also inhibit clotting.[53] Formation of a firm clot is beneficial because it leads to a gradual release of peptides and fatty acids as the clot

Table 22-6 *Toxic mineral concentrations in milk replacers*

Mineral	Concentration causing problems (mg/kg or ppm)	Signs
Copper*	200	Reduced growth
	1000	Hemolytic crisis
		Death
Iron	5000	Reduced growth
		Decreased feed intake
Selenium	10	Reduced growth
Zinc	700	Pneumonia
		Ocular problems
		Diarrhea
		Anorexia

*Toxic levels are highly dependent on levels of molybdenum, sulfate, and other minerals in ration.

is digested in the abomasum.[44,53] This should result in better digestion and less diarrhea. Unfortunately, many milk replacers fail this test. Limited studies indicate that clotting by itself is not essential to normal dietary digestibility,[45] presumably because the small intestine compensates for reduced abomasal function. However, clottability of milk replacer powder may be important if calves are challenged by infectious agents or other adverse conditions. Poor clottability may also indicate that badly heat-damaged milk protein or nonmilk protein was used in the replacer.

Experience with a particular product is important. Milk replacer quality is determined not only by the ingredients but also by the method by which they are prepared. Details of the type of process used to prepare skim milk powders and the methods used to disperse fat are often not available. Methods of preparation and sources of ingredients may also vary from time to time. Calf performance is the best indicator of overall quality.

CONCLUSION

Most systems of calf rearing have been made to work. The primary objective of liquid feeding for calves intended for replacement heifers and beef is to wean a healthy calf. This is most easily accomplished when whole cow's milk is fed. Feeding preserved colostrum and waste milk is the next most successful method utilized, but there are opportunities for spread of pathogens, particularly *Salmonella*. Milk replacers based on all-milk products are acceptable. High-fat, acidified products work better. Milk replacers containing fish and vegetable products should be reserved for calves older than 3 weeks, and even then they will be inferior to milk-based products. The rule of thumb for feeding liquid diets is to offer a solution containing 9% to 13% dry matter at 10% of body weight. All calves should have access to ad lib water. Calf starter and good quality hay should be offered from 10 days of age.

REFERENCES

1. Akinyele IO and Harshbarger KE: J Dairy Sci 66:825, 1983.
2. Anderson KL, Nagaraja TG, and Morrill JL: J Dairy Sci 70:1000, 1987.
3. Andren A and Bjorck L: Acta Physiol Scand 126:419, 1986.
4. Arave CW, Mickelsen CH, and Walters JL: J Dairy Sci 68:923, 1985.
5. Boggs DL and others: J Anim Sci 51:550, 1980.
6. Bomba A and others: Z Vet Med Praha 34:141, 1989.
7. Daniels LB and others: J Dairy Sci 60:992, 1977.
8. Dawson DP and others: J Dairy Sci 71:1301, 1988.
9. Fallon RJ and Harte FJ: Ann Rech Vet 14:473, 1983.
10. Friend TH, Dellmeier GR, and Gbur EE: J Anim Sci 60:1095, 1985.
11. Graham TW and others: J Am Vet Med Assoc 190:1296, 1987.
12. Griebel PJ, Schoonderwoerd M, and Babiuk LA: Can J Vet Res 51:428, 1987.
13. Guilloteau P and others: Reprod Nutr Dev 26:717, 1986.
14. Guilloteau P and others: Br J Nutr 55:571, 1986.
15. Hand MS, Hunt E, and Phillips RW: Vet Clin North Am (Food Anim Pract) 1:589, 1985.
16. Hill KJ, Noakes DE, and Lowe RA: In Phillipson AT, editor: Physiology of digestion and metabolism in the ruminant, New-Castle-upon-Tyne, 1970, Oriel Press, Ltd, pp 166-179.
17. Holland RE, Herdt TH, and Refsal KR: Am J Vet Res 47:2020, 1986.
18. James RE, McGilliard ML, and Hartman DA: J Dairy Sci 67:908, 1984.
19. Jenkins KJ and Bona A: J Dairy Sci 70:2091, 1987.
20. Jenkins KJ and Hidiroglou M: J Dairy Sci 69:1865, 1986.
21. Jenkins KJ and Hidiroglou M: J Dairy Sci 70:2349, 1987.
22. Jenkins KJ and Hidiroglou M: J Dairy Sci 72:150, 1989.
23. Jenkins KJ, Kramer JK, and Emmons DB: J Dairy Sci 69:447, 1986.
24. Jenkins KJ and others: J Dairy Sci 68:669, 1985.
25. Jenness R: In Larson BL and Smith VR, editors: Lactation: a comprehensive treatise, vol 3, Nutrition and biochemistry of milk, New York, 1974, Academic Press.
26. Jenny BF, O'Dell GD, and Johnson MG: J Dairy Sci 60:453, 1977.
27. Keith EA and others: J Dairy Sci 66:833, 1983.
28. Kertz AF: J Dairy Sci 60:1006, 1977.
29. Kertz AF, Reutzel LF, and Mahoney JH: J Dairy Sci 67:2964, 1984.
30. Keusenhoff R and Piatkowski B: Arch Tierernahr 33:179, 1983.
31. Keys JE, Pearson RE, and Weinland BT: J Dairy Sci 62:1408, 1979.
32. Keys JE, Pearson RE, and Weinland BT: J Dairy Sci 63:1123, 1980.
33. Khorasani GR and others: J Anim Sci 67:1634, 1989.
34. Kopelman IJ and Cogan U: J Dairy Sci 59:196, 1976.
35. Laiblin VC, Koberg J, and Hofmann W: Berl Munch Tierarztl Wochenschr 102:236, 1989.
36. Mylrea PJ: Res Vet Sci 7:407, 1966.
37. Nocek JE, Braund DG, and Warner RG: J Dairy Sci 67:319, 1984.

38. Nunes do Prado I and others: Reprod Nutr Dev 28 (suppl 1):157, 1988.
39. Nunes do Prado I and others: Reprod Nutr Dev 29:425, 1989.
40. Nunes do Prado I and others: Reprod Nutr Dev 29:413, 1989.
41. Odde KG, Kiracofe GH, and Schalles RR: J Anim Sci 61:307, 1985.
42. Otterby DE, Dutton RE, and Foley JA: J Dairy Sci 60:73, 1977.
43. Otterby DE, Johnson DG, and Polzin HW: J Dairy Sci 59:2001, 1976.
44. Pelissier JP and others: Reprod Nutr Dev 23:161, 1983.
45. Petit HV, Ivan M, and Brisson GJ: J Anim Sci 66:986, 1988.
46. Polzin HW, Otterby DE, and Johnson DG: J Dairy Sci 60:224, 1977.
47. Rawson RE and others: Can J Vet Res 53:268, 1989.
48. Reddy PG and others: J Dairy Sci 68:2259, 1985.
49. Reddy PG and others: J Dairy Sci 69:164, 1986.
50. Rindsig RB: J Dairy Sci 59:1293, 1976.
51. Rindsig RB and Bodoh GW: J Dairy Sci 60:79, 1977.
52. Roy JHB: The calf, ed 4, London, 1980, Butterworth.
53. Roy JHB: In Batt RM and Lawrence TLJ, editors: Function and dysfunction of the small intestine, proceedings of the second George Durrant memorial symposium, Liverpool, England, 1984, Liverpool University Press.
54. Roy JHB: The calf, vol 1, Management of health, ed 5, London, 1990, Butterworth.
55. Schoonderwoerd M: In Smith BP, editor: Large animal internal medicine, St Louis, 1990, Mosby–Year Book.
56. Schoonderwoerd M and Misra V: J Dairy Sci 72:157, 1989.
57. Sedgman CA and others: Br J Nutr 54:219, 1985.
58. Sedgman CA, Roy JH, and Thomas J: Br J Nutr 53:673, 1985.
59. Seegraber FJ and Morrill JL: J Dairy Sci 69:460, 1986.
60. Silva AG and others: J Dairy Sci 69:1387, 1986.
61. Steen RWJ: Anim Prod, in press.
62. Veen WA, Veling J, and Van der Aar PJ: Arch Tierernahr 39:515, 1989.
63. Webster, J: Calf husbandry, health and welfare, London, 1984, Collins.
64. Wijayasinghe MS, Smith NE, and Baldwin RL: J Dairy Sci 67:2949, 1984.
65. Wise GH, Anderson GW, and Linnerud AC: J Dairy Sci 67:1983, 1984.
66. Wright KL and others: J Dairy Sci 72:1002, 1989.
67. Youanes YD and Herdt TH: Am J Vet Res 48:719, 1987.
68. Yu Yu, Stone JB, and Wilson MR: J Dairy Sci 59:936, 1976.
69. Yvon M and others: Reprod Nutr Dev 25:495, 1985.
70. Yvon M and others: Reprod Nutr Dev 26:705, 1986.

Nutrition of Dairy Replacement Heifers

WALTER M. GUTERBOCK

HEIFER RAISING AS PART OF THE DAIRY ENTERPRISE

Dairying is the conversion of feed, labor, and capital inputs into salable products: milk and animals. Milk is the main source of income to dairy farmers. High-producing cows convert feed and the other inputs into milk more efficiently than low producers. The fixed cost (overhead) of keeping a high-producing cow and paying for her maintenance feed are a smaller proportion of the total expense than they are for a low producer, so proportionately more of the outlay can go for profitable production. Similarly, a dairy that is full spreads its fixed costs over a greater number of productive animals than a dairy with unused capacity.

In order to maintain maximal efficient production, it is necessary to remove unproductive animals from the herd, such as cows that are no longer able to produce enough milk to pay the fixed costs of keeping them, that could not conceive in time to allow a reasonable dry period, or those with pendulous udders, chronic diseases, injuries, or locomotor problems. A few cows die on the farm. Culled and dead cows must be replaced to keep the number of milking animals at its maximal efficient level. On well-managed dairies the replacement rate is generally about 30% per year.[11]

Replacement cost is one of the major expenses of dairying. In most surveys it comes close to labor cost and is usually the third highest expense after feed and labor. It may account for 7% to 10% of gross receipts.[11,21]

A well-managed replacement program allows genetic improvement of the herd because the heifers are the offspring of carefully selected bulls. A good flow of heifers also allows discretionary culling for low production or other undesirable traits and eventual expansion of the herd. A dairy farmer who is looking for room for freshening heifers is more likely to cull marginal cows. A dairy farmer who does not have a good flow of heifers will continue to milk the undesirable cows, and the herd cannot grow. Well-run replacement programs can produce surplus heifers that can be sold to increase the revenue of the farm.

A well-run heifer program is also a source of pride and satisfaction to a dairy farmer. A poor one can be a source of great frustration and wasted time. Heifer programs are also often the means by which the younger generation on the farm gets involved with the running of the enterprise.

A well-managed heifer program allows rapid growth of the animals. With rapid growth, fewer animals need to be held in inventory, less capital is tied up in heifers, and less space is needed to house them[47] (Table 23-1).

Heifer programs often are managed as an afterthought. Heifer facilities may be crowded. Because there is no easily monitored index of performance like daily milk yield, it is a tempting area in which to skimp on feed quality. Little attempt is made to account for feeds, especially forages, that are fed to heifers. Raising young calves can be extremely frustrating, especially when they are sick, and many dairy farmers dislike it. They may delegate the responsibility to

Table 23-1 *Effect of age at first calving on number of heifers needed per 100 milking cows*

Replacement rate (%)	Age at first calving		
	24 months	30 months	36 months
20	40	50	60
25	50	62	75
30	60	75	90

From data in Thicket B and others: Calf rearing, Ipswich, UK, 1986, Farming Press, Ltd.

poorly paid and poorly supervised laborers.

The right person in charge of the heifer raising, especially of the young calves, is probably more important than the type of housing, method of feeding, or even ration fed. The author has seen the most improbable systems work well in the right hands and other situations with good facilities and feeds that do not work because no one cares. Good heifer raisers really love their calves, and calves seem to respond to caring human contact. Women and older children often make good calf rearers. If no one on the farm cares about heifer raising, purchasing replacements should be recommended.

For a number of reasons it is possible to dairy very successfully using only purchased replacements. Equivalent performance can often be obtained with purchased or home-raised heifers, although the reasons are unclear. One might be that dairy farmers who raise replacements tend to milk them all, whereas those who purchase heifers are more selective. Another might be that the genetics of the purchased replacements are superior or equivalent to the herd's. Feeding and management deficiencies may erase the genetic differences between purchased and home-raised heifers. Purchased replacements may have been raised by a specialist, whereas problems in the dairy farmer's raising program may lead to poor performance in the home-raised animals.

The true costs of raising replacements are often hidden. Few dairy farmers carefully account for all of the feed, labor, facilities, and veterinary costs of raising heifers. Many of these expenses may be blended in with the expenses of the milking herd. For example, interest, depreciation, and property tax on the part of the facility that is used for heifers may not be separated. Facilities used to house heifers take up room that potentially could be used for milking cows. Accounting methods also vary. For example, income from cull cows should offset some of the heifer-raising expense. Depreciation on the milking herd may be counted as a part of replacement expense, even though it is not a cash outlay.

On many dairies the herd of replacement heifers is almost equal to the milking herd in size (Table 23-1). Also, when a heifer-raising program is begun, the dairy faces the double cost of raising heifers while still purchasing replacements until the first raised heifers freshen. Nevertheless, a replacement-raising program can have favorable tax consequences by allowing reinvestment of profits in animals and a certain amount of tax avoidance or postponement.

A custom calf raiser may be the answer for the dairy manager who does not want to raise calves or is unable to. In areas where almost all feed is purchased and land is expensive, it often is advantageous to send heifers to areas with cheap feed rather than to pay to haul feed in, as long as the custom raiser does a good job. However, a custom raiser often removes the heifer replacement program from the oversight of the dairy's veterinary and nutritional consultants. Few custom raisers seem to be able to do an excellent job of feeding, record keeping, and breeding. Dairy farmers should be especially careful about sending heifers to beef feedlots that may lack the facilities and experience to manage the breeding program. It is important that dairy managers visit their heifers often to be sure that they are being fed and bred correctly.

GOALS OF A HEIFER REPLACEMENT PROGRAM

It is currently commonly accepted that Holstein heifers should freshen at 520 to 550 kg (1150 to 1200 lb) at 24 months of age, with a withers height of about 133 cm (52 inches). Table 23-2 gives suggested weights and heart girths at different ages to attain that goal.[30] Heinrichs and Hargrove[17] found that both height and weight

Table 23-2 *Growth guide for dairy heifers*

Age (months)	Holstein Girth (cm)	Holstein Weight kg	Holstein Weight lb	Jersey Girth (cm)	Jersey Weight kg	Jersey Weight lb
Birth	79	44	97	61	25	55
2	94	73	161	81	46	101
4	109	124	273	97	82	181
6	127	180	397	112	126	278
9	145	254	560	132	185	408
12	157	324	714	142	236	520
15	165	365	805	150	265	584
18	173	415	915	155	300	661
21	180	465	1025	163	336	741
24	188	522	1151	168	373	822

Based on data in the Penn State dairy reference manual, University Park, 1980, the Pennsylvania State University.

of freshening heifers were positively correlated with herd average milk yields. It should be noted, however, that the mean milk yield for their high group was 8055 kg (17,720 lb). It is not clear if bigger heifers are needed to attain herd averages on the order of 11,000 kg (24,000 lb).

Experimental work on ideal size, weight, and age for freshening heifers has been confounded by a number of factors. Because higher-producing (and presumably better-managed) herds also tend to have better heifer programs, their heifers tend to be bigger and younger at calving.[17] The better production may be the result of superior management, not just bigger heifers. Age, weight, and height are highly correlated,[17] so it is difficult to separate these effects.

Several trends emerge from the literature. Mature size of animals is under genetic control, and by the fourth lactation groups of cows that were grossly underfed or overfed as calves are about equal in size.[39,46] As age at first calving increases, milk yield in the first lactation increases but total lifetime milk decreases.[12,22] It is not clear whether decreased first lactation production in younger heifers is due to the effect of size or age. Smaller heifers have to put more feed nutrients into growth during first lactation than those closer to their mature size. Small heifers have more dystocia[46] and more reproductive problems than bigger ones. The optimal age at first calving for total lifetime production is 22.5 to 23.5 months.[12,22] Profit per day of herdlife is maximized at an age at first calving of 24 to 25 months.[12,17] Age at first calving influences lifetime production and profitability more in herds with high culling rates and lower cow longevity. Overall, it appears that the desirable range for age at first calving is 21 to 25 months.[8]

Heifers reach puberty when they reach about 50% of their weight at first calving (about 275 kg or 600 lb in Holsteins), regardless of the age at which they attain that weight.[5,10,34] Because the gestation period is fixed, to decrease age at first calving the strategy must be to accelerate growth to breeding size. It is well established that attempts to accelerate this growth too much reduce milk yield.[14,44,46] Mammary tissue growth has an allometric phase, in which the mammary gland grows at a greater rate than the body, during the period immediately before and after puberty.[42] Overfeeding during this allometric phase decreases serum growth hormone levels and formation of mammary parenchyma.[40,41] Fat deposition in the udder is increased.[40] When heifers were fed 60% or 140% of requirements,[21] the underfed group gave more milk per unit of metabolic body size than the overfed and normal groups. It appears clear that mild underfeeding does less harm to future productivity than gross overfeeding in the period around puberty.

A veterinarian or nutritional consultant should not blindly embrace the 24-month, 525-kg (1150 lb) goal. Some dairy farmers can exceed this goal and bring in productive heifers at 21 or 22 months. By growing the heifers rapidly in the period before puberty, they can take advantage of the greater feed efficiency of the younger animals. By decreasing age at first calving, they tie up less capital in animals and facilities and may lower total feed costs. Nevertheless, they may pay more for the higher-quality feeds needed to accelerate growth. As already stated, mild underfeeding does no harm to future productivity. A dairy farmer with cheap feed and spacious facilities may be better off to grow heifers more slowly with less purchased feed. Dairies that are expanding or that have a high culling rate and a relatively young herd, as well as purebred breeders trying to make impressive records, may desire bigger, older heifers at freshening that produce more milk in the first lactation. Dairy farmers with a longer-term perspective may want to maximize lifetime milk by increasing growth rate and decreasing age at first calving. In most cases, accelerated growth reduces the final cost of raising the heifer.

The consultant should evaluate the goals, facilities, and feed resources to determine the optimal age at first calving and growth rate goals. The cost of inputs to accelerate heifer growth should be weighed against the benefits of lower age at first calving.

The average growth rate required to attain 525 kg (1150 lb) at 730 days is 0.66 kg/day (1.5 lb/day). To breed heifers at 14 to 15 months of age at 360 kg (795 lb) requires 0.74 kg/day (1.6 lb/day) gain for 14 months (for Holstein calves with a birth weight of 45 kg or 100 lb). During gestation, 0.6 kg/day (1.3 lb/day) would be required. Overfattening during the period around puberty should be avoided. In Danish work, production in first lactation was reduced when rates of gain around puberty exceeded 0.6 kg/day (1.3 lb/day), but the experiment was done with animals of much shorter stature than American Holsteins.[18] In the North American situation, rate of gain in this critical period should not exceed 0.8 kg/day (1.8 lb/day). Condition score should be maintained at 3.0 to 3.5. Withers height should be monitored as well as body weight to help ensure that weight goals are not met by overfattening. Feeding strategies to maximize frame growth have not been well defined.

The most profitable strategy may be to try for maximal growth (0.8 to 0.9 kg/day or 1.8 to 2 lb/day) during the first 4 months of life to take advantage of the greater feed conversion efficiency of the young calf. This period of rapid growth would then be followed by a period of 0.6 to 0.7 kg/day (1.3 to 1.5 lb/day) gain around puberty and 0.6 kg/day (1.3 lb/day) during gestation. Gain is more efficient in younger calves because a lower percentage of their feed energy is used for maintenance[47] (Table 23-3). More research is needed that incorporates body condition scoring (Chapter 26) and withers height so as to separate the effects of frame growth from those of fattening.

Table 23-3 *Percentage of total energy needed for maintenance at various live weights and rates of gain*

Live weight, kg (lb)	Rate of gain	
	1.0 kg/day (2.2 lb/day) (%)	0.5 kg/day (1.1 lb/day) (%)
50 (110)	30	36
100 (220)	45	59
300 (660)	51	68
500 (1100)	55	71

From data in Thicket B and others: Calf rearing, Ipswich, UK, 1986, Farming Press, Ltd.

MONITORING THE REPLACEMENT HEIFER PROGRAM

One of the most important contributions that a production medicine program can make to a dairy farm is monitoring heifer growth. Because it is often physically difficult to weigh animals, weight tapes are commonly used to estimate body weight. They seem to be accurate enough for this purpose. Withers height should also be measured and tracked.

One effective means of monitoring is to measure a cross-section of the heifers on the farm.

Nutrition of Dairy Replacement Heifers

On a small farm this group would include all the heifers. On a large one it may only be a sample of heifers of all ages. Age can be graphed against height and weight and compared to the standard. It is helpful to have preprinted graph forms (Fig. 23-1) that can be used to chart this information. A prerequisite for this program to work is that heifer calves be individually identified and their birth dates recorded.

The range of calving ages and weights should also be examined. Means alone may not give an accurate representation of the heifers. A wide variation may be a sign of overcrowded facilities that never allow the smaller, weaker animals to catch up.

A monitoring program should include an accounting of the costs to raise heifers. This requires a certain amount of extra record keeping, especially to record forage consumption. It also may require changing the accounting system somewhat to separate drugs, feeds, and labor used for the heifer program from those used for the milking herd. Accurate knowledge of the costs helps in goal setting and in evaluating proposed improvements in the program.

Calf morbidity and mortality should be monitored. Mortality is relatively easy to track. Death loss of heifers should be less than 2% from day 2 of life to calving. In general, deaths during the first week are due to periparturient events rather than to deficiencies in the calf-raising program. Stratifying deaths by age is helpful in suggesting

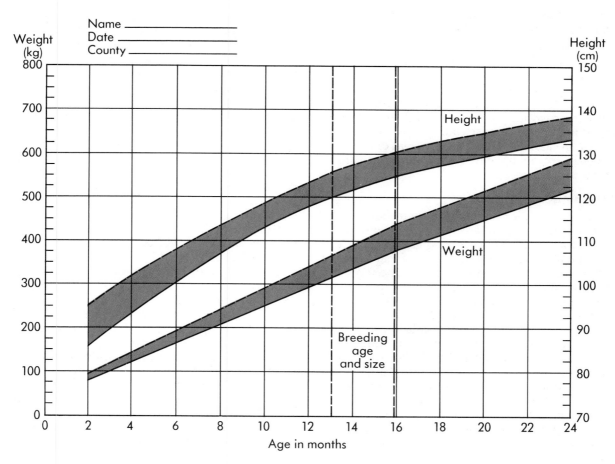

Fig. 23-1 Holstein calf and heifer growth chart.

management changes to decrease mortality.

It is much harder to get calf raisers to record morbidity and treatments. Preprinted cards stapled to the calf hutches have been used successfully to do this. Letter codes are used to denote the conditions treated and the medications used. The calf raiser can then record these data without keeping lists and files. Excessive morbidity should lead the dairy farmer and consultant to evaluate calf nutrition, housing, and sanitation. It is far more productive to focus on management and nutrition deficiencies than to chase etiologic agents. Healthy, vigorous calves that are well fed and kept in a good environment should not get sick except when weather stress is truly overwhelming.

MANAGEMENT AND HOUSING AND THEIR RELATION TO NUTRITION

Accessibility of feed is as important in assuring adequate intake as the ration that is being fed. Weather, overcrowding, and disease can modify feed requirements and also affect accessibility. Poor accessibility is often a result of poor design or overcrowding of facilities, and these factors predispose to disease and decrease growth rates. In calf raising, as in other areas of livestock husbandry, management and nutrition cannot be separated.

The preweaning period

The right people can make indoor facilities work. However, the consultant should recommend the most forgiving system, that is, least likely to lead to a breakdown in animal health if management errors or weather changes occur. The most forgiving calf-housing system is the individual outdoor hutch.[26,50] Outdoor hutches allow greater space between calves and thus decrease the pathogen load and minimize calf-to-calf contact that can spread disease. However hutches are less convenient and comfortable for the people who must take care of the calves during inclement weather. They may fail under these situations if calf rearers are not dedicated.

Group feeding of calves on milk can be successful. Group housing and ad lib feeding of acidified milk replacers can save labor in calf feeding. Grouped calves fed ad lib tend to consume more milk replacer and less dry starter ration and water than calves fed twice daily.[35] Some report[47,53] that calves on the ad lib system gain more than calves that were limit-fed milk replacer, but this result is not consistent.[35] Ad lib–fed calves are fatter and bigger at weaning than limit-fed calves. However, they consume less grain and have a harder time adjusting to weaning.[35,47] Housing and bedding of ad lib–fed calves must accommodate the copious amounts of urine (and, potentially, ammonia fumes) produced.

The design and placement of calf hutches must be adapted to local conditions, and design and placement that are adequate for one season may be inappropriate for another. For example, in the summer in central California, calves need maximal air movement and protection from heat and sun. This means that hutches should have relatively open sides and a high shade over them, and the hutch opening should face north so that the sun does not shine in under the roof. In the winter, by contrast, the calf needs protection from cold and damp. This would mandate a hutch that is closed on three sides and that is not in the shade to allow solar warming. The hutch should face south to admit sunlight. A hutch with gaps between the boards on the sides helps cool calves and maximize air circulation in the summer but does not allow the calf's body heat to warm the inside of the hutch in the winter. Long hutches with fully closed sides and deep bedding are necessary in the north-central United States and Canada but not in areas with mild winters.

The calf hutch should be designed to keep the calf dry. It should be on a deep bed of round pebbles to allow urine and water to drain away from the calf or on wooden slats or expanded metal. If the hutch is on a solid surface, it must be bedded deeply to allow drainage and to provide insulation. Bedding must be changed to keep the calf dry. The hutch should be designed to minimize drafts, especially if it has a slatted floor or no bedding is used. In some areas this may require putting up a windbreak. Calves can withstand great cold if they can stay dry and out of drafts. Feed and water should be accessible but protected from fecal soiling. The grain bucket should be protected from getting wet. The greater the distance between hutches, the better.

Successful calf raisers have used both buckets and bottles to feed milk. Suckling from a nipple bottle probably stimulates esophageal groove closure better than drinking from a pail.[52] This means that less milk will putrefy in the rumen of the very young calf. Milk-feeding equipment must be kept clean. Buckets are easier to wash but are perhaps less likely to be washed because they are cumbersome to transport and handle. If raisers feed milk in buckets, they are likely to fill the buckets on the hutches and wash them infrequently. Milk in buckets is vulnerable to contamination by flies and feces. Bottle washing can easily be automated on a large operation. Nevertheless, buckets last longer than plastic nipple bottles, and there is no nipple replacement expense. In the end, the raiser must make the choice based on facilities and convenience.

Milk and colostrum should be handled so as to minimize bacterial growth. On large dairy farms, warm hospital (mastitis- and antibiotic-treated) milk may sit in unrefrigerated cans in hot weather for a whole day before it is fed. It is dramatic to take a sample of such milk and show the owner that the bacteria count has increased to over 10 million colony-forming units (CFU) per ml! Calf-feeding schedules and milking schedules should be coordinated to minimize the incubation time. If possible, calf milk should be refrigerated. A drawback is that this practice encourages dilution by hot water to warm the milk.

One of the aims of calf management is to encourage consumption of dry feeds and establish a functional rumen as early as possible. Calves first offered starter concentrate rations at 28 days gained less in 56 days than those first offered starter at 14 days.[51] Calf starter should be offered to calves after day 7, although some calves eat a bit before this. Young calves should be offered only what they will consume in a day. Calves reject starter that is dusty, weathered, wet, or sour. Excess or spoiled starter from the grain buckets should be removed, fed to weaned calves or cows, and replaced with fresh grain. Calves seem to prefer coarse mixes of rolled ingredients and pellets to powdery rations.[47] Many successful raisers use complete pelleted rations. Calves given a choice consumed pellets and mixed dry feed equally.[38] Pellets eliminate selection of ingredients by the calves and are easy to handle. The disadvantage of pellets is that they become very dusty when they are weathered or after they have become wet. It is also difficult to monitor the quality of ingredients in a pellet. Pellets should not be too big or too hard for baby calves to eat.

Good calf raisers encourage grain consumption by putting a bit of starter in the calf's mouth. Calves also seem to like a bit of milk replacer powder sprinkled on the grain. The powder is sweet, and as the calves lick it up they also get some starter.

Clean water should be available free choice to calves after the first week of life. The calf must be able to drink water if it is to consume grain. It has been estimated that a calf needs 6.5 kg of total (consumed and feed) water per kg of dry matter intake (6.5 pints/lb DM).[31]

Weaning

Weaning generally takes place at 45 to 60 days of age. Earlier weaning has its adherents, but the author has not seen it work very well. Again, the right person can probably make it work, but it is certainly not for everyone. In general, the later the calves are weaned, the better they adjust to the stress. Nevertheless, there seems little advantage of leaving calves in hutches after 60 days. They get too big for the hutches, they produce copious amounts of feces, and the buckets may not hold enough grain to meet their needs.

Weaning should be managed so that the calves face as few stresses at one time as possible. It is a mistake to stop milk feeding, remove the calves from the hutches, dehorn, castrate, and vaccinate on the same day. Vaccinating, dehorning, and castrating should be done either before weaning (between days 21 and 45) or after weaning. Milk feeding should be stopped several days before the calves are to be removed from the hutches. Milk should not be stopped unless the calf is consuming 1 kg (2 lb) or more of starter a day.

A common mistake in calf management is to put newly weaned calves into a group with older ones and to feed them excessive forage. Newly weaned calves should be put in small, uniform groups, ideally 8 to 12 calves per group. The pen should allow at least 3 m^2 (3.6 yd^2) per calf. As a group grows, the smaller members of it can be sorted out and put with younger calves so they can compete more effectively. Newly weaned calves should be fed the same starter they ate in the hutches. They should be fed from a raised trough inside the pen because they are not used to eating through stanchions or cables. Self-feeders can be used successfully, but they must be cleaned frequently. There should be adequate feeder space so that all calves can eat at once. Clean water should be available. Water troughs should be long and shallow so that all calves have easy access to water and the water stays fresh. Large water troughs designed for mature cows do not work for calves because the water gets stale and full of algae.

Calves that are newly removed from hutches and put into groups are often exposed to coccidial oocysts in the manure in the pen. Overcrowding and muddy conditions will increase the exposure. Periodic corral cleaning and the addition of a coccidiostat to the grower grain mix will help prevent coccidiosis. Calves respond to the addition of a coccidiostat to the grain even when clinical signs of coccidiosis are not seen.

Older weaned heifers

As heifers get older, group size can be increased. The group size should be adjusted to the facilities and to the flow of heifers through the facility. The key is to ensure that heifers within a group are uniform in size. This minimizes the effects of competition for feed on the smaller animals. Smaller herds need smaller groups to maintain uniformity. Overcrowding leads to wider disparities in size, increased risk of coccidiosis, and increased respiratory disease. Heifers are much better off in the open with minimal shelter than in poorly ventilated portions of dairy barns that are shared with mature cows.

NUTRITION DURING THE FIRST THREE WEEKS OF LIFE
Basic physiology

Timely feeding of adequate quantities of high-quality colostrum is essential to the future well-being of the calf (Chapter 21). The calf is dependent on the antibodies in colostrum for immunity to disease until its own immune system becomes functional. Assuring that the colostrum program is working should be the first step in any consultation on calf-rearing problems.

Because the standard calf nipple bottle in the United States is 1.9 L (4 pints), calves are commonly fed 3.8 L (1 U.S. gal) of milk or replacer per day in two feedings. This is not enough to meet the energy requirements of the calf for maintenance and growth. In cold weather, this situation gets even worse. If a 20% fat milk replacer is diluted at the rate of 120 g/L of liquid (1 lb/gal), it yields a product that is 2.4% fat, compared to 3.5% to 3.7% for whole milk. The whole milk requirement for maintenance of a 45-kg (100 lb) calf is calculated to be 3.3 L (7 pints) per day. For 450 g (1 lb) of growth per day, 5 L (11 pints) or about 600 g (1.3 lb) of dry matter are required.[31] It is not surprising that some calves lose weight and body condition during the first 2 weeks of life when dry feed consumption is low.

Rumen function develops in response to solid feed consumption. The rumen can be functional as early as 2 weeks and is functional in all calves by 6 to 8 weeks of age if they are offered solid feed.[28] The rumen develops earlier in calves that are limit-fed milk than in those fed milk ad lib because solid feed consumption is higher in the limit-fed calves.

Milk and milk substitutes

Milk and milk substitutes may be placed in a hierarchy of desirability as calf feeds. They are described here from most desirable to least.

Colostrum. Timely feeding of high-quality, first-milking colostrum is paramount to the future health of the calf. Undiluted colostrum is too rich for routine feeding after 2 days of age. Feeding of undiluted colostrum after 5 days of age leads to an increased incidence of scours.[28] Colostrum is the main source of fat-soluble vitamins to the baby calf. Colostrum may be diluted 1:1 or 2:1 with water and used as a routine feed for calves after day 2. There may be some benefit from the local action of immunoglobulins in the gut.

On large dairies it is important to motivate milkers to separate true first-milking colostrum from later collections. It is also important that colostrum be fed as soon as possible after milking or cooled promptly if it is to be saved. Bacterial quality of colostrum on large dairy farms in warm climates is often very poor.

Fresh cow milk. This is the secretion from cows in the first 3 days of lactation. It is intermediate between whole milk and colostrum in composition. It does not contain enough immunoglobulins to confer adequate passive immunity to calves, but it may contain enough to have some local antigen-binding activity in the gut. It should be fed to the youngest calves on the farm, following the colostrum feeding.

Whole milk. Whole milk is the product for which the calf's digestive system was designed[13] and is the standard by which milk replacers should be judged. It is the ideal feed for the first 3 weeks of life, which is the period of greatest danger of scours. After 3 weeks the calves can be fed milk replacer until weaned. Greater calf health and decreased treatment cost for young calves fed whole milk help to compensate for the cost advantage of milk replacer.

Mastitis milk. Milk from mastitic cows that cannot be sold and milk from cows on antibiotics often is fed to calves. Mastitic milk is lower in nutrients than whole milk and often is highly contaminated with bacteria. Like colostrum, it should not be allowed to incubate between milking and feeding. Most dairy farmers cannot bear to discard this milk and instead feed it to calves. If possible, it should be fed only to older calves.

Return milk. This milk consists of whole milk and other dairy products that have been returned from stores. The containers are crushed and the product recovered. Often it will contain yogurt, soft cheeses, chocolate milk, and other products in addition to milk. Because of the way it is handled, it is often high in bacteria and may contain sucrose. It is used successfully by raisers but is probably best fed to calves over 3 weeks old.

Milk replacer. Calves on milk replacer tend to scour more than those on whole milk.[31] The digestibility of milk replacer nutrients is dependent on the protein source and the processing of the milk protein. Milk protein–based replacers are preferable to those containing vegetable proteins (Chapter 22). In the author's opinion, milk replacers are better used to dilute and extend the whole milk supply than to replace whole milk completely.

Quantity of milk. Newborn calves should be fed 8% to 10% of their body weight of whole milk or its equivalent.[18] For a 45-kg (100 lb) Holstein calf, this works out to about 4.5 L (1.2 gal). Feeding 5 L (1.3 gal) per day supports 460 g (1 lb) per day of growth.[31] One study[37] showed the best weight gains and no additional scouring when calves were fed 5 L (1.3 gal) of milk daily. Feeding large quantities of milk to healthy calves does not seem to cause scours in research trials,[18] although the belief persists among calf raisers that it does. The quantity of milk should be adjusted to the size of the calf, as larger calves have larger maintenance requirements than smaller ones. Twice-a-day feeding leads to better daily gains than once-a-day feeding.[38] Calves that are consuming starter can be switched successfully to once-a-day feeding at 3 weeks to a month of age.

The quantity of milk fed should not be increased as calves get older. Holding milk intake constant encourages consumption of dry feed. Calves fed greater quantities of milk tend to consume less starter and have a more difficult time at weaning.[35]

Cold weather increases the maintenance energy requirements of calves. The strategies for increasing caloric intake include increasing the energy density, palatability, or both of the calf starter; increasing the frequency of liquid feeding; and increasing the nutrients in the liquid feed. Of these, increasing the frequency of feeding is the most successful.[37,43] Increasing the dry matter content of the liquid feed above 15% decreases feed efficiency.[31] The need to increase the caloric intake of the liquid feed is greatest for calves under a month of age because their enzyme systems are not well adapted to digesting calf starter. Older calves can increase caloric intake by eating more starter.

Calf starter

In order to meet the protein requirements of the young calf (assuming that 3.8 L of milk, or 1 gal, will be fed daily), calf starter mixes should supply at least 18% crude protein.[47] Nonprotein nitrogen should not be used in calf starter because the rumen is not fully functional in very young calves.[29] Crude fiber should be 6% to 8%. Crude fiber below 6% may predispose calves to bloat. Further requirements are shown in National Research Council (NRC) tables in the appendix.

Calves under 3 weeks of age may benefit from the use of a prestarter that contains milk proteins and lactose because their digestive enzymes are adapted to milk digestion.

Starter may be pelleted or consist of steam-rolled grains and pellets containing protein and the vitamin-mineral mix. Calves do not like finely ground or dusty feed. If the starter is a mix of grains and pellets, it also should contain 7% to 10% molasses to prevent dustiness and increase palatability. If the starter is to be handled in bulk with tanks and augers, the molasses content may have to be reduced to allow the product to flow.

Cottonseed meal should be used with caution as an ingredient in calf starter (p. 272).

NUTRITION FROM THREE WEEKS TO WEANING

This is a period of increasing starter consumption and decreasing dependence on milk. Concentrate intake increases from about 0.5 kg (1 lb) at 3 weeks of age to 3 kg (6.6 lb) at 12 weeks of age. Concentrate consumption increases rapidly when milk feeding is stopped.[47]

Milk feeding may be stopped as early as 30 days of age, although it is more common to continue for about 55 days. In one study[6] comparing weaning at 35 days to weaning at 70 days, the early group had lower weight gains and consumed more starter feed. In the author's experience, weaning at about 55 days and removal of the calves from the hutches at 60 days is a very forgiving system. Excellent calf raisers should be able to wean earlier.

After 3 weeks of age, labor may be saved by feeding a more concentrated milk replacer product once a day. This is commonly done by feeding one 1.9 L (2 quart) bottle daily, containing milk replacer mixed at 150% normal strength. This mixture would exceed the 15% dry matter recommendation for maximum digestibility.

Another commonly used method to encourage grain consumption is to reduce milk feeding by half after day 35 to 40 of age, preparatory to weaning at day 55. This would mean once-a-day feeding of the normal milk replacer or whole milk. This method can also work well and save some labor. Milk feeding should not be reduced for calves that are not consuming calf starter.

Hay may be fed to calves before weaning, but it is optional.

NUTRITION FROM WEANING TO BREEDING

Newly weaned calves should continue to get calf starter. Two to 3 weeks after removal from the calf hutches, they may be fed a grower mix containing 16% crude protein. It can be made using the same ingredients in the starter in different proportions.

The most common feeding error after weaning is to overfeed forages. Corn silage cannot meet the protein requirements of young calves, and alfalfa hay cannot meet their energy requirements. To achieve optimal growth rates, some concentrates should be fed to heifers up to breeding age (see appendix table for requirements). Heifers that are overfed forages develop a potbellied look (haybelly) and have very thin flesh covering over the spine and transverse processes of the vertebrae. The consultant should palpate the backs of heifers with long winter coats.

There is apparently no benefit to feeding rumen bypass protein to heifers in this age group.[23]

NUTRITION FROM BREEDING TO CALVING

Heifers may be fed straight forage during this period, provided that the basic nutrient requirements are met (see Appendix). The crude protein level of the total ration should not be less than 12%. Corn silage alone does not meet this requirement.

Feeding concentrates to heifers during the 2 or 3 weeks before calving to adapt the rumen flora to concentrates is a sound nutritional practice but may lead to excessive udder edema. Affected heifers may be treated with furosemide before calving. Salt should not be fed to heifers close to parturition. In some cases dairy farmers may elect to keep heifers separate from cows until after parturition and not feed them concentrates before calving.

Heifers should freshen with a body condition score of 3.5.

ADDITIVES USED IN REPLACEMENT HEIFER FEEDING

Sodium bicarbonate

Sodium bicarbonate may be used as a rumen buffer at up to 2% of the ration dry matter. One study showed a small but nonsignificant increase in average daily gain and feed efficiency at about 2% bicarbonate.[15] Sodium bicarbonate is relatively expensive and displaces ingredients that provide energy and protein to the calf. It is probably preferable to provide adequate digestible fiber to the calf to maintain rumen pH in the normal range rather than to supplement with bicarbonate.[47]

Decoquinate

Decoquinate (Deccox, Rhone Poulenc, Inc.) is a coccidiostat. The recommended feeding rate is 0.5 mg/kg (0.5 ppm) body weight. Decoquinate reduces coccidial oocyst shedding and the clinical signs of coccidiosis. It also may stimulate the immune system of calves, or at least counteract a possible immunosuppression caused by coccidia.[36] Reduction of diarrhea also leads to a decrease in respiratory disease. Decoquinate was shown to increase weight gains (but not height) significantly in Holstein calves.[7,16] It may be fed to calves under 180 kg (400 lb) live weight.

Lasalocid

Lasalocid (Bovatec, Hoffman LaRoche) is an ionophore coccidiostat that is recommended for use in dairy heifers over 180 kg (400 lb). It is as effective as decoquinate and monensin in the prevention of coccidiosis in artificially infected calves. As an ionophore it increases the yield of propionate in the rumen fermentation and thereby increases feed efficiency. Unlike monensin, it is approved for heifers under 180 kg (400 lb) and does not appear to affect the palatability of feed. It may be fed at a level of 60 to 200 mg per head per day in at least 500 g (1 lb) of feed or in a free-choice, self-limiting supplemental feed.[27]

Monensin

Monensin (Rumensin, Elanco) is an ionophore that is approved for use in dairy heifers over 180 kg (400 lb) live weight. It may be fed at a rate of 50 to 200 mg per head per day, depending on weight. The dose should be reduced by half when it is first fed because of its effect on palatability. It is reported[19] to increase average daily gain of heifers by 6% to 14%. It does not appear to have any deleterious effect on reproduction.[1] Monensin is as effective against coccidia as lasalocid and decoquinate.[9] It also has been reported to reduce the incidence of legume-induced bloat.[19]

Tetracyclines

Oxytetracycline added to milk fed to calves at a dose of 83 mg per head per day improved weight gain, increased starter consumption, and decreased diarrhea.[25] Chlortetracycline (Aureomycin Soluble Powder, Cyanamid) also may be added to milk at the rate of 2.2 mg/kg (1 mg/lb) live weight per day. It increases daily gain[45] by about 10%. Additional gains may be obtained by adding chlortetracycline to the calf starter at the rate of 25 to 70 mg per head per day (or 5 kg of Aureomycin crumbles per tonne [10 lb/ton] of feed).

Tetracycline-sulfonamide combinations may effectively be added to calf starter and grower mixes. They may also be top-dressed during periods when unfavorable weather or other stresses might precipitate outbreaks of respiratory disease. The sulfamethazine in the product helps to control coccidiosis, although it is not labeled as a coccidiostat.[49]

SPECIFIC IMPORTANT NUTRIENTS IN CALF FEEDS

Fat-soluble vitamins

The calf is born with no liver reserves of vitamins A, D, or E and depends on the colostrum as a source.[31] Because proper absorption of these vitamins requires a normal intestinal epithelium, calves with gastrointestinal infections may not be able to absorb them from the milk replacer. The minimum amount of vitamin A required to prevent an increase in cerebrospinal fluid pressure in calves fed purified diets is 4200 international units (IU) per 100 kg of live weight (1900 IU/100 lb).[31] The minimum amount of vitamin D required for growth, well-being, and proper bone calcification is about 660 USP units/100 kg live weight per day (300 USP units/100 lb).[2] Calves that eat 1.5 kg (3 lb) of alfalfa hay or silage per 100 kg of live weight and that have access to sunshine do not need additional vitamin D. The use of cod liver oil as a source of vitamins A and D exacerbates vitamin E deficiency (muscular dystrophy) because of the high concentration of unsaturated fatty acids in cod liver oil.[48]

Vitamin E (α-tocopherol) is required by the calf as an antioxidant. In addition to preventing muscular dystrophy, vitamin E supplementation also stimulates the immune system of calves.[33] The vitamin E requirement of calves is increased by stress, by high concentrations of polyunsaturated fatty acids in feed, and by rapid growth. The presence of fish protein in the diet also appears to increase the requirement.[24] The muscles of the pharynx are among those affected earliest by muscular dystrophy. Difficulty in swallowing and aspiration pneumonia are early signs of the deficiency.[3,37] A daily dose of 148 mg of d-α-tocopherol prevented the deficiency symptoms in one study.[37]

Other vitamins

Thiamin, riboflavin, niacin, pyridoxine, pantothenic acid, biotin, choline, folic acid, inositol, and para-aminobenzoic acid have all been shown to be required by the calf[31] (Table 23-4). Calves fed cows' milk do not require supplementation.

Minerals

Calcium. The calcium requirement of calves increases with increased growth rates.[31] When milk replacer containing nonmilk (and therefore less digestible) ingredients is fed, the calcium requirement may be higher because of decreased digestibility. Liquid diets containing skim milk powder and added fats may tie up calcium in soaps and therefore require higher calcium levels than whole milk.[31]

Magnesium. Hypomagnesemic tetany may occur in calves after several weeks on an all-milk diet.[4] Not much difference in the growth rate of calves may be seen until the clinical signs appear. It therefore is important that calf starter and milk replacers contain adequate magnesium (Table 23-4).

Salt. Salt should be available free choice to all but preparturient heifers, in which salt should be restricted in order to help prevent udder edema. Feeding potassium chloride instead of sodium chloride is probably of no benefit in preventing udder edema.[32]

Selenium. Calves obtain selenium from the dam in utero. Injecting the calf or supplementing the calf starter does not significantly affect blood selenium levels.[20] It would appear that the most rational way to provide selenium to the young calf is to provide it to the dry cow.

Cottonseed products

Cottonseed contains gossypol, which is toxic to nonruminants. Whole cottonseeds contain more gossypol than cottonseed meal. The amount of gossypol in cottonseed meal varies widely, depending on the variety of cotton, growing conditions, and the amount of heating and protein

Table 23-4 *Minimum vitamin and mineral requirements of preruminant calves*

Vitamin A	4200 IU/100 kg body weight (bw)
Vitamin D	660 IU/100 kg bw
Thiamin	95 µg/kg bw
Riboflavin	35-45 µg/kg bw
Niacin	26 mg/100 kg bw
Pyridoxine	65 mg/100 kg bw
Pantothenic acid	19.5 mg/100 kg bw
B_{12} (Cyanocobalamine)	20-40 µg/kg dry matter consumed
Biotin	0.19 mg/100 kg bw
Choline	2.6 g/100 kg bw
Folic acid	15 µg/kg bw
Inositol	5.2 mg/kg bw
Para-aminobenzoic acid	260 µg/kg bw
Dietary calcium	6.7-18.3 g/day, depending on growth rate
Phosphorus	3.0 g/day
Magnesium	0.4-0.78 g/day
Iron	100 mg/day
Copper	10 mg/kg dry matter consumed
Cobalt	0.1 mg/kg dry matter consumed
Zinc	50 mg/kg dry matter consumed

From Radostits OM and Bell JM: Can J Anim Sci 50:405, 1970.

binding that occurs during the extraction of the oil from the seeds. Between 1986 and 1988, several severe incidents of gossypol poisoning of calves occurred in the United States.

The exact tolerance of calves for gossypol and the ages at which they are most susceptible to poisoning have not been well defined. At present, the safe level of gossypol in the complete feed of preruminant calves is assumed to be 100 ppm, a figure based on work in swine.

If cottonseed meal is to be fed to calves, every load should be tested for gossypol content by an experienced laboratory. Heating, binding to iron in the feed or rumen, and protein binding inactivate gossypol. Under no circumstances should whole cottonseed be fed to preruminant calves.

REFERENCES

1. Baile CA and others: J Dairy Sci 65:1941, 1982.
2. Behdel SI and others: P Agricul Exp Sta Bull 364:1, 1938.
3. Blaxter KL and others: Br J Nutr 6:125, 1952.
4. Blaxter KL and Rook JAF: J Comp Pathol 64:157, 1954.
5. Crichton JA and others: Anim Prod 2:159, 1970.
6. DePeters EJ and others: J Dairy Sci 69:181, 1986.
7. Ducharme GA: J Dairy Sci 72(suppl 1):242, 1989.
8. Foldager J and Sejrsen K: Proceedings of the 12th World Congress on the Diseases of Cattle 1:451, 1982.
9. Foreyt WJ and others: Am J Vet Res 47:2031, 1986.
10. Gardner RW and others: J Dairy Sci 60:1941, 1977.
11. Genske and Mulder: Dairy income and expense averages for California dairy clients, 1988, 1989.
12. Gill GS and Allair FR: J Dairy Sci 59:1131, 1976.
13. Guilloteau P and others: Reprod Nutr Dev 21(6A):885, 1981.
14. Hansson A: Proc Br Soc Anim Prod 20:51, 1956.
15. Hart SP and Polan CE: J Dairy Sci 67:2356, 1984.
16. Heinrichs AJ and Bush GJ: J Anim Sci 67(Suppl 1):414, 1989.
17. Heinrichs AJ and Hargrove GL: J Dairy Sci 70:653, 1987.
18. Huber JT and others: J Dairy Sci 67:2957, 1984.
19. Hutjens MF: Dairy Herd Management 23(12):32, 1986.
20. Kincaid RL and Hodgson AS: J Dairy Sci 72:259, 1989.
21. Klingborg DJ and others: Financial herd health assessment and monitoring. Paper presented at Am Assn Bov Pract seminar, 1988.
22. Lin CY and others: J Dairy Sci 69:760, 1986.
23. Mantysaari PE and others: J Dairy Sci 72:2107, 1989.
24. Michel RL and others: J Dairy Sci 55:498, 1972.
25. Morrill JL and others: J Dairy Sci 60:1105, 1977.
26. Murley WR and Culvahouse EW: J Dairy Sci 41:977, 1958.
27. National Research Council: Nutrient requirement of dairy cattle, Washington, DC, 1989, National Academy of Sciences.
28. Nocek JE and others: J Dairy Sci 67:319, 1984.
29. Otterby DE and Linn JG: J Dairy Sci 64:1365, 1981.
30. Penn State dairy reference manual, University Park, 1980, Pennsylvania State University.
31. Radostits OM and Bell JM: Can J Anim Sci 50:405, 1970.
32. Randall WE and others: J Dairy Sci 57:472, 1974.
33. Reddy PG and others: J Dairy Sci 69:164, 1986.
34. Reid JT and others: Cornell Univ Agricul Exp Sta Bull 987, 1964.
35. Richards AL and others: J Dairy Sci 71:2193, 1988.
36. Roth JA and others: Am J Vet Res 50:1250, 1989.
37. Safford JW: Am J Vet Res 15:373, 1954.
38. Schingoethe DJ and others: J Dairy Sci 69:1663, 1986.
39. Schultz LH: J Dairy Sci 52:1321, 1969.
40. Sejrsen K and others: J Dairy Sci 65:793, 1982.
41. Sejrsen K and others: J Dairy Sci 66:845, 1983.
42. Sinha YN and Tucker HA: J Dairy Sci 52:507, 1969.
43. Stewart GD and Schingoethe DJ: J Dairy Sci 67:598, 1984.
44. Swanson EW: J Dairy Sci 43:377, 1960.
45. Swanson EW and Hazelwood BP: Tennessee farm and home science progress report 45, Knoxville, 1963, University of Tennessee.
46. Swanson EW and Hinton SA: J Dairy Sci 47:267, 1964.
47. Thickett B and others: Calf rearing, Ipswich, UK, 1986, Farming Press, Ltd.
48. Thompson JR and others: J Dairy Sci 66:1119, 1983.
49. Todd AC and Thacher J: Vet Med Small Anim Clin 68:527, 1973.
50. Warnick VD and others: J Dairy Sci 60:947, 1977.
51. Weiss WP and others: J Dairy Sci 66:1101, 1983.
52. Wise GH and others: J Dairy Sci 1983, 1984.
53. Woodford ST and others: J Dairy Sci 70:888, 1987.

Feeding the Dairy Cow for Optimal Production

J. T. HUBER

For high milk production, the dairy farmer needs cows of high production potential. This has been achieved principally by genetic progress, which averages advances of 1% to 2% per year. Recent developments such as embryo transfer and bovine somatotropin (BST) treatment greatly accelerate that rate. Estimates are that 10 to 15 years of genetic progress might be accomplished by injection of BST.

Good herds have sound calf- and heifer-raising programs that allow cows to be physiologically prepared for high milk yields at first calving. Specifics of such a program are (1) low calf mortality; (2) rapid growth that avoids overfattening prepubertal heifers, which diminishes the mass of mammary secretory tissue; and (3) breeding to calve at about 24 months of age. Weights at calving should be 540 kg (1200 lb) or more (for large breeds). To achieve this size, an average growth rate of about 725 g/day (1.6 lb/day) from birth to 24 months is required.

In addition to good nutrition, herds that achieve high milk yields have sound practices in other critical management areas such as reproduction, mastitis control, proper milking technique, and herd health. The barrel-stave concept aptly applies to milk production and management; that is, milk yields will be limited to the level of the most severe management problem. This must be corrected before benefit from other improvements can be realized.

ENERGY NEEDS

Mismanagement of energy feeding in dairy rations can be costly. Overfeeding during late lactation and dry periods causes excessive fattening and subsequent metabolic embarrassment. Undernutrition in early lactation, when the potential for milk production is greatest, is another common mistake made on dairy farms. Source of energy also affects milk yield and composition. This discussion deals mainly with energy needs at the different stages of lactation and systems for supplying that energy.

Early lactation

For the first 5 to 8 weeks postpartum, propensity for milk production by the cow is high and desire to eat low.[24] This imbalance often results in an energy deficiency that can be partially compensated for with body fat reserves, but a severe energy deficit results in decreased milk yields, poor breeding performance, decreased milk protein concentrations, and retarded growth in younger animals (Fig. 24-1).

To achieve a rapid increase in energy intake after calving, cows should be preconditioned to some grain during the last 2 weeks of their dry period. By 2 weeks postpartum, high-producing cows can be eating 60% to 65% of their total ration as concentrate. Forage quality must be high during this period. Added fat as whole oil seed (e.g., cottonseed, soybean) plus some supplemental fat can alleviate the energy deficit, but to avoid reduced cellulose digestibility, fat should not exceed about 6% of the ration dry matter (DM).

The diet fed in early lactation should be palatable, of high quality, and one to which the cow is adjusted. The faster a fresh cow is encouraged to consume high levels of energy without incurring digestive upsets, the greater will be her

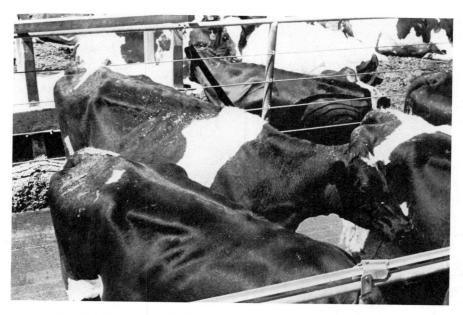

Fig. 24-1 Extremely thin lactating cows in a California dairy herd.

lactational performance. One system suggested for building up energy intake of fresh cows fed in individual mangers is to offer concentrate at about 5 kg/day (10 lb/day) just after calving and increase the amount by about 1 kg/day (2 lb/day) until levels commensurate with needs (12 to 15 kg/day or 25 to 35 lb/day) are reached. For most cows, it will take 10 to 14 days to build up to the desired concentrate intake.

Adjusting intakes of individual cows under conditions of group feeding often is difficult, so close control of concentrate levels fed to early lactation animals housed in groups is imperative. Grouping according to production level, stage of lactation, or both is often needed to deliver energy accurately. Also, younger cows may need to be grouped separately from older ones because they may not be able to meet their needs for both growth and lactation when in competition with larger, older cows. In herds where physical grouping is difficult, extra energy for high producers and young cows can be provided through special feeding units. Further discussion of dividing cows for optimum performance follows.

Feeding to achieve as high a peak in milk production as possible should be encouraged. English workers[10] proposed that for each kg increase in peak milk yield, 200 kg more milk for the total lactation was attained. Many cows attain peak milk production (50 to 60 kg/day or 100 to 130 lb/day) while feed intakes lag requirements; to survive, they incur an energy deficit that must be met by mobilization of body fat. Fortunately, the dairy cow in early lactation has great capacity for redirecting body energy to milk. Energy balance studies have shown that a deficit equal to 3 kg (7 lb) of body fat per day may be incurred by cows with very high production during the first 3 to 4 weeks of lactation.

Practices for maintaining high feed intakes in early lactation cows are as follows:

1. Provide an adequate balance of necessary nutrients (especially protein, minerals, and vitamins).
2. Ensure that fiber is sufficient to maintain optimal rumen function (20% to 22% ADF on a DM basis). The minimum acceptable ADF depends on type and particle size of forage fed. Average particle size of forages

should exceed 1 cm, and some long hay is often beneficial, especially in "off feed" and "digestive upset" situations. Dairy cows in Israel receive as high as 80% concentrate, but much of this so-called concentrate comes from by-products such as grain brans, citrus peels and pulp, and brewer's grains that contain liberal amounts of hemicellulose and pectin. If corn silage is the main forage (with limited hay), concentrate should not exceed 50% of the total DM, but 60% to 65% concentrate may be fed with long-stem hay. High-grain, low-fiber rations result in erratic consumption, decreased milk fat, and an increased incidence of displaced abomasa.

3. Make the ration chewy and sweet. Cows prefer cracked and rolled grains or pellets to finely ground meal, which cakes on their noses and mouths. They also tend to eat larger quantities of a concentrate when molasses is added at 3% to 5%. Figure 24-2 shows happy, hungry cows during experiments at the University of Arizona in Tucson.

Mid to late lactation

As milk production decreases during mid lactation (14 to 28 weeks), energy concentration of the diet may be decreased, depending on the production and body condition of the cows. Body weights should start to increase (500 to 750 g/day or 1 to 5 lb/day) to replenish the tissue lost in early lactation. Top producers (more than 35 kg/day or 77 lb/day at 20 weeks postpartum) should be kept on the high-energy diets during this period.

As the cow moves into late lactation (29 to 44 weeks), ration energy density can be lowered. The grain content can be decreased to 30% to 40%, depending on forage quality and the production level of cows. Body weight gains should be 900 g (2 lb) or more per day during this period. Again, top producers should be fed diets of sufficiently high energy density to meet needs for milk and replenishment of tissue. Forage quality should be high during the whole lactation, but if low-quality forages must be fed, they should be reserved for late-lactation and dry cows. Body fat replenishment should take place during mid and late lactation and not while cows are dry

Fig. 24-2 Cows eating from gate feeders at the University of Arizona Dairy Cattle Center.

because the efficiency of conversion of feed to milk is about 50% higher for fat accumulated when milking than when dry.[49]

Body condition

High-producing cows lose tissue during early lactation when milk yields are greatest and appetite lowest. Too rapid a loss in weight causes metabolic disturbances that seriously impair the productivity and profitability of the cow for the entire lactation. However, a moderate amount of body fat mobilization for milk synthesis is desirable. Large cows can safely lose 100 kg (220 lb) of body fat during the first 70 days of lactation. This would amount to about 15% of their postcalving weight and should furnish enough energy for production of approximately 650 kg (1430 lb) of milk. Body weight changes, however, do not accurately reflect tissue energy loss because of changes in gut fill and shifts in body composition (by water filling adipose space).

If cows are allowed ad libitum consumption, intakes usually are maximized at 40 to 50 days postpartum, and cows reach an energy equilibrium shortly thereafter. As milk production decreases, feeding excess energy relative to needs replenishes lost tissue.

Extra allotment of energy usually should not continue into the dry period because fat accumulated by nonlactating cows is less efficiently converted to milk.[49] Moreover, there is greater danger of excessive fattening if dry cows are allowed to overconsume energy. Extremely fat cows are highly susceptible to ketosis, parturient paresis, mastitis, metritis, displaced abomasum, and poor appetite. The "fat cow syndrome," characterized by periparturient ketosis and fatty livers and kidneys, results in decreased resistance to pathogenic challenges.[50] Table 24-1 lists metabolic diseases encountered in a herd of cows with the fat cow syndrome. Approximately 1 year previous to this problem, a reproductive difficulty occurred in this herd that increased the nonlactating period of many cows. Some of these cows became excessively fat, partly because they were dry for longer than normal periods. Also, the energy fed to the cows was about 50% above that recommended by the NRC.[52] It is desirable that large breeds of cows gain about 70 kg (140 lb) for calf and placental growth during the dry period. Typical rations for dry cows are given in Table 24-2.

Accurate assessment of the body condition of a cow is often difficult, and, as mentioned, body weight changes do not always indicate gain or loss in body fat. The experienced dairy farmer who observes the cows day after day is often the best judge of when a cow is becoming "too fat." Animals so heavy that ribs are not visible and hipbones barely protrude above the plane of the back usually suffer postparturient problems. Cows that are dry for long periods of time—because of a miscalculated breeding date or early cessation of lactation—usually become

Table 24-1 *Disease conditions and losses from the fat cow syndrome in a 600-cow Holstein-Friesian dairy herd during a 4-month period before and after treatment and preventive procedures were initiated*

		Disease condition			Losses		
Time	No.	Parturient Paresis (%)	Ketosis (%)	Retained fetal membranes (%)	Mastitis (%)	Sold (%)	Died (%)
Before prevention* (2/1-5/31)	120	5	38	62	6	3	25
After prevention* (6/1-9/30)	120	2	5	13	2	2	3

*Prevention consisted of decreasing intakes of energy and other nutrients from about 150% to 100% of the recommended allowance. The white blood cell count from eight cows diagnosed to have "fat cow syndrome" averaged just 2800/mm³ (2.8 × 10⁹ cells/L) compared with 8100 cells/mm³ (8.1 × 10⁹ cells/L) for normal cows.
Adapted from data presented in Morrow DA and others: J Am Vet Med Assoc 174:161, 1979.

Table 24-2 *Typical rations for dry cows based on different forages (daily allotment)*

Requirement: 14 Mcal NE, crude protein (CP) 1 kg/day

Types of ration*

  Corn silage (25 kg or 55 lb at 33% DM) plus 0.45 kg or 1 lb 50% CP soybean meal

  Sorghum silage (30 kg or 65 lb at 33% DM) plus 0.7 kg or 1.5 lb soybean meal

  Sorghum silage treated with ammonia or urea (32 kg or 70 lb at 33% DM)

  Alfalfa hay† (12 kg or 26 lb)—protein sufficient

  Alfalfa hay (6 kg or 13 lb) plus corn silage (13.5 kg or 30 lb)

  Grass hay† (13.5 kg [30 lb])—may need supplemental protein depending on protein content of hay

  Pasture—protein sufficient

---

*Mineral supplementation varies with the ration fed, but all should include trace mineralized salt, and the Ca:P ratio should be about 1.5:1.

†If forage quality is low, then a limited amount of grain is recommended. Alfalfa hay is high in calcium and may predispose to milk fever.

DM = Dry matter

From Huber JT: In Howard JL, editor: Current veterinary therapy, ed 2, Philadelphia, 1986, WB Saunders.

excessively fat unless energy intake is restricted. Figure 24-3 shows an excessively fat cow.

Regular evaluation of body condition of cows during the complete lactation and dry periods is used by dairy farmers and veterinarians to assess the adequacy of the feeding regimen.[23] Rating cows on a scale of 1 to 5 (from very thin to very fat) is the system that has been widely adopted[64] (Chapter 26).

Grouping of cows for maximum milk production

Group feeding of forages and concentrates has often caused problems in furnishing the individual cow her nutrient needs. Each cow has specific nutrient requirements related to milk production, body size, age, stage of lactation, and other factors. Because of imbalanced rations, social behavior, availability of feed, bunk space, and a variety of other reasons, it is doubtful that many cows receive their exact requirement. This problem can be reduced by splitting the herd into groups with similar nutritional requirements.

The practice of grouping cows according to stage of lactation and milk production has worked successfully for many dairies. Often cows not confirmed pregnant, which may re-

Fig. 24-3 An excessively fat cow in a Michigan dairy herd.

quire estrus detection and insemination, are placed in one group. This group usually includes all cows up to 100 to 200 days after calving. Other cows are divided according to milk production into two or three groups. Dry cows and springing heifers (heifers in late pregnancy) usually comprise an additional group. Where facilities permit (in larger herds), first- and second-calf heifers should be separated from mature cows to allow extra feed for growth needs. Nutrients for the different groups are allotted according to the group needs. Generally, energy and protein are supplied at about 10% in excess of the group's average requirement to accommodate the needs at the different stages of lactation. Complete rations fed to the groups vary from 0 to 60% concentrate to meet nutrient needs at the different stages of lactation. Feeding grain to late-lactation and dry cows is often unprofitable. However, these rations should be balanced for protein, vitamins, and minerals.

Where facilities do not permit physical separation of cows, magnet feeders, electronic gates, or transponder feeding units have allowed consumption of extra energy and other necessary nutrients by high producers and young cows. Magnet feeders are usually the simplest and most economical of the three and are widely used on dairy farms in the central United States (Fig. 24-4). Computerized feeding systems that electronically control the quantity of feeds allowed each cow depending on her needs are used on some farms. One advantage of computerized feeders over magnet systems is that release of feed is timed so that the cow consumes only her own feed. In the future a large expansion of these practices may occur in smaller herds (up to 150 cows) using loose housing systems. Many dairies have boosted milk yields and profits by providing extra feed to high producers. However, if the systems are not well managed by controlling which cows have access to them, they can prove quite unprofitable.

Forages for lactating cows

Type and quality of forage can often make the difference between profit and loss in a dairy feeding program and are the variables over which farmers usually can exert greatest control. Low-quality forages are often unpalatable and

Fig. 24-4 Magnet feeder dispenses feed when activated by the magnet attached to the cow's neck chain.
From Huber JT: In Howard JL, editor: Current Veterinary Therapy, Philadelphia, 1986, WB Saunders.

of reduced nutrient value, principally in content of energy, protein, and minerals.

Legume forages are preferable to grasses for sustaining high milk production because of their more rapid rate of cell wall digestion and greater buffering capacity. Studies have shown that legume quality was more effective in maintaining milk yields in early lactation than concentrate level (Fig. 24-5).

Because starch accumulates in corn kernels with advanced maturity, organic matter digestibility of the corn plant changes very little between the tassel and hard dough stages, but nutrient yield is almost doubled during this time (Fig. 24-6). Thus, corn should be harvested for silage at approximately the time when starch deposition in the kernels is complete. An easy method for detecting this stage is the appearance of a small black layer where the kernel is at-

Fig. 24-5 Dairy cows in California eating high-quality alfalfa hay.

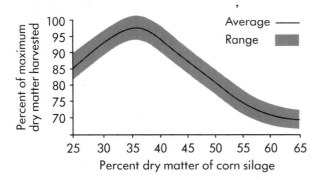

Fig. 24-6 Effect of stage of maturity of corn silage on dry matter harvested per acre (summary of research conducted at Michigan, Indiana, and USDA).
From Huber JT: In Howard JL, editor: Current Veterinary Therapy, Philadelphia, 1986, WB Saunders.

tached to the cob. Corn harvested at this maturity possesses 33% to 37% dry matter, which is ideal for ensiling in most silo structures. Fortuitously, corn silage harvested at hard dough (about 35% dry matter) is also more palatable for cattle compared with that harvested earlier (25% dry matter) or later (45% dry matter), and this results in higher milk yields in controlled studies[35] (Table 24-3). Figure 24-7 portrays the ideal time for harvest of corn silage.

The particle size of forages can exert a profound effect on utilization by cows. Forages cut too fine (whether grass, legume, or grains harvested as hay or silage) exit the rumen rapidly and cause a shift in rumen fermentation toward propionate and a decreased acetate-to-propionate ratio. This condition results in depressed butterfat percentages and a higher incidence of displaced abomasa. A minimum average size of 1 cm (0.4 inch) for forage particles should avoid most of these problems, but minimum length may need to be greater for denser materials. Silages chopped to achieve an average particle size of 8 to 12 mm (0.3 to 0.5 in) are desirable. Nevertheless, silage particles that are too large result in poor packing and a greater chance of spoilage in the silo.

PROTEIN NEEDS, NITROGEN METABOLISM, AND NONPROTEIN NITROGEN
Protein sources

Top-producing cows yield more than 350 kg (770 lb) of milk protein per lactation, which equals

Feeding the Dairy Cow for Optimal Production

Table 24-3 *Influence of maturity on the feeding value of corn silage*

Maturity	DM of silage (%)	Total DM as ears (%)	Milk yield (kg/day)	Silage DM* intake (% BW)	TDN in silage (% DM)
Soft dough	25	37	17.2	1.95	68.2
Medium dough	30	47	18.4	2.13	68.4
Hard dough	33	51	19.1	2.30	68.0

*Cows were fed only corn silage (ad libitum) and soybean meal (1 kg/9.2 kg milk).
Adapted from data in Huber JT, Graf GC, and Engel RW: J Dairy Sci 48:1121, 1965.

Fig. 24-7 A physiologically mature ear of corn at approximately the correct stage to cut for silage.

the amount in about 7 steers ready for market. Milk protein is synthesized mainly from blood amino acids that have two sources. One is rumen microbial protein, and the other is rumen bypass protein; each supplies about 50% of the total used for milk protein synthesis in high-producing cows. Diets formulated with insufficient amino acids for intestinal absorption limit milk yields. This occurs when the balance of rumen-degradable and rumen-bypass (undegradable) protein is improper. The degradable fraction supplies nitrogen for microbial growth. An adequate supply of rumen NH_3 is also necessary for optimal digestion of fiber and other nutrients in the rumen.[36]

Protein concentration

Much research is still needed to define accurately the protein needs for lactation in terms of both quantity and quality. It is thought that ruminal protein degradation limits protein absorption and milk production. Experiments over the past 15 years have shown that increasing crude protein for high producers from 12% to 18% results in increased milk yields, which could often be explained by higher feed intakes.[36] Table 24-4 lists the results of three studies[19,31,44] on protein level in early lactation. Raising protein from 8% to 15% increases diet digestibility and causes cows to consume more dry matter, but what stimulates feed intake at protein levels higher than 14% to 15% is not clearly understood. Increasing dietary protein may not always be economical, particularly when the cost of the extra protein and increased feed are considered.[57] There is a diminishing milk yield and economic response as protein concentration in the diet increases (Table 24-5).

The new NRC guidelines recommend 12% to 18% crude protein for the 700-kg (1550-lb) cow yielding 12 to 60 kg (25 to 130 lb) milk per day (Table 24-6). Because of lower dry matter intakes, the percent protein for early lactation

Table 24-4 *Milk yields and intakes in response to protein in high-yielding dairy cows*

Location	Protein (% DM)	Milk yield (kg/day)	DM intake (kg/day)
Kansas State University	13.0	24.8	18.4
8 cows/treatment	15.0	28.4	19.6
1-21 weeks postpartum	17.0	27.9	19.0
Michigan State University	11.3	31.6	16.5
12 cows/treatment	14.5	33.6	19.1
3-13 weeks postpartum	17.5	34.6	21.3
New Hampshire State University	11.1	27.2	17.5
8 cows/treatment	13.7	30.6	18.3
7-23 weeks postpartum	15.7	34.1	18.7
	19.2	33.3	18.9

From Huber JT: In Howard JL, editor: Current veterinary therapy, ed 2, Philadelphia, 1986, WB Saunders.

Table 24-5 *Economics of increasing protein concentration in the diet dry matter (DM) by 2% from different basal protein levels*

Final protein (% diet dry matter)	Change in			Feed cost—soybean meal at		
	Milk yield (kg/day)	Intake of DM (kg/day)	Milk value (cents/day)	$220/tonne (cents/day)	$275/tonne (cents/day)	$330/tonne (cents/day)
10	3.4	2.0	98	39	47	52
12	1.7	.7	51	21	28	36
14	1.1	.3	31	16	24	31
16	.7	.2	20	15	22	30

Prices used were milk at 29 cents/kg (13 cents/lb) and grain at 13.2 cents/kg (6 cents/lb).
Adapted from data presented in Satter LD, Whitlow LW, and Santos KA: Proc Distiller's Feed Research Council 34:77, 1979.

Table 24-6 *Suggested energy crude protein (CP) and undegradable intake protein (UIP) concentrations for lactating cows (700 kg or 1500 lb body weight producing 3.5% fat milk)*

kg/day lb/day	Milk yield kg/day					Early lactation 2-3 wk
	12 26	24 52	36 78	48 104	60 130	
Dry matter intake (DMI), kg/day	14.9	19.3	23.3	26.7	31.5	AMAF*
NEl intake, Mcal/day	21.1	29.4	37.2	45.9	54.1	AMAF
TDN intake, kg/day	9.3	12.9	16.5	20.0	23.7	AMAF
Crude protein, kg/day	1.83	2.72	3.75	4.35	5.26	—
% of DMI	12.2	14.1	16.1	16.3	16.7	—
NRC recommendation, %	12	15	16	17	18	19
% of CP as UIP	37	35	36	35	34	37

As much as feasible without causing digestive disturbances.
Adapted from National Research Council: Nutrient requirements for dairy cattle, Washington DC, 1989, National Academy of Sciences, National Academy Press.

cows (0 to 3 weeks postpartum) is 19%. These recommendations are presented as g/day, suggesting a knowledge of protein requirements that is more precise than the "percent of dry matter" given in the earlier NRC.

Assumptions for deriving the 1989 NRC[52] absorbed protein values were extracted mainly from the NRC's *Ruminant Nitrogen Usage*,[51] with some changes based on the judgment of the Dairy NRC Committee. Suggested amounts of rumen UIP (undegradable intake protein, i.e., rumen bypass protein, escape protein) and DIP (degradable intake protein) are given. The percent of total protein recommended as UIP varies from 34% to 37%, with an average of about 35% (Table 24-6). It is surprising that recommended UIP was not increased with higher levels of milk production; most studies have shown greatest response to UIP in high-yielding cows.

A serious deficiency in UIP recommendations is the limited and often variable information available on UIP concentrations in feedstuffs.[52] However, the recognition of a need for UIP in lactating cows under certain conditions is an improvement over the 1978 NRC. It is to be hoped that more accurate data on UIP values in feedstuffs will be generated, thus allowing more valid calculation of UIP in dairy cow diets.

Another beneficial feature of the 1989 dairy NRC is a computer program on a disk that provides a printout of nutrient recommendations for the feeding and management conditions that are specified (Table 24-7).

Rumen degraded versus undegraded protein

To maintain high milk production in most cows, considerable protein must escape rumen degradation without limiting rumen ammonia concentrations.

Sources high in rumen undegradable protein include extruded or heated oilseed meals or oilseeds (which are also high in fat), brewer's grains, distiller's grains, corn gluten meal, fish meal, blood meal, and feather meal. Little is known about their relative value for milk production when fed for rumen bypass. Generally, vegetable proteins are low in essential amino acids. However, after mixing with microbial and endogenous proteins in the rumen and small intestine, there is a "complementary effect" that raises the biological value of the protein mixture to a level higher than the single sources. More research is needed on dietary bypass protein and the amino acid complement available for absorption from the small intestine. A recent study by King and associates[42] showed a remarkably high correlation between the amino acid profile of diet protein and that delivered to the duodenum when three protein sources were compared (blood, corn gluten, and cottonseed meals). The extraction of amino acids by the mammary gland differed with the protein source, but serum amino acid concentrations did not.

Synchronization of protein and energy degradation in the rumen

Several recent reports have shown that the rate of energy breakdown in the rumen affects efficiency of protein utilization by cows. In one study,[25] we obtained higher milk production in early lactation cows from a barley–cottonseed meal (fast starch and fast protein release) diet than a barley–brewer's grain (fast-slow), milo–cottonseed meal (slow-fast), or milo–brewer's grains (slow-slow) diet (Table 24-8). From NRC tables,[51] the NEl and percent protein values for the rations were equal, but we suggest that important interactions occur when grains and protein sources of differing rumen degradabilities are mixed.

Nonprotein nitrogen (NPN)

Ruminants possess the unique ability to meet a large percentage of their nitrogen requirements with simple nitrogenous compounds such as urea, ammonia, biuret, and others. It was recently estimated that feeding NPN to ruminants saves U.S. producers more than $500 million annually. When oilseed meal costs are high and grain costs low, savings are greater (Table 24-9). Dairy farmers who replace supplemental natural protein with NPN can increase their profit by $40 to $80 per cow annually.

A high level of management is necessary for successful incorporation of NPN into dairy cattle rations. Failure to follow recommended practices might reduce milk yields and make NPN unprofitable. Rations ideal for maximal NPN use are low in protein and high in energy, such as

Table 24-7 *Computer printout of nutrient intake and requirements for a lactating cow*

NRC Dairy (1989) Requirements calculated on 8/1/89 at 16:38
Pregnant or lactating cattle

Energy concentration fed/NRC assumed is	1.000			
Live weight in kg is	600.			
Milk production in kg is	45.			
Milk fat test % is	3.50			
Number of days pregnant is	45.			
Lactation number is	4			
Proportional feed NEl/required NEl is	1.00			
Weight change in lactation is	.000	kg	or	4.13 % LW
NEl needed is	42.59	Mcal	or	1.72 Mcal/kg
ME needed is	71.68	Mcal	or	2.89 Mcal/kg
DE needed is	82.00	Mcal	or	3.31 Mcal/kg
Baseline TDN needed is	18.60	kg	or	75.10 % DM
Crude protein intake needed is	4054.	G	or	16.37 % DM
Undegraded intake protein needed is	1372.	G	or	5.54 % DM
Degraded intake protein needed is	2579.	G	or	10.42 % DM
Intake protein (IP) needed is	3951.	G	or	15.96 % DM
Calcium needed is	158.0	G	or	.638 % DM
Phosphorus needed is	99.6	G	or	.402 % DM
Vitamin A needed is	45600.	IU	or	1841. IU/kg
Vitamin D needed is	18000.	IU	or	727. IU/kg
Undegraded intake protein in IP is				34.72 % IP

LW = Live weight; DM = dry matter
From the National Research Council: Nutrient requirements for dairy cattle, Washington, DC, 1989, National Academy of Sciences, National Academy Press.

Table 24-8 *Influence of protein and starch degradability on milk production and components*

	Diets				
Item	B-CSM	B-BDG	M-CSM	M-BDG	Standard error
Number of cows	8	8	8	8	1.75
Pretreatment milk (kg/d)	32.2	34.0	33.1	33.3	.70
Milk (kg/d*)	37.4[a]	34.9[b]	34.2[b]	34.6[b]	.70
3.5% FCM (kg/d)	34.3[a]	31.4[b]	33.6[a]	34.8[a]	.57
Milk fat (%)	3.1[b]	2.9[b]	3.4[a]	3.6[a]	.11
Milk fat (kg/d)	1.20	1.01	1.20	1.24	.25
Milk protein (%)	2.9	3.0	3.0	2.8	.12
Milk protein (kg/d)	1.10	1.0	1.02	1.00	.31
Milk/CP intake (kg/kg)	8.70	8.51	8.54	8.43	.74
Milk/starch intake (kg/kg)	4.50	4.91	4.62	5.96	.56
Milk/DMI (kg/kg)	1.50	1.50	1.41	1.45	.37

*Means with different letter superscripts differ ($p < 0.05$).
B, Barley; M, milo; CSM, cottonseed meal; BDG, brewer's dried grains; FCM, fat corrected milk; DMI, dry matter intake.
Adapted from data in Herrera-Saldana R and Huber JT: J Dairy Sci 72:1477, 1989.

Table 24-9 *Savings in feed cost (per cow daily) from substituting 1.6 kg (3.5 lb) shelled corn and 0.23 kg (0.5 lb) urea (26 cents/kg) for 1.8 kg (4 lb) soybean meal*

Price of shelled corn ($)		Price of soybean meal ($/tonne)				
(per bu)	(per tonne)	180	225	270	310	355
				Savings, cents per cow daily		
2.00	80	14	22	30	37	47
2.50	100	10	19	27	34	42
3.00	120	7	15	23	31	39
3.50	140	4	12	20	27	36
4.00	160	1	9	17	24	32
4.50	180	-2	6	14	21	29

To calculate savings per year for a herd, multiply the daily savings by the number of cows and then by 310 days. For example, a 23¢ per cow daily savings ($3 corn and $270 soybean meal) would result in $14,260 annual savings for a 200-cow herd.

those using liberal quantities of grain and grain silages, typical of midwestern and eastern U.S. rations.

Response of lactating cows to NPN

A common practice for feeding lactating cows in many dairy herds in the midwestern United States is to provide a basal ration of 11% to 13% natural protein and add NPN to increase total crude protein to 14% to 16%. There has been some controversy over feeding NPN-containing rations to lactating cows, particularly in the early stages of lactation (up to 4 months). Some workers have suggested that NPN is contraindicated when the crude protein requirement exceeds 11% to 12%.[57,65] However, several experiments show milk yields of early lactation cows[36,45] fed urea-containing rations of 15% to 17% crude protein to be just as high (32.4 kg/day) as those of cows fed equal protein from all natural sources (32.6 kg/day) and significantly higher than negative control rations (30.6 kg/day) containing 12% to 13% crude protein. Milk yields of lactating cows from several Michigan and New Hampshire experiments[32,36] also showed NPN equal to soybean meal for supplementing rations up to 14% to 15% crude protein.

In conclusion, present data do not warrant exclusion of NPN from rations for early lactation cows. Utilization of nitrogen from a limited amount of supplemental nonprotein sources is as apparently efficient as that from natural protein sources in rations containing up to about 16% crude protein. Consideration of the ratio of rumen degradable to undegradable protein also is necessary. Naturally, NPN should be used judiciously; if lowered energy consumption results from feeding NPN at critical times (early lactation), it should be withdrawn.

Limits for and methods of feeding NPN

The maximum level at which NPN is profitable depends largely on the feeding system used. When it is added as urea to concentrate fed twice daily, intake should not exceed about 200 g (7 oz)/day (as urea or urea equivalent). Adding NPN to complete rations, silages, gelatinized grain (Starea), or to dehydrated alfalfa during pelletization (Dehy-100) better synchronizes ammonia release with rumen energy availability for microbial protein synthesis.[36] These improved methods of NPN delivery have allowed greater quantities of NPN to be consumed without depressing milk yields or decreasing intakes. The approximate limit for adding urea to dairy concentrate is 1.5% (or equivalent as nitrogen from other NPN sources), to complete rations 1%, and to silage 2% of the silage dry matter. For Starea or Dehy-100, the approximate limit is 300 g of urea equivalent per cow daily.

Liquid supplements

Liquid feeds furnish a convenient method to supply cows with needed protein, vitamins, and minerals. They are palatable and convenient to handle. Liquid supplements contain 10% to 20% urea with up to 25% of the nitrogen as true protein (derived from corn, fish, or other solubles). Liquids are mixed with forages, concentrates, or complete feeds prior to feeding. Mixing NPN-containing liquid supplements with total mixed rations is one of the preferred methods of feeding these materials. Liquids are also consumed free choice from lick wheels. With the lick wheel, a key problem is overconsumption when it is initially offered to animals. After several days, however, most cattle regulate their intake of liquid to a level commensurate with their needs.[7]

Urea-treated silage is one of the preferred methods of incorporating NPN into rations for lactating cows. Comparisons of isonitrogenous rations containing urea-treated and control silages have shown slightly higher milk yields for the urea corn silages.[36] However, neither urea nor ammonia should be added to silages harvested in excess of 45% dry matter.

Ammonia addition to grain silages has been extensively investigated and is now commonly used for silage treatment in the United States and Canada (Fig. 24-8). A key incentive for developing the ammonia treatment methods was that ammonia nitrogen usually costs only about 60% as much as urea nitrogen.

Unlike the bound ammonia in ammoniated industrial wastes, which is largely unavailable for microbial use, most of the ammonia in ammonia-treated silage is present as the free ammonium salts of silage organic acids. These acids act to keep the ammonia from escaping into the atmosphere before feeding.

Ammonia treatment of silage inhibits initial plant respiration and proteolysis, thereby conserving silage energy and natural protein.[36] This protein-sparing action of the ammonia was probably responsible for superior results with ammonia-treated silages compared with urea-treated silages.[15,32] Ammonia treatment also protects the silage from mold and yeast damage (with accompanying heating) after exposure to air. This is of particular importance when silage is fed to cattle in feed bunks. Prevention of secondary fermentation is beneficial in warm weather during feeding of silages stored in silos with large exposed surfaces. A third benefit of treating silages with ammonia is preservation of

Fig. 24-8 A cold-flow box for addition of anhydrous ammonia to silage.

ensiled energy. Studies by Beltsville workers[63] showed that 5% dry matter and 8% energy were gained by adding ammonia to corn silage.

Reproductive performance of cows not impaired by NPN

Some veterinarians and dairy farmers have suggested that feeding rations containing NPN lowers conception rates of cattle. To test this claim, 85,000 individual lactation records from Michigan DHIA herds (from 1965 to 1969) were analyzed for length of calving interval.[56] Also, feeding practices were documented carefully through an interview with individual dairy farmers. The data show that absolutely no decrease in reproductive efficiency resulted from feeding NPN. Cows receiving no NPN (45% of total cows) had an average calving interval of 380 days, which was identical to that of cows

Table 24-10 *Relationship of urea intake to calving interval, percent of cows sold as sterile, and milk yields in Michigan DHIA herds, 1965-69*

Urea intake (g/day)	None	1-370	1-60	61-120	121-180	181+
Average urea intake (g/day)	0	79.8	36.1	90.5	146.7	219.5
Herd-year observations (no.)	1442	1715	760	653	219	83
Calving interval (days)*	380.4	379.9	379.4	380.7	379.8	377.8
Cows sold as sterile (%)	2.15	2.40	2.40	2.60	2.60	1.71
Milk per cow (kg/lactation)	5932	5867	5676	5893	5981	5977

*Calculated by dividing the total number of individual cow observations for each herd by the total calving interval for each herd.
Adapted from data in Ryder WL, Hillman D, and Huber JT: J Dairy Sci 55:1290, 1972.

fed NPN (55% of total cows) for at least 3 of the 5 years (Table 24-10). The pattern did not change as NPN in the rations increased. In fact, herds fed more than 120 g of urea (or urea equivalent) per day produced slightly more milk and were slightly lower in calving interval than those fed less urea, but the differences were not significant. The percentage of cows sold from herds because of infertility was about 2% and showed no relationship to the feeding of NPN.

Heifers fed high levels of urea have shown decreased reproductive performance in some studies,[21] but not in others.[15] Even though NPN feeding per se does not decrease reproductive efficiency in cows, poor dietary management resulting in high blood or tissue ammonia concentrations or an amino acid imbalance may impair normal development of spermatozoa, ova, or survival of early embryos.[13] These effects have also been observed at high levels of dietary and rumen degradable protein. They can be exacerbated by low energy concentrations.

Need for adaptation to NPN rations

Finnish workers showed that it took 2 months to fully adapt dairy cows to an all-NPN ration.[62] Three to 6 weeks is needed for adjustment to lower levels of NPN.[48] Enhancement of urea synthesis by the liver, which minimizes ammonia toxicity to tissues and facilitates urea recycling into the rumen, is the primary mechanism through which adaptation to NPN occurs. At recommended feeding levels, a gradual increase in the NPN-containing feed for 14 days is probably sufficient time for adaptation. An easy system would be to feed one third of the total target amount of NPN during the first week, two thirds for the second week, and a full feed by the third week. With lick wheels, time allowed for animals to lick might be restricted initially and then gradually increased.

Combining bypass protein and NPN for high milk production

Dairy cow diets that combine rumen bypass protein (to provide amino acids for high milk yields) and NPN (ammonia or urea) for rumen microbial protein synthesis have resulted in greater profits than conventional diets. Ammonia-treated corn silage and heat-treated soybean meal (to achieve increased UIP) combined in a diet yield the highest milk production and greatest income over feed costs (Table 24-11).[44] Tailoring of protein supplements (by combining NPN and bypass sources in early lactation cows) also is productive and can yield greater profit.[20,28] This increases profitability without decreasing productivity. In a South African study,[20] addition of 1.6% urea (DM basis) to the total diet decreased UIP to about 35% of the total crude protein. The authors recommended a maximum of 1% of the total DM as urea for high-producing cows. This level (about 200 g/day) corresponds well with earlier recommendations.[36]

Conclusion

The 1989 NRC recommendations for supplying protein to lactating cows are a marked improvement over the 1978 NRC. The absorbed protein system was developed from sounder theory and newer research information. More studies are needed to expand the usefulness of rumen UIP

Table 24-11 *Milk production, intake, and economic evaluation of different protein regimens for early lactation cows*

Corn silage/soybean meal	Protein type*		
	Normal/normal	Normal/heated	Ammoniated/heated
Milk production (kg/day)	34.1	35.3	35.6
Dry matter intake (kg/day)	20.4	19.3	20.2
Cost of feed ($/day)†	3.33	3.17	3.01
Milk income ($/day)‡	9.85	10.11	10.26
Income over feed cost ($/day)	6.52	6.94	7.25

*Each protein type represents a mean of 24 cows; 12 on 14.5% and 12 on 17.5% CP. Treatments were from 3 to 13 weeks of lactation.
†Feed costs ($/tonne): hay, 66; corn silage (CS), 24; corn silage (AS), 26; high moisture ear corn (85% DM), 94; soybean meal (NS), 330; soybean meal (HS), 440.
‡Milk cost: 28.6¢/kg.
Adapted from Kung L Jr and Huber JT: J Dairy Sci 66:227, 1983.

values, but they are likely to be forthcoming. Incorporation of NPN into diets for early, mid-, and late-lactation cows will be feasible under the 1989 NRC system[52] and usually will provide an avenue for increased profits to dairies. The amount of NPN allowable is largely dependent on the amount of undegraded protein in the basal ration. In early lactation, protein supplements especially high in undegraded protein might be combined with NPN for highest yields and greatest income over feed costs as long as guidelines for use of rumen undegradable protein are followed.

SPECIAL CONSIDERATIONS OF MINERAL AND VITAMIN NEEDS

There is much confusion regarding the feeding of minerals to dairy cattle. Farmers are continually bombarded by numerous mineral salesmen whose product is claimed to be superior to all others. A problem often encountered in the field is that the same mineral is sold regardless of the type of ration the farmer is feeding. Need for supplemental minerals differs greatly depending on the other ration ingredients.

Calcium, phosphorus, and vitamin D

There may not be one ideal ratio of calcium to phosphorus for all dairy cows. The optimum ratio varies according to the function of the cow (whether milking or dry) and the type of feeds consumed. A ratio of 1.4:1 is calculated from NRC standards. This ratio is probably satisfactory for the dry cow but too narrow for lactating animals. Ratios less than 1:1 and greater than 2.5:1 tend to increase the incidence of milk fever. High-calcium diets prior to calving predispose to milk fever.

Phosphorus also can be present in excess. Primiparous Holstein cows fed phosphorus at 138% of the NRC allowance for the first 12 weeks of lactation yielded an average of 2.7 kg (6 lb) less milk per day for the entire lactation than control cows fed 98% of the suggested requirement.[11] The reason for decreased milk production resulting from high phosphorus intake is not known, but these data should discourage the indiscriminate addition of several phosphorus sources to diets for dairy cows.

Heavy supplementation of vitamin D_3 on a routine basis is not desirable, but feeding massive doses (10 to 20 million IU per day) for 7 days before calving alleviated milk fever in aged cows.[26] However, feeding such a high dose of vitamin D for longer than 7 days is dangerous. It is difficult to predict calving dates close enough for the treatment to be effective.

Many questions relating to calcium and phosphorus metabolism are still unanswered, but current recommendations are to provide phosphorus at 100% of the allowance recommended by the NRC standard (from 0.24 to 0.48% of ration DM as phosphorus depending on milk yield). Calcium intake would then be calculated so as to provide about a 1.7:1 ratio of calcium to phosphorus for milking cows and a 1.5:1 ratio for dry cows. This intake roughly corresponds to NRC allowances. In rations where corn silage constitutes the main forage, both calcium and phosphorus should be supplemented, whereas only phosphorus is needed when legumes are the main forage.

Sulfur (S)

When NPN is used in ruminant diets, there is often a decrease in natural protein supplements that contain S amino acids. Also, heavier feeding of grain and grass silages with less legume hay decreases S intake by cows.

Sulfur needs of cattle are often expressed in relation to the nitrogen content of the diet. Diets for lactating dairy cows containing 17% crude protein should furnish 0.25% S or an S:N ratio of about 1:10. When grain and grain silages are the main components, supplemental S is often needed. Addition to about 2.7 kg/tonne (5.5 lb/ton or 0.27%) of calcium or sodium sulfate to dairy concentrate should correct any possible S deficiency of high-grain and NPN-containing diets. Precautions should be made not to exceed 0.35% S in the diet, or feed intakes might decrease. When legume hay is fed in liberal amounts, supplemental S is not needed.

Selenium and vitamin E

Excessive selenium (3 to 5 ppm) accumulates in some plants in areas of high soil content and has been toxic to livestock. Conversely, soils in large regions of the United States are deficient in selenium, and cattle diets formulated from feeds from such areas require supplementation. Recommended concentrations[52] for dairy cattle diets are 0.3 ppm.

A problem related to low selenium is increased incidence of retained placentas. Vitamin E–selenium injections before the expected date of calving have effectively decreased the incidence of retained placentas in experiments and among herds fed diets that were borderline in selenium, vitamin E, or both.[41]

Fluoride

It was generally thought that dairy cattle tolerate up to 30 ppm fluoride without adverse effects. However, damaged teeth (mottled and broken) and bone lesions have occurred in cattle that consumed phosphates that furnished 15 to 20 ppm fluoride in the total diet.[30] Cornell workers[43] also reported fluoride toxicity in cattle on Cornwall Island in animals exposed to less than 40 ppm for prolonged periods (3 months of age to adult). Fluoride dietary intakes of less than 20 ppm are mandatory in some states. This often requires special processing of phosphate supplements that render them more expensive. Until research clearly establishes toxic levels in cattle, caution is recommended because bone damage from fluoride is not reversible.

How to feed minerals

Feeding minerals free choice has long been practiced by dairy farmers. It was thought that if the correct minerals were available to cows, they would consume minerals according to their needs. However, studies in New York[17] and Minnesota[53] showed no relationship between the amount of minerals needed by cattle and their free-choice consumption.

Minerals should be force-fed by addition to the concentrate, forage, or total mixed rations. These major diet components are consumed in relation to the animals' needs; thus, problems of mineral shortages or excesses to individual cows would be less likely to occur than when they are fed free choice. Silages can be ideal carriers of many minerals needed in dairy cow diets; in some cases, minerals (such as $CaCO_3$) exert a beneficial effect upon fermentation. When cattle are grazing, there is usually no good way to force-feed minerals, so offering them free choice to cattle on pasture is still recommended.

RECENT ADVANCES AND FEEDING IMPLICATIONS

Nutrition of bovine somatotropin (BST)–treated cows

Nutritional regimens that increase nutrient intake, improve efficiency of nutrient utilization, improve digestibilities, or improve the milieu of absorbed nutrients should better allow cows to express a BST response, so long as that regimen does not exert a specific inhibitory effect on action of the hormone.[3] Figures 24-9 and 24-10 illustrate the powerful effects on milk yields and energy intakes that resulted from BST injection. Maximum response to BST in early lactation requires high energy intake in order to sustain the extra production and to avoid excessive tissue fat mobilization. In most long-term studies with BST, energy concentrations of early lactation diets have been as high as possible without causing milk fat depression. In trials involving about 285 complete lactations (some cows for four consecutive lactations) and a number of partial lactations, we commenced BST injections (14-day intervals) at 60 days after parturition.[37] Cows should be in good body condition when injection starts, although this is often difficult to achieve at 60 days postpartum. The need for good body condition at calving (probably 4 on a scale of 1 to 5) and a palatable, high-energy diet during early lactation are obvious. The feeding regimen for our studies has been as follows:[37]

1. Feed a diet of 40% concentrate for 30 days prepartum, or longer, if warranted by poor body condition.
2. After parturition, feed a 60:40 concentrate (C):forage (F) diet (containing 15% whole cottonseed or about 5% fat and approximately 1.63 Mcal/kg or 0.74 Mcal/lb) until cows average less than 32 kg/day (70 lb/day) milk for 1 week and have a body condition in excess of 3.
3. Between 23 and 32 kg milk/day (50 lb to 70 lb/day), a 50:50 C:F diet is fed, and whole cottonseed is dropped to 11%.
4. At milk yields less than 23 kg/day (50 lb/day) until the end of lactation, cows are

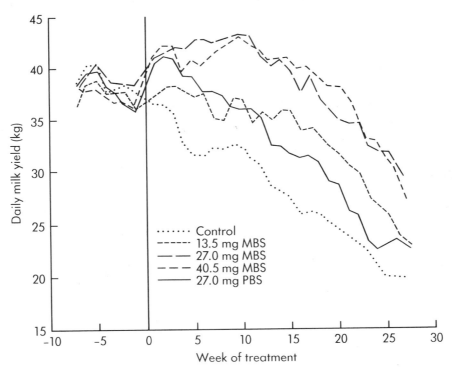

Fig. 24-9 Effect of exogenous bovine somatotropin on mild yield. Treatments commenced at week 0 (84 ± 10 days postpartum) and continued for 27 weeks. Milk production data (unadjusted) represent weekly averages. Daily dose of methionyl bovine somatotropin (MBS) and pituitary bovine somatotropin (PBS) as indicated.
From Bauman DE and others: J Dairy Sci 68:1352, 1985.

decreased to 40:60 C:F, but body condition is considered before dropping to this ration (with 7% whole cottonseed).

Our experience with this feeding schedule has been favorable. We have avoided excessive fattening of prepartum cows while allowing lactating cows to express near-maximum response to BST. The incidence of ketosis or other metabolic disorders has been negligible and similar for both BST-treated and control groups. The cow in Fig. 24-11 averaged more than 13,000 kg (28,000 lb) milk for each of three consecutive lactations while receiving a biweekly injection of 500 mg BST after 60 days postpartum.

The response of cows to BST is 1.1 kg/day (2.4 lb/day) more on 17% versus 14% crude protein diets, and 1.6 kg/day (3.5 lb/day) more when undegradable versus degradable protein was fed (40% versus 33% UIP)[47] (Table 24-12). Soybean and canola meal fed as the main protein supplements gave similar responses to BST. There was no depression in thyroid function in cows fed a combination of canola meal and BST. BST treatment increases protein needs due to increased milk yields, so recommended protein levels and degradabilities in accordance with NRC[51] recommendations for higher production are suggested.

Effects of BST and buffers are apparently additive in that both increase output of energy in milk, but the mechanisms are different. BST has a stronger influence on milk yield, and buffers

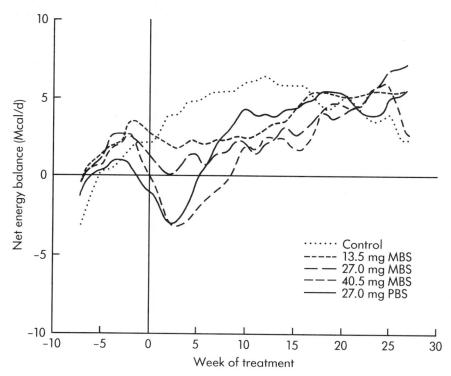

Fig. 24-10 Effect of exogenous bovine somatotropin on net energy balance (Mc/day). Treatment period same as in Fig. 24-9. Net energy balances (unadjusted) represent 3-week rolling averages. Daily dose of methionyl bovine somatotropin (MBS) and pituitary bovine somatotropin (PBS) as indicated.
From Bauman DE and others: J Dairy Sci 68:1352, 1985.

Fig. 24-11 This University of Arizona dairy cow (#899) produced a total of 40,000 kg of milk from three consecutive lactations on BST injections.

Table 24-12 *Effect of protein level and rumen degradability on response of cows to bovine somatotropin*

	No BST LP, LU†	+ BST*			
		LP, LU	LP, HU	HP, LU	HP, HU
Milk (kg/d)	29.8	32.2	33.8	33.3	35.1
Fat (%)	3.23	3.09	3.25	3.16	3.29
Protein (%)	3.23	3.17	3.14	3.19	3.14
Total solids	12.19	11.84	12.30	12.10	12.26
3.5% FCM (kg/d)	28.1	29.3	32.0	31.0	33.2
DMI (kg/d)	21.2	21.6	21.8	20.7	21.9
Average body weight (kg)	584	579	587	579	586

*640 mg BST injected at 28-d intervals for 112 d (4 injections), 8 cows/treatment.
†*LP*, 14%; *HP*, 17% crude protein; *LU*, 33% undegradable intake protein; *HU*, 40% undegradable intake protein; DMI = dry matter intake; 3.5% FCM = milk yield corrected to a 3.5% fat content.
From McGuffey RK, Green HB, and Basson RP: J Dairy Sci 71:120, 1988.

have a stronger influence on fat percentage[14] (Table 24-13).

Added fat (Megalac-ML) was tested in cows with and without injection of 50 mg/day of BST.[58] Even though there was no effect of fat alone on milk production, the increased energy contribution from fat enhanced the cows' response to BST (37.6 kg/day or 82.9 lb/day) on added fat alone versus 40.3 kg/day or 88.8 lb/day for added fat plus BST). A fertile field for

Table 24-13 *Influence of bovine somatotropin (BST) and sodium bicarbonate alone (NaHCO₃) and in combination on response of lactating dairy cows*

	Treatment*			
	Control	BST	NaHCO$_3$	BST + NaHCO$_3$
Milk (kg/day)	33.7	37.6	31.9	37.4
Fat (%)	2.97	2.82	4.14	3.73

*8 cows/treatment; 50 mg BST administered daily; NaHCO$_3$ was 1.2% of diet dry matter.
Adapted from data presented in Chalupa and others: J Dairy Sci 68(suppl 1):143, 1985.

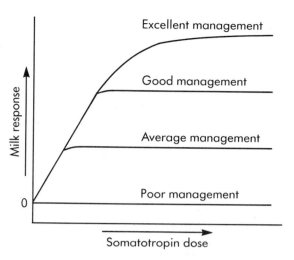

Fig. 24-12 Relationship of somatotropin dose to milk response in lactating cows.
Reprinted with permission from Dr. Dale Bauman, Cornell University, Ithaca, NY.

future research is delineating the effects of other feed adjuncts in BST-treated cows. The basic premise might be that anything enhancing diet intake, digestibility, or overall utilization of nutrients combines well with the hormone.

Lower producers and cows in mid and late lactation respond to BST as well as early lactation high producers[60] (Fig. 24-12). However, good management and sound nutrition are extremely important for optimum response to BST.[3] Sound, effective programs in reproduction, herd health, milking, housing, cow handling, and replacement raising are vital if maximum profits from the hormone treatment are to be realized.

Summary. Nutrient needs of BST-treated cows are greater because of increased milk yields but do not differ from recommended guidelines. Energy density and rumen undegraded protein intake may be critical in limiting response to BST. Supplements providing increased dietary fat and high-quality undegradable protein could aid in meeting nutritional needs of BST-treated cows. Improved palatability of diets might also enhance BST response. In no way is BST treatment a replacement for good management, and shoddy practices negate the efficacy of BST.

Nutritional management of heat-stressed cows

Heat stress severely limits production and reproduction of dairy cows. This sections deals with feeding systems that might alleviate the detrimental effects of heat stress on dairy cows (Fig. 24-13).

Energy concentration of diets. Body heat production and rectal temperatures are higher on high-forage compared to high-concentrate diets. Increased heat increment has been associated with higher concentrations of acetate in the rumen of cows eating high forage.[61] Often cows voluntarily limit their forage consumption during hot weather, even to the extent of severely lowering the butterfat content of milk. Addition of buffers (about 1% sodium bicarbonate, NaHCO$_3$; potassium bicarbonate, KHCO$_3$; or sodium sesquioxide) alleviates milk fat depression and maintains healthier rumen fermentation during periods of heat stress when forage consumption is lower than 1% of body weight.

Added fat (from whole cottonseed, soybeans, other oilseeds, or commercial fat sources currently available) has benefited heat-stressed cows on low-forage intakes. However, for good rumen function, total fat intake should not exceed 6% of the ration dry matter (see Chapter 25).

Fermentation of higher-quality compared to low-quality forages produces less heat, because of differences in fiber content. A minimum fiber (about 21% acid detergent fiber, ADF) is rec-

Fig. 24-13 Cows suffering severe heat stress during the summer in Central Arizona.

Fig. 24-14 Evaporative cooling for dairy cows at the University of Arizona Dairy Cattle Center.

ommended even at hot ambient temperatures in order to sustain good rumen function and avoid milk fat depression.

Feed intake. Cooling systems probably exert their major commercial benefit through increasing cows' intake (Fig. 24-14). After cows are cooled in holding pens (before or after milking), they eat for a longer time when returned to corrals than cows that are not cooled. Moreover, cows that are pen cooled during hot weather approach the feed manger more frequently than those not cooled. Misting cows at the manger also substantially increases feed intake and milk production.

Minerals: sodium (Na) and potassium (K). A need for considerably higher levels of Na and K for lactating cows than indicated by NRC recommendations during hot weather has been reported.[4,18] Raising dietary Na from 0.18 to 0.55% of DM resulted in about a 10% increase in milk yields. The increased dietary requirement of K in heat-stressed cows was attributed to greater excretion of K in sweat during hot weather. Potassium intake is also decreased because less forage is eaten in hot weather. Positive responses in milk yields have been obtained in cows fed as much as 1.5% K in the dry matter, as compared to the NRC recommendations of 0.87% (Table 24-14).

Level and type of protein. High-protein diets of high rumen degradability were detrimental to cows subjected to hot summer temperatures. Three trials in Arizona showed decreased milk yields and feed intakes when cows were fed 19% protein of high rumen degradability. Lower rumen degradability protein, as well as the two 16%-protein diets (of normal or high degradability), did not decrease milk production[29] (Table 24-15).

Fungal culture. Studies at the University of Arizona[33] have shown reduced rectal temperatures and respiratory rates as well as increased milk yields in cows fed a fungal culture from *Aspergillus oryzae* (Amaferm) (Table 24-16). A 12% increase in conception rates was observed for fungal-fed cows compared to controls. Confirmation of improved reproduction and its relationship to heat stress alleviation due to feeding fungal cultures awaits further investigation.

Table 24-14 *Summary of relative increases in daily milk yield with increasing dietary potassium or sodium in complete mixed diets*

Experiment*	Dietary level (%) lower to higher	Milk yield % increases
Potassium		
1	0.66 to 1.08	+4.6
	0.66 to 1.64	+2.6
	1.08 to 1.64	no change
2	1.00 to 1.50	+2.8
3	1.30 to 1.80	+4.2
4	1.07 to 1.51	+3.4
5	1.07 to 1.58	no change
6	1.14 to 1.58	no change
Sodium		
7	0.20 to 0.43	+3.6
8	0.18 to 0.55	+9.6
	0.18 to 0.88	+10.8
	0.55 to 0.88	no change
9	0.28 to 0.47	+3.4
10	0.16 to 0.42	+9.2
11	0.24 to 0.62	no change

*Experiments 1, 2, 3, 4, 7, 8, and 9 were during warm weather, and 5, 6, 10, and 11 were during cool weather. Adapted from data in Beede DK: Proc SW Nutr and Mgmt Conf, Tempe, AZ, 1987.

Summary. Diets high in grain and low in forage reduce heat stress for lactating cows because of their lower heat of digestion. However, milk fat is depressed and digestive disorders increase when forage intake is severely decreased. Buffers, supplemental fat, or both allow high-concentrate diets to be fed without many of the undesirable effects. By-product feeds (e.g., almond hulls or citrus pulp) might also aid in keeping milk and milk fat at acceptable levels during heat stress. Cooling systems improve feed consumption, milk production, and reproductive performance. Milk yields and feed intakes are decreased in heat-stressed cows fed diets high in protein of high rumen degradability. Milk yields were higher in heat-stressed cows when Na and K in the diet were increased above NRC recommendations. Finally, feeding

Table 24-15 *Influence of protein level and degradability on performance of lactating cows at hot and moderate temperatures*

Protein level (% of DM)	19	29	16	16
Rumen degradability (%)	60	41	59	45
Hot environment (60 cows)*				
Milk (kg/d)	26.9	28.9	28.5	28.4
3.5% FCM (kg/d)	23.6[a]	26.6[b]	26.2[b]	27.0[b]
DM intake (kg/d)	21.5[c]	21.9[c]	23.3[d]	23.1[d]
Milk fat (%)	2.72	3.04	3.01	2.95
Milk protein (%)	3.04	3.04	3.13	3.11
Moderate environment (60 cows)†				
Milk (kg/d)	36.6[a]	35.0[ab]	34.1[b]	36.0[ab]
3.5% FCM (kg/d)	34.7[a]	31.8[b]	32.3[ab]	32.4[ab]
Milk fat (%)	3.11[a]	2.89[b]	3.04[a]	2.89[b]
Milk protein (%)	2.89	2.94	2.92	2.96

Means not showing a common letter superscript are different; for a and b, $p < 0.05$; for c and d, $p < 0.10$.
3.5% FCM, milk yield corrected to 3.5% fat.
DM = Dry matter
*Adapted from data in Higginbotham GE, Torabi M, and Huber JT.[29]
†Adapted from data in Higginbotham GE and others.[27]

a fungal culture during periods of heat stress may increase milk yields and feed intakes while decreasing rectal temperatures.

Additives for lactating cows

Of the several feed additives recently reported to benefit lactating cows, some are rumen fermentation agents or metabolite potentiators, and others are legitimate nutrients administered in excess of what normally might be considered as the requirement. This section summarizes effects on performance, mode of action (when known), and profitability (under current U.S. conditions) of several feed additives for lactating cows.

Isoacids. Mixtures of four organic acids (namely, valeric, isovaleric, isobutyric, and 2-methyl butyric), three of which are precursors to the branched-chain amino acids, have been shown to stimulate growth of cellulolytic organisms in rumen cultures.[2] They increase milk production 5% to 10% when incorporated as ammonium or calcium salts into dairy cattle rations at about 90 g/day.[54] Efficiency of conversion of feed to milk has usually increased because of the higher milk yields and little change in feed intake. Most studies with isoacids have been with diets high in silage (particularly corn silage) and grain. More controlled studies are needed under conditions where pasture and legume hay are the primary sources of forage.

The principal investigator and inventor of isoacids as a feed additive is Bob Cook (Michigan State University), who commenced studies in the area more than 25 years ago.[16] For the past 15 years, many university and field trials have tested isoacids (under the sponsorship of Eastman Chemicals Co.) in several thousand lactating cows with generally positive responses (Fig. 24-15).

In calculating profits from feeding isoacids, the assumption is an average 6% (1.6 kg or 3.5 lb/day) increase in milk yields on a 27 kg (60 lb)/day herd average, with no change in butterfat percent (milk protein percent was lower in some studies) or feed intake. The increased milk is worth about 42 cents/day (26 cents/kg milk). Cost of the isoacids ranges between 20 and 25 cents/day, so income over costs (assuming the 6% response) would be about 20 cents/day.

Table 24-16 *Effect of feeding* Aspergillus oryzae *(AO) extract on milk yields and rectal temperatures*

Study	No. cows	Milk yields (kg/day)		Rectal temperature (°C)		Remarks
		Control	AO	Control	AO	
Huber et al., 1985*	48	18.5	20.3†	39.39	38.78	Normal concentrates
	48	19.3	18.7	39.72	39.22	Low concentrates
Marcus et al., 1986*	205	28.7	30.2	39.67	39.44	Commercial herd
Huber et al., 1986‡	24	22.6	23.5	39.17	39.00	Mid lactation
Gomez-Alarcon et al., 1987‡	46	38.1	40.0§	39.90	39.30	Early lactation

*90 g Vitaferm per day.
†Significantly higher ($p < 0.05$).
‡3 g Amaferm per day.
§Significantly higher ($p < 0.08$).
Adapted from data in Huber JT: Proc Pacific NW Nutr Conf, Portland, OR, 1987.

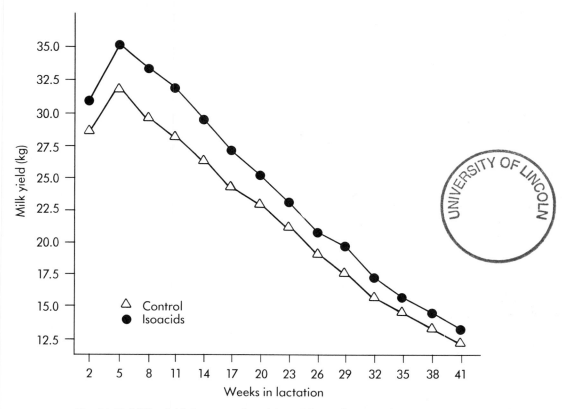

Fig. 24-15 Milk yield for control and isoacid supplemented cows. From Pappas AM and others: J Dairy Sci 67:276, 1984.

Most studies suggest greater responses in early rather than mid or late lactation, so feeding after 6 months postpartum might be contraindicated.

Fungal additives

The normal flora of the rumen includes anaerobic fungi that occupy a special niche in degradation of lignocellulose bonds.[1] During recent years, viable extracts of fungal spores have been fed to livestock. The most common of these have been cultures of *Aspergillus oryzae* and *Saccharomyces cerevisiae*. Rumen studies suggest that numbers of cellulolytic bacteria are enhanced, digestibility of celluose is increased, and synthesis of microbial protein is greater when these cultures are included in diets for cattle. Additional observations include less diurnal variability in rumen pH, ammonia and VFAs and reduced rectal temperatures and respiration rates during heat stress. There is more rapid conversion of lactic acid to propionic and acetic in in vitro studies.

In 12 direct comparisons, cows fed *A. oryzae* outyielded control cows by an average of 1.5 kg/day (3.3 lb/day) (Table 24-17). The response was nearly double for comparisons made in cows in early compared to mid or late lactation. In four trials with *S. cerevisiae* (three with cows and one with goats), fat-corrected milk yields were increased by about the same magnitude as in the *Aspergillus* trials. However, *Saccharomyces* has a greater effect on butterfat content, whereas *Aspergillus* increased milk yields with little effect on fat percent.

The profit forecasts for fungal additives assume increased milk production and higher feed intakes of about 5% each with a cost for the fungi of 5 cents/day. (Many of the cultures are mixed with vitamins, minerals, and the like, but this cost applies only to the fungal cultures.) Taking into account the higher feed cost and increased milk flow, income over costs amounts to 12 to 17 cents/day. Greater profits are realized by feeding fungal cultures to cows in early lactation.

Rumen buffers. Large and frequent dietary changes occur during early lactation and often result in decreased dry matter intakes and "off feed" problems. Buffer compounds such as sodium bicarbonate, potassium carbonate, potassium bicarbonate, and magnesium oxide stabilize the rumen environment and buffer acid loads, with a resultant improvement in intake, milk production, and milk fat percent[9,22] (Table 24-17).

Cows fed diets high in grain silages and concentrates are more responsive to buffers than those fed plentiful legume and grass forages. Generally, the period for effective buffering is during the first 8 to 10 weeks postpartum. How-

Table 24-17 *Effect of buffer treatment on feed intake, milk production, fat test, and body weight change in the first 8 weeks postpartum*

Parameter	Ration			
	Control	1.5% NaHCO$_3$	0.8% MgO	0.8% MgO plus 1.5% NaHCO$_3$
Dry matter intake (kg/day)	18.5^a	20.7^b	19.2^{ab}	20.6^b
Milk production (kg/day)	34.5^a	36.1^{ab}	34.9^a	38.3^b
Fat test (%)	3.80	3.96	3.62	4.05
Kg fat/day*	1.24^a	1.38^{ab}	1.22^a	1.52^b
Fat-corrected milk (kg/day)*	32.5^a	35.1^{ab}	32.3^a	39.1^b
Body weight change (kg) (week 1—week 8)	−10.1	+5.7	−20.2	−32.7

Means in the same row with different letter superscripts are different ($p < 0.05$).
*Adjusted by covariance for previous 305-day mature-equivalent lactation average.
Adapted from data in Erdman RA and others: J Dairy Sci 63:923, 1980.

ever, buffers might be beneficial during the total lactation for cows fed high-concentrate diets that depress milk fat. Buffers might also be more effective in maintaining rumen stability when large quantities of grains containing starch of rapid rumen degradability such as barley, wheat, oats, or flaked sorghum are fed. Starch in rolled sorghum and corn is more slowly degraded in the rumen and results in less acid production.

Addition of sodium bicarbonate and potassium carbonate should not exceed 1.3% of the ration DM, and the limit for MgO is 0.6% to 0.8%. Too much buffer or nonuniformity of mixing decreases feed intakes. Silages are good vehicles to mask the taste of buffers if a total mixed ration is not possible. The silage acids apparently neutralize the unpalatable taste of buffers. During heat stress, buffers with both sodium and potassium can supply the increased need for these minerals. However, the diet also must be balanced for chloride.

The profit opportunity with rumen buffers is quite favorable, assuming increases in milk yields of 1 kg/day, milk fat increases of 0.3% and dry matter intake increases of 2 kg (4 lb)/day are obtained. Cows averaging 34 kg (75 lb)/day of milk in early lactation would show a profit of 20 to 25 cents/day if the cost of buffer was 12 cents/day.

Niacin. This vitamin functions in two coenzyme systems in the body that are important in energy metabolism. When added at 6 g/day for lactating cows (about 20 times the estimated requirement), niacin has generally increased milk yields in natural protein diets, but negative effects were shown when NPN comprised a substantial part of the supplemental protein. Alleviation of ketosis was reported when niacin was added at 12 g/day to diets for early lactation cows. The exact mechanisms of action are still unclear. Niacin stimulated the rumen protozoal population in one study, but overall effects on rumen microbes and their relationship to performance need further elucidation. A good review on supplemental niacin for lactating cows was given by Hutjens.[38]

Assuming a 1 to 1.5 kg (2 to 3 lb)/day increase in milk yields and a 0.5 kg (1 lb)/day increase in feed intake from feeding 6 g niacin daily, the profit from added niacin (which costs 5 to 7 cents/day) would be 5 to 15 cents/day.

Methionine hydroxy analogue (MHA). This precursor to the amino acid methionine increases milk fat percent in cows fed normal diets.[55] Many subsequent studies have con-

Table 24-18 *Influence of methionine hydroxy analogue (MHA) on milk and milk fat production*

Item	Milk (kg/day)	Fat (kg/day)	Fat (%)	Fat-corrected milk (3.5%) (kg/day)
Early postpartum cows (1-12 days postpartum)				
Control (24 cows)	27.9	0.794[a]	2.85[a]	25.0[a]
MHA (23 cows)	28.6	1.031[b]	3.63[b]	29.1[b]
Later postpartum cows (17-102 days postpartum)				
Control (24 cows)	23.2	0.672	2.82[a]	21.0
MHA (29 cows)	22.0	0.722	3.29[b]	21.2
All cows				
Control	25.6	0.735[a]	2.84[a]	22.9
MHA	24.9	0.858[b]	3.44[b]	24.7

Means for treatment comparisons sharing a different letter superscript differ ($p < 0.05$).
Adapted from data in Huber JT and others: J Dairy Sci 67:2525, 1984.

firmed a consistent increase in milk fat, but the mechanism of action has not been clarified.

Because of degradation in the rumen, some controversy exists as to whether feeding MHA results in increased blood methionine. Apparently, MHA is more resistant to rumen breakdown than dl-methionine. One study[34] showed increased blood methionine from feeding about 28 g/day (1 oz/day) of MHA to well-managed dairy cows (normal feeding levels are 25 to 30 g or 1 oz/day). In a herd traditionally low in milk fat, MHA increased fat content 0.8% in early lactation cows and slightly increased milk yields (Table 24-18). Early lactation cows gave the greatest MHA response, but increases in milk fat have been observed during mid and late lactation.

Considering profit from using MHA, an average increase of 0.3% in milk fat for a herd producing 27 kg (60 lb) of milk per day would raise the value of milk about 30 cents/day per cow. The estimated cost of MHA is 10 to 15 cents/day, so income over costs might be projected at 15 to 20 cents/day.

β-Carotene. German workers discovered sizable amounts of β-carotene in the corpora lutea of dairy cows.[46] They linked high levels of plasma β-carotene to improved reproductive performance, characterized by decreased days open, fewer services per conception, and a reduced incidence of silent heats and cystic ovaries. Trials in the United States have been inconclusive. There was no benefit to supplementation of normal cows with 300 mg β-carotene/day.[5,9] However, a Virginia study using cows with poor reproductive performance showed that feeding 600 mg β-carotene per day shortened the open period (116 versus 186 days).[6] This amount of β-carotene (600 mg daily) is six times that recommended to supply vitamin A needs of the cow. The mechanism of the β-carotene effect on reproduction has not been established, and further controlled research is needed to show efficacy and mode of action.

Protected methionine (or other amino acids). Many nutritionists assume that the amino acid that most limits milk yields is methionine because of its low concentrations in many plant proteins and in rumen microbial protein. Successful systems for protecting amino acids from rumen degradation have been developed. However, studies to date with protected methionine in lactating cows produced equivocal benefits. South Dakota workers[12,39] showed substantial increases in milk yields when the additive was fed with normal soybean meal. No effect was shown when soybean meal or whole soybeans were heat-treated to increase escape of protein to the small intestine. Other studies have shown increased percent of milk protein from supplementing protected methionine and lysine.[59]

More research is needed on protected amino acids to determine the dietary and management conditions under which they are beneficial. Some diets will benefit from additional methionine, whereas others lack lysine or other amino

Table 24-19 *Maximum tolerable dietary levels of certain elements*

Element	Maximum tolerable level mg/kg or ppm
Aluminum	1,000*
Arsenic	
Inorganic	50
Organic	100
Bromine	200
Cadmium	0.5†
Fluoride	40‡
Lead	30†
Mercury	2†
Molybdenum	10§
Nickel	50
Vanadium	50

*As soluble salts of high bioavailability. Higher levels of less soluble forms found in natural substances can be tolerated.
†Levels are based on human food residue considerations.
‡As sodium fluoride or fluorides of similar toxicity. The maximum safe level of fluoride for growing heifers and bulls is lower than for other dairy cattle. Somewhat higher levels are tolerated when fluorine is from less available sources such as phosphates (see text). Morphological lesions in cattle teeth may be seen when dietary fluoride for the young exceeds 20 ppm, but a relationship between the lesions caused by fluoride levels below the maximum tolerable levels and animal performance has not been established.
§Toxicity related to the dietary level of copper.
Adapted from data in National Research Council: Nutrient requirements for dairy cattle, Washington, DC, 1989, National Academy of Sciences, National Academy Press.

acids. If we assume a 5% to 7% increase in milk yields from supplementing protected methionine (which some studies show), then income over costs (assuming a cost of 20 to 25 cents/day), after taking into account extra feed, would be 10 to 15 cents/day.

Zinc Methionine. Two studies[66,67] involving 162 cows compared the feeding of 2 to 4 g/day of zinc methionine to a control ration. Treated groups averaged 1 to 2 kg/day (2 to 4 lb/day) more milk with no change in fat, protein, or lactose content. There was also a 20% to 50% decrease in somatic cell counts in milk. Superior zinc absorption and greater immune response have also been reported with zinc methionine compared to zinc oxide.

No feed intake or supplement cost data were reported for either study, so it is difficult to assess profitability of the additive. However, with the magnitude of response reported in the Colorado study, a profit might be assumed.

Conclusion. It would be naive to assume that all feed additives are beneficial in all situations. I have conservatively presented some available response data and have attempted to clarify conditions under which an additive might or might not give a response. A further question is which additives are complementary and which are not. Very few data are available on the interactions between additives. Generally the milk yield response reported for most additives is 5% to 10%. Unless there is an improvement in nutrient utilization (through digestion, absorption, or another mechanism) or a greater efficiency of conversion of absorbed nutrients to milk (by repartitioning the milieu of available metabolites), higher feed intakes would accompany the greater milk yields and should be factored into any benefit to cost estimate.

NUTRITIONAL REQUIREMENTS FOR DAIRY COWS

Tables in the Appendix summarize the 1989 NRC[52] recommendations for content of nutrients to be furnished to the various classes of dairy cattle at different production levels and stages. When using these tables, close attention should be paid to the footnotes, which clarify conditions

Fig. 24-16 High-producing cows at feeding time at the University of Arizona Dairy Cattle Center.

for provision of several of the nutrients.

Estimates of ruminal undegradability of a number of feeds are given in the Appendix. As mentioned in the footnote, these data were compiled by the NRC authors[51] from many sources. Table 24-19 gives maximum levels of minerals suggested to be tolerable for dairy cattle.[52]

CONCLUSION

Successful feeding of the modern commercial dairy herd is both an art and a science. Knowledge of the normal and abnormal behaviors of cows and rapid response to their changing needs are imperative. Provision of balanced diets at different phases of the lactation and dry periods requires great skill. Excessive energy, protein, minerals, or vitamins can be just as costly and detrimental to animals as deficiencies of these nutrients. Correct timing of nutrition is essential if desired results are to be realized. Figure 24-16 shows high-producing, early lactation cows in excellent health and body and mammary condition.

REFERENCES

1. Akin DE, Gordon GLR, and Hogan JP: Appl Environ Microbiol 46:738, 1983.
2. Allison MJ and others: J Bacteriol 83:184, 1961.
3. Bauman DE and others: J Dairy Sci 68:1352, 1985.
4. Beede DK: Proceedings of the Southwest Nutrition and Management Conference, Tempe, AZ, 1987.
5. Bindas EM and others: J Dairy Sci 67:1249, 1984.
6. Bindas EM and others: J Dairy Sci 67:2978, 1984.
7. Braund DG: University Press of Florida, 1979.
8. Braund DG and others: Pat 4,188,153, US Patent Office, Washington, DC, 1978.
9. Bremel DH and others: J Dairy Sci 65:178, 1982.
10. Broster WH and Strickland ME: ADAS Q Rev 26:87, 1977.
11. Carstairs JA: PhD thesis, Michigan State University, 1978.
12. Casper DP and others: J Dairy Sci 70:321, 1987.
13. Chalupa W and Ferguson JD: Proceedings of the Minnesota Nutrition Conference, Bloomington, 1989.
14. Chalupa W and others: J Dairy Sci 68 (suppl 1):143, 1985.
15. Clark JH: University of Illinois data, personal communication, 1977.
16. Cook RM: Dairy in-service file 320:38, East Lansing, 1983, Department of Animal Science, Michigan State University.
17. Coppock CE, Everett RW, and Merrill WG: J Dairy Sci 55:245, 1972.
18. Coppock CE and West JW: Proceedings of the Southwest Nutrition and Management Conference, Tempe, AZ, 1987.
19. Edward JS, Bartley EE, and Dayton AD: J Dairy Sci 63:243, 1980.
20. Erasmus LJ and others: S Afr J Anim Sci 16:169, 1986.
21. Erb RE and others: J Dairy Sci 59:656, 1976.
22. Erdman RA and others: J Dairy Sci 63:923, 1980.
23. Ferguson JD and Otta KA: Proceedings of the Cornell Nutrition Conference, Syracuse, NY, 1989.
24. Flatt WP, Coppock CE, and Moore LA: Proceedings of the third symposium on energy metabolism, European Association of Animal Products, Publ 11:121, 1965.
25. Herrera-Saldana R and Huber JT: J Dairy Sci 72:1477, 1989.
26. Hibbs JW and Conrad HR: J Dairy Sci 43:1124, 1960.
27. Higginbotham GE and others: J Dairy Sci 72:1818, 1989.
28. Higginbotham GE and others: J Dairy Sci 67(suppl 1):12, 1984.
29. Higginbotham GE, Torabi M, and Huber JT: J Dairy Sci 72:2554, 1989.
30. Hillman D, Bolenbaugh D, and Convey EM: Michigan State University Farm Science Research Report 365, 1978.
31. Holter JB, Byrne JA, and Schwab CG: J Dairy Sci 65:1175, 1982.
32. Holter JB, Colovos NR, and Urban Jr WE: J Dairy Sci 51:1403, 1968.
33. Huber JT: Proceedings of the Pacific Northwest Nutrition Conference, Portland, OR, 1987.
34. Huber JT and others: J Dairy Sci 67:2525, 1984.
35. Huber JT, Graf GC, and Engel RW: J Dairy Sci 48:1121, 1965.
36. Huber JT and Kung L Jr: J Dairy Sci 64:1176, 1981.
37. Huber JT and others: J Dairy Sci 71(suppl 1):207, 1988.
38. Hutjens MF: Anim Nutr Health, p 23, April 1987.
39. Illg DJ, Sommerfeldt JL, and Shingoethe DJ: J Dairy Sci 70:620, 1987.
40. Jorgensen NA: In: Dairy science handbook, vol 6, Clovis, CA, 1972, Agriservices Foundation.
41. Julien WE, Conrad HR, and Moxon AL: J Dairy Sci 59:1960, 1976.
42. King KJ and others: J Dairy Sci 71(suppl 1):159, 1988.
43. Krook L and Maylin GA: Cornell Vet 6:8:1, 1979.
44. Kung Jr L and Huber JT: J Dairy Sci 66:227, 1983.
45. Kwan K and others: J Dairy Sci 60:1706, 1977.
46. Lotthammer KH: Paper presented at Roche Symposium, London, 1978.
47. McGuffey RK, Green HB, and Basson RP: J Dairy Sci 71(suppl 1):120, 1988.
48. McLaren GA and others: J Anim Sci 18:1319, 1959.
49. Moe PW, Tyrrel HF, and Flatt WP: J Dairy Sci 54:548, 1971.
50. Morrow DA and others: J Am Vet Med Assoc 174:161, 1979.
51. National Research Council: Ruminant nitrogen usage, Washington, DC, 1985, National Academy of Science.
52. National Research Council: Nutrient requirements for dairy cattle, Washington, DC, 1989, National Academy of Science.

53. Pamp DE, Goodrich RD, and Meiske JC: J Anim Sci 45:1458, 1977.
54. Pappas AM and others: J Dairy Sci 67:276, 1984.
55. Patton RA, McCarthy RD, and Griel LC Jr: J Dairy Sci 53:776, 1970.
56. Ryder WL, Hillman D, and Huber JT: J Dairy Sci 55:1290, 1972.
57. Satter LD, Whitlow LW, and Santos KA: Proceedings of the Distiller's Feed Research Council 34:77, 1979.
58. Schneider PL, Vecchiarelli B, and Chalupa W: J Dairy Sci 70(suppl 1):177, 1987.
59. Schwab C: Paper presented at Rhone-Poulenc Animal Nutrition Technical Symposium, Fresno, CA, 1989.
60. Sullivan JL and others: J Dairy Sci 71(suppl 1):207, 1988.
61. Tyrell HF, Reynolds PJ, and Moe PW: J Anim Sci 48:598, 1979.
62. Virtanen AI: Science 153:1603, 1966.
63. Waldo DR and others: Abstracts of the 72nd meeting of ASAS, 1980.
64. Wildman EE and others: J Dairy Sci 65:495, 1982.
65. Wohlt JE and Clark JH: J Dairy Sci 61:902, 1978.
66. ZinPro Corporation Technical Bulletin, TBD-8601, 1986.
67. ZinPro Corporation Technical Bulletin, TBD-8802, 1988.

25

The Role of Dietary Fat in the Productivity and Health of Dairy Cows

WILLIAM CHALUPA

Concentrations of nutrients needed in rations are determined by animal requirements and by feed intake. As the genetic potential for milk production increases, it is more difficult to formulate rations with the required concentrations of energy. For example, cows often produce in excess of 45 kg (100 lb)/day milk within the first few weeks of the lactation cycle. During this time, feed intake is submaximal and rations need to contain 1.75 to 1.85 Mcal/kg (about 0.8 Mcal/lb) net energy to meet requirements for production and maintenance.

Energy density of rations can be increased by replacing forages with grains. This, however, can lead to a multitude of digestive and metabolic problems like acidosis, rumen indigestion, bloat, reduced fiber digestibility, secretion of milk with low concentrations of fat, lameness, and liver damage (p. 110).[40]

Substitution of fat for a grain is a method for increasing energy density without compromising fiber content.[61,64] Fats have more than twice the energy density of grain, so they can be used to boost ration energy density by replacing grain, leaving the fiber portion intact. So that fiber concentrations are not compromised, rations to support more than 30 to 35 kg/day (70 lb/day) of milk need to contain supplemental fat.

Until recently, fat and high-fat feed ingredients were not routinely used in the formulation of rations for lactating dairy cows. In fact, feed composition tables did not include concentrations of fat in commonly used ingredients.[52] In 1989 the National Research Council (NRC) acknowledged the value of fats and high-fat feed ingredients in provision of energy for high levels of milk production.[54] Because certain fats can impair fermentative digestion, NRC emphasized that fats used in rations for lactating dairy cows should be inert in the rumen.[54]

Dietary fats, like proteins and carbohydrates, are subjected to fermentative digestion brought about by bacterial and protozoal enzymes in the reticulorumen and hydrolytic digestion by enzymes secreted into the small intestine (Fig. 25-1). Successful use of supplemental fat depends on knowing how the level of fat and the form of fat affect fermentative and hydrolytic digestion.

DIETARY LIPIDS

Ruminant rations formulated from forages, cereal grains, and fat-extracted seeds contain about 3% fat (dry matter basis). The fat content of grains is 2% to 4%. The residual fat in meals prepared from oilseeds (e.g., soybean, cottonseed, peanut, or canola) depends upon the method used to extract fat. Solvent extraction of fat yields protein meals with about 2% fat. Removal of fat by the expeller process results in protein meals with 4% to 5% fat.[54] Forages contain 2% to 4% fat. Most of the fat is in leaf chloroplasts, with linoleic, linolenic, and oleic acids being the main fatty acids. Whereas most of the fats in feed ingredients are triglycerides, there

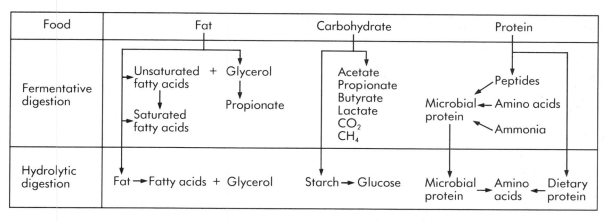

Fig. 25-1 Outline of fermentative and hydrolytic digestion in ruminants.

are also waxes and glycosyl, phospholipids, and other complex lipids.[51]

RUMEN FERMENTATION

Dietary triglycerides, galactoglycerides, and phospholipids are rapidly hydrolyzed by microbial and plant lipases to yield free fatty acids. Glycerol and galactose are fermented to propionate. Free fatty acids are adsorbed onto feed particles and ruminal bacteria.[21,51] Eighteen-carbon fatty acids are hydrogenated to stearic acid or mono-unsaturated fatty acids. Biohydrogenation takes place only on free fatty acids that have been adsorbed onto particulate matter. A number of protozoa and bacteria can hydrogenate unsaturated fatty acids but *Butyrivibrio fibrinosolvens* may be of major importance.[21,51] Because of extensive biohydrogenation of dietary fatty acids in the rumen, milk and tissue fats in ruminants contain high proportions of saturated fatty acids.

Only 3% to 5% unprotected dietary fat is tolerated by ruminal microorganisms.[61,64] The mechanisms by which long-chain fatty acids affect fermentation involve adsorption onto bacteria and onto feed particles.[33,34] Adsorption onto bacteria can impair nutrient uptake; coating of feed particles can decrease digestibility.

Ruminal microorganisms contain about 10% lipid.[67] This lipid arises from incorporation of dietary long-chain fatty acids into cellular components of rumen bacteria and from bacterial synthesis of fatty acids using glucose and acetate as carbon sources.

INTESTINAL DIGESTION AND ABSORPTION

Ruminants are more efficient than nonruminants in absorbing long-chain fatty acids from the small intestine.[51,78] There is an inverse relationship between fatty acid melting point and fatty acid digestibility in ruminants.[77] High melting point may interfere with emulsification and lipolysis of fat in the small intestine.[37,45] Thus absorption decreases as fatty acid chain length and saturation increase. Unsaturated fatty acids are more than 90% absorbed in sheep compared to 55% to 65% absorption for saturated fatty acids.[72] Highly hydrogenated tallow (59% stearic acid and 25% palmitic acid in the concentrate) is poorly absorbed[50] (Table 25-1).

Fatty acid profile, amount of fat, and form of fat affect absorption of palmitic and stearic acids from the small intestine. Intestinal absorption of stearic acid (a high-melting-point fatty acid) decreases as the proportion of the fatty acid in the concentrate increases (Table 25-1). Calcium salts of fatty acids are insoluble in the rumen. As their pK_a is between 4 and 6, they dissociate to free calcium and fatty acids in the acidic environment of the abomasum.[60,80] The dissociated fatty acids are efficiently absorbed from the small intestine.[30,70] Absorption of palmitic and stearic acids from animal fat is greater when the dietary fatty acids are in the form of calcium salts (palm oil

Table 25-1 *Absorption of palmitic and stearic acids from the small intestine*

Measurement	Palm oil		Animal fat		
	CSFA	FFA	CSFA	Tallow	Saturated*
Fat intake (kg/day)					
Total	1.47	1.38	1.17	1.47	1.61
Concentrate	1.09	1.00	0.77	1.13	0.85
Fatty acids in concentrate (% wt)					
Palmitic	43.0	44.1	24.1	23.6	24.5
Stearic	4.5	4.6	13.6	14.7	58.7
Duodenal fatty acid absorption (%)					
Palmitic	73.3	77.8	76.2	65.4	53.1
Stearic	80.3	82.6	64.4	47.1	37.4

CSFA, calcium salts of fatty acids; FFA, free fatty acids.
*Hydrogenated animal fat.
Data from Moller PD: Futterfette in der Tierernahrung, 1988.

or animal fat) than in the form of triglycerides (tallow) (Table 25-1). Because dispersion of fatty acids in micelles is a prerequisite for absorption, particle size of dietary fat may affect absorption.[76]

The picture that emerges is that ruminally inert fat supplements should not be in the form of completely hydrogenated triglycerides. They resist hydrolysis in the rumen and in the small intestine.[37,45] The high melting point of stearic acid may interfere with emulsification of fat in the small intestine. Oleic acid may aid emulsification in the small intestine of fat supplements that contain high proportions of stearic acid.

Mucosal cells of the small intestine synthesize triglycerides, phospholipids, cholesterol, and cholesterol esters.[51] Triglycerides can be synthesized either by the α-glycerophosphate or by the monoacylglycerol pathways. Under normal dietary conditions, most triglycerides are synthesized by the α-glycerophosphate pathway but the monoglyceride pathway is important when the flow of lipid to the small intestine is increased by adding fat to ruminant rations.[51] With apoproteins synthesized in mucosal cells, the synthesized lipid components are packaged in the endoplasmic reticulum and Golgi apparatus to form chylomicrons and very-low-density lipoproteins (VLDL). In ruminants, most of the newly synthesized triglycerides are incorporated into VLDL rather than into chylomicrons. Chylomicrons and VLDL are transported via the lymphatic system to venous blood.

SYNTHESIS OF FAT

Lactating cows have two major sites of lipid synthesis: the mammary gland and adipose tissue. In adipose tissue, most of the fatty acids are synthesized from acetate.[24] About 50% of the fatty acids in cow's milk (short-chain fatty acids of 4 to 12 carbon atoms, most of the myristic acid, and about half of the palmitic acid) are synthesized in the mammary gland from acetate and β-hydroxybutyrate. The 18 carbon acids and the other half of the palmitic acid are derived from triglycerides and free fatty acids in blood.[5,78]

Following parturition, dairy cows are normally in negative energy balance. This causes mobilization of fat from adipose tissue, which can accumulate in the liver to produce fatty liver or can be converted to ketones to produce ketosis. Nevertheless, dietary long-chain fatty acids normally are not metabolized in the liver of ruminants but instead may be used for synthesis of milk fat, deposited in adipose tissue, or oxidized for energy.[61] Thus, supplementing rations of lactating cows with long-chain fatty acids not only reduces negative balances of energy but also can decrease the risk of ketosis.[43,43]

RUMEN BYPASS FATS

Initially, fat was protected by coating with protein that was then treated with formaldehyde so that it was not degraded in the rumen. This product did not become commercially viable because of high cost of manufacture, difficulty in producing a consistently reliable product and the lack of governmental approval for use of formaldehyde.[61]

Progress has been made in developing dry fats that are easy to handle and do not interfere with ruminal fermentation. To be of benefit, however, protected fats must be not only inert in the rumen but also efficiently absorbed from the small intestine.

Ruminal inertness can be achieved either by forming calcium salts of long-chain fatty acids or by having a high proportion of saturated long-chain fatty acids such as palmitic and stearic acids.[13,17,61,64] Both approaches reduce solubility in the rumen.

Degree of unsaturation affects the pK_a of calcium salts of fatty acids and hence the amount of dissociation (Table 25-2). For example, the pK_a of calcium salts of soya fatty acids is 5.6. Following feeding, ruminal pH of high-producing cows may be below 6. If ruminal pH decreased to 5.5, 50% of soya fatty acids would be in the free acid form. By contrast, the pK_a of calcium salts of stearic acid, tallow fatty acids, and palm oil fatty acid distillate (Megalac) is 4.5 to 4.6. At ruminal pH of 5.5, only about 10% of these fatty acids would be in the form of free acids. These small proportions of free fatty acids are not likely to have adverse effects on ruminal fermentation.

Highly saturated fatty acids in the form of triglycerides are also insoluble in the small intestine and so may have poor intestinal digestibility.[37]

Commercial products

Critical evaluation of nutritional interventions such as supplemental fat requires continuous experiments for at least the first half of lactation and preferably over the entire lactation cycle. This is demonstrated by experiments that show nutritional interventions often produce an immediate response (response during the period of the nutritional intervention) and a residual response (response after the nutritional intervention is withdrawn).[6]

Despite the interest in ruminally inert fats, only fat supplements that rely on a high proportion of palmitic and stearic acids in the form of free acids* and calcium salts of palm oil fatty acid distillate† have been evaluated for ruminal inertness, intestinal digestibility, milk production, and reproduction. Partial hydrogenation of triglycerides‡ yields a product that does not interfere with intestinal digestion,[65] but detailed information on ruminal inertness and effects on productivity is not currently complete.

Fat supplements that contain high proportions of palmitic and stearic acids. Products such as Dairy Fat Prills and Energy Booster 100, which contain 47% palmitic acid, 36% stearic acid, 14% oleic acid, and 1% to 2% each of lauric, myristic, and 20-carbon fatty acids,[49] may not be completely inert in the rumen, due to small decreases of ruminal acetate-propionate ratios in vitro and in vivo.[16,29,32,69] This probably reflects the oleic acid in the fat supplement. The small effects on ruminal fermentation, however,

Table 25-2 *Effect of ruminal pH on dissociation of calcium salts of long-chain fatty acids*

Source of fatty acids	pK_a	Rumen pH			
		6.5	6.0	5.5	5.0
		Dissociation (%)			
Soya	5.6	11	28	51	80
Stearic	4.5	1	3	9	24
Tallow	4.5	1	3	9	24
Palm oil distillate*	4.6	1	4	11	28

*Megalac (Church and Dwight Co., Inc., Princeton, NJ). Based on pK_a values reported by Sukhija PS and Palmquist DL: J Dairy Sci 73:1784, 1990.

---

*Dairy Fat Prills, BP Nutrition, Wincham, Northwich, UK; Energy Booster 100, Milk Specialties Co, Dundee, IL, USA.
†Megalac, Volac Ltd, Orwell, Royston, Herst, UK; Church and Dwight Co, Inc., Princeton, NJ, USA.
‡Alifet, Alifet USA, Cincinnati, OH, USA.

Table 25-3 *Responses of cows to calcium salts of long-chain fatty acids*

	Response to fat supplementation			
	3.5% Fat-corrected milk (kg/day)	Milk (kg/day)	Fat (%)	Protein (%)
Multiparous cows (10 comparisons)				
Mean	2.64	2.40	0.05	−0.16
Standard deviation	1.14	0.75	0.22	0.10
Probability	<0.01	<0.01	>0.10	<0.05
Primiparous cows (3 comparisons)				
Mean	1.07	1.17	−0.02	−0.03
Standard deviation	0.82	1.04	0.14	0.09
Probability	>0.10	>0.10	>0.10	>0.10

Based on data in references 3, 22, 26, 27, 28, 68. Cows were supplemented with 0.45 to 0.57 kg/day Megalac for the first 150 days of lactation.

should not be a problem unless high levels are fed. Digestibility of energy from fatty acids in such supplements is about 85%.[32]

Three comparisons are available on the effects of fat supplements high in palmitic and stearic acids on productivity. In commercial dairies in southeastern Pennsylvania,[29] cows were supplemented with 0.45 kg (1 lb)/day Dairy Fat Prills. Yield of 3.5% fat-corrected milk was increased 0.9 kg/day (2 lb/day) in multiparous cows and 0.03 kg/day (0.07 lb/day) in primiparous animals. Conception rate, was also improved. At the University of Wisconsin,[71] multiparous cows received about 1 kg/day (2.2 lb/day) Energy Booster 100. In cows supplemented with fat and either 0 or 12 g/day niacin, yield of 3.5% fat-corrected milk increased 3 and 3.7 kg/day (6.6 and 8.2 lb/day), respectively.

In the three comparisons, increased yield was mainly the result of increased milk production, as concentration of fat in milk increased only 0.03%. A minor decrease in milk protein of 0.02% was reported.

Calcium salts of fatty acids. Supplements such as Megalac with 44% palmitic acid, 5% stearic acid, 40% oleic acid, 9.5% linoleic acid, and 1.5% myristic acid[19] are apparently completely inert in the rumen.[14,32,69]

Fatty acids in Megalac have digestibilities in lactating cows of 85% to 94%.[1,30,32,70] The net energy value of Megalac, as determined in the open-circuit respiration chambers at the USDA in Beltsville, is 6.52 Mcal/kg (2.96 Mcal/lb).[2] For comparative purposes, NRC states that the net energy content of animal and vegetable fat is 5.84 Mcal/kg (2.65 Mcal/lb).[54]

Responses of lactating cows to calcium salts of long-chain fatty acids in 10 long-term comparisons are summarized in Table 25-3. When supplemented with 0.45 to 0.57 kg/day (1 to 1.25 lb/day) Megalac, multiparous cows produced an additional 2.6 kg/day (5.7 lb/day) 3.5% fat-corrected milk. Most of the increase was accounted for by increased milk yield (2.4 kg/day), as concentration of fat in milk increased only 0.05%. Response of primiparous cows was less than that observed in multiparous cows. Concentration of protein in milk decreased slightly in multiparous cows but not in primiparous animals.

STRATEGIC USE OF SUPPLEMENTAL FAT

Supplemental fat is especially valuable during the early stages of the lactation cycle when cows fail to consume sufficient feed for high yields of milk and thus are in negative energy balance. Whether supplemental fat increases energy intake depends on the effects on feed intake. The production response to supplemental fat depends upon how the increased energy intake is partitioned. Supplemental fat may increase milk yield, alter milk composition, or reduce negative energy balance and thus lead to improved reproductive performance. Ration, animal, and environmental factors affect the partitioning of energy.

Responses to supplemental fat

Feed intake. Dry matter intake must be maintained if supplemental fat is to increase energy intake. In some, but not all, studies,[60] dietary fat reduced feed intake. This may be secondary to inhibition of rumen fermentation,[41] to slowing of intestinal motility,[55] or to taste and palatability factors.[62] Feed intake is not affected if fats do not interfere with ruminal fermentation and if precautions are taken to allow cows to adjust to new flavors and odors. This can be done by introducing fats gradually rather than abruptly at calving or by allowing cows access to fats prior to parturition.[62]

Milk yield. In the 1940s investigators at Cornell[44,48] found that adding 3% to 4% fat to concentrates increased milk production 2% to 10%; the response was found to be highly significant when the Cornell data was pooled.[64] Cows respond to higher levels of protected than unprotected fat (Fig. 25-2).

Supplementation with ruminally inert forms of fat increases yield of fat-corrected milk primarily by increasing milk production. Glucose utilization in the mammary gland for lactose synthesis is a major determinant of milk volume.[43] Increased milk volume following fat supplementation may reflect sparing of glucose from metabolism by other pathways. Glucose is metabolized via the pentose pathway to provide NADPH needed for synthesis of short-chain and medium-chain fatty acids. Increased extraction of long-chain fatty acids from blood by the mammary gland decreases synthesis of short-chain and medium-chain fatty acids by feedback inhibition of long-chain fatty acids (via acyl CoA) on acetyl CoA carboxylase. Glucose thus spared could be diverted to other milk synthetic processes.[64]

Potential responses to supplemental fat can be estimated from the difference between the net energy value of the fat and the net energy value of the ingredient replaced by fat. For example, Megalac contains 6.52 Mcal/kg net energy.[2] It usually replaces grains in rations. Corn contains 1.96 Mcal/kg net energy.[54] Thus replacement of 0.5 kg corn with 0.5 kg Megalac results in an increase of 2.3 Mcal, enough energy for an extra 3.3 kg/day of 3.5% fat-corrected milk. Experimentally, the actual response was 2.6 kg/day, or 80% of the potential response (Table 25-3). Energy not partitioned to milk production can be expected to spare mobilization of

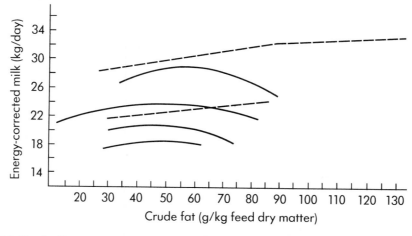

Fig. 25-2 Production responses to protected (– – –) and unprotected (———) dietary fat. From Østegaard V and others: Beretning fra statens husdyrbrugs forsøg, no. 508, Copenhagen, 1981.

energy from adipose tissue and decrease negative energy balance.

Multiparous cows supplemented with 0.45 kg/day of Megalac for the first 150 days of the lactation cycle produced an additional 3.8 kg/day of 3.5% fat-corrected milk.[27] After Megalac was removed from rations,[11] multiparous cows continued to produce more milk (2.5 kg/day of 3.5% fat-corrected milk) so that over the entire lactation, cows fed 0.45 kg/day (1 lb/day) Megalac for the first 5 months of lactation produced 3.0 kg/day (6.6 lb/day) more 3.5% fat-corrected milk than cows that were not supplemented with Megalac. Economic benefits of supplemental fat, therefore, should be evaluated on the basis of milk yield during the entire lactation and not only on the production increases during the period of feeding supplemental fat.

Milk composition. Although dietary fat may increase milk fat percentages that are depressed because of poor nutrition or management, it probably will not significantly increase milk fat percentage above an animal's genetic potential.[79]

Increased dietary fat often decreases the concentration of protein in milk.[61] The effect is specific for casein synthesis.[23,61] Palmquist and Moser postulated a mechanism involving insulin because insulin resistance was observed when protected soybean oil was fed.[66] However, lower concentrations of protein in milk associated with fat supplementation may be due to decreased synthesis of microbial protein in the rumen.[35,36]

Pricing of milk according to amounts of carrier (water, lactose, and minerals), fat, and protein has rekindled interest in factors affecting milk composition.[81] Depressions in concentration of protein in milk caused by feeding fat have been alleviated by supplementation with niacin,[26,35,36] formulation of rations to contain higher levels of undegraded intake protein,[27,28] and by supplementation with rumen-protected amino acids.[18]

Reproduction. Parturition is accompanied by changes in lipid, carbohydrate, protein, and mineral metabolism to provide nutrients for milk synthesis (Table 25-4). The intense postpartum drive for milk secretion has a higher metabolic priority than reproduction.[74]

In the past 25 years, milk production of cows in the United States has increased at the rate of about 100 kg (220 lb)/year. During this period, fertility has decreased. Erb and Smith concluded that days open and services per conception increase by 5.9 days and 0.1 services, respectively, per 1000 kg (2200 lb) milk.[25] However, the conception rate of virgin heifers has not declined in the last 25 years.[9] This suggests that the intense genetic selection for high milk yields has not resulted in selection for cows with lower fertility.

Feeding strategies that do not provide sufficient nutrients for high yields of milk cause the cow to draw on reserves of body nutrients. First ovulation usually occurs at 17 to 42 days after parturition.[9] Whereas days to first ovulation are related to milk yield, stronger correlations are found with energy balance during the early postpartum period.[8] The effect of high milk yield on conception rate is, therefore, primarily a reflec-

Table 25-4 *Metabolic changes associated with lactogenesis in ruminants*

Physiological function	Metabolic change	Tissues involved
Milk synthesis	Increased use of nutrients	Mammary
Lipid metabolism	Increased lipolysis	Adipose
	Decreased lipogenesis	
Glucose metabolism	Increased gluconeogenesis	Liver
	Increased glycogenolysis	
	Decreased use of glucose	Body tissues (general)
	Increased use of lipid	
Protein metabolism	Use of protein reserves	Muscle
Mineral metabolism	Increased absorption and mobilization of calcium	Kidney, liver, gut, bone

From Bauman DE and Currie J: Dairy Sci 63:1514, 1980.

tion of inadequate intake of energy to meet requirements. Negative energy balance appears to interfere with the ability of the hypothalamo-hypophyseal axis to develop pulsatile luteinizing hormone secretion patterns necessary for ovarian follicular development and ovulation. Additionally, the ovary may be less responsive to gonadotropin stimulation because of hypoglycemia and hypoinsulinemia during the early stages of the lactation cycle.[9]

Whether supplemental fat improves reproduction depends upon how much of the additional energy is used for increased milk production. Where supplementation causes moderate increases in milk yield (i.e., 1 to 2 kg [2 to 4 lb]/day), the conception rate improves by 20%.[29,73] When supplementation with ruminally inert fat increased milk yield by 3 to 4 kg/day (7 to 9 lb/day), conception rate was not improved. However, conception rate was maintained even though production increased.[27]

Factors affecting responses to supplemental fat

Ration. The protein-energy ratio is important in optimizing production and efficiency.[67] The ability of protein to regulate the partition of energy from rations and mobilization of energy from body fat was demonstrated by Orskov and co-workers.[57] Responses to supplemental fat must be considered in terms of the balance between energy-yielding and amino acid–yielding nutrients.

When there is an excess of energy-yielding nutrients in relation to amino acid–yielding nutrients, excess energy-yielding nutrients spare the mobilization of body fat and lessen negative energy balance. When amino acid– and energy-yielding nutrients are adequate, milk production can be fueled without mobilization of tissue stores. When amino acid–yielding nutrients are high in relation to energy-yielding nutrients, milk yield increases but negative energy balance is exacerbated. When formulating rations, it is important to recognize that the amount of protein and its degradability in the rumen can be used to direct the partition of energy in fat supplements.

Because fat is minimally fermented in the rumen, it cannot provide energy for growth of ruminal microorganisms.[56] Thus rations containing supplemental fat yield less ruminally synthesized microbial protein and should be formulated to contain more rumen undegradable protein. NRC estimated the proportions of undegraded and degraded intake protein needed in rations using data from diets that contained about 3% fat.[53,54] Based on the relationship between net energy intake and flow of bacterial crude protein to the small intestine developed by NRC,[53] an additional 72 g of undegraded intake protein should be provided per Mcal net energy from fat concentrations above 3% fat in rations. Thus, if a ration contains 0.5 kg (1.1 lb) Megalac (3.3 Mcal net energy from added fat), 238 g (0.52 lb) of degraded intake protein should be replaced with undegraded intake protein.

Animal. Logically, cows with the genetic potential for high levels of milk production should show the greatest responses to supplemental fat.

A major difficulty at the farm level is identifying whether genetic potential or management and nutrition are limiting productivity. Artificial insemination provides dairy farmers with semen from sires of high and superior genetic potential for milk production. In the United States, average production of Dairy Herd Improvement Association (DHIA; an organization that emphasizes optimal management and nutrition) herds is about 8000 kg (17,600 lb)/year.[46] Many herds produce less milk; management and nutrition and not genetics may, therefore, be limiting production. Thus, use of supplemental fat should not be limited to high-producing herds (i.e., in excess of 9000 kg [20,000 lb]/year). Lower-producing herds should be challenged with supplemental fat to determine whether they have the genetic potential for higher productivity.

Parity may affect the response to dietary fat.[47] The response of milk yield in primiparous cows is variable (see Table 25-2). Primiparous cows require nutrients not only for maintenance and milk production but also for growth. In general, the drive to attain mature body size (growth) has a higher metabolic priority than the drive to produce milk. Thus primiparous cows that are closer to mature body size are more likely to respond in terms of milk yield to nutritional interventions like supplemental fat than cows that must achieve considerable growth during their first lactation. Clearly, the impact of parity on responses to ruminally inert forms of fat needs further investigation.

Environment. Supplementation with a mixture of palmitic, stearic, and oleic acids in the form of free acids (Energy Booster 100) increased the yield of 3.5% fat-corrected milk[71] of cows in Wisconsin that calved during the warm season (April 2 through August 2) but not of cows that calved during the cool season (November 1

Table 25-5 *Example of a least-cost ration for maintenance (650-kg or 1450-lb cow) and production of 45 kg/day or 100 lb/day milk with 3.7% fat*

Ingredients	% Dry matter	
Alfalfa hay	25.0	
Corn silage	25.0	
Corn grain	20.8	
Cane molasses	2.0	
Whole cottonseed	10.0	
Soybean meal (48% CP)	0.9	
Distiller's dried grains w/solubles	9.0	
Blood meal	2.9	
Urea	0.4	
Megalac	1.8	
Monosodium phosphate	0.6	
Sodium bicarbonate	0.7	
Magnesium oxide	0.2	
Trace mineralized salt	0.5	
Selenium premix	0.1	
Vitamin premix	0.1	
Analyses (dry matter basis)	**Amount**	**Recommendation***
Net energy (Mcal/kg)	1.74	1.73
Protein (% dry matter)		
Crude (intake)	17.6	17.6
Undegraded	7.6	7.8
Degraded	10.0	9.8
Soluble	4.9	4.9
Carbohydrate (% dry matter)		
Total	68.5	
Acid detergent fiber	20.4	20.2
Neutral detergent fiber	32.3	28.5
Nonfiber	36.2	30.0
Fat (% dry matter)		
Total	7.0	7.0
Basal ingredients	2.4	3.0
Whole cottonseed plus distiller's grain	2.8	2.0
Basal ingredients plus cottonseed and distiller's grain	5.2	5.0
Ruminally inert	1.8	2.0

*Recommendations for energy and protein are from NRC.[53,54] Soluble protein recommendation is 50% of degraded protein; neutral detergent fiber recommendation is 1.1% of body weight; nonfiber carbohydrate recommendation (minimum) is 30% of ration dry matter.

through April 1). Responses of cows that calved during the warm season to supplemental fat was largely the result of higher energy intake.

RATION FORMULATION
Amount of dietary fat

The amount of fat to include in rations for lactating cows is not constant and should be determined by formulating for caloric density and fiber concentrations. Limits to the quantity of fat that can be absorbed from the small intestine and the effects of fat on metabolic efficiency set the upper limit of total fat in rations. Effects on ruminal fermentation determine the kinds of fat that are used.

Data summarized by Palmquist suggests that lactating cows can digest and absorb about 1 kg (2.2 lb) of fat per day.[61] Maximum energetic efficiency of the lactating cow occurs when 15% to 20% of the metabolizable energy (equivalent to 7% to 8% of ration dry matter) is from fat.[7,42,61] Palmquist suggests that the upper limit of dietary fat is equivalent to the amount of fat in milk (i.e., 40 kg/day or 88 lb/day of milk with 3.5% fat equals 1.4 kg or 3 lb dietary fat).[63]

The source of fat is extremely important. Fat in rations for lactating cows is provided by base feed ingredients such as forages, grains, and oilseed proteins (fraction 1); whole oilseeds and by-product proteins like distiller's grains (fraction 2); and ruminally inert fats (fraction 3). Fat in base feed ingredients yields rations that contain about 3% fat. Whole oilseeds such as soybeans and cottonseeds often are plentiful and easily incorporated into total mixed diets.[20,59,75] The seed coat may provide some degree of protection and fat in whole seeds usually has minimal effects on ruminal fermentation.[15] Whole oilseeds can be used to increase ration fat from the base level of 3% to 5%. When base feed ingredients provide less than 3% fat in ration dry matter, additional whole oilseeds may be used so the fat in fractions 1 and 2 is 5% of ration dry matter. In the United States, there is an increased use of tallow in rations for lactating cows, but research to define amounts of tallow that can be fed is not available. High-quality tallows that are low in unsaturated fatty acids (i.e., derived from ruminant animals) probably can replace some of the fraction 2 fat provided by soybeans and cottonseed. Ruminally inert forms of fat should be used to increase ration fat from 5% to 7%.

Sample ration

The least-cost ration in Table 25-5 was formulated with an interactive microcomputer program[31] using the previous guidelines for amounts and types of fat and protein. The neutral detergent fiber (NDF) recommendation was set at 1.1% of body weight. The minimum nonfiber carbohydrate (NFC, storage carbohydrates) recommendation was 30% of ration dry matter. To meet these carbohydrate guidelines, the diet was formulated to contain 50% forage (alfalfa hay and corn silage). So that the required energy density could be achieved without compromising fiber intake, supplemental fat was provided by whole cottonseed and Megalac. Distiller's dried grains with solubles and blood meal were

Table 25-6 *Example of a supplement for high-producing cows that contains rumen bypass fat and rumen bypass protein*

Ingredients	% Dry matter
Soybean meal (48% CP)	14
Fish meal	10
Blood meal	16
Corn grain	34
Megalac	20
Cane molasses	5
Sodium bicarbonate	0.7
Salt	0.3
Analyses (dry matter basis)	**Amount**
Net energy (Mcal/kg)	2.82
Protein (% dry matter)	
Crude (intake)	32.2
Undegraded	21.7
Soluble	4.3
Carbohydrate (% dry matter)	
Total	37.4
Acid detergent fiber	2.0
Neutral detergent fiber	4.1
Nonfiber	33.3
Fat (% dry matter)	21.3

used as sources of protein resistant to ruminal degradation. Urea was needed to provide sufficient soluble protein. Because the formulation attempted to meet recommendations for 11 ration components, it is not surprising that some deviations from recommendations occurred.

To ensure that the protein-to-energy ratio is maintained when rations contain additional fat, it often is advantageous to provide a supplement that contains both bypass fat and bypass protein. Many combinations of feed ingredients can provide proper balances of fat and protein. Feeding 1 kg/day (2.2 lb/day) of the sample supplement in Table 25-6 provides net energy and crude protein for production of 4 kg/day (8.8 lb/day) of 3.5% fat corrected milk. There is sufficient nonfat net energy to allow the assimilation of rumen degraded protein in the supplement to microbial crude protein.

REFERENCES

1. Andrew SM and others: J Dairy Sci 72:2227, 1989.
2. Andrew SM and others: J Dairy Sci 73(suppl 1):191, 1990.
3. Baker JG and others: J Dairy Sci 72(suppl 1):483, 1989.
4. Bauman DE and Currie: J Dairy Sci 63:1514, 1980.
5. Beitz DC and Allen RS: In Swenson MJ, editor: Dukes physiology of domestic animals, Ithaca, NY, 1984, Cornell University Press.
6. Brooster WH: In Haresign W and Cole DJA, editors: Recent developments in ruminant nutrition, vol 2, London, 1988, Butterworth.
7. Brumby PE and others: J Agric Sci (Camb) 91:151, 1978.
8. Butler WR, Everett RW, and Coppock CE: J Anim Sci 53:742, 1981.
9. Butler WR and Smith RD: J Dairy Sci 71:767, 1989.
10. Chalupa W: In Haresign W and Cole DJA, editors: Recent developments in ruminant nutrition, ed 2, London, 1988, Butterworth.
11. Chalupa W: Proceedings of the Georgia Dairy Management Conference, Athens, 1989, University of Georgia.
12. Chalupa W and Ferguson JD: J Dairy Sci 73(suppl 1):244, 1990.
13. Chalupa W and others: J Dairy Sci 67:1439, 1984.
14. Chalupa W and others: J Dairy Sci 68(suppl 1):110, 1985.
15. Chalupa W and others: J Dairy Sci 68(suppl 1):115, 1985.
16. Chalupa W and others: J Dairy Sci 69(suppl 1):213, 1986.
17. Chalupa W and others: J Dairy Sci 69:1293, 1986.
18. Chow JM, DePeters EJ, and Baldwin RL: J Dairy Sci 73:1051, 1990.
19. Church and Dwight Co, Inc, Princeton, NJ, 1988.
20. Coppock CE, Lanham JK, and Horner JI: Anim Feed Sci Tech 18:89, 1987.
21. Czerkawski JW and Clapperton JL: In Wiseman J, editor: Fats in animal nutrition, Boston, 1984, Butterworth.
22. Downer JV and others: J Dairy Sci 70(suppl 1):221, 1987.
23. Dunkley WL, Smith NE, and Franke AA: J Dairy Sci 60:1863, 1977.
24. Emery RS: In Ruckebusch Y and Thivend P, editors: Digestive physiology: metabolism in ruminants, Lancaster, Eng, 1980, MTP Press Ltd.
25. Erb HN and Smith RD: In Veterinary clinics of North America, bovine reproduction, Philadelphia, 1987, WB Saunders.
26. Erickson PS, Murphy MR, and Clark JH: J Dairy Sci 72(suppl 1):483, 1989.
27. Ferguson JD and others: J Dairy Sci 71(suppl 1):242, 1988.
28. Ferguson JD and others: J Dairy Sci 72(suppl 1):415, 1989.
29. Ferguson JD and others: J Dairy Sci 73:2684, 1990.
30. Filley SJ and others: J Dairy Sci 70(suppl 1):221, 1987.
31. Galligan DT and others: J Dairy Sci 72(suppl 1):445, 1989.
32. Grummer RR: J Dairy Sci 71:117, 1988.
33. Harfoot CG: Prog Lipid Res 17:21, 1978.
34. Harfoot CG and others: J Appl Bacteriol 37:663, 1974.
35. Horner JL and others: J Dairy Sci 69:3087, 1986.
36. Horner JL and others: J Dairy Sci 71:1239, 1988.
37. Jenkins TC and Jenny BF: J Dairy Sci 72:2316, 1989.
38. Jenkins TC and Palmquist DL: J Dairy Sci 64(suppl 1):129, 1981.
39. Jenkins TC and Palmquist DL: J Dairy Sci 67:978, 1984.
40. Kesler EM and Spahr SL: J Dairy Sci 47:1122, 1964.
41. Kowalczyk J and others: Br J Nutr 37:251, 1977.
42. Kronfeld DS: Adv Anim Physiol Anim Nutr 7:5, 1976.
43. Kronfeld DS: J Dairy Sci 65:2204, 1982.
44. Loosli JK, Maynard LA, and Lucas HL: Cornell University Agricul Exp Sta Memoir 265, 1944.
45. MacLeod GK and Buchanan-Smith JG: J Anim Sci 35:890, 1972.
46. Majeskie JL: National cooperative dairy herd improvement program handbook, Columbus, OH, 1988.
47. Mattias JE and others: J Dairy Sci 65(suppl 1):151, 1982.
48. Maynard LA, Loosli JK, and McCay CM: Cornell University Agric Expt Sta Bull 753, 1941.
49. Milk Specialties Co, Dundee, IL, 1988.
50. Moller PD: Futterfette in der Tierernahrung, 1988.
51. Moore JH and Christie WW: In Wiseman J, editor: Fats in animal nutrition, Boston, 1984, Butterworth.
52. National Research Council: Nutrient requirements of dairy cattle, ed 6, Washington, DC, 1978, National Academy of Science.
53. National Research Council: Ruminant nitrogen usage, Washington, DC, 1985, National Academy of Science.
54. National Research Council: Nutrient requirements of dairy cattle, ed 6, Washington, DC, 1989, National Academy of Science.
55. Nicholson T and Omer Brit SA: J Nutr 50:141, 1983.
56. Nocek JE and Russell JB: J Dairy Sci 71:2070, 1988.
57. Orskov ER, Reid CW, Tait and CAG: Anim Prod 45:345, 1987.

58. Ostergaard V and others: Berentning fra statens husdyrbrugs forsog, Copenhagen, 1981.
59. Owen FG, Larson LL, and Lowery SR: J Dairy Sci 1985.
60. Palmquist DL: Can J Anim Sci 64(suppl 1):240, 1984.
61. Palmquist DL: In Wiseman J, ed: Fats in animal nutrition, Boston, 1984, Butterworth.
62. Palmquist DL: Proceedings of the 23rd Pacific Northwest animal nutrition conference, Pacific Northwest Grain and Feed Association, Portland, OR, 1988.
63. Palmquist DL: Personal communication, 1988.
64. Palmquist DL and Jenkins TC: J Dairy Sci 63:1, 1980.
65. Palmquist DL, Kelby A, and Kinsey DJ: J Dairy Sci 72(suppl 1):572, 1989.
66. Palmquist DL and Moser EA: J Dairy Sci 64:789, 1981.
67. Preston TR and Leng RL: Matching ruminant production systems with available resources in the tropics, sub-tropics, Armidale, New South Wales, Australia, 1987, Penambul Books.
68. Robb EJ and Chalupa W: J Dairy Sci 70(suppl 1):220, 1987.
69. Schauff DJ and Clark JH: J Dairy Sci 72:917, 1989.
70. Schneider P and others: J Dairy Sci 71:2143, 1988.
71. Skaar TC and others: J Dairy Sci 72:2028, 1989.
72. Sklan D and others: J Dairy Sci 68:1667, 1985.
73. Sklan D and others: J Dairy Res 56:675, 1989.
74. Smidt D and Farries E: Curr Top Vet Med Anim Sci 20:358, 1982.
75. Smith NE: In Garnsworthy PC, editor: Nutrition and lactation in the dairy cow, Boston, 1988, Butterworth.
76. Steele W: J Dairy Sci 66:520, 1983.
77. Steele W and Moore JH: J Dairy Res 35:371, 1968.
78. Storry JE: In Cole DJA and Haresign W, editors: Recent developments in animal nutrition, Boston, 1988, Butterworth.
79. Storry JE and others: J Dairy Res 41:465, 1974.
80. Sukhija PS and Palmquist DL: J Dairy Sci 73:1784, 1990.
81. Thomas PC and Martin PA: In Garnsworthy PC, editor: Nutrition and lactation in the dairy cow, London, 1988, Butterworth.

Body Condition Scoring: Use and Application

RONALD L. TERRA

BODY CONDITION SCORING: WHY?

With the current situation of decreasing financial return in the dairy industry, dairy farmers must maximize their profit. The goal of a nutritional consultant is to feed the herd so as to maximize milk production while maintaining optimal health and reproductive efficiency. An overriding concern is how to do this as inexpensively as possible. To achieve this goal, the consultant must evaluate the efficacy of the recommendations that are given to the dairy farmer. The variables to be measured are as follows:

1. *Milk production* is the measure of greatest concern to the producer.
2. *Feed costs* are easily determined. They can be followed on a monthly basis and compared to the costs of preceding months either on an absolute basis or relative to the milk produced.
3. *Reproductive efficiency* is measured by analyzing the information obtained from fertility examinations.
4. *Health status* is more subjective. Clinical illness is easily noted, and the economic effects of that illness can be measured, for example, as decreased milk production or increased drug costs. However, subclinical illness is more subtle both in terms of economic loss and in identification. A means to readily identify the presence of subclinical disorders is valuable to the consulting nutritionist.

Body condition scoring is used to evaluate the presence of subclinical disorders and assess the response of the cattle to the nutritional program. It involves estimation of the muscle and fat cover of animals and is based on the principle that during negative energy balance flesh melts away from bone. Unlike body weight, condition scores are largely unaffected by frame size.

The primary components that determine body condition score are *nutrition* and *production.* Increasing production without the necessary increase in the plane of nutrition decreases body condition. Maintaining the nutritional plane in the face of decreasing production increases body condition. Health status also affects body condition score. Subclinical illness may decrease body condition, which, in fact, can be the only sign seen if milk production and food intake are not markedly depressed. Subtle decreases in body condition can be missed, which reinforces the importance of an ongoing condition-scoring program. Differences in condition score between animals within production groups should be noted by the veterinarian. Steps can then be taken to identify the problem and remedy the situation.

Body condition can affect health status and level of production. With the increase in milk production by dairy cattle over the last decade, cows are more susceptible to metabolic disorders and infectious diseases. Most of them occur periparturiently, and body condition can influence the incidence and severity of these maladies.[3,10,15,18] Overconditioning before or during the dry period is associated with higher than desired body condition scores and contributes to the incidence of fat cow syndrome.[10] Other research indicates that fatness—or high body condition score—results in milk fat depression,[8] decreased milk yield,[6] and reproductive inefficiency.[13,17] Lowered body condition score also decreases milk production and reproductive efficiency.[8,13,17]

The overall body condition score derived from observation of a group of dairy cattle indicates the level of nutrition, production, and overall health status. If the average body condition score is considered optimal and one animal shows a marked deviation, then that animal should be singled out for further evaluation.[17] Body condition scoring also evaluates how well the nutritional program is working. If the average body condition score of a group of cattle is considered optimal and production is at the desired level, then the nutritional program is probably optimal. However, if production is suffering and body condition score is not optimal, then the nutritional program should be investigated. In general, body condition reflects the interaction of nutrition, health, and production.

BODY CONDITION SCORING: WHEN?

The high levels of milk production achieved by modern dairy cows increase their vulnerability to mismanagement. This emphasizes the need for optimal conditions at each stage of the lactation cycle. Body condition must be optimal during each lactation period to ensure the health of the animal, to maximize production, and to help maximize income over production costs.

The timing of body scoring is arbitrary. A particular design should be developed and followed so that similar groups at different times of the year or under different nutritional schemes can be compared reliably. I find it convenient to score cattle at the time of feed-level changes. Other valuable times to conduct scoring are at freshening, peak lactation, mid lactation, and drying off.

At freshening, the cow should be at her highest level of condition. The increased fat deposits can be used in early lactation to support milk production. Problems occur in cows that are overconditioned and predisposed to metabolic disorders. On the 5-point system discussed later in this chapter, an optimal score at parturition is about 3.5 but varies slightly between breeds.[18] Scoring the cow at this time allows the nutritionist to evaluate the feeding program in the dry period and during late lactation. If the cow entered the dry pen in adequate condition and freshens while either underconditioned or overconditioned, then the dry cow and late lactation rations need to be reevaluated. The condition score at this time also serves as a reference point for the rest of lactation. Normally, the high-producing cow loses condition in early lactation because of her genetic and hormonal drives for milk production. The dairy cow mobilizes up to 50 kg of fat in early lactation to accommodate the deficit between energy lost in milk production and energy intake in feed.[1,14] Milk production peaks at 40 to 60 days postpartum, and the body condition score at this time should be approximately 2.5.[18] A precipitous (greater than 1 point) drop in condition score over this time span is cause for concern, as it predisposes to decreased lactational and reproductive efficiency.[4,14]

From peak of lactation to the start of the dry period, the cow should be fed to regain condition that was lost during early lactation. The goal over the last two thirds of the lactation is to attain a condition score of 3.5 at dry off.[18] Evaluation of the cattle at certain points during this time period allows the nutritionist to determine the adequacy of the rations. In large dairies, scoring cattle at the time cows are moved between milking strings has several benefits.

1. If a cow is due to be moved down a production string, yet is below the condition score of the rest of the string she is leaving, then she should be allowed to remain behind in the higher-producing string a while longer. The higher plane of nutrition in the high-producing string allows the cow to gain weight as her milk production falls and dietary energy is diverted to body fat synthesis. The opposite scenario—that of an overconditioned cow being sent to a lower production string for weight loss prior to dry off—can be seen occasionally.
2. The average condition score for the group at the time of string change allows evaluation of the ration that was fed to the string. It also serves as a reference point for the string the cattle are joining.

Keeping records of the cattle scored enables the

scorer to attempt to score the same animals at the next string change, whether those cows are being moved or not. This will tell whether the cattle are responding favorably or adversely to the ration changes.

Finally, an effort should be made to score the cows at dry off. By comparing the condition scores at dry off versus freshening, the dry cow ration can be evaluated. It is best to change body condition of cows during late lactation. The body condition of thin cows can be improved, but the time period is short. Slimming fat cows in late gestation can precipitate metabolic problems. Therefore, the condition score at dry off is the final data point needed to evaluate the overall nutritional package offered to the dairy.

BODY CONDITION SCORING: HOW?

Numerous body condition scoring systems have been developed or adapted for use in dairy cattle.* Some systems are based on 8 or 10 points,[2,7,11] but the one most commonly used in North America is the 5-point system described here.[3,18] The original method involved external palpation of the cows to assess the amount of subcutaneous fat and muscle present;[18] this has since been adapted to allow scoring by visual inspection only.[3] A chart aids in the process (Table 26-1; Fig. 26-1). This adaptation is a significant improvement over the palpation method, as the animals do not have to be restrained to be scored. This method is preferred when large numbers of freely moving cattle are assessed.

In the scoring process, the cow is initially observed from a distance to give the assessor a general impression. Then the individual body parts are scored and the data recorded. The parts analyzed are the lumbar and sacral areas from the back and sides. Palpation of the amount of fat covering is also useful[3,9,16,18] (Table 26-1; Figs. 26-1 and 26-2). Finally, an overall body condition score is assigned, based on an average of the scores assigned to the individual body areas. With practice, the results can be calculated mentally, and only the final score and cow number are written down.

---

*References 2, 3, 7, 9, 11, 16, 18.

Body condition scores reflect the nutrient reserves of the dairy cow. Researchers have shown that the condition score is related to live weight change, body weight to wither height ratio, and the proportion of fat in that weight change. The overall levels of body water, protein, ash, and body energy also are related to body score.[18,19] A breed difference in the partitioning of fat among the various fat deposits has been observed. This results in differences in the proportion of total body fat at the same body condition score.[19] Because of all these factors, it is difficult to predict accurately the live weight change associated with changes in condition score. Published estimations can be used to approximate the amount of energy increase necessary to yield a change of one point in the condition score.[5] As the condition score and body weight of the animal change, a change in maintenance requirement also occurs. Using estimates of body weight change and the nutritional requirements necessary for maintenance, milk production, and weight gain, the amount of feed that would increase body condition score by one point can be determined.

As an example, in Holsteins a live weight change of approximately 50 kg is needed to cause an increase of one point in the body condition score. Assuming that this condition score change occurs between peak production (60 days postpartum) and dry off (305 days postpartum), then the cow needs to gain 0.2 kg per day. The energy density of 1 kg of weight gain is approximately 6 Mcal; thus an extra $6 \times 0.2 = 1.2$ Mcal of NE_g is needed daily to support this weight gain.[12] This additional energy could be provided by feeding another 2.7 kg of corn silage or 1.1 kg of dehydrated sugar beet pulp daily. Depending on the costs of the feedstuffs used, this weight gain would cost approximately 6 to 8 cents per day. Theoretically this increased level of feeding would result in the desired increase in body condition score from 2.5 at peak lactation to 3.5 at dry off. However, slight miscalculations in the amounts of feedstuffs mixed in the ration, in the amounts fed, or in the energy density could result in unwanted consequences. Therefore, constant monitoring is important.

Table 26-1 *Body condition scoring in dairy cattle*

Area	1	2	3	4	5
General	Emaciated Frame protruding	Moderate condition Frame obvious	Good condition Frame and covering well balanced	Fat Covering more obvious than frame	Obese Severe overconditioning
Lumbar area					
Vertebral spinous processes	Sharp ends Little flesh	Easily discernible	Smooth ridge Individual processes not visible	Not discernible Covering almost flat	Buried in fat
Transition between spinous and transverse processes	Concave	Slightly concave	Smooth slope or concave	Smooth slope	Convex
Transverse processes	Half length visible	Third length visible	Quarter length visible	Not discernible Smooth, rounded edge	Buried in fat
Transition between transverse processes and paralumbar fossae	Prominent shelf, gaunt	Overhanging shelf	Slight shelf	None	Bulging
Palpation	Transverse processes and vertebral bodies feel sharp	Transverse processes feel sharp	Apply pressure to feel transverse processes Smooth	No bones palpable	No bones palpable
Sacral area					
Sacral vertebra	Individual vertebrae distinct Extremely sharp No tissue cover	Individual vertebrae not visible Prominent	Smooth covering	Smooth covering	Smooth covering
Tuber coxae (hooks) Tuber ischii (pins)	Severe depression	Very sunken	Smooth	Rounded with fat	Surrounded by fat
Between tuber coxae and tuber ischii (gluteal muscle)		Very depressed	Depression Little fat deposition	Slight depression	Flat
Between tuber coxae	Severely depressed No flesh	Very depressed	Moderate depression	Flat	Rounded
Tailhead to tuber ischii	Deep V-shaped cavity under tail	U-shaped cavity under tail	Shallow cavity under tailhead	Slight depression	Folds of fatty tissue under tailhead
Palpation	No fatty tissue Skin drawn tight over pelvis	Some fatty tissue	Can feel pelvis with slight pressure; fatty tissue over whole area	Bones difficult to feel	Bones difficult to feel

Adapted from Edmonson AJ and others: J Dairy Sci 72:68, 1989.

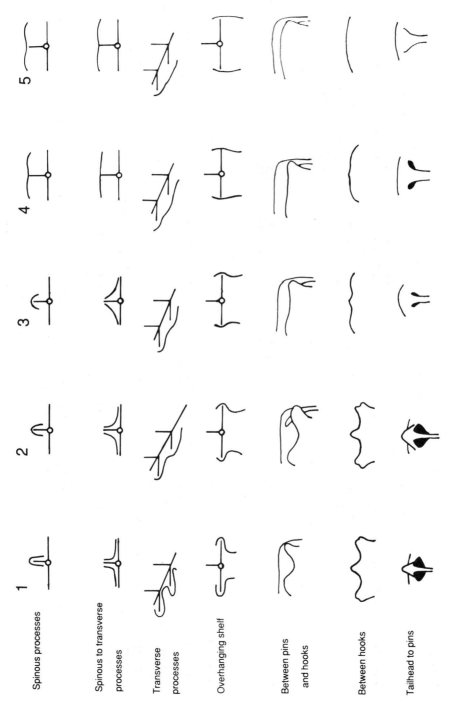

Fig. 26-1 Schematic of anatomical conformation at various condition scores. From Edmonson AJ and others: J Dairy Sci 72:68, 1989.

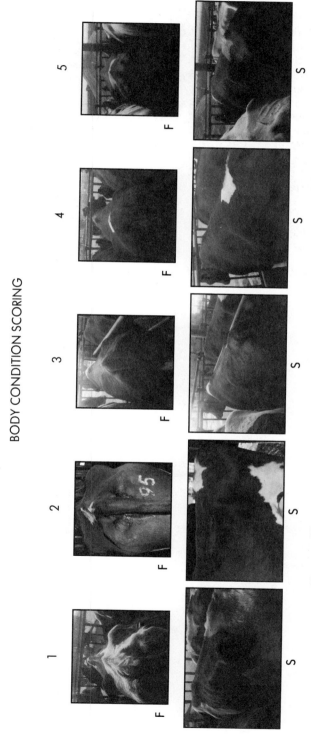

Fig. 26-2 Photographs of cows at body condition scores of 1-5. **1F** and **1S**, score 1; **2F** and **2S**, score 2; **3F** and **3S**, score 3; **4F** and **4S**, score 4; **5F** and **5S**, score 5.

CONCLUSION

This chapter has described the importance of condition scoring in dairy cattle, the timing of scoring, and the method with which to score. This is only one tool in the delivery of a sound nutritional consultation program to the dairy farmer. In implementing a condition scoring program, the nutritionist is ensuring that subtle changes produced in response to changes in the level of nutrition will be recognized. Once recognized, management steps can be recommended either to alleviate a detrimental effect or to maintain a beneficial one.

REFERENCES

1. Bauman DE and Currie WB: J Dairy Sci 63:1514, 1980.
2. Earle DF: Aust Dep Agric J (Victoria) 74:228, 1976.
3. Edmonson AJ and others: J Dairy Sci 72:68, 1989.
4. Ferguson JD: Paper presented at Monsanto Technical Symposium, Syracuse, 1980.
5. Frood MJ and Croxton D: Anim Prod 27:285, 1978.
6. Garnsworthy PC and Topps JH: Anim Prod 35:113, 1982.
7. Grainger C and McGowan AA: N Z Soc Anim Prod occasional publication 8:134, 1982.
8. Jaquette RD and others: J Dairy Sci 71:2123, 1988.
9. Maas J: In Smith BP, editor: Large animal internal medicine, St Louis, 1991, The CV Mosby Co.
10. Morrow DA: J Dairy Sci 59:1625, 1976.
11. Mulvany P: Handout 4468, Natl Inst Res Dairying, Reading, UK, 1981, Shinfield.
12. National Research Council: Nutrient requirements of dairy cattle, ed 6, Washington, DC, 1988, National Academy of Science.
13. Richards MW and others: J Anim Sci 62:300, 1986.
14. Samuels WA: Feedstuffs 62(7):13, 1990.
15. Troutt HF: Proc Am Assoc Bov Pract 6:68, 1974.
16. Veterinary aspects of dairy cattle: nutrition and housing. In Blood DC and Radostits OM, editors: Herd health, 1982.
17. Weaver LD: Proc Am Assoc Bov Pract 17:82, 1984.
18. Wildman EE and others: J Dairy Sci 65:495, 1982.
19. Wright IA and Russel AJF: Anim Prod 38:23, 1984.

Nutritional Problems Encountered in Dairy Practice

JAMES D. FERGUSON

The most frequently encountered health problems in dairy herds may be broadly categorized as infectious (mastitis, metritis), metabolic (ketosis, retained fetal membranes, milk fever, ruminal acidosis, displaced abomasum), physical (feet and leg problems, dystocia), and reproductive (anestrous, cystic ovaries, infertility) disorders.[7,8,14,17] One disorder may predispose to another, and these categories are not mutually exclusive. Grouping disorders into these categories assists in determining if hygiene, feed management, physical environment, or cow attributes need to be improved (Table 27-1). Incidence rates are under 10% for most of these problems,[14,17] although in individual herds incidence rates may approach or exceed 50%. Nutritional risk factors have been identified for many of these disorders (see Table 27-1). Nonetheless, correcting underlying nutritional imbalances often does not eliminate all cases within a herd. Many of these disorders occur as a complex, have no single underlying etiology, and are associated with mismanagement more often than malnutrition alone. The nutritional component may predispose but not be sufficient to cause disease by itself.

Herd problems associated with nutrition tend to have multiple etiologies. Additionally, utilization of body tissue reserves may mask presentation of problems associated with dietary imbalances. The relationships between nutrition, disease, productivity, and reproductive performance are complex and confounded by physiological state (growth, lactation, gestation). The practitioner must integrate information on nutrient imbalances, feed management, environment, and body condition to analyze herd problems.

Production occurs as a cycle and can be divided into dry, periparturient, and early, middle, and late lactation periods. Nutrient requirements and dietary intake vary throughout the production cycle and may not change contemporaneously. Changes in nutrient requirement may precede alterations in intake by 4 to 8 weeks. Body stores provide nutrients when intake does not meet requirements. The use of fat stores is quite apparent, as dairy cows decline in body condition in early lactation. Utilization of other body reserves is not as apparent. For example, calcium release from bone stores to maintain plasma calcium concentration at parturition is only apparent when the system fails and parturient paresis occurs. Depletion and repletion of body nutrient reserves occur throughout the production cycle. The extent of body store depletion and repletion is related to the degree of asynchrony between nutrient requirement and nutrient intake.

For nutrients with low body reserves, diet must consistently meet requirements. Dietary deficiencies are poorly tolerated for these nutrients. For nutrients with appreciable stores, nutritional programs need to ensure adequate reserves are available at critical periods.

There are critical periods during the production cycle when inappropriate nutritional management is likely to result in reduced productivity, increased metabolic and reproductive disorders, or both. Critical times include the periparturient and early lactation periods. However, late lactation and dry cow diets may influ-

Table 27-1 *Incidence and association of health problems and risk factors in dairy herds*

Metabolic	Infectious	Physical	Reproductive
Ketosis (2.1%)	**Mastitis (14.1%)**	**Feet/legs (3.8%)**	**Cystic ovaries (7.9%)**
↓ Fiber	↓ Hygiene	Environment	↓ Manganese
NSC	↓ Selenium	Stalls/bedding	↓ Selenium
Protein	↓ Vitamin E	↓ Hygiene	↓ Vitamin E
↓ Fat	↓ Vitamin A	↓ Copper	↑ Metritis
↑ Body condition at calving		↓ Zinc	
		↓ Selenium	
		↑ Body condition at calving	
		↑ Laminitis	
Milk fever (4.3%)	**Metritis (9.6%)**	**Dystocia (8.5%)**	**Anestrus (4.9%)**
↑ Calcium	↑ Body condition at calving	↑ Age	↓ Energy
↑ Phosphorous	↑ Retained fetal membrane	Bull	↓ Zinc
↓ Magnesium	↑ Dystocia	↑ Milk fever	↓ Selenium
↑ Energy	↑ Pyometra	Condition loss	↓ Cobalt
↑ Body condition at calving	↓ Hygiene	Body condition at calving	↓ Iodine
			↓ Manganese
			↓ Copper
			↑ Protein
Retained fetal membranes (9.4%)		**Downer cow (0.4%)**	**Infertility (14%)**
↓ Selenium		↑ Milk fever	↓ Energy
↓ Vitamin E		Dystocia	↓ Protein
↓ Vitamin A		Grass tetany	↓ Copper
		Injury	↓ Zinc
			↓ Selenium
			↓ Carotene
			↓ Manganese
			Condition loss
Fatty liver			**Abortion (1.3%)**
NSC			↓ Iodine
↑ Body condition at calving			↓ Vitamin A
			Toxins
Grass tetany			
↓ Magnesium			
Ruminal acidosis			
↑ NSC			
↓ Fiber			
↓ Fiber length			
Left displaced abomasum (1.2%)			
Chopped versus long forage			
Physical form			
↑ Ketosis			
Body condition			

NSC, nonstructural carbohydrate.
↑ Increased risk of health problem with increasing level of predisposing factor.
↓ Increased risk of health problem with decreasing level of predisposing factor.
Incidence reports reported by Milian-Suazo F, Erb HN, and Smith RD: Prev Vet Med 7:19, 1989.

ence problems during the periparturient and early lactation periods by affecting body reserves.

DRY PERIOD

Milk yield is maximized in the subsequent lactation when cows have dry periods of 40 to 70 days. This period is associated with mammary gland involution and regeneration. Fetal growth is maximal and nutrient needs of the conceptus are high. At the end of lactation, dry matter intake may be 2% to 2.5% of body weight; intake decreases to 1.5% to 2.0% of body weight by the second to third dry week. Prior to calving, dry matter intake sharply declines for a brief period. Feeding programs that are adequate early in the dry period may not be adequate closer to term because of decreases in dry matter intake and increases in requirements. Ideally, dry cows should be managed as two groups to permit ration changes to adapt to conceptus requirement and reduced dry matter intake as parturition approaches. Rations need to be adjusted about 3 weeks prior to calving to prevent periparturient complications and body condition loss and to prepare for impending parturition.

Dry parturient complications: dystocia, mastitis, metritis, retained fetal membranes, and milk fever (parturient paresis)

Periparturient disorders tend to occur as a complex. For example, cows with milk fever are at increased risk for dystocia, retained fetal membranes, and metritis (Tables 27-2 and 27-3).[4-8,14] Periparturient disorders increase the risk for postpartum metabolic and reproductive disorders, reproductive failure, and culling.[4-8,14] Successful treatment of primary periparturient disorders does not eliminate the risk of future complications. A major benefit of preventing periparturient disorders is a reduced incidence of postpartum metabolic and reproductive problems.[4-8,14]

Periparturient problems of particular concern include overconditioning (fat cows), parturient paresis (milk fever), retained fetal membranes, metritis, and dystocia.[4-8,18,19] Parturient paresis, retained fetal membranes, and mastitis are the

Table 27-2 *Odds ratios showing the associations between periparturient disorders, metabolic disorders, and reproductive function*

Primary disorder	Secondary disorder							
	Milk fever	Dystocia	RFM	Metritis	Mastitis	Cystic ovaries	LDA	Ketosis
Milk fever*		2.8	2.0	1.6	8.1		3.4	8.9
		4.2	3.2		5.4		+	+
Dystocia*			4.0	3.0				
				4.9				
				3.5				
RFM*				5.8	5.7		+	16.4
Metritis*						1.7	+	+
Age†		+		+		+		
Milk yield†	+							+
Over condition at dry off or calving†	+		+	+		2.5	+	+

RFM, retained fetal membranes; LDA, left displaced abomasum.
*Numbers represent odds ratios. Different values come from separate studies. Values > 1 indicate the magnitude of increased risk for the secondary disorder if the primary disorder occurs in a cow. For example, cows with retained fetal membranes are 5.8 times more likely to develop metritis than cows which cleanse normally.
† + indicates a predisposing effect.
From Curtis CR and associates[4] and Erb HN and Grohn YT.[8]

Table 27-3 *Reproductive performance and parturient disorders in multiparous cows*

Disorder	Reproductive measures			
	Days to first service	Total number services	Conception at first service	Culling
Dystocia				3.7*
Metritis	4.4†		0.6*	
Cystic ovary	10†	0.4†	0.7*	1.5*
Retained fetal membranes	5†			
Age, years†		0.04†	−‡	
Mature equivalent, 305-day milk yield (kg)	2.8†	135		−‡

*Values are odds ratios; for example, a cow with metritis has only 0.6 of the chance of becoming pregnant at first service as a cow without this problem.
†Deviation from herd mean. For example, metritis increases the interval to first service by 4.4 days.
‡ − indicates a preventive effect.
Adapted from Erb HM and others, J Dairy Sci, 68:3337, 1985.

major periparturient problems contributing to postpartum complications.[14,17] Dry cow nutritional risk factors for periparturient complications include excessive intakes of energy, protein, calcium, and phosphorus; inadequate intakes of magnesium, selenium, and vitamins A, D, and E; and body condition loss during the dry period.

Parturient paresis. Parturient paresis (milk fever) is a major cause of economic losses, both from direct treatment costs and from subsequent complications.[4-8] The incidence of parturient paresis is about 9% in the United States for all dairy cows;[13] however, individual herds may experience incidence rates of 30% to 50%. Treatment with intravenous calcium solutions often is initially successful, but relapse is common, and secondary postpartum complications are not eliminated by treatment. Therefore, prevention is preferred to treatment of milk fever.

Risk of parturient paresis increases with age and level of milk yield and is further influenced by breed and dry cow nutrition.[1,16] Risk of parturient paresis is higher in Channel Island than Friesian breeds. Risk of parturient paresis increases with excessive intakes of calcium and phosphorus during the dry period. Cows in the fourth or later lactation are at higher risk than younger animals, particularly if they are consuming high-calcium dry cow diets. Herd incidence is influenced by breed, diet, age of cows, and overall management. Thus, incidence may differ in herds fed on similar forages and grains.

Prevention of milk fever with parenteral vitamin D (10 million units IM, 1 week before calving; repeat once if the cow does not calve in 8 days) or oral megadoses of vitamin D works by enhancing gastrointestinal absorption of calcium. Therefore, use of vitamin D as a preventative is most beneficial in herds feeding on high-calcium dry cow diets. However, in cows that develop milk fever despite vitamin D prophylaxis, the problem is often more severe and prolonged than in naturally occurring cases[16] because vitamin D analogues repress rather than augment the natural homeostatic system to maintain plasma calcium. The increased intestinal absorption of calcium inhibits parathyroid hormone (PTH) secretion, necessary for bone calcium resorption. The normal endogenous 1,25-$(OH)_2$D response to hypocalcemia may also be inhibited. Additionally, when hypercalcemic effects of vitamin D analogues wane, cows experience hypocalcemia for 24 to 72 hours until endogenous homeostatic controls return to normal. This may occur 2 to 3 weeks postpartum and results in clinical signs of hypocalcemia. Because of these problems and because repeated administration of vitamin D is toxic, vitamin D prophylaxis is best reserved for cows with an accurate calving date and a history of previous milk fever.

Diets containing more than 60 to 80 g of calcium provide more calcium than needed by cows at the end of gestation.[22] Excess calcium suppresses homeostatic mechanisms to the point that full activation of these processes may require several days when milk secretion begins.[24,25] Low-calcium (less than 60 g/day) diets during the dry period help keep calcium homeostatic mechanisms active by stimulating bone mobilization and gut absorption. Diets very low in calcium (less than 30 g/day) virtually eliminate parturient paresis but may be associated with other problems.[1,12]

Forages compose the bulk of dry cow diets and thus are extremely important as a source of calcium and other nutrients. Ideally, dry cow forage should be palatable, of moderate quality, high in fiber, and low in calcium, protein, and total cations. Legume forages are high in cations, protein, and calcium. Corn silage is high in energy. Dry cow diets based primarily on corn silage or alfalfa hay or haylage are not appropriate for dry cows. Grass hay is high in fiber and low in calcium and protein; it is therefore ideally suited to meet dry cow nutritional needs. Grass hay or haylage, corn stover, or straw, alone or in combination with corn silage, provide an excellent forage base for dry cow feeding. Many farms lack a supply of grass hay or haylage and do not feed dry cows appropriately. However, nutritional advisors should encourage the use of low-calcium forages.

Anion-cation balance has received considerable attention as a method to prevent milk fever.[1,12,22] Diets high in cations (Na + K + Ca + Mg) are associated with increased incidence of milk fever.[2] Increasing the anion concentration (Cl + P + S) of the ration by adding ammonium chloride can reduce the incidence of milk fever.[2,23] The mechanism by which high-anion diets reduce parturient paresis is not clear but appears to be related to induced metabolic acidosis, which enhances intestinal absorption and bone resorption of calcium.[9,10] The incidence of parturient paresis in cows consuming high-anion diets is similar to incidence rates in cows consuming low-calcium diets (less than 35 g/day).

Use of ammonium sulfate and ammonium chloride (100 g [3 oz] each daily for 3 weeks prior to calving) to prevent milk fever by providing a highly anionic diet may reduce palatability and increase the cost of feeding dry cows. It may be cheaper and more appropriate to provide low-calcium forages than to alter diets by use of ammonium salts. The appropriate strategy has to be determined for each farm. Calcium containing mineral preparations should be avoided in dry cow diets.

Retained fetal membranes. Retained fetal membranes are a common parturient problem in dairy cows.[4,7,17,20] Incidence rates average about 7% in the United States but can range from less than 2% to more than 50% on individual farms.[20] Mortality is low, but cows with retained fetal membranes are at risk for metritis, ketosis, and LDA (see Tables 27-2 and 27-3).[7,8] The benefits of selenium and vitamin E in preventing retained fetal membranes are well recognized.[15,27] However, retained fetal membranes can have multiple etiologies, and increasing Se and vitamin E above recommended levels should not be relied on as a universal preventative. It is of interest that cows with retained fetal membranes are at higher risk of milk fever at subsequent calving.[7,8] This suggests that hypocalcemia that is not severe enough to induce paresis may inhibit uterine contraction and produce retained fetal membranes. Thus strategies to reduce parturient paresis may reduce the incidence of retained fetal membranes.

Placental expulsion is related to placental maturation over the last 1 to 2 months of gestation.[20] The junction of maternal crypts and fetal microvilli must mature and break down for placental disruption. Chemotaxis of leukocytes into the placentome also are required for loosening.[20] Additionally the mechanical process of uterine contraction and placental hypovolemia following rupture of the cord are important in placental expulsion.[20] Retention is associated with a disturbance in the loosening mechanism or with uterine inertia.[20]

Dry cow nutrition and management cannot be ignored in the prevention of retained fetal membranes. Curtis and co-workers[5] identified higher intakes of protein and energy during the dry period as protective against retained fetal membranes and ketosis (Table 27-4). However, excessive feeding of grain and protein is not to be encouraged, as it may lead to the fat cow syndrome (obesity, dystocia, fatty liver, ketosis, death).[18,19] Dry cows need to be fed to meet requirements throughout the dry period. Underfeeding may reduce maternal reserves of protein and energy and minimize reserves available after calving for lactation.

When retained fetal membranes exceed 10% of calvings, selenium and vitamin E status should be examined, as should the entire dry cow program. The selenium status should be evaluated by taking blood samples from cows fresh less than 30 days. Whole blood selenium should be above 0.2 µg/ml and plasma selenium above 0.07 µg/ml.[27] If blood concentrations of selenium are adequate, oversupplementation is not beneficial. If blood samples are equivocal, liver and kidney tissue concentrations may better reflect body stores.

Management

Often dry cows are neglected—out of sight, out of mind. Dry cows are fed the worst forages on the farm, and intakes are not monitored closely.

Dry cows should be housed separately from lactating cows to ensure they are fed appropriately for their needs. In a stanchion barn this may require separating dry cows at one end of the barn or dividing manger space from neighboring cows. Diets should be fed to maintain body condition throughout the dry period. Live weight increases due to fetal growth and accumulation of fluids of pregnancy, but body condition should be stable. Dry cows should be managed as two groups—those dry for the first 4 to 6 weeks and those less than 3 weeks from calving. This is necessary because changes in fetal requirement, dry matter intake, and impending parturition alter feeding management late in the dry period. The last 2 to 3 weeks of the dry period are particularly important for preventing milk fever and other parturient disorders.

Table 27-4 *Association between periparturient disorders and dry period nutrition*

Dry period nutrition*	Secondary disorder				
	Milk fever	Dystocia	RFM	Mastitis	Ketosis
Protein					
1 (142%)					
2 (180%)			0.4		0.8
3 (224%)			0.7		0.2
Energy					
1 (131%)					
2 (150%)		0.4			0.3
3 (178%)		0.2			0.4
Phosphorus					
1 (122%)					
2 (174%)	0.9				
3 (226%)	1.6				
Calcium	+				
Selenium			−		
Vitamin E			−		

Numbers represent odds ratios. Different values come from separate studies. Values greater than 1 indicate increased risk for the disorder; values less than 1 indicate decreased risk. For continuous variables, + indicates an increased risk; − indicates a decreased risk.
*Nutritional intakes are divided into three groups according to level of intake. The figures in parentheses show the median intake for the group expressed as a percent of the NRC recommended intake. Odds ratios compare groups 2 and 3 to 1. For example, cows with very high phosphorus intakes (group median 226% of NRC recommendations) are 1.6 times more likely to develop milk fever than cows with intakes close to NRC recommended levels (group median 122%).

Dry cows should be fed long hay at up to 1% of body weight, to maintain rumen function. Ideally, the hay should be a grass variety. Corn silage may be included in the ration at a rate of 9 to 15 kg/head/day. If corn silage is fed, no grain may be needed in the early dry cow ration. Vitamins A, D, and E and trace minerals (particularly Se) should be provided either as injections at dry off or as daily supplements in small amounts of grain. Grain fed at 1.4 kg (3 lb)/head/day maintains rumen adaptation to grain diets. Crude protein concentration of the total diet should be at least 12%. White salt or a trace mineral salt block should be available free choice, but daily intakes should be less than 30 g (1 oz)/head/day. Daily calcium should be less than 60 g (2 oz)/day and phosphorus less than 40 g (1.5 oz)/day. The ration should contain adequate amounts of magnesium (0.16% of ration dry matter).

Three weeks prior to calving, dry cows should move to a "close-up" (to parturition) group. Calcium homeostatic mechanisms may require 2 weeks to be activated prior to calving. Because gestation may vary by 1 week, entrance into the close-up group 3 weeks prior to expected calving ensures that all cows are adapted for at least 2 weeks before calving. Grain should be fed at a daily intake of 3 to 4 kg/head/day (6 to 9 lb/head/day). Forage should consist of grass hay alone or in combination with corn silage. Daily calcium and phosphorus intake should be less than 40 g (1.5 oz)/head/day. Grain should not be overfed, as high intakes depress forage intake and reduce rumen fill. Niacin may be included in the diet (6 g/head/day) to help prevent ketosis and fatty liver problems if cows are overconditioned.

Sensible dry cow nutrition and suitable body condition and hygiene at calving minimize periparturient complications.

LACTATION
Rumen acidosis and milk fat depression

Following the periparturient period, cows commence the next critical phase of the production cycle. During the first 100 days of lactation, cows are at risk for the majority of metabolic, physical and health disorders. Metabolic complications include ketosis, displaced abomasum, rumen acidosis and laminitis, fatty liver, and milk fat depression. Difficulties during this period may be associated with feeding and management problems in the dry or the early lactation period.

Milk energy output increases rapidly after calving and exceeds energy intake from feed. To maximize energy intake, rations are often formulated to contain 50% to 60% of dry matter as grain. As a consequence, fiber concentration is marginal. Additionally, forages often are fermented and chopped, reducing the physical size of the fiber particles. Long-stem fiber is needed to form a mat in the rumen (the rumen pack) to trap food particles and increase the retention of food particles within the rumen. In addition, the rumen pack stimulates chewing, salivation, and rumination. As a result of marginal fiber intakes, early lactation cows are at risk for rumen acidosis (p. 110). This reduces dry matter intake, resulting in excessive body condition loss. Other complications include laminitis, milk fat depression, digestive upsets, diarrhea, ketosis, displaced abomasum, and even death.

Excessive grain feeding predisposes to rumen acidosis, which may depress feed intake. However, rations too high in forage lower the dietary energy concentration and may depress total energy intake. Either problem may place a constraint on total energy intake and increase negative energy balance and the rate and extent of condition loss postpartum. If fat mobilization is excessive, fatty liver and ketosis may occur, which may further compromise intake. Poor condition also reduces fertility.

Dry matter consumption is dependent upon palatable diets and a healthy rumen. Rumen microbes require structural (fiber) and nonstructural carbohydrate, protein, and minerals to maintain a balanced pattern of fermentation. Ideally, substrates for fermentation should be available on a continuous basis so that large fluctuations in the rumen environment do not occur. Feed should be available free choice throughout the day.

Although able to utilize ammonia and urea as sole nitrogen sources, the rumen ecosystem is more efficient when some protein is provided in the form of peptides and amino acids. The proportion of nonstructural carbohydrate in the ration affects the efficiency of protein utilization in the rumen. Diets need to be balanced in rumen supply of protein as well as nonstructural and structural carbohydrates. Diets need to contain at least 19% acid detergent fiber and 28% to 30% neutral detergent fiber to ensure adequate fiber in the ration. Chop length should be at least 0.75 cm (¼ to ⅜ inch) for fermented feeds.

Cows tend to eat multiple small meals a day. This provides a continuous supply of substrate to the rumen and minimizes fluctuations of rumen metabolites. For most efficient nutrient utilization, fluctuation of rumen substrate should be minimal. The degree of fluctuation of rumen metabolites depends on the type of feeds in the ration and the rate and extent of rumen fermentation. Rapidly fermentable feeds fed infrequently cause large fluctuations in rumen substrates; slowly fermentable feeds result in less fluctuation. However, in either situation minimal fluctuations occur when cows are fed frequently. This may mean feeding grain three to four times a day in a stanchion barn, and two to three times a day in group housing. Less frequent feed offerings may result in pulsatile intakes, with large fluctuation in rumen metabolites and less efficient rumen digestion.[26] Wide swings in rumen environment, particularly pH, predispose to digestive upsets and reduced feed intake.

Slug feeding may occur in situations where bunk space per cow is limited, eating time is reduced by excessive milking time, if the feed bunk is empty for long periods during the day or night, or if grain is fed only twice a day in parlor feeding systems. These conditions produce pulsatile feed delivery.[26]

Imbalances in carbohydrate sources (nonstructural versus structural), finely chopped forages, low-forage diets, and pulsatile slug feeding can lead to rumen acidosis. Severity of signs may depend on fermentability of feeds and the overall chemical composition of the ration. Although overt signs of disease may not be apparent, when less than 50% of cows are chewing their cud, and excessive amounts of fiber and grain are present in the manure, forage may be chopped too fine, fiber concentration may be too low in the total diet, or feed delivery may be a problem.

Two basic systems are utilized to deliver feed to lactating cows. Total mixed rations (TMR) combine all feed ingredients in a blended mix so that cows consume forage and grain together. On many farms a cafeteria-style feeding system offers forages separately from grains. Combinations of these two systems may be used if computer or parlor feeders are used on the farm. Each system has disadvantages.

Total mixed rations make it difficult to feed hay and forage in long physical form to dairy cows. Careful regulation of the chopping process at harvest and added dietary buffers are important to prevent rumen upsets when total mixed rations are fed. Excessive mixing when blending TMR diets may further reduce particle size and should also be monitored. Additionally, feed may not be available ad libitum. This may create a situation similar to slug feeding and result in rumen upsets and butterfat depression.

In cafeteria-style feeding, rapid increases in grain intakes postpartum may depress forage consumption and produce rumen upsets. Additionally, sequence of feed offering may influence rumen function, particularly when rapidly fermented feeds are fed.[26] Forage should be fed prior to grain to slow the passage rate and to provide chewing and buffering prior to grain feeding.

If all goes well in the postpartum period, dry matter intake should meet the needs of production around week 8 of lactation. This will be associated with small changes in body condition. By monitoring body condition, adequacy of dry and lactation rations may be assessed.

REFERENCES

1. Allen WM and Sansom BF: J Vet Pharmacol Ther 8:19, 1985.
2. Block E: J Dairy Sci 67:2939, 1984.
3. Butler-Hogg BW, Wood JD, and Bines JA: J Agricul Sci (Camb) 104:519, 1985.
4. Curtis CR and others: J Am Vet Med Assoc 183:559, 1983.
5. Curtis CR and others: J Dairy Sci 68:2347, 1985.
6. Curtis CR and others: J Dairy Sci 67:817, 1984.
7. Erb HN and others: J Dairy Sci 68:3337, 1985.
8. Erb HN and Grohn YT: J Dairy Sci 71:2557, 1988.
9. Fredeen AH, Depeters EJ, and Baldwin RL: J Anim Sci 66:159, 1988.
10. Fredeen AH, Depeters EJ, and Baldwin RL: J Anim Sci 66:174, 1988.
11. Garnsworthy PC: In PC Garnsworthy, editor: Nutrition and lactation in the dairy cow, London, 1988, Butterworth.
12. Gerloff BJ: Vet Clin North Am (Food Anim Pract) 4:379, 1988.

13. Goff JP, Kehrli ME Jr, and Horst RL: J Dairy Sci 72:1182, 1989.
14. Grohn YT and others: J Dairy Sci 72:1876, 1989.
15. Harrison JH, Hancock DD, and Conrad HR: J Dairy Sci 67:123, 1984.
16. Littledike ET and Goff J: J Anim Sci 65:1727, 1987.
17. Milian-Suazo F, Erb HN, and Smith RD: Prev Vet Med 7:19, 1989.
18. Morrow DA, Tyrell HF, and Trimberger GW: J Amer Vet Med Assoc 155:1946, 1969.
19. Morrow DA: J Dairy Sci 59:1625, 1976.
20. Paisley LG and others: Theriogenology 25:353, 1986.
21. Perkins B: Production, reproduction, health and liver function following overconditioning in dairy cattle, doctoral thesis, Ithaca, NY, 1985, Cornell University.
22. National Research Council: Nutrient requirements of dairy cows, ed 6, Washington, DC, 1989, National Academy Press.
23. Oetzel GR and others: J Dairy Sci 71:3302, 1988.
24. Ramberg CF: Fed Proc 33(2):183, 1974.
25. Ramberg Jr CF and others: Am J Physiol 219:1166, 1970.
26. Robinson PH: J Dairy Sci 72:1197, 1989.
27. Smith L and Hogan JS: Large Animal Veterinarian, pp 20-24, Nov-Dec 1988.
28. Wildman EE and others: J Dairy Sci 65:495, 1982.
29. Wright IA and Russel AJF: Anim Prod 38:23, 1984.

28

Choosing a Computerized Ration-Balancing Program

DAVID T. GALLIGAN

FUNDAMENTAL COMPONENTS

The tedious calculations involved in ration evaluation and formulation have been greatly eased by development of interactive computer programs. Many ration formulation and evaluation programs are now available for use on microcomputers. Although there is considerable variation in performance, user friendliness, and cost, all share basic fundamental components. These fundamental components are (1) nutrient requirement determination, (2) feed reference libraries, (3) ration evaluation capabilities, (4) ration formulation techniques, (5) economic evaluations, and (6) printed outputs (feeding instructions). In this chapter each of these components is reviewed, emphasizing functional attributes beneficial for use in a veterinary nutritional consulting environment.

Nutrient requirements

The nutrient requirement section of most programs is based on nutrient requirements determined by scientific bodies such as the National Research Council *Nutrient Requirements of Dairy Cattle*.[5] Generally, requirements are expressed as the daily quantity of nutrient required (lb of crude protein, gm of calcium, Mcals of net energy of lactation) or as a nutrient density (quantity of nutrient per lb of dry matter intake). Most computer programs have the user enter certain parameters about the cow (body weight, milk production level, milk fat, physiological status, weight change) that are then used in equations to determine requirements (Fig. 28-1). Requirements should be able to be displayed and printed, along with ration evaluations and formulations.

Programs that permit the user to edit default levels allow greater flexibility in formulating rations where nutrient requirements are unique to a feeding situation. For example, it often is desirable to increase the bypass protein requirement in rations when feeding high levels of fats.

Some programs use linear programming techniques to formulate rations. This type of program should allow the user to specify and modify maximum and minimum tolerable ranges for each nutrient in a ration. This is particularly important when infeasible solutions are obtained because of requirement constraints that need modification.

Last, the capability to create customized nutrient fields can allow users to investigate effects that new nutrient fractions might have on the ration evaluation or formulation process.

Feed reference library

A feed analysis reference library is usually available or can be created with most programs. Many program libraries have nutrient analysis of common feed ingredients based on reported NRC values. The ability to combine multiple

Information entered by operator

COW PARAMETERS

BODY WEIGHT	1400 lb
MILK	80 lb
FAT TEST	3.7 %
WEIGHT CHANGE	0 lb/DAY

Output calculated by computer program

NUTRIENT REQUIREMENTS

	lb DM	Mcal NE	CP	BP	ADF	Ca	P
			\----------% / lb-------------				
% DRY MATTER	100	0.82	17.8	6.5	22.1	0.75	0.48
MINIMUM LEVEL	44.0	36.1	7.9	2.9	9.7	0.332	0.212
USER SPECIFIED MIN.	0	0	0	0	0	0	0
%CHANGE	-5	0	0	0	0	0	0
NRC REQUIREMENT	46.3	36.1	7.9	2.9	9.7	0.332	0.212
%CHANGE	0	10	10	10	10	40	40
USER SPECIFIED MAX.	0	0	0	0	0	0	0
MAXIMUM LEVEL	46.3	39.7	8.6	3.2	10.7	.464	.297

CP = Crude protein
BP = Bypass protein

Fig. 28-1 Example of computer output for nutrient requirements. From DairyLP.[2,4]

feed reference libraries is helpful in customizing feed libraries for individual producers.

Feed libraries should allow the user to store, delete, and edit ingredient analysis unique to an individual herd in order to facilitate evaluation and reformulation of rations. Nutrient compositions of ingredient mixes created by the user should be calculated by the program and stored in the feed library as well. The feed library capacity should permit a sufficient number of feeds and mixes to be conveniently stored (100 or more is recommended). The feed bank should store the following information about each ingredient: net energy of lactation, crude protein, bypass protein, acid detergent fiber, calcium, phophorus, fat, neutral detergent fiber, sodium, nonfiber carbohydrate, magnesium, potassium, sulfur, iron, copper, cobalt, manganese, zinc, chloride, vitamin A, vitamin D, vitamin E, and selenium. Dry matter content, cost, and feed description also should be maintained for each ingredient.

Ration evaluation

Ration evaluation is an important functional capacity of a computer ration formulation package and is essential for herd problem solving. Within this environment, the user can specify rations being fed, and the program then will calculate each ration's total nutrient content. Hence evaluation is of a ration not created by the program itself but one entered by the user. The ability to enter ingredient amounts on an as-fed basis simplifies this process because most producers record as-fed amounts. Furthermore, the capability to make mixes of feed ingredients and to calculate their nutrient compositions allows evaluation of particular mixes used in the feeding system.

It is important that the nutrient content of evaluated rations be expressed as (1) total nutrient weight or Mcals, (2) nutrient density, (3) percent NRC requirement met by the ration, and (4) percent NRC requirement met by the ration adjusted to dry matter intake (Fig. 28-2).

Ration calculated by program for a cow

FEEDS	lb/day
1 Haylage	30.0
2 High Moisture corn	30.0
3 Distillers Grain	2.0
4 Soybean Meal	2.0
5 Whole Cotton Seeds	6.0
6 Dicalcium Phosphate	.43
7 Limestone	.20

Nutrient content in the ration

	DM lb	NE Mcal	CP lb	BP lb	AF lb	Ca lb	P lb
TOTALS NUTRIENTS	43.96	36.7	6.97	2.19	5.78	0.332	0.212
RATION DENSITY %DM		0.84	15.9	5.0	13.1	0.754	0.482
% NRC		95	102	89	76	59	100	100
NRC DMI CORRECTED	100	107	94	80	63	105	105

CP = Crude protein
BP = Bypass protein
AF = Acid detergent fiber

Fig. 28-2 Example of computer-generated ration formulation. Adapted from DairyLP.[2,4]

These evaluations allow one to distinguish nutrient deficiencies or excesses due to inappropriate nutrient delivery or density.

To evaluate feeding situations in which a variety of feeds are fed, it is advantageous to have the capacity to change feed intake levels easily to simulate intakes. For example, when forage intakes are not known, it is hard to evaluate their energy and nutrient contribution to the ration. By simulating a range of possible intakes, potential nutrient deficiencies and excesses may be revealed. Programs that run in an interactive mode allow this to be done relatively easily.

Ration formulation

Ration formulation can be performed under a variety of methods. Linear programming techniques have been used frequently in the feed industry.[1] These programs formulate with the objective of maximizing either least cost or profit margin. The linear program finds the combination of feed ingredients within the constraints (e.g., nutrient and ingredient) that is minimum cost (total feed cost) or maximum profit (milk revenue–feed cost), depending on the objective. Many programs also allow the user to constrain feed ingredients to certain feeding ranges (minimum, fixed level, or maximum). For example, it may be advantageous to constrain corn silage to a level that ensures inventories are sufficient for the entire year. Other programs allow the user to specify ratios of feed ingredients to be maintained during the formulation process.

Other options available with some packages include inventory control and parametric analysis.[7] Parametric analysis traces out how the ration changes as the coefficients or parameters (feed prices or nutrient requirement levels) are changed.

Linear programming techniques have also been developed to formulate rations across periods of time.[3] These approaches take advantage of seasonal changes in feed prices by recommending appropriate use of home grown forages during the year to minimize total feed cost over the year. Currently, these methodologies are limited to mainframe systems because of the size of these linear programs.

Algebraic methods have also been developed to deal with the complexity of ration balancing and the need for operator intervention and understanding.[2,4] These techniques allow the user to enter the most economically efficient feeds (generally forages) in fixed amounts. Exact amounts entered depend on availability, nutrient density, and the feed management system. Nutrients from these fixed ingredients are summed, and remaining nutrient deficits (energy and protein) are then balanced with selected variable feeds (usually concentrate ingredients). Algebraic simultaneous equations are used to solve these problems.

To increase the flexibility and economic efficiency of this technique, algorithms have been developed to calculate the substitution feed values[6] so economically efficient feeds can be selected and used in the formulation process to maximum advantage. Solutions from these methods often can approximate a least-cost ration.

Economic evaluation

The economic evaluation component is perhaps the most important part of the program package. Many approaches have been taken in this area.

Linear programming techniques are very efficient in that if a solution exists (i.e., is feasible and thus meets all constraints) it also will be the lowest in cost possible (or maximum in profit). These packages also have the ability to specify the price range over which a feed will remain in the ration by calculating the reduced costs. Reduced costs tell the change in price needed for a feed ingredient to be included in a ration and hence can be used to calculate the opportunity price. It is important to realize that the calculated reduced cost of a feed ingredient is based on the assumption that all other costs and constraints are constant (Fig. 28-3).

Marginal nutrient values also are calculated and indicate the change in ration cost from increasing or decreasing a nutrient requirement level by one unit. Hence sensitivity of the total ration cost to marginal changes in each nutrient is known.

Ratios of substitution values:market price of feed ingredients also are used to rank the economic efficiency of feed ingredients.[2] These values are based on the assumption that the economic value of a feed is a sum of the economic values of the individual nutrients in feed.[6] It is important to realize that these methods ignore the economic value of nutrient "density" and merely calculate the cost necessary to supply total nutrient amounts.

Outputs

Printed outputs are an important feature of a ration formulation and evaluation program. This is the component that communicates feeding instructions directly to the producer. They should have a format that ensures that the ration can be administered efficiently.

The program should have the capability of printing necessary instruction sheets for the variety of feed delivery systems commonly found in practice. Stanchion sheets (grain tapes) are required where individual production levels are fed. In general, these sheets are in table format with milk production level on one axis and feed ingredients on the other. The table is filled with amounts of each ingredient to be fed for each production level (Fig. 28-4).

Total mixed ration instruction sheets are required for group feeding systems. These sheets specify the group that the ration is for and then list feeds and their amounts (either on a per head or per group basis) to be fed per day along with the ration nutrient composition (Fig. 28-5). Flexible programs allow the user to specify the group size and report feeds to be fed on an as-fed or dry matter basis. By reporting feed amounts on a dry matter basis, the producer can easily adjust feed levels according to changes in moisture content of ingredients. Some programs have batch-mixing outputs that display incremental load weights as feeds are added to a mixer in a particular order.

Instructions for making mixes of ingredients is an essential printout and is used in almost all feeding management systems. Adaptable programs (1) allow specification of desired total weight of the mix, (2) list component ingredient amounts and prices, and (3) list the nutrient composition of the mix.

Dual Cost Analysis

Nutrients	Marginal value
Dry matter	$ 0
Net energy	$ .10[a]
Crude protein	$ .28
Bypass protein	$ 0[b]
Calcium	$ 0
Phosphorus	$ 0

[a] Ration cost will increase by $.10 if net energy requirement is increased by one unit.

[b] Ration cost will not change if the bypass protein requirement is increased by one unit.

Reduced Cost Analysis

Name	Current $/Ton	Solution lb/day	Opportunity Price for inclusion $/Ton
HAYLAGE 1	30	0.0	28[a]
HAYLAGE 2	35	71.1	35
HIGH MOISTURE CORN	80	25.8	80
CORN SILAGE	25	0.0	15
DISTILLERS GRAINS	180	0.0	161
SOYBEAN MEAL 48%	320	1.8	320
WHOLE COTTON SEEDS	200	0.0	186
LIMESTONE	50	0.0	50
DICALCIUM PHOSPHATE	324	0.0	324

[a] The cost of HAYLAGE 1 will have to be reduced from $30/ton to $28/ton to be included in the ration.

Fig. 28-3 Computer output showing dual cost (marginal values of nutrients) and reduced cost information for ration.

Fig. 28-4 Computer-generated instruction sheet for stanchion barn feeding system. Adapted from DairyLP.[2,4]

GENERAL RECOMMENDATIONS lb/d[a]
WHITE SALT FREE CHOICE

MILK PRODN	Protein/ Grain Mix	High Moisture Corn	Hay	Corn Silage
100	15.4	31.3	10.0	35.0
95	14.5	29.8	10.0	35.0
90	13.5	28.2	10.0	35.0
85	12.6	26.6	10.0	35.0
80	11.6	25.1	10.0	35.0
75	10.7	23.5	10.0	35.0
70	9.8	21.9	10.0	35.0
65	8.8	20.4	10.0	35.0
60	7.9	18.8	10.0	35.0
55	6.9	17.2	10.0	35.0
50	6.0	15.6	10.0	35.0
45	5.0	14.1	10.0	35.0
40	4.1	12.5	10.0	35.0
35	3.1	10.9	10.0	35.0
30	2.2	9.4	10.0	35.0

```
                    TOTAL MIXED RATION^a
                  WHITE SALT FREE CHOICE
        Milk:     65                          Group:    LOW HERD
                  Feed     %            % DM     -      COW #     +
    DESCRIPTION   LB/DAY   DM   LB DM   MIX    0.05^b    50     0.05^b
    CORN SILAGE   29.0    30.2   8.8   20.55  1377.5  1450.0  1522.5
    HAYLAGE       34.0    41.8  14.2   33.35  1615.0  1700.0  1785.0
    LOW GRP MIN.   0.3   100     0.3    0.70    14.3    15.0    15.8
    HMEC          19.8    69.8  13.8   32.49   942.5   992.1  1041.7
    SOYBEAN HULLS  5.6    90     5.1   11.89   267.4   281.5   295.6
    LIMESTONE      0.339 100     0.3    0.79    16.1    16.9    17.8
    DICALCIUM      0.099  96     0.1    0.22     4.7     5.0     5.2
    ***************************************************************
        TOTALS    89.2         42.6   100     4237    4460    4683
```

	DM^c	AF^d			DM^c	AF^d	
DM	100	48	%	MG	0.3	0.1	%
NE	0.765	0.366	MC/LB	K	1.0	0.5	%
CP^e	15.9	7.6	%	S	0.2	0.1	%
BP^f	4.8	2.3	%	FE	50.0	24.0	PPM
SOL-P^g	5.8	2.8	%	CO	0.1	0.05	PPM
ADF^h	19.2	9.2	%	CU	10	5	PPM
NDF^i	31.6	15.1	%	MN	40	19	PPM
NFC^j	3.4	1.6	%	ZN	40	19	PPM
FAT	0.1	0.1	%	NA	.18	0.1	PPM
CA	0.66	0.32	%	CL	.25	0.1	PPM
P	0.42	0.20	%	DCAB^k	30	14.5	MEQ/100G
Selenium	0.29	0.14	PPM				
Vitamin A	1411	674	IU/LB				
Vitamin D	141	67	IU/LB				
Vitamin E	15	7	IU/LB				

a Adapted from DairyLP (2,4)
b 5% Tolerance limits
c Dry matter basis
d Crude protein
e Bypass protein
f Acid detergent fiber
g Soluble protein
h Acid detergent fiber
i Neutral detergent fiber
j Non fiber carbohydrate
k Dietary cation-anion balance

Fig. 28-5 Computer-generated instruction sheet for total mixed ration system for a group of 60 cows producing a mean of 65 lb milk per day. Adapted from DairyLP.[2,4]

Greater flexibility is allowed in programs where the user can customize outputs for particular feeding systems. Several programs allow the user to enter individualized comments or notes on printout sheets.

DOCUMENTATION AND USER FRIENDLINESS

Documentation and support vary considerably with programs. All programs should have accompanying operational manuals. Good documentation greatly facilitates the user friendliness of any program. In a survey of features deemed most important by program users, ease of use was ranked first.[7]

Some programs have on-line help facilities or tutorial programs to familiarize the user with the software. In general, programs driven by menus versus direct commands are easier to learn. After the program is mastered, however, experienced users desire a direct command access system to increase speed of use.

Because dairy nutrition is constantly progressing and new techniques being explored, it is important that developers of programs realize the need to update the software periodically.

REFERENCES

1. Black JR and Hlubik J: J Dairy Sci 63:1366, 1980.
2. Galligan DT and others: J Dairy Sci 69:1656, 1986.
3. Galligan DT and others: J Dairy Sci 72 (suppl 1) 1989.
4. Galligan DT and others: J Dairy Sci 72 (suppl 1) 1989.
5. National Research Council: Nutrient requirements of dairy cattle, Washington, DC, 1989, National Academy of Science.
6. Peterson WE: J Dairy Sci 15:293, 1932.
7. Weaver LD, Olivas MA, and Gallard JC: J Dairy Sci 71:1104, 1988.

SECTION C Goats

29

The Goat Industry: Feeding for Optimal Production

KATHERINE BRETZLAFF, GEORGE HAENLEIN, ED HUSTON

THE GOAT INDUSTRY
Dairy goats

There are approximately 1 million dairy goats in the United States.[8] Dairy goats can be found throughout the United States, although there is a very high concentration on the West Coast, especially in California.

There are 6 recognized breeds of dairy goats in the United States, the Nubian, Alpine, Saanen, Toggenburg, Oberhasli, and American LaMancha breeds (Fig. 29-1). Nubians and Alpines are the most popular, comprising more than 50% of goats registered with the American Dairy Goat Association.[8] Average to good dairy goats should produce 900 to 1800 kg (2000 to 4000 lb) of milk in a 305-day lactation. High-producing dairy goats can achieve records in the 2300 to 2700 kg (5000 to 6000 lb) range. There are breed variations in milk production, with Nubians sacrificing some volume for their typically higher milk fat contents (4% to 6%). At the present time, Oberhaslis produce less than the other U.S. dairy breeds.

Dairy goats are larger in general than hair goats or meat (Spanish) goats. Mature does (female goats) of the Nubian, Saanen, and Alpine breeds should weigh at least 60 kg (135 lb) and not uncommonly weigh 70 kg (150 lb) or more. LaManchas, Oberhaslis, and Toggenburgs may be slightly smaller.

Dairy goat operations vary markedly. Most producers have relatively small herds of 20 or fewer animals. Dairy goats may be kept for home milk consumption, for 4-H or other agricultural projects, for breeding and show stock, for companion animals, as pack animals, or for commercial milk production. Unwanted buck kids are slaughtered or sold for meat. As would be expected with U.S. or European dairy animals, most dairy goats are kept in intensive systems where they are in relatively confined areas for at least part of the day and are fed some processed feeds. The dairy goat "industry" is not very commercialized in the United States. Goat milk sells in specialty markets to adults and infants who are allergic to cow milk, have trouble digesting cow milk, or have gastric ulcers and other ill-defined intestinal disorders. The composition of goat milk is different in several aspects from that of cow milk. For example, it has a greater proportion of medium- and short-chained fatty acids, lacks the agglutinating protein that causes fat globule clustering and rapid separation of cream, contains smaller micelles of casein (which result in higher digestibility) and has very little α_{s1}-casein relative to cow milk.[6,13,14] There are other, more minor differences.[7] The production cost of goat milk at retail is two to three times that of cow milk, which will probably always restrict the volume of goat milk sales in

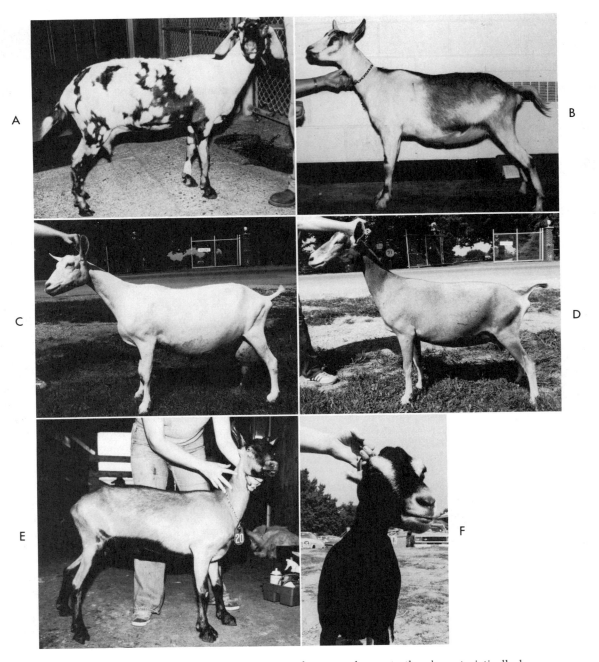

Fig. 29-1 Dairy goats. **A,** Nubian goats can be any color; note the characteristically long ears. **B,** Alpine goats can be any color. **C,** Saanen goat, all white. **D,** Toggenburg goat, chocolate brown with white markings. **E,** Oberhasli goat, brown with black markings. **F,** LaMancha goat. The body can be any color; note the short ears.

the United States. Nevertheless, there is a strong demand for goat cheeses, milk, and other goat products among various ethnic groups; in many areas this demand represents a market exceeding supplies.

Because of their affectionate personalities, dairy goats are companion animals as well as food-producing animals. Therefore, economic considerations do not always dictate management decisions. Purebred breeders are interested in selling breeding stock and want to feed optimally to establish outstanding production records with their animals. Some of them purchase the highest-quality feedstuffs available. However, many commercial producers are interested in optimal nutrition at the lowest cost.

One of the main challenges of working with dairy goats is the variety of backgrounds and goals of the owners, who are often relatively new to agricultural enterprises and do not have strong experience with feedstuffs and ruminant nutrition. Often such owners have difficulty finding a veterinarian or nutritionist who is knowledgeable about dairy goats. Thus, they rely on advice from other dairy goat owners who may have had experiences that have led to conclusions that are less than nutritionally or economically sound.

Angora (hair) goats

There were reported to be more than 2.3 million Angora goats in the United States in 1988[18] (Fig. 29-2). The majority (approximately 2 million) were in Texas, with only one other state (New Mexico) having more than 100,000 animals. Arizona had the third largest number of Angoras and Michigan had the fifth. The Angora industry is small but growing rapidly in Oklahoma (number four), Missouri, Ohio, and Kansas.

Receipts for mohair account for 85% of the income from Angora goats. The average clip per Angora goat for most states in 1988 generally was about 3.5 kg (7.5 to 8 lb) of mohair.[18] Exceptions were New Mexico and Arizona, which reported average clips of 2.5 and 2.1 kg (5.5 and 4.7 lb), respectively. Angora goats are usually shorn twice a year. Differences in mohair production are associated with genetic potential and with varying levels of nutrition. Maintenance of animals on rangelands without supplemental

Fig. 29-2 Angora (hair) goat (white). Compliments of the Mohair Council of America.

feed may result in reduced hair growth primarily because of a lack of adequate levels of protein.

There is great variation in the size of Angora goats because of variable nutritional status during their growing years. Mature does frequently weigh between 40 and 55 kg (90 and 120 lb), but underdeveloped females may mature at under 36 kg (80 lb). For optimum reproductive efficiency, young females should weigh at least 30 kg (60 lb) at first breeding.

Angora goats traditionally are managed extensively on rangeland. Ranchers vary in the extent to which they supplement their Angoras. Angora herds in states outside Texas, New Mexico, and Arizona are generally more intensively raised and have increased mohair yield and improved kid survival. In these situations the goats frequently are kept at least part of the year on improved pastures and are provided some kind of shelter for kidding.

Mohair production varies with the genetic potential of the goat, age, and nutrition. Finer hair (small diameter) is more valuable and comes from young animals and animals under poor nutritional conditions. Although animals under nutritional stress produce finer hair, this factor cannot be abused or the animals will suffer from low mohair production and have poor reproductive efficiency.

The average market price for mohair reached all-time highs from 1978 through 1984, peaking at $11.20/kg ($5.10/lb) in 1979. Prices have declined each year since 1984, the average being $4.16/kg ($1.89/lb) in 1988. Angora producers also receive an incentive payment from the U.S. Department of Agriculture in any year in which the average market price is less than an established support price.[21] Incentive payments for mohair, as for wool, are provided entirely by tariffs placed on imported wool and mohair products and not from domestic taxes. Texas produces 85% to 90% of U.S. mohair, which is approximately 25% to 30% of world production. The vast majority of U.S. mohair is exported and processed in other countries, primarily in Europe. The volatility of market prices is a result of the international nature of the mohair industry.

Most Angora goats are considered to be worth what they could bring at an auction barn. Animal value varies from year to year, depending on the market price for mohair and the age and condition of the goat, but typically ranges between $30 and $50. This usually means that health problems are approached from a whole-herd point of view. Breeding stock is the exception, with young breeding males bringing as much as several hundred dollars or more at production sales or through private treaty.

The efficiency of reproduction in Angoras is poor compared to dairy goats, which commonly produce two or three offspring per kidding. Because of nutritional limitations, vulnerability to climatic extremes, poor mothering instincts, and predators, Angora kid crops weaned are frequently only 50% to 70% or even less. In areas where Angoras are more intensively raised, kidding occurs in sheds. Kids are watched to be sure they nurse and are left with the dam until bonding has occurred. With proper nutrition and supervised kidding, weaning percentages per doe can exceed 100% (i.e., more than one kid per doe).

Cashmere goats are growing rapidly in popularity in the United States. *Cashmere* actually refers to a fine fiber produced by goats, rather than to a specific breed of goat. Any breed of goat may contain individuals that produce cashmere-quality fiber as their natural downy undercoat. Although this fiber is worth considerably more per kg than mohair, even a goat that has been bred for cashmere production typically produces only 0.5 kg or less of cashmere fiber per year. Therefore, from a commercial view of return per animal, at current prices cashmere goats may not be any more profitable than mohair goats.[20] However, the scarcity of animals and the sudden high demand for them has driven up prices for breeding stock. Instead of purchasing purebred animals, some producers are selecting Spanish goats that have a natural downy undercoat and breeding them with imported cashmere semen. This results in an upgrading program that allows harvesting of the extra cashmere commodity from already existing meat goat resources. Time will tell whether U.S. cashmere production will become a viable industry.

Spanish (meat) goats

There were estimated to be 500,000 meat goats in the United States in 1982 (Fig. 29-3), the majority of them in Texas.[3] These goats are often used for brush control as well as for meat production. Goat meat is widely used in the southwestern United States among people of Latin American and Mediterranean extraction. Kid (cabrito) and processed goat meat (sausage or chili meat) are the most popular commodities. Kid is leaner than lamb, so considering today's

Fig. 29-3 Spanish (meat) goats can be any color.

health interests it might have a growing market if properly promoted and distributed.

Spanish goats are typically managed under extensive conditions. Of all types of goats, Spanish goats are the least likely to receive supplemental feed or health care. Animal value varies with the market and type of animal. Large muttons and bucks might bring $50 to $75, breeding does $30 to $60, and kids $5 to $40, depending on size. They are typically more similar to Angora goats than to dairy goats in body size. However, Spanish goats generally are more resilient to factors causing reproductive failure in Angoras. Higher twinning and survival rates result in up to double the kid crop expected for Angoras.

COMMON FEEDSTUFFS
Forages

Although goats have some unique characteristics that should be considered in a detailed discussion of goat nutrition, the general principles of feeding ruminants apply to goats. The most significant unique factor is a goat's instinct to browse more and graze less than sheep or cattle. Because of this tendency, goats that have browse available to them ingest a diet that varies nutritionally from that consumed voluntarily by other domesticated ruminants. Goats select the most nutritious parts of plants, using their mobile upper lip to bite off the youngest, growing parts of a variety of plants. Goats also are extremely agile and can climb or stand on their hind legs to select their diet. Plants consumed include grasses, forbs (weeds), and browse (woody-stemmed plants). Forbs and browse typically exceed grasses in their content of protein, phosphorus, and calcium. When goats select seeds, nuts, and fruits of plants, they are ingesting more highly concentrated sources of digestible energy than are found in grasses alone.

Because of their different eating habits and preferences, goats can complement sheep or cattle under extensive grazing systems. However, if areas become overgrazed, the opportunity for goats to select their diet diminishes, and deficiencies can occur. Seasonal changes in type and quality of available vegetation should also be considered to determine the amount of supplementation that may be needed. For example, Angora goats maintained on open rangeland are commonly supplemented with a high-protein concentrate late in gestation or during lactation. Energy supplementation also may be necessary for does in mid to late gestation, which usually occurs in the winter when good-quality forage may be scarce.

Dairy goats are often kept under circumstances where they have access to minimal or no browse, but instead are confined and fed similarly to dairy cows. Some dairy goats are pastured part of the day, but available pastures vary markedly. Even if browse is available, dairy goats cannot sustain high milk production on browse and pasture alone. Because of the seasonal variation in pasture quantity and quality, many producers use pastures primarily as a source of exercise for their dairy goats instead of as a significant source of nutrition. Hay is by far the most common harvested forage used for U.S. dairy goats (at least during part of the year), and supplementation with concentrates is almost universal.

Good-quality mixed pastures that contain alfalfa or clover and a palatable grass can provide the bulk of a dairy goat's needs. Pastures need to be properly fertilized and clipped to obtain optimal production, however, and should not be overgrazed. Because the quality of available forage varies with season, adjustments in supplemental feeding are necessary for maximal production. Many dairy goat owners do not have the resources to maintain ideal pasture conditions for their goats. Sometimes small overgrazed areas that have become overgrown with unpalatable grass and weeds are considered "pastures" that look adequate yet supply few of the animals' needs. However, the nutritive value of forages and browse available needs to be considered when calculating the overall ration. Lush pastures in spring may contain so much water that a high-producing dairy goat cannot consume enough to meet her needs. Lush alfalfa pastures carry the risk of bloat, and rapidly growing grass or small grain pastures can predispose goats to grass tetany or nitrate poisoning. For these reasons it is beneficial to offer pastured goats free access to hay.

Angora and Spanish goats in Texas commonly are maintained on native rangeland that is ideally suited for browsing animals, provided it has not been overstocked and the variety of available plants has not been reduced by overgrazing. Spanish goats may be superior to Angoras in utilizing pastures containing a large proportion of browse. As the number of goats on a pasture increases, the ability of individuals to select the most nutritious parts of the plants decreases, and the quality of ingested diets declines. Season also affects the quantity and quality of forage available. The feeding value of forage can decrease markedly during the winter.[12] If animal productivity is to be maintained under these conditions, supplementation must be increased. Stocking rates as well as the amount of supplementation may need to be changed markedly from year to year because of variation in rainfall. Generally, six goats are considered equivalent to one cow when calculating stocking rates. Five to ten goats per hectare (two to four goats per acre) may be appropriate for some native brushy pastures as found in parts of southern Oklahoma, compared to 25 goats per hectare (10 goats per acre) of wheat pasture or 37 goats per hectare (15 goats per acre) of alfalfa pasture.[17] In more intensive grazing of small grain pastures, as in some Angora operations, grass tetany may be a problem.

Hay is the most common harvested forage fed to goats. The same considerations of quality apply as to other ruminants. Usually goats favor leafy hays. They are known for wasting hay because they select leaves and seed heads first and then refuse to eat what falls through the feeder to the ground if it becomes contaminated with manure. Goats can be forced to eat poorer-quality parts of hay or to eat off the ground by not offering alternative feeds; however, feed intake and production may decrease drastically. Some goats on a high-grain ration, however, may seem to prefer stemmy hay if it is not moldy. Consumption of good-quality forage is the basis of a successful feeding program, so some feed refusal is tolerated with high-producing dairy goats. It is very important to construct goat feeders so that leafy parts of hay do not fall on the ground and goats cannot stand in the hay and contaminate it with manure.

Silage and haylage are not commonly used for goats in the United States, partly because of the small scale of most goat operations. Goats are safely fed silage in other countries and usually learn to like it after an initial learning period. As with any changes in feeds, introduction to silage should be done gradually.

Root crops and garden produce frequently are fed to goats as treats or as seasonally significant parts of the diet. Goats can adjust to a wide variety of by-products, as can other ruminants. Provision of variety to a goat diet enhances interest in eating and increases overall dry matter intake. Care should be taken to consider the potential of some feeds such as turnips and onions to cause off-flavors in milk. These feeds should be fed to lactating animals immediately after milking to prolong the time between consumption and the subsequent milking.

Concentrates

Cereal grains such as corn, sorghum (milo), oats, barley, and wheat are commonly used components of concentrate mixtures for goats. Grain digestion is enhanced by some type of processing such as flaking, rolling, or cracking, but it is not essential. Grains should not be ground, however, as goats pick out the larger pieces of feed and leave the finer ones.

Concentrate mixtures for goats should contain a protein source such as cottonseed meal, linseed meal, soybean meal, or peanut meal. Pelleted concentrates are ideal because goats may pick out the coarser cereal grains and leave the finer meals in loose grain mixes. Molasses may be included in concentrate mixtures for goats at approximately 5% of the ration to reduce dustiness and improve palatability, but it is not essential. Some goats do not prefer molasses-flavored feeds.

Prepared feeds

Commercially available feeds for dairy cattle, horses, or sheep are commonly used for dairy goats. Use of these feeds relieves goat owners of the need to monitor feed quality as when home grown feeds are mixed with purchased grains and by-products.

There are companies that specialize in supplying nutritional products specifically for dairy goats. Most of those products are supplements, but some are suitable for part or all of a goat's daily nutrient requirements, including roughage from grazing, browsing, hay, silage, straw, leaves, and by-products.

Horse feeds can be very satisfactory for goats as long as the products are not high in molasses. They may be used for backyard goats kept as pets or used for brush control and as a total ration on commercial farms. Little additional feeding is needed except for high producing dairy goats. Pelleted horse feeds (e.g. "maintenance pellets") with high fiber (25% to 28%) and low protein (12% to 14%) contents make an excellent complete goat ration. They are well balanced for average producers, cause no overeating problems, and avoid waste associated with loose hay. The copper content of horse feeds, contrary to popular opinion, is not too high for goats, only for sheep. The calcium content of horse feeds also is acceptable for female goats because of the female's high calcium requirements when milking and growing. Only bucks may be at risk with urinary calculi if the ration has high calcium and phosphorus contents.

Angora goats maintained on rangeland pastures are often fed a high-protein supplement, depending on range conditions. These commercially available supplements are available as pelleted feeds that are hand-fed or as self-limiting meals or blocks. For the self-limiting feed, salt is often included at 10% to 30% of the mix depending on the desired consumption level.[5] A good water source should be readily available, especially if using salt to limit intake.

For neonates, there are commercially available goat milk replacers that produce excellent results and are preferred over other types of milk replacers.* A colostrum replacement for kids is also commercially available.† Lamb milk replacers, which have elevated fat content, need to be diluted to prevent scours in goat kids. Calf milk replacers work well in the hands of some producers, whereas others feel their kids do not thrive on them. Usually kids perform satisfactorily on calf milk replacers if they are not overfed and the fat contents are not too high.

Supplements

Concentrate mixtures frequently contain a source of calcium or phosphorus as needed to balance the forage part of the ration. Supplemental phosphorus and perhaps calcium is needed for grass-fed goats but is not necessary for alfalfa-fed goats or goats grazing brushy pastures.

Trace mineralized salt can be added to balance the ration and is frequently included in commercial feeds. Some dairy goat producers purchase minerals specifically produced for dairy goats and supply them cafeteria-style to their animals.* These may include a number of products such as beet pulp, dried brewer's yeast, carotene, kelp meal, wheat germ oil, and diatomaceous earth, in addition to minerals and vitamins. Buffers such as sodium bicarbonate or magnesium oxide also are used as with dairy cattle to prevent milk fat depression. Buffers in diets that have a large proportion of concentrates are advisable. Although individual producers may recommend certain products or methods of supplementation, there are few data from controlled studies to support the expense of using some of these items. Unfortunately, there are few data from actual goat studies on many aspects of nutrition. Most of the supplements marketed for dairy goats do not appear to be harmful, so use will depend on the producer's economic and production goals. Supplements produced with the dairy goat in mind can provide an appropriate source of minerals for a variety of purposes. In general, a goat ration balanced in a way similar to a dairy cow ration is adequate.

In Angora-producing areas, minerals formulated for Angoras are mixed by some mills. These often are supplemented with sulfur to

---

*Dairy Goat Nutrition, Kansas City, MO 64113; Nutritional Research Associates, Inc., South Whitley, IN 46787; Land O'Lakes, Minneapolis, MN 55413. The authors intend no endorsement by listings in these footnotes.
†Dairy Goat Nutrition, Kansas City, MO 64113.

*Dairy Goat Nutrition, Kansas City, MO 64113; Nutritional Research Associates, Inc., South Whitley, IN 46787.

meet the demand for hair growth. Supplemental sulfur also compensates for impaired absorption associated with the ingestion of the tannin-containing plants common on the southwestern ranges.[16]

FEEDING FOR OPTIMAL PRODUCTION IN DAIRY, MEAT AND HAIR GOATS

There is a National Research Council (NRC) publication on the nutrient requirements of goats that contains tables concerning the specific needs of goats in different phases of production.[16] This publication was based on a review of the world literature concerning goat nutrition. Detailed information on feeding goats is limited in many respects, and more recent research concerns nutrition of goats in tropical regions, where the goat is a comparatively more important species than in the United States. Goats in these regions are often kept under extensive management principally for production of meat. Recommendations for feeding dairy goats in temperate climates often are based on older research with goats or on recommendations for dairy cattle.

In the calculation of nutrient requirements for goat rations, however, goats should not be considered small cows. A dairy goat weighs one tenth of what a dairy cow weighs but has nutrient requirements (and production potential) that are one seventh to one eighth those of the cow.[10] Another aspect of goat nutrition that varies from cow nutrition is the ability of the dairy goat to consume as much as 5% to 8% of body weight in dry matter daily. The caprine digestive tract also appears to vary from that of a bovine or ovine animal in that the retention time of ingested food is less than in the other domesticated ruminants. Low-quality feed may pass through the digestive tract of goats with less digestion than in cows. However, goats perform better than cattle or sheep when they can choose combinations of more highly digestible plant parts. These feeds also move through the digestive system more rapidly in goats and allow room for greater intakes.[11] Although this implies that goats need a higher-quality diet than do sheep or cattle, this is not necessarily the case. Because of their ability to consume a wide variety of plant parts, goats can survive in some areas where sheep or cattle would starve.

When matching nutritional requirements of goats to nutritional supply, there are four critical periods: (1) postweaning body development, (2) breeding period, (3) late gestation (and mid gestation in Angoras), and (4) perinatal period. These periods are especially critical in Angora goats under extensive management systems. These are times when nutritional requirements increase and supplemental nutrients, if needed, help animals reach their production potential. Supplemental nutrition to range goats at these times is worth the expense because of increased reproductive efficiency and decreased kid losses.

Maintenance

The NRC tables begin with a basic recommendation for maintenance. To these requirements are added those associated with activity used to acquire the diet. The energy expended by rangeland goats while browsing can be high. One recent study suggested the maintenance requirements of 2-year-old Angora wethers may be underestimated in the NRC recommendations by 17%.[2]

Maintenance requirements increase dramatically in Angora goats after shearing if changes in temperature occur at this time. "Freeze loss" is an extremely important aspect of Angora production. The risk is greatest when freshly shorn goats are held off feed for some reason and then not given time to fill up before the sudden onset of cold, wet weather. Freshly shorn goats should not be turned out late in the day if there is a risk of a sudden weather change during the night. Some producers shelter their Angoras for 4 to 6 weeks after shearing. Freeze losses can occur after summer shearing in August as easily as after shearing in February. The effects of wind and rain are as important as the temperature itself. Mortality with freeze loss can approach 100%. Although optimal nutrition cannot necessarily overcome the effects of exposure, adequate dietary energy can make a difference under marginal conditions.

Breeding

Flushing refers to provision of an increased plane of nutrition to females beginning several weeks prior to the introduction of males for breeding and continuing for several weeks thereafter. The purpose is to increase the ovulation rate at the time of breeding and decrease the incidence of early embryonic death postbreeding. Dairy goats or other goats in good flesh may not respond to flushing. Range goats in marginal body condition may benefit from being moved to a fresh pasture or being fed ¼ to ½ kg (½ to 1 lb) of corn or other energy source per head per day during this period of time.

Gestation

Requirements for early pregnancy are minimal, and no special nutritional allowances for this period are made. Additional requirements for late pregnancy are important, due to the rapid growth of the fetus(es) during the last 6 weeks of gestation. Undernutrition during this time period can result in pregnancy toxemia in dairy goats. Chronic undernutrition during mid gestation (days 90 to 110) leads to stress abortions in Angora goats, especially young does. Underfed females also have weaker offspring and are not as good at mothering, resulting in increased neonatal losses. Overfeeding prior to late gestation is not uncommon in dairy goats; it can be as detrimental as underfeeding by predisposing to fatty livers and pregnancy toxemia.

Pregnant dairy goats should be evaluated 2 months prior to their due date. If they are thin, they should be offered ad libitum forage and approximately 0.5 kg (1 lb) of concentrate per day. If they are fat, they should be offered only forage, with zero to ¹⁄₁₀ kg (zero to ⅕ lb) of concentrate. As they enter the last 6 weeks of gestation, the amount of concentrate offered should increase gradually up to ¾ kg (1.5 lb) per doe per day. Forage of very high quality should also be offered to encourage dry matter intake and stimulate rumen function and size. Dry matter intake becomes progressively reduced in late gestation in overconditioned dairy goats that may carry twins or triplets in addition to excessive abdominal fat that compete with the rumen for space. This makes the digestibility of feeds critical. There should not be an attempt to compensate for poor hay quality by merely increasing the amount of concentrate, as this can predispose to acidosis, especially in fat goats. Also, the rumens of goats that are fed too much concentrate in the dry period do not function well during early lactation. Provision of a high-quality forage during late gestation is critical for optimal production in dairy goats.

The body condition of goats can be evaluated in the same manner as for sheep or cows; however, lower body scores than in dairy cows are desirable. Goats have a tendency to lay down internal fat and to appear thinner than other domesticated ruminants when they are actually in good body condition.

Angora and other goats raised on extensive systems usually raise their own kids. Angoras are notoriously poor mothers that abandon their kids if frightened or disturbed at kidding time. Proper feeding of pregnant does helps to produce strong kids that get up and nurse soon after birth. Angoras that kid in good body condition also produce more milk during lactation. Kid losses due to starvation and predation are two reasons why weaning rates are so low in many Angora herds.

Lactation

General principles of feeding lactating dairy goats are similar to those of feeding dairy cows. High-producing goats are in a negative energy balance in early lactation and need to be fed high-quality forages and concentrates. Changes in amounts of concentrates need to be made gradually to allow for adaptation of ruminal microbes. The quantity of concentrates fed varies with the quality of forage and production level of the goat. The NRC tables provide an allowance for each kg of milk at various percentages of fat. Dairy goats in peak lactation are encouraged to consume a maximum amount of dry matter. Even then, they lose weight from using body reserves to produce milk. Once milk production begins to decline and they are in a positive energy balance later in lactation, the level of concentrate supplementation should be decreased as necessary to prevent accumulation of body fat. It should be realized, however, that goats at this time must be fed not only to produce milk but also to regain weight lost during early lactation, assuming they were not too fat at kidding.

Feeding high levels of concentrates to dairy

goats causes acidosis just as in dairy cows. Feeding buffers such as sodium bicarbonate aids in controlling this condition.

Growth

Newborn kids need to suckle colostrum as soon after birth as is practical. Dairy goat kids are often removed from the dam at birth and hand-fed colostrum. The colostrum may be heat-treated (1 hour at 55° C [131° F]) if a control program for caprine arthritis-encephalitis (CAE) is used. About 125 ml (4 oz) of colostrum are adequate for a first feeding. During the first few days of life, frequent feedings (four to six times daily) are often used to raise small, weak kids. Tube feeding 30 to 60 g (1 to 2 oz) of colostrum every 2 hours helps to stimulate kids that are too weak to stand and nurse on their own. Total intake of colostrum per day for the first 2 to 3 days should be approximately 250 ml (8 oz) for small kids weighing 1.5 to 2.0 kg (3 to 4 lb), and up to 600 ml (20 oz) for the larger kids weighing 3.5 to 4.5 kg (8 to 10 lb). Larger volumes should be divided into multiple small feedings and may cause scours.

After the first 24 to 48 hours, kids can be switched from colostrum to goat milk, cow milk, or a commercially available milk replacer. Milk replacers made for goat kids are available and are preferred to calf or lamb milk replacers. Lamb milk replacers are too high in fat and can result in scouring if not diluted. Milk replacers for kids should contain 16% to 24% fat and 20% to 28% milk-based protein on a dry matter basis.[15] Young kids can be switched to three and then two times per day feedings by the third week of age or earlier, if necessary. The rate at which kids are switched to twice daily feeding varies with management goals. Multiple small feedings are best to reduce the incidence of bloat but may not be practical on many farms. Daily intake should gradually increase to 1.1 to 1.4 L (1.1 to 1.5 qt) of milk by approximately 2 weeks of age.[4] Early consumption of forage and grain must be encouraged. Consumption of forage is essential to develop volume, muscular tone, and absorbing papillae of the rumen. Therefore, attention should be paid to what forage kids like to eat so they can be provided with plenty of that type of feed. Provision of fresh cut succulent plants may enhance forage intake.

Kids can be weaned early (4 weeks of age) but weaning at 6 to 10 weeks is more typical, with some producers feeding milk up to 3 months of age. Kids have less weaning shock if they weigh 9.0 to 11 kg (20 to 25 lb) at weaning. Kids can be weaned gradually by reducing the volume of milk fed, decreasing the number of daily feedings, and by watering down the milk. Reduction in milk consumed can begin at 3 to 4 weeks of age.

Weaning is a shock to kids, and growth frequently is reduced or stopped at this time. The effect is more pronounced in young kids (less than 1 month) and in kids that were on ad libitum milk and therefore were ingesting less solid feed. This is why young kids should be offered high-quality hay and a high-protein concentrate soon after birth. Kids suffer minimal weaning shock if they are consuming ¼ to ½ kg (½ to 1 lb) of a 16% protein concentrate daily in addition to available forage.

Growth during the first month depends to a large degree on the birth weight of the kid and weight gain during the first week.[15] Small kids tend to have a lower growth rate. Growth is correlated with dry matter intake and especially with energy intake. Ad libitum feeding of milk improves weight gain but can result in increased storage of fat.

Daily weight gains of dairy goat kids in French experiments were almost constant during the first 12 weeks of life at approximately 170 g (6 oz) per day (excluding the weaning period at 5 to 6 weeks of age).[15] From weeks 12 to 30, average daily gain gradually decreased to 75 g (2.5 oz).

After 3 months of age, growth slows down. Growth rates are affected by the amount of energy ingested and therefore are maximized by feeding some concentrates. Dairy doelings are frequently bred the first breeding season after they are born in order to have them kid at approximately 1 year of age. Pregnancy during the first year of life slows the growth rate during that year. If adequate nutrition is provided, however, the doeling will eventually reach full adult size.

Angora kids raised on rangeland after weaning often are not large enough (25 to 30 kg or

60 to 65 lb) to breed until the second breeding season after they are born (18 months of age). Lifetime productivity is correlated closely with body size at first breeding.[19] Female Angoras should approximately double in size during the first 12 months after weaning. If creep-fed prior to weaning, they can grow fast enough to breed their first year and carry kids to term. Creep feeds for Angora kids include a high-quality forage and a 15% to 16% crude protein concentrate mix. During this period of growth, young does should be protected from internal parasites, including coccidia. Addition of a coccidiostat to the feed is a good idea if kids congregate in small areas. Monensin at 20 g/tonne (20 ppm) or decoquinate at 0.5 mg/kg (22.7 mg/100 lb) body weight are both approved for goats to control coccidiosis.

Angora and Spanish kids often are left to nurse their mothers until 4 to 6 months of age. Leaving them with their dams beyond 4 months of age is of little nutritional benefit, as Angora does produce little milk after 100 days of lactation. If male kids have not been castrated, they can be fertile and breed by this time. Therefore, sexes should be separated by 4 months of age. However, many Angora kids are not weaned and separated until 6 months of age for practical reasons.

Fiber production

Angora goats have been bred for many years to produce hair fiber at the expense of other physiological functions such as growth and reproduction. Hair growth is a nutrient drain, especially for protein. Increasing dietary protein up to as much as 23% has been shown to enhance mohair production but also increases fiber diameter.[1] The gain in overall production and health of the animals usually compensates for the slightly reduced value of coarser hair. However, the economics of feeding a high level of protein must be considered. Sulfur, an important component of cystine and methionine, which are necessary for hair growth, is frequently supplemented to Angora goats as well.

Ultimately, feeds and feeding levels are economic choices that vary with availability, price, and alternatives. Table 29-1 is an example of alternatives for Angora goats on typical Texas

Table 29-1 *Example range supplements and feeding levels for various classes of angora goats*

	Mixes			Corn	Cottonseed meal
	1	2	3		
Example feeds (%)					
Ingredients					
Cottonseed meal	75	50	25	—	100
Corn	22	46	70	97	—
Dicalcium phosphate	—	1	2	3	—
Molasses	3	3	3	—	—
	100	100	100	100	100
Crude protein %	33	25	16	9	42
Feeding levels (kg/day)					
Kids and yearlings					
Summer	0.18	0.24	0.34	0.67	0.19
Fall/winter	0.34	0.45	0.64	1.25	0.35
Developing billies					
Summer	0.24	0.23	0.32	0.63	0.25
Fall/winter	0.59	0.57	0.81	1.58	0.62
Breeding does (Sept 1-Nov 1)	0.24	0.23	0.32	0.63	0.25
Pregnant does (Dec 1-Mar 15)	0.4	0.39	0.36	0.71	0.42

Table 29-2 *Sample protein supplement for goats (percent of supplement)*

	Total protein content		
	14%	16%	18%
Corn	37	35	32
Oats	37	35	32
Wheat bran	16	14	15
Oil meal (soybean, etc.)	9	15	20
Dicalcium phosphate	0.5	0.5	0.5
Trace mineral salt	0.5	0.5	0.5

rangelands. Table 29-2 shows the composition of a simple grain mix made to contain different percentages of protein.

REFERENCES

1. Calhoun MC and Lupton CJ: Presentation at Sheep and Goat Field Day, Texas A&M Research and Extension Center, San Angelo, 1987.
2. Calhoun MC and others: Texas Agricul Exp Sta, PR-4589, 1988.
3. Council for Agricultural Science and Technology, Report 94, 1982.
4. Furber H: Dairy goat production, Guelph, Ontario, 1985, Independent Study Division, University of Guelph.
5. Gray JA and Groff JL: Texas Agricul Exp Sta Bull B-926.
6. Haenlein GFW: Int Goat and Sheep Res 1:173, 1980.
7. Haenlein GFW: J Dairy Sci 63:1729, 1980.
8. Haenlein GFW: J Dairy Sci 64:1288, 1981.
9. Haenlein GFW: Proceedings of the fourth international conference on goats, Brasilia, 1987.
10. Haenlein GFW and Devendra C: International Foundation for Science provisional report 14, Workshop on small ruminant research in the tropics, Ethiopia, 1983.
11. Huston JE: J Dairy Sci 61:988, 1978.
12. Huston JE, Shelton M, and Ellis WC: Texas Agricul Exp Sta Bull B-1105, 1971.
13. Jenness R: J Dairy Sci 63:1605, 1980.
14. Maree HP: Dairy Goat J 56(5):62, 1978.
15. Morand-Fehr P: In: C Gall, editor: Goat Production, London, 1981, Academic Press.
16. National Research Council: Nutrient requirements of goats: Angora, dairy, and meat goats in temperate and tropical countries, Washington, DC, 1981, National Academy of Science.
17. Pinkerton F: Proceedings of the 4th Annual Field Day, American Institute for Goat Research, 1989.
18. Ranch Magazine 70(9):53, 1989.
19. Shelton M: Texas Agricul Exp Sta Bull MP-496, 1961.
20. Shelton M and Lupton CJ: Proceedings of the 4th Annual Field Day, American Institute for Goat Research, 1989.
21. USDA Agricultural Stability and Conservation Service: Mohair Commodity Fact Sheet, June 1988.

Common Nutritional Problems
Feeding the Sick Goat

KATHERINE BRETZLAFF, GEORGE HAENLEIN, ED HUSTON

NUTRITIONAL DISEASES
Deficiencies

Energy. Severe energy limitations at the time of breeding result in reduced conception rates (failure to cycle) and ovulation rates. In Angora goats, energy deficiencies are common and result in mid-term stress abortions in young does. These can be minimized by not breeding young goats until they are approximately two thirds of their expected adult body weight. In the Angora, uniformity (continuity) of dietary intake appears to be of equal importance to the level of nutrition. Mid-term abortions frequently occur following an event that disrupts normal feeding behavior or patterns.

The most common disease associated with energy deficiency in dairy goats is pregnancy toxemia. This occurs in goats that are in late gestation and usually carrying multiple fetuses. Goat fetuses achieve 80% of their growth during the last 6 weeks of gestation and put a tremendous demand for glucose on the dam. If ingested energy is not sufficient to maintain plasma glucose levels, body fat reserves are mobilized, resulting in ketosis. Very thin and very fat goats are predisposed to the condition. It can be triggered by a sudden decrease in feed intake during late pregnancy or can be due to chronic malnutrition. The primary clinical signs are of central nervous system dysfunction, including dullness, depression, blindness, and eventually recumbency, coma, and death. Once clinical signs are evident, the chances of successful treatment are reduced. Early cases can be treated with oral propylene glycol (175 to 250 ml [6 to 8 oz]) twice daily for 2 days and, if necessary, intravenous glucose (50 ml of 50% dextrose slowly).

Prevention of pregnancy toxemia is much preferred to treatment. Goats should not be fat as they approach late gestation; however, it is too late to put a fat goat on a diet at dry off. They should be started on a small amount of concentrate 6 weeks prior to kidding, with the amount gradually increasing to 0.5 to 0.7 kg (1 to 1.5 lb) daily by the time of kidding, regardless of body condition.

Fiber. As in dairy cows, reduced fiber levels in the diet of lactating dairy goats can lead to milk fat depression. The goat diet should contain at least 16% crude fiber, and forage should not be finely ground. Pelleted roughages may not contain adequate particle length. Goats on lush, growing pasture with a low fiber content should be offered some hay prior to grazing. Goats on high-concentrate diets may not eat enough forage to prevent milk fat depression. In this case, the addition of a buffer reduces the depression. Levels recommended for dairy cows are 1% of the grain ration as sodium bicarbonate, sodium sesquicarbonate, or magnesium oxide, or sodium bentonite at 5%.[11] Some dairy goat producers use free-choice feeding of bicarbonates.

A commercially available buffer for dairy goats can be top-dressed, mixed into the grain, or offered free choice.*

Calcium. Hypocalcemia or milk fever is not as common in dairy goats as it is in dairy cows. Dairy goats with hypocalcemia usually do not show the recumbency typical of dairy cows. More typically the does are somewhat off feed and off milk production and may show signs of smooth muscle stasis such as bloat, constipation, or weak labor at kidding. Slow intravenous treatment with 50 to 100 ml of 23% calcium borogluconate usually provides a noticeable response within an hour. Prevention guidelines are similar to those in dairy cattle and include reduction of calcium intake prior to kidding and supplementation with vitamin D. Reduction of calcium intake is primarily accomplished by reducing or eliminating the consumption of alfalfa hay. Recognition of the association of hypocalcemia with a fixed cation excess in the diet of:

$$mEq\ (Na + K - Cl)/100\ g\ diet\ dry\ matter > 50$$

rather than simply a consequence of an excess of calcium prepartum may lead to more sophisticated management of dairy goats to prevent this condition in problem herds.[3]

Magnesium. As in other ruminants, hypomagnesemia (grass tetany) can occur in goats grazing early, lush pastures, especially those heavily fertilized with potassium and nitrogen. It has been reported in Angora goats grazing on winter wheat or other cereal grain pastures. Clinical signs are accelerated breathing, possibly hyperexcitability, and tremors followed by convulsions, coma, and death. Treatment is by intravenous administration of calcium preparations which contain magnesium. Prevention includes feeding hay before turning animals out on pasture and limiting the amount of time on pasture until the animals become accustomed. Feeding of a free-choice mineral mix containing 8% magnesium prior to and during the time animals are on lush pastures is also helpful. An example of such a mineral mix is 400 kg calcium carbonate, 150 kg magnesium oxide, 50 kg dehydrated molasses, and 200 kg trace mineral salt, added to 200 kg ground corn or other concentrate (total 1000 kg).[10]

Sulfur. The most relevant sign of sulfur deficiency is reduced mohair production by Angora goats. The sulfur-containing amino acids cysteine and methionine are present in high proportion in mohair. Sulfur requirements are increased by the consumption of tannic acid–containing plants (such as shin oak) or with diets containing nonprotein nitrogen (urea). Sodium sulfate and ammonium sulfate are used most commonly for ration formulation at 0.16% to 0.32% of the diet.[12]

Selenium. White muscle disease is a commonly recognized syndrome associated with selenium deficiency in kids. Young kids may become progressively weaker yet continue to drink milk even if recumbent. Alternatively, kids may show stiffness after exercise or even a sudden death syndrome. At necropsy, cardiac or skeletal muscles may show the white "chicken meat" coloration. In adults, selenium deficiency can produce rather obscure infertility problems (early embryonic death and return to estrus) or postpartum problems with retained fetal membranes and failure of uterine involution leading to metritis.

Treatment and prevention of selenium deficiency require injection or oral administration of selenium. Because of the association between selenium and vitamin E, the administration of both substances is routine. Injection of does with a commercially available selenium–vitamin E preparation 30 days prior to kidding can be used as a preventive measure, but inclusion of 0.20 ppm selenium in the ration allows a more consistent delivery of the mineral.[8] In deficient areas, higher levels of supplementation may be necessary. Blood selenium levels in a representative sampling of animals should be done. The maximum tolerable level of selenium in the feed for dairy cattle is 2 ppm.[11] There are no data on goats, but it is assumed that levels safe for cattle or sheep are acceptable for goats.

Iron. Kids have minimal stores of iron at birth, and goat milk has a low iron content. Therefore, if anemia is diagnosed as a problem, kids can be injected with 150 mg iron dextran at 2- to 3-week intervals. Caution is advised as

---

*Dairy Goat Nutrition, Kansas City, MO 64113.

some injectable iron products are toxic to large animals. For animals consuming solid feed, ferrous sulfate or ferric citrate may be added at 0.03% of the diet.[6] Iron deficiency is rarely a problem in mature animals.[12]

Iodine. Goiter is overdiagnosed in dairy goats, especially in young Nubians, which normally have a large thymus gland that is often mistaken to be enlarged thyroids. If these kids are otherwise healthy, they do not need to be treated for iodine deficiency. The enlarged thymus will become less noticeable as the animal matures.

Iodine deficiency is primarily diagnosed by the birth of stillborn or weak kids that have true goiters and are hairless or have short, fuzzy hair. Affected kids can be treated with 20 mg potassium iodine orally or 3 to 5 drops of Lugols Iodine in the milk daily for a week.[9] Does in late pregnancy can be drenched with 280 mg potassium iodide.[9] Iodine deficiency can easily be prevented by provision of iodized salt (0.007% to 0.01% iodine).[5,8] Goitrogenic feeds (e.g., cabbage, soybeans, turnips, kale, and rape) can increase the dietary requirement for iodine. Some commercial dairy goat products contain kelp as a source of iodine. The advantages of these compounds over iodized salt have not been demonstrated conclusively.

Zinc. Parakeratosis with hair loss on the head and limbs, unthriftiness, lameness due to fissures on the feet, small testicles, and reduced libido can be associated with zinc deficiency.[15] Males seem to be more sensitive to marginal dietary levels of zinc. Minimum levels (10 ppm) in the ration should be provided to prevent zinc deficiency.[6] Levels as high as 40 ppm are recommended for dairy cattle.[11] Diagnosis is enhanced by observation of response within 2 weeks to oral supplementation with 1 g zinc sulfate daily.[15]

Copper and molybdenum. Copper, molybdenum, and sulfur metabolism interact in goats as in other species. Addition of copper to the diet alleviates signs of molybdenum toxicity, whereas copper deficiency is enhanced by excessive intakes of molybdenum, sulfur, or cadmium.[6,10] Copper deficiency is reported to be likely when pasture diets contain less than 5 ppm copper, when molybdenum exceeds 1 ppm, or when sulfur exceeds 2000 ppm.[9] Clinical signs include unthriftiness, chronic diarrhea, and microcytic anemia in adults; enzootic ataxia may be seen in young kids but is less commonly reported than in lambs. Copper deficiencies are most easily prevented by the addition of copper to the diet (0.5% copper sulfate in the mineral mix).[5]

Thiamine. Polioencephalomalacia (cerebrocortical necrosis) is associated with consumption of feeds containing thiamine antagonists or thiaminase activity. It is also commonly associated with large amounts of grain in the diet. Animals show central nervous system signs, including opisthotonus and paddling, and are blind. Early treatment with intravenous (5 to 10 mg per kg) followed by intramuscular (5 to 10 mg per kg every 6 hours) thiamine hydrochloride can produce dramatic results. Blindness may persist but shows gradual improvement over a period of several days. Animals being fed amprolium for prevention of coccidiosis have been reported to be more susceptible to polioencephalomalacia. Thiamine should be supplemented in the diets of these animals. Otherwise, prevention involves removing the source of the offending feed, whether it be moldy feed or plants that contain antithiamine activity (bracken fern).

Toxicities (excesses)

Phosphorus. Urinary calculi occur primarily in young wethers but can occur in intact males. The condition often is associated with feeding of a high-concentrate diet with a 1:1 calcium-phosphorus ratio instead of the recommended 2:1 ratio.[10] Urinary calculi are a combination of calcium, magnesium, and ammonium phosphates under these conditions. Less commonly, depending on the region and the types of plants being grazed, calculi may be of the silicate ($SiO_2 \cdot H_2O$), calcite ($CaCO_3$) or weddellite ($CaC_2O_4 \cdot 2H_2O$) type. Clinical signs consist of straining to urinate, dribbling urine, and kicking the abdomen. The calculi most frequently lodge in the urethral process or at the level of the sigmoid flexure.

Treatment may require removal of the urethral process, urethrotomy, or urethrostomy.

Acidification of the urine is frequently attempted by oral dosing of 10 g of ammonium chloride (for a 30 kg or 65 lb kid) daily for a week in an attempt to dissolve the additional calculi that are frequently present.

Prevention involves reduction of the percentage of concentrates in the diet if possible or provision of at least a 2:1 calcium-phosphorus ratio in the total ration. Increasing the sodium chloride content of the ration to 4% to 5% may promote water intake and diuresis. Although salt-stimulated polydipsia might be assumed to prevent precipitation by reduction of the concentration of the ions in question, the benefit of salt in the diet may actually be due to an ionic effect rather than a dilutional effect.[10] Addition of ammonium chloride at 2% of the concentrate ration to acidify the urine also may be beneficial. A clean supply of water, preferably warmed to some extent in cold climates, should be available.

Grazing of subterranean clover has been reported to predispose to formation of calcite calculi because of the high calcium and protein content.[9] This may explain the predisposition to urinary calculi of some dairy bucks that are fed high-quality alfalfa. In general, it is advisable not to feed bucks any more alfalfa or concentrate than necessary to maintain good body condition. Good-quality grass hays are preferred when possible.

Copper. Goats do not appear to be as sensitive to copper toxicity as sheep but can be adversely affected if oversupplemented, exposed to copper sulfate footbaths, fed rations with high levels (200 ppm) of copper intended for swine, or given access to plants causing liver damage (*Heliotropium* spp.).[1] Acute toxicity results in gastroenteritis with blue-green diarrhea. Chronic toxicity results in a hemolytic crisis during a period of stress, with hemoglobinuria and jaundice. There is little information on this condition in goats, with treatment protocols generally following those used for sheep. These include the use of D-penicillamine (52 mg/kg or 20 mg/lb body weight for 6 days) to enhance urinary excretion of copper[10] and blood transfusions. Prevention requires consideration of molybdenum and sulfate levels in the ration with the molybdenum-copper ratio not to drop below 1:10.[10]

Other preventive measures used for sheep would probably be useful in goats.[10]

Grain. High-concentrate feeding predisposes goats to a number of problems, including urethral calculi, acidosis, and enterotoxemia. These conditions are primarily a problem in dairy goats or young goats of other types that are being fed to enhance growth. In enterotoxemia (overeating disease), sudden changes in feed change the environment in the small intestine, allowing an overgrowth of *Clostridium perfringens* and the release of the prototoxin, which is converted to the epsilon toxin by trypsin.[10] The toxin is absorbed, resulting in the characteristic central nervous system signs and sudden death. Prevention includes vaccination as well as management to ensure that goats do not have access to abrupt changes in the type, quality, or amount of feed. Any changes in goat rations should occur gradually over a 2-week period and forage-grain ratios should not be less than 40:60. Feeding sunflower seeds has been proposed as a way to introduce more fiber into the diet of dairy goats without sacrificing energy, mineral, and protein. This in turn could reduce the likelihood of enterotoxemia.[7]

Acidosis can occur in lactating dairy goats and in kids that are consuming a high proportion of concentrates in their diet. The syndrome is similar to that in dairy cows and can be controlled by the use of buffers supplied in the ration or free choice. A buffer for dairy goats is commercially available.*

FEEDING THE SICK GOAT
Neonates

Young kids that are born weak and do not try to get up to nurse within 2 hours should be assisted. The teat plug from the dam's teats should be cleared, and the kid should be assisted to nurse. If the kid is to receive heat-treated colostrum, it should be offered from a bottle with a hole in the nipple that is small enough to prevent the kid from aspirating milk. If the dam's colostrum is not available, the kid can be fed colostrum that was saved from another doe, frozen, and then thawed; cow colostrum; or a com-

---

*Dairy Goat Nutrition, Kansas City, MO 64113. No endorsement intended.

mercially available colostrum substitute.* As a last resort, a home-made colostrum substitute can be made from 0.75 L (26 oz) of goat milk or canned milk diluted in half with water to which has been added 2.4 ml (½ teaspoon) of cod liver oil or castor oil, 15 ml (1 tablespoon) glucose or honey, and 1 beaten egg yolk.[4,13] This emergency formula provides minimal amounts of immunoglobulins. If it is used, extra precautions should be taken to isolate the kid from potential sources of infection. Kids that refuse to suckle can be tubed with a no. 14 French catheter that has been marked for the desired length (nose to the last rib) to be passed. Proper placement of the feeding tube can be checked by palpating for the tube next to the trachea and trying to aspirate air from the tube. If properly placed, no air should be aspirated.

Kids with diarrhea are managed with the same principles as calves with scours. The underlying causes (e.g., too much milk, viral or bacterial infection, parasitism, or change in feed) should be determined and measures taken to correct them. Dehydration should be corrected with electrolytes, and acidosis countered with bicarbonate. Dehydrated kids should receive a balanced electrolyte solution, 20 to 40 ml/kg body weight if mildly dehydrated (3% to 5%), 40 to 80 ml/kg if moderately dehydrated (5% to 8%), or 60 to 120 ml/kg if severely dehydrated (8% to 12%).[2] Additional bicarbonate may be needed with severe diarrhea (2 to 4 mmol (mEq)/kg body weight, 4 to 6 mmol (mEq)/kg body weight, or 6 to 8 mmol (mEq)/kg body weight if mildly, moderately, or severely dehydrated, respectively). Oral administration is easiest if the kid is not severely dehydrated.

Adults

Adult goats have the same problem as other ruminants when they have been off feed and rumen microbial populations have changed. It is important to identify the cause of the loss of appetite as well as correct any temporary deficits that have occurred. Correction of dehydration and acidosis can be approached as in other ruminant species. Oral administration of fluids or medications by stomach tube is feasible in goats, although dairy goats that are not extremely ill may resist the intubation procedure. Transfaunation from the rumen of another goat may be helpful for a goat that has been off feed. The use of microbial culture products to reinoculate the rumen may be useful, although it has not been demonstrated conclusively in goats.

Goats that are marginally ill can sometimes be tempted to eat by offering them some freshly cut browse or other succulents.

Angora goats raised on rangelands should not be handled during late gestation because of their susceptibility to stress abortions. Medications needed at this time are best administered in the feed or water if at all possible.

REFERENCES

1. Baxendell SA: The diagnosis of the diseases of goats, Vade Mecum Series B, No. 9, Sydney, Australia, 1988, University of Sydney Post-Graduate Foundation in Veterinary Science.
2. Blackwell TE: Vet Clin North Am 5(3):557, 1983.
3. Fredeen AH, DePeters EJ, and Baldwin RL: J Anim Sci 66:159, 1988.
4. Furber H: Dairy goat production, Guelph, Ontario, 1985, Independent Study Division, University of Guelph.
5. Guss S: Management and diseases of dairy goats, Scottsdale, Ariz, 1977, Dairy Goat Journal Publishing Corp.
6. Haenlein GFW: J Dairy Sci 63:1729, 1980.
7. Haenlein GFW: Dairy Goat J 60(8):22, 1982.
8. Haenlein GFW: Proceedings of the fourth international conference on goats, Brasilia, 1987.
9. Howe PA: Diseases of goats, Vade Mecum Series, No. 5, Sydney, Australia, 1984, Univeristy of Sydney Post-Graduate Foundation in Veterinary Science.
10. Kimberling CV: Jensen and Swift's Diseases of Sheep, ed 3, Philadelphia, 1988, Lea & Febiger.
11. Linn JG and others: University of Illinois Extension Circular M-1183, 1988.
12. National Research Council: Nutrient requirements of goats: Angora, dairy, and meat goats in temperate and tropical countries. Washington, DC, 1981, National Academy of Science.
13. Simmons P: Raising sheep the modern way, Charlotte, VT, 1983, Garden Way Publishing.
14. Smith BP: Large animal internal medicine, St Louis, 1990, Mosby–Year Book.
15. Smith MC: Vet Clin North Am 5(3):449, 1983.

---

*Dairy Goat Nutrition, Kansas City, MO 64113. No endorsement intended.

SECTION D Llamas

31

The Llama Industry: Feeding Systems and Special Feeding Requirements

LYNN R. HOVDA, SUSAN SHAFTOE

THE LLAMA INDUSTRY

The llama industry is one of the most diverse of all the large animal species. Owners range from backyard hobby farmers with little or no experience to those accustomed to feeding large groups of animals.

Only recently have the nutritional demands of llamas begun to be accurately evaluated.[7,20,24] Prior to these studies, the majority of information on llama nutrition was based on personal observations.

FEEDING SYSTEMS

Llamas can be raised either on pasture or in confinement.[2,5,13] There is no recommended number of llamas per hectare of land, although 7 to 9 llamas per hectare (3 to 4 per acre) seems to be the average.[12,13] This number varies widely according to the amount of rainfall and nature of the pasture. Depending on the severity of the winter and quality of the pasture, hay and supplemental grain may be required.

Many different confinement systems can be utilized. In large operations, llamas are often grouped according to sex, size, body condition, and stage of gestation.[7,9] Smaller operations may separate only according to sex.[13] Regardless of the system used, llamas should be fed from feeders to decrease the ingestion of parasites.[3] A variety of feeding devices, including bunks, racks, and stall feeders, can be used. Bunk feeders have the advantages of mobility, decreased waste, and ease of cleaning; they are suitable for both hay and grain.[3] Although most llamas are fed twice a day, once-a-day feeding is a successful alternative.[19]

All llamas should have access to fresh, free-choice water.[5,13,19] Automatic waterers, stock tanks, and natural sources such as streams, ponds, and lakes are acceptable means of providing water. When a natural water source is used, it should be examined frequently for contamination. In northern climates, heaters may be required in the winter to prevent the water from freezing. A normal, healthy llama will drink about 5% to 8% of body weight each day.[19,20] This amount varies, depending on stage of gestation, activity level, environmental temperature, and moisture present in feedstuffs.[5]

Mineral supplements should be offered in elevated off-ground feeders and protected from the weather.[3] A loose mixture is preferable, as llamas do not use salt blocks as well as other animals.[13,20] Alternatively, a pelletized mixture may be added to a small amount of grain daily or every other day.[13] This method may be better as it assures that each llama has received the correct amount of a mineral mixture.

ANATOMY OF THE GASTROINTESTINAL TRACT

Mature llamas are modified ruminants and are anatomically different from true ruminants.[19,26] As opposed to the four-compartment stomach in ruminants, the llama has only three compartments.[7,20,26] The function of these compartments, however, is similar to those of the true ruminant.[7,20,26]

The first two compartments (C1 and C2) comprise approximately 90% of the total stomach volume and function primarily as mixing units and fermentation vats.[20] Smaller individual saccules (approximately 2 to 3 cm in diameter) are also present in both C1 and C2.[20,26] These saccules appear to be unique to the llama stomach and function both as miniature mixing units and as a source of bicarbonate secretion.[7,26] Stratified squamous epithelium covers the surface area of C1 and C2, and glandular mucus-secreting epithelium lines the saccules.[20,26]

The third compartment (C3) makes up the remaining 10% of total stomach volume.[20] Gastric and pyloric glands are present in the terminal portion of C3 and appear to be histologically similar to the abomasum of true ruminants.[20,26] Secretion of digestive fluids such as hydrochloric acid is thought to be the primary function of the third compartment.[7] Glandular epithelium lines the surface area of C3.[20,26]

The motility of the llama stomach is different from that of the true ruminant.[27] The stomach of a healthy, ruminating llama contracts from five to eight times per minute, followed by a pause of 1 minute.[7] During feeding, the rate increases, the pause disappears, and the rhythm becomes more regular.[7]

The contraction pattern of C1 and C2 is such that the second compartment contracts first.[7] This is followed by the caudal sac of C1 contracting in a cranial direction and then the cranial sac in a caudal direction. The saccules present in the walls of C1 and C2 empty their contents into the main compartment during contraction.[7,26]

With few exceptions, the remainder of the gastrointestinal tract does not differ markedly from that of the true ruminant. The liver is positioned entirely on the right side and has a fimbriated caudal border.[7,8] No gallbladder is present.[7,8] The diameter of the spiral colon decreases distally and is a possible site of obstruction.[9]

PROTEIN, ENERGY, AND FIBER REQUIREMENTS

As with all species, protein and its breakdown products, amino acids, are required for growth, cellular repair, and enzyme production. Protein fed in excess of nutritional demands is not stored in the body, but rather converted to glucose, other energy substrates, and ammonia. The ammonia is converted to urea by hepatocytes and excreted by the kidneys. Excess protein is unlikely to be a problem in healthy llamas, but would be detrimental to llamas with compromised renal function and potentially could contribute to hepatoencephalopathy in the presence of severe liver damage.[20]

It appears that healthy, normal llamas can be maintained on a diet containing 10% protein.[17,18] Growing, undernourished, working, and lactating llamas, as well as those in late gestation, do well on a 12% to 14% protein diet.[18] Creep feed offered to suckling llamas should begin at 16% protein and then slowly decrease to 12%.[5,20]

The digestible energy requirement in kcal for maintenance in adult llamas is $75 \times W$ in $kg^{0.75}$.[20] Growth, cold weather, gestation, and lactation further increase energy requirements. It is interesting to note that the maintenance energy requirements in kcal for sheep are $119 \times W$ in $kg^{0.75}$.[20] If llamas are fed at the same nutritional plane as sheep, they deposit the excess energy as fat.

The dry matter intake of llamas is currently estimated to be 2% of the body weight per day.[17,20] Llamas are much more efficient than sheep in utilizing poor quality feedstuffs. This may be attributed to prolonged retention of particulate matter in the gastrointestinal tract.[24]

Llamas appear to require a minimum of 25% fiber in their diet.[20] The provision of inadequate fiber has been associated with rumen atony and the development of gastric and duodenal ulcers.

COMMONLY USED FEEDSTUFFS

Llamas have frequently been described as "cafeteria-style" eaters, browsing on a smorgasbord of forages and grains when offered.[3,12,13,20] Depending upon the feedstuffs available, variable amounts of protein, fat, carbohydrates, vitamins and minerals are ingested.

Forages

Llamas are generally grazing animals, preferring tall, coarse grass over the richer legumes.[3,20] They will, however, browse on anything in their pasture including trees and shrubs.[13]

The majority of energy should be supplied by forages.[4,22] Usually forages supply all the energy requirements of a healthy llama. When demands are increased over maintenance, however, dry matter intake available from forages may not meet energy requirements.

A wide variety of hays are used, including various grasses (7% to 10% crude protein), oat (9% to 10% crude protein), and alfalfa (12% to 18% crude protein).[4,20,22] Leafy, green, fresh-smelling hay is preferred.[4]

Forages, especially coarse and stemmy hays, provide the majority of fiber in the llama diet.[20] Fiber is composed mainly of lignins and cellulose.[4] These substances stimulate the lining of the forestomachs and cause appropriate motility, tone, and secretions. Bacteria and protozoa in the forestomachs aid in the degradation of cellulose.[28]

Alfalfa hay is used successfully in many well-managed llama herds with no complications.[5,9] However, the high protein intake is an expensive source of energy.[20] Generally, if alfalfa is used, a mixture of alfalfa and grass hay (25% legume and 75% grass) is all that is necessary. This results in a protein concentration of 10% to 12%, which under normal situations is more than adequate.[5]

Cereal grains

Corn, oats, wheat, and barley are commonly used cereal grains.[4] They may be used to provide supplemental energy when it is required. Cereal grains provide an average of 9% to 10% protein and limited amounts of fiber.[4] They are normally higher in phosphorus than in calcium,[4,22] so the calcium-phosphorus ratio must be carefully adjusted if large amounts of grain are used.

Other supplements

Many other products are used as supplements. Soybean meal (38% to 42% protein), whole cottonseed (19% to 20% protein), and malt pellets (25% protein) are frequently used as additional sources of protein.[20] Fat is very calorie dense and is added to the diet when required. Vegetable oils, such as corn, soybean, or safflower oil, are excellent fat sources.[4,20] Whole cottonseed is also high in fat and is frequently used in rations for packing llamas when a small volume of feed is required.[1,20]

Prepared mixes

Many commercially available equine, bovine, caprine, and ovine prepared mixes have been fed to llamas. They are generally a mixture of cereal grains and alfalfa pellets sweetened with molasses. These mixes vary widely in protein, energy, and mineral content, depending upon the product and concentration used. Purina Mills has prepared mixes* designed specifically for llamas.[10]

Mineral supplements

Currently, there are no mineral supplements formulated specifically for llamas. Calcium, phosphorus, and salt are perhaps the areas of greatest concern in llamas. Iodized salt should be fed unless the diet is already high in iodine.[19] Calcium and phosphorus are normally considered together, as the presence of one in the ration influences the requirement for the other. The ratio of calcium to phosphorus in the llama diet should be between 1.2:1 and 1.6:1, with the ratio not exceeding 2:1.[18] If grass or oat hay is the primary forage offered, calcium concentration will be deficient and will have to be supplemented daily.[4,20,22] Alfalfa hay provides more than adequate amounts of calcium, but phosphorus will need to be supplemented.[4,20,22]

Mineral intake is ultimately dependent upon the amounts in the soil in which plants are grown, and micronutrients may be present in abnormally low quantities. Trace minerals such as selenium, copper, zinc, manganese, iron, and cobalt often are added to the salt mixture. Of these, the requirement for selenium is best documented, and 1 to 1.5 mg per llama per day often is added to the mineral mix in selenium-deficient areas.[20]

---

*Llama Chow, Purina Mills, Inc., St. Louis, Mo.

FEEDING REQUIREMENTS
Neonates and suckling llamas

Neonates are considered to be infant llamas less than 14 days old, and suckling llamas are between 14 days and weaning.[23] Newborn llamas weigh between 7 and 14 kg (16 to 30 lb).[7,14,23] Newborns less than 7 kg (16 lb) are premature or dysmature.[7,23] The expected growth rate for both neonatal and suckling llamas is 0.2 to 0.5 kg (0.5 to 1 lb)/day.[15,18] Weight gain tends to taper somewhat at 2 to 3 months, when 0.2 to 0.3 kg (0.5 to 0.7 lb)/day is expected.[15]

Neonatal llamas generally begin to pick at solid foods within the first 1 to 2 weeks of life.[14,20] They graze along with their dams but do not adequately utilize hay and grain until 2 or 3 months of age.[20] After that time, suckling llamas that are not gaining adequate weight should be supplemented with a grain mixture (16% to 18% protein).[5,20] Small amounts of a grain mixture should be introduced into the diet of suckling llamas 1 to 2 months prior to weaning. A creep feeder should be used to prevent the dam from ingesting the grain.

Weanling and yearling llamas

Suckling llamas are normally weaned between 5 and 6 months of age.[5,14] Young, growing llamas appear to do well on a 12% protein diet.[5,20] Special attention to the calcium-phosphorus ratio and possibly to copper and zinc levels is required during this period.

Orphans

Llamas may be orphaned at birth or at any time during the suckling period. Young llamas failing to thrive may be deliberately removed from the dam and raised as orphans. When a neonatal or suckling infant is losing weight or has failed to gain weight for 2 weeks, it should be separated from the dam and raised as an orphan. Physical separation of the infant and mother is important, as most infant llamas will continue to suck on a dry teat and starve before they will accept a supplementary bottle. It is not desirable to bottle-feed male orphans, but it has been done successfully without development of the behavioral disorder called "berserk male syndrome."[2,14]

As with true ruminants, the ingestion of colostrum at birth is essential.[11] If llama colostrum is not available, goat or mare's colostrum should be used.[14,18] Goat's colostrum is preferred as it is most similar in composition to the llama's and because goats frequently are immunized against clostridial toxins known to cause disease in infant llamas. Newborns should receive 10% of their body weight in good-quality colostrum within the first 12 to 18 hours of life.[18,23] It may be given in multiple feedings, but ideally they should receive half the calculated amount within the first 1 to 2 hours.[23] Not only is colostrum the sole source of immunoglobulins but it also provides necessary fat-soluble vitamins (A, D, and E) and acts as a valuable energy supply.

Orphans may be bottle-fed goat's milk, lamb or kid milk replacer, or whole cow's milk.[18,23] Goat's milk appears to be the best, although some orphans do not like the taste of frozen goat's milk and prefer only fresh. They should receive a minimum of 10% of their body weight per day. The exact amount necessary for maintenance and growth has not been determined, but many infants have received up to 16% of their body weight per day without ill effects.

Bottle-fed llamas seem to prefer nipples designed for human infants rather than lamb or calf nipples. Extremely small llamas suckle well on nipples designed for premature human infants or flutter-valve nipples. Most orphaned llamas will not drink milk from a pail.[25]

Orphaned llamas that cannot or will not suckle may be fed through an indwelling nasogastric tube in the distal esophagus (Fig. 31-1).[23] Alternatively, they may be placed on parenteral nutrition (Figs. 31-2 and 31-3).[16] Both methods have been used successfully to maintain the protein and calorie balance until the infant is suckling well.

Pregnant and lactating llamas

More problems have been reported in overfed, excessively conditioned pregnant llamas than in malnourished ones.[2,13,20] During the first 8 months of gestation, no changes need to be made in the diet. Depending upon the weight and body condition, a higher protein grain mixture (14% to 16%) may be necessary in the last 2 to 3 months.[5,20] It should be slowly introduced into the diet until the llama is receiving 0.2 to 0.5 kg (0.5 to 1 lb) twice a day.

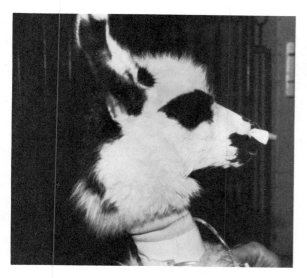

Fig. 31-1 Indwelling nasogastric feeding tube in neonatal llama.

Lactating llamas may quickly lose body condition, especially if they are producing large amounts of milk or nursing twins. A decreased appetite is often present during this time, and the llama may not compete adequately for feed if group housed.[5] In large groups or if weight loss occurs, they may need to be separated and fed 0.5 to 1 kg (1 to 2 lb) of grain twice a day.[5] Weekly weighing helps to determine the response to feeding and prevent continuing losses or excessive weight gains.

As the time of weaning approaches, the grain mixture should be decreased.[20] Once the infant is weaned, the dam should receive no grain. This results in a more rapid decrease in the milk supply and helps to prevent the incidence of mastitis.

Working llamas

Working llamas need to be supplemented with a grain mixture. They may require as much as 50% more energy than their more sedentary counterparts.[1,20] The amount given depends upon the work load and available forage and should be determined by body condition of the individual working llama. Whole cottonseed is calorie dense and is generally mixed 50:50 with

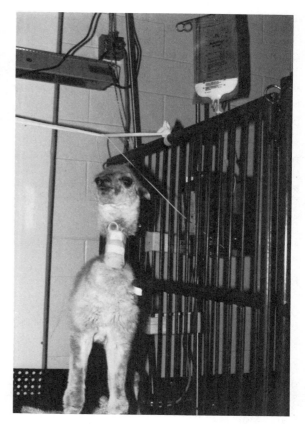

Fig. 31-2 Neonatal llama receiving parenteral nutrition.

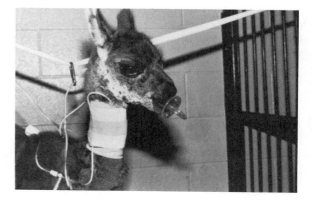

Fig. 31-3 Human infant pacifier used to stimulate sucking in neonatal llama receiving parenteral nutrition.

a grain mixture when packing, thus decreasing the weight of the feed carried.[1]

Geriatric llamas

The lifespan of a llama is between 15 and 22 years.[7] As llamas age, they are susceptible to the same age-related disorders as other species. The appetite may become more capricious, weight loss may occur, and water intake may decrease. Weight loss may be secondary to parasitism, abnormal dentition, or systemic diseases rather than to malnutrition. A thorough physical examination as well as feed analysis can help to determine the source of the weight loss. The addition of table salt (1 to 2 teaspoonful two times a day) to the mineral or grain mixture appears to increase water consumption in those llamas which do not drink well.

Hospitalized adult llamas

Hospitalized adult llamas—regardless of the disease process—frequently do not eat well. Offering a wide variety of hays, cereal grains topped with molasses or a sweet syrup, and premixed feeds helps to stimulate appetite. Llamas are by nature very social animals, and the presence of another llama in the same or an adjacent stall may increase the desire to eat.

Show animals

Contrary to popular belief, animals groomed for the show ring do not require excess grain. The primary problem associated with excessive grain feeding is obesity.[18,20] It may be necessary, however, to carry personal water supplies, especially for those llamas not accustomed to chlorinated water.

CONCLUSION

Llamas are, by nature, hardy animals that thrive on low-quality forages. This has recently been well documented by two controlled studies. Growing, pregnant, lactating, working, and geriatric llamas, however, may require special supplementation.

REFERENCES

1. Biggs S: Llama life, p 15, Summer 1988.
2. Casey C: Llamas 3:109, 1989.
3. Ebel SL: Vet Clin North Am (Food Anim Pract) 5:1, 1989.
4. Ensminger ME and Olentine CG, editors: Feeds and nutrition, Clovis, CA, 1978, Ensminger Publishing.
5. Fenimore RL: Scientific proceedings, Western States Veterinary Conference, 1988, pp 311-316.
6. Fenimore RL: Scientific proceedings, Western States Veterinary Conference, 1988, pp 317-323.
7. Fowler ME: Zoo and wild animal medicine, ed 2, Philadelphia, 1986, WB Saunders.
8. Fowler ME: In Scientific proceedings, Western States Veterinary Conference, 1988, pp 334-342.
9. Fowler ME: Llamas 3:15, 1989.
10. Freeman M and Greene N: Llamas 3:73, 1989.
11. Garmendia AE and others: Am J Vet Res 48:1472, 1987.
12. Gavin R: Country Journal 16:62, 1988.
13. Hart R: Living with llamas, Ashland, Ore, 1988, Tuniper Ridge Press.
14. Herriges S: 3L Llama 19:82, 1983.
15. Herriges S and Smith B: Personal communication, 1988, Bend, Oregon.
16. Hovda LR, McGuirk SM, and Lunn DP: J Am Vet Med Assoc (in press).
17. Johnson LR: Llama Life, p 24, Winter 1988.
18. Johnson LW: Scientific proceedings, Western States Veterinary Conference, 1988, pp 358-368.
19. Johnson LW: Scientific proceedings, Western States Veterinary Conference, 1988, pp 330-333.
20. Johnson LW: Vet Clin North Am (Food Anim Pract) 5:37, 1989.
21. Oldfield JE: Llama World 1:10, 1983.
22. National Research Council: Nutrient requirements of sheep, ed 6, Washington, DC, 1985, National Academy of Science.
23. Paul-Murphy J: Vet Clin North Am (Food Anim Pract) 5:183, 1989.
24. San Martin HF: Doctoral thesis, Lubbock, 1987, College of Agriculture, Texas Tech University.
25. Sharp J: Llamas 2:69, 1988.
26. Vallenas A, Cummings JF, and Munnell JF: J Morphol 134:399, 1971.
27. Vallenas A and Steven CE: Am J Physiol 220:275, 1971.
28. Vallenas A and Steven CE: Cornell Vet 61:239, 1971.

Common Nutritional Problems in Llamas

LYNN R. HOVDA, SUSAN SHAFTOE

JUVENILE LLAMAS
Selenium–vitamin E deficiency
Young llamas have been reported to be affected with nutritional muscular dystrophy (white muscle disease),[6,15] and selenium deficiency has been proposed as the cause (Fig. 32-1).[17] Normal blood selenium concentrations have not been established for young llamas. Many veterinarians, therefore, routinely recommend supplemental injections of selenium and vitamin E in those areas known to be selenium deficient.[7] Newborn llamas should receive 0.25 to 1.0 mg per animal of selenium using L-Se* (equivalent to 0.25 mg selenium and 68 IU vitamin E per ml), and weanlings should receive 1 to 3 mg of selenium using Bo-Se* (equivalent to 1 mg selenium and 68 IU vitamin E per ml).[7] E-Se* (equivalent to 2.5 mg selenium and 68 IU vitamin E per ml) is a product recommended for horses and generally is not used in juvenile llamas.[7] Selenium injections may be given either intramuscularly (IM) or subcutaneously (SC). Depending on soil selenium concentrations, injections may be repeated every 3 to 4 months.[17] Llamas showing clinical signs of white muscle disease should receive 0.13 mg/kg BW (0.06 mg/lb) selenium and 7 IU/kg BW (3 IU/lb) vitamin E SC or IM.[17] Usually, only one injection is necessary, and a response is seen in 3 to 5 days.

Enterotoxemia
Infants nursing heavily milking dams are susceptible to the development of enterotoxemia from *Clostridium perfringens* C and D.[6,8] Of these, type C is most commonly seen in the nursing infant.[19] Affected infants are depressed and anorexic and suddenly develop bloody diarrhea within the first 3 weeks of life. Sometimes death may occur without any signs developing. Treatment is generally nonproductive. Prevention consists of vaccinating the dam with a C and D toxoid 8 and 4 weeks before parturition.[6,8]

Nutritional diarrhea
Bottle-fed infants may develop diarrhea as their gastrointestinal tracts adapt to an alternate milk source.[10] Normally all that is required are smaller, more frequent feedings of the same amount of milk or the temporary reduction of milk and supplementation with an oral electrolyte solution. A kaolin-pectin mixture 30 to 60 ml (1 to 2 ounces, four times a day) may be added to the milk for 1 to 2 days if indicated. Rarely is it necessary to remove the infant completely from milk.

Constipation
Bottle-fed infants that are undergoing a change in feeds may develop constipation, as can suckling llamas that are just beginning to browse.[10,17] In suckling llamas, the forestomachs cannot adequately process roughage until 2 to 3 months.[13,17] The poorly digested material is passed into the intestinal tract, resulting in hard, dry fecal material. Constipated llamas often remain bright and alert and nurse intermittently. They strain and vocalize as they make frequent attempts to pass the fecal material. Treatment consists of a warm, soap water or mineral oil

---

*Schering Plough Corporation, Animal Health Division, Kenilworth, NJ.

Fig. 32-1 Juvenile llamas with white muscle disease may require a sling until they are able to stand unassisted.

enema. If constipation is refractory to treatment, intravenous fluids may be required. Oral mineral oil (90 to 120 ml or 3 to 4 oz) or a laxative such as psyllium hydrocolloid may be added to the diet for 2 to 3 days to prevent a recurrence.[17]

Gastrointestinal ulceration

Neonatal or suckling llamas that are stressed or receiving excessive amounts of nonsteroidal antiinflammatory agents and orphaned llamas are prone to the development of gastric or duodenal ulcers.[6,17] Early clinical signs may be subtle and include anorexia, depression, and mild abdominal pain.[17] More severe ulcers may be accompanied by bruxism, salivation, and severe abdominal pain.[17] Cimetidine* (5 mg/kg [2.3 mg/lb]) intravenously has been utilized as a therapeutic agent.[6,17] A kaolin-pectin mixture may decrease abdominal pain in llamas suspected of having ulcer disease.[17] Antacids such as magnesium and aluminum hydroxide have been suggested but need to be given frequently to be effective (5 to 15 ml every 2 hours).

Angular limb deformities

Reports of infant llamas with crooked legs have been increasing. Genetic predisposition, trauma, conformation, lack of exercise, rapid growth rate, endocrine imbalance, and nutrition all have been suggested as underlying factors in the development of this disease state.[13,20] It appears that rations improperly balanced for calcium and phosphorus and those containing excessive amounts of protein play a role.[14,20] The onset of developmental orthopedic diseases in horses has recently been shown to be correlated with abnormal levels of copper and zinc.[14] This may be a factor for infant llamas as well. Low levels of magnesium, sulfate, and vitamins A and D may also play a role.[20]

Micronutrient deficiencies

Copper, zinc, iron, iodine, and cobalt deficiencies have all been implicated as causing disease in young, rapidly growing llamas.[13] A syndrome similar to enzootic ataxia in lambs has been associated with copper deficiency in both young and mature llamas.[16] Copper deficiency may result in a poor hair coat with excessive shedding and the development of depigmented, stringy wool fibers.[13] Zinc deficiencies have been associated with a dermatosis similar to that found in other species.[13] Iron deficiency has been reported in two yearling llamas.[13]

MATURE LLAMAS
Stomach atony

A wide variety of disease processes can cause atony of the stomach in a llama.[6,8,17] These include bloat, traumatic reticuloperitonitis, grain overload, vagal indigestion, simple indigestion, and gastric and duodenal ulcers. The first four rarely occur in llamas.[6,8] This may be due to the unique anatomical features of the llama fore-

---

*Tagamet, SmithKline and French Laboratories.

stomachs or to the rather fastidious appetite of the mature llama.[17]

Indigestion of unknown origin (simple indigestion), however, has been observed in a number of mature llamas.[17] These llamas may present as dull, anorexic, and dehydrated, with diarrhea or constipation and decreased forestomach motility. Analysis of the first compartment fluid shows an increase or decrease in the pH (normal is 6.6 ± 0.27) and a decrease in the number of live, motile microorganisms.[2,17] Conservative therapy may be all that is required. Treatment generally includes good-quality hay, correction of pH abnormalities, and transfaunation of stomach contents from a normal llama. Bovine or ovine rumen liquor may be used if llama contents are not available.[6,17] Intravenous fluids may be necessary in dehydrated llamas who will not drink.

Gastric ulcers have been reported in mature llamas. Although the exact etiology remains unknown, they have been associated with decreased quantities of fiber and increased amounts of grain in the diet, stress, and the use of nonsteroidal antiinflammatory drugs.[5,17] Cimetidine (5 mg/kg [2.3 ml/lb] intravenously every 6 hours) has been used successfully, as has the introduction of a good-quality grass hay in the diet.[6,17]

Esophageal choke

Choke, although rare, does occur in llamas.[9] Llamas receiving pelletized feed or alfalfa cubes appear to be more susceptible.[12,17] Clinical signs, including bloat, are similar to those of any choked ruminant. Therapy includes resolution of the bloat and attempts to move the food material into the forestomachs with a stomach tube.[17]

Obesity

More mature llamas are overweight than underweight. Obesity has been implicated in heat stress and infertility in both males and females.[1,13,18] Pregnant llamas who are obese have more problems with dystocias, premature infants, and stillbirths.[13] Milk production may be decreased in overweight llamas, especially primiparous llamas.[11]

Most llama owners agree that it is exceedingly difficult to put a llama on a diet. Obese llamas are best separated from the group and fed on an individual basis until they have reached their desired weight.[3,4] Separation may be all that is required in aggressive llamas who are consuming more than their required quantity of food. More frequently, however, the amount of food offered must be decreased or exercise increased. Complete starvation is not recommended, as llamas have been shown to develop ketosis.[17] Weekly weighing helps to monitor response to the diet. The mature male should weigh between 160 and 240 kg (350 to 530 lb) and the female between 110 and 190 kg (240 to 420 lb).[6] If a scale is not available, careful palpation of the ribs and thoracic vertebrae is beneficial.[4] The pelvis, even in obese llamas, is frequently bony and should not be used to assess body condition.[4]

Vitamin E–selenium deficiency

Adult llamas are susceptible to selenium deficiencies. It is associated with poor reproductive performance in adults.[13] In areas known to be selenium deficient, selenium should be added to the mineral mix based on ration analysis. Alternatively, it may be supplemented as a prophylactic injection. Adult llamas should receive 5 to 10 mg of selenium and 70 to 140 IU vitamin E per animal using Mu-Se (equivalent to 5 mg of selenium and 68 IU of vitamin E per ml).*[6,7]

Other micronutrient deficiencies

All of the micronutrient deficiencies responsible for producing disease in juvenile llamas can cause disease in mature llamas. If a zinc, copper, or magnesium deficiency is suspected, blood level analysis may be helpful. In most instances, however, blood levels are at best inexact and normal values are not well established for llamas. Feeding history, feed analysis, and soil analysis are required to diagnose and properly correct the deficiency.

CONCLUSION

Diseases related to overnutrition and undernutrition are, for the most part, rare in llamas. Obesity is perhaps the most common and has far-reaching complications. Careful attention to the

---

*Schering Plough Corporation, Animal Health Division, Kenilworth, NJ.

environment, ration analysis, and judicious feeding protocols will avoid the majority of nutritionally related diseases.

REFERENCES

1. Casey C: Llamas 3:109, 1989.
2. Engelhardt WV, Ali KE, and Wipper E: J Comp Physiol 132:337, 1979.
3. Fenimore RL: Scientific proceedings, Western States Veterinary Conference, 1988, pp 311-316.
4. Fenimore RL: Scientific proceedings, Western States Veterinary Conference, 1988, pp 317-323.
5. Fowler ME: 3L Llama 27:45, 1985.
6. Fowler ME: Zoo and wild animal medicine, ed 2, WB Saunders, 1986, Philadelphia.
7. Fowler ME: Llamas 31:37, 1986.
8. Fowler ME: Scientific proceedings, Western States Veterinary Conference, 1988, pp 334-342.
9. Goldsmith B and Fowler ME: 3L Llama 21:3, 1984.
10. Herriges S: 3L Llama 19:82, 1983.
11. Herriges S and Smith B: Personal communication, 1988, Bend, Oregon.
12. Johnson LR: Llama Life, p 24, Winter 1988.
13. Johnson LW: Vet Clin North Am (Food Anim Pract) 5:37, 1989.
14. Knight DA and Reed S: In Proceedings of the ACVIM, 1987, pp 495-505.
15. Oldfield JE: Llama World 1:10, 1983.
16. Palmer AC and others: Vet Rec 107:10, 1980.
17. Smith JA: Vet Clin North Am (Food Anim Pract) 5:101, 1989.
18. Strain MG and Strain SS: Vet Med 82:494, 1988.
19. Thedford TR and Johnson LW: Vet Clin North Am (Food Anim Pract) 5:145, 1989.
20. Turner AS: In Proceedings of the llama workshop for veterinarians, Fort Collins, 1987, Colorado State University.

SECTION E Sheep

33

Feeding Sheep for Optimal Production

MILLARD CALHOUN

Sheep are raised under a variety of conditions in the United States. In the western states, flocks tend to be large, and sheep graze in large pastures or on extensive range areas. In the Pacific states and in the midwestern, eastern, and southeastern states, flocks tend to be smaller, and the sheep enterprise is generally a part of a larger farming operation. Regardless of the sheep enterprise (range, farm flock, or confinement), the largest cost associated with sheep production is the cost of providing feed. A unique advantage of sheep is their ability to produce wool and a market-finished lamb entirely on pasture or conserved forages. However, sheep also make efficient use of grains and by-product feedstuffs by converting these materials into high-quality animal products (meat, wool, hides, and milk) for human use.

Feeding sheep for optimal production requires knowledge of the nutritional requirements of the animal, the nutrient composition of available feedstuffs, and how these feeds can be combined to meet the requirements. The purpose of this chapter is to provide current information on the nutritional requirements of sheep, to make practical recommendations for the use of feedstuffs and feeding practices to minimize nutrition- and disease-related problems, and to provide diets needed for optimal production.

The nutrients of primary concern or importance in sheep nutrition are water, energy, protein, and minerals (salt, calcium, phosphorus, magnesium, sulfur, copper, selenium, and iodine are the most critical), and vitamins (A, D, and E are of primary concern). Except in the very young lamb prior to development of normal rumen function, rumen microorganisms make all necessary B vitamins and vitamin K. With the possible exceptions of niacin and thiamin (vitamin B_1) in some feeding situations, the amounts of B vitamins and vitamin K produced in the rumen are adequate to meet the animals' requirements.[27]

Appendix Table A-17 gives the daily requirements for dry matter, energy, crude protein, calcium, phosphorus, vitamin A, and vitamin E, and Appendix Table A-18 gives information for the same nutrients expressed in terms of the concentration in the diet. Tables 33-1 and 33-2 provide information on macro and micro (trace) mineral requirements and toxic levels, respectively, in sheep.

WATER

Water is the most critical nutrient in feeding sheep. It plays a vital role in body processes and is essential for the health and well-being of animals. Sheep deprived of water stop eating, and survival is threatened. Adequate water is essential for sheep to excrete excess substances such as oxalates, ammonia, and magnesium ammonium phosphates that otherwise would cause

Table 33-1 *Macromineral requirements of sheep (percentage of diet dry matter)*

Nutrient	Requirement
Sodium	0.09-0.18
Chlorine	—
Calcium	0.20-0.82
Phosphorus	0.16-0.38
Magnesium	0.12-0.18
Potassium	0.5-0.8
Sulfur	0.14-0.26

Table 33-2 *Micromineral requirements of sheep and maximum tolerable levels (ppm or mg/kg of diet dry matter)*

Nutrient	Requirement	Maximum tolerable level*
Iodine	0.1-0.8†	50
Iron	30-50	500
Copper	7-11‡	25§
Molybdenum	0.5	10§
Cobalt	0.1-0.2	10
Manganese	20-40	1000
Zinc	20-33	750
Selenium	0.1-0.2	2
Fluorine	—	60-150

*NRC (1980).[25]
†High level for pregnancy and lactation in diets not containing goitrogens; should be increased if diets contain goitrogens.
‡Requirement when dietary molybdenum concentrations are less than 1 mg per kg dry matter.
§Lower levels may be toxic under some circumstances.

urinary calculi. Providing an adequate supply of good-quality, clean, fresh water is basic to achieving optimal production. In most situations, good-quality water is readily available at minimal cost. However, in the western states and in the semiarid Southwest, water quality and availability are often serious problems. Also, in some areas, contamination resulting from industrial activities can be a problem, and in farming areas there are concerns with pesticide, herbicide, and nitrate contamination of waters.

Water is available to sheep in the form of free water (water in ponds, streams, and water tanks and as dew or snow), water contained in feed, and metabolic water (water obtained from oxidation of nutrients in feed). The amount of water required varies considerably, depending on the physiological functions being supported (maintenance, growth, reproduction, and lactation), ambient temperature, wool covering, amount of feed consumed, and the composition of the feed.[27] Voluntary water consumption is two or three times dry matter intake and increases with high-protein and salt-containing feeds. Pregnancy, lactation, and elevated ambient temperatures also increase water intake. Water intake starts to increase by the third month of gestation and increases further with the number of fetuses. Restricting water predisposes ewes to pregnancy toxemia in late gestation. To ensure optimal performance, ewes must have an unlimited supply of good-quality water readily available during the critical periods of late gestation and early lactation. However, excessive water in some foods can adversely affect performance because of limitation of dry matter intake. In some cases silages and succulent pastures are low in dry matter content and require supplementation with energy-dense feeds to enable productive classes of sheep to achieve optimal performance. Nonpregnant, nonlactating sheep may go for weeks without drinking water when consuming feeds of high moisture content.

There is considerable variation in the mineral composition of water, regardless of whether the source is groundwater or surface water in streams, ponds, or lakes. Because of this, it is advisable for a producer to determine the minerals present in the water available to sheep. Although mineral content is variable, the contribution from water usually is ignored because the amounts present generally do not provide more than a small percentage of the minerals required for sheep. However, in some cases, the amounts of salts present (chlorides, sulfates, carbonates, and bicarbonates of sodium, calcium, and magnesium) are sufficiently large to necessitate adjustments in the amounts of these minerals added to diets or provided free choice. In extreme cases the amounts of salt (NaCl) and sulfates contained in water may be sufficient to adversely affect animal performance. In these instances the salt and sulfate content of the diet may contribute to toxicity problems. This is of particular concern when salt additions to supplements are used to control feed intake.

When NaCl is the predominant salt present, the upper safe limit for salt in drinking waters for sheep appears to be about 13,000 ppm.[24] When magnesium chloride provides more than 1000 ppm of the salts present, feed consumption and live weight gains are decreased. In contrast, sodium sulfate up to 5000 ppm, calcium chloride up to 3000 ppm, and mixtures of sodium carbonate and bicarbonate up to 4100 ppm in solutions with NaCl to give a total salt concentration of 13,000 ppm did not affect general health, feed consumption, or wool production.

Guidelines established by the National Academy of Sciences[24] for use of saline waters with sheep are (1) less than 5000 ppm, water can be safely used for all classes of sheep; (2) 5000 to 7000 ppm, water can be used with reasonable safety for sheep, but caution should be used when feeding water approaching the higher limit to pregnant and lactating sheep; (3) 7000 to 10,000 ppm, water may be suitable for maintenance of older sheep, but considerable risk may exist in using them for young, pregnant, or lactating animals or for any animals subjected to heavy heat stress or water loss; and (4) more than 10,000 ppm, not recommended for any use. Alkalinities and nitrates also should be measured in highly saline waters, as both can detract from the suitability of water for sheep.

ENERGY

The most common limiting nutrient for sheep is energy. The critical periods are the last 4 to 6 weeks of gestation and the first 6 to 8 weeks of lactation. Energy is also a critical nutrient for young, rapidly growing lambs. Overgrazing of pastures, utilization of low-quality pastures or harvested forages, and drought all contribute to the problem of energy deficiency. Quite often energy shortages are accompanied by deficiencies of other nutrients (protein, minerals, vitamins). The consequences of inadequate energy intake are decreased growth, reproductive performance, and milk and wool production and increased mortality related to decreased resistance to infectious and parasitic diseases.

The provision of sufficient feed to meet energy requirements is the major cost associated with feeding sheep. Whenever possible, energy should be provided from the cheapest source available. Consideration should be given to the costs of transportation and feeding and the amounts of feed wasted. The major sources of energy for sheep are native range and pasture plants, improved pastures, hay, silage, grains, and by-product feedstuffs.

Overfeeding is generally wasteful of feed resources. In some grazing situations, however, the maximal utilization of excess feed is a good management practice. During periods of above-average rainfall in arid regions, feeding available range vegetation reduces requirements for supplemental feed when vegetation is not abundant. A number of practices have been used effectively to control intake and prevent overfeeding, including hand-feeding just enough to maintain body weight, restricting access to pastures or feed bunks to the minimum time necessary for sheep to eat the desired amount of feed, and adding intake limiters, such as salt, to supplemental feeds offered free choice to grazing animals.

Total digestible nutrients (TDN), digestible energy (DE), metabolizable energy (ME), and net energy for maintenance (NEm) and gain (NEg) are used widely for expressing the energy requirements of sheep and the energy values of feeds. (The energy requirements for various classes of sheep and levels of production are given in Appendix Table A-17.) The terms commonly used for all classes of sheep in the United States are TDN and DE. To a lesser extent NEm and NEg are used to express the energy value of feeds and energy requirements of growing-finishing lambs fed in feedlots. The current National Research Council (NRC) publication for sheep presents net energy requirements separately for growing lambs of small, medium, and large mature weight genotypes (Table 33-3).

PROTEIN

In the current NRC publication for sheep, crude protein was used to express animal requirements and the protein value of feeds. The crude protein system is used in this chapter. In many instances it is advisable to correct crude protein values for the amount of protein contained in the acid detergent fraction (ADF) of a feed be-

Table 33-3 *Net energy requirements for lambs of small, medium, and large* mature weight genotypes (Mcal/d)*

Body weight† (kg):	10	20	25	30	35	40	45	50
NEm requirements:	0.315	0.530	0.626	0.718	0.806	0.891	0.973	1.053
Daily gain† (g)				NEg requirements				
Small mature weight lambs								
100	0.178	0.300	0.354	0.406	0.456	0.504	0.551	0.596
150	0.267	0.450	0.532	0.610	0.684	0.756	0.826	0.894
200	0.357	0.600	0.708	0.812	0.912	1.008	1.102	1.192
250	0.446	0.750	0.886	1.016	1.140	1.261	1.377	1.490
300	0.535	0.900	1.064	1.219	1.368	1.513	1.652	1.788
Medium mature weight lambs								
100	0.155	0.261	0.309	0.354	0.397	0.439	0.480	0.519
150	0.233	0.392	0.463	0.531	0.596	0.658	0.719	0.778
200	0.310	0.522	0.618	0.708	0.794	0.878	0.960	1.038
250	0.388	0.653	0.771	0.884	0.993	1.097	1.199	1.297
300	0.466	0.784	0.926	1.062	1.191	1.316	1.438	1.557
350	0.543	0.914	1.080	1.238	1.390	1.536	1.678	1.816
400	0.621	1.044	1.234	1.415	1.589	1.756	1.918	2.076
Large mature weight lambs								
100	0.132	0.221	0.262	0.300	0.337	0.372	0.407	0.439
150	0.197	0.332	0.392	0.450	0.505	0.558	0.610	0.660
200	0.263	0.442	0.524	0.600	0.674	0.744	0.813	0.880
250	0.329	0.553	0.654	0.750	0.842	0.930	1.016	1.099
300	0.394	0.663	0.785	0.900	1.010	1.116	1.220	1.320
350	0.461	0.775	0.916	1.050	1.179	1.303	1.423	1.540
400	0.526	0.885	1.046	1.200	1.347	1.489	1.626	1.760
450	0.592	0.996	1.177	1.350	1.515	1.675	1.830	1.980

*Approximate mature ram weights of 95 kg, 115 kg, and 135 kg, respectively.
†Weights and gains include fill.
From National Research Council: Nutrient requirements of sheep, Washington, DC, 1985, National Academy of Sciences.

cause the protein associated with the ADF fraction is assumed to be unavailable to the animal. The correction is made as follows:

Available crude protein =
crude protein − (ADF nitrogen × 6.25)

Newer methods for expressing the protein requirements of ruminants and the protein value of feeds based on partitioning protein into fractions that are degraded and undegraded in the rumen have been proposed[28] but generally are not used by sheep producers. This may change when values become available for the rumen degradation of proteins of various feedstuffs under a wide range of feeding conditions, and when software programs to make the necessary calculations for sheep are written for personal computers. With the exception of very young lambs, the amount of protein in sheep rations is much more important than protein quality.

Urea can be beneficially used whenever ruminal ammonia levels are below the optimum required for growth of rumen microorganisms. Ammonia levels in the rumen depend on the level of protein in the diet, the extent to which dietary protein is broken down in the rumen, and the availability of dietary energy to support growth of the microbial population. When these factors are known, the usefulness of urea in the diet can be estimated.

Table 33-4 *Urea usefulness with various feed intakes, TDN levels, and ruminal digestions of dietary protein*

TDN %	Percent dietary protein escaping digestion in the rumen											
	20	20	20	30	30	30	40	40	40	50	50	50
	Daily TDN Intake, (kg)											
	0.23	0.45	0.68	0.23	0.45	0.68	0.23	0.45	0.68	0.23	0.45	0.68
	Percent dietary protein above which urea is useless											
55	8.2	8.7	8.9	9.2	9.8	10.0	10.4	11.1	11.3	12.0	12.8	13.0
60	9.0	9.5	9.7	10.0	10.7	10.9	11.4	12.1	12.3	13.1	13.9	14.2
65	9.7	10.3	10.5	10.9	11.6	11.8	12.3	13.1	13.4	14.2	15.1	15.4
70	10.5	11.1	11.4	11.7	12.4	12.7	13.3	14.1	14.4	15.3	16.3	16.6
75	11.2	11.9	12.2	12.5	13.3	13.6	14.2	15.1	15.4	16.4	17.4	17.8
80	12.0	12.7	13.0	13.4	14.2	14.5	15.2	16.1	16.4	17.5	18.6	19.0
85	12.7	13.5	13.8	14.2	15.1	15.4	16.1	17.1	17.5	18.6	19.8	20.1

All values are on a dry matter basis.
From the National Academy of Science: Ruminant nitrogen usage 1985, Washington, DC, National Academy of Sciences. Used with permission.

Table 33-4 shows the effects of TDN and the percentage of dietary protein escaping degradation in the rumen on the usefulness of urea in sheep diets. The values indicate dietary protein levels above which the addition of urea would be useless. The values need to be adjusted for the moisture content of the feed to give crude protein percentages on an as-fed basis. The usefulness of urea increases as the energy content of the diet increases and also as the percentage of dietary protein escaping digestion in the rumen increases.

Plant proteins, such as cottonseed meal and soybean meal, are more effective than urea in supporting animal growth and production. However, response is generally quite satisfactory when urea provides a portion of the protein requirements for animals fed medium- and high-energy diets, for example, growing-finishing lambs. The situation with respect to grazing animals is less clear but depends on the energy content and the amount of forage available and its protein content. As in other ruminants, sheep consuming low-quality forages (less than 50% TDN) containing more than 6% to 7% protein would probably not benefit from urea supplementation unless the supplement contained an energy source such as grain. During the winter months, mature range forage often contains as little as 3% to 5% protein. Supplementation of this forage with an energy supplement and urea would most likely increase voluntary forage intake and improve the nutritional status of the animal to a level adequate to maintain body condition during the early stages of pregnancy. Urea supplementation under such circumstances most likely would be unsatisfactory for ewes during the last one third of pregnancy or early lactation.

The use of urea requires some general considerations. Urea can be used at about 1% of the total ration or not more than 3% of the concentrate portion of the ration but should not be used to replace more than about one third of the total protein in the diet. A reasonable level for use of urea in range supplements would be to replace 25% to 35% of the natural protein, or no more than 5% to 7% crude protein equivalent from urea in a typical 20% crude protein range supplement. Urea should not be used in rations for very young lambs or in creep rations.

Efficient use of urea by rumen microorganisms requires 2 to 3 weeks for full adaptation. It is used more efficiently when consumed in small amounts at frequent intervals. Thus urea is most useful when fed in weaned lamb rations or in self-fed or salt-limited ewe rations. It is not as efficiently used, for example, with cubes fed one to three times per week. Sulfur is essential for microbial protein synthesis. Diets low in protein that could benefit from urea often are low in sulfur. This can be corrected by adding 0.1% sulfur to complete diets or by including sulfur in a free-choice loose mineral or mineral block offered to grazing animals. Because of the uncertainty of the actual protein value of urea in feeds, it is unwise to use urea supplements unless they are priced substantially lower than supplements containing all-natural protein.

An important point to remember when feeding urea to sheep is that high levels can be toxic. Urea should be thoroughly mixed with the diet, using formulations that prevent separation of ingredients. Sudden high intakes by hungry animals or as a result of separating out urea in a feeder can cause toxicity and death. Because of this, the recommended levels and feeding instructions should be closely followed.

A number of plant proteins derived as by-products during the commercial extraction of vegetable oils are widely used as protein supplements for sheep. They contain 35% to 45% crude protein and include soybean meal, cottonseed meal, sunflower meal, linseed meal, canola meal and peanut meal. In general these are excellent protein sources. Soybean meal and cottonseed meal are the most widely used, but the others can be substituted on an equivalent protein basis when available at a competitive price. The corresponding oilseeds can also be used and have the added advantage of providing fat, which increases the energy content of the diet. However, high levels of fat can have a negative effect on fiber digestion in sheep.

Lambs younger than 8 weeks should not be given access to cottonseed, and cottonseed meals should be used with caution. All the gossypol in cottonseed is free gossypol and toxic, whereas gossypol in cottonseed meal exists in both bound and free forms. Bound gossypol is assumed to be nontoxic. The amount of free gossypol present depends on the oil extraction process used to produce the cottonseed meal. Mature sheep can detoxify free gossypol, but young lambs without functional rumens are susceptible to gossypol poisoning and should not be fed cottonseed or cottonseed meal until they have been consuming dry feed for several weeks and are 8 to 12 weeks old. After 8 weeks, lambs can be safely fed cottonseed meal processed by the expeller process or the expander modification of the direct solvent process, but caution should still be used in feeding direct solvent cottonseed meal or cottonseed. Mature sheep can be fed cottonseed meal regardless of the processing method, but cottonseed should probably be restricted to 0.5% of live weight, particularly if fed for extended periods.

Numerous methods are available to provide supplemental protein to grazing sheep. Animals can be hand-fed daily or at less frequent intervals, or they may be self-fed. When hand-feeding grain, it is best to have adequate trough space so that all animals can eat simultaneously. Pellets or cubes can be spread out on the ground. Protein blocks in varying degrees of hardness are commercially available and can be strategically placed to improve range or pasture utilization. Harder blocks are consumed more slowly. Although protein blocks save labor, they are more expensive than alternative protein supplements. Consumption is also regulated by calculating the needs of the flock for a week and distributing the required amount of feed weekly. Salt also is used to limit voluntary intake of a self-fed protein supplement. It is generally necessary to adjust the salt level (between 10% and 25% of the supplement) to obtain the desired level of intake. Trace mineralized salt should never be used for this purpose, and water should be readily available because of the increased water intake associated with increased salt consumption.

NUTRITION OF THE EWE

Feeding the ewe for optimal production requires knowledge of nutritional requirements during the critical periods of early growth and development, breeding, gestation, lactation, and recovery. Information contained in Appendix Tables A-17 and A-18 is based on research data from experimental animals maintained in a moderate environment and with minimal activity. Environmental conditions such as temperature, humidity, and wind velocity affect energy requirements, and the length and density of fleece are known to moderate the effects of climate.[26] Because of the moderating influence of fleece covering, sheep are more tolerant of climatic extremes than other farm animals.

Body condition scoring is a simple but useful procedure that can help producers make management decisions about the amounts of feed required by a ewe flock to optimize performance. The advantage of body condition scoring is that the procedure can be learned and used with ease. However, the procedure is subjective, and it must be remembered that the rate of change in body condition is too slow to indicate a short-term change in nutrient intake.

Scoring is done by using the hand to feel for the fullness of muscling and fat cover over and around the vertebrae in the loin region (Fig. 33-1). Sheep condition scores range from 0 to 5, with a 0 representing the thinnest and 5 representing the fattest animal. Typically more than 90% of the sheep in a flock fall within the scores 2, 3, and 4, and often 70% to 80% of the animals are covered by a range of two scores.

Sheep scoring 0 and 1 have no fat and very limited muscle energy reserves. These scores describe stages of severe emaciation. These sheep need individual care and feeding. Sheep scoring 2 and 2.5 are thrifty yet are in need of extra nutrition for body weight gain prior to breeding and to allow an additional 15 to 20 kg (30 to 45 lb) gain through gestation. Sheep scoring 3.5 are in above-average flesh but not fat enough to impair productivity. Sheep scoring 4 are moderately fat. Sheep scoring 5 are very fat. Ewes scor-

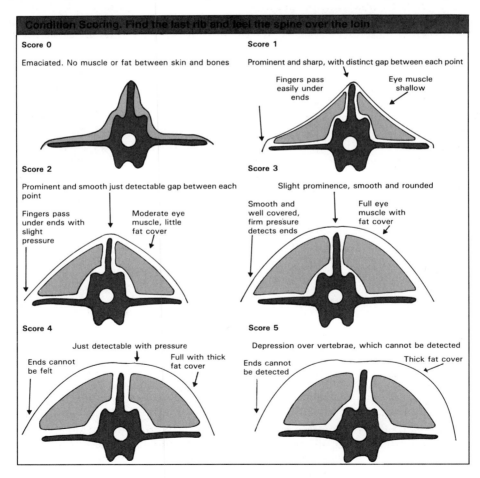

Fig. 33-1 Body condition scoring in sheep.
Reproduced with permission from Hindson J: In Pract, 11:152, 1989.

ing 4 and 5 should be fed to lose weight during gestation and lactation and hold condition during breeding and late gestation.

Ideally, ewes should range from condition scores of 2.5 to 3.5 during the year. A score of 2.5 at the end of lactation and 3.5 at lambing is recommended.

MAINTENANCE

Ewe weight and body condition score normally fluctuate throughout the year, generally reaching minimum values during lactation. Subsequently, during the time from weaning the lamb until the beginning of the next breeding period, it is often desirable for the ewe to be fed to regain body condition. Usually this can be accomplished by providing suitable grazing without the need for supplemental feeds.

The nutritional requirements for maintenance given in Appendix Table A-18 are for ewes in moderate condition. Thin ewes should be fed at the next higher weight category until condition improves. Excessive body conditions can adversely affect reproductive performance, and it is uneconomical to allow ewes to become fat during this period. Fat ewes should be fed to gradually lose condition during the dry period. Where grazing is lush, it may be necessary to limit grazing time in order to prevent ewes from becoming overly fat.

When ewes are being bred to lamb more frequently than once a year (accelerated lambing programs), lambs are early weaned, and the ewe may not have adequate time to recover body condition. The usual procedure is to breed the ewe to lamb once a year. With this schedule the ewe has a period of 3 to 4 months to recover weight lost during lactation.

Although feeding for optimal production is the usual objective, occasionally it becomes necessary to feed for survival of the breeding flock. Short-term survival feeding would be the situation during a winter storm. Long-term survival feeding occurs during an extended or severe drought. Australian scientists have demonstrated that dry ewes have a remarkable ability to survive on limited rations over an extended period of time. Mature sheep in reasonable condition can survive a 30% loss in body weight over a 6-month period. During a drought, as long as there is adequate dry forage available, the primary nutrient to provide is protein. When forages become limited, the primary ingredient to provide is energy. In a survival feeding situation, ewes can probably survive on about 60% of the NRC 1985[27] recommended levels for dry, mature ewes.

REPRODUCTION

Feeding the ewe flock to achieve optimal reproductive performance requires not only an adequate supply of feed (dry matter) but also that the feed provided contains the required nutrients. It is not unusual to find ewes confronted with an ample supply of low-quality harvested forage or mature forage in grazing situations that does not meet the protein needs of late gestation. In such circumstances, lamb birth weights are below normal, lambs are less vigorous at birth, and survivability is decreased. Similar results can be observed if energy is severely restricted, as can an increased incidence of pregnancy toxemia. Ewes also should not be allowed to become overly fat, because reproductive performance is adversely affected by obesity.

The critical periods during the reproduction cycle are from 2 to 3 weeks prior to breeding to 3 to 4 weeks after breeding and again during the last 4 to 6 weeks of gestation. The need for additional feed during these periods is reflected in the feeding levels given in Appendix Tables A-17 and A-18. These recommended nutrient requirements vary with maturity, body weight, stage of gestation, and lambing rate. Ewe lambs need to be fed at a higher rate than mature ewes during gestation to allow for continued growth and development.

Providing additional feed to ewes just prior to and during breeding is called *flushing*. The NRC[27] suggests increasing feed intake about 50% over maintenance during the period from 2 weeks before breeding to the first 3 weeks of breeding. This procedure is widely practiced and is accomplished by providing ewes with fresh pasture, supplemental harvested forage, or up to 0.45 kg (1 lb) of grain per ewe daily. This is done in an attempt to increase ovulation rate and consequently the number of lambs born, although it is not always effective. Thin ewes that have not recovered from the weight

Table 33-5 *Ewe rations appropriate at various stages of production (amounts to feed each ewe daily)**

Stage of production	Ration no.	Alfalfa hay, midbloom (kg)	Corn silage, mature (kg)	Alfalfa haylage (kg)	Corn stover, mature (kg)	Corn grain, dent (kg)	Soybean meal, 44% CP (kg)
Maintenance	1	1.36	—	—	—	—	—
	2	—	2.72	—	—	—	0.09
	3	—	—	2.72	—	—	—
Gestation (first 15 weeks)	1	1.59	—	—	—	—	—
	2	—	3.63	—	—	—	0.09
	3	—	—	3.18	—	0.09	—
	4	0.45	—	—	0.91	0.23	—
	5	—	—	—	1.13	0.23	0.18
Gestation (last 4 weeks; 130%-150% lambing rate expected)	1	1.59	—	—	—	0.34	—
	2	—	3.18	—	—	0.34	0.18
	3	—	—	3.18	—	0.34	—
	4	0.79	—	—	0.91	0.34	—
	5	—	—	—	1.36	0.34	0.27
Gestation (last 4 weeks; 180%-225% lambing rate expected)	1	1.59	—	—	—	0.57	—
	2	—	3.18	—	—	0.45	0.23
	3	—	—	3.18	—	0.57	—
	4	0.91	—	—	0.91	0.45	—
	5	—	—	—	1.36	0.57	0.27
Lactation (first 6 to 8 weeks suckling singles)	1	1.81	—	—	—	0.91	—
	2	—	4.54	—	—	0.45	0.34
	3	0.45	3.63	—	—	0.50	0.27
	4	—	—	3.63	—	0.91	—
Lactation (first 6 to 8 weeks suckling twins)	1	2.27	—	—	—	0.91	—
	2	—	4.54	—	—	0.45	0.57
	3	0.45	3.63	—	—	0.50	0.45
	4	—	—	3.63	—	1.02	0.11

Feeding directions

1. These rations are designed to be hand-fed at least once daily. The amounts given are required per ewe daily.
2. During maintenance and the first 15 weeks of gestation, each ewe should have 36 cm (14 inches) of bunk feed space. In late gestation (last 4 weeks) and during lactation, bunk space should be increased to 40-46 cm (16 to 18 inches).
3. Some of these rations are deficient in calcium or phosphorus; therefore, a supplement containing 50% trace mineral salt (for sheep) and 50% dicalcium phosphate should be fed free choice. The consumption of 23 g per sheep per day of this mixture provides the amounts of calcium and phosphorus needed for maintenance and the first 15 weeks of gestation; 45 g per day provides that needed for late gestation and lactation. Vitamins A and E should be added to the salt-mineral mix when sheep are fed the wheat straw and corn stover rations.
4. Ohio State University researchers have added 10 kg urea, 5 kg limestone, and 2 kg dicalcium phosphate to a tonne of corn silage (20 lb urea, 10 lb limestone, 4 lb dicalcium phosphate per ton) at time of ensiling or at feeding time. They found this provided adequate nutrition for the gestating ewe. Their recent research indicates the addition of 2.5 kg/tonne (5 lb/ton) of sodium sulfate would be particularly beneficial.

*These rations were formulated to meet the requirements of a 70-kg ewe in average body condition. Depending on actual size of the ewe and body condition score the amounts fed can be adjusted.

Table 33-6 *Complete diets for gestation and lactation*

		Gestation diets				Lactation diets			
		1	2	3	4	1	2	3	4
		Proportion (kg/tonne or lb/1000 lb)				Proportion (kg/tonne or lb/1000 lb)			
Ingredient									
Corn grain (dent yellow)		200	—	—	75	287.5	—	—	236.1
Alfalfa hay (mature)		715	—	640	—	592.5	—	500	—
Corn silage (mature)		—	964	—	—	—	922.5	—	—
Corn and cob meal		—	—	300	—	—	—	365	—
Alfalfa haylage		—	—	—	925	—	—	—	771.1
Soybean meal (solvent 44% CP)		—	30	—	—	62.5	67.5	75	—
Molasses (cane)		75.0	—	50	—	50	—	50	—
Calcium carbonate		—	4	—	—	—	5	—	—
Dicalcium phosphate		5	—	5	—	2.5	—	5	—
Trace mineral salt (sheep)		5	2.5	5	—	5	5	5	—
Nutritional content (%)									
Dry matter		87.2	38.0	87.2	48.2	87.5	40.2	87.2	51.4
TDN	As fed	51.8	26.5	52.4	30.1	56.6	28.3	56.5	34.2
	DM basis	59.4	69.8	60.1	62.5	64.7	70.4	64.8	66.5
Crude protein	As fed	10.2	4.1	9.8	7.7	12.2	5.7	12.1	7.8
	DM basis	11.7	10.8	11.3	16.0	14.0	14.1	13.9	15.2
Calcium	As fed	0.92	0.24	0.84	0.63	0.74	0.28	0.72	0.58
	DM basis	1.06	0.64	0.96	1.30	0.84	0.71	0.82	1.12
Phosphorus	As fed	0.26	0.10	0.27	0.13	0.26	0.12	0.30	0.14
	DM basis	0.30	0.25	0.31	0.28	0.30	0.29	0.35	0.28

loss of the previous lactation benefit the most from flushing. However, flushing during the peak of the breeding season, when the ovulation rate is highest, may not be advantageous. It is also not profitable to supplement ewes that are in average body condition (2.5 to 3) at the beginning of the breeding season.

It is desirable to have ewes in average body condition (2.5 to 3) when they are bred and to avoid sudden changes in feeding level during the 30 days after breeding. Evidence indicates this minimizes embryo and early fetal losses. It is recommended that animals be kept at a near maintenance level of feeding in the first month of pregnancy. If the animals have been flushed with additional feed, then feeding levels should be gradually reduced after breeding.

The period from 3 to 4 weeks after breeding until the last 4 to 6 weeks of gestation nutrition is not as critical as the immediate time before and after breeding. Body condition of the ewe should be the primary consideration in feeding during this time. Ewes in thin condition (less than 2) should be fed to allow them to achieve a body condition score of about 3 at lambing. Ewes that are fat (more than 3.5) can be allowed to lose 5% to 7% of their weight during this period. The desired goal is a body condition score of about 3 when the ewe lambs.

Table 33-5 provides a list of ewe rations appropriate at various stages of reproduction. The rations shown should be hand-fed to each ewe at least once daily. Table 33-6 lists complete diets for gestation and lactation.

ARTIFICIAL REARING OF LAMBS

Lambs can be raised successfully on milk replacer. This becomes necessary with orphaned lambs and in cases where the ewe refuses to claim her lambs or is providing an inadequate supply of milk. Although it is preferable to graft or foster these lambs (often called *bummer* lambs) onto a ewe that has adequate milk and is nursing only one lamb, quite often the only resource is to raise them artificially. This can be accomplished successfully with one of the good commercial lamb milk replacers.[29,35]

To minimize death losses, primarily from pneumonia but also from other diseases, it is essential that lambs receive colostrum within the first 12 to 24 hours of life. They should be given 180 to 240 ml (6 to 8 oz) of colostrum from another ewe or from a cow.[3] The sooner the decision is made to raise a lamb on milk replacer, the better. It is easier to train a newborn lamb to nurse a nipple than one that has nursed its dam for a period of days.[2,33] If lambs are taken from the ewe for a short period (4 to 6 hours) before feeding, then they will be hungry and more likely to accept a bottle.

The manufacturer's instructions should be followed in mixing and feeding lambs milk replacer. During the first 1 to 2 days, lambs are fed warm milk replacer. Subsequently, they can be provided with cool milk on an ad libitum basis. Each lamb consumes 0.9 to 1.4 L (1 to 1.5 qt) of liquid milk replacer per day. Lamb weight at 30 days should be about 11.4 to 14 kg (25 to 30 lb). In addition to the milk replacer, clean, fresh water should be provided free choice and a palatable creep feed should be started after the first week. Liquid milk replacer can be discontinued after 4 to 5 weeks and the lambs weaned onto a creep ration. During this period, good-quality hay should be provided in limited quantities.

Because lamb milk replacers are relatively expensive, researchers have examined the effects of alternative, lower-cost formulations[16,33] and use of other commercially available alternatives, such as a high-quality calf milk replacer or cow's milk.[17] Based on this research, it is generally recommended that lamb milk replacers contain 25% to 30% fat and 20% to 25% milk protein on a dry matter basis. Milk protein can be partially replaced by protein from other sources, but lamb performance is generally greater when skim milk powder (cow) is used.[33] A wide range of fats can be used, but rapeseed oil containing high levels of erucic acid should not be fed.[18,33] Under good management, satisfactory performance has also been reported with high-quality calf milk replacer (20% protein from spray-dried milk products and 20% fat) or even cow's milk used alone or blended with lamb milk replacer. Gains achieved with whole cow's milk or calf milk replacers are about 90% of those achieved with high-quality lamb milk replacers.[2,17] Probably the best approach to minimize the cost of artificial rearing is to wean lambs as soon as possible onto dry feed.

EARLY WEANING

Early weaning has been defined as the removal of the lamb from its milk supply before the normal weaning age is reached.[15] Obviously, normal weaning age varies with the production system. In grazing situations, lambs generally are weaned at about 12 to 16 weeks. Lambs raised artificially are weaned when they are 4 to 6 weeks old.

There are many good reasons for early weaning. When pasture resources are limited, lambs can be weaned at 6 to 8 weeks and placed on feed in drylots to reduce pressures on available forage that should be used to maintain the ewe flock.[38] For lambs artificially reared, milk replacers are more expensive and require more labor and greater attention to management than feeding a high-quality dry diet. Early weaning is also useful in accelerated lambing systems to ensure high conception and ovulation rates in the ewes.[15]

Early weaning has been accomplished successfully in artificially reared lambs at 10 to 14 days of age,[23] but inadequate voluntary water intake immediately after weaning in these young lambs[34] suggests that weaning at 21 to 28 days is safer.[19,23,30] The important factors determining the success of early weaning are (1) consumption of a dry diet prior to weaning, (2) voluntary water intake at time of weaning, and (3) postweaning consumption of a high-quality diet. Birth weight of the lamb and live weight gain prior to weaning are also important.[15,41] Ewe milk production is important in determining lamb gains[40] (see section on Lactation in this chapter).

The response of lambs to early weaning, regardless of the production system, is related to the stage of functional development of the rumen at weaning. Very little rumen development occurs until lambs begin to eat dry feed; thus, consumption of dry feed (pasture, conserved forage, ewe feed, or a creep feed) prior to weaning is essential to minimize the setback in performance that occurs when lambs are weaned. When lambs are to be early weaned, it is important to introduce a ration or feeding system that assures early intake of sufficient dry feed to encourage microbial digestion and rumen development. A grain ration with some roughage that is highly palatable, such as alfalfa, may be the best preweaning ration to assure optimal performance upon early weaning.

Suckling lambs obtain water from milk and may not voluntarily drink water in sufficient amounts to maintain normal rumen function. This can be especially critical at weaning. Recently weaned lambs should be closely observed to ensure adequate water intake, and a conveniently placed source of clean, fresh water

Table 33-7 *Lamb creep diets*

						Creep diets			
						Diet 5[4]		Diet 6[4]	
		Diet 1 (%)	Diet 2 (%)	Diet 3 (%)	Diet 4 (%)	Complete	Supplement	Complete	Supplement
Ingredient									
Corn grain (dent yellow)		51.5	61.0	—	—	85.0	—	—	—
Sorghum grain (milo)		—	—	53.0	65.0	—	—	80.0	—
Alfalfa hay (mid bloom)		20.0	20.0	15.0	15.0	—	—	—	—
Soybean meal (solvent 44% CP)		22.0	12.5	—	—	13.0	86.7	—	—
Cottonseed meal (solvent 41% CP)		—	—	25.0	13.0	—	—	17.0	85.0
Molasses (cane)		5.0	5.0	5.0	5.0	—	—	—	—
Trace mineral salt (sheep)		0.5	0.5	0.5	0.5	—	—	2.5	12.5
Calcium carbonate		1.0	1.0	1.5	1.5	1.5	10.0	0.5	2.5
Ammonium chloride		—	—	—	—	0.5	3.3	—	—
Vitamin A				1102 IU/kg			11,023 IU/kg	1102 IU/kg	8267 IU/kg
Vitamin E				22 IU/kg			220 IU/kg	22 IU/kg	165 IU/kg
Nutritional content (%)									
TDN	As fed	72.4	73.1	68.7	69.7	79.2	64.5	72.5	57.3
	DM basis	80.5	81.2	76.3	77.4	88.0	71.7	80.6	63.7
Crude protein	As fed	17.9	14.5	18.3	14.5	14.3	43.7	15.8	38.6
	DM basis	19.9	16.1	20.3	16.1	15.9	48.6	17.6	42.9
Calcium	As fed	0.76	0.73	0.86	0.85	0.63	4.00	1.01	4.74
	DM basis	0.84	0.81	0.96	0.94	0.70	4.44	1.12	5.27
Phosphorus	As fed	0.32	0.29	0.45	0.36	0.31	0.54	0.40	0.85
	DM basis	0.36	0.32	0.50	0.40	0.34	0.60	0.45	0.94

Feeding directions

1. A trace mineral salt mixture formulated for sheep should be used.
2. An antibiotic should be added to these diets. Chlortetracycline (Aureomycin) can be used at levels of 20 to 50 g/tonne (ppm) or oxytetracycline (Terramycin) at 10 to 20 g/tonne (ppm).
3. Lasalocid (Bovatec) can be used at 20 to 30 g/tonne (ppm) for control of coccidiosis, but it is not cleared by the Food and Drug Administration for use in combination with either of the above antibiotics.
4. Creep diets 5 and 6 can be fed as complete feeds or the supplement portion can be pelleted and mixed with whole grain (85% corn and 15% supplement for diet 5; 80% milo and 20% supplement for diet 6).

should be provided. Postweaning diets, whether pasture or complete feed, should be of a high quality. Young lambs do not have sufficient rumen development to handle poor-quality roughages. Lambs should be weaned onto high-quality forage, either pasture or hay, and/or a palatable, high-energy (3 Mcal DE per kg), high-protein (20% CP) complete feed. The particular postweaning diet depends on the production system, but the younger the lamb, the more important is diet quality. Sample diets for rapidly growing early-weaned lambs are given in Table 33-7.

Many equally suitable diets could be prepared by using other ingredients; for example, barley and wheat could be substituted for the corn or milo. However, wheat should constitute no more than half of the grain portion of the ration. Oats should not be used for very young lambs but can be worked into the diet of lambs with fully functional rumens.[30]

Depending on price and availability, canola meal, sunflower meal, peanut meal, and fish meal could be used to replace soybean meal and cottonseed meal. Cottonseed meal used in diets for early-weaned lambs (less than 6 to 8 weeks old) should contain no more than 2 g/kg (0.2%) of free gossypol and be restricted to 20% of the diet. In general, this means that cottonseed meal manufactured by the expeller process, the pre-press solvent process, or the expander modification of the direct solvent process would be safe to use.[8] The only concern is with using direct solvent cottonseed meal containing more than 0.2% free gossypol. The reported range of values for cottonseed meal produced by the direct solvent process is 0.1% to 0.5%.[5]

Lamb performance on soybean meal is slightly superior to cottonseed meal.[12,21] The use of 2% to 3% menhaden fish meal in combination with either cottonseed meal or soybean meal has improved performance compared with cottonseed meal or soybean meal alone.[1,12] Performance improvement with fish meal has been attributed to increased bypass of high-quality protein and greater availability and absorption of amino acids.[4] A nonprotein nitrogen source (urea) has been effectively used in high-energy diets once the rumen becomes fully functional.[19]

An antibiotic such as chlortetracycline or oxytetracycline often is included in weaning rations.

In situations where coccidiosis is a concern, lasalocid can be used at 30 g per tonne; however, it is not approved for use in combination with either of the above antibiotics. Thirty g of lasalocid per tonne is the highest level approved. Although this level aids in the control of coccidiosis, it may not provide complete protection for lambs that are severely stressed and challenged with large numbers of coccidial oocysts.[20]

CREEP FEEDING LAMBS

The practice of providing suckling lambs access to feed in an area that excludes the ewe is called *creep feeding*. Simple, palatable creep feeds that give excellent performance can be formulated from grains and a protein supplement.[22] Ingredients to use in a particular area depend on availability and price, but corn, milo, oats, barley, alfalfa hay or meal, and soybean and cottonseed meals are widely used and readily accepted by lambs. The value of creep feeding depends upon the system of production and the management program followed. Creep feeding is most likely to be profitable when lamb prices are relatively high and whenever the nutritional requirements of the lamb are not being met by milk and natural forage, such as when pastures begin to decline from overgrazing or drought and when ewes are nursing twins or triplets.[39]

Creeps (barriers that exclude ewes but allow access to lambs) may be used in pasture systems to allow lambs to use the better-quality pasture. Similar barriers may be used in either drylot or pasture situations to exclude ewes and allow lambs access to a creep ration. Creep rations may be introduced at any time but are particularly useful in accustoming lambs to a dry feed. Creep feeding is an essential part of any early weaning or accelerated lambing program.

Regardless of the production program, creep feeding generally is not economical unless the feeding system and the ration used assure an adequate intake of feed to stimulate growth. If the intake of a creep ration does not average 0.225 kg (0.5 lb) per day from 20 days of age to weaning, then no increase in lamb performance is realized from creep feeding. Examples of lamb creep diets based on either corn and soybean meal or milo and cottonseed meal are shown in Table 33-7.

Table 33-8 Growing-finishing diets for lambs

	Diets using corn and soybean meal*					Diets using milo and cottonseed meal*				
	1	2	3	4	5	1	2	3	4	5
Ingredient										
Corn grain (dent yellow)	310	415	517.5	630	732.5	—	—	—	—	—
Sorghum grain (milo)	—	—	—	—	—	195	327.5	462.5	607.5	737.4
Alfalfa hay (mature)	550	450	350	250	150	150	150	150	150	150
Cottonseed hulls	—	—	—	—	—	400	300	200	100	—
Soybean meal (solvent 44% CP)	70	65	60	55	50	—	—	—	—	—
Cottonseed meal (solvent 41% CP)	—	—	—	—	—	175	140	105	70	40
Molasses (cane)	60	60	60	50	50	60	60	60	50	50
Calcium carbonate	—	—	2.5	5	7.5	10	12.5	12.5	12.5	12.5
Trace mineral salt (sheep)	5	5	5	5	5	5	5	5	5	5
Ammonium chloride	5	5	5	5	5	5	5	5	5	5
Nutritional content										
Dry matter (%)	87.5	87.5	87.5	87.6	87.6	89.0	88.9	88.7	88.7	88.5
TDN (%)	65.9	70.1	74.0	78.1	82.0	60.4	64.3	68.5	72.7	76.8
Net energy for maintenance (Mcal per kg)	1.54	1.65	1.76	1.90	2.01	1.39	1.50	1.61	1.72	1.83
Net energy for gain (Mcal per kg)	0.772	0.90	1.01	1.15	1.28	0.64	0.75	0.88	1.01	1.15
Crude protein (%)	15.0	14.5	14.0	13.5	13.0	15.1	14.5	14.0	13.5	13.1
Protein bypass (%)	37.0	38.6	40.1	42.0	43.5	39.7	42.3	45.2	48.5	51.2
Calcium (%)	0.75	0.62	0.61	0.59	0.58	0.76	0.85	0.83	0.81	0.79
Phosphorus (%)	0.25	0.26	0.27	0.27	0.28	0.33	0.33	0.32	0.32	0.32

Feeding directions

1. These diets can be fed once daily in troughs or bunks if there is capacity for a day's feed. They can also be self-fed if the feeders are designed to handle such feed without bridging.
2. Offering lambs a good-quality hay for 1 to 3 days along with diets 1 or 2 (provided free choice) can be used to start lambs on feed.
3. About 7.5 cm (3 inches) of self-feeder or trough space must be provided per lamb for self-feeding and about 30 cm (12 inches) if hand-fed.
4. Gradually adapt the lambs to the higher-energy diets by allowing 4 to 7 days on a diet before switching to the diet with the next higher energy level.
5. Uniform mixing is essential.
6. Lambs must not be allowed to be without feed for even a short period of time.
7. Use the mineral and vitamin mixture for lamb diets in Table 33-9 with all these diets.

*Values are kg/tonne or lb/1000 lb.

FEEDING PROGRAMS AND DIETS FOR GROWING-FINISHING LAMBS

An important advantage of sheep is that they can be raised to market weight and finish completely on forage diets. Growing-finishing sheep also make effective use of a wide variety of harvested feeds, grains, crop residues, and plant by-products. Because of this versatility, numerous feeding programs are used to prepare lambs for market. Regardless of the program, the major consideration is price and availability of feedstuffs and the cost per unit of live weight gain.

Each year a portion of lambs coming off spring and summer pastures in the range states are sent directly to market. Actual numbers depend on availability of forage, as influenced by growing conditions, and in some cases the use of creep diets or supplemental feeds. Lambs that are not ready for slaughter quite often are placed in the large feedlots located in Texas, Colorado, New Mexico, and California and finished on high-grain diets. They also may be placed on a field of harvested alfalfa, sorghum or Sudan grass, small grains (wheat, oats, or barley), beet tops, or crucifers (kale, rape, turnips) to graze (aftermath grazing) as long as these are available and then moved to feedlots.[13,14]

Examples of the types of rations that have been successfully used for growing finishing lambs in feedlots are given in Table 33-8. One series of diets uses corn as the energy source and soybean meal as the protein source; the other uses milo as the energy source and cottonseed meal as the protein source. These diets increase in energy value (grain content) as they go from 1 to 5 in each series. They are suitable for starting all classes of lambs on feed and carrying them to market weight and finish. Lambs must be switched gradually from forage to high-grain diets to minimize problems with acidosis and enterotoxemia. Therefore, they should be started on either diet 1 or 2 and then gradually switched to the higher-energy diets 3, 4, and 5 by allowing a 4- to 7-day adjustment period on each diet before changing to the next higher energy level.

These diets were formulated to be nutritionally adequate and balanced for calcium-phosphorus and nitrogen-sulfur ratios. A trace mineral, vitamin, and salt mixture designed for use in complete sheep feeds and similar in composition to that shown in Table 33-9 should be included in the diets. It is essential that the calcium-phosphorus ratio be at least 2:1 to minimize problems with urinary calculi. Ammonium chloride was added to these diets at 0.5% to acidify the urine and reduce problems with urinary calculi. Ammonium sulfate could also have been used for the same purpose at a level of about 0.75%. Either lasalocid (for prevention and control of coccidiosis) or an approved antibiotic (chlortetracycline or oxytetracyline) could also have been added.

Many other feed ingredients can be used. Barley can be substituted for all or part of the corn or milo. Generally, wheat can replace up to half of the corn or milo in the diet. Dried beet pulp (up to 30% of the total grain portion of the ration) has been shown to be equal to corn in feed value and can be substituted for grain in these diets. Sheep use the nutrients available in whole grains very efficiently and grinding, dry rolling, steam flaking, or micronizing does not improve their digestibility.[9,10,31] The only justification for processing grains for sheep would be to alter the physical characteristics so that sorting of ration ingredients is minimized.

Many different roughages can be used successfully in lamb finishing diets. The choice of roughage depends on cost and availability and

Table 33-9 *Mineral and vitamin mixture for lamb diets*

Ingredient	kg/tonne (lb/1000 lb)	Contribution* to complete diet
Salt, plain fine mixing	864.81	0.43% salt
Sulfur, elemental	100	0.05% S
Cobalt carbonate ($CoCO_3$)	0.040	0.1 ppm Co
Ethylenedaimine dihydroiodide (EDDI)	0.050	0.2 ppm I
Manganese oxide (MnO)	5.15	20 ppm Mn
Zinc oxide (ZnO)	5.15	20 ppm Zn
Vitamin A†	8.8	1323 IU/kg
Vitamin E‡	16	22 IU/kg

*Contribution to the complete diet when the mineral and vitamin mixture is added to 1 complete lamb diet at 5 kg/tonne or 10 lb/ton.
†Contains 30,000 IU of vitamin A per g.
‡Contains 276 IU of vitamin E per g.

equipment for handling, processing, and feeding. Alfalfa hay is widely used; in the Southwest, however, plant residues such as cottonseed hulls and peanut hulls are effectively used to prevent digestive disturbances when high-concentrate rations are fed. When roughages make up a significant portion of the finishing ration (20% to 50%), the energy value of the roughage is important and needs to be considered in order to purchase the most economical source of energy. In minimum roughage (5% to 15%) rations, the physical characteristics of the roughage are more important than the energy value. Under such conditions alfalfa hay, cottonseed hulls, and peanut hulls work equally well.[7]

Pelleting high-roughage finishing diets generally increases feed intake and improves performance; however, the advantages usually do not warrant the additional cost of pelleting. There are no advantages to pelleting high-grain diets.

Net energy values for maintenance and gain can be used to estimate live weight gain for given levels of feed intake. The percentage of dietary crude protein bypassing the rumen can be used to indicate the possibility of using urea to supply a portion of the crude protein requirements. Urea may offer an opportunity for decreasing the cost of gains in high-concentrate rations. Rations containing moderate to high levels of roughage are less suitable with urea than with a natural protein source such as soybean or cottonseed meal.

Feed additives and implants

Feed additives, whether for nutritional (improvement in rate of gain and feed efficiency) or medicinal uses (control or prevention of specific diseases) are regulated by the U.S. Food and Drug Administration (FDA). Because clearances for feed additives are subject to change and new additives are occasionally cleared, knowledge of the regulations is essential to ensure that usage of a particular additive is legal. An excellent source for this information is the *Feed Additive Compendium*.* The following additives are currently approved by the FDA for use in sheep feeds: chlortetracycline (Aureomycin), 22 to 55 mg/kg (10 to 25 mg/lb) of feed; oxytetracycline (Terramycin), 11 to 22 mg/kg (5 to 10 mg/lb) of feed; lasalocid (Bovatec), 22 to 33 mg/kg (10 to 15 mg/lb) of feed; and Neomycin Base, 77 to 154 mg/kg (35 to 70 mg/lb) of feed. Each of these drugs must be used alone in diets, as no combinations are cleared for use.

Responses to antibiotics are variable and are markedly affected by differences in management and the amount of stress to which animals are subjected.[32] For example, Rumsey and coworkers[36] in a summary of three lamb feeding trials reported responses varying from −2.7% to +25% in live weight gain and from +8% to −30.9% in feed requirements for gain in lambs fed 25 mg of chlortetracycline per kg of feed, and Calhoun and Shelton[11] reported a 14.1% improvement in live weight gain and a 19% reduction in feed requirements for gain when lambs were fed 55 mg of chlortetracycline per kg of feed. Although improvements in rate of gain and feed efficiency ranging up to 40% have been observed when chlortetracycline or oxytetracycline was added to rations, the average response is in the range of 5% to 10%.

There is some evidence that antibiotics help reduce the incidence of enterotoxemia.[32] Chlortetracycline can be used at the level of 22 mg/kg (10 mg/lb) of feed for this purpose, and oxytetracycline can be used at a level of 25 mg per lamb per day; however, it would be unwise to depend solely upon antibiotics to control enterotoxemia. Vaccination with *Clostridium perfringens* type D or types C and D is the preferred method of controlling enterotoxemia. The primary response to feeding diets containing lasalocid is an improvement in the feed efficiency and control of coccidiosis. Lasalocid is approved for control of coccidiosis in sheep fed in confinement. The average improvement in gains and feed efficiency, with all types of diets, is about 7% when lasalocid is included at levels from 22 to 33 mg/kg (10 to 15 mg/lb).

Ammonium chloride or ammonium sulfate has been widely used in sheep diets to help prevent urinary calculi problems. The levels generally used are 0.5% for ammonium chloride and 0.75% for ammonium sulfate. Ammonium chloride is slightly more effective for this purpose

---

*Published annually. Available by subscription from Miller Publishing Co., 12400 Whitewater Drive, Suite 160, Minnetonka, MI 55343 (612-931-0211).

than ammonium sulfate; however, the addition of 0.5% of ammonium sulfate also adds 0.12% sulfur, which is an important consideration in diets requiring supplemental sulfur.

Zeranol (Ralgro) is the only implant approved for use with sheep. It is safe and effective when properly used. Generally, the response is greater with wethers than with ewe lambs and when the potential for growth is greater. When growth is restricted by nutrition (available feed) or stress (such as extremes in temperature), the response to zeranol is less. Because some estrogenic activity is associated with zeranol, sheep kept for breeding should not be implanted. Zeranol has little effect on carcass characteristics other than carcass weight. It has been shown to improve gains of suckling lambs under pasture range[37] and drylot conditions. It has also been shown to improve rate of gain and feed efficiency in growing-finishing lambs fed a wide range of diets. In feedlot trials with more than 6000 lambs, the average improvement in live weight gain was 15% and the average improvement in feed requirements for gain was 11%.[6]

REFERENCES

1. Adam AI: Doctoral dissertation, Ithaca, NY, 1982, Cornell University.
2. Ainsworth L and others: Tech Bull 11e, Ottawa, 1987, Agriculture Canada.
3. Al-Jawad AB and Lees JL: Anim Prod 40:123, 1985.
4. Beermann DH and others: J Anim Sci 62:370, 1986.
5. Berardi LC and Goldblatt LA: In Liener IE, editor: Toxic constituents of plant foodstuffs, New York, 1980, Academic Press.
6. Brown RG: J Am Vet Med Assoc 157:1537, 1980.
7. Calhoun MC: Prog Rep 3389, Texas Agricultural Experiment Station, 1976.
8. Calhoun MC and others: Prog Rep 4790, Texas Agricultural Experiment Station, 1990.
9. Calhoun MC and Shelton M: Prog Rep 2753, Texas Agricultural Experimental Station, 1970.
10. Calhoun MC and Shelton M: Prog Rep 2910, Texas Agricultural Experimental Station, 1971.
11. Calhoun MC and Shelton M: J Anim Sci 37:1433, 1973.
12. Calhoun MC and others: Prog Rep 4792, Texas Agricultural Experiment Station, 1990.
13. Chappell GLM: Proceedings of the 20th sheep profit day, Prog Rep 318, Lexington, 1989, University of Kentucky.
14. Fitzgerald S: In Haresign W, editor: Sheep production, Boston, 1983, Butterworths.
15. Glimp HA: In Proceedings of the Iowa State University sheep nutrition and feeding symposium, 1968.
16. Heaney DP, Shrestha JNB, and Peters HF: Can J Anim Sci 62:837, 1982.
17. Heaney DP, Shrestha JNB, and Peters HF: Can J Anim Sci 62:1135, 1982.
18. Heaney DP, Shrestha JNB, and Peters HF: Can J Anim Sci 62:1241, 1982.
19. Heaney DP, Shrestha JNB, and Peters HF: Can J Anim Sci 63:631, 1983.
20. Horton GMJ and Stockdale PHG: Am J Vet Res 42:433, 1981.
21. Huston JE and Shelton M: J Anim Sci 32:334, 1971.
22. Jordan RM and Hanke HE: SID Res Digest 4(1):24, 1987.
23. Magee BH, Lane SF, and Gillet D: SID Res J 4(2):26, 1988.
24. National Academy of Science: Nutrients and toxic substances in water for livestock and poultry, Washington, DC, 1974, NAS.
25. National Research Council: Mineral tolerance of domestic animals, Washington, DC, 1980, National Academy Press.
26. National Research Council: Effect of environment on nutrient requirements of domestic animals, Washington, DC, 1981, National Academy Press.
27. National Research Council: Nutrient requirements of sheep, Washington, DC, 1985, National Academy Press.
28. National Research Council: Ruminant nitrogen usage, Washington, DC, 1985, National Academy Press.
29. Nutrition. In Scott GE, editor: SID: the sheepman's production handbook, Denver, 1986, Abegg Printing.
30. Orskov ER: In Haresign W, editor: Sheep production, Boston, 1983, Butterworths.
31. Orskov ER, Fraser C, and McHattie I: Anim Prod 18:85, 1974.
32. Ott EA: In Proceedings of the Iowa State University sheep nutrition and feeding symposium, 1968.
33. Penning PD: In The management and diseases of sheep, Slough, England, 1979, Commonwealth Agricultural Bureaux.
34. Pond WG and others: J Anim Sci 55:1284, 1982.
35. Ross CV: Sheep production and management, Englewood Cliffs, NJ, 1989, Prentice Hall.
36. Rumsey TS and others: J Anim Sci 54:1040, 1982.
37. Snowder GD, Matthews NJ, and Matthews DH: SID Res J 5(2):5, 1989.
38. Thomas VM, Ayers E, and Kott RW: SID Res Digest 3(2):13, 1987.
39. Thomas VM and Kott RE: SID Res Dig 5(2):1, 1989.
40. Torres-Hernandez G and Hohenboken W: J Anim Sci 50:597, 1980.
41. Walker DM and Hunt SG: Aust J Agricul Res 32:89, 1981.

Nutritional Diseases in Sheep

JAMES G. MORRIS

Nutritional disease is a major cause of suboptimal production in sheep. The objective of this chapter is to briefly cover the major nutritional diseases of sheep which are of economic importance.

Three criteria should be used for the diagnosis of nutritional imbalances in sheep:
1. Evidence that the diet contains deficient or excessive amounts of nutrient
2. Exhibition of clinical signs compatible with a deficiency or excess of the nutrient
3. Supplementation of the diet with the missing essential nutrient or deletion of excess should reverse the clinical signs.

1. A simple deficiency of an essential nutrient may be evident from a low concentration of the nutrient in the diet. However, deficiencies can result from the chemical form of the nutrient being unavailable to the animal or because substances in the diet interfere with absorption of the nutrient. Excessive amounts of a given substance may cause either direct toxicity (e.g., copper) or indirect deficiencies by interference with other nutrients (e.g., molybdenum). The requirements of the animal may also be modified such that a "normal" diet may be inadequate. For example, ewes in late pregnancy have a higher demand for nutrients than dry ewes, and lambs with heavy intestinal parasite loads may have reduced absorption of nutrients from the gastrointestinal tract.

2. Experimental studies on nutrient deficiencies and toxicities have provided clinical values and necropsy findings that are typical of that deficiency. Under practical conditions, they often are modified by other factors such as deficiencies or excesses of other nutrients, length of time the animal has been exposed to the diet, individual variation, and environmental conditions.

3. Reversal of the clinical signs of the disease following supplementation of the diet with the nutrient suspected to be lacking is presumptive evidence of a nutrient deficiency. Reversal following deletion of a toxic nutrient is often slower and more difficult to interpret. Also, sheep can make "spontaneous" recoveries due to changes in weather, pasture species grazed, or physiological conditions (e.g., lambing). Therefore, control groups of comparable sheep are necessary to give validity to a response from alteration of the diet.

DEFICIENCIES OF ENERGY AND PROTEIN

Most of the nutrients absorbed from the feed are oxidized to meet the body's demand for energy to support metabolic processes. Although energy per se is not a nutrient, it is convenient to sum the contribution of all oxidized substrates under the general term of *energy*. A deficiency of energy is the most common nutrient deficiency that limits production in sheep. Inadequate intake of energy may result from deficits in feed availability, feed quantity, the inability of the animals to ingest and absorb nutrients from the feed, or a combination of these factors.

At low levels of forage availability, the intake of grazing sheep is directly related to the amount of forage present. Sheep have only a limited ability to extend their time of grazing or to increase their rate of biting while grazing in order to compensate for low forage availability. At higher levels of forage availability, feed intake is limited not by the amount of forage that can be harvested but by the rate at which the rumen can process the forage. Once the rumen of a sheep has reached normal capacity, the rate of feed intake equals the rate of material leaving the rumen. Particles of feed leave the rumen when they attain a critical particle size to pass through the omasum. Particle size reduction is achieved by a combination of mastication, particularly during rumination, and digestion by the microbes. Although time spent ruminating is related to the amount of cell wall ingested, sheep are apparently unable to increase rumination time above about 10 hours a day.

Deficiency of protein

As plants mature, the protein content declines, so sheep grazing mature forage often have a low intake of protein. Sheep are selective grazers and attempt to select a diet of higher quality than that of the average of all available plants ("on offer"), but this generally results in a reduction of total dry matter intake. Even with selective grazing, the protein content of mature plants may fall to levels where the availability of nitrogen and amino acids may limit microbial growth and decrease the rate of ruminal digestion of food, which in turn depresses dry matter and energy intake.

A deficiency of dietary protein in sheep with a functional rumen is commonly associated with a deficiency of energy. Apparent digestibility of energy and dry matter, voluntary feed intake, and protein concentration of forages are generally correlated. Low levels of dietary protein per se may reduce microbial activity in the rumen and cause a decreased appetite and a lowered feed intake. Parasitic burdens in sheep also depress appetite. The combined effects of protein loss via the parasites and reduced intake of protein frequently result in hypoproteinemia and the development of "bottle jaw" (edema in the ventral mandible).

Deficiency of energy

Many factors contribute to sheep having an inadequate intake of energy-yielding substrates. These factors can be grouped into those associated with forage (inadequate quantity or quality of forage) and those associated with animal factors that prevent normal intake of forage. In young lambs the quantity of forage that can be processed by the rumen, even if of high quality, is inadequate to sustain reasonable rates of growth. Older sheep may be unable to harvest adequate forage because of missing, broken, or defective teeth. Defective teeth may be caused by high intake of fluorine or inadequate intake of calcium in the diet as lambs. Sheep grazing abrasive forages high in silica or closely grazing plants growing on sandy soils may also suffer abnormally rapid wear of the teeth. Adverse weather conditions such as low temperatures may inhibit grazing, especially if accompanied by rain and wind, which destroy the thermal insulation of the fleece. Falls of snow that cover the forage also may greatly reduce grazing.

An uncomplicated deficit in energy intake by sheep leads to nonspecific clinical signs of depressed rate of gain, lowered fertility and overall reproductive rate, and higher lamb mortality. Potential underlying causes of the depressed energy intake should be investigated and corrected, if possible, rather than merely providing more feed.

METABOLIC DISTURBANCES IN SHEEP

The demand for nutrients, particularly energy, is greatest in the lactating ewe and in the ewe in the latter stages of pregnancy when carrying multiple fetuses. During lactation the energy- or protein-deficient ewe is able to reduce milk production. If the diet is inadequate in late gestation, however, she is unable to reduce the fetal demand for nutrients. Therefore, metabolic disease is of greatest economic importance in the pregnant ewe.

Ovine ketosis, pregnancy toxemia (twin lamb disease)

Ovine ketosis is a disease particularly of ewes carrying multiple fetuses in the late stages of pregnancy. In this disease, the demand of the fetuses for glucose exceeds the ability of the ewe's body to supply glucose. The fetal lamb is able to maintain very low plasma glucose levels (about 0.4 mmol/L or 8 mg/dl), which results in a marked concentration gradient across the placenta and efficient glucose transfer. In addition, the sheep fetus maintains a relatively high concentration of fructose in plasma (4 to 5.5 mmol/L or 80 to 100 mg per dl) that is derived from maternal glucose. Even though the concentration of fructose is greater than glucose in fetal plasma, glucose is the main energy substrate of the fetus. Therefore, there is a considerable drain of glucose from the ewe.

Although the high demand of the fetus for glucose is the major factor initiating the disease, the precipitating factor leading to clinical signs of ketosis is probably the adrenal response to hypoglycemia. Once the hypoglycemia has been sustained for a long period, an irreversible hypoglycemic encephalopathy ensues.

Etiology. The primary predisposing cause of ovine ketosis is undernutrition in the late stages of pregnancy that leads to a failure to absorb an adequate amount of glucogenic precursors. The cause of the undernutrition may be a lack of feed, an uneven distribution of feed among greedy and shy feeders, or stresses leading to periods of starvation. Stresses, including transport, storms that interfere with grazing, exposure to cold prevailing winds, or disease conditions (e.g., pneumonia, foot rot), can precipitate an outbreak in susceptible ewes. Ewes that are overfat are more susceptible than thin ewes.

Clinical signs. The early stage of ovine ketosis generally lasts 2 to 5 days. Affected ewes appear listless and do not eat. They walk aimlessly or remain in one position, often for hours, separated from the main flock. They often stand with the head held high in a star-gazing position and appear to be alert. Vision is impaired, however, as indicated by a poor or absent menace reflex. Ewes do not show the usual reaction to moving objects such as dogs. They are reluctant to move and, when forced to do so, often stumble into objects or knuckle-walk on their hind legs. Other behavioral anomalies include grinding of teeth, continual lapping of water, and head-pressing.

In later stages of the disease, ewes become drowsy. Some ewes lie quietly; others may show interspersed periods of seizure activity involving the muscles of the head, twitching of the lips, salivation, and movements of muscles of the cervical region that result in lateral deviation or dorsiflexion of the head. The muscle tremors eventually spread over the whole body, and the ewe goes into tonic-clonic convulsions and falls to the ground. Rarely after such a convulsive period will the ewe arise only to repeat the process a number of times before dying. The ground around ewes dying of pregnancy toxemia generally shows evidence of struggling.

The disease is characterized by hypoglycemia, ketonemia, ketonuria, and a large increase in the plasma free fatty acids. The blood glucose level of affected sheep falls from the normal of 3 mmol/L (50 to 60 mg/dl) to levels as low as 1 mmol/L (20 mg/dl) or less. Prolonged hypoglycemia leads to irreversible encephalopathy. Acidosis and dehydration may accompany the ketonuria. Mortality in ewes in the absence of treatment has been reported to be 90%. Necropsy findings include pale, fatty liver (up to 35% fat on a dry weight basis) and kidneys, enlarged adrenal glands, and pulmonary changes associated with recumbency prior to death.

Diagnosis. Differential diagnosis should exclude acute hypocalcemia before lambing and, less frequently, hypomagnesemia. Ewes with hypocalcemia generally respond in a short period to calcium therapy.

Treatment. Early recognition and treatment of the disease in the pregnant ewe flock is important. Mild cases have a high recovery rate and ewes that go on to produce live lambs. In severe cases, however, there can be a high mortality in both ewes and lambs despite treatment. Once ewes exhibit advanced CNS signs or are comatose, irreversible hypoglycemic encephalopathy has occurred, and no treatment is effective. Treatment in mildly affected ewes is aimed at providing sufficient glucose for the ewe until she resumes eating or gives birth. Oral glucose is ineffective, as it is metabolized to volatile fatty acids in the rumen. Daily intravenous glucose (1 to 3 L of a 5% solution) or 120 ml (4 oz) of oral glycerol or propylene glycol (1,2 propane diol but *not* ethylene glycol) also causes a temporary rise in blood glucose. Oral sodium propionate (50 to 100 g in water) is also effective. Acidosis is also a frequent problem of ewes with pregnancy toxemia and should be treated with intravenous bicarbonate. Following successful treatment, the ewe needs to be given feed, preferably a palatable ration containing about 25% grain, to prevent relapse.

A cesarean section to remove the lambs can be very helpful in severely affected ewes. Some clinical experience with the use of glucocorticoids in sheep has been disappointing, which is in contrast with their efficacy in treatment of dairy cows with ketosis. Large doses of dexamethasone (25 mg/ewe) may induce abortion within 48 hours in ewes that survive this long. If the ewe is in late gestation, a live lamb may be recovered. If ewes fail to respond to intravenous glucose and oral propylene glycol in 24 hours, ewes should either be treated with dexamethasone or have the lamb(s) taken by cesarean section.

Prevention. Ewes should be kept on a rising plane of nutrition, especially during the last 6 weeks of pregnancy. Obese ewes carrying multiple fetuses are more susceptible because excess abdominal fat results in a depression of feed intake. Avoid stresses that may cause a temporary lack of feed, such as shearing, trucking, or moving ewes to another location. Provision of shelter during storms and supplementary feeding can mitigate against adverse weather conditions. The incidence of the disease can be reduced by avoiding breeds that have a high propensity for multiple births (e.g., Finn) and not flushing ewes before and at mating, which increases the number of multiple births. The most economical approach is better nutrition in late pregnancy; provide grain 0.25 to 1 kg (0.5 to 2 lb) per head per day, depending upon forage quality and climate and stage.

Parturient paresis in ewes (acute hypocalcemia)

Parturient paresis is a metabolic disturbance of ewes that occurs in either late pregnancy or early lactation and is associated with hypocalcemia, hyperexcitability, ataxia, paresis, coma, and death. The exact cause of the disease is unknown, but it is a result of defective mobilization of calcium from body reserves and possibly a concurrent failure to absorb the normal amount of calcium from the gut.

Etiology. Normally there is only a sporadic incidence of the disease in ewes. Outbreaks of the disease occur when ewes are subjected to stresses such as being transported over long distances, forced into strenuous exercise, or suddenly deprived of feed. The period of highest susceptibility is 6 weeks before to 10 weeks following lambing.

Clinical signs. Early signs of hypocalcemia include mild hyperexcitability, muscle tremors, and a stilted gait. The ewe becomes depressed, which leads to recumbency with the head extended on the ground and the hind legs either under the body or extended. The disease is associated with a markedly depressed or absent corneal reflex, a weak pulse, and ruminal atony that results in tympany. Ruminal contents may be regurgitated and present in the mouth and nostrils. Death generally ensues in 6 to 12 hours in untreated animals.

Diagnosis. When an outbreak of hypocalcemia occurs, clinical history and the absence of urinary ketones usually exclude pregnancy toxemia from the differential diagnosis. In pregnancy toxemia the disease is restricted to pregnant ewes, the course of the disease is longer, and ketone bodies are present in urine. Treatment with calcium usually gives a rapid and dramatic response in cases of acute hypocalcemia but only temporary or no response in cases of hypoglycemia. Total serum calcium concentration is depressed in affected ewes to below 2.5 mmol/L (less than 5 mg/dl).

Treatment. Treatment involves intravenous administration of 100 ml calcium borogluconate solution (20%), with an optional 25 to 50 ml given subcutaneously. Cardiac output and blood flow to most organs is reduced about 50% in sheep during hypocalcemia. Heart rate should be monitored, especially if the solution is administered rapidly. Irregularities in beats or an increase in heart rate signals a need for diminution or cessation of the rate of infusion. Recovery is generally rapid, the ewe being apparently normal within 30 minutes after therapy.

Prevention. The prevention of stress such as transportation or sudden deprivation of food in the susceptible period of gestation is the best preventive measure. The role of high calcium diets in the last month before parturition in the induction of acute hypocalcemia in ewes has not been investigated. However, experience with dairy cattle indicates that a relatively low intake of calcium in the last 5 weeks of pregnancy may reduce the incidence of hypocalcemia. Low levels of magnesium in the diet may also be a complicating factor.

Hypomagnesemic tetany (lactation tetany, grass tetany)

Hypomagnesemic tetany is a metabolic disturbance occurring most commonly in adult ewes and characterized by hypomagnesemia, clonic-tonic muscular spasms, convulsions, and ultimately death due to respiratory failure.

Etiology. Hypomagnesemia occurs in all breeds and classes of sheep, but ewes within the first month of lambing are most likely to be affected. Many factors may interact to predispose sheep to an outbreak of lactation tetany, but the consistent biochemical aberration in all cases is hypomagnesemia. The hypomagnesemia is frequently associated with hypocalcemia, and the fall in serum calcium may be the precipitating factor. A short period of starvation in lactating ewes is capable of inducing an appreciable fall in both serum calcium and magnesium. This may be the mechanism responsible for the production of transit tetany.

Hypomagnesemia occurs under three main conditions: (1) when sheep are turned out in the spring onto young, lush grass pasture after being fed indoors on hay and concentrates in winter, (2) when sheep graze on young grown cereal crops such as wheat, and (3) sheep are overwintered on feeds (particularly hays) low in magnesium. A fourth condition, the feeding of lambs on all-milk diets for several months, also can produce hypomagnesemia.

Clinical signs. Affected ewes exhibit hyperesthesia, especially to natural stimuli such as lambs nursing, dogs barking, or other noise. Some ewes show an apparent lack of ability to move; others exhibit a stiff-limbed gait due to uncoordinated muscle movements. Often muscular fasciculations are present, especially of the facial muscles. Respiration rate is generally elevated, but temperature is normal at this stage of the disease. As the condition progresses, ewes collapse and go into tetanic spasms with all four limbs extended. During these spasms, body temperature is often elevated. Death can be rapid, without the coma typical of hypocalcemia.

Diagnosis. Plasma magnesium levels of less than 0.2 mmol/L (0.5 mg/dl) in live ewes constitute a positive diagnosis. Magnesium levels in the blood of dead sheep are always elevated over premortem levels and are not of diagnostic value. Differential diagnosis requires exclusion of hypocalcemia of lactating ewes and clostridial diseases. Hypocalcemia is associated with depressed serum calcium levels and generally near-normal serum magnesium levels, and it responds to calcium borogluconate therapy. Enterotoxemias can often be excluded through history, clinical findings, and postmortem findings, but in particular the presence or absence of specific toxins by means of a neutralization test differentiates between the two diseases.

Prevention. Prevention is based on supplying magnesium supplements at the time the sheep are exposed to low-magnesium forages. Magnesium supplements in general are unpalatable, and a carrier such as molasses, grain, or a combination of both has to be used to induce sheep to consume adequate amounts of the supplement. Magnesium oxide (7 g/day per ewe) in a grain mixture for confined sheep or a dehydrated molasses block may be used to supplement sheep on range. Magnesium sulfate in liquid molasses supplements is also employed. Supplements should be fed well before the projected period when an outbreak is likely to occur, as sheep do not actively seek magnesium supplements. The efficiency of delivering adequate amounts of magnesium by free-choice supplements is frequently low.

Excessive levels of magnesium intake also should be avoided. Magnesium salts can promote osmotic diarrhea and predispose to struvite calculi (magnesium ammonium phosphate). Diets fortified with magnesium should not be fed to male entire or castrated sheep, as these diets may predispose to the formation of urinary calculi that cause blockage of the urethra and death.

Treatment. The success of treatment depends on early diagnosis and intravenous administration of a combined magnesium and calcium solution such as calcium borogluconate–magnesium chloride or gluconate solution. An additional dose of the solution is often given by subcutaneous injection. Recovery after treatment is generally rapid, but, as with hypocalcemia, relapses are not infrequent. Treated animals should be kept under observation. The prognosis for sheep that require retreatment is generally poor.

Copper deficiency

Etiology. Ruminants are particularly susceptible to copper deficiency because metabolism of anaerobic bacteria in the rumen leaves less than 10% of the naturally occurring copper compounds in a form that is available for absorption. The presence of molybdenum and sulfate in the diet further reduces the availability of copper by the formation of thiomolybdate compounds that reduce absorption of copper and enhance the excretion of copper by the kidney.

The efficiency of copper uptake from feed varies with the breed of sheep. Some sheep maintain normal copper status when consuming feedstuffs that are seriously inadequate for other breeds. These sheep with high efficiency of copper uptake, however, develop copper poisoning on pastures that are healthy for normal sheep.

A distinction frequently is made between simple copper deficiency induced by a low concentration of copper in the diet and "conditioned" copper deficiency induced by the presence of factors that decrease absorption of copper or increase the requirement for copper by enhanced excretion of the mineral. Simple copper deficiency occurs infrequently in animals consuming natural diets, as copper is an essential element for plants and the ovine requirement is low. The adequacy of copper in a diet depends not only on the level of copper in the diet but also on the concentration of other minerals, particularly molybdenum, sulfate, calcium, and zinc, and protein. Areas of conditioned copper deficiency of grazing sheep occur in every major continent. In the United States, California, the Pacific Northwest, and southern Florida are problem areas.

Clinical signs. In adult wool breeds of sheep, some of the earliest signs of copper deficiency involve changes in the fleece. Copper-deficient sheep produce wool that lacks the normal crimp, a condition known in Australia as "steely wool." Steely wool has a lower tensile strength than normal wool because of fewer disulfide linkages in the wool and more free sulfydryl groups. In black-wooled sheep, copper deficiency produces a loss of normal melanin pigment from the fleece and causes a color change to a red or dirty gray. Copper-deficient adult sheep may also have microcytic hypochromic anemia.

In lambs, the clinical signs and pathology of copper deficiency are largely neurological, and the condition is often referred to as *swayback*. In young lambs, swayback may be evident at birth (neonatal or congenital ataxia) or may not appear until the lamb is up to 3 months of age (delayed ataxia). In the congenital form, severely affected lambs are stillborn or, if born alive, are small, dull, and depressed, with flaccid limbs, nystagmus, and depressed corneal and pupillary reflexes. Mildly affected lambs appear alert and have normal reflexes but posterior weakness and incoordinated gait. When delayed ataxia occurs in lambs, they exhibit progressive posterior weakness, incoordination, and paralysis. In older lambs the condition may not be evident until the lambs are herded or forced to run, at which time they exhibit lateral oscillations or swayback accompanied by a hopping gait.

On necropsy, the brains of lambs with severe congenital swayback frequently have gross lesions including cavitation of the white matter and hydrocephalus. In less severe cases, only microscopic lesions are evident, including chromatolysis in the brain stem and bilateral demyelination of the spinal cord. The brains and cords from lambs with delayed swayback show the presence of lesions in the large neurons of the red and vestibular nuclei, the reticular formation, and the ventral horns of the cord.

In contrast to young lambs, copper deficiency in older lambs does not produce neural lesions. These lambs, however, show growth retardation often referred to as *ill thrift*. Such lambs have emaciated carcasses at necropsy and hemosiderin deposits in the liver and spleen because of an inability to mobilize iron. Microcytic hypochromic anemia, fractures of the long bones, and osteoporosis may also be present. A number of other pathological changes relating to copper metabolism, including defective collagen formation that leads to weak bones and defective elastin formation, occur less commonly in sheep than in other species.

Diagnosis. Clinical diagnosis of swayback is based on the characteristic ataxia of the newborn or young lambs. Diagnosis is confirmed by the presence of lesions in the brain and spinal cord on histological examination of necropsy specimens and by analysis of liver and blood for copper. Values in blood of less than 9.5 μmol/L (0.6 μg/ml) and of liver of less than 1.25 mmol/kg dry matter (80 mg/kg dry matter) indicate low copper status. Analysis of feed or forage for copper is of limited value unless the concentrations of molybdenum and sulfur are measured. Depressed superoxide dismutase activity of erythrocyte is a useful index of copper deficiency in sheep.

Prevention. The damage to the CNS of lambs with swayback and to the wool fibers of copper-deficient sheep is irreversible. Emphasis should be placed on prevention of copper deficiency rather than on treatment. Copper deficiency can be corrected by increasing the copper content of the natural forage, administering copper to individual animals, or supplementing the flock. Because of the dangers of toxicity, copper should be supplemented only when there is clear evidence of a deficiency.

Fertilizer application. The application rate of fertilizers needed to raise the copper status of pastures depends on the soil type. As copper is an essential mineral element for plants, some pastures show a response in plant growth following application of copper-containing fertilizers.

Treatment of individual animals. Oral dosing of ewes with 1 g of bluestone (copper sulfate heptahydrate; $CuSO_4.5H_2O$) at 8 and 4 weeks before lambing is an effective preventive measure for swayback. As placental transfer of copper is low and milk contains little copper, lambs usually require further dosing with 6.25 mg/kg of body weight to prevent the delayed form of swayback. The efficiency of absorption of copper by preruminant lambs is much higher than by adult sheep.

Most copper salts are irritant to tissues when administered by the subcutaneous or intramuscular routes. Commercial formulations containing copper combined with amino acids, such as glycine or methionine, or complexed with EDTA are available. The copper in these preparations is translocated slowly from the injection site and frequently results in local tissue reactions in which the copper salt is sequestered. These copper salt preparations have low toxicity but variable efficacy. In contrast, another copper salt, diethylamine oxyquinoline sulfate, is rapidly translocated and gives no tissue reaction but can cause acute toxicity at levels close to the required amount of copper.

Three slow-release copper-containing preparations have been marketed for oral use in sheep. "Bullets" containing cupric oxide and a carrier are designed to be retained in the rumen-reticulum because of a high specific gravity. The bullets slowly release copper while in the rumen-reticulum. Copper oxide needles or wires that are administered in a gelatin capsule containing either 2 g (for lambs) or 4 g (adult sheep) copper also provide an effective method of increasing copper status without associated toxicity. The short copper oxide needles become lodged in the abomasal plicae and release cupric chloride, which is absorbed. Copper glass beads that slowly erode in the rumen are available in some countries.

Mineral mixtures offered free choice using salt as an attractant and containing 0.5% to 1% $CuSO_4.5H_2O$ have been used. These, however, have the serious shortcoming of variable intake, which may result in ineffective supplementation or toxicity.

Chronic copper toxicity

Chronic copper toxicity is a disease induced by prolonged consumption of a diet that results in continual storage of copper in the liver. Although development of the disease is slow, the hemolytic crisis that represents the full expression of the disease develops rapidly.

The ovine liver has a remarkable capacity to store copper. The amount of copper stored depends on not only the level of dietary copper but also the concentration of other dietary constituents. Higher than normal levels of molybdenum (1 to 2 mg/kg or ppm) in the presence of sulfate, calcium, and zinc as well as other elements interact to reduce copper storage. A dietary concentration of about 5 mg/kg (ppm) copper is normally necessary to provide the copper requirements of sheep. However, if the level is slightly increased (e.g., 8 to 10 mg/kg or ppm copper) and the concentration of molybdenum is extremely low, toxicity may develop. The reason for sheep being more sensitive to copper toxicity than most other animals is that sheep show a linear response in liver copper concentration over a range of dietary intakes of copper. In contrast, animals such as the laboratory rat exhibit little change in liver copper concentration over a wide range of copper intakes (10 to 900 μg copper per day). Only after copper intake exceeds 1000 μg/day does liver copper increase rapidly in rats.

Molybdenum and sulfate play major roles in determination of copper storage through the action of thiomolybdates formed in the rumen. Thiomolybdates reduce absorption of copper from the gastrointestinal tract of sheep. They also affect systemic copper metabolism by changing the distribution of copper in plasma and reduction of the availability of copper to metabolic sites in the body. In addition, thiomolybdates cause a release of copper from short-term storage and an increase in biliary and urinary excretion of copper.

Chronic copper toxicity occurs in Australia in sheep on pastures of normal to low copper and extremely low molybdenum. However, these sheep often have sustained liver damage from ingestion of hepatotoxins (e.g., pyrrolizidine alkaloids), which also adversely affects hepatic storage.

Clinical signs and diagnosis. Sheep with excessive copper storage in the liver show a rise in plasma aspartate aminotransferase (AAT), indicative of central lobular necrosis, when liver concentrations of copper exceed 12 mmol/kg (750 μg/g) liver dry matter. They do not exhibit clinical signs until the hemolytic crisis occurs. The crisis is precipitated by the excessive release of copper from the liver into the blood. The concentration of copper in plasma may rise to 20 times normal. The high concentration of copper is taken up by the erythrocytes and results in methemoglobin formation, Heinz body production, and hemolysis. At this time there is a release of AAT from the erythrocytes that further increases plasma concentration. Other liver enzymes also elevated in plasma include glutamic and sorbitol dehydrogenase and arginase.

At the time of the crisis, animals become anemic and weak, and there is a rise in the respiration rate. Kidney function declines due to hemoglobin nephropathy. Serum urea nitrogen and creatinine increase. Most animals die after a short period of recumbency.

On necropsy there is a generalized jaundice, a friable yellow-orange-brown liver, bronze-black kidneys, and hemoglobinuria. Histopathological examination of the liver shows centrilobular degeneration. Copper levels of the blood and liver are greatly elevated. Liver levels of 1000 to 3000 mg/kg (ppm) dry matter are common.

Prevention and treatment. Once the hemolytic crisis occurs, treatment is ineffective. In cases where toxicity has occurred, the concentration of copper in the body can be reduced by oral dosing with ammonium molybdate and sodium sulfate. Experimentally, thiomolybdates have been shown to be effective in depleting sheep of copper. The addition of ammonium molybdate to the drinking water at a rate of 20 g/L also has been effective. Sheep should be prevented from consuming sources of feed or mineral supplements high in copper. Horses, cattle and swine are much more tolerant to excess copper in the diet than sheep. Mineral supplements prepared for cattle or rations prepared for swine or horses should not be fed to sheep until it is ascertained that they do not contain excessive levels of copper.

SUMMARY

Sheep are particularly sensitive to energy deficiencies in late gestation. Special attention should be given to the provision of adequate feed before and after lambing. Ketosis and hypocalcemia should be treated as soon as possible to reduce losses. Ewes grazing low-magnesium pastures should be given magnesium supplements, but in male sheep excesses should be avoided because of a predisposition for urolithasis. Both toxicity and deficiency of copper are not uncommon in sheep, which are more sensitive to toxicity than are cattle or swine. Feed concentrations of copper, sulfur, and molybdenum should be monitored closely.

SELECTED REFERENCES

Blood DC and Radostits OM: Veterinary medicine, ed 7, London, 1989, Baillière Tindall.

Branzanji AAH and Daniel RCW: Effect of hypocalcemia on blood flow distribution in sheep. Res Vet Sci 42:92-95, 1987.

Ford EJ: Pregnancy toxemia. In WB Martin, editor: Diseases of sheep, Oxford, 1983, Blackwell Scientific Publications.

Kimberling CV: Jensen and Swift's diseases of sheep, ed 3 (rev), Philadelphia, 1988, Lea & Febiger.

Lindsay SD and Petherick DO: Dynamic biochemistry of animal production, Amsterdam, 1983, Elsevier.

Mosdøl G: Hypocalcemia in the ewe. Nord Vet Med 33:310-326.

National Research Council: Nutrient requirements of sheep, ed 6 (rev), 1985.

Russel AJF: Deficiencies of macro-elements in mineral metabolism. In WB Martin editor: Diseases of sheep, Oxford, 1983, Blackwell Scientific Publications.

Suttle NF and Linklater KA: Disorders related to trace element deficiencies. In WB Martin, editor: Diseases of sheep, Oxford, 1983, Blackwell Scientific Publications.

SECTION F Alimentation of Clinically Ill Ruminants

35
Nutritional Support for Sick Cattle and Calves

JONATHAN M. NAYLOR

Although some evidence suggests that short-term anorexia may aid recovery from disease, long periods of poor feed intake reduce immune function, wound healing, and tissue repair. These deleterious effects are due mainly to protein calorie malnutrition. However, poor water and electrolyte intake often accompanies aphagia. The most pressing problem in aphagic animals is often to prevent dehydration and maintain electrolyte balance.

CATTLE
Water and electrolytes

Cattle have a well-developed salt drive that can be harnessed to help maintain water and electrolyte balance. A simple method of providing sick cattle with water and electrolytes is to offer the choice between plain water and water fortified with salt and other electrolytes. Table 35-1 shows a mix that can be offered to support cattle with fluid losses such as those caused by abomasal displacements.

Although this mix works well in some situations, it cannot be relied upon entirely. Some animals, particularly those with severe disease, refuse the electrolyte mix. Thus hydration status and water and electrolyte intake should be monitored carefully in sick animals. If the animal does not correct its fluid deficit voluntarily, electrolyte solution can be administered by stomach tube, providing rumen motility is still present. Severely compromised animals with poor rumen motility should be treated intravenously using appropriate solutions.

Improving voluntary feed intake

The usual method of coaxing sick ruminants to eat is to offer a cafeteria of foods. Fresh grass is very palatable and is often the last feed to be refused. Adding molasses to grain may improve intake; some cattle prefer silage. It is important to feed small quantities at a time to keep the feed fresh and to avoid gorging on highly palatable feeds.

A variety of pharmacological methods have been tried to improve feed intake. Some success

Table 35-1 *Oral electrolyte solution for sick cattle*

Constituent	Amount (g)
Sodium chloride, NaCl	90
Potassium chloride, KCl	30

Dissolve in 20 L water. Final composition is Na^+ 77 mmol/L, K^+ 20 mmol/L, Cl^- 97 mmol/L. Offer free choice along with plain water.

Table 35-2 *Composition of various oral electrolyte solutions designed to be fed to diarrheic calves*

Product (manufacturer)	Glucose (mmol/L)	Glycine (mmol/L)	Sodium (mmol/L)	Potassium (mmol/L)	Chloride (mmol/L)
Revibe (Langford)	120	40	120	20	50
Hydra-Lyte (Vet-A-Mix)	368	16	85	30	45
Revive (Tech America)	High	Present	120	25	85
Hydra (Vetrepharm)	60	60	100	15	75
Electro-Plus A (Pitman-Moore)	60	60	100	15	75
Life-Guard H.E. (Norden Labs)	446	33	113	26	58
Biolyte (Upjohn)	398	0	142	24	80
Life-Guard (Norden Labs)	166	36	105	26	51
Electrolyte Powder (Rogar/STB)	110	0	139	10	101
DiaProof (Diamond Scientific)	80*	0	75*	15*	45*
Calf-Lyte (Austin Lab)	139	47	80	19	89
Ion Aid (Syntex)	117	107	76	24	76
Resorb (Beecham)	120	44	78	17	78

Minor differences in concentration exist between the same product used in different countries depending on whether or not the fluid is dissolved in quarts or liters of water.
*Values are based on our analyses and have been rounded; packet to packet variation exists. DiaProof also contains a gelling agent.
Note that for sodium, potassium, and chloride, mmol/L and mEq/L are identical.
Table modified from Vet Clin North Am (Food Anim Pract) 6:60, 1990.

has been claimed for the administration of B vitamins. In the past compounds such as ginger and even arsenic were administered to "stimulate appetite." Nonsteroidal antiinflammatory agents such as flunixin meglumine may improve feed intake by alleviating pain and counteracting some of the effects of endotoxin. They may be useful, but one should be wary of possible adverse side effects (e.g., an increased tendency to abomasal ulceration) and of masking signs indicating a deterioration in the animal's condition.

Tube feeding, transfaunation

Sometimes sick cattle are transfaunated, which involves collection of rumen contents from a healthy cow and straining out the solid matter (with gauze or a large, bowl-shaped kitchen sieve). Some hospitals maintain special donor cows fitted with a rumen fistula for rumen fluid collection. An alternative source is from cattle slaughtered at an abattoir. Precautions should be taken to prevent the spread of pathogenic organisms (such as *Salmonella*) from the donor to the recipient cow. If rumen contents are to be transported long-distance, a sealed container should be used to maintain an anaerobic atmosphere. The liquid is administered into the rumen of the sick cow using a stomach tube. Typically 1 to 2 L (1 to 2 qt) of rumen fluid are given. Often nutrients are administered at the same time; ground alfalfa meal (about 1 kg or 2 lb), brewers yeast (about 250 g or 0.5 lb), glycerol (250 to 500 ml or 0.5 to 1 pint), and 20 L (5 gallons) of electrolyte solution might follow the liquid into the rumen. A good bilge pump is useful for pumping the mixture into the rumen, and long-sleeved plastic gloves help prevent spread of odor. Transfaunation is thought to help restore the rumen flora following a period of anorexia or toxic insult such as grain overload. The hope is that this, in turn, will improve feed intake.

SUPPORT OF THE SICK CALF
Water and electrolytes

A considerable amount of research has been performed in the area of oral fluid therapy in calves.

Table 35-2 *Composition of various oral electrolyte solutions designed to be fed to diarrheic calves—cont'd*

Alkalinizing ability (mEq/L)	Bicarbonate (mEq/L)	Citrate (mEq/L)	Acetate (mEq/L)	Citric acid (mEq/L)	Comments
72	0	0	80	8	High-energy products
70	0	10	70	0	High-energy products
60	0	0	60	Present	
40	0	40	0	0	Lower in glucose and alkalizing ability. Citrate interferes with milk clotting
40	0	40	0	0	Lower in glucose and alkalizing ability. Citrate interferes with milk clotting
80	80	0	0	0	High energy. Bicarbonate can interfere with milk clotting and digestion
85	85	0	0	0	High energy. Bicarbonate can interfere with milk clotting and digestion
80	80	0	0	0	High energy. Bicarbonate can interfere with milk clotting and digestion
48	48	0	0	0	Bicarbonate can interfere with milk clotting and digestion
55*	Present	Present	0	0	Bicarbonate can interfere with milk clotting and digestion
6	0	6	0	0	No significant alkalizing ability
0	0	0	0	0	No significant alkalizing ability
0	0	1	0	1	No significant alkalizing ability

It has been spurred by the realization that dehydration and acidosis are the major causes of death in calves with diarrhea.

In general, oral electrolyte solutions should contain sodium (60 to 120 mmol/L), potassium, and chloride to replace losses. Severely sick diarrheic calves are often acidotic, and an alkalinizing agent should be present if the calf is diarrheic. Both bicarbonate and metabolizable bases are used as alkalinizing agents. Metabolizable bases include lactate, acetate, propionate, citrate, and gluconate. They work because hydrogen ions are consumed as the bases are metabolized to glucose or to water and carbon dioxide.[8] Some work suggests that bicarbonate- or citrate- containing solutions interfere with milk digestion; however, the magnitude of this effect is quite variable.[5,9] Glucose, glycine, or both are usually added to increase the rate of sodium and water absorption. Absorption is probably at its peak when the sum of glucose plus glycine in the solution is 150 to 200 mmol/L.[8] A summary of various oral electrolyte solutions is shown (Table 35-2).

Voluntary food intake

Calves that have a suck reflex can usually be coaxed to suck fluid from a bottle; bucket-fed calves may prefer their food in a bucket. There is some doubt regarding the optimum diet for a diarrheic calf. On the one hand, a considerable body of evidence clearly shows that digestion and absorption of nutrients from the gut is suboptimal in diarrheic calves.[12-14] On the other hand, the gut has a considerable reserve capacity.[7] Experimental studies show that calves can gain weight at a normal rate on appropriate diets (whole cow's milk plus appropriate oral electrolyte solutions), despite the presence of severe diarrhea.[5,9]

There is little point in holding diarrheic calves that are bright and alert off feed. They should be supplemented with oral electrolyte solutions and carefully monitored. Diarrheic calves that are in good body condition but depressed and reluctant to suck are generally held off feed while they are rehydrated with oral electrolyte solutions. As they become stronger and regain their suck reflex, milk is gradually reintroduced. The author usually withholds milk from depressed calves for about 24 hours (day 1) and then offers milk in small quantities if the calf is willing to suck. A total of 5% of body weight as whole cow's milk divided among two to four feeds is offered on day 2, and this amount is gradually increased to a total of 10% of body weight per day as milk over the next 2 to 3 days. Oral electrolyte solutions are given throughout the diarrheic period. These should be fed separately from milk to avoid interfering with clotting of milk in the abomasum. Bicarbonate- and citrate-containing solutions can interfere with clotting of the milk in the abomasum[2] and may reduce digestibility of milk, even when fed in separate feeds.[5] Acetate-containing solutions allow correction of acidosis without interfering with milk digestion. Milk replacers are usually inferior in digestibility to milk and are not as suitable for diarrheic patients.[3,4,6] The ideal source of milk for a sick calf is a cow. Suckled calves may benefit from milk intake because they take several small meals throughout the day[1,10] rather than the two large meals that are the norm in bucket-feeding operations.

If the calf will suck, feeding from a nipple pail or nipple bucket offers the best chance of stimulating the esophageal groove to close. This promotes rapid passage of milk into the abomasum. Tube feeding of electrolytes is acceptable, provided the calf is strong enough to stand and still has some suck reflex. In very weak calves, the gut becomes atonic; fluid deposited in the rumen may ferment and cause bloat.

Diarrheic calves may be in poor body condition when they are presented for treatment. Sometimes this is due to poor feeding practices, inadequate milk supply from the dam, or overzealous withholding or dilution of milk by the farmer. These calves should be gradually reintroduced to milk. Calves that have a poor tolerance for milk as evidenced by a deterioration in their condition when on milk, as well as calves that refuse milk, should be offered a high-energy oral electrolyte product to help preserve body condition. High-energy oral electrolyte preparations generally supply about 70% to 80% of maintenance energy requirements when they are fed at the rate of 2 L (2 qt) per feed three times a day.

Intravenous feeding (Chapter 36) may give better results than high-energy electrolyte solutions in those cases where this option is economically viable.[11]

CONCLUSION

The field of nutritional support for sick cattle is in its infancy. Problems of poor feed intake in adults are usually countered by offering a cafeteria of feeds and allowing the cow to choose. Grass is particularly palatable. Diarrheic calves should be supplemented with oral electrolytes solution. Although maldigestion is well documented, many calves are still able to benefit from milk feeding. Whole cow's milk is the ideal diet; some citrate- and bicarbonate-containing oral electrolyte solutions interfere with milk digestion.

REFERENCES

1. Boggs DL and others: J Anim Sci 51:550, 1980.
2. Bywater RJ: Vet Rec 107:549, 1980.
3. Dalton RG, Fisher EW, and McIntyre WIM: Vet Rec 72:1186, 1960.
4. Fallon RJ and Harte FJ: Ann Rech Vet 14:473, 1983.
5. Heath SE and others: Can J Vet Res 53:477, 1989.
6. Kertz AF: J Dairy Sci 60:1006, 1977.
7. Mylrea PJ: Res Vet Sci 7:407, 1966.
8. Naylor JM: Vet Clin North Am (Food Anim Pract) 6:51, 1990.
9. Naylor JM and others: Can Vet J (in press).
10. Odde KG, Kiracofe GH, and Schalles RR: J Anim Sci 61:307, 1985.
11. Sweeney RW and others: Proceedings of the 6th Annual Veterinary Medical Forum, American College Veterinary Internal Medicine 6:735, 1988.
12. Woode GN, Smith C, and Dennis MJ: Vet Rec 102:340, 1978.
13. Woode GN and others: J Clin Microbiol 18:358, 1983.
14. Youanes YD and Herdt TH: Am J Vet Res 48:719, 1987.

36

Alimentation of the Clinically Ill Neonatal Ruminant

LYNN R. HOVDA, SHEILA M. McGUIRK

Neonatal ruminants are in a more precarious nutritional situation than adults. Decreased body mass, increased metabolic rate, low fat stores, and poor feeding techniques all combine to place the neonate at a greater risk. Neonatal ruminants are susceptible to a wide variety of diarrheal diseases, each of which has specific effects on the gastrointestinal tract. Diarrhea of several days duration frequently results in malnutrition.[5] Just the increased metabolic demands of an illness can cause a neonatal ruminant to enter into a negative energy balance.

The economic value of the ruminant often plays an important role in determining the therapy chosen. Enteral products such as milk, milk replacers, and oral electrolyte solutions are cost-effective but may not provide the support required. The majority of commercially available enteral products designed for human beings have not shown to date any therapeutic or economic advantages when used in neonatal ruminants. The cost of parenteral nutrition has decreased over the past 5 to 10 years, but it still remains an expensive option. When the total value of some neonatal ruminants is considered, however, this amount may be affordable.

Frequently, clinically ill neonatal ruminants can be managed on the premises. This requires a warm, dry environment, provision of adequate nursing care, and good veterinary supervision. When these animals fail to respond to therapy or require more intensive nursing support, they should be transported to a larger referral center.

ENERGY, PROTEIN, AND FAT REQUIREMENTS

Daily energy requirements for maintenance in calves have been determined to be about 50 kcal/kg (25 kcal/lb) body weight (BW). Additional kcal needed for growth have been determined to be about 300 kcal/100 g gain in BW.[4,6,7,27] For calves weighing less than 35 to 40 kg (80 to 90 lb) and for lambs, kids, and neonatal llamas, maintenance energy requirements in kcal should be calculated using Kleiber's formula ($140 \times kg\ BW^{0.75}$).[31] This is more reliable, particularly at a low body weight.

The provision of energy is important because critically ill ruminants may quickly enter into a negative energy balance. This is due to their small amounts of white adipose tissue and to an apparent inability to mobilize triglycerides with subsequent low plasma or serum concentrations of free fatty acids. Adipose tissue serves as a storage depot and becomes a primary energy source in periods of fasting.[24] Both brown and white adipose tissue may be found in neonatal animals.[36] Brown adipose tissue is involved in nonshivering thermogenesis, producing heat and contributing to cold resistance.[1,36] White adipose tissue is similar to that in adult ruminants and acts as an energy source.[36] The majority of adipose tissue in newborn calves is brown, comprising 2% of the total body weight.[1] Recently, one group has shown that perirenal fat, a major fat depot in calves, can be used as an energy source.[36] Disease often puts additional demands on the already limited energy supplies. Even simple abnormalities such as temperature elevations assume importance, as each Celsius degree increase over the basal temperature is associated with a 12% increase in energy expenditure.[33]

The protein requirements for maintenance and growth in neonatal ruminants are influenced by the biological value of the protein.[6,32] When the biological value is 100, the formula ADN = 65.3 + 32.2(G) (where ADN is mg of apparent digestible nitrogen required per kg BW per day, 65.3 is the endogenous nitrogen production in mg per kg BW per day, and 32.2G is the mg nitrogen retention associated with 1 g body weight gain × G, the growth rate in g per kg BW per day) may be used.[6] As the biological value decreases, protein requirements increase proportionately.[6] Alternatively, the daily protein requirement in calves can be estimated as 0.6 g nitrogen per kg BW per day.[3] The requirement for dietary essential amino acids in calves has been shown to be similar to those of human beings, dogs, and rats.[32]

Dietary fatty acid requirements for neonatal ruminants have not been determined.[32] A syndrome resembling fatty acid deficiency has been produced in newborn calves fed a fat-free diet for 5 weeks. The introduction of lard or hydrogenated cocoa butter into the diet reversed the signs and symptoms.[10]

ENTERAL SUPPORT

Oral fluid therapy is often the first and most appropriate method for supporting ill neonatal ruminants. Milk, milk replacers, oral electrolyte solutions, or commercially available enteral solutions may be used to provide maintenance fluid and electrolyte requirements, correct dehydration, and compensate for ongoing losses. The quantity and quality of nutritional support provided depend upon the individual product used, as milk replacers and oral electrolyte solutions contain wide variations in energy (Chapter 35).

Milk

Whole milk is still considered to be the best form of nutrition for nondiarrheic neonatal ruminants that are able to suckle.[4,37] It is an energy-rich source of carbohydrates, proteins, and fat in forms that are easily digested by the neonatal GI tract. Although the appropriate dam's milk is desirable, in some instances it may not be readily available. Neonatal llamas have been raised successfully on both whole goat's and whole cow's milk; lambs tolerate whole cow's milk.

Depressed lactase activity has been demonstrated in calves who have been diarrheic for several days.[8] Possibly they have a reduced ability to digest milk; this can aggravate their diarrhea. Lactaid,* a commercially available lactase enzyme, can be added to the milk to enhance digestion. Calves, goats, and neonatal llamas have been observed to benefit from the use of Lactaid. However, diarrheic calves fed milk without Lactaid may gain weight at normal rates, indicating considerable reserve enzymatic capacity in the gut in some cases.

Milk replacers

A good-quality milk replacer may be used if milk is not available. Currently, milk replacers are available for the calf, kid, and lamb but not for the neonatal llama. Lamb milk replacer has been successfully used in the healthy llama but has not been adequately evaluated in the clinically ill neonate.

COMMERCIALLY AVAILABLE ENTERAL SOLUTIONS

Recently, interest has developed in the use of commercially available enteral solutions manufactured for human consumption. Most of these products are convenient, easy to use, calorie rich, and provide adequate amounts of carbohydrates, fats, and proteins. Over 75 enteral products are currently available for human use, varying both in cost and composition. They can be classified as blenderized, milk based, lactose free, elemental (predigested), protein component based, fat component based, carbohydrate component based, and specialized.[9,25,30] Of these, the two most commonly used by human beings are lactose free and elemental.[30]

The majority of enteral solutions used in human therapy contain high molecular weight protein, fat, and carbohydrates (primarily glucose,

---

*Lactaid Inc., Pleasantville, NJ.

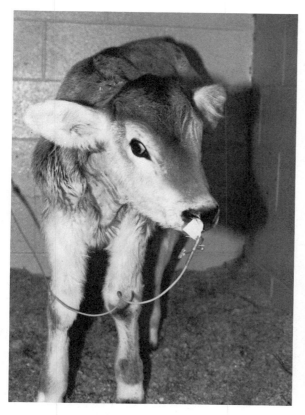

Fig. 36-1 Nasogastric feeding tube sutured into calf's nostril.

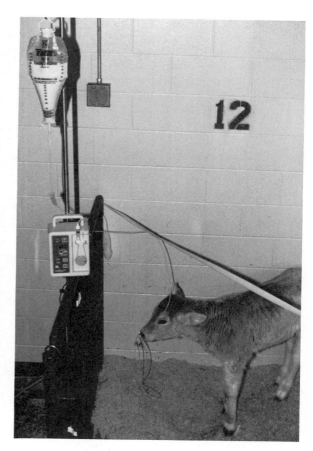

Fig. 36-2 Calf receiving continuous enteral nutrition with indwelling nasogastric tube and pump.

sucrose, maltose, and fructose).[30] These solutions are mainly nonlactose containing, isosmolar and require digestion. In human beings, they are well tolerated with a minimum of complications. The most commonly reported human adverse side effects include nausea, abdominal distension and diarrhea.[21,30,41] When used in neonatal calves this type of product frequently has resulted in diarrhea, due to the use of starches and sugars which calves are unable to digest.

Elemental diets are used when digestion is impaired or when a low residue diet is required.[30,41] Elemental diets generally are hyperosmotic and contain amino acids or a mixture of short chain peptides and amino acids as the protein source, oligosaccharides and monosaccharides as the carbohydrate source, and fat primarily in the form of medium chain triglycerides. The contents are present in a predigested form which enhances absorption. Reported human complications of elemental diets include unpalatability, diarrhea and dehydration, constipation, hyperglycemia, pancreatic acinar atrophy, alteration of gastrointestinal flora and decreased gastric emptying.[21,30,41] In the author's experience calves tolerate this formula somewhat better than the lactose free, although soft feces have developed in some cases.

In calves that are able to suckle, formula is administered via a nipple bottle. Calves that are unable or unwilling to suckle can have a nasogastric tube placed in the esophagus and the solution continually infused with an enteral pump (Figs. 36-1 and 36-2). These tubes are small and their placement should be verified with endoscopy or radiographs if there is any question regarding the positioning.[40] The slow, continuous delivery provided by the pumps may help to increase absorption and decrease side effects such as diarrhea, abdominal distension, and regurgitation associated with bolus feeding.

Medium chain triglycerides (MCT Oil*) are a fat component based formula which has been used successfully in calves and kids. It is composed of 100% coconut oil and provides 8.3 kcal/g. The mixture of medium chain saturated fatty acids (6 to 12 carbon atoms) is hydrolyzed and absorbed more readily than long chain triglycerides especially when lipase is decreased.[30] The fatty acids produced by hydrolysis are not incorporated into chylomicrons, but instead are transported to the liver by the portal vein system. MCT Oil should be gradually introduced into the diet to avoid the side effects of nausea and diarrhea. Doses of 15 to 30 cc four times a day have been tolerated by calves and 2.5 to 5 cc four times a day by kids with minimal complications. To prevent aspiration, MCT Oil should be mixed with milk and administered through a nipple bottle.

PARENTERAL NUTRITION

Parenteral nutrition has been widely used in malnourished and traumatized human beings for the past 20 years.[22] More recently, the use has been reported in puppies and dogs,[33] cats,[33] foals and horses,[16,17] calves,[3,18] and neonatal llamas (Fig. 36-3).[19] Parenteral nutrition is classified as either total parenteral nutrition, partial parenteral nutrition, or hyperalimentation.[3,22,33] In total parenteral nutrition nothing is given orally. An intravenous solution containing water, carbohydrates, protein, fat, electrolytes, vitamins and minerals provides enough nutrients for the maintenance of body weight and metabolic functions.[3,22] Partial parenteral nutrition refers to the use of both parenteral and enteral routes of administration. When nutrients are supplied in excess of those required for maintenance, the term *hyperalimentation* is used.[3,22]

Indications for parenteral nutrition are nu-

---
*Mead Johnson, Evansville, Ind.

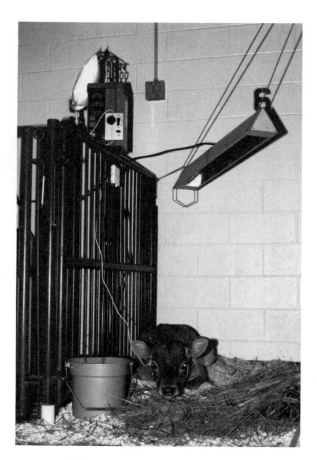

Fig. 36-3 Calf receiving parenteral nutrition.

merous and widespread. The single, dominant factor in determining if parenteral nutrition is needed is the inability to obtain adequate nutrients by any other route.[22,33] In human infants, parenteral nutrition often is used if 4 or 5 days of inadequate oral intake have occurred.[22] Low birth weight infants or infants with prior nutritional depletion may require parenteral nutrition after only 2-3 days.[22] In neonatal ruminants, parenteral nutrition is primarily used to provide supportive therapy for chronic, persistent diarrhea, congenital gastrointestinal anomalies, preoperative and postoperative malnourishment, failure to thrive, and in animals that are unwilling or unable to tolerate enteral feedings.

Partial parenteral nutrition is used more frequently than total parenteral nutrition in neonatal ruminants, especially in the therapy of diarrhea. As soon as they are able to suckle, neonates are give 1% to 2% of their body weight as milk or milk replacer divided into frequent daily feedings. In dogs, rats, and human infants, the use of small amounts of oral nutrients in addition to parenteral nutrition prevents the development of abnormal gastrointestinal tract changes.[12] These oral nutrients provide for adaptive increases in carbohydrate digestion and enzyme activity and stimulate epithelial cell regeneration.[12] In addition, they may also prevent the recurrence of diarrhea when the transition is made to enteral nutrition.[3]

In diarrheic neonatal ruminants, parenteral nutrition helps to stop the perpetuation of diarrhea caused by oral intake.[14] It bypasses the need for digestion and absorption from the damaged GI tract and allows time for the intestinal mucosa to regenerate and for disaccharide and trypsin activities to increase. It also provides nutrients necessary for mucosal repair.

A base solution containing dextrose, soybean or safflower oil, and amino acids is used to provide both protein and energy sources.[3,33] It is important to provide enough calories as dextrose and soybean oil to prevent the utilization of amino acids as an energy source and thereby allow them to be used as building blocks for growth and repair.[22] In human beings, this is accomplished by keeping the ratio of nonprotein calories to nitrogen in the range of 135:1 to 185:1.[3,22]

Both dextrose and soybean or safflower oil are used as energy sources. Glucose has the advantages of lower cost, better tolerance by neonates, and being the normal source of energy.[44] Soybean oil is very calorie dense and provides more than 2.5 times the calories of dextrose on an equal volume basis (9 kcal/gm versus 3.4 kcal/gm).[43] It does not promote insulin release, is isosmotic with plasma, and, in stressed states, may be preferentially utilized over dextrose.[43]

Currently, most parenteral nutrition solutions used in human beings derive 30% to 50% of the calories from soybean or safflower oil and the remainder from dextrose.[22,29] The majority of neonatal ruminants appear to be able to tolerate this amount of fat without adverse side effects.

The use of a balanced solution of carbohydrates and fats as a calorie source in parenteral nutrition helps to avoid water retention, fatty infiltration of the liver from an excess carbohydrate load, and worsening respiratory compromise, if it is present.[29,44]

A variety of different synthetic crystalline L-amino acid solutions are available. The amino acid pattern and concentration vary and may be similar to human serum or to egg protein.[26] Crystalline L-amino acid solutions provide all 20 essential and nonessential amino acids.[22,26] Specialized amino acid solutions for use in renal (Nephramine*) and liver disease (Heptamine*) are available but have not been evaluated in neonatal ruminants.[30,33]

Protein requirements for human infants vary, but in general parenteral nutrition should supply 2 to 3 g protein per kg per day.[22,26] Current estimates for calves are 3.75 g protein (0.6 g nitrogen) per kg body weight per day, considerably higher than human infant recommendations.[3,18] Most neonatal ruminants, however, have shown adequate growth when supplied with 2.5 to 3 g protein per kg per day. Concentrations greater than 4 g protein per kg per day in human infants are associated with significant side effects including azotemia, hyperaminoaciduria, hyperaminoacidemia, and hyperammonemia.[22]

In human infants and small animals, maintenance water requirements are normally calculated from body surface area.[33] In calves, daily water requirements for maintenance have been estimated to be 50 ml/kg body weight.[3,18] Fluid requirements can be increased by a number of disease conditions, including fever, diarrhea, and third space losses.[33] If the parenteral nutrition solution does not provide adequate water, lactated or acetated Ringer's solution may be administered to compensate for the difference.

Electrolytes, vitamins, and trace elements may be added to the base solution as required. In diarrheic calves, potassium chloride (15 to 25 mmol/L or mEq/L) and sodium bicarbonate (5 to 15 mmol/L or mEq/L) are generally the only electrolytes added to the solution. The amount added varies according to serum concentration

---

*American McGaw, Irvine, Calif.

and response to therapy. Calcium, phosphorus, and magnesium become important in prolonged use.[28,33] Multiple vitamins containing B vitamins, ascorbic acid, and fat-soluble vitamins are added daily. Unless parenteral nutrition is maintained for longer than 14 days, trace minerals are not normally added. Zinc may be the possible exception to this, as hypozincemia can develop in only 1 week and occurs more rapidly with amino acid infusions and diarrhea in young, rapidly growing animals.[20,28,35] Zinc deficiencies are manifested by alopecia, dermatitis, mental depression, diarrhea, and growth retardation.[28,35] A selenium injection may be necessary if the neonate is from a selenium-deficient area, as parenteral nutrition solutions do not contain selenium.[22,30,35]

A variety of different concentrations of solutions can be used. Two solutions have been used effectively in calves, kids, and infant llamas. Solution I contains 3.4% amino acids, 20% dextrose, and 2% soybean oil (a nonprotein calorie–nitrogen ratio of 165:1) and provides 1.036 kcal/ml. Neonates are started on this solution at an administration rate of 60 ml/kg body weight/day (30 ml/lb/day). If they tolerate it well and do not show evidence of lipemia or hyperglycemia, the amount of soybean oil and flow rate are increased. Solution II contains a final concentration of 2.8% amino acids, 16.6% dextrose, and 3.3% soybean oil. This solution contains a nonprotein calorie–nitrogen ratio of 205:1, provides 1.047 kcal/ml, and is administered at 70 ml/kg BW (32 ml/lb BW) per day. Neonates are maintained at this rate for a minimum of 5 days. When they have resolved their disease process and have shown adequate growth, the parenteral nutrition solution is tapered slowly, and the amount of milk or milk replacer in the diet increased accordingly.

Recent technical advances have allowed fats, dextrose, and amino acids to be mixed together in one container.[2,42] These admixtures are stable and compatible and do not result in an increased risk of contamination (Fig. 36-4).[2,42]

In neonatal ruminants and llamas, the parenteral nutrition solution is infused through a 14 or 16 gauge 9 or 13 cm (3½- or 5¼-inch) catheter placed in the external jugular vein or via a longer catheter placed in the superior vena cava.[3,18,19,33] To minimize the incidence of catheter-related sepsis and thrombophlebitis, the catheter is inserted using aseptic technique and used only for delivery of the parenteral nutrition solution.[33,34] The site is monitored three or four times a day for evidence of warmth or pain but is otherwise not disturbed. The catheter is left in place as long as there is no evidence of phlebitis or thrombosis or for a maximum of 4 to 5 days and then changed to the contralateral vein.[19] The parenteral nutrition solution and administration set are changed every 24 hours. To maintain a constant flow rate, all solutions are administered with an infusion pump.

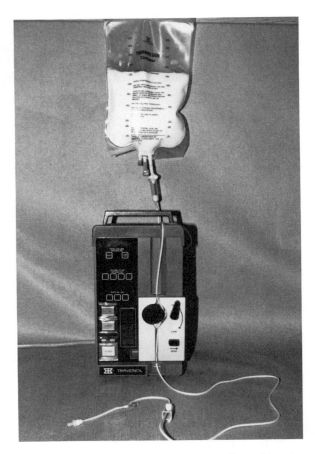

Fig. 36-4 Delivery system for parenteral nutrition incorporates all components in one container and utilizes an infusion pump.

Neonates maintained on parenteral nutrition are monitored closely. Daily weight, hematocrit, and plasma protein concentrations are obtained to aid in the assessment of hydration. Weight gain in excess of 0.25 to 0.5 kg (½ to 1 lb)/day is generally associated with overhydration.[3,19,33] The vital signs are monitored four times a day, and any unexplained temperature rise is suggestive of impending septicemia.[33] Serum is monitored once daily for evidence of lipemia or hyperglycemia; the urine is monitored once daily for the presence of glucose or excessive protein. Serum electrolyte and creatinine concentrations are determined at the initiation of therapy and daily until they have stabilized. Biochemical profiles including liver enzymes should be obtained at the onset of therapy and then weekly.

The most commonly reported complications of parenteral nutrition are thrombophlebitis, thrombus formation, and sepsis.[30,33,34] The majority of solutions used for parenteral nutrition are hyperosmotic (1200 to 1700 mOsm/l) and irritating to peripheral veins. The jugular vein and superior vena cava have larger diameters and faster flows and tolerate this osmolality well if the catheter is carefully placed and maintained.[26,34] The incidence of thrombophlebitis is closely related to catheter selection.[30,39] In general, shorter, thinner, and softer catheters are less thrombogenic.[30,39] Various catheter materials have been used to decrease thrombogenicity. Most catheters currently are derived from plastic monomers (polyethylene, polyvinylchloride), fluorinated hydrocarbons (polytetrafluoroethylene), or silicone rubber.[30,33,39] When the degree of catheter softness is considered, silicone rubber, polyethylene, and polyurethane all have similar thrombogenicities.[30,39]

Venous thrombosis is related to a number of factors including catheter placement and composition, infections, preexisting coagulopathies, and irritating solutions.[30,33,34] Shortly after introduction of a catheter, platelet-mediated fibrin deposition occurs on the external catheter surface.[30,33] When the catheter is removed, any associated thrombi are dislodged and remain in the venous circulation until they reach the small pulmonary vasculature.[33]

The potential for sepsis with parenteral nutrition is high because solutions with large concentrations of dextrose and fat provide excellent growth media for bacteria, yeast, and fungi. In human beings, bacterial infections are associated with common skin contaminants such as *Staphylococcus epidermidis*, enterococcus, and gram-negative bacteria.[33,39] Fungal infections are usually due to *Candida* spp.[33] Catheter-related sepsis occurs shortly after insertion of the catheter. Unexplained fever, leukocytosis, hyperglycemia, and chills are signs associated with sepsis. If catheter-related sepsis is suspected, the catheter should be removed and the tip cultured. Blood cultures may help to identify the offending organism, although fungal organisms are more difficult to culture than bacteria. Usually all that is required is removal of the catheter.[30,33,34,39] If septic thrombophlebitis or deep-seated fungal infections are suspected, the appropriate antimicrobial therapy should be instituted.

Major metabolic complications include overhydration and underhydration, hyperglycemia and hypoglycemia, electrolyte deficiencies, and hematological and biochemical abnormalities.[23,33,34,38] Daily monitoring allows early and rapid correction of most of these problems. Less common complications in human beings are polymyopathies, fat overload syndrome, metabolic bone disease, and sudden death.[34,38]

Two of the most common metabolic complications in neonatal ruminants are hyperglycemia and hypokalemia. Hyperglycemia can occur secondary to glucose intolerance, excessive administration rates, exogenous glucocorticoid administration, or sepsis.[33,34,38] A slow increase in the rate of glucose administration and concentration allows time for endogenous insulin levels to rise and helps to avoid hyperglycemia. Neonates that have been malnourished for extended periods of time may require lower administration rates until an adequate insulin response occurs. Infant llamas normally have higher baseline serum glucose concentrations and therefore higher levels while receiving parenteral nutrition.

If parenteral nutrition infusions are suddenly stopped, a rebound hypoglycemia may develop within 30 minutes.[33] When parental nutrition is interrupted, peripheral infusion of a solution of 5% dextrose in water prevents hypoglycemia until it can be reinstituted. The use of an intravenous pump or controller prevents rapid swings in administration rates that may cause hyperglycemia or hypoglycemia (see Fig. 36-4). Neonatal llamas and kids appear to be especially susceptible to rebound hypoglycemia.

Hypokalemia may occur for a variety of reasons. High concentrations of infused glucose stimulate insulin secretion, resulting in intracellular shifts of potassium, as does the correction of preexisting metabolic acidosis.[26] Decreased oral intake and losses through the kidney and gastrointestinal tract also contribute to the development of hypokalemia.[16,28] Until the neonate has resolved the disease process, daily serum potassium determinations are necessary to supply adequate amounts of potassium.

Neonatal ruminants and llamas maintained on parenteral nutrition tolerate it well, often becoming bright and alert after only a few hours of therapy. Parenteral nutrition has been shown to be beneficial in reversing weight loss, cachexia, and electrolyte and acid-base abnormalities associated with clinical illness. Complications are few and can be avoided with close and early monitoring.

REFERENCES

1. Alexander G, Bennett JW, and Gemmell RT: J Physiol 224:223, 1975.
2. Ang SA, Canham JE, and Daly JM: JPEN 11:23, 1987.
3. Baker JW and Lippert AC: Comp Cont Educ 9:F71, 1987.
4. Blaxter KL and Wood WA: Br J Nutr 5:55, 1951.
5. Bowie MD, Barbezat GO, and Hansen JDL: Am J Clin Nutr 20:89, 1967.
6. Brisson GJ, Cunningham JM, and Haskell SR: Can J Anim Sci 37:157, 1975.
7. Bryant JM and others: J Dairy Sci 50:1645, 1967.
8. Bywater RJ and Penhale WJ: Res Vet Sci 10:591, 1969.
9. Chernoff R: Am J Hosp Pharm 37:65, 1980.
10. Cunningham HM and Loosli JK: J Dairy Sci 37:453, 1954.
11. Dollar AM and Porter JG: Nature 179:1299, 1957.
12. Dunn L and others: J Pediatr 112:622, 1988.
13. Fettmman MJ and others: J Am Vet Med Assoc 188:397, 1986.
14. Gunn R and others: Can Med Assoc J 117:357, 1977.
15. Hallback DA and others: Gastroenterology 74:683, 1978.
16. Hansen TO: In Proc Am Assoc Eq Pract 32:153, 1986.
17. Hansen TO, White NA, and Kemp DT: Am J Vet Res 49:122, 1988.
18. Hoffsis GF and others: J Am Vet Med Assoc 171:677, 1977.
19. Hovda LR, McGuirk SM, and Lunn DP: J Am Vet Med Assoc 196:319-322.
20. Huber JT: J Dairy Sci 52:1303, 1969.
21. Kaminski MV and Pinchcofsky GD: Proceedings of the international symposium on parenteral and enteral nutrition, 1983.
22. Kerner JA: Manual of pediatric nutrition, New York, 1983, J Wiley & Sons.
23. Kien CL and Chusid MJ: JPEN 3:468, 1979.
24. Leat WF and Cox RW: In Lawrence TJ, editor: Growth in animals, London, 1980, Butterworth.
25. Letsou A: Nutr Supp Ser 7:12, 1987.
26. Lorch V and Lay SA: Pediatr Clin North Am 24:547, 1977.
27. Medina M and others: Comp Cont Ed 5:S148-S155, 1983.
28. Nutr Supp Ser 2:8, 1982.
29. Pelham LD: Am J Hosp Pharm 38:198, 1981.
30. Phillips GD and Odgers CL: Parenteral and enteral nutrition: a practical guide, ed 3, Edinburgh, 1986, Churchill Livingstone.
31. Phillips RW: Vet Clin North Am (Food Anim Pract) 1:541, 1985.
32. Radostits OM and Bell JM: Can J Anim Sci 50:405, 1970.
33. Raffe MR: In Slatter DH, editor: Textbook of small animal surgery, vol 1, Philadelphia, 1985, WB Saunders.
34. Reimer SL, Michener WM, and Steiger E: Pediatr Clin North Am 27:647, 1980.
35. Rudman D and Williams PJ: Nutr Rev 43:1, 1985.
36. Schoonderwoerd M and others: Can Vet J 27:365, 1986.
37. Schugel IM: Proc Am Assoc Bov Pract, 1974.
38. Seashore JH: Surg Clin North Am 60:1239, 1980.
39. Spurlock SL and Spurlock GH: In Proceedings of the ACVIM, 1988.
40. Theodore AC and others: Chest 86:931, 1984.
41. Todd E: In Grant A and Todd E, editors: Enteral and parenteral nutrition, ed 2, Oxford, 1987, Blackwell Scientific Publications.
42. Vasilakis A and Apelgren KN: JPEN 12:356, 1988.
43. Wolfe BM and Ney DM: In Rombeau JL and Caldwell MD, editors: Parenteral nutrition, Philadelphia, 1986, WB Saunders.
44. Wolfe RR: In Rombeau JL and Caldwell MD, editors: Parenteral nutrition, Philadelphia, 1986, WB Saunders.

Nonruminants

SECTION A Horse

37

The Horse Industry and Feeding Horses for Optimal Reproduction and Growth

H. F. HINTZ, SARAH L. RALSTON

THE HORSE INDUSTRY

The horse and mule population of the United States reached a peak number of nearly 27 million shortly before the United States entered World War I. Then, with the increased use of the internal combustion engine–powered vehicles, the number of horses and mules gradually declined, although not as precipitously as might be imagined. Even as recently as 1955, horses and mules outnumbered tractors on farms. By 1960, however, the U.S. Department of Agriculture reported only 3 million horses and mules; this was the last year they were counted.

People predicted that the horse would become a novelty, but the increased mechanization brought with it increased leisure time and more money that could be spent on recreation. The horse population consequently rebounded; by the 1970s some authorities estimated the country had 8 to 10 million horses. The equine industry once again became a powerful economic force. In 1988 the American Horse Council estimated that horses contribute approximately $15 billion annually to the U.S. economy and that horse owners account for roughly $13 billion in annual investment and maintenance expenditures.[1] Horse sports draw more than 110 million spectators each year, with 70 million attending racetracks. The wagering on horse races exceeds $13 billion annually.

The greatest use of horses remains recreational. More than 27 million people ride horses each year, with 54% riding on a regular basis and 46% riding for occasional recreation.

Table 37-1 outlines the number of annual registrations recorded for the various breed registries, an indication of the changes in the horse population. The annual registrations reached a peak in 1984.

It is difficult to predict the future of the horse industry. Currently, the signs seem to indicate that the recent decline has reached its nadir and the population has stabilized.

PRINCIPLES OF FEEDING FOR OPTIMAL PERFORMANCE AND GROWTH

The horse is a nonruminant herbivore. As in the ruminant, the bacteria in the digestive tract digest fiber and produce volatile fatty acids that the horse can use for energy. Unlike the ruminant, the equine bacterial population cannot be relied upon to provide substantial amounts of amino acids and water-soluble vitamins because the site of fermentation is beyond the small intestine, which is where these nutrients are primarily absorbed.

Table 37-1 Registration figures for major American light horse breeds, 1960 to 1989

Breed	1960	1968	1975	1980	1984	1985	1986	1987	1988	1989
Appaloosa	4,052	12,389	20,175	25,384	17,384	16,189	14,551	12,589	12,317	10,746
Anglo and Half Arab	2,200	9,800	11,351	14,257	11,400	10,099	6,907	NA	6,688	4,974
Arabian	1,610	6,980	15,000	19,725	29,175	30,004	28,283	26,240	24,578	21,723
Hackney	459	656	1,015	595	733	744	791	621	866	779
Morgan	1,069	2,134	3,400	4,537	5,411	4,538	4,329	3,803	3,526	2,238
National Horse Show	0	0	0	0	782	856	927	964	1,036	903
Paint	0	2,390	5,287	9,654	14,673	12,692	11,273	15,518	14,929	14,390
Palomino	657	1,262	1,539	1,548	3,594	1,200	1,500	1,746	1,747	2,080
Paso Fino	0	0	380	645	1,155	1,335	1,323	1,249	1,464	1,453
Quarter Horse	35,507	65,326	97,179	137,090	169,675	157,360	153,773	147,007	128,352	123,294
Saddlebred	2,329	3,589	4,064	3,879	3,986	4,353	4,363	3,918	3,811	3,708
Standardbred	6,413	10,682	12,830	15,219	19,795	18,384	17,637	17,579	17,393	16,896
Tennessee Walker	2,623	8,493	6,591	6,847	7,610	7,633	7,900	8,000	8,400	8,847
Thoroughbred	12,901	23,201	29,225	39,367	47,288	50,382	49,700	50,885	38,854*	49,000
Total	69,820	146,902	208,036	278,747	332,661	315,769	303,257	290,119	263,961	261,031

*Estimated or incomplete figures
Courtesy American Horse Council (1989), Horse Industry Directory, Washington, DC.

Table 37-2 *Expected feed consumption by horses (% body weight)**

	Forage	Concentrate	Total
Mature horses			
Maintenance	1.5-2.0	0-0.5	1.5-2.0
Mares, late gestation	1.0-1.5	0.5-1.0	1.5-2.0
Mares, early lactation	1.0-2.0	1.0-2.0	2.0-3.0
Mares, late lactation	1.0-2.0	0.5-1.5	2.0-2.5
Working horses			
Light work	1.0-2.0	0.5-1.0	1.5-2.5
Moderate work	1.0-2.0	0.75-1.5	1.75-2.5
Intense work	0.75-1.5	1.0-2.0	2.0-3.0
Young horses			
Nursing foal, 3 months	0	1.0-2.0	2.5-3.5†
Weanling foal, 6 months	0.5-1.0	1.5-3.0	2.0-3.5
Yearling foal, 12 months	1.0-1.5	1.0-2.0	2.0-3.0
Long yearling, 18 months	1.0-1.5	1.0-1.5	2.0-2.5
Two year old, 24 months	1.0-1.5	1.0-1.5	1.75-2.5

*Air-dry feed (about 90% DM).
†Includes estimated milk dry matter.
Adapted from National Research Council: Nutrient requirements of horses, ed 5, Washington, DC, 1989, National Academy of Sciences, National Academy Press.

The horse usually requires some dietary fiber to maintain normal digestive function and to prevent vices such as wood chewing; the amount is not clearly established. Horses have been fed diets providing only 0.5 kg of hay per 100 kg of body weight (0.5% BW) without apparent adverse effect. However, the rule most generally quoted is 1 kg of hay (or roughage equivalent) per 100 kg (1%) of body weight. Excessive amounts of fibrous feeds can limit energy intake by filling the digestive tract. Examples of hay-grain ratios for various classes of horses are shown in Table 37-2.

The success of any feeding program depends on good management, including routine dental care[2] and parasite control.[28] Horses should be observed closely and frequently. Because even experienced horse owners can be fooled about the extent of weight change in a horse, routine weighing (or estimating weight with tapes placed around the heartgirth) is an excellent management aid. However, weight tapes may not be reliable for use on mares in late pregnancy. Body score systems such as those developed by Henneke and co-workers[12] and Carroll and Huntington[7] are very useful for the evaluation of body condition.

If horses are group fed, the feeding facility should allow all animals access to their share. Head partitions can be used to help prevent fighting.[14]

Animals must be provided adequate amounts of water and be fed the amounts of energy necessary to achieve proper body condition. The ration must be balanced for protein, minerals, and vitamins. Complete feeding recommendations can be found in *Nutrient Requirements of Horses*[22] (see Table A-19 and A-20 in the Appendix).

FEEDS FOR HORSES

Any of several grains and roughages can be fed to horses as long as the nutritional characteristics of the feedstuffs are considered in ration formulation.

Grains

Horse owners usually prefer oats as the grain for horses. Oats contain higher concentrations of fiber and protein and a lower digestible energy content than does corn. Thus, oats are usually considered to be safer than corn. However, oats vary greatly in quality and are usually more costly per Mcal of digestible energy than corn. The production of oats has decreased in recent years, and prices are likely to remain high.

Oats do not have to be processed for horses that do not have dental problems. Oats average 30% hulls, but there is great variation in the proportion of hulls, and feed value varies. Poor oats may be 50% hulls, whereas high-quality oats may be only 25% hulls and thus yield a much lower fiber content.

Corn is often used as an energy source and is usually the most economical grain in most parts of North America. It can be fed whole, cracked, or as ear corn. Corn has a higher energy concentration than oats. Contrary to popular belief, corn is not a "heating feed" when fed properly. The heat increment per unit of digestible energy of corn is actually less than that of oats or hay. Moldy corn can cause leukoencephalomalacia. Mold is most likely to occur in processed (cracked, flaked) corn.

Barley is commonly fed to horses in the western United States and Canada. The energy concentration of barley is greater than that of oats but less than that of corn. Protein is similar to oats. Flaked or rolled barley is an excellent source of energy to horses.

Milo (sorghum grain) must be ground, crimped, or rolled because the small, hard whole kernels cannot be efficiently digested by horses. Some varieties of milo contain high levels of tannic acid and are not palatable.

Wheat must also be processed, but fine grinding should be avoided because of dust problems. Wheat is usually too expensive to feed to horses but, when the price is right, can be successfully incorporated into horse rations.

Forages

Hay. The key to simplified horse feeding is good-quality hay. Every horse owner should make an effort to learn to evaluate hay. Hay provides energy, but the energy concentration is lower than in grains. Hay usually cannot supply all the energy needs of hardworking horses, lactating mares, and rapidly growing foals, which need grain in addition to hay. The lower energy concentration in hay limits the total energy intake.

When buying hay, it is important to consider the nutritive value in relation to cost rather than the kind of hay. High-quality hay that costs more per tonne is often a better buy than cheaper, poor-quality hay. One important factor when evaluating hay is age at harvesting. Young plants contain more digestible energy and nutrients per kg than older plants. As hay matures, the content of indigestible material increases.

Other criteria of good-quality hay are (1) freedom from mold, dust, and weeds; (2) lack of excessive weathering; (3) leafiness and lack of stems; and (4) green color.

There are two general classes of hay: legume and grass. Legume hay contains more digestible energy, calcium, protein, and vitamin A activity than grass hay harvested at the same stage of maturity. Horses should be gradually changed from grass hay to alfalfa because an abrupt change could cause digestive upsets and diarrhea. Horses fed alfalfa may urinate more and require more water, and there may be a stronger smell of ammonia in the barn because alfalfa contains more nitrogen than grass hay. The urine also contains more crystalline sediment because of the high calcium content of alfalfa. However, contrary to popular belief, alfalfa hay does not harm normal kidneys. Some alfalfa hay, however, may contain blister beetles, which are lethal to horses.[25]

Bird's-foot trefoil and clover hays such as alsike, crimson, and red have nutritional characteristics similar to alfalfa hay when properly harvested. Some horses are not particularly fond of bird's-foot trefoil hay and may select other hay when given a choice. Moldy red clover hay may cause excessive salivation (slobbering disease). Moldy sweet clover hay may induce vitamin K deficiency because of an anti–vitamin K factor in the mold. Moldy hay may cause colic and trigger chronic obstructive pulmonary disease or other allergic reactions (heaves).

Timothy hay is the favorite hay of many horsemen. Good-quality timothy is an excellent hay for horses, but the grain mixture for brood mares and young horses must contain more protein and calcium than when legume hay is fed. Other grasses such as bromegrass, orchard

Table 37-3 *Digestible energy content of some feeds for horses*

Feed	Digestible energy* (Mcal/kg)
Alfalfa hay	
Early bloom	2.25
Mid bloom	2.10
Full bloom	2.00
Alfalfa meal	2.10
Barley	3.26
Barley straw	1.45
Beet pulp	2.35
Bermuda grass hay	
15-28 days' growth	1.92
29-42 days' growth	1.95
43-56 days' growth	1.75
Bromegrass hay	
Mid bloom	1.90
Mature	1.60
Corn	3.40
Oats	2.85
Orchard grass hay	
Early bloom	1.95
Late bloom	1.70
Sorghum grain	3.35
Soybean meal (44% protein)	3.2
Timothy hay	
Early bloom	2.00
Mid bloom	1.80
Full bloom	1.70
Late bloom	1.60
Wheat grain	3.40
Wheat bran	2.90

*Air-dry (90% dry matter) values.

Table 37-4 *Potential feeding problems for horses*

Feed	Problem
Alfalfa containing blister beetles	Cantharidin toxicosis, colic, stranguria, hematuria
Alsike clover	Photosensitization, liver cirrhosis
Feeds designed for other species and containing certain ionophores and antibiotics	Salinomycin poisoning, Monensin poisoning, Lincomycin poisoning
Kleingrass	Chronic hepatitis and fibrosis, positive drug test
Legumes infested with *Rhizoctonia legumenicola*	Slaframine intoxication ("slobbering")
Moldy corn (*Fusarium moniliforme*)	Leukoencephalomalacia, colic, ataxia
Moldy sweet clover	Vitamin K deficiency leading to hemorrhage
Reed canary grass	Hordenine in urine (disqualification in competitive events in some states)
Moldy ryegrass	"Ryegrass staggers," trembling, hypersensitivity, spasms due to lolitrem B
Sprouted barley	Hordenine in urine (disqualification from competition in some states)
Sudan grass or Sudan hybrids	Cystitis
Tall fescue infested with *Acremonium coenophealum*	Agalactia, Prolonged gestation, Abortion, Thickened placenta

grass, coastal Bermuda grass, and bluegrass can have a nutrient composition similar to that of timothy if properly harvested and also can make good horse hay. Cereal hays also can be used for horses. For example, oat hay is a popular feed in some areas.

Silage. Ensiling is an effective means of preserving forage for horses if properly harvested feeds are stored in an airtight chamber (silo) where they can undergo fermentation. Either hay crop silage or corn silage can be a very economical source of nutrients for horses, replacing one third to one half the hay usually fed, but silage must be free of mold. Spoiled silage can be lethal to horses due to botulism. Unfortunately, it is difficult to produce mold-free silage, and thus it is generally not recommended that horses be fed silage, although it is commonly fed in Europe and Scandinavia (colder climates).

The digestible energy content of some feeds for horses is given in Table 37-3. More values can be found in the NRC.[22] Some feedstuffs should be avoided (Table 37-4).

FEEDING PROGRAMS
Feeding the mature horse at maintenance

The feeding program for maintenance of the average mature horse can be very simple. Good-quality hay consumed at a rate of 1.5 to 2 kg of hay per 100 kg (1.5-2%) of body weight, water, and trace-mineralized salt fed free choice should supply all the needed nutrients. Additional mineral supplements may be needed if the hay was grown on soil lacking in minerals such as selenium, copper, or phosphorus. If the climate is harsh, energy requirements could be increased, and some grain is needed. The preceding recommendation also assumes reasonable parasite control and no dental problems. Some nervous horses may require additional energy.

Many horses at maintenance are fed excess amounts of energy, leading to obesity, a higher incidence of founder, lipomas, and a waste of feed and money.

The protein requirement for the horse at maintenance is only 40 g per Mcal of digestible energy (DE) per day.[22] The DE requirement for a 500 kg (1100 lb) horse is 16.4 Mcal/day; thus a 500 kg horse would need 656 g of protein per day. If hay containing 2 Mcal/kg was fed, the horse would need 8.2 kg (18 lb) of feed per day (which is within the recommended 1.5% to 2% of body weight; 7.5 to 10 kg). The hay would need to contain only 8% protein (0.656 kg ÷ 8.2 kg) to meet protein requirements.

Feeding the brood mare

Energy. In the past, overfeeding of the mare often was thought to be a major cause of reproductive problems, but recent studies suggest that underfeeding is more likely to cause problems than is overfeeding.[12] Kubiak and associates[19] concluded that nonpregnant mares should be brought into the breeding season in a moderately fat condition (not less than 15% body fat) and then kept in a positive energy balance to attain maximum reproductive performance.

Recent studies with women demonstrate that when body fat content becomes too low, estrogen production decreases and sterility results.[9] Thus women athletes such as long-distance runners and bodybuilders often stop menstruating.[9] This amenorrhea can be reversed by increasing the fat content of the body. Some of the difficulty in getttting mares recently arriving from the racetrack in foal possibly could be related to lack of body fat.

Inadequate dietary energy can cause loss of body condition in pregnant mares, prolonged gestation, and decreased fertility at rebreeding. Theoretically, somewhat smaller foals may be produced, but the mare apparently protects the fetus during periods of energy deprivation. For example, Jordan[15] reported that fat pregnant pony mares fed to lose 20% of their body weight did not have smaller foals than mares fed energy to maintain weight. No differences in birth weights of foals from quarter horse mares that were fed two different levels of energy were noted, even though one group of mares lost an average of 22.5 kg (50 lb).[4] Banach and Evans[3] concluded that energy restriction during the last 90 days of gestation did not decrease birth weight. Of course, severe or prolonged energy restriction can cause foals to be small or aborted.[29]

The increase in energy requirements during pregnancy is significant only during the last third of gestation. The energy intake should be gradually increased by 12% to 20% during that period. It is important to observe the mare and increase the energy according to desired body condition. It has been recommended that thoroughbred and quarter horse mares gain 70 to 90 kg (150 to 200 lb) during gestation.[24]

The energy requirement of the lactating mare depends on the amount of milk produced. Some mares may require almost twice as much energy during lactation as when they are open.

Protein. Protein is needed for maintenance of the mare, for development of the fetus, and for milk production. The National Research Council[22] suggests 7% dietary protein would be adequate for maintenance. Average- to good-quality forage should contain at least 8% protein, and most grains contain at least 9% protein. Thus the open mare is not likely to need a protein supplement when good forage is fed. The pregnant mare (in the last third of gestation) requires a diet containing 9% to 10% protein. If legume or grass-legume hays are fed, a protein supplement is not necessary during that period. If grass hay or cereal hay is fed with grain, the grain mixture may need to contain at least 12% protein.

The lactating mare may require about 13% protein in the total ration. If a grass hay or cereal hay is fed, the grain mixture may need to contain 16% to 18% protein. If a good-quality legume or grass-legume hay is fed, a grain mixture containing 12% protein should provide adequate intake of protein.

Minerals. Calcium and phosphorus are essential for a sound skeleton. The ration of the open mare should contain at least 0.2% calcium and 0.15% phosphorus and that of pregnant or lactating mare 0.4% to 0.45% calcium and 0.3% phosphorus.[22]

The mineral content of the hay varies according to many factors, such as species, soil type, and age of plant at harvest. Good-quality grass hay could be expected to contain at least 0.35% calcium and 0.2% phosphorus. Legumes such as alfalfa may contain 1% to 1.5% calcium and 0.2% to 0.25% phosphorus. Thus calcium and phosphorus problems would not be expected when the open mare is fed hay of reasonable quality. Pregnant and lactating mares may require a calcium supplement because their calcium requirements are greater and more grain (which contains very low levels of calcium) is fed to meet energy demands. When grass hay is fed, the grain mixture should usually contain at least 0.55% calcium. The addition of 1% dicalcium phosphate and 1% limestone to the grain mixture may supply adequate levels of calcium and phosphorus, but the diet should be analyzed. When legume hay is fed, a calcium supplement may not be needed, but a phosphorus supplement may be necessary.

These guidelines for energy, protein, calcium, and phosphorus are based on average analyses of feedstuffs. More accurate formulations can be made if the nutritive content of the feedstuff is determined rather than calculated. Many laboratories analyze feedstuffs for reasonable fees. Local cooperative extension agents can be contacted for further information on the testing of feedstuffs.

Crops grown in many parts of the United States such as the Northeast, East Coast, Northwest, and parts of California contain low levels of selenium. A low intake of selenium may cause decreased fertility in mares and white muscle disease in foals. The requirement for selenium for mares has been estimated to be 1 to 2 mg/day. Daily intakes of 50 mg or greater can be toxic. Horses fed toxic levels may slough hoofs, lose their manes and tails, and eventually become paralyzed and die. Thus caution must be exercised when using selenium supplements. Fortunately, many feed companies in the low-selenium areas add selenium to horse feeds, and selenium supplements are then not needed if a commercial feed is used.

Pregnant mares fed iodine-deficient diets may have weak or dead foals with enlarged thyroid glands (goiter). The foals may be born hairless. Iodine can be easily supplied by providing free-choice trace-mineralized salt. As with selenium, excess iodine can be toxic. Foals of mares fed high levels of iodine (more than 40 mg/day) may have signs of toxicosis, including enlarged thyroid, weakness, and death. Thus the signs of toxicosis may be similar to the signs of deficiency.

There has been considerable interest in the copper nutrition of horses. Copper deficiency in pregnant mares may contribute to the appearance of orthopedic disease in foals. Further studies are needed to determine the optimal level of copper needed in the diet of the mare. It has been suggested that the grain mixture may need to contain up to 30 mg/kg (ppm) copper,[10,16] although NRC[22] recommends a minimum of 10 ppm copper in the total diet. Many hays contain 6 ppm or less. Zinc deficiency also may contribute to the development of orthopedic disease in foals, whereas in other instances very high levels of zinc have been claimed to induce copper deficiency. Again, further studies are needed, but it appears that a grain mixture containing 75 to 100 mg/kg (ppm) zinc would provide adequate levels of zinc for the pregnant mare.

Vitamins. Grains are poor sources of several of the vitamins, but vitamin deficiencies usually would not be expected in the mare if good-quality hay is fed. However, hay that has been stored for more than 2 years or that has been extensively damaged by weather may lack in vitamin activity, particularly that of vitamin A. In such cases a vitamin supplement may be useful if a fortified commercial feed is not used.

Feeding the foal

The feeding program for a young horse can be broken into two parts: suckling and weanling.

Suckling. The amount of nutrients received by the suckling foal is determined by the amount and composition of the mare's milk and by the intake of creep feed, grain stolen from the mare's feed box, and forage (hay, pasture, or both). The amount of grain a foal can steal from the mare's feed box depends on how much grain the mare is fed and on the temperament of the mare, but the manager can control the amount and composition of the creep feed. Creep feeding may be useful because there is great variation in the milking ability of mares and not all mares are benevolent and readily share their feed. It is thought that foals that readily consume grain prior to weaning have less stress at weaning.

Several different grains can be used in creep feeds, but corn, oats, and barley are recommended over wheat and rye. The grains should be processed, crimped, or steam rolled to improve digestibility. Soybean meal is often used as the protein source because it is a richer source of lysine than other vegetable proteins, such as linseed meal and cottonseed meal, and is more economical than animal proteins such as dried skim milk. A creep feed should contain at least 14% to 16% protein, 0.8% calcium, 0.55% phosphorus, and 30 mg/kg (ppm) copper.

The amount of creep feed is controversial. For example, Lewis[20] suggests that the foal should not be fed grain separately from the mares until the foal is at least 2 months and preferably 3 months old. He suggested that the intake should be limited to 0.5% to 0.75% of body weight per day so that the incidence of epiphysitis would be minimized. A foal with an expected mature weight of 500 kg (1100 lb) might weigh 160 kg (350 lb) at 3 months of age. Therefore, the 3-month-old foal would be fed 0.8 to 1.1 kg/day. Others feel that creep feeding can be started at 1 month of age and that a rule of thumb of 0.45 kg (1 lb) of feed per month of age per day (up to 12 months of age) could be used.[23] Thus, a 3-month-old foal would be fed 1.35 kg/day. According to the Lewis's guidelines, a 4-month-old foal weighing 190 kg would be fed 0.9 to 1.4 kg/day (about 2.5 lb/day), whereas the latter rule of thumb would indicate 1.8 kg/day (4 lb/day). Of course, the important consideration is the body condition and growth of the foal. The eye of the master is the most important factor in the determination of how much grain a foal should receive.

Weanling. The total ration for a weanling should contain at least 14% protein, 0.7% calcium, and 0.45% phosphorus.

The amino acid content is important. The lysine content should be at least 2.1 gm/Mcal DE.[22] Other amino acids may also be critical, but requirements have not been studied. Trace minerals should also be considered. Concentrations of 30 to 50 mg/kg (ppm) copper, 75 to 100 mg/kg (ppm) zinc, and 0.2 mg/kg (ppm) selenium are not unreasonable for grain mixtures fed to weanlings unless, of course, the forage contains high concentrations of these minerals.

The number one problem of raising weanlings is probably skeletal abnormalities. *Developmental orthopedic disease* is a term that is used to describe a "package" of skeletal problems including osteochondrosis, physitis (ephiphysitis), and flexural deformities.

The etiology of developmental orthopedic disease is complex. Some of the factors that have been suggested are (1) genetic capacity for a rapid growth rate, as physitis is much more commonly found in foals that are growing rapidly than in slower growing animals, and (2) genetic predisposition for conformation that causes greater stress on the skeleton, for example, toe-in conformation and upright conformation in fetlock joints that have been associated with physitis.

Deficiencies or excesses of many nutrients such as vitamin A, vitamin D, iodine, magnesium, and manganese might also be involved in orthopedic problems. The nutritional situations that have received the most attention recently are overfeeding, particularly of carbohydrates; calcium deficiency or excess; and copper deficiency.

Overfeeding. Overfeeding has long been considered by some authors to be harmful to young horses. In 1856 Dadd[8] wrote, "Many thousands of horses die in consequence of being too well or rather injudiciously fed. Men who prepare horses for the market attempt to get them into condition without regard to their general health." Forty-two years later, Henry[13] stated, "In no other way can colts be ruined so surely and so permanently as by liberal feeding and close confinement."

Glade[11] conducted a series of experiments in which the effect of diet on endocrine response was examined. High intakes of carbohydrate caused significant changes in serum insulin, thyroxin, and triiodothyronine. Growth plates from the animals fed the highest intakes of soluble carbohydrate had thicker reserve and hypertrophic zones. The hypertrophic cartilage had lost the usual columnar organization of lacunae and remained unpenetrated by metaphyseal capillaries. These changes were indicative of retarded cartilage maturation and therefore could lead to osteochondrosis. Glade concluded that chronic overfeeding caused an endocrine involvement that predisposed the development of osteochondrosis. Thompson and co-workers[26] studied the incidence of radiographic bone aberrations in foals gaining more or less than 1.1 kg/day (2.4 lb/day) when suckling and more or less than 0.7 kg/day (1.5 lb/day) after weaning. They reported an increased incidence of aberrations in the high-gaining group and that the distal bones of the leg were more affected than the more proximal bones. They concluded that there was a link between above-average weight gains and the onset of physitis.

By contrast, some feel that growing horses cannot be fed too much if the diet is balanced.[27] The NRC[22] suggests that weanlings 6 months of age with an estimated mature weight of 500 kg be fed about 3 to 3.5 kg (about 7 lb) of grain and 2 to 2.5 kg (about 5 lb) of hay. At 12 months of age, intake might be 3 kg of grain and 3 kg of hay. As with the suckling, close observation of the youngster is needed to determine the amount of feed actually needed.

Calcium. Calcium has received a great deal of attention in both human and equine nutrition in recent years. Both overfeeding[18] and underfeeding[17] of calcium can cause bone problems. There is, however, considerable difference in opinion among authors as to what constitutes overfeeding and underfeeding. The NRC[22] recommends that foals with an expected weight of 500 kg should receive 34 g of calcium per day, about 0.6% of the ration when conventional feeds are used. To illustrate the confusion that exists, some authors think that the level of 0.6% is too low, whereas others believe it is too high.[18] I feel that a range of 0.6% to 0.8% is certainly reasonable.

However, diets containing 0.6 to 0.8% calcium are adequate only when the calcium in the diet is efficiently utilized. Some vegetables and subtropical grasses may contain oxalic acid that binds calcium so that animals cannot use it effectively. Grasses that may contain oxalic acid include buffel, kikuyu, pangola, Rhodes, and purple pigeon grass.

Recent studies from Canada indicate that the calcium in cereal hays is not used efficiently because of oxalate binding. The calcium in legumes such as alfalfa and in supplements such as limestone and dicalcium phosphate usually are utilized effectively by horses.

Copper. As mentioned earlier, copper has received considerable attention recently. Skeletal abnormalities due to copper deficiency have been produced experimentally in foals.[5,6] Copper deficiency may cause decreased activity of the bone-forming cells and therefore thin bone cortices. Copper is also required for normal activity of the enzyme lysyloxidase, which is necessary for normal skeletal development.

There is, however, controversy as to the amount of copper required to prevent skeletal problems in horses. Some recommend that weanlings be fed 150 to 175 mg of copper daily, which is equivalent to a dietary concentration of 30 mg/kg (ppm) when conventional diets are used.[10] This level is safe for horses but would be toxic to sheep.

Other workers, however, reported that they found no problems in foals fed diets containing only 10 mg/kg (ppm) copper.[22] The differences between the studies might be due to several factors, such as genetic influences and interactions with other nutrients in the diet.

Copper supplementation may reduce the incidence of bone disease, but it is not a panacea. Some farms have a high incidence of developmental orthopedic disease in their foals even though the foals have been fed 30 mg/kg (ppm) of copper or more in the diet. Then again, some farm managers who have increased the copper intake, although usually in addition to other dietary changes, have reported a significant decrease in the incidence of developmental orthopedic disease. Thus it seems reasonable to feed a level of at least 30 ppm to foals in herds or breeds predisposed to the problem.

Selenium. Selenium is not normally associated with skeletal problems. It is, however, required for normal muscle development. Deficiency can cause white muscle disease. Some feel that muscular problems could lead to conformational problems that could lead to skeletal problems. Foals fed only grains and forages grown in many areas of the eastern United States may need a selenium supplement. Selenium supplements should be added very carefully as little as 5 ppm is toxic.

SUMMARY

The important factor in designing diets for horses is that the ration should supply the necessary nutrients. The most practical ration varies among regions. In some areas it may be more sensible to use milo rather than corn, or barley rather than oats. A horse owner who wants to formulate a ration is best off working with local informed people such as county extension specialists or university personnel. Most horse owners have one to three horses. For small numbers of horses, it may be more practical to buy a properly prepared commercial feed rather than to formulate a mixture. A good feed company makes certain that the trace nutrients are adequately added and mixed. The commercial feed can also contain useful economical by-products that are not readily available to those mixing small amounts of feed. The company might also provide other services such as forage analysis. Commercial feeds designed for other species should be used with extreme caution. Many horses have died from eating beef, swine, or poultry rations containing certain antibiotics and ionophores (see Chapter 39) because of the large variation in the tolerance among species.

REFERENCES

1. American Horse Council: Horse industry directory, Washington, DC, 1989, American Horse Council.
2. Baker GJ: Comp Cont Ed 4:S507, 1982.
3. Banach M and Evans JW: Proceedings of the Equine Nutr Physiol Society, 1981.
4. Breuer LH: Proceedings of the 4th Equine Nutr Physiol Society symposium, 1975.
5. Bridges CH and Harris ED: JAVMA 193:215, 1988.
6. Bridges CH and others: JAVMA 188:173, 1984.
7. Carroll CL and Huntington PJ: Equine Vet J 20:41, 1988.
8. Dadd GH: Am Vet J 1:434, 1856.
9. Frisch RE: Sci Am 258(3):88, 1988.
10. Gabel AA and others: Proceedings of the American Association of Equine Practice, 1987.
11. Glade M: Proceedings of the American Association of Equine Practice, 1987.
12. Henneke DR, Potter GD, and Kreider JL: Equine Vet J 15:371, 1983.
13. Henry WA: Feeds and feeding, Madison, WI, 1898, Henry Publications.
14. Holmes LN, Song G, and Price EO: Appl Anim Behav Sci 19:179, 1987.
15. Jordan RM: Proceedings of the 7th Equine Nutr Physiol Society symposium, 1981.
16. Knight DE and others: Proceedings of the 1985 American Association of Equine Practice convention, 1985.
17. Krook L and Lowe JE: Pathol Vet (suppl):1, 1964.
18. Krook L and Maylin G: Cornell Vet 78 (suppl 2):1, 1988.
19. Kubiak JR and others: Theriogenology 28:587, 1987.
20. Lewis L: In Stashak TS, editor: Adam's lameness in horses, ed 4, Philadelphia, 1987, Lea & Febiger.
21. Masri MD, Olcott BM, and Kornagay WR: Proceedings of the 33rd annual meeting of the American Association of Equine Practice, New Orleans, 1987.
22. National Research Council: Nutrient requirements of horses, Washington, DC, 1989, National Academy of Science.
23. Ott EA: Personal communication, 1988.
24. Rich G: Proceedings of the advanced equine management short course, Fort Collins, 1987, Colorado State University.
25. Schoeb RR and Panciera RJ: Vet Pathol 16:18, 1979.
26. Thompson KN, Jackson SG, and Rooney JR: Proceedings of the 10th Equine Nutr Physiol Society symposium, 1987.
27. Tyznik W: Personal communication, 1989.
28. Uhlinger C: Proceedings of the 33rd annual meeting of the American Association Equine Practice, New Orleans, 1987.
29. Van Niekerk CH, Morgenthal JC, and Starke CJ: J South Afr Vet Assoc 54:65, 1983.

38

Performance Horse Nutrition

SARAH L. RALSTON

Diets that maintain body condition and performance are as varied as the types of competition expected of horses (Table 38-1). However, all horses need water, energy, fiber, protein, and at least minimal levels of vitamins and minerals to maintain the desired condition for performance.

NUTRIENTS
Energy sources

The hardworking horse needs more energy than the average horse on pasture. Carbohydrates and fats are the most concentrated and efficient sources of calories for horses; fiber is the safest source.[7] Substances used directly in energy pathways are: glycogen and glucose from carbohydrate or amino acid metabolism, long-chain fatty acids and glycerol from lipid metabolism, and volatile fatty acids from large intestine fermentation of fiber and carbohydrates. The energy pathways utilized are either aerobic (requiring the presence of oxygen) or anaerobic (utilizing no oxygen). Fuels and by-products for aerobic and anaerobic work differ. Aerobic sources of energy include long-chain fatty acids, volatile fatty acids, and glucose. Metabolic by-products of aerobic metabolism are primarily carbon dioxide, water, and variable amounts of heat (Table 38-2). Only glucose and glycogen are utilized in anaerobic work, which generates heat and lactic acid as the major by-products. Diet influences the energy sources utilized during work.[8,9] Most feeds contain variable amounts of carbohydrate, fat, and fiber. The choice of diet should be based on the type of performance expected of the horse and availability and economics of feeds available. Nutrients emphasized depend on the percent aerobic versus anaerobic work expected of the horse.

Grains, fresh green grasses, and legume (alfalfa or clover) hays are excellent sources of carbohydrates. Grass hay contains variable levels of available carbohydrates, depending on its quality, but is not generally considered to be a good source of soluble carbohydrate.

Grains and hays normally contain only 2% to 5% fat. Oils (seed or animal liquids that are virtually 100% fat) contain twice the amount of energy as carbohydrate or protein. Although vegetable oils are readily digested and utilized by horses,[12,13] it is not recommended that a horse's diet contain more than 10% fat.

The digestion of fiber may provide a significant amount of energy if a horse is fed only grass and hay.[7] It is not as efficient a source of energy as carbohydrates or oils. High-fiber forages are, however, the primary source of energy to which the horse was adapted over evolutionary time. The resultant volatile fatty acids provide a "cool" (minimum heat generated in muscle) source of energy for aerobic work. A horse's diet should consist of at least 50% of a good source of fiber by weight. The best sources of fiber are roughages.

Protein is used as an energy source only if carbohydrates, fats, and fiber are deficient *or* if there is excess protein in the diet. The utilization of protein for energy has a higher heat increment than an equal amount of sugar, fat, or volatile fatty acids (see Table 38-2). This may cause a

Table 38-1 *Examples of type of work, body type, and primary energy sources theoretically utilized by performance horses*

Type of competition	Primary type of work	Body type	Primary energy sources
Flat racing (thoroughbred, quarter horse), gymkhana	Anaerobic (speed)	Lean, minimal fat	Glycogen, glucose
Harness racing	Anaerobic, aerobic (speed, stamina)	Lean, some fat reserves	Glycogen, glucose (fat, volatile fatty acids)
Show jumping, steeplechase, rodeo, polo	Anaerobic, aerobic (speed, strength, stamina)	Muscular, moderate fat reserves	Glycogen, glucose, fat, volatile fatty acids
3-day event, dressage, distance riding	Aerobic, anaerobic (stamina, ± some strength, speed)	Lean, muscular but adequate fat reserves	Fat, volatile fatty acids (glycogen, glucose)

Table 38-2 *Relative heat increments for nutrient sources and metabolic by-products*

Nutrient	Relative heat increments
Glucose (aerobic)	Moderate
Fat	Low
Volatile fatty acids	Low*
Glycogen/glucose (anaerobic)	Moderate
Protein	High

*Heat increment in body tissues low; heat of fermentation generated in intestines fairly high.

relative increase in sweating and possibly, during prolonged exercise or high heat and humidity, elevation of body temperature.[15]

Protein

The absolute need for protein is slightly increased in the hardworking horse.[7] However, the percent of protein in the diet need not be above maintenance (8% to 12%). The horse will be eating more to meet its energy needs and thereby taking in more protein. Commonly used feeds that are high (above 14%) in protein include alfalfa, clover, soybean meal, and brewer's grains.

Electrolytes and water

The important electrolytes are potassium, sodium, chloride, magnesium, and calcium. If the horse sweats for a prolonged period of time during training or competition, it will need to ingest more water and electrolytes than normal to replace its losses.[4,10,11] Grasses and legumes (alfalfa, clover) are high in potassium. Legumes are also high in calcium and magnesium. Most grains are deficient in all of the major electrolytes. Both hays and grains have low levels of sodium (less than 0.05%). Performance horses in particular should have free access to sodium chloride. Additional electrolytes (see box) are not usually necessary except after prolonged sweating (more than 1 hour) during training or competition if an otherwise balanced diet with at least 50% hay or pasture is fed.[6,10,14]

Water requirements are not adequately met by any horse feed. A separate source of clean, ice-free water is necessary for optimal performance. A horse fed only dry hay and grain needs more water than a horse on lush pasture. Water consumption may increase to as high as 30 to 40 L/day in hardworking horses. The only time water intake should be restricted is if the animal is very hot, the water is very cold, or the horse is not going to continue to work after drinking. Under these circumstances the horse

> **Home-Made Electrolyte Mix**
>
> 500 g plain salt (NaCl)
> 400 g "lite" salt (NaCl/KCl mixture)
> 60 g $CaCO_3$ (or limestone)
> May be fed free choice *but* watch that horses do not overconsume (more than 230 g/day). The "lite" salt contains dextrose and may encourage overconsumption. Horses should be given 30 to 60 g/hr of strenuous work.

should be offered 2 to 3 L of water every 15 to 30 minutes until it is no longer thirsty or it is completely cooled.

Under conditions of prolonged sweating, such as endurance races or 2 to 3 hours of intense training, significant amounts of water and electrolytes are lost. If not replaced by allowing the horse to drink, graze, or consume 60 to 90 g (2 to 3 oz) of electrolyte supplement (see box) every 2 to 3 hours, severe losses may result, necessitating intravenous therapy. Clinical problems associated with acute electrolyte and water losses includes synchronous diaphragmatic flutter, exertional myopathies, cardiac arrhythmias, and ileus.[1]

Vitamins and minerals

Stress and high energy expenditure increase the levels of B vitamin and possibly ascorbic acid and vitamin E required in an athlete's diet.[3,7] The requirements for vitamin A and D, however, are not altered by strenuous exercise.[4] High levels of these (more than 10 times in excess of need) may be detrimental to the animal and should be avoided.[7]

Fresh grass, alfalfa, and grass hays contain moderate to high levels of vitamins A, D, E, and B-complex. Brewer's grains and brewer's yeast are fairly good sources of B vitamins. There are a multitude of vitamin-mineral supplements on the market, and the labels must be read carefully. Many contain near-toxic levels of vitamin A (more than 100,000 IU/dose) and D (more than 10,000 IU/dose). Trace mineral needs (i.e., copper, iron, zinc, and iodine) are not known to be increased by exercise. Supplementation of these may interfere with absorption and metabolism of other minerals and is not recommended unless a known dietary deficit exists.

SAMPLE PERFORMANCE DIETS

This section is prefaced with a warning: These are guidelines only. Each horse is an individual. The horse's performance and body condition must be evaluated on a weekly basis and rations adjusted accordingly. Body condition may be assessed in a fashion similar to that described for beef cattle, where the relative condition and depth of fat and muscle on the neck, withers, loin, tailhead, and ribs are each scored on a scale of 1 to 9, with 1 being emaciated and 9 extremely obese (Table 38-3 and Fig. 38-1).

Racehorses, gymkhana: speed only

Horses in this category include the "flat" racers of both thoroughbred and quarter horse breeding. As a rule, these animals are not worked for prolonged periods (longer than 60 minutes) during training, and their performances involve strenuous, anaerobic work of less than 2 to 5 minutes. The ideal diet for such animals is relatively high in rapid energy (grain) and low in bulk (forage). Supplemental vitamins have not been proven to be of benefit, but in theory more B vitamins than normally provided by a hay and grain diet may be needed due to stress.[7] Water and mineral requirements are probably not increased above maintenance. Since many race horses are less than 3 years old and therefore still growing, they will need higher protein and minerals (see Chapter 37) than their mature counterparts.

In general, horses should not be fed more grain than hay, and no more than one well-balanced vitamin supplement should be used. Beware of excessive grain (more than 50% of total ration), which potentially leads to a relative deficit of calcium in relation to phosphorus. High-grain diets may also predispose the horse to colic and laminitis.

Polo, harness racing, show jumping, steeplechase: speed and/or strength over distance

These horses require strength and speed but also need a certain amount of stamina. They also have increased energy requirements but because

Table 38-3 *Body scoring system for horses*

Condition (score)	Neck	Withers	Loin	Tailhead	Ribs
Emaciated (1)	Bones visible	No fat	No fat or muscle	No fat, prominent hipbones	Protruding
Very thin (2)	Bones barely visible	Accentuated, minimal fat	Spinous and transverse processes prominent	Prominent hipbones	Prominent
Thin (3)	Thin, flat musculature	Accentuated, some fat	Spinous processes prominent	Prominent but hipbones barely visible	Easily discernible
Moderately thin (4)	Some fat	Smooth	Slight ridge evident	Hips covered, some fat	Faintly visible
Moderate (5)	Blends smoothly with body	Smooth, some fat buildup	Level	Some fat at tailhead	Not visible but easily palpable
Moderately fleshy (6)	Fat palpable	Fat buildup pronounced	Rounded	Fat soft at tailhead	Fat soft over ribs
Fleshy (7)	Fat visible	Rounded, firm	Slight crease	Rounded	Ribs barely palpable
Fat (8)	Thickened	Soft, rounded fat	Crease evident	Flabby	Hard to feel
Obese (9)	Bulging fat	Bulging fat	Crease obvious	Bulging fat	Patchy fat pads

Each area should be scored individually and then the scores averaged to determine the overall condition score of the horse. This will allow for breed differences in conformation (e.g., high, prominent withers are common in thoroughbreds).
Modified from Henneke and others: Equine Vet J 15:371, 1983.

Fig. 38-1 Horses of different body condition scores. For most purposes scores of 5 to 7 are ideal. Note the crease in the back of the grade 9 horse. Note also that grades 2, 4, and 5 are represented by the same horse at various stages in its life.

Table 38-4 *Supplements for a mature, 500-kg performance horse*

Maintenance levels only needed (probably not necessary if on balanced diet)*

	Concentration in 60-g dose	Maximum recommended amount per dose of supplement
Sodium	8-10%	10-12 g
Copper	0.09%	70 mg
Zinc	0.30%	300 mg
Iron	0.30%	300 mg
Manganese	0.30%	300 mg
Selenium†	0.001%	1 mg
Vitamin A†	25,000 IU	25,000-50,000 IU
Vitamin D†	2,000 IU	2500-50,000 IU
Vitamin K†	Not needed	Not needed

Increased amounts needed

	Sources	Amount
Energy	Grains, vegetable oil	As needed to maintain body weight
Electrolytes	Salt, electrolyte mix (see box on p. 419)	30 to 60 g/hour strenuous work
B Vitamins	Grass, alfalfa	Free choice
	Brewer's yeast	60 to 90 g/day
	Various commercial supplements	As recommended by manufacturer
Vitamin E	Grass, alfalfa	Free choice
		Up to 1000 IU vitamin E/day in supplements

*Amounts given provide 100% of the recommended daily allowance for 400- to 500-kg horses.
†Toxic at 10 times recommended dose.
Author thanks James Kenney for assistance in preparation of this table.

of more aerobic phases of work, can utilize fiber sources efficiently. They should be fed diets containing at least moderate levels of fiber (16% to 20%). Because of their longer periods of training and performance, they have an increased need for water and electrolytes that is caused by sweat losses. Dietary protein of 8% to 12% and perhaps a B vitamin supplement are recommended. Standardbred harness horses are included here due to longer training sessions and races than are common in thoroughbred or quarter horse racing.

Three-day events, dressage, competitive trail rides, and endurance races: stamina, some strength and speed

In competition, endurance horses cover long distances at relatively slow speeds for 100 to 200 km/day (60 to 125 miles/day). They are trained daily for prolonged periods of time over a variety of terrains. Dressage and three-day event horses train intensively for hours at a time. All work is primarily under aerobic conditions, with occasional bursts of speed requiring anaerobic energy utilization. They demand more water, electrolytes, and B vitamins than do horses working over shorter distances and time periods. Their diets should be relatively low in protein (8% to 10% protein on a dry matter basis), with emphasis on fiber and energy sources. Fats such as vegetable oil may be added to the diet for extra energy (no more than 10% of total diet) (Table 38-4).

Draft or carriage horses

Draft animals require tremendous strength and muscle bulk but not much speed. In these animals, a diet of moderate energy, moderate protein (up to 12% on a dry matter basis), and high fiber would be adequate. If they perform over

Table 38-5 *Sample diets (amounts fed per day)*

Contents	Race horse	Distance/endurance competition
Grain (kg)	3.5-5.5	0-3.5
Hay (kg)	7-9	9-12 (preferably grass or grass-legume mix)
Others	60 g of a B vitamin–mineral mix	60 g trace mineral salt
	Free access to trace mineral salt and water	B vitamin supplement
		Free access to plain salt (white) and water

For electrolytes, see home-made electrolyte mix in box on p. 419.

prolonged periods of time during training or during performance, they have an increased need for water and electrolytes.

SUMMARY

Any animal under stress has an increased need for B vitamins. All animals that are worked hard need more available energy but not necessarily more protein in their diets. The need for water and electrolytes is increased in any animal that is worked for prolonged periods of time and sweats profusely. Energy and protein needs usually are adequately met by a balanced diet of hay and grain. Supplements should *not* be necessary unless poor-quality hay or forage is fed (Tables 38-4 and 38-5) or animals are highly stressed. Body condition and performance are the best determinants of adequacy of diet. Proper diet only permits a horse to reach its genetic potential for performance within the limits of its training. Nutritional supplements do not compensate for poor conformation, lack of natural ability, or improper training.

The condition of the performance horse is in the "eye of the owner," who must monitor the horse's condition closely and adjust its rations either up or down to allow for its individuality.

REFERENCES

1. Carlson OP: Comp Cont Educ 7:S542, 1986.
2. Hintz HR: Horse nutrition: a practical guide, New York, 1983, Arco Publishing.
3. Hintz HF: Equine Sports Med 5(3):1, 1986.
4. Hintz HF: Equine Sports Med 5(3):4, 1986.
5. Meyer H: Proceedings of the 8th Equine Nutr Physiol symposium, Lexington, Ky, 1983.
6. Meyer H, Guner C, and Lindner A: Proceedings of the 9th Equine Nutr Physiol symposium, Michigan State University, 1985.
7. National Research Council: Nutrient requirements of horses, Washington, DC, 1989, National Academy of Science.
8. Pagan JD and others: Proceedings of the 10th Equine Nutr Physiol symposium, Fort Collins, Co, June 11-13, 1987, pp 425-430.
9. Pagan JD and others: Proceedings of the 10th Equine Nutr Physiol symposium, Fort Collins, Co, June 11-13, 1987, pp 431-435.
10. Ralston SL: Cornell Vet 78:53, 1988.
11. Ralston SL and Larson K: J Eq Vet Sci 9:13, 1989.
12. Rich VA: Proceedings of the 8th Equine Nutr Physiol symposium, Lexington, Ky, 1983.
13. Rich GA, Fontenot JP, and Meacham TN: Proceedings of the 7th Equine Nutr Physiol Society, Warrenton, Va, 1981.
14. Sanders DE: JAVMA 188(9):912, 1986.
15. Slade LM and Lewis LD: Proceedings of the 4th Equine Nutr Physiol symposium, 1975.
16. Winter LD and Hintz HF: Proceedings of the 7th Equine Nutr Physiol symposium, Warrenton, VA, 1981.

Nutritional and Feed-Induced Diseases in Horses

SARAH L. RALSTON

Horses often differ from other agricultural animals in their sensitivity to various toxins, feed additives, and nutrient imbalances. For example, horses are much more resistant to copper, molybdenum, and urea toxicity than are ruminants but are extremely sensitive to ionophores such as monensin that are commonly included in cattle rations. This chapter covers metabolic problems that may be caused or exacerbated by improper diet and specific nutritional diseases in the equine animal.

METABOLIC DISORDERS WITH MULTIPLE ETIOLOGIES
Colic

Colic is a generic term for abdominal pain. There are many causes of colic, many of which have little or no connection to nutrition (Table 39-1). Dietary problems, however, can cause colic. If there is a high incidence of colic in a horse herd, the diet should be carefully evaluated, as should the worming program and dentition of the horses.

Impaction of the large intestine is most commonly associated with reduced water intake or feeds with very low digestibility; however, ingestion of foreign materials such as rubber fencing also may cause the problem.[3] Impactions also may result from a sudden change from highly digestible feeds (high-quality legume hay or complete pelleted feeds) to low-digestibility, high-fiber feeds (poor-quality grass hay, straw). The impaction is caused by the reduced water content of fecal material, reduced rate of passage (allowing greater water resorption in the intestine), or increased fecal bulk and particle size. Fecal masses are often palpable per rectum. Impactions usually respond to aggressive fluid therapy in conjunction with intragastric administration of mineral oil and surfactants such as dioctyl sulfylsuccinate. In most cases analgesics are necessary to keep the animal quiet and comfortable until the impaction has been resolved. Light exercise (hand walking) stimulates propulsive gastrointestinal activity and is recommended. Surgery is not often necessary.

It has been suggested that feeds that have a tendency to swell when wet should be thoroughly soaked before feeding (beet pulp, wheat bran, linseeds) to prevent gastric distention and impaction. Gastric impactions have been reported in ponies and horses fed large amounts of cracked corn, persimmon seeds, or mesquite beans.[3,11] Concentrates, especially pelleted feeds, fed in large amounts (more than 50% of the total diet) also have been associated with a high incidence of colic, especially gastric impactions leading to rupture.[7] If horses are fed large amounts of concentrates, it is best to divide the daily ration into three or four separate feedings.

Spasmodic colics are common in stressed horses fed high levels of concentrates (more than 50% of total ration). They usually resolve within 24 hours with medical management only. Gas colic, due to overfermentation of feed in the large intestine, is seen in horses receiving high grain or lush forage diets.

Sand colic is most effectively treated with prevention of sand ingestion, if possible. However, this is impossible for horses pastured on sandy soils. Administration of 0.5 kg (1 lb) psyllium hydrolytic mucilloid preparations twice a day in

Table 39-1 *Classification of colics in horses*

Category	Potential cause(s)	
	Nutritional	Other
Gastric impaction	High intake of cracked corn or foreign objects Low-digestibility feed	Poor dentition Parasitic damage to gut
Large intestine impaction	Large amounts of pelleted feeds Sudden switch from low-fiber to high-fiber diet Inadequate water Ingestion of foreign materials or sand	
Obstructive	Enterolithiasis	Parasitism Torsion Incarceration of bowel
Spasmodic	High-concentrate diet with altered motility in gastrointestinal system	Stress Altered gut motility
Sand	Ingestion of sand	
Parasitic		Parasitic damage to bowel wall or gut circulation
Other	A wide variety of dietary toxins	*Salmonella* Duodenitis Hepatitis Nephritis Urinary obstruction Colitis X

the concentrate rations of such horses has been effective in eliminating sand from the gastrointestinal tract.[2] The efficacy of wheat bran, which is commonly used, is not proven. Diagnosis in acute cases is sometimes difficult because 81% of sand impactions are not palpable per rectum.[29] Diagnosis is based on history of exposure to sand, documentation of sand in feces, or radiographic evidence of sand on the gastrointestinal tract.[2]

Colic occurs in horses regardless of diet, environment, worming program, or dentition. However, to reduce the incidence of colic, feed good-quality hay or pasture, free-choice potable water (see Chapter 6) and only as much grain or concentrate as necessary to maintain body weight. Do not feed concentrates to excited or exhausted horses or less than 1 hour before strenuous exercise (race or other competition) or stress (long trailer transport).

Laminitis

Laminitis refers to inflammation of the sensitive laminae of the hoof that results in severe pain and lameness. As in colic there are both nutritional and nondietary causes of the metabolic disorders leading to this problem. Laminitis may be caused by severe concussion ("road founder") but is most commonly associated with systemic disease caused by acute toxic (e.g., retained placentas, sepsis, salmonella) or nutritional insults.

The most common nutritional etiology of laminitis is sudden access to a highly fermentable feed (high-soluble carbohydrate, low fiber) consumed in large quantities by a horse or pony adapted to a less digestible diet. The classic examples are horses gaining access to a grain bin when they have been accustomed to limited grain rations and ponies turned out on lush pasture in the spring. The excessive ingestion of soluble carbohydrates overwhelms small intestine digestion and causes delivery of large amounts of highly fermentable material to the cecum and large colon. The resident microbial population, adapted to a diet higher in cellulose and lignin, produces more volatile fatty acids and lactic acid than usual. They draw fluid into the gut and may damage the intestinal lining. Endotoxins are released from dying bacterial cells that, upon absorption from the intestine, contribute to the hemodynamic alterations observed. The time frame from exposure to the feed to the appearance of clinical signs is important. If the insult is sufficiently large, the release of endotoxins occurs within 12 hours and may cause a fatal colic syndrome (colitis X).[8] More commonly only the pulse, blood pressure, and rectal temperature begin to increase 12 to 16 hours after the overload was ingested. Clinical signs of pain and lameness appear 20 to 24 hours after exposure to the feed. If overfermentation of the ingested feed can be prevented or reduced within the first 12 hours, laminitis may be prevented or at least reduced in severity.

It is important to treat all grain overloads aggressively before clinical signs appear. Early treatment includes administration of mineral oil, mild laxatives, antiinflammatory drugs, and perhaps antihistamines. The use of steroids is contraindicated.[30] Nonsteroidal antiinflammatory drugs should be used with caution if renal function is impaired. Mild exercise before clinical signs appear may increase the rate of passage and benefit the horse. Once metabolic signs (increased heart rate, respiration, or rectal temperature) appear, however, exercise should be discontinued and the horse confined to a stall with sand or deep sawdust bedding to reduce strain on the weakened laminae.

Obese ponies and horses are considered to be more prone to develop laminitis than are lean animals. Whether metabolic changes associated with obesity and laminitis (relative glucose intolerance and hyperinsulinemia) are reversible with weight loss has yet to be determined.[5,12] The contribution of relative glucose intolerance to the development of laminitic changes in horses also has not yet been determined. The clinical correlation between obesity and laminitis, however, is strong enough to warrant severe cautions against maintenance of horses or ponies in obese body condition.

Hyperlipemia and hyperlipidemia

Mobilization of lipid stores during starvation in excess of the body's ability to utilize the triglycerides is a potential problem in both ponies and horses. Fatty acids released by adipose tissue are taken up by the liver, and some are converted to triglycerides. Triglycerides are released back into the bloodstream. Triglycerides may be inadequately cleared from the blood during periods of stress or fasting, especially in ponies and obese animals. Two distinct syndromes exist.

1. *Hyperlipidemia* is characteristic of anorexic, clinically ill horses. It is rarely life-threatening and appears to reflect the presence of severe negative energy balance. Serum triglycerides are elevated but under 500 mg/dl.
2. *Hyperlipemia* is seen most commonly in obese ponies, especially in mares in the immediate periparturient period. It usually is preceded by a period of aphagia. Serum triglycerides are in excess of 500 mg/dl, and fat accumulates in the liver. *Mortality is high, approaching 80% without treatment.* Fatty infiltration of the liver and ultimately the kidneys is characteristic of this syndrome.

In both syndromes, relative insulin resistance has been documented. It is considered to be a predisposing factor in the development of hyperlipemia.[5,20]

Azoturia and rhabdomyolysis

Rhabdomyolysis, characterized by azoturia, myoglobinuria, and severe muscle cramping ("tying up") is a common problem in horses. Classically it was associated with feeding a high-concentrate diet to physically fit horses during a brief (1 to 2 days) period of rest, followed by a period of moderate to intense exercise ("Monday morning disease"). It originally was most common in heavily muscled draft horses. In recent years, however, rhabdomyolysis has been recognized in a wide variety of horses under highly variable circumstances. Two distinct syndromes have been recognized, although other categories probably exist.

The first and more classic rhabdomyolysis occurs relatively early during strenuous exercise (first 10 to 30 minutes) usually in horses that are physically fit. Theoretically these horses have increased muscular stores of glycogen, especially after a brief period of rest. If blood flow to the muscles is inadequate to support aerobic respiration in the cells, glycogen is mobilized in large quantities for anaerobic production of energy. The main by-products of anaerobic respiration are lactic acid and heat, both of which, if not removed by adequate blood flow, are toxic to the cell membranes and cause rhabdomyolysis. In vitamin E– and selenium-deficient animals, the cells are even more susceptible to oxidative damage, increasing the risk of rhabdomyolysis. Clinically these animals suffer severe, acute muscular cramping as the primary clinical sign. If sufficient muscle breakdown occurs, myoglobin and other protein by-products appear in the urine and cause a dark, cloudy color. Severe rhabdomyolysis may precipitate acute renal failure due to the kidneys' inability to excrete the sudden large load of solutes. This type of "tying up" is best treated with analgesics, muscle relaxants, and possibly intravenous fluids (to maintain renal function). Some horses are acidotic, but treatment with bicarbonate is controversial. Emphasis should be placed on keeping the muscle masses warm and increasing blood flow to the peripheral tissues. Preventive measures include:

1. Cut grain rations at least in half during days of rest.
2. Provide slow, gradual warm-up periods (walking at least 10 minutes on a loose rein) during the first training session after a brief (1- to 4-day) respite.
3. In some cases, supplementation with sodium bicarbonate (20 to 30 g per day in the grain) vitamin E, or selenium (only if proven to be deficient by ration analysis) may be beneficial.

The other type of tying up might be more appropriately termed *exertional rhabdomyolysis*. It appears most commonly in horses that have worked for several hours (e.g., endurance horses or 3-day event horses after roads and track), are dehydrated, and are suffering variable degrees of electrolyte imbalance. The onset of severe muscular spasm is often preceded by sudden cooling of the major muscle masses, either when cold water is poured over them or a hot, exhausted horse is allowed to stand in a cool breeze. Poor muscle blood flow may be the cause of this type of problem. Treatment in these horses should emphasize not only restoration of blood flow to the muscles by warming them but also replenishment of body water and electrolytes by aggressive intravenous administration of isotonic electrolyte solutions. Because of the usual presence of hypovolemia and dehydration, vasodilators are contraindicated in these animals, as is the use of bicarbonate. Nonsteroidal antiinflammatory drugs should be used with caution because of dehydration. Unlike horses that tie up early in exercise, exertional rhabdomyolysis victims are usually alkalotic instead of acidotic. Administration of bicarbonate would worsen their physiological status.

Prevention should focus upon maintenance of hydration and replenishment of electrolyte losses during prolonged exercise (see Chapter 38 on performance horse nutrition). A hot, tired horse should never have cold water poured over the major muscle masses and should be protected from cold breezes while cooling out. To rapidly reduce body temperature, it is safer to apply cold water to the lower extremities, head, abdomen, and jugular groove.

SPECIFIC DEFICIENCY AND TOXICITY SYNDROMES
Nutritional secondary hyperparathyroidism

Nutritional secondary hyperparathyroidism (NSH) also is called fibrous osteodystrophy, bran disease, miller's disease, bighead, or, in Portuguese, *cara enchada*. It is caused by a relative deficiency of calcium. This may be due to diet, such as feeding excessive amounts of bran (high in phosphorus), or induced by interference with the absorption of calcium (e.g., high oxalate content).[15] The result is the resorption of calcium (with concomitant release of phosphorus) from bone in the attempt to maintain blood calcium levels within normal limits (2.7 to 3.2 mmol/L or 10.9 to 12.8 mg/dl). This results in demineralization of skeletal bone, starting with bones subjected to the least stress: facial and ribs. The bones become soft and swollen because of fibrous tissue proliferation, thus the name *big head*. These changes may be preceded by a period of shifting leg lameness. There is also a high incidence of pathological fractures. Supplementation with calcium and reduction in the phosphorus and oxalate intake can reverse the disease, but the facial bones usually do not remodel. Prevention is to avoid overfeeding bran, grain, or high-oxalate grasses (buffel, setaria) to horses and to provide a calcium supplement when necessary.

Vitamin K toxicity

Vitamin K is frequently administered to race horses as an empiric preventative of nasopulmonary bleeding during racing activity. The active synthetic form of vitamin K, menadione (K_3) may be toxic to horses at manufacturers' recommended dosages.[23] Horses have been documented to develop acute renal failure following the parenteral administration of K_3 (1000 to 2500 mg in 300 to 500 kg horses).[23] Clinical signs were observed within 6 to 48 hours after injection in most horses. The animals initially experienced a period of depression followed by signs of renal colic with stranguria. Hematuria, azotemia, hyponatremia, hypochloremia, hyperkalemia and hypercalcemia were characteristic clinicopathological findings. Postmortem examinations revealed varying degrees of tubular nephrosis. Dehydration may exacerbate the effects of vitamin K_3 on the kidney. It is not recommended that synthetic vitamin K be administered to horses unless a deficiency is documented, and in that case K_1 is better. Vitamin K deficiency is unlikely to occur in horses.

Vitamin D toxicity

Vitamin D (D_2 and D_3) is known to be toxic in excess in a variety of species. In horses, however, the precursor form, ergocalciferol, given in doses up to 22,000 IU/kg body weight per day for 21 days did not cause overt signs of toxicosis. Daily doses of 47,200 IU of ergocalciferol per kg body weight caused depression and limb stiffness and reluctance to move within 6 days. Anorexia, weight loss, progressive weakness, and recumbency followed. All horses given vitamin D_2 in excess of needs were hyperphosphatemic. Blood calcium concentrations tended to be normal but fluctuated widely in a 24-hour period. Vitamin D toxicity also causes mineralization of vessel endothelium, endocardium, and skeletal muscles.[10]

Certain plants contain substances that have potent vitamin D–like activity and may cause signs associated with vitamin D toxicosis if ingested by horses. These plants include *Cestrum diurnum* (day-blooming jessamine) and *Solanum* species found in the subtropical regions of the world. The major difference between the signs observed with the plant-induced toxicities and the signs described for vitamin D toxicity is the consistent presence of hypercalcemia in affected animals.[16]

Vitamin E and selenium

Syndromes associated with vitamin E and selenium deficiency in horses include white muscle disease[18] and steatitis in foals,[10] reproductive failure in mares,[25] and possibly azoturia in performance horses.[18] Reduced immune competence has been documented in ponies fed selenium-deficient diets.[14] Grains and forages produced in some northwestern, northeastern, and Atlantic seaboard states may be very low in selenium (see Fig. 17-2 on p. 220).

White muscle disease in foals has been documented in both vitamin E– and selenium-deficient individuals. It is characterized by aphagia; stiff, stilted gaits with a peculiar hopping gait in the rear legs; and weakness. Upon necropsy, mineralization of the major muscle masses is evident; mottled white patches of muscle feel gritty when cut with a knife, hence the name of the syndrome.

Selenium deficiency in the presence of adequate vitamin E resulted in generalized steatitis in a foal.[11] In this case fat in the ligamentum nuchae, omentum, and subcutaneous tissue was grossly tan and necrotic in appearance, with evidence of mineralization.

There is apparently little transplacental transfer of selenium and vitamin E in mares. Foals depend heavily on mare's milk to maintain adequate stores. Supplementation of lactating mares increases milk content of selenium and presumably vitamin E. Diets high in fat increase the dietary requirements for vitamin E, although the extent has not yet been determined in this species. Stress and disease may increase the dietary requirement, especially for vitamin E.[14,21]

Diagnosis of vitamin E or selenium deficits in horses may be initiated by determination of dietary and serum levels of tocopherol (vitamin E) or selenium. Red blood cell glutathione peroxidase is indicative of selenium status in horses but is slow to respond to supplementation and very labile under storage. Plasma selenium concentrations are not as reliable as serum concentrations. In general, determination of the selenium content of the ration and serum selenium concentrations are the best diagnostic tests. Serum selenium levels lower than 8 μg/dl serum are suggestive of a deficiency. Diet vitamin E analyses are not readily available.

Dietary selenium of 0.02 mg/kg (ppm) was inadequate to maintain normal immune function in ponies.[14] Dietary concentrations as high as 2.2 mg/kg (ppm) resulted in adequate immune function and plasma concentrations. It is not recommended that selenium be supplemented to horses unless a true deficiency is proven to exist. This trace mineral may have toxic effects at only 10 times recommended levels.

Serum tocopherol varies dramatically over a 24-hour period, and horses with supposedly low levels are frequently clinically normal. Serum tocopherol lower than 0.8 μg tocopherol per dl serum has been found in horses considered clinically deficient in this vitamin. Although serum tocopherol concentrations appear to be dependent on serum lipid content in other species, this is not true in horses.[6] In general, several samples should be taken at at least 12-hour intervals from more than one individual in the herd to determine dietary adequacy.[6] Empiric supplementation of stressed horses or of those consuming high levels of dietary fat with 100 to 500 IU alpha tocopherol on a daily basis is recommended. Single large doses are not as effective at preventing deficiencies.[25]

Iron

Iron overload is probably more likely in horses than a true deficiency status. True deficiency may be diagnosed only by the concomitant existence of low levels of both serum iron and ferritin.[28] Iron-deficiency anemia has been documented in foals but may be a normal physiological status, especially in view of the existence of neonatal anemias in most other species studied to date. Indeed, in swine, oral administration of iron increased the susceptibility to enterotoxic colibacillosis.[13]

In 1983 an "outbreak" of hepatic disease occurred in neonatal foals throughout the United States. After exhaustive epidemiological investigations, it was traced to the oral administration of a paste containing ferrous fumarate (16 mg/kg body weight) in the first few days of life. The syndrome was characterized by jaundice, liver atrophy, bile duct hyperplasia, lobular necrosis, intrahepatic cholestasis, and varying degrees of periportal fibrosis.[13] It is recommended that foals not be given iron supplements in the first 3 days of life.

TOXIC FEED, CONTAMINANTS, AND ADDITIVES

Some feeds and plants can be problematic for horses due to molds and fungi associated with them (Table 39-2). It should be noted that the feeds listed in Table 39-2 are all quite acceptable as fodder if they are *not* moldly or infested with the problematic fungus. Many ornamental plants are extremely toxic. The plants listed in Table 39-3 are known to be toxic, but the list is by no means exhaustive. Lawn and garden clippings should *never* be fed to horses.

Blister beetles

There are many species in the genus *Epicauta*, commonly known as blister beetles. These beetles contain cantharidin in varying amounts, according to species and sex.[4] Blister beetles are attracted to the blooms on alfalfa plants. When alfalfa fields are cut late (a large number of flowers) in the midwest, southwest, and mountain regions of the United States, there is a high probability that blister beetles will be on the hay when it is cut. If the modern technique of crimping the hay as it is cut is used, many beetles may be crushed and baled with the hay. Unfortunately cantharidin is extremely heat and storage resistant. The amount of cantharidin contained in only five or six beetles of some species is sufficient to kill a full-grown horse.[4] Clinical signs include colic, playing in water, anorexia, and acute death.[26] Cantharidin can be assayed in the gastrointestinal contents, urine, or blood of the affected animals. Metabolic alterations observed include profound hypocalcemia and hypomagnesemia.[27] To prevent blister beetle toxicosis, only alfalfa grown in unaffected regions, certified to be free from contamination or cut before the bloom stage, should be fed.

Table 39-2 *Problematic molds and fungi associated with various feeds*

Name (reference)	Feed associated	Clinical signs
Acremonium coenophealum	Tall fescue	Agalactia Tough placentas Weak newborn Prolonged gestation Abortion
Fusarium moniliforme	Corn	Leukoencephalomalacia Colic, ataxia, blindness
Rhizoctonia leguminicola	Red clover	Slaframine slobbers Excessive salivation Weight loss
Dicoumarol	Sweet clover hay (moldy)	Vitamin K deficit Internal hemorrhage
Penicillium spp.	Ryegrass, fresh or dried	Ryegrass staggers Excitability, spastic ataxia, tetany if disturbed

Table 39-3 *Toxic or problematic feeds and plants*

Feed	Clinical syndrome and signs	Toxic principle
Acorns and oak leaves	Colic, hemorrhagic diarrhea, tenesmus, mucosal edema of gut, ulcerative enterocolitis, nephrosis, high gallic acid in urine	Tannic acid
Alsike clover	Photosensitization Liver cirrhosis	
Black walnut	Ventral limb edema Laminitis	Juglone?
Bracken fern Horsetail	B vitamin deficit Central nervous system signs	Thiaminase
Cestrum spp. (Wild Jasmine)	Hypercalcemia, calcinosis, kyphosis, weight loss	Vitamin D activity
Larkspur *Solanum Malacoxylon*	Excitability, abnormal gait Constipation	Polycyclic diterpene

Locoweeds
  Astragulus spp.
  Oxytropis spp.
	Neurologic signs
	Selenium toxicosis
	Abortion
		Pyrrolizidine alkaloid
		Selenium

Kleingrass Panic grass	Hepatotoxicosis Secondary photosensitization, sporadic occurrence	Saponins? Fungus?
Pyrrolizidine alkaloid plants Hound's-tongue (*Cynoglossum* spp.)	Weight loss, icterus, photosensitization, hepatic encephalopathy	Pyrrolizidine alkaloids (locoine Swainsonine)

Continued.

Table 39-3 *Toxic or problematic feeds and plants—cont'd*

Feed	Clinical syndrome and signs	Toxic principle
Pyrrolizidine alkaloid plants—cont'd		
Ragwort/groundsel (*Senecio* spp.)	Hepatic fibrosis, biliary hyperplasia, megalocytic hepatopathy	
Tarweed/fiddle-neck (*Amsinckia* spp.)		
Crotolaria spp.		
Heliotropium spp.		
Comfrey		
Photosensitization plants	Sloughing of nonpigmented skin	Photoreactive pigments
Buckwheat		
St.-John's-wort		
Red maple leaves (wilted or dried only)	Nephrosis, anemia Cyanosis, dyspnea Generalized hemolysis Methemoglobinemia	Some form of oxidant
Reed canary grass Sprouted barley	Positive urine drug test	Hordenine
Selenium toxic plants	Selenium toxicity *Acute:* acute diarrhea, incoordination, fever, death	

Paintbrush (*Castilleja* spp.)
  Snakeweed (*Gutierrezia* spp.)
  Astragalus spp.
  Prince's plume (*Stanleya* spp.)
  Woody aster, goldenweed
  Gumweed (*Grindelia* spp.)
  Saltbrush (*Atriplex* spp.)
	Chronic: sloughing mane and tail, lameness, horizontal hoof cracks, weight loss, poor fertility	Selenium

Nightshade (*Solanum* spp.)	Intense gastrointestinal and CNS disturbance Sudden death	Alkaloid
Sagebrush	Incoordination and ataxia when excited or restrained	?
White snakeroot	Stiff gait, sweating, subnormal body temperature, muscle tremors, constipation, dysphagia, increased respiration rate, prostration, death	Tremetol
Sudan grass and Sudan hybrids	Cystitis, dysuria Urinary incontinence Posterior weakness Incoordination	Cyanide
Yellowstar thistle, Russian knapweed (*Centaurea* spp.)	Nigropallidal encephalomalacia, inability to prehend feed (appetite is normal, they can swallow)	

H. F. Hintz contributed to the preparation of this table.

Carboxylic ionophores

Monensin, narasin, and lasalocid are all carboxylic inophore compounds commonly included in cattle and chicken rations as growth promotants and coccidiostats. These compounds transport metal ions through biological membranes and are produced by *Streptomyces* spp. Only 2 to 3 mg monensin per kg body weight will kill a horse.[1] Lasalocid is less toxic than monensin,[9] but all ionophores should be excluded from horse and pony rations. Chicken and cattle rations may contain as much as 125 g lasalocid per tonne (125 ppm) or 121 g monensin per tonne (121 ppm). Do not permit horses access to cattle or chicken feed. Horses may develop clinical signs of depression, ataxia, paresis, colic, or rhabdomyolysis within 6 to 12 hours of ingesting feed contaminated with monensin.[17] Heart rate is dramatically increased; the animals may sweat profusely and refuse further feed. Horses receiving high doses usually die in 48 to 72 hours. Animals that survive longer than 72 hours may suffer long-term deterioration of cardiac and renal function.[19] Clinical metabolic alterations observed include hypocalcemia, hyperphosphatemia, hyperglycemia, and increased total bilirubin. Accidental contamination of horse feeds with monensin has resulted in several outbreaks of toxicity. Monensin or lasalocid toxicity should be suspected when there is an outbreak of the signs listed here that are associated with a new batch of mixed feed or when horses have access to cattle or chicken feed.

Antibiotics

Antibiotics are not commonly included in horse rations. However, rations formulated for food animals which contain lincomycin,[22] salinomycin,[24] or tetracycline may result in colic, diarrhea, or other signs of toxicosis in horses.

REFERENCES

1. Amend JF and others: Proceedings of the 31st meeting of the American Association of Equine Practice, 1985.
2. Bertone JJ and others: J Am Vet Med Assoc 193:1409, 1988.
3. Bohanon TC: J Am Vet Med Assoc 192:365, 1988.
4. Capinera JL, Gardner DR, and Stermitz FR: J Econ Entomol 78:1052, 1985.
5. Coffman JR and Colles CM: Can J Comp Med 47:347, 1983.
6. Craig AM and others: Am J Vet Res 50:1527, 1989.
7. Fontaine M and deFaucompret P: Practique Vet Equine XII:10, 1980.
8. Green EM, Sprouse RF, and Garner HE: Proceedings of the 2nd colic research symposium, 1983.
9. Hanson LJ, Eisenbeis HG, and Givens SV: Am J Vet Res 42:456, 1981.
10. Harrington DD: J Am Vet Med Assoc 180:867, 1982.
11. Honnas CM and Schumacher J: J Am Vet Med Assoc 187:501, 1985.
12. Jeffcott LB and others: Equine Vet J 18:97, 1986.
13. Kadis S and others: Am J Vet Res 45:255, 1984.
14. Knight DA and Tyznik WJ: J Anim Sci 68:1311, 1990.
15. Krook L and Lowe JE: Path Vet (suppl):1, 1964.
16. Krook L and others: Cornell Vet 65:26, 1975.
17. Matsuoka T: JAVMA 169:1098, 1976.
18. Maylin GE, Rubin DS, and Lein: Cornell Vet 70:272, 1980.
19. Muylle E and others: Equine Vet J 13:107, 1981.
20. Naylor JM: In Robinson NE, editor: Current therapy in equine medicine, Philadelphia, 1983, WB Saunders.
21. Nockels CF: Fed Proc 38:2134, 1979.
22. Rainsbeck MF and Osweiler GD: J Am Vet Med Assoc 179:362, 1981.
23. Rebhun WC and others: J Am Vet Med Assoc 184:1237, 1984.
24. Rollinson J, Taylor FGR, and Chesney J: Vet Rec 121:126, 1987.
25. Roneus BO and others: Equine Vet J 18(1):51, 1986.
26. Schoeb TR and Panciera RJ: J Am Vet Med Assoc 173:75, 1978.
27. Shawley RV and Rolf LL: Proceedings of the 2nd colic research symposium, 1983.
28. Smith JE and others: J Am Vet Med Assoc 188:285, 1986.
29. Specht TE and Colahan PT: Proceedings of the 2nd colic research symposium, 1983.
30. Yelle M: Equine Vet J 18:156, 1986.

Feeding Sick Horses

SARAH L. RALSTON, JONATHAN M. NAYLOR

Horses are unique among the large animals in that many, if not most, are regarded as pets. Although large horse operations usually are as concerned with economic profitability as food animal producers, most horse owners have fewer than five animals. Horse owners often are willing to incur large medical expenses to prolong the life of a critically ill or aged horse, which is regarded as a pet or even family member rather than an economic unit. Whereas the nutritional management of cattle, sheep, swine, and most goats is driven by economics, nutrition of horses is often based on a more anthropomorphic approach by the owners and is fraught with tradition and nonscientific theory.

Because of this interest in individual survival and well-being, the nutritional needs of horses with chronic or potentially debilitating diseases need to be addressed.[10,25,35,40,44] Despite many interspecific similarities in basic physiological responses that make comparisons possible,[47] there are also variations in metabolic responses and dietary requirements that alter recommendations for a given species relative to others.[34,37,40] Many of the recommendations in this chapter are extrapolated from data from other species and therefore must be regarded as theoretical until definitive equine data are available.

The following sections deal with nutritional recommendations for sick horses. These recommendations are summarized in Table 40-1; details of common feed supplements are found in Table 40-2.

TRAUMA, SEPSIS, AND BURNS

The goals of nutritional support of traumatized, burned, or septic horses are to (1) minimize the need for mobilization of body tissues for energy and essential amino acids and (2) maximize wound healing and immune competence. These goals should be accomplished without overloading an already compromised system.

Although glucose is the predominant energy source in normal tissue metabolism, utilization of glucose is altered in the septic, traumatized, or severely burned animal (see Chapter 11). Prolonged administration of glucose-containing fluids without exogenous sources of other nutrients such as amino acids and vitamins has been associated with fatty infiltration of hepatocytes in people.[22,47] In humans it is recommended that the rate of administration of glucose should not exceed 5 mg per kg of body weight per minute.[47] This would translate to a maximum of 3 L of 5% dextrose per hour in a 500-kg horse, which would provide approximately 13.5 Mcal/day, only 2.9 Mcal less than the estimated maintenance requirement.[34] Water-soluble vitamins, vitamin E, and carnitine, however, are needed for utilization of long-chain fatty acids in cells.[22,47] To minimize the danger of hepatic lipidosis, a source of protein, water-soluble vitamins, vitamin E, and perhaps lipid should be provided enterally or parenterally when glucose is infused at high rates for more than 24 hours.[22,25]

Fatty acids and ketones are metabolized preferentially by most tissues in simple starvation, but utilization may be reduced in septic or traumatized animals, despite suboptimal feed intakes.[47] Plasma triglycerides and fatty acids are increased in healthy horses fasted for longer than 2 days but, unlike humans, ketosis is not apparent.[1] Clinically ill, anorexic horses develop hyperlipidemia or hyperlipemia instead of ketosis.[37] Ingestion of adequate amounts of a bal-

Table 40-1 *Dietary recommendations for clinically ill horses*

Clinical problem	Nutritional objectives	Diet ingredients (% of total or daily amount)*
Trauma, sepsis, burns	Increased† protein, energy, vitamins B, C, and E Increased zinc?	Mixed legume hay or alfalfa pellets (50-100%) Grain (0-50%) Vegetable oil (100-400 ml) B vitamin supplement Vitamin E (500-1000 IU) Zinc? (200-400 mg)
Chronic diarrhea	Increased water, electrolytes	Free-choice water, salt B vitamin supplement Cecal flora transplant?
With weight loss	Increased energy, protein	Grain mixes (25-50%) Good quality hay (0-75%) Vegetable oil (100 ml)
No weight loss	Increased fiber	Grass hay (100%) Pasture, not lush (100%)
Extensive large bowel resection	Increased carbohydrates, protein, water, B vitamins, and electrolytes	Alfalfa hay or pellets (50-100%) Complete feed pellets (25-100%) Grain mixes (0-50%) B vitamin supplement Vegetable oil (100-400 ml) Phosphate (5-10 g) Cecal flora transfaunation?
Small bowel resection; malabsorption		
Immediately after resection (1-10 days)	Increased digestibility Decreased bulk	Liquid diet or slurry of pelleted feed (25-100% of needs) Soy or linseed meal (100-400 g)
Long-term management	Increased soluble fiber and B vitamins Decreased carbohydrates	Alfalfa or mixed legume hay (75-100%) Vegetable oil (100-400 ml) B vitamin supplement Processed grain (0.5-2 kg for 450-kg horse)
Enterolithiasis	Decreased cecal pH, calcium, phosphorus, and magnesium	Grass hay (0-50%) Grain (0-50%) Complete pelleted feed (0-100%) Apple cider vinegar (if on hay) 110 ml
Renal disease	Increased water and salt Decreased protein, calcium, and phosphorus	Grass hay (50-100%) Corn grain (0-50%) Flaked milo (0-25%) Free-choice salt and water
Hepatic disease	Increased simple sugars, B vitamins, and short-branched-chain amino acids Decreased aromatic amino acids	Grass hay (50-75%) Corn grain (25-50%) Flaked milo (0-25%) Linseed meal (50-200 g) B vitamin supplement Ascorbic acid (10 g) Vitamin E (500 IU)

*Assume free-choice water and trace mineral salt unless otherwise noted.
†All increases and decreases are relative to normal maintenance and vary with the severity of the disease.

Table 40-1 *Dietary recommendations for clinically ill horses—cont'd*

Clinical problem	Nutritional objectives	Diet ingredients (% of total or daily amount)*
Cardiac disease	Decreased salt	Grass hay (75-100%) Corn, oats, barley (0-25%) No salt if edema present
Geriatric (older than 20 years)	Increased protein and phosphorus Decreased fiber Calcium:phosphorus <3:1 but >1:1	Grass hay (50%) Grain mixes (0-50%) Pelleted feeds (12%-14% protein) (slurries) (0-100%) Linseed or soybean meal (200-400 g) B vitamin supplement Ascorbic acid (10 g) Phosphate (5 g) Vegetable oil (50-100 ml)
Rectal or vaginal surgery	Low bulk, laxative diet	Alfalfa pellets (50-100%) Complete pelleted feed (50-100%) Fresh green grass (0-100%) No hay

*Assume free-choice water and trace mineral salt unless otherwise noted.
†All increases and decreases are relative to normal maintenance and vary with the severity of the disease.

Table 40-2 *Supplements for horses*

Supplement	Comments
Protein	
Linseed meal	30-40% protein, excellent short branched chain:aromatic amino acid ratio, low lysine, high arginine
Soybean meal	40-44% protein, high lysine
Corn gluten meal	30-36% protein, low lysine, high in short branched chain amino acids
Energy	
Grains	Wide variety used (see text)
Vegetable oil	Usually soybean derived, cheaper than other forms of oil
Corn oil	Palatable, expensive
Wheat germ oil	Expensive

anced diet results in rapid resolution of hyperlipidemia in most horses.[37]

No studies of the energy and protein requirements of septic, burned, or traumatized horses have been published. Basal digestible energy (DE) requirements, however, may be approximated by the requirements of a horse in a metabolism stall:[34]

$$DE (Mcal/day) = 0.975 + 0.021 (body\ weight, kg)$$

To estimate the requirements of a clinically ill horse confined to a stall, the basal energy requirement should be multiplied by a fudge factor to reflect changes in metabolic rate associated with disease (see Table 11-1, p 141). For force-feeding via enteral or parenteral routes, it is better to underestimate energy requirements than to provide excess calories. Therefore, conservative use of the fudge factors, which are based primarily on human and rat data, is recommended. If the horse is willing and able to eat on its own, however, it should be safe to offer amounts of feed that meet or exceed the NRC recommendations for maintenance (approximately double the basal DE).

Free-choice access to good-quality grass, legume hay, or both and to salt and water usually meets the needs of moderately stressed horses (e.g., carpal chip removal, cervical lacerations). However, severely injured or septic animals may have reduced voluntary intake or be unable to consume enough to meet their needs. Energy-dense supplements such as grain mixes and vegetable oil are recommended for horses that are willing to eat but losing condition. Maximizing feed intake in horses that are unwilling to eat is discussed later in this chapter.

The estimated Mcal DE should be used to calculate protein requirements. The maintenance crude protein requirement of horses is 40 g crude protein per Mcal DE consumed[34] (maintenance nitrogen intake as digestible protein is 1 g N per 350 kcal DE).[34] Because of the enhanced protein needs in trauma and sepsis, supplementation of protein above this level should not be a problem unless the horse is suffering concurrent hepatic or renal failure. In liquid formulas such as Osmolite and Ensure, which are sometimes used in sick horses, the protein is assumed to be digestible protein. The amount of protein usually is given in terms of g N per kcal DE, and most formulas contain 1 g N per 120 to 150 kcal DE (approximately 45 g crude protein per Mcal).

All dietary changes, including refeeding after a prolonged (longer than 2-day) period of fasting, should be made slowly. Only a quarter of the estimated amount of feed, divided into several small feedings, should be given on the first day. If no adverse reactions are noted, the amount can be increased to half on the next day and to full feeding by day 5. Adverse reactions to sudden dietary change in horses, regardless of the route of feeding, include diarrhea, impactions, colic, and laminitis. Laminitis and diarrhea are particularly likely when diets high in simple sugars are fed to horses. If laminitis or colic occur during introduction of the diet, the amount fed should be reduced or stopped completely, and the nutritional recommendations reevaluated. Soft feces may be caused by highly purified diets and are tolerable unless associated with fever or dehydration. If the feces are small, dry, and hard, additional bulk (wheat bran, Metamucil, or alfalfa pellets), water, and mineral oil should be added to the diet.

Supplementation of vitamin C does not harm and may benefit stressed, chronically ill horses. Horses do not require supplemental vitamin C under normal circumstances because of their adequate hepatic synthesis of ascorbic acid from glucose.[34] The vitamin is mobilized from stores in the adrenal gland in periods of stress.[5,50] Old age and the stress of disease or transportation result in lower plasma levels of ascorbic acid in horses.[5,50] Whether these lower levels reflect true depletion of body stores is unknown at this time. However, chronic stress or disease theoretically may deplete body stores and reduce hepatic synthesis because of competition for glucose by glucose obligate tissues (red and white blood cells, neural tissue, bone marrow). Ten to 20 g of ascorbic acid (20 to 40 g of ascorbyl palmitate) is recommended, preferably given in the feed or enteral slurry. Ascorbyl palmitate results in the greatest increases in plasma vitamin C,[50] but ascorbic acid powder also may be effective.[42]

Based on data from other species, nutrients of concern also include the water-soluble vitamins, especially thiamin, folic acid, pantothenic acid, and niacin.[47] Biotin, cyanocobalamin (B_{12}), and fat-soluble vitamin K, which are synthesized by bacteria in the large intestine and stored in the body, would be of concern only if there had been prolonged use of oral antibiotics[46,47] or large bowel dysfunction. It is important to note that the injectable form of vitamin K (menadione) may cause renal damage, especially in dehydrated horses. Good-quality hay and fresh grass are excellent sources of B vitamins. Brewer's yeast has been overrated as a source of B vitamins in general but is still a good, safe supplement for debilitated or stressed horses. If the horse is not eating on its own, injectable B-complex solutions may be used.

The vitamin E recommendation for healthy horses is 500 mg per 500-kg horse per day.[34] More vitamin E (750 to 1500 mg α tocopherol per 500 kg per day) may be required to maximize wound healing and immune competence in a severely ill horse. There is enough vitamin A and D stored in a horse's body to last 4 to 6 months without further intake. It is not likely that these fat-soluble, potentially toxic vitamins will need to be supplemented, but if a deficiency is suspected up to 50,000 IU vitamin A and 5000 IU vitamin D may be added to the daily ration.

Most clinically ill horses do not need supplemental iron. Despite low blood iron and anemia, horses in a clinical survey did not have low body stores of iron.[49] Indeed, 10 horses in that study had indications of iron overdoses. Iron is sequestered in body tissues in response to bacterial infections.[29] The resultant hypoferremia is an apparent defense against bacterial infections and should not be overridden by exogenous supplementation with iron.[29,49] Unless a horse has suffered prolonged, severe blood loss or has been on a diet documented to be low in iron for more than 2 or 3 months, iron should not be supplemented. If iron supplementation is necessary, 400 to 600 mg per 500-kg horse per day would meet the normal requirements.

Stress and disease also result in decreased blood zinc concentrations.[4,19,23] Severely burned or traumatized horses may benefit from 200 to 400 mg zinc per horse per day. Excessive supplementation, however, should be avoided.

GASTROINTESTINAL DYSFUNCTION
Diarrhea

Diarrhea is primarily a large bowel disease in the horse.[32] It may be due to either malabsorption, disrupted bacterial fermentation, or enhanced rate of passage. There is also a strong correlation in other species between hypoalbuminemia and the development of diarrhea.[12,39]

Based on recent data in other species,[3,47] it is not recommended that horses with diarrhea be held off feed unless abnormal gastric emptying, gastric reflux, or bloating is a problem.[10,35] Enteral provision of energy and protein appears to be necessary to maximize gastrointestinal integrity and healing ability. Diarrhea can be induced by fasting alone in horses. Glutamine, aspartate, and fatty acids are the primary energy sources for the enterocyte, glucose being absorbed rapidly for utilization in other tissues.[47] The traditional approach to the management of the diarrheic horse has been to withdraw grain and feed only hay. The fiber in the hay may stimulate intestinal segmental motility (which slows passage of digesta) and adds bulk and form to the feces. This approach is acceptable in horses that are maintaining their body weight. In horses that are losing condition, more intensive nutritional care is needed. Small bowel function may be normal and allow nutritional balance to be achieved despite ongoing large bowel losses.[32] Low-residue, relatively high-protein (1 g nitrogen/120 to 150 kcal DE) liquid diets (Equical, Osmolite, Ensure)* have been used to maximize absorption from the small bowel and minimize large bowel stimulation but may be associated with an exacerbation of diarrhea. If the animal can tolerate bulk in its diet, diets of grain (1 to 2 kg or 2 to 5 lb/day) and good-quality hay free choice or 1.5% to 2.5% body weight as complete pelleted feed will maximize digestion in the small bowel relative to grass hay alone. If abnormal bacterial fermentation is suspected, oral administration of cecal and colonic fluid obtained from a horse with normal gastrointestinal function (preferably *Salmonella* negative), yogurt, or a commercial bacterial inoculant may help.[3]

Gastrointestinal resection

Surgical resection of devitalized intestine is more common than it was 10 years ago. Surgical resection of large portions of the equine large intestine is compatible with long-term survival and even reproduction and athletic performance if appropriate diets are fed (Fig. 40-1).[6-8,45,52]

Immediate postoperative period (0 to 14 days). Postoperative fasting with or without parenteral alimentation often is recommended to provide "bowel rest." However, enteral alimentation has been shown to enhance bowel adaptation after resection and may reduce the incidence of postoperative gram negative infections in humans.[47] Following large bowel resection, horses usually are able to consume high-quality, long-stem hay within 12 to 24 hours of surgery without adverse effects. Horses with small bowel disease often are routinely fasted for 1 to 4 days postoperatively. This is in part due to concerns about potential disruption of

---
*Ross Laboratories, Columbus, Ohio

Fig. 40-1 This mare produced a healthy foal 9 months after resection of 95% of the left and right colons. She maintained good body condition throughout her lactation on a diet of alfalfa hay and grain.
Photo courtesy of Al Kelminster, Colorado State University.

suture lines and abnormal motility patterns. Diets of long-stem hay and grain cannot be recommended immediately postoperatively, but liquid diets may be used to provide at least some nutritional support. Small amounts (0.25 to 1 kg or 0.5 to 2 lb per feeding) of complete pelleted feed or alfalfa pellets offered at 2 to 6 hour intervals may also be tolerated. These diets must be introduced slowly; less than a quarter of the estimated amount required, divided into several small feedings, should be fed on the first day. The amounts fed can be gradually increased over the next 2 to 3 days until estimated needs are met.

Up to 3 weeks after typhlectomy, colon resection, or large bowel ischemia without resection, there is decreased apparent digestion of protein, fiber, and phosphorus.[21,33,45] Therefore, the immediate postoperative diet after large bowel resection should be relatively high in protein (12% to 14%) and low in fiber (less than 26%). Alfalfa pellets, complete feed pellets, or high-quality alfalfa hay frequently are recommended. The increased loss of phosphorus should not be of concern in the short-term management of these horses.

Long-term management

Large colon resection. After resection of only the left colon or cecum, horses appear to suffer few long-term alterations in digestive ability or nutritional requirements[33,45] and are able to maintain adequate body condition on even low-quality (6% to 8% protein) grass hay fed free choice.[45] However, for more than 6 months after extensive resection of both the left and right colons, horses have reduced apparent digestion of fiber and protein and are in negative phosphorus balance when fed diets of 2% to 2.5% of their body weight in pelleted alfalfa or mixed hay.[8,21] They are, however, able to maintain normal blood chemistry and body weight.[6,8] For performance, advanced pregnancy, or lactation, additional grain is required to maintain body weight. There is increased fecal water loss that suggests that greater than normal water intake may be necessary.[8] The long-term consequences of negative phosphorus balance remain to be determined but could possibly lead to weakened bones and, in growing horses, developmental orthopedic disease.[34] Supplementation with a calcium-phosphorus salt mix to bring the total dietary phosphorus concentration to 0.45% with a calcium-phosphorus ratio of 1.5 to 2.0:1 may help prevent overt phosphorus deficiency.

Small bowel resection, malabsorption. Following resection of less than 50% of the small intestine, ponies are able to maintain adequate body condition when fed 1.5% of their body weight in complete pelleted feed.[52] However, ponies fed the same diet after more than 60% resection of the small intestine rapidly lose weight and become debilitated.[52] A mature stallion with fibrosis in the cranial duodenum and jejunum resulting in malabsorption (roughly equivalent to more than 50% of the small bowel being nonfunctional) was maintained in adequate body condition when fed alfalfa hay free choice, 1 kg (2 lb) grain, and 0.5 L (2 cups) of vegetable oil per day. It is important to note that this horse lost weight when fed large amounts of grain or grass hay. Increased concentrate-forage ratios decrease digestion of feeds in the large intestine and should be avoided in cases of small intestinal malabsorption or resection. Instead, digestion in the large intestine should be emphasized with easily fermentable roughages such as alfalfa as the primary sources of energy. Addition of fat to the diet may also be beneficial. Calcium, which is absorbed primarily in the proximal small intestine, may need to be supplemented.

Enterolithiasis

Horses prone to the development of enteroliths may benefit from a diet that consists primarily of complete pelleted feed or the addition of apple cider vinegar to a diet of two parts forage to one part grain.[27] The aim is to reduce cecal pH to less than 6.5, which will prevent the formation of the calcium-phosphorus-magnesium stones. Addition of 110 ml (0.2 lb) apple cider vinegar per feeding of grain reduced cecal pH effectively in ponies fed timothy hay and grain.[27] It is important to note, however, that reduced cecal pH also may be associated with an increase in wood-chewing activity.

Rectal lacerations or vaginal surgery

Pelleted feeds result in feces that are more easily passed because of relatively small particle size and normal to high water content. Complete pelleted feeds and pelleted alfalfa, therefore, are the feeds of choice for animals following recto-vaginal surgery, where the major concern is to avoid strain on the suture lines. Another good choice is highly digestible fresh grass when it is available. If fecal contamination is of major concern, reduction of total fecal volume may be possible by feeding limited amounts. It is not recommended that horses be starved to reduce fecal volume. It takes at least 3 to 4 days of total starvation to empty the equine gastrointestinal tract, at which point the animal would be immunosuppressed and have reduced wound healing.[36]

GERIATRICS

Geriatric horses (above 20 years old) have special nutritional needs. Many have poor dentition (Figs. 40-2 and 40-3). In one study 10 of 13 old horses had pituitary tumors.[44] All old horses with tumors, whether or not exhibiting the classic clinical signs of Cushing's disease (hirsutism, polydipsia, polyuria and hyperglycemia), had reduced responsiveness to insulin and dexamethasone suppression of cortisol secretion.[43] These changes are associated with an increased susceptibility to laminitis. Old horses may also have reduced plasma concentrations of vitamin

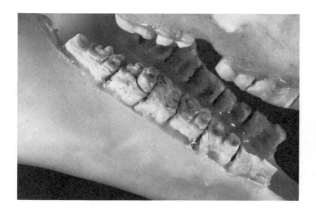

Fig. 40-2 "Wave mouth" is not uncommon in aged horses. If a molar is missing from the upper or lower arcade, the opposing teeth continue to erupt, resulting in protrusions as seen here. These must be regularly filed down ("floated") to allow normal mastication.
Photo courtesy of Al Kelminster, Colorado State University.

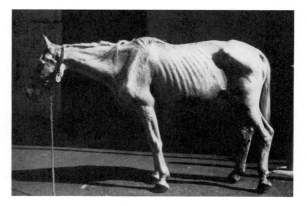

Fig. 40-3 Result of putting a horse with wave mouth out on pasture. This horse was 35 years old and had extreme wave mouth. When given proper dental care and fed a pelleted diet, he gained weight and overall strength.
Photo courtesy of Al Kelminster, Colorado State University.

C relative to younger animals fed the same diet.[44] Apparent digestion of protein, fiber, and phosphorus is reduced in horses above 20 years old.[41]

In general, old horses probably do not need special dietary manipulations until there is loss of body condition that cannot be accounted for by a treatable disease process. In the "failing" older horse, a diet formulated to meet the needs of a weanling would be the most appropriate because the nutritional concerns are basically the same: protein 14% to 16%, Ca-P ratio greater than 1.5:1, but phosphorus in increased amounts relative to adult maintenance, low fiber, and high digestibility. Avoid feeding geriatric horses dry pelleted feeds for aged horses appear to be more prone to choke. If dentition is poor, pelleted feed soaked in water to make a soup consistency may be used to maximize intake without danger of choke. Efficacy of supplementation with ascorbyl palmitate has not been tested, but moderate amounts (10 to 20 g/horse/day) probably would not harm an older horse. Energy supplements of choice would be vegetable oils or pelleted feeds soaked in water. Due to the potential for reduced renal function in the older horse, access to legumes should be limited. Soybean or linseed meals may be used if a protein supplement is necessary.

CHRONIC OBSTRUCTIVE PULMONARY DISEASE (COPD)

Horses with COPD (heaves) are greatly affected by nutritional management. Indeed, management of nutrition and housing offers the best route for long-term treatment of horses with heaves. The majority of COPD horses are allergic to molds associated with hay and bedding. These horses are best managed by moving them to a corral, paddock, or pasture, where natural ventilation reduces exposure to allergens. A three-sided shed can be used for shelter from rain and horse blankets for additional insulation. It takes about 3 to 4 weeks for a horse to recover completely from heaves when placed in this type of environment. Medication (bronchodilators, antihistamines, corticosteroids) in the acute phase of the disease may improve the speed of recovery and make the horse more comfortable.

If horses have to be fed preserved feeds, they should be offered at ground level to reduce dust in the air. Moldy hay contains the most allergens (Fig. 40-4). Molds are particularly likely to grow when hays are baled wet at 35% to 50% moisture, which leads to heating and molding in the bale. Hays baled at 20% to 30% water are of moderate quality, and hays baled at less than 20% moisture are usually the cleanest.[15,16] If hay is to be fed indoors, then it should be soaked by immersion for 12 hours before feeding. This can be accomplished by placing the hay in a barrel of water. If long-stemmed hay is not tolerated, cubed or pelleted hay may be used. Hay particles in cubed hays are longer than those in pellets and may be a better source of roughage than pelleted feeds. Grain that has been stored carefully should be low in molds, but it also can be soaked if dust is a problem.

Bedding can contain molds. The cleanest beddings are wood shavings and shredded paper that are removed daily.[15]

Because management of horses with COPD is aimed at improving air quality, it is important that all horses in the same air space be placed on hypoallergenic management. If affected horses must be stabled, they should be moved to well-ventilated corner stalls away from areas where hay is handled. Horses in adjacent stalls should be managed as if they were affected too to improve the overall air quality.

HYPERKALEMIC PERIODIC PARESIS

This recently recognized genetic metabolic disease of Quarter horses can be partially controlled by nutritional management. The disease is characterized by episodes of muscle fasciculation, spasm, and weakness accompanied by hyperkalemia. The most common recommendations are to avoid high potassium feeds and to increase the intake of alkali. Molasses is particularly high in potassium and is a constituent of most "sweet feeds," which therefore should be avoided. Hays vary in potassium content. Coastal grass hays are lowest but may not be available in inland regions. Timothy hay is lower than brome or alfalfa hays. Owners of affected horses may wish to analyze different sources of hay before making long-term commitments to purchase. To further prevent hyperkalemia, the

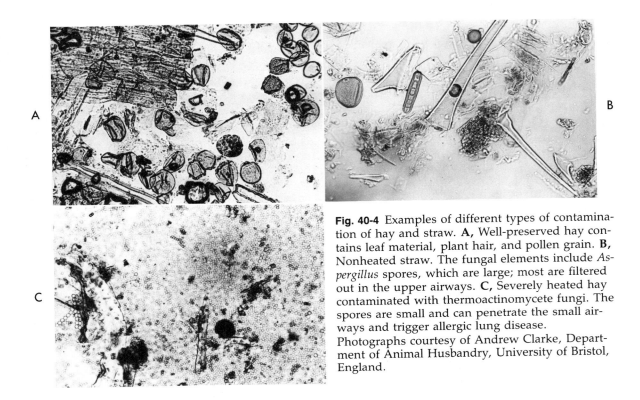

Fig. 40-4 Examples of different types of contamination of hay and straw. **A,** Well-preserved hay contains leaf material, plant hair, and pollen grain. **B,** Nonheated straw. The fungal elements include *Aspergillus* spores, which are large; most are filtered out in the upper airways. **C,** Severely heated hay contaminated with thermoactinomycete fungi. The spores are small and can penetrate the small airways and trigger allergic lung disease.
Photographs courtesy of Andrew Clarke, Department of Animal Husbandry, University of Bristol, England.

diet can be made more alkaline by adding baking soda (30 g or 1 oz fed twice daily) to the food.

CRACKED HOOVES

There has long been interest in the nutritional prevention of cracked hooves. However, many traditional "cures" have been found to be ineffective. For example, gelatin, a substance made from collagen, was once recommended. Even though both collagen and hoof wall are rich in cross-linked disulfide bridges, controlled trials showed that gelatin does not improve hoof quality.[13] Recent attempts at nutritionally altering hoof quality have been more successful. In pigs there is clear evidence that biotin supplementation improves the strength of hoof wall horn and decreases the incidence of lameness.[28,30,48] The evidence in horses is less rigorous, but some horses with weak, crumbly hooves appear to benefit from biotin supplementation. Other signs noted in biotin-responsive horses have been ridging of the hoof wall and soreness. It takes about a year for the hoof to grow out completely, and supplementation is needed for at least 6 to 9 months. Doses of 15 to 30 mg of biotin per horse per day are used.[17]

HEPATIC FAILURE

The liver is of vital importance to the metabolic regulation of nutrient utilization. It is a major site of vitamin storage and synthesis, gluconeogenesis, and metabolism of a variety of toxins and drugs. In severe liver disease the primary concern is to provide adequate amounts of energy and protein in amounts and forms that minimize the need for hepatic metabolism and maximize the potential for regeneration.[31,38] Simple sugars such as glucose reduce the need for gluconeogenesis and lipolysis and, to a lesser degree, reduce muscle catabolism and urea production. Horses with alkaloid-induced hepatic failure have reduced plasma concentrations of branched-chain amino acids (BCAA: leucine, isoleucine, and valine) relative to an increase in aromatic amino acid (tyrosine and phenylalanine) concentration. Aromatic amino acids and tryptophan have been implicated in development of hepatoencephalopathy.[47] If the ratio of the sum of the three BCAA to the sum of the aromatic amino acids in the plasma is less than 3:1, horses may be at risk of developing hepatoencephalopathy. Diets and supplements should be selected that minimize excess protein and aromatic amino acid intake (see Tables 40-1 and 40-2). Supplementation with BCAA may be beneficial in cases of acute hepatic failure. Addition of arginine, an amino acid essential for the clearance of ammonia via the urea cycle, has been recommended because of the apparent contribution of ammonia toxicity to the development of hepatoencephalopathy.[4] Protein restriction to levels below maintenance should not be necessary unless hepatoencephalopathy develops. Inappropriate restriction of protein and energy intake may impede healing and regeneration of damaged hepatic tissue.[47] Other nutrients of concern if hepatic damage is severe or the course of disease is prolonged include requirements for fat-soluble vitamins A and E and the water-soluble vitamins. These may be supplemented orally or parenterally with the exception of vitamin E. Vitamin E should be given orally because the currently available parenteral selenium–vitamin E combination preparations would have to be given in amounts that would induce selenium toxicity in order to provide adequate vitamin E (500 to 1000 IU) to a mature horse.

Feeds that have succeeded in maintaining horses with severe liver disease include grass or oat hay, corn, soybean meal, and milo. The diet should consist of at least 50% hay to maintain large bowel function. Corn should comprise 25% to 50% of the total diet, with milo, soybean meal, or linseed meal added as necessary to maintain body condition and weight. Concentrate rations should be divided into three or more separate feedings.

RENAL DISEASE

Restriction of protein and phosphorus in the diet of animals with chronic renal failure improves the quality of life and may retard progression of the disease.[31] Horses with chronic renal failure may benefit from such restrictions but are unique in the need for additional restriction of dietary calcium intake.[7] Urinary concentrations of calcium reflect the dietary intake, for in horses the kidney is a major organ of excretion of this mineral.[14] In horses with reduced renal function, dietary calcium fed in excess of their needs results in hypercalcemia and may be life-threatening.[7] A diet of at least 50% grass or oat hay, grain consisting primarily of corn, and free-choice water is recommended for horses in renal failure. Salt should be provided free choice and vegetable oil added to the grain as needed for maintenance of body weight. Protein status should be monitored carefully to prevent excessive protein restriction. If plasma protein drops significantly, casein, linseed or soybean meal may be added to the ration in amounts necessary to maintain normal plasma protein concentrations. Feeds to be avoided are clover and alfalfa, because of their high calcium content, and wheat bran, because of its high phosphorus content.

CARDIAC DISEASE

Cardiac dysfunctions such as atrial fibrillation and valvular insufficiency are not uncommon in horses. The goals of nutritional management of animals with chronic heart disease are to maintain lean body weight and reduce fluid accumulation.[31] This may be accomplished by feeding maintenance amounts of a standard diet and reducing salt intake. If ventral edema or ascites are present, salt intake is easily restricted by feeding a diet of hay or pasture, grain mixes without added salt, and free-choice water. The salt blocks traditionally offered free choice should be removed. Most hays, pasture grases, and unprocessed grains are naturally low in salt (less than 0.05%). However, many commercial grain mixes and pelleted feeds may have salt added in excess of 1%. If potassium-sparing drugs such as captopril or spironolactone are used, sodium intake should not be restricted. If diuretics such as furosemide are used, plasma potassium levels should be monitored on a regular basis to detect hypokalemia. However, as long as the horse is eating well and at least 50% of its total diet consists of grass or alfalfa hay, potassium intake should be adequate. If potassium supplementation is needed, it is easily accomplished by adding 5% to 10% molasses to the grain mix. Molasses contains 4.7% potassium as fed.

NUTRITIONAL EVALUATION

Because the energy requirements of clinically ill horses are estimated with data from other species, response to nutritional support must be carefully monitored. Body weight and condition score (see Chapter 38) is useful for long-term evaluations and should be recorded at least once a week. If large animal scales are not available, a weight tape may be used. Weight tapes must be placed in exactly the same place around the girth of a horse to reflect weight changes over time. Moving the measurement site only a few centimeters cranially or caudally results in an apparent gain or loss of 25 to 50 kg (50 to 100 lb). Loss of more than 10% of body weight or reduction in condition score of more than 1 point in a relatively short (1 to 2 weeks) period of time is an indication that additional nutritional support is necessary. In the short term, however, body condition and weight do not correlate well with overall nutritional status. Weight especially may be affected by hydration status and intestinal fill. By the time significant changes are detected, the animal may be severely compromised.[9,20,26,35,36]

The adequacy of protein and energy delivery potentially may be estimated by measuring blood triglycerides, albumin, and total protein.[20,47] In one study all clinically ill horses and ponies with hyperlipidemia[37] had severe hypophagia in the days or weeks previous to sampling. However, not all of the clinically ill horses with histories of hypophagia had hyperlipidemia. In a controlled study,[36] severe nutritional restriction was needed before serum lipids increased. Increased blood triglycerides have also been reported in horses with pituitary adenomas but normal feed intake.[37] Hyperlipidemia (more than 500 mg triglycerides per dl serum) indicates the need for immediate nutritional supplementation in horses that do not have clinical signs of pituitary adenomas. Normal or reduced blood triglycerides, however, do not necessarily imply that nutritional needs are being met.

Based on data from other species,[47] plasma protein lower than 40 g/L (4 g/dl) or albumin lower than 30 g/L (3 g/dl) may indicate that, for whatever reason, a horse is receiving inadequate protein to meet its needs.[9,47] Hypoalbuminemia also may be the result of protein-losing enteropathies and hepatic or renal disease but should be corrected, if possible, even under these conditions. Fibronectin, retinol-binding protein, prealbumin, and transferrin all are used as sensitive measures of nutritional status in human medicine[47] but have not been critically evaluated in this context in horses. Serum or plasma concentrations of vitamins and minerals also have been used to evaluate the nutritional status of animals,[2,47] but it is not known if circulating levels actually reflect whole body status for vitamins, especially in clinically ill horses. Plasma levels of certain trace minerals vary with intake, and deviations from normal ranges are considered to reflect dietary imbalances. However, inflammation alters copper, iron, and zinc concentrations irrespective of total body status or need (Chapter 4). Acute changes of plasma electrolytes may or may not reflect whole body needs. For example, blood levels of potassium and phosphorus are affected by the animal's acid-base status and should be evaluated with caution in acidotic or alkalotic horses.[42]

Hair trace mineral analysis has been recommended for the diagnosis of various trace mineral deficiencies.[18] However, hair mineral content reflects only long-term imbalances, and there are few, if any, good correlations between mineral intake and hair content of that mineral.[53]

MODES OF ALIMENTATION
Oral

If a horse is able to eat, it should be encouraged to do so. Feeding small amounts at frequent intervals and offering a variety of feeds (e.g., hays, grain mixes, bran mashes, carrots, apples) increases the amount ingested. Fresh grass is often particularly palatable; whole grains are preferred to rolled grains. If the horse is mobile and grassy areas are available (even dried-out grasses in winter), 20 to 30 minutes of grazing several times a day often dramatically improve intake and overall nutrient status. Sick horses may have unusual food preferences and may eat poor-quality hay while refusing grain or lush hay. Unpalatable medications should not be put into the feed. Although not proven scientifically, addition of small amounts of vinegar, licorice, or molasses to the feed may increase intake. It has been noted clinically that parenteral injection of B vitamins often increases feed intake in severely stressed horses.

Several categories of drugs may be used to stimulate feed intake in hypophagic or anorexic horses. Analgesics and antiinflammatory drugs reduce the anorexia associated with pain and fever.[40] Diazepam has been used to stimulate appetite in severely anorexic horses but should be used with caution. A very low dose (10 mg/500 kg horse) is given intravenously when the horse is in a quiet environment with feed in front of it. The effect is immediate and lasts only 15 to 20 minutes. Repeated doses may be used, although responses are not consistent over time. Diazepam requires hepatic metabolism and should be used with extreme caution in horses with hepatic dysfunction. Excessive tranquilization and ataxia have been noted and are particularly likely in horses that are severely depressed or debilitated prior to diazepam administration. Lower doses (2.5 to 5 mg/500 kg) should be tried initially in such animals.

Anabolic steroids increase intake over time but do not have an immediate effect. They are often used in convalescent horses but should be given only once and at the recommended dosage. Repeated doses or doses higher than the recommended amounts may adversely affect the reproductive performance of mares and stallions.

Feeds should be placed in containers that are easy for the horse to reach. For example, a horse with severe burns on the dorsal aspect of its neck should be fed from a rack at chest height instead of on the ground or from a rack above its head.

If the horse is recumbent, getting it into a sling to support it on its feet often results in improved intake (Fig. 40-5) and general attitude.

Nasogastric and esophagostomy

Nasogastric feeding usually is reserved for horses that are unwilling or unable to eat for

Fig. 40-5 Horse in a sling.
Photo courtesy of Al Kelminster, Colorado State University.

periods of more than 3 days. There is some evidence that short-term anorexia associated with acute infections is beneficial; however, long-term starvation compromises immune function and recovery. Despite problems associated with long-term extraoral feeding, it can be lifesaving in horses facing prolonged periods when they are unable to eat. In other circumstances it may help to maintain condition and speed recovery.

Horses have been maintained successfully on tube feed for months. Slurries of pelleted feed, alfalfa meal, or liquid formulas may be delivered via nasogastric tube or esophagostomy.[24] Liquid diets, Equical, Ensure, and Osmolite* (the latter two designed for use in humans) have been fed successfully to horses when offered in less than maintenance amounts. Some preparations, however, are associated with a high incidence of diarrhea when fed at maintenance levels without additional fiber sources.[51]

A semipurified diet designed for horses that has satisfactorily maintained horses for several weeks is shown in Tables 40-3 and 40-4. This diet should not be fed to horses with historical or clinical evidence of laminitis because it is rich in soluble carbohydrates. The inclusion of alfalfa meal in the ration greatly reduces the incidence of problems with diarrhea. The electrolyte mix-

---

*Ross Laboratories, Columbus, Ohio 43216.

Table 40-3 *Tube feeding regimen for a 450-kg horse*

	Day						
	1	2	3	4	5	6	7
Electrolyte mixture* (g)	210	210	210	210	210	210	210
Water (L)	21	21	21	21	21	21	21
Dextrose (g)	300	400	500	600	700	800	900
Casein (g)	300	450	600	750	900	900	900
Dehydrated alfalfa meal (g)	2000	2000	2000	2000	2000	2000	2000
Megacalories/day (DE)†	7.4	8.4	9.4	10.4	11.4	11.8	12.2

*See Table 40-4.
†Digestible energy.
The above allowances should be divided into three feeds daily.
Maintenance requirements for a 450-kg horse are 15 Mcal of DE and 580 g of crude protein.

Table 40-4 *Maintenance oral electrolyte mixture (1 day's requirements for a 450-kg horse)*

Substance	Amount
Sodium chloride, NaCl	10 g
Sodium bicarbonate, $NaHCO_3$	15 g
Potassium chloride, KCl	75 g
Potassium phosphate (dibasic anhydrous) K_2HPO_4	60 g
Calcium chloride, $CaCl_2 \cdot 2H_2O$	45 g
Magnesium oxide, MgO	24 g

ture is rich in potassium and low in sodium, which is typical of maintenance electrolyte mixtures and normal horse diets. If hyponatremia develops, 15 g of additional sodium chloride may be added to the ration. Routine supplementation of salt is not recommended. Whenever a clinician takes over the long-term feeding of a sick horse, it is important to evaluate the horse's clinical status and blood chemistry routinely.

Nasogastric tubes may be passed at each feeding or left in place, depending on the disposition of the horse. However, pressure necrosis of nasal passages is a significant problem if tubes are used for more than a week to 10 days. If prolonged (longer than 1 or 2 weeks) periods of tube feeding are anticipated, an esophagostomy[32] may be the method of choice. Complications associated with placing an indwelling feeding tube directly into the esophagus include electrolyte imbalances due to chronic loss of saliva; infections, both local and spread along fascial planes to the thorax; stricture or fistula formation after removal of the tube; and damage to the vagus or recurrent laryngeal nerves, jugular vein, or carotid artery. Owners should be warned of the risks.

A mature horse's stomach capacity is only 20 to 40 L, and most horses tolerate a volume of only 7 to 10 L by stomach tube. Fluids and small particles empty fairly rapidly (30 to 60 minutes). Small, frequent feedings are therefore recommended.

Intravenous alimentation

Guidelines given for intravenous alimentation of neonates (Chapter 36) may be used in mature horses. Parenteral supplementation should be considered only for horses in which enteral alimentation is contraindicated for more than 2 to 4 days (e.g., anterior enteritis, gastric reflux, after small intestinal resection). Although long-term maintenance on parenteral nutrition may be possible,[25] total parenteral support is prohibitively expensive in most cases. Partial alimentation by providing as much as the owner can afford per day, however, is preferable to total starvation.[11]

REFERENCES

1. Baetz AL and Pearson JE: Am J Vet Res 33:1941, 1972.
2. Baker H and others: Am J Vet Res 47(7):1468, 1986.
3. Baker JC and Ames TR: Equine Vet J 19(4):342, 1987.
4. Barbul A: J Parenter Enter Nutr 10(2):227, 1986.
5. Baucus K and others: J Anim Sci (in press).
6. Bertone AL, Ralston SL, and Stashak TS: Am J Vet Res 50(9):1621, 1989.
7. Bertone JJ and others: J Am Vet Med Assoc 191:565, 1987.
8. Bertone AL, Van Soest PJ, and Stashak TS: Am J Vet Res 50(2):253, 1989.
9. Blumenstock FA: In Jenks JS, editor: Nutrient effects on immune function, American Society of Parenteral and Enteral Nutrition eleventh clinical congress, New Orleans, 1987.
10. Booth AJ and Naylor JM: J Am Vet Med Assoc 191(1):62, 1987.
11. Brennan MF and others: J Parenter Enter Nutr 10(5):446, 1986.
12. Brinson RR, Curtis WD, and Singh M: J Am Coll Nutr 6(6):517, 1987.
13. Butler KD and Hintz HF: J Anim Sci 44:257, 1977.
14. Caple IW, Doake PA, and Ellis PG: Aust Vet J 58:125, 1982.
15. Clarke A: In Practice 9:196, 1987.
16. Clarke AF and Madelin T: Equine Vet J 19:442, 1987.
17. Comben N, Clark RJ, and Sutherland DJB: Vet Rec 115:642, 1984.
18. Coombs DK: J Anim Sci 65:1753, 1987.
19. Cossack ZT and Prasad AS: Nutr Res 7:1161, 1987.
20. Dominioni L and Dionizi R: J Parenter Enter Nutr 11(5):708, 1987.
21. Ducharme NG and others: Proc Am Coll Vet Surg 50, 1984 (abstract).
22. Dudrick SJ and Rhoads JE: In Sabison CO, editor: Davis-Christopher textbook of surgery: the biological basis of modern surgical practice, ed 2, vol 1, Philadelphia, 1981, WB Saunders.
23. Fisher GL: Am J Vet Res 38(7):935, 1977.
24. Freeman DE and Naylor JN: Vet Med Assoc 172(3):314, 1978.
25. Hansen TO, White NA, and Kemp DT: Am J Vet Res 49(1):122, 1988.
26. Haydock DA and Hill GL: J Parenter Enter Nutr 10:550, 1986.
27. Hintz HF and others: Proceedings of the eleventh Equine Nutr Physiol symposium, Stillwater, Ok, May 18-20, 1989.
28. Johnston AM and Penny RHC: Vet Rec 125:130, 1989.
29. Kadis S, Udeze FA, and Polanco J: Am J Vet Res 45(2):255, 1984.
30. Kempson SA, Currie RJW, and Johnston AM: Vet Rec 124:37, 1989.
31. Lewis LD, Morris MM, and Hand MH: Small animal clinical nutrition, Topeka, KS, 1988, Mark Morris Associates.
32. Merrit AM: Proc Am Assoc Equine Pract 21:401, 1975.
33. Meyer H and others: Am Assoc Equine Pract Newsletter 3:30, 1981 (abstract).
34. National Research Council: Nutrient requirements of horses, ed 5, Washington, DC, 1989, National Academy of Science.
35. Naylor JM, Freeman DE, and Kronfeld DS: Comp Cont Educ 6(2):S93, 1984.
36. Naylor JM and Kenyon SJ: Proc Am Assoc Eq Pract 24:505, 1978.
37. Naylor JM, Kronfeld DS, and Achland H: Am J Vet Res 41:899, 1980.
38. Pearson EG: Comp Cont Educ 4(3):S114, 1982.
39. Pietsch JB: Surg Forum 36:153, 1985.
40. Ralston SL: Comp Cont Educ 10(3):356, 1988.
41. Ralston SL: Equine Vet Sci 9(4):203, 1989.
42. Ralston SL and Larson K: Equine Vet Sci 9(1):13, 1989.
43. Ralston SL, Nockels CF, and Squires EL: Am J Vet Res 49(8):1387, 1988.
44. Ralston SL: Vet Clin North Am 6:339, 1990.
45. Ralston SL, Sullins KE, and Stashak TS: Am J Vet Res 47(10):2290, 1987.
46. Rao SSC and others: Gastroenterology 94:928, 1988.
47. Rombeau JL and Caldwell MD: Enteral nutrition and tube feeding, vol 1, Philadelphia, 1984, WB Saunders.
48. Simmins PH and Brooks PH: Vet Rec 122:431, 1988.
49. Smith JE and others: J Am Vet Med Assoc 188(3):285, 1986.
50. Snow DH and Frigg M: Proceedings of the 9th Equine Society symposium, Nutr Physiol, 1987.
51. Sweeney RW: Proceedings 7th annual Veterinary Medical Forum 7:455, 1989.
52. Tate LP and others: Am J Vet Res 44(7):1187, 1983.
53. Wells L and others: Proceedings of the 10th Equine Nutr Physiol Society symposium, 1987.

SECTION B Swine

41

The Swine Industry

JOHN F. PATIENCE

Like most other sectors of agriculture, pork production is experiencing major changes, not only in management practices but also in the fundamental structure of the industry. These changes reflect developments in nutrition, quantitative genetics, housing, and herd health. The information age is also making its presence felt by encouraging new approaches to record keeping, financial management, and business strategies.

Exciting developments in biotechnology also appear on the horizon. As might be expected with any new technology that has the potential to bring about fundamental change, however, there are signs of resistance from many quarters. Consequently, it is now clear that the impact of biotechnology will be determined not only by technical ability and economics but also by political and social will.

The North American consumer continues to change; factors such as cost, convenience, ethnic tradition, religion, health, and a multitude of others influence food-purchasing decisions. The world market for pork is changing as a result of an improved standard of living in Third World nations, an increase in the influence of animal rights activists, and changing world trade patterns.

However, the dynamics of the pork industry are not likely to worry the pig. It has demonstrated over more than 7000 years of domestication that it is a very adaptable animal, capable of thriving under a wide range of circumstances. Because modern agriculture established carefully defined conditions that optimize productivity and efficiency, the innate adaptability of domesticated animals often receives less consideration than perhaps it should. In the case of the pig, adjusting to change is no more foreign than eating truffles or hunting wild game, two of its former and perhaps more notorious occupations.

HISTORY

The pig was domesticated some 5000 to 7000 years ago. The modern pig, *Sus scrofa domesticus*, is descended from the European wild boar, *Sus scrofa scrofa*. Toward the end of the eighteenth century, the European pig was improved by the introduction of the Chinese pig, *Sus vittatus,* and possibly the Siamese pig, *Sus indicus*. Thus, the pig as we now know it is derived from more than one wild species and represents the results of selection over hundreds of years.[10]

The pig adapted well following its move to North America. The pig was first introduced to North America by Columbus in 1493 to the island nation Haiti. It is believed to have been first brought to the mainland by Cortés in 1521. The Portuguese introduced the pig to Canada in 1553.[10]

Initially, the pig fulfilled the role of scavenger, living off garbage in New World settlements, roots, berries, and any other morsels that suited the pig's omnivorous tendencies. Later, the pig continued in this role and was used to dispose of skim milk provided by the sale of butterfat. With the establishment of corn as the mainstay of crop production throughout much of North America and wheat and barley in more arid areas, the pig served as a means to convert excess grain production into cash. This was particularly useful when other markets, notably export sales, failed to take all the grain available. In this manner, the pig became known as the "mortgage lifter" and established itself on farms throughout the continent. The development of soybean meal, initially a by-product of the soybean oil industry, as a very effective supplement to cereal grains in pig diets consolidated the position of the pig in North American agriculture. Although corn and soybean meal—or wheat plus barley and soybean meal—remain the standard diet for the pig, some pork producers use alternative ingredients to reduce their cost of production without compromising performance. Examples include pulse crops, such as peas, lentils, and faba beans; oilseed meals, such as canola and sunflower (dehulled); animal by-products, such as blood, feather, and meat and bone meals; and distilling and milling by-products. It is apparent that the diet of the pig will continue to change as the use of synthetic amino acids and supplemental fats increases.

The pig has fulfilled other roles besides that of provision of meat for the family larder. It has, at various times and in various cultures, been the focus of religious worship or scorn, used as a hunting "dog" when the real thing was banned by royal decree, and trained to seek buried delicacies such as truffles.

Whatever its role, the pig has survived and even excelled. Indeed, history teaches us another lesson—that some so-called new technology is in fact centuries old. For example, the superiority of boars versus castrates was recognized as far back as the sixteenth century, when Fitzherbert wrote, "Then say how many swine thou art able to keep: then let them be boars and sows all, and no hogs."[4] In another example, the use of slatted floors was recorded as early as the mid-1800s, as were self-feeders and central heating.[2] The influence of disinfection, ventilation, and freedom from drafts in maintaining animal health was similarly noted more than 200 years ago![19]

THE INDUSTRY TODAY

Pork production is distributed throughout much of North America, although the north central region of the United States continues to dominate, representing 81% of total U.S. production (Table 41-1).[13] In Canada, production is divided equally among Quebec, Ontario, and the two prairie provinces, Alberta and Manitoba, with the remaining three provinces contributing about 13% of the total (Table 41-2).[1]

Per capita consumption of pork increased during the 1960s but has remained essentially static since that time. Concurrently, beef consumption in the United States and Canada also remained fairly constant, but poultry consumption has increased (Table 41-3). This is a major concern to the industry and is being addressed by increased advertising and greater attention to meat quality.

PRODUCTION SYSTEMS

The raising of pigs can be divided into two functional parts: breeding-farrowing and feeding out. The former includes breeding, gestation, farrowing and lactation. Offspring are weaned at 21 to 35 days of age, at which time the dams are either culled for slaughter or returned to the breeding herd for remating. The weaned pigs are placed in specialized housing, called *nurseries*, in which temperature, airflow, and humidity are controlled to minimize stresses on the young pig. Particular attention is paid to avoiding antagonistic social interaction at this time because disturbance appears to be particularly trying to the newly weaned pig.

After 4 to 6 weeks, the pigs are transferred to grow-out facilities until they reach market weight of about 100 kg (220 lb). On some farms, called *farrow-to-finish operations*, both the breeding and grow-out activities are combined. However, some specialize in one or the other to take advantage of labor, financial, housing, management, or business opportunities that may be available locally.

Housing options vary throughout North America because of economics, weather conditions, and the production system preferred by the owner. For example, gestation housing may be either outside drylots, inside group housing, or individual gestation stalls. Outside drylots are

The Swine Industry

Table 41-1 U.S. hog production by state in 1987

State	Hogs (thousands)	State	Hogs (thousands)
Iowa	23,386	Maryland	332
Illinois	8855	Oklahoma	274
Indiana	7202	Florida	264
Minnesota	7146	Arizona	251
Nebraska	6607	California	247
Missouri	4702	Oregon	197
North Carolina	4223	New York	195
Ohio	3601	Idaho	123
South Dakota	2672	Delaware	120
Kansas	2371	Washington	91
Wisconsin	2231	Louisiana	75
Michigan	1987	Hawaii	66
Georgia	1879	New Jersey	59
Kentucky	1375	New Mexico	50
Pennsylvania	1334	West Virginia	48
Tennessee	995	Massachusetts	39
Arkansas	890	Wyoming	38
Texas	841	Utah	37
Virginia	765	Nevada	26
South Carolina	614	Maine	14
North Dakota	533	New Hampshire	11
Alabama	514	Vermont	11
Mississippi	430	Connecticut	7
Montana	398	Rhode Island	5
Colorado	333	Alaska	1
		Total	88,465

Adapted from Schoeff R: Feed Management 39(10):40, 1988.

Table 41-2 Canadian pork production by province, 1988

Province	Production (1000 head)
British Columbia	337
Alberta	1835
Saskatchewan	969
Manitoba	1774
Ontario	4579
Quebec	4626
Atlantic provinces	582
Total	14,702

Adapted from Canada Livestock & Meat Trade Report, Agriculture Canada, Ottawa, 1989.

becoming less popular because of increased feed costs in cold climates and difficulty in maintaining accurate breeding records. Other disadvantages of drylots are the inability to control parasites, exposure to disease-causing organisms, and problems associated with interactions among sows that often lead to poor weight control and physical injury to "subordinate" animals. Moving sows indoors has helped to reduce feed costs and aid disease control, but the problems of antagonistic social interactions in group housing remain. Consequently, many farms have adopted individual housing of dry sows in stalls. This facilitates record keeping, allows more precise feeding and thus better control of body weight, reduces injuries, and generally simplifies management. However, sow stalls

Table 41-3 *Average annual per capita meat consumption (kg) in Canada and the United States*

Year	Pork*		Beef*		Poultry†	
	Canada	U.S.	Canada	U.S.	Canada	U.S.
1970-74	28.0	28.2	41.4	38.0	18.6	22.2
1975-79	25.9	25.3	46.8	39.8	19.5	24.4
1980-84	29.2	28.6	39.8	35.1	21.5	28.7
1985-87	28.1	27.2	38.8	34.9	24.8	33.2

*Based on cold carcass weight.
†Based on eviscerated weight (Canada) or ready-to-cook (U.S.) chicken plus turkey.
Adapted from Robbins LG: Handbook of food expenditures, prices and consumption, Publ 81/1, Agriculture Canada, Ottawa, 1988.

have generated criticism, largely because of restrictions on movement that prevent sows from turning around. A recent development has been the electronic sow feeder; it allows computer-controlled feeding of sows in a central feeder while maintaining freedom of movement during the remainder of the day. Early experience has been mixed, although certain advantages appear clear, including greater freedom of movement for the sow and very precise feed intake and weight control with a minimum of human intervention. However, management is more difficult because the health and general activity of individual animals are more difficult to monitor. Also, problems with vulva biting persist, at least in some designs. Perhaps more experience with this system will overcome these difficulties.

Lactation housing is almost always indoors, in farrowing crates that protect the young piglet from being crushed by its much larger mother. All-in-all-out rooms normally are employed. In the all-in-all-out system, rooms are emptied following use and thoroughly cleaned, disinfected, and preheated before the next group of animals are brought in. In consequence, death loss in piglets and illness in sows from infectious disease are reduced. Although the all-in-all-out system increases construction costs, the benefits are sufficient to reduce net operating costs.

Weanling facilities also are operated on an all-in-all-out basis to facilitate disease control and provide a better physical environment for the newly weaned piglet. For example, when pigs first enter the room, the temperature can be increased to about 30° C (86° F) and gradually reduced as the young pig matures. In barns operated on a continuous basis, room temperature must be kept at a compromise temperature that is cold for the young pig and hot for the older pig.

Typically, weanling pens are sized to house a single litter (about 1.5 m by 1.5 m or 5 feet by 5 feet) to eliminate the need to mix pigs at the time of weaning. The number of pens per room is dictated by the number of litters weaned per week; ideally, the room should be filled over a maximum of 1 week. In larger units, operators have the option of having larger rooms that fill on a weekly basis or twice as many smaller rooms that fill every 3 to 4 days. For example, a 200-sow unit is expected to farrow up to 450 litters per year and wean about 4200 to 4600 pigs per year. This is equivalent to almost nine litters per week. In this case, each nursery room would contain nine pens, measuring 1.5 m by 1.5 m to accommodate 10 pigs per pen. The number of rooms depends on the amount of time each litter spends in the nursery. If pigs are weaned at 4 weeks of age and remain in the nursery for 5 weeks, then a minimum of 6 rooms is required because one room fills each week and an additional room is in clean-up. Some operators add an additional room for sickly or slow-growing pigs. If such a room is not available, some pigs are forced into the grower barn before they are physically or physiologically ready, and their growth is further compromised.

When pigs are about 9 weeks of age and

weigh about 23 kg (50 lb), they are moved into the grow-out barn to enter the final phase of pork production. Although it is now common in North America to feed a single diet from 23 kg (50 lb) to market weight (100 kg [220 lb]), there is little doubt that more specialized feeding programs, with two to four different diets, will become the norm in the future. Similarly, barrows and gilts generally are housed and fed together; in the future, they may be fed different diets and managed differently to maximize the profitability of the two sexes.

The following is a description of a typical grower barn that is common in many parts of the North American continent. Because there is a wide range in climate and therefore housing needs, however, variations from this fully confined, artificially ventilated design exist. For example, in the warmer climates, curtain wall barns are more common for cost and engineering reasons.

The grower barn typically is long and narrow, with a single row of pens on either side of a central alley. Pens tend to be about twice as long as they are wide, with about one third or more of the floor area slatted. Some owners prefer fully slatted floors. There are advantages to both systems. Pens should be at least 2.1 m (7 ft) wide to permit normal traffic patterns around the feeder, with solid concrete partitions between pens to minimize drafts. The front door of the pen also should be solid for the same reason. The pen partition over the slatted floor area should be perforated to permit animal contact and encourage defecation in the slatted area.

The waterer should be located over the slats and the feeder located along the pen partition adjacent to the slatted area. This design keeps the sleeping area, located near the door of the pen, free of activity associated with eating or drinking. A typical grower pen is 2.1 m wide and 4.5 m long (7 ft by 15 ft), with a holding capacity of about 13 pigs up to 100 kg (220 lb). Wider and longer pens increase capacity proportionately. However, pens should not be too large because large numbers of pigs in a single pen appear to increase the incidence of behavioral problems associated with an unstable social hierarchy.

The all-in-all-out concept is gaining favor in the grow-out barns, nurseries, and farrowing units. The rooms can be disinfected between lots and then preheated before repopulation to eliminate chilling of the new occupants. In warmer climates, the benefits of temperature regulation may not be as important, but the ability to break disease cycles is critical in many circumstances.

STRUCTURE OF THE INDUSTRY

In the past, grain farmers could enter or leave the hog markets almost at will because of low housing requirements and the relatively short (compared to cattle) reproductive cycle of pigs. The cyclical nature of pork production meant that high pork prices, relative to grain prices, enticed farmers to feed hogs. The expanded production resulted in oversupply, prices fell, and many producers abandoned pigs. This reduced supplies and strengthened prices, and the cycle was repeated. Between 1945 and 1965, a relatively predictable 4-year cycle of production and pricing was seen. In the period from 1965 until 1985, two cycle lengths, of 3 and 7 years, have become more apparent,[11] and it is difficult to predict future events. Shonkwiler and Spreen[14] suggest that the changing nature of the industry, with fewer corn belt mixed farms and more larger, specialized units, is largely responsible for the change.

Typically, the hog cycle has been influenced by the grain markets or, more precisely, the hog:corn or hog:barley ratio. These are the ratios of pork and grain prices and reflect in part the profitability of pork production. The relevance of the hog:corn and hog:barley ratios is likely to diminish because a two-price grain structure appears to be evolving. Thus corn and the coarse grains can have two "values": the higher export price, which is the primary market, and the local price, related to livestock feeding.

The trend toward larger, more specialized units is clearly evident (Table 41-4). Between 1964 and 1978, units producing more than 1000 pigs per year increased their share of the U.S. market from less than 8% to almost a third. At the same time, the number of pork-producing farms declined from 2.1 million in 1950 to less than half a million in 1974 and about 333,000 in 1987. The average production per farm increased from 31 to 178 to 266 pigs for the same time

Table 41-4 *Percentage distribution by size of farms and sales in 1964 and 1978 in the United States*

Size category (annual hog/pig sales)	1964 Farms	1964 Sales	1978 Farms	1978 Sales
1-99	68	23	60	10
100-199	17	23	15	10
200-499	12	33	16	24
500-999	2	13	6	22
1000+	1	8	3	34

Adapted from Wilson PN and Eidman VR: Amer J Agric Econ 67:279, 1985.

periods.[13,18] In Canada, the number of production units declined from 154,000 in 1966 to 60,000 in 1976, and average production per farm rose from 44 to 128 pigs during the same period.[5] Although smaller units, generally associated with mixed farming, may survive in some form, it is clear that both economic and social forces favor larger units (Table 41-4). Orr[9] reported that larger farms tend to obtain better performance and are more profitable (Table 41-5). Other studies have reached essentially the same conclusion.[16] Data obtained from Kentucky and Minnesota farms revealed that units producing more than 2000 pigs per year weaned almost 1 more pig per litter than units with sales of fewer than 200 pigs annually.[18]

In some parts of North America, the industry is vertically integrated, meaning that off-farm production activity (e.g., feed manufacturing, slaughter, and processing) is tied to the actual raising of pigs. These activities tend to be geographically localized, such as in the southeastern United States or the Canadian provinces of Quebec and Manitoba. Some of these units control half a million hogs or more and take financial advantage from linking two or more phases of pork production and processing.

In Canada, marketing boards dominate the marketing of live hogs. These organizations, owned by the pork producers, serve to centralize the selling of live animals to the meat-processing industry. Although their business practices vary among the provinces, they all seek to maximize the competitive environment surrounding the selling process. Unlike many dairy and some poultry marketing boards, they do not regulate production levels or prices. However, they do regulate the selling process, and they attempt to improve the competitive position of the farmer in pig sales. Marketing boards also tend to be advocates for pork producers and interact with government and other agencies on issues of interest to their membership.

Table 41-5 *Effect of farm size on productivity and profitability in the United States*

Item	Sow herd size <100	Sow herd size >300
No. of farms	41	27
Pigs sold/sow/year	15.6	16.5
Average selling price	112	116
kg feed/kg pork	3.78	3.69
Feed cost (cents/kg)	11.77	11.97
Total cost, $	35.23	33.38
Profit ($/pig sold)	35.60	43.32

Adapted from Orr D: Feed Management 39(5):24, 1988.

The industrial concentration of the meat-processing industry attracts considerable attention. In the United States, the size of beef-processing units has increased considerably in the past 15 years, but there has been little change in pork, where the four largest firms still slaughter about 35% of total annual production.[17] In Canada, the concentration of the pork-processing industry is about the same as that in the United States. Other industries, such as petroleum, pulp and paper, and automobile manufacturing, have higher levels of industrial concentration (Table 41-6).[3]

INDUSTRY GOALS

The major goal of the pork industry is to maintain profitability in the face of changing economic circumstances. Production efficiency plays a major role in profitability so that increasing attention is focused on record keeping, financial management, and the adoption of relevant and effective technology.

Performance standards within the industry vary widely. Well-managed units are capable of weaning about 2.3 to 2.4 litters per sow per year, with litter size averaging 9.5 to 10 pigs or more. Although some units exceed these figures, most fall well short, weaning perhaps 16 pigs or less per sow per year. However, it is clear that weaning 23 to 25 pigs per sow per year is an achievable goal, and the long-term financial viability of units producing less is suspect.

Feed conversions vary considerably among farms. Whereas some use as little as 2.7 kg (6 lb) of feed per kg of pig gain (including requirements to maintain the breeding herd), others use 3.5 kg (7.7 lb) or more. The nature of the diet has a major impact on feed efficiency. However, nonnutritional factors, especially those related to animal health, also play a major role in feed utilization efficiency. Many units are upgrading their health status, as it is recognized that common respiratory and gastrointestinal diseases, as well as internal and external parasites, cost the industry an enormous amount of money. Minimum-disease herds are becoming more common, although the difficulty of maintaining such status in areas of dense pig population remains a major problem.

Production efficiency, and thus the ability to withstand financial stress, varies widely among individual farms. The financial success of a production unit does not depend directly on production, but rather on the ability to handle financial commitments. However, it is clear that in the long term higher-producing units are likely to outlast their less efficient neighbors.

Table 41-6 *Concentration of major manufacturing industries in Canada, 1985*

	Percent of sales by	
Industry	4 largest enterprises	8 largest enterprises
Automobile manufacturing	95	99
Steel	78	88
Petroleum	64	89
Meat slaughter and processing	36	49
Pulp and paper	33	53

Adapted from Canadian Meat Council; Canada's meat processing industry—background information, 1988.

Whether this translates into larger units totally displacing smaller independents, as has occurred in the poultry industry, will depend on many factors, including productivity, availability of financing, odor control, and ability to attract capable staff.

Overall meat consumption is under pressure, following a steady increase since the mid-1930s. For example, from 1930 until 1971, per capita consumption of meat grew at an average annual rate of 1.4%; since then, growth has slowed to less than 0.4%.[15]

The industry continues to seek improvement in pork quality and acceptability to the consumer. In Canada, the market hog grading system offers a financial incentive to produce leaner carcasses. Herds producing better-quality carcasses receive a 7% bonus over average herds and as much as 12% more than the poorer herds. This translates to $9 to $20 per pig, depending on market prices. Based on measurements of individual carcass weight versus backfat thickness, the grading system assigns an arbitrary index (baseline = 100) based on predicted lean tissue yield. During the period from 1969 until 1977, the proportion of graded carcasses indexing above 100 increased from less than 50% to more than 65%. Jones[8] reported that between 1967 and 1981 the Canadian industry achieved a 12% increase in total lean yield.

In the United States, carcass grading is less common. A recent survey conducted by the Uni-

Table 41-7 *Nutrient composition of modern meat products average of common cuts (lean only)*

Species	Calories (kcal)	Protein (g)	Fat (g)	Iron (mg)	Thiamine (mg)
Beef	209	29.0	9.5	2.8	0.10
Lamb	193	27.9	8.2	2.0	0.15
Pork	220	28.9	10.6	1.8	0.67
Veal	228	27.1	12.4	3.3	0.08

Adapted from Canadian Nutrient File, Health and Welfare Canada, based on surveys in 1987 (beef and pork) and 1963 (lamb and veal). Pork data does not include side ribs, which contain 25% fat.

versity of Wisconsin–Madison revealed that only about 30% of all pigs are purchased on a carcass merit system.[7] However, this is twice the level of 3 years ago, and continued growth is expected. Indeed, leaner meat is an absolute necessity if the U.S. industry is to progress at the same rate as that in other pork-producing nations.

The benefits to the consumer of carcass grading have been very apparent. A recent survey of commercial pork cuts in Canada revealed that the pork available today is much more nutritious than that described in older reference books on nutrient composition. The results of this survey indicate that pork is competitive in quality with other red meats (Table 41-7). A major challenge of the pig industry is to ensure that this new information reaches the consumer and consulting professionals who are in a position to influence the eating patterns of the human population.

REFERENCES

1. Agriculture Canada: Canada livestock and meat trade report, Ottawa, 1989, Agriculture Canada.
2. Caird J: English agriculture in 1850-51, London, 1852, Longmans.
3. Canadian Meat Council: Canada's meat processing industry—background information, 1988.
4. Fitzherbert A: Husbandrye, London, 1523, Richard Pynson (Quoted by Fuller, 1985).
5. Fredeen HT and Harmon BG: J Anim Sci 57(suppl 2):100, 1983.
6. Fuller MF: In Cole DJA and Haresign W, eidtors: Pig nutrition, London, 1985, Butterworths.
7. Hog Farm Management 25(10):40, 1988.
8. Jones SDM: Can J Anim Sci 66:23, 1986.
9. Orr D: Feed Management 39(5):24, 1988.
10. Patience JF: Western Hog J 8(2):12, 1986.
11. Purcell JC and Sullivan GD: J Agribusiness, p 53, February 1986.
12. Robbins LG: Handbook of food expenditures, prices and consumption, Publ 88/1, Ottawa, 1988, Agriculture Canada.
13. Schoeff R: Feed Management 39(10):40, 1988.
14. Shonkwiler JS and Spreen TH: South J Agricul Econ 18:227, 1986.
15. Trapp JN: Current Farm Economics 58(3):3, 1985.
16. Van Arsdall RN and Nelson KE: USDA Tech Bull 1712, Economic Research Service, 1985, Washington, DC, Government Printing Office.
17. Ward CE: Feedstuffs 61(4):1, 1989.
18. Wilson PN and Eidman VR: Am J Agricul Econ 67:279, 1985.
19. Youatt W: The pig, London, 1847.

42

Feeding the Gilt and Sow for Optimal Production

FRANK X. AHERNE

Reproductive efficiency generally is measured in terms of the number of pigs weaned per sow per year and is influenced by the number of pigs born and weaned per litter and the number of litters produced per sow per year. A further criterion of reproductive efficiency is the number of litters produced per sow per lifetime, which is influenced by culling rate. For a meaningful between-herd comparison of number of pigs weaned per sow per year, it is necessary that gilts be considered to be part of the breeding herd from the time they are selected for breeding and are withheld from market. When this is done, reproductive efficiency is maximized by mating gilts early and keeping them producing large, healthy litters for as long as possible. The data presented in Table 42-1 are representative of the average level and ranges in sow productivity of swine units.[135] As can be seen, there is a very considerable range in production efficiency; the potential for improvement is great.

There have been numerous reviews on the effects of nutrition on sow reproductive efficiency.* However, many sow experiments were undertaken with insufficient accuracy, replication, or duration to allow detection of even a 10% difference in sow productivity or evaluation of long-term effects on breeding regularity.[35] A further criticism is that the experiments often were conducted on discrete segments of the breeding cycle (i.e., pregnancy, lactation, or postweaning period) and therefore could not observe the important interrelationships among the various phases. Furthermore, little effort was made in the majority of the experiments to explain the nature of any observed nutritional effects on sow reproductive efficiency.

In the sixties and early seventies, most gilts had considerable fat reserves when they were bred. It was suggested that these body reserves could be exploited during successive reproductive cycles without affecting subsequent fertility, provided the sow made a positive weight gain of 10 to 11 kg (22 to 25 lb) during each parity.[36,37,85,105] However, on some farms low feed intakes over successive parities were associated with delayed return to estrus, especially in first-litter sows.[90] Whittemore and co-workers[134] showed that it was possible for sows to gain 22 kg (48 lb) body weight over two successive parities and still lose 6 to 8 mm (0.25 to .3 inch) of backfat. Also, between 1972 and 1987 the average backfat thickness of gilts tested on-farm decreased very significantly.[3] These lean gilts with 14 to 15 mm (about .6 inch) of backfat at time of breeding cannot sustain a 6 to 8 mm (0.25 inch) loss of backfat over two parities without suffering a decrease in reproductive performance. It is now generally recognized that the long-term reproduction of the sow is best served by minimizing weight and fat loss in lactation.[21] If body condition is conserved during lactation, then only small amounts of weight need to be restored during the following pregnancy. There is now ample evidence that this strategy is both achievable and desirable in most modern swine units.

This chapter examines the effects of nutrition

---

*References 1, 3, 19, 21, 35, 66, 102.

Table 42-1 *Typical sow productivity figures*

	Mean	Ranges
Number born alive per litter	10.1	8.5-12.3
% Born dead per litter	7.0	0.1-14.0
% Preweaning mortality	18.7	7.4-30.1
Number weaned per litter	8.2	5.9-9.9
Weaned per sow per year	16.8	9.1-19.4
Litters per sow per year	2.1	1.8-2.5
Nonproductive sow days	50.8	14.1-116.3

Wilson MR and others: J Anim Sci 62:576, 1986.

on the prolificacy of gilts and sows. *Prolificacy* is defined in terms of pigs weaned per sow per year and per breeding lifetime. For simplicity, phases of the sows' breeding cycle are dealt with separately, but the possible effects of nutrition in one period on any other period are stressed.

GILT NUTRITION

Survey data have shown that an average of 30% to 45% of sows are culled each year and replaced by gilts. Gilts produce smaller litters than sows, and therefore a high culling rate has a significant impact on the number of pigs weaned per breeding female. It therefore is important that the gilts' reproductive potential be fully expressed. Among the factors that may affect a gilt's reproductive performance are age, weight and body condition at mating, ovulation rate, and the ability to be rebred after weaning her litter.

Age at puberty

Fewer than 1% of gilts, housed and managed with commercial pigs, reach puberty by market weights of 90 kg (200 lb). However, puberty can be induced in gilts by relocating them to another pen, mixing them, and exposing them to a mature boar.[72] The mean age at puberty of nonstimulated gilts is about 200 days, with a range of 135 to 250 days.[29,64] For stimulated gilts the variation in age at puberty is equally large, but the mean age at puberty is 30 to 40 days younger than for noninduced gilts.

The effects of feed, energy, and protein on age at puberty. In general, a restriction of feed or energy intake to 60% to 70% of ad libitum intake delays the onset of puberty by about 9 days, but less severe restriction during the feeder period (50 to 90 kg or 110 to 200 lb) does not influence age at puberty.[6,26,71] Gilts may be more sensitive to restriction during the grower period (20 to 60 kg or 45 to 130 lb).[47] However, it would appear that conditions of management and nutrition that support commercially acceptable growth rates during rearing are likely to be adequate for the proper development of the gilt and its subsequent reproductive performance.[19]

In the majority of cases, moderate protein restriction does not influence age at puberty.[26] However, severe protein restriction or amino acid imbalance does delay the onset of puberty.[51,54,65,132] Environmental factors and management practices are likely to have a much greater influence on age at puberty of the gilt than nutrition.

Ovulation rate

The primary constraint on litter size in swine is ovulation rate.[17,70] It has been suggested that when puberty occurs at a young age and lower weight, ovulation rate increases with successive estrous periods; thus delaying mating until the second or third estrus increases the ovulation rate and embryo survival.[7,71] However, beyond a minimum age and weight, neither age, weight, nor number of estrus periods experienced influences ovulation rate or litter size.[74,108] Increasing feed or energy intake (flushing) of restricted fed gilts for 11 to 14 days before mating increases the ovulation rate,[6,15,26,64] but only to the levels achieved by ad libitum fed gilts (Table 42-2). Therefore, the ovulation rate is maximized in gilts by ad libitum feeding during the growing-finishing phase and up to the time of mating.

At protein levels between 12.5 and 16%, the source and level of protein in the diet have little effect on ovulation rate.[51,139]

Pregnancy and conception rate

Pregnancy rate is defined as the proportion of gilts selected for breeding that become pregnant or farrow, and *conception rate* is the proportion of selected gilts *mated* (i.e. have a normal estrus)

Table 42-2 *Influence of flushing on ovulation rate*

	Plane of nutrition				
	Restricted	Flushed	Restricted	Flushed	Ad lib
Prepubertal feed (kg/day)	2.0		2.4		Ad libitum
Postpubertal feed (kg/day)	2.0	2.4	2.4	2.8	Ad libitum
1st ovulation rate	11.2		12.6		13.3
2d ovulation rate	12.1	13.5	13.5	13.4	13.7

Data from the University of Alberta, 1986.

Table 42-3 *Influence of feeding level on embryo survival in gilts*

Period	Number of trials	Energy intake (Mcal/day)	Number of surviving embryos	Embryo survival (%)
Prepubertal	19	8.5	9.8	69.7*
	46	5.5	10.0	77.5
Premating	36	9.2	10.1	73.2†
	31	5.2	9.7	78.3

*$p < 0.01$.
†$p < 0.04$.
Adapted from Den Hartog LA and van Kempen GJM: Neth J Agric Sci 28:211, 1980.

that become pregnant or farrow.[29] In gilts fed 1.5 kg (3.3 lb) of feed per day for 35 days before mating, conception rate is not affected, but pregnancy rate is lower than that in gilts fed a more adequate level (2.25 to 3.0 kg/day or 5 to 6.6 lb/day).[29] In an extensive series of experiments, Den Hartog and van Kempen[26] showed that high or low levels of feed intake up to puberty or during the estrous cycle did not significantly affect conception rates (80.5% to 88%).

Embryo survival

A high level of feeding during rearing or the premating period is not advantageous. Increased embryo mortality compensates for any increase in ovulation rate[6,26] (Table 42-3).

Body composition and reproduction

Currently, a major question in the swine industry is whether age, weight, and backfat thickness at the time of original mating influence long-term sow prolificacy. With induction of puberty and ad libitum feeding, today's gilts may reach puberty at 80 to 90 kg (175 to 200 lb), with 10 to 12 mm (about 0.44 inch) of backfat. If maintained on ad libitum feeding, these gilts should weigh 95 to 105 kg (210 to 230 lb) and have 13 to 16 mm (0.5 to 0.6 inch) backfat at second estrus and weigh 110 to 120 kg (240 to 265 lb) and have 15 to 16 mm (about 0.6 inch) backfat at third estrus.

Farm survey data and some research data suggest that gilts that are bred at younger ages and lighter weights with lower backfat depths have higher subsequent culling rates.[58] In contrast, it has also been demonstrated that with good nutrition and management these lean gilts can maintain their original backfat depths and remain productive for equally as long as gilts bred at heavier weights with greater backfat depths. However, an as yet undetermined minimum fat

Table 42-4 *Nutrient requirements of developing gilts being reared for breeding and allowed feed ad libitum*

Intake levels*	Body weight (kg)	
	20-50	50-110
Energy concentration (kcal ME/kg diet)	3255	3260
Crude protein (%)	16	15
Nutrient†		
Lysine (%)	0.80	0.70
Calcium (%)	0.65	0.55
Phosphorus, total (%)	0.55	0.45
Phosphorus, available (%)	0.28	0.20

*Percentages in this table are on an *as-fed* basis.
†Sufficient data are not available to indicate that requirements for other nutrients are different from those of growing-finishing pigs.
Reproduced with permission from National Research Council: Nutrient requirements of swine, ed 9, Washington, DC, 1988, National Academy of Sciences, National Academy Press.

cover, perhaps as low as 10 mm (0.4 inch) backfat,[133] is required for successful reproduction, and breeding very lean gilts with 15 mm (0.6 inch) backfat or less allows little margin for error in either nutrition or management over the entire breeding cycle. Therefore it is recommended that genetically lean gilts bred at less than 120 kg (265 lb) live weight be fed ad libitum during the growing-finishing period and up to the time of breeding. The minimum nutrient content of their diet should be that recommended by the National Research Council (NRC)[102] and presented in Table 42-4. Such a feeding program maximizes the gilts' body weight and condition at time of breeding and may increase their ovulation rate, litter size, and lifetime in the herd.[68,69]

SOW NUTRITION
Gestation

Nutrient requirements and daily feed allowances can be expressed as a fraction of the diet (Table 42-5) or in terms of amount of feed to be ingested (Table 42-6). The basis on which the energy and feed intakes of sows of different weights were calculated is shown in Table 42-7. A 12% protein diet containing 3.34 Mcal DE per kg (1.5 Mcal/lb) and 0.43% lysine is sufficient for gestating sows and gilts.[102] The suggested daily feed requirements range from 1.8 to 2 kg (4 to 4.4 lb)/day, depending on the weight of the sow and assuming a total weight gain during gestation of 45 kg (100 lb) (25 kg [55 lb] of maternal weight gain plus 20 kg [45 lb] increase in weight due to the products of conception). These recommended feeding levels also assume individual feeding, a corn and soybean based diet and comfortable environmental conditions. For every 1° C (1.8° F) below 19 to 20° C (66 to 68° F) for individually housed sows or 14 to 15° C (57 to 59° F) for group-housed sows, the daily feed requirement must be increased 40 to 60 g (0.9 to 0.1 lb)/sow, depending on the condition of the sow. An approximate guide to the feeding of a gestating sow is to provide 1% of the sow's weight plus 0.5 kg (1.1 lb) if fed a corn-soybean meal diet or plus 0.7 kg (1.5 lb) if fed a barley-wheat-soybean-based diet. (Table 42-8 gives an example of a diet that can be used.) Thus, a 140-kg (310 lb) sow would get 1% of her weight or 1.4 kg (3.1 lb), plus 0.5 kg (1.1 lb) of a corn-soybean meal diet or 0.7 kg (1.5 lb) of a barley-wheat-based diet, for a total of 1.9 or 2.1 kg (4.2 to 4.6 lb)/day (Table 42-7). It has been suggested that gilts should gain 30 kg (65 lb) body weight in their first parity, 25 kg (55 lb) in their second parity, and 20 kg (45 lb) in each subsequent parity. In large-scale swine units, sows should be weighed at breeding, before farrowing, and at weaning. Backfat measurements also could be taken at service, each month throughout gestation, at farrowing, and at weaning. These backfat readings would provide an excellent guide as to whether the feeding and management program is adequate. As mentioned previously, the objective is to conserve body condition, and therefore sows should gain 1 to 2 mm (about 0.06 inch) of backfat in gestation and not lose more than 1 to 2 mm (about 0.06 inch) during lactation.

Feed intake in gestation is highly correlated with live weight gain of sows ($r = 0.71$) and reasonably correlated with piglet birth weight ($r = 0.46$).[37] For each additional kg of feed consumed per day, pig birth weight increases by 20 g (0.7 oz) and 35 to 50 g (1.25 to 1.8 oz) for gilt and sow litters, respectively.[24,61] Feed intake dur-

Table 42-5 *Nutrient requirements of breeding swine*

Composition of diets*	Bred gilts, sows, and adult boars	Lactating gilts and sows
Digestible energy (kcal/kg diet)	3340	3340
Metabolizable energy (kcal/kg diet)	3210	3210
Crude protein (%)	12	13
Nutrient		
Indispensable amino acids (%)		
Arginine	0.00	0.40
Histidine	0.15	0.25
Isoleucine	0.30	0.39
Leucine	0.30	0.48
Lysine	0.43	0.60
Methionine + cystine	0.23	0.36
Phenylalanine + tyrosine	0.45	0.70
Threonine	0.30	0.43
Tryptophan	0.09	0.12
Valine	0.32	0.60
Linoleic acid (%)	0.1	0.1
Mineral elements		
Calcium (%)	0.75	0.75
Phosphorus, total (%)	0.60	0.60
Phosphorus, available (%)	0.35	0.35
Sodium (%)	0.15	0.20
Chlorine (%)	0.12	0.16
Magnesium (%)	0.04	0.04
Potassium (%)	0.20	0.20
Copper (mg)	5.00	5.00
Iodine (mg)	0.14	0.14
Iron (mg)	80.00	80.00
Manganese (mg)	10.00	10.00
Selenium (mg)	0.15	0.15
Zinc (mg)	50.00	50.00
Vitamins		
Vitamin A (IU)	4000	2000
Vitamin D (IU)	200	200
Vitamin E (IU)	22	22
Vitamin K (menadione) (mg)	0.50	0.50
Biotin (mg)	0.20	0.20
Choline (g)	1.25	1.00
Folacin (mg)	0.30	0.30
Niacin, available (mg)	10.00	10.00
Pantothenic acid (mg)	12.00	12.00
Riboflavin (mg)	3.75	3.75
Thiamin (mg)	1.00	1.00
Vitamin B_6 (mg)	1.00	1.00
Vitamin B_{12} (µg)	15.00	15.00

*These requirements are based upon corn-soybean meal diets, feed intakes, and performance levels listed in Tables 42-6, 42-7, and 42-8. Percentages are on an *as-fed* basis.
Reproduced with permission from National Research Council: Nutrient requirements of swine, ed 9, Washington, DC, 1988, National Academy of Sciences, National Academy Press.

Table 42-6 *Daily nutrient intakes and requirements of intermediate-weight breeding animals*

	Bred gilts, sows, and adult boars	Lactating gilts and sows
Mean gestation or farrowing weight (kg)	162.5	165.0
Intake and performance levels		
Daily feed intake (kg)	1.9	5.3
Digestible energy (Mcal/day)	6.3	17.7
Metabolizable energy (Mcal/day)	6.1	17.0
Crude protein (g/day)	228	689
Nutrients		
Indispensable amino acids (g)		
Arginine	0.0	21.2
Histidine	2.8	13.2
Isoleucine	5.7	20.7
Leucine	5.7	25.4
Lysine	8.2	31.8
Methionine + cystine	4.4	19.1
Phenylalanine + tyrosine	8.6	37.1
Threonine	5.7	22.8
Tryptophan	1.7	6.4
Valine	6.1	31.8
Linoleic acid (g)	1.9	5.3
Mineral elements		
Calcium (g)	14.2	39.8
Phosphorus, total (g)	11.4	31.8
Phosphorus, available (g)	6.6	18.6
Sodium (g)	2.8	10.6
Chlorine (g)	2.3	8.5
Magnesium (g)	0.8	2.1
Potassium (g)	3.8	10.6
Copper (mg)	9.5	26.5
Iodine (mg)	0.3	0.7
Iron (mg)	152	424
Manganese (mg)	19	53
Selenium (mg)	0.3	0.8
Zinc (mg)	95	265
Vitamins		
Vitamin A (IU)	7600	10,600
Vitamin D (IU)	380	1060
Vitamin E (IU)	42	117
Vitamin K (menadione) (mg)	1.0	2.6
Biotin (mg)	0.4	1.1
Choline (g)	2.4	5.3
Folacin (mg)	0.6	1.6
Niacin, available (mg)	19.0	53.0
Pantothenic acid (mg)	22.8	63.6
Riboflavin (mg)	7.1	19.9
Thiamin (mg)	1.9	5.3
Vitamin B_6 (mg)	1.9	5.3
Vitamin B_{12} (μg)	28.5	79.5

Nutrient concentrations are expressed on an as-fed basis.
Reproduced with permission from National Research Council: Nutrient requirements of swine, ed 9, Washington, DC, 1988, National Academy of Sciences, National Academy Press.

Table 42-7 *Daily energy and feed requirements of pregnant gilts and sows*

	Weight (kg) of gilts and sows at mating*		
	120	140	160
Mean gestation weight (kg)†	142.5	162.5	182.5
Energy required (Mcal DE per day)			
Maintenance‡	4.53	5.00	5.47
Gestation weight gain§	1.29	1.29	1.29
Total	5.82	6.29	6.76
Feed required per day (kg)‖	1.8	1.9	2.0

*Requirements are based on a 25-kg maternal weight gain plus 20-kg increase in weight due to the products of conception; the total weight gain is 45 kg (100 lb).
†Mean gestation weight is weight at mating + (total weight gain/2).
‡The animal's daily maintenance requirement is 110 kcal of DE per $kg^{0.75}$.
§The gestation weight gain is 1.10 Mcal of DE per day for maternal weight gain plus 0.19 Mcal of DE per day for conceptus gain.
‖The feed required per day is based on a corn-soybean meal diet containing 3.34 Mcal of DE per kg (1.5 Mcal/lb). Reproduced with permission from National Research Council: Nutrient requirements of swine, ed 9, Washington, DC, 1988, National Academy of Sciences, National Academy Press.

Table 42-8 *Proportional composition of sow diets (per 1000 kg)*

	Pregnant	Lactating
Barley	640*	564
Wheat	220	220
Soybean meal (46.5% crude protein)	100	146
Fat	—	30
Salt	5	5
Calcium phosphate	15	15
Limestone	10	10
Mineral-vitamin premix	10	10
	1000	1000

*If values are taken as kg makes 1 metric tonne, if used as lb makes ½ a U.S. ton.

ing gestation is poorly correlated with litter size at birth ($r = 0.14$), but increasing sow feed intake in late gestation does result in 0.3 more pigs at 21 days postpartum.[52,120,133,137] The effect of feed intake on culling rate has been studied in sows fed from 1.7 to 2.3 kg (3.7 to 5.1 lb)/day throughout gestation and an average of 6.0 kg (13 lb)/day during lactation.[133] Sows fed higher levels were considerably heavier, and their culling rates were significantly lower. It also has been shown that overfeeding sows in gestation (8.84 Mcal DE per day) does not affect reproductive performance over three parities, but culling rate increases due to lameness and failure to exhibit estrus.[55] Thus, the response to energy intake above 6.0 Mcal DE per day during pregnancy is usually small[1,62,82] and is probably beneficial only in the last week of pregnancy.

Protein. Although sow weight responds to increased protein intake in gestation up to levels of 300 g (0.7 lb)/day (about 13% of the diet), no improvement in reproductive performance or birth weight is apparent beyond a daily intake of approximately 140 g (0.3 lb), supplying 8 to 10 g (0.3 to 0.35 oz) lysine.[1,21,56,57,91] Speer[120] concluded that feeding diets very low in protein (0.5% to 2.0%), protein-free diets, or even periods of complete inanition did not adversely affect embryo or fetal development or litter size but did reduce birth weight. These data support the NRC[102] recommendations as outlined in Tables 42-5 to 42-7.

Pattern of feed intake. In general, level of feeding in early gestation has shown little, if any, effect on litter size at birth, possibly because of the equalizing effect of fetal mortality due to uterine crowding and placental insufficiency later in gestation[30,73,127] (Table 42-9).

Several large-scale studies have reported that increasing sow feed intake in late gestation increases piglet birth weight by 30 to 50 g (1.1 to 1.8 oz).[24,63] In general, this increase in birth weight had only a small influence on piglet growth and survival. In the interest of simplicity, therefore, substantial increases in sow feed intake in late gestation are not recommended. This also may reduce complications with sow constipation and agalactia. The pattern of feed intake during gestation is less important than the total amount fed, and constant levels of feed intake give results as good as systems in which

Table 42-9 *Influence of energy level during rearing, the estrous cycle, and postmating on reproductive performance*

	Energy intake		
Sequence of treatments	Low/low/low	Low/high/low	High/high/high
Number of trials	26	14	15
Number of ova shed (calculated)	12.6	13.9	14.0
Embryonic survival (%)	78.7	77.6	70.3
Number of surviving embryos	9.9	10.75	9.84

Note that there is a significant decrease in embryo survival on the high/high/high energy intake and that the final number of embryos is similar.
Data adapted from Den Hartog LA and van Kempen GJM: Neth J Agric Sci 28:211, 1980.

Table 42-10 *Overall effects of adding fat to sow diets on milk composition and piglet growth*

	Control	Added	Fat benefit
Pigs born alive	10.0	9.9	−0.1
Pigs weaned	8.1	8.4	+0.3
Survival (%)	82.0	84.6	+2.6
Average birth weight (kg)	1.41	1.39	−0.02
Average 21-day weight (kg)	5.57	5.66	+0.09
Fat in colostrum (%)	7.3	9.1	+1.8
Fat in milk (%)	9.1	10.1	+1.0

Data are summarized from 31 experiments from Moser BD and Lewis AJ: Feedstuffs 59(9):36, 1980.

feed intake was varied in early or late gestation.[34] However, for sows fed 2 kg (4.4 lb) or less per day throughout gestation, a 20% increase in feed intake from day 109 of gestation until farrowing reduces or prevents loss of backfat by the sow without significantly reducing feed intake during lactation.

Fat and carbohydrate supplementation in late gestation. The effects of fat supplementation on milk composition and piglet performance have been extensively reviewed.[97,110,116] Fat supplementation increased litter size at weaning by 0.3 pigs, mainly because of the increased survival of light weight (less than 0.9 kg or 2 lb) pigs (see Table 42-10). Although the fat content of the colostrum and milk increases, there is little, if any, effect on piglet growth. Glycogen and fat content of the liver and skeletal muscle of piglets are marginally increased by fat supplementation in late gestation. In order to obtain the greatest response to added fat in terms of improved piglet survival, sows should consume at least 1 kg (2.2 lb) of fat in the week before farrowing.[110] Recent experiments have confirmed that manipulating a sow's diet in late gestation by adding fat, butanediol, fructose, or starch has only minor effects on reproductive performance, pig survival, or litter growth.[20,84]

Laxatives. Constipation in sows at the time of farrowing is a concern among pork producers and is frequently associated with reduced milk yield in early lactation. The use of laxatives (wheat bran, high levels of potassium chloride) is widespread in prefarrowing diets. However, research does not indicate that laxatives improve sow performance (Table 42-11). More recently, a large-scale cooperative trial in the United States confirmed that there were no obvious benefits from the inclusion of 0.75% potassium chloride in prefarrowing diets.[103]

Table 42-11 *Effects of laxative diets on the performance of lactating sows*

Item	Treatment	
	Corn-soybean meal	Plus laxative*
Pigs born alive per litter	9.62	9.59
Pigs weaned per litter	7.99	7.98
Survival (%)	83.1	83.2
Weaning weight (kg)	5.05	4.99

*5% to 15% wheat bran, 5 to 15% alfalfa meal, 5% beet pulp, or 0.75% KCl fed 3 to 7 days before farrowing and for 2 to 4 weeks of lactation.
Summary of 5 experiments, 612 letters, from data in Cromwell GL: Il Pork Ind Conf, 1980.

Lactation

The nutrient and energy requirements of the lactating sow depends on the sow's weight, milk yield and composition, and the change in body condition targeted for the lactation (Tables 42-5, 42-6, and 42-12). Daily feed intakes of 4.4 to 6.1 kg (9.7 to 13.5 lb) of a corn-soybean meal diet containing 13% protein and 0.6% lysine allow the production of average milk yields of 5 to 7 kg (11 to 15 lb)/day and minimal loss in sow weight or condition throughout lactation.[102] Sows producing larger milk yields than these require additional feed.

Estimates of milk yield can be made by assuming each g of litter weight gain up to 21 days of age requires about 4 g of milk.[87,104] Each kg of milk produced requires approximately 0.6 kg of a corn-soybean meal diet or 0.65 kg of a barley-wheat–based diet (0.6 or 0.65 lb diet per lb milk).

The maintenance requirement of the sow can be estimated as 1% of her body weight. Thus, a 165-kg (360 lb) sow producing 6.25 kg (13.8 lb) of milk per day and fed a corn-soybean meal diet would require 1.65 kg (3.6 lb) of feed for maintenance plus 6.25 × 600 g or 3.75 kg (8.27 lb) of feed for milk production, for a total of 5.40 kg (11.9 lb) per day, which is in close agreement with recommended levels (Table 42-12).[102]

Feed intake in lactation. The NRC[101] suggests that a lactating sow, given voluntary access to feed containing an adequate supply of nutrients, will consume sufficient nutrients to maintain body weight and support milk production. However, many sows do not consume or obtain adequate feed levels during lactation.[101] Approximately 40% of first-litter gilts, 50% of second-parity sows, and 70% or more of third and greater parity sows do not consume the levels of feed recommended by NRC[102] (Table 42-13). Many sows mobilize their body reserves, both protein and fat, to maintain milk production.[48,67,113,114] However, there are limits to this process; thin sows produce significantly less milk than sows in good body condition.[117,131] Our work has demonstrated that underfeeding lactating sows for more than one parity significantly reduces milk yield.

The amount of weight, fat, or both a sow may lose before it affects her milk yield and its composition or it is delayed in return to estrus is still a matter of debate. Whittemore and co-workers[133] suggest that the lower threshold for backfat is 10 mm, but there is considerable variation among sows. Moreover, the minimum level of backfat should be an academic question because the objective in swine production is to minimize weight and condition loss during lactation. The more the sow eats in lactation, the more milk she produces, the less weight she loses, and the heavier the litter she weans. Therefore, it is important to maximize sow feed intake during lactation.

Factors affecting feed intake. Among the more important factors affecting feed intake during lactation are appetite, sow feed intake during gestation, farrowing barn temperature, and system of feeding.

Appetite. There is some evidence that modern lean types of sows eat less than their counterparts of 20 years ago.[1] Therefore, traditional systems of feeding may not be appropriate for such sows. The appetite of the sow also is influenced by her milk yield, which in turn is influenced by the size and vigor of the litter.

Gestation-lactation feed intake interactions. Feed intake in lactation is greatly influenced by feed intake in gestation, but the reason for this is unknown. The more the sow eats during gestation, the less she eats during lactation (Table 42-14). It is important that producers re-

Table 42-12 *Daily energy and feed requirements of lactating gilts and sows*

	Weight of lactating gilts and sows at postfarrowing (kg)		
Intake and performance levels	145	165	185
Milk yield (kg)	5.0	6.25	7.5
Energy required (Mcal DE per day)			
Maintenance*	4.5	5.0	5.5
Milk production†	10.0	12.5	15.0
Total	14.5	17.5	20.5
Feed required per day (kg)‡	4.4	5.3	6.1

*The animal's daily maintenance requirement is 110 kcal of DE per $kg^{0.75}$.
†Milk production requires 2 Mcal of DE per kg of milk.
‡The feed required per day is based on a corn-soybean meal diet containing 3.34 Mcal of DE per kg.
Reproduced with permission from National Research Council: Nutrient requirements of swine, ed 9, Washington, DC, 1988, National Academy of Sciences, National Academy Press.

Table 42-13 *Variation in feed intake during different lactations of sows fed to appetite*

	Parity			
Feed per day (kg)*	1	2	3	4
Less than 3.0	8†	1	3	4
3.1-4.0	32	7	4	4
4.1-5.0	45	43	32	28
5.1-6.0	15	41	52	54
More than 6.0	0	8	9	13
Mean feed per day (kg)	4.2	5.0	5.1	5.3

*Barley-wheat–based diets containing 3.0 Mcal DE per kg.
†Percentage of total sows at this parity.
From Lynch PB: Proc British Society Anim Prod, Leeds, 1988.

Table 42-14 *Effect of feed intake and weight gain during gestation on feed intake and weight loss in lactation*

	Plane of nutrition during gestation		
	High	Medium	Low
Gestation			
Feed per day (kg)	2.64	2.0	1.5
Weight gain (kg)	67	48	30
Lactation			
Feed per day (kg)	3.40	4.46	4.90
Weight loss (kg)	30.7	15.8	3.6

From data in Mullan BP: Doctoral thesis, 1987, Department of Animal Science, University of Western Australia.

sist the temptation to overfeed sows during gestation. This relationship between gestation and lactation feed intake also exposes a disadvantage of group-feeding pregnant sows: the more aggressive sows are likely to be overfed.

Barn temperature. The temperature of the environment has a significant effect on sow feed intake. Voluntary sow feed intake declines by 12% when the temperature of the farrowing barn increases from 21° C to 27° C (70 to 80° F).[88] The daily depression in average feed intake amounted to about 0.1 kg/1° C (0.12 lb/1° F) rise in barn temperature above 25° C (77° F). Sows with drip coolers consume more feed in hot conditions than control sows. Cooling the heads of lactating sows with a jet of cooled air has been shown to increase sow feed intake when barn temperatures exceed 28° C (82° F), but not at lower temperatures.[88,92,122]

Systems of feeding. It is a common practice for pork producers to increase feed intake of lactating sows gradually over the first 3 to 7 days of lactation. In general, restricted feeding increases sow weight loss and can increase baby pig mortality by increasing the activity of the sow.[40,99,121] Many producers now allow sows to feed to appetite immediately after farrowing.

The sows are fed three times per day, with the major meal in the morning, and have access to fresh feed at all times. On these farms there is no evidence of increased problems with constipation, milk scours, agalactia, or sows going off feed.

Energy intake. There is ample evidence that energy intakes of less than 12 Mcal DE per day for first-litter gilts increase weight and condition loss, reduce weaning weights, increase the interval from weaning to estrus, reduce the percentage of sows in estrus within 10 days of weaning, and reduce conception rate, embryo survival, and subsequent litter size.[67-69,113-115] The effects of restricted energy intake are less obvious in multiparous sows[100,113,114] at least for one parity, but continued underfeeding of a lactating sow increases weight loss, decreases milk production, and delays return to estrus. These data emphasize the importance of ensuring an adequate energy intake by lactating gilts and sows.

Fat supplementation. The results of several experiments indicate that the digestible energy intake of a lactating sow can be increased by supplementation of her diet with fat.[22,98] Adding fats or oil to the diet of the lactating sow may decrease feed intake by 5% but results in increased energy intake, especially in hot weather.[22,118] It is advisable to increase the protein content of the diet by 1% to 2% to compensate for possible reductions in feed intake.

Protein and amino acid intake. The data available[10,56,57,91,103] support recommendations that the diets of lactating sows contain 13% protein and 0.6% lysine.[102] In some of these studies, levels of dietary protein above 13% slightly decreased sow weight loss during lactation and increased litter weight. There is also some evidence that low protein intake during gestation may be compensated for by adequate protein intake during lactation.[91] Recently, King and Dunkin[67] have substantiated NRC[102] recommendations that intakes of 700 to 750 g (about 1.6 lb) protein and 37 g (1.3 oz) lysine per day are needed to maximize sow performance.

Calcium and phosphorus requirements. Because sow's milk is rich in calcium and phosphorus, the sow's requirement for these minerals is very high during lactation. Usually, calcium and phosphorus are drawn from bones during lactation to meet the needs for milk synthesis. This causes no problems unless the demineralization is excessive. Although culling sows for feet and leg problems is a serious problem in many herds, the cause of such lameness is generally osteochondrosis or feet problems rather than bone fractures caused by calcium or phosphorus deficiencies.

There is no evidence available that lactating sows require more calcium and phosphorus than the levels suggested by NRC.[102] Feed intake may be reduced by increasing calcium levels to 0.9% and phosphorus levels to 0.7 percent.

Sow weight loss. There is still some controversy as to whether the fat or the protein component of lactation weight loss in sows is responsible for delayed return to estrus.[66,100,115] It is likely that excessive loss of either body protein or fat reduces sow reproductive performance. However, this is not a consistent finding.[32,74] It appears likely that the amount of body reserves at farrowing and weaning is the critical factor affecting sow reproductive performance, rather than the amount of tissue mobilized during lactation. Thus, a small weight loss in a thin sow may be more serious than a large weight loss in a well-conditioned sow.[46]

Postweaning

The major objectives of nutrition during the postweaning period are to shorten the interval to conception, to synchronize the onset of estrus, and to maximize ovulation and conception rates. In the past, recommendations have included a "drying off" period for at least 24 hours postweaning, during which the sow receives no food or water. It has been suggested that this is an effective means of shortening and synchronizing the interval to estrus.[13] More recent evidence indicates no benefit from "drying off" sows when conception rates are greater than 80%,[5,128] and it may even be detrimental with some management systems. This system of management may be effective where conception rates are less than 70%.

Increasing the level of feed postweaning has been reported to shorten the interval to service in primiparous sows (gilts)[14,69] and to increase the number of such sows exhibiting estrus within 10 days of weaning.[25,50,128] It is possible that primiparous and multiparous sows respond differently to postweaning feed intake.[14] It appears that gilts or sows that have lost excessive weight or body condition during lactation respond to high levels of feed during the postweaning period. The postweaning performance of sows in good body condition is not influenced by level of feeding after weaning.

The information on the influence of postweaning nutrition on ovulation rate at the first estrus and on subsequent litter size is contentious. For a sow the normal interval from weaning to mating varies but may be as low as 4 to 5 days, which corresponds to the follicular phase of a normal estrous cycle. There is little evidence that high-level feeding between weaning and breeding affects either the ovulation rate or subsequent litter size of sows. Although increasing the level of feed does not affect ovulation rate at the first postweaning estrus, it does increase ovulation rate at the second postweaning estrus.[28] Lodge and Hardy[86] did report an increased litter size in flushed sows (9 versus 10.8), but this may have been because the control sows had a low mean litter size: flushing seems to have brought low litter size back to "normal" rather than to have caused an increase above what was to be expected. Indeed, with larger litter sizes in the control groups, various authors have failed to confirm a stimulatory effect of increased feed intake on ovulation rate.[50,128]

Increasing postweaning feed levels for primiparous sows has been reported to improve conception rates,[14] although this is not confirmed by results from older sows.[16,27] However, the interaction between lactation and postweaning feed levels needs to be determined.

The level of feeding in the interval between weaning and breeding is unlikely to improve reproductive performance except where it reverses a reduction in performance due to poor nutrition in lactation.[112]

SOW'S MILK
Milk yield

The function of the sow is to produce large litters of viable pigs and then to produce sufficient milk to ensure their survival and rapid growth. The most important factor affecting the milk yield of the sow are her genetic potential to produce milk, body condition, nutrient and energy intake, and the demands made by the litter. Estimates of milk yield have been made by weighing the pigs immediately before and after nursing; using oxytocin to induce milk letdown and then milking the sow by hand or machine, and, more recently, using deuterium oxide tracer techniques.

In general, estimates of the milk yield of sows have ranged from 5 to 10 kg (11 to 22 lb)/day.[81] Gilts produce less milk than sows, and milk yield increases up to the third parity after which it plateaus and then declines at the fifth or sixth parity. Milk yield increases gradually postpartum and reaches a peak at about day 21 of lactation and starts to decline in the fifth week of lactation.[104]

In swine the efficiency of milk production (milk energy per unit of feed energy) is about 65%.[1,130] The conversion efficiency of milk energy into piglet body energy varies with milk yield and ranges from 36% to 58%.[129] As the size and vigor of a litter increases, total milk yield increases, but milk intake per piglet generally is lower in larger litters.

By the fourth week of lactation the daily nutrient requirement of the piglets exceeds the amount supplied by the milk. Supplemental feed (creep feed) must be supplied if piglet weights at weaning (4 weeks of age or later) are to be maximized.

Milk composition

The composition of sows' milk is very variable (Table 42-15).[13,104,129] Factors contributing to this variability include nutrient and energy intake of the sow in gestation and lactation and also the stage of lactation.[104]

The addition of fat to the diet of a lactating sow usually increases the dry matter and fat content of her milk but may not increase the weight gain of the litter.[18,20,80,97,110] This may be because milk protein content limits piglet growth.[104] The primary advantages of fat supplementation of the lactating sow's diet are increased piglet survival and decreased sow weight loss.

Reduced feed or energy intake by the sow during lactation reduces milk yield but increases the dry matter, protein, and fat content of the milk. It thus has no significant effect on litter

Table 42-15 Composition of sow milk

	Mean value	Range of values
Dry matter (%)	19.4	17.1-25.8
Fat (%)	7.2	3.5-10.5
Protein (%)	6.1	4.4-9.7
Lactose (%)	4.8	2.0-6.0
Ash (%)	0.96	0.78-1.30
Calcium (%)	0.21	0.12-0.35
Phosphorus (%)	0.14	0.10-0.19

From Bowland JP: In Bustad LK, McClellan RG, and Burns MP, editors: Swine in biomedical research, Seattle, 1966, Frayn Printing Co.

Table 42-16 Changes in gross composition of sow colostrum and milk

	Hours after parturition			
	0	6	12	15-24
Dry matter (%)	30.2	26.6	20.8	19.6
Fat (%)	7.2	7.8	7.2	7.7
Protein (%)	18.9	15.1	10.2	7.2
Lactose (%)	2.5	2.9	3.4	3.7
Ash (%)	0.63	0.62	0.63	0.66
Calcium (%)	0.05	0.05	0.06	0.07
Phosphorus (%)	0.11	0.11	0.11	0.12

Data from Perrin DR: J Dairy Res 22:103, 1955.

Table 42-17 Pig weights and estimates of milk and creep feed intake

		Intake (g per pig per day)	
Age (days)	Weight (kg)	Creep feed	Milk
Birth	1.43	—	—
4	2.37	—	537
14	3.45	—	617
21	4.65	21	686
28	6.30	42	800

growth rate over one or two parities. However, continuation of the reduced feed intake over more than two parities adversely affects milk yield, milk composition, and litter weight at weaning. Sows in good condition can mobilize body reserves to maintain energy output in milk; thin sows cannot and therefore produce less milk than sows in good body condition.[81,117]

The composition of colostrum changes rapidly during the first 12 hours postpartum (Table 42-16). The dry matter of colostrum decreases by approximately 10% within 12 hours of parturition, primarily because of a 9% to 10% reduction in its protein content. The composition of colostral protein also changes dramatically: the gamma globulin fraction declines from approximately 8.6% at parturition to 2% at 12 hours postpartum.[12] After the first 48 hours postpartum and throughout the remainder of lactation, the composition of a sow's milk changes little.

CREEP FEEDING

Creep feed commonly is introduced to piglets when they are 10 to 14 days old in order to maintain rapid growth when the sow's milk yield declines. Creep feeding accustoms the baby pig to eat dry feed and may result in gut flora and enzyme adaptations that minimize growth reduction and incidence of diarrhea after weaning.[9] However, several studies have shown that creep feed is of little benefit to pigs weaned at 3 or 4 weeks of age.[83,101,106,125] In these studies creep feed intake was 20 g/day or less during the third week of lactation (Table 42-17). This level of feed intake did not significantly influence piglet weight at weaning, nor did it influence feed intake, feed utilization, or piglet performance in the postweaning period.[83,106,107,125]

It also has been suggested that small preweaning creep feed intakes may promote postweaning diarrhea by providing dietary antigens that precipitate a hypersensitivity response.[93-95,123] Enteric hypersensitivity reactions are accompanied by intestinal tissue damage, which reduces nutrient absorption and enzyme activity[123] and predisposes the intestinal tract to *E. coli* infection. Several approaches may be taken to overcome this potential problem: later weaning, "abrupt" weaning, or antibiotic supplementation of the postweaning diet. *Abrupt weaning* is the term given to a system whereby piglets are not given creep feed while suckling.

Late weaning produces immunological tolerance to dietary antigens prior to weaning. English and associates[42] have suggested that a total preweaning creep feed intake of at least 400 g (about 1 lb) is necessary to produce tolerance and prevent postweaning diarrhea. In the third and fourth weeks of lactation, creep feed intake for pigs fed a semicomplex diet is approximately 20 and 40 g (0.7 and 1.4 oz) per pig per day, respectively.[106] Thus, by 4 weeks of life piglets should have consumed more than 400 g (0.9 lb) of creep feed. Assuming a milk:weight gain ratio of 4:1, it can be calculated from the data in Table 42-17 that the creep feed:weight gain ratio of the pig in the fourth week of lactation is approximately 1.24:1 and that creep feed intake is contributing approximately 17% of the piglet weight gain. In the fifth and sixth weeks of lactation, creep feed intake is 110 and 240 g (0.25 and 0.53 lb) per pig per day and contributes 33% and 50% of the pig's nutrient intake, respectively. In some studies creep feed intake did not significantly influence weaning weight (mean = 7.2 kg or 15.9 lb) at 5 weeks,[106] whereas in other studies creep feeding did significantly affect weight at weaning (mean = 9.4 kg or 20.7 lb).[2] There is a tremendous variation in the creep feed intake per lactation of different litters (0.4 to 22.8 kg or 1 to 50 lb) weaned at 5 weeks of age.[2] Creep feed intake before weaning did not influence growth rate or feed intake in the postweaning period.

Creep feed composition

During the first 2 to 3 weeks of life, the digestive system of the baby pig is adapted primarily to digest milk. If a suckling pig is to eat and utilize a creep feed, the feed must be palatable and digestible. In the formulation of creep diets, these criteria must be met at a reasonable cost.

Because the suckling pig meets most of its amino acids requirements from sow's milk, creep feeds containing 14% protein support the same level of performance as diets containing 22% protein.[53] However, if the creep feed is going to be fed for the first 2 weeks after weaning, then it must contain 18% to 20% protein and a minimum of 1.15% lysine.[102] Postweaning pig performance is superior when the pigs are fed the same diet before and after weaning, rather than being changed from a creep to a starter diet. Creep feeding simple, semicomplex, or complex diets to pigs weaned at 3 or 4 weeks of age did not significantly affect their weight at weaning or their feed intake or performance after weaning.[106,125] Therefore, the cost-effectiveness of creep feeding pigs weaned before 4 weeks of age could be questioned. For fast-growing pigs weaned at 4 or 5 weeks of age, creep feeding increases weight at weaning but does not influence growth rate or feed intake after weaning. These results suggest that feeding a semicomplex diet similar to the one shown in Table 42-18 would give the most economic response for pigs weaned between 3 and 5 weeks of age. Recently, Taylor and co-workers have suggested that, even for pigs weaned at 3 weeks of age, complex starter diets could be replaced by cheaper, simple diets with minimal loss in performance.[126]

The ingredients used in formulating creep starter diets for pigs in North America are corn, wheat, barley, oat groats, soybean meal, fish meal, whey, skim milk powder, and fats or oils. The degree of complexity of the diet depends on the age of the pig, farmer preference, and eco-

Table 42-18 *Proportional composition of creep and starter diet*

	Proportion
Barley	177.0
Wheat	180.0
Oat groats	200.0
Soybean meal (46.5% CP)	156.0
Whey (70% lactose)	150.0
Fish meal (60% CP)	65.0
Fat and oil	40.0
Dicalcium phosphate	1.6
Limestone	6.0
Salt	0.4
Lysine HCl	2.0
Antibiotic	2.0
Mineral-vitamin premix	20.0
	1000.0

*If values are taken as kg, this makes 1 metric tonne of feed. If values are used as lb, this makes ½ U.S. ton of feed.

nomics. (An example is shown in Table 42-18.) Oat groats are commonly included at levels of up to 20% of the diet. Although the level of whey added to creep and starter diets varies considerably because of cost, the best response to whey is obtained at levels of 15% to 20% of the diet (Table 42-19). It has been suggested that the whey used be edible-grade quality and contain a minimum of 70% of lactose.[4]

Although most creep and starter diets contain flavors or aroma modifiers, few experiments have shown their inclusion to improve feed intake or growth rate consistently.[76,125] The response to antibiotics in creep feeds is generally small under conditions of good husbandry.[138] Because creep feed is often continued as the starter diet after weaning, however, it is desirable to include an antibiotic in creep feed. Reviews suggest that a tetracycline-sulfamethazine-penicillin combination and mecadox give satisfactory responses in weaner diets.[60]

Most creep and starter diets contain 3% to 5% added fat or oil. Speer[119] reported that baby pigs show a preference for creep feeds containing fat. Several studies have shown that when the energy-protein ratio is kept constant, the addition of fat to the diet significantly improves pig performance.[2,38]

Water

Water should be readily available to the suckling pig at all times. Several studies have shown that creep feed intake is reduced if water is not available. Nipple drinkers lead to more water waste but are easier to keep clean.

Table 42-19 *Effect of level of whey in starter diets on piglet performance*

	Level of whey (%)		
	0	10	20
Average daily gain (kg)			
0-14 days	.11	.12	.16
Average daily feed intake (kg)			
0-14 days	.23	.22	.26
Gain: feed	.35	.54	.61

Data from Allee GL and Hines PH: J Anim Sci 35:210, 1972 (abstract).

PIGLET MORTALITY

Baby pig mortality is still a major limitation to the number of pigs weaned per sow per year and is a major cause of reduced efficiency and profitability in swine production.[45,49] The average baby pig mortality in North America is approximately 25%, with a range of 10% to 48%.[39,135]

Stillbirths

Stillbirths account for 7.0% of all pigs born, with a range of 1% to 14.0% (Table 42-1). In many cases no differentiation is made between genuine stillbirths and pigs found dead at first observation of the litter. It has been estimated that 90% of all stillborn pigs die between the start of farrowing and actual birth. The main cause of stillbirths is anoxia from premature breakage of the umbilical chord. In most herds 40% to 50% of the sows produce all the stillborn pigs, with half of these sows having more than one stillborn pig per litter. It is likely that sows with a high stillbirth rate in one litter are likely to have stillborn pigs in subsequent litters. Some of the factors or conditions that may contribute to stillbirths are anemia in sows (less than 9 g hemoglobin per 100 ml blood), thin or fat sows, old age (six parities or more), high farrowing barn temperature, and small (fewer than five pigs) or large (more than nine pigs) litters.

In sows identified as producing stillborn pigs, farrowings could be induced by prostaglandin injections to allow more accurate prediction of the onset of farrowing and thereby allow the farrowing to be attended. Studies have shown that attending farrowing and assisting pigs by drying them off and directing them to a teat increases piglet survival by 7%.[125]

Preweaning piglet mortality

It has been estimated that 18.5% of liveborn pigs die before weaning (Table 42-20). Among the more important factors contributing to piglet mortality are larger litter size, low birth weight, chilling, poor sanitation, and poor farrowing facilities. However, diarrhea, crushing, starvation, or a combination of these factors are the actual causes of death.[127] It has been shown that approximately 50% of preweaning deaths occur within 3 days of parturition, and most of this

Table 42-20 *Relationship of birth weight to survival of suckling pigs*

Pig weight (kg)	Survival (%)
0.45	16
0.68	39
0.91	59
1.14	74
1.36	86
1.59	95

Data from a 10,000-pig study by Hall and others: J Anim Sci 1(suppl):388, 1984.

loss is in the first 24 hours after birth.[41,124]

The piglet at birth has an environmental temperature requirement of 32° C to 34° C (90 to 93° F). This high temperature requirement, together with a poor hair coat, poor fat and glycogen stores, large surface area, and poor temperature-regulation capabilities, makes the baby pig very susceptible to chilling.[41] At an environmental temperature of 19° C (66° F), the newborn pig uses all of the energy from its milk intake just to maintain its body temperature.

Piglet birth weight and survival tend to decrease as litter size increases, with maximum survival occurring in litters of six to nine pigs.[31,49] However, the greatest number of pigs weaned generally is obtained from litters of 11 to 12 liveborn pigs. Therefore, large litters at birth are still paramount to obtaining more pigs at weaning. The survival rate of piglets increases as birth weight increases, at least up to birth weights of 1.36 kg (3 lb) (see Table 42-20). Pigs lighter than 0.9 kg (2 lb) at birth have only a 40% chance of survival unless they receive some supplemental feeding. Edwards and associates[33] have reported that 15% of all pigs born weigh less than 1 kg (2.2 lb) at birth and that these pigs accounted for 61% of the mortality within the first 2 days of parturition. High variation in birth weights also tends to decrease baby pig survival.[43]

Because of the known relationship between birth weight and piglet survival there have been many attempts to increase birth weight by manipulating the diet of sow in late gestation. More promising prospects for decreasing baby pig mortality are cross-fostering, providing more heat in the creep area, and providing surplus and small pigs with supplemental feeding and artificial rearing.

Because of the importance of starvation as a cause of baby pig mortality, there has been considerable interest over the years in cross-fostering and supplemental feeding of surplus or underprivileged pigs.[39,44] Several studies have shown that the survival of pigs of less than 1 kg (2 lb) birth weight can be significantly improved by dosing them with colostrum or sow milk replacer.[43] Winship and co-workers[136] reported a survival rate of 84% in pigs dosed with 15 ml of colostrum and 11 feeds of 25 ml of sow milk replacer compared to 43% survival in untreated lightweight pigs. Dosing pigs with oil during the first 48 hours postpartum did not improve piglet survival rate or weight gain.[111]

Artificial rearing. Numerous reports have described systems of artificially rearing surplus or lightweight pigs.[8,77,78] In general, the results have been variable, and systems of artificially rearing piglets have not been widely adopted at the farm level, mainly because of cost, inconvenience, and ineffectiveness. This situation is likely to change significantly in the future as improved delivery systems and long-life liquid milk supplements are developed and marketed. For artificially raised pigs, growth rate is not as important as survival. Pigs gain physiological maturity with increased age and size; if pigs are kept alive for 2 weeks on an artificial rearing system, they will be mature enough to go onto a complex creep feed. It is very important not to overfeed such pigs. A system of frequent, very short feedings is more effective than a system in which the pigs are fed ad libitum. The industry cannot continue to allow lightweight pigs and surplus pigs to die as they have in the past. It is likely that new, dry and liquid milk replacers will be developed and the improved systems used to supplement the nutrient intake of suckling and newly weaned pigs.[11,79] Mortality in a litter is highly correlated with 3-day weight gain, suggesting that lack of milk intake during the first 3 days of life is a major cause of death. Low milk yield in early lactation is normal for most sows, and very low milk yields are commonly the result of fever, lack of appetite, inadequate hormonal stimulation of milk production, and disease. Any system that could supplement the nutrient intake of the litter, especially of the lightweight pigs, during the first days of lactation would significantly reduce baby pig mortality and increase litter weights at weaning. Taverner and associates[125] reported that providing a liquid supplement to suckling pigs increased weaning weights of pigs by 2 kg (4.4 lb) over those given a dry creep feed. Furthermore, when a flavor was added to the milk, pigs consumed more of the supplement and were 18% heavier at weaning than pigs given only plain milk supplement and 27% heavier than pigs given the dry creep feed. Knudson and co-workers[75] also reported an improvement in litter weight at weaning of pigs offered a liquid milk supplement.

REFERENCES

1. Agriculture Research Council: The nutrient requirement of pigs, Slough, 1981, Commonwealth Agricultural Bureaux.
2. Aherne FX, Danielsen V, and Neilsen HE: Acta Agric Scand 32:155, 1982.
3. Aherne FX and Kirkwood RN: J Reprod Fertil 33(suppl):169, 1985.
4. Allee GL and Hines PH: J Anim Sci 35:210, 1972 (abstract).
5. Allrich RF and others: J Anim Sci 48:359, 1979.
6. Anderson LL and Melampy RM: In Cole DJA, editor: Pig production, London, 1972, Butterworth.
7. Archibong AE, England DC, and Stormshak F: J Anim Sci 64:474, 1987.
8. Armstrong WD and Clawson AJ: J Anim Sci 50:377, 1980.
9. Aumaitre A: World Review of Animal Production 8:54, 1972.
10. Baker DH and Speer VC: J Anim Sci 57(suppl 2):284, 1983.
11. Bark LJ and English PR: J Anim Sci 65(suppl 1):107, 1987.
12. Bourne FJ: Anim Prod 11:337, 1969.
13. Bowland JP: In Bustad LK, McClellan RG, and Burns MP, editors: Swine in biomedical research, Seattle, 1966, Frayn Printing Co.
14. Brooks PH and Cole DJA: In Swan H and Lewis D, editors: Proceedings of the conference for feed manufacturers, University of Nottingham, London, 1972.
15. Brooks PH and Cole DJA: Livestock Production Science 1:7, 1974.
16. Brooks PH and others: Anim Prod 20:407, 1975.
17. Christenson RK: J Anim Sci 63:1280, 1986.
18. Cieslak DG, Leibbrandt VD, and Benevenga NJ: J Anim Sci 57:954, 1983.
19. Close WH and Cole DJA: Livestock Production Science 15:39, 1986.
20. Coffey MT, Yates JA, and Combs GE: J Anim Sci 65:1249, 1987.
21. Cole DJA: In Cole DJA, and Foxcroft GR, editors: Control of pig reproduction, London, 1982, Butterworth.
22. Cox NM and others: J Anim Sci 56:21, 1983.
23. Cromwell GL: Illinois Pork Industry Conference, 1980.
24. Cromwell GL and others: J Anim Sci 67:3, 1989.
25. Den Hartog LA and van der Steen HAM: Neth J Agric Sci 29:285, 1980.
26. Den Hartog LA and van Kempen GJM: Neth J Agric Sci 28:211, 1980.
27. Dyck GW: Can J Anim Sci 52:570, 1972.
28. Dyck GW: Can J Anim Sci 54:277, 1974.
29. Dyck GW: Can J Anim Sci 68:1, 1988.
30. Dyck GW, Palmer WM, and Simeraks S: Can J Anim Sci 60:877, 1980.
31. Dyck GW and Swierstra EE: Can J Anim Sci 67:543, 1987.
32. Eastham PR, Smith WC, and Whittemore CT: Anim Proc 46:71, 1988.
33. Edwards SA, Malken SJ, and Spechter HH: Anim Prod 42:470, 1986.
34. Elsey FWH: In Falconer, editor: Lactation, London, 1971, Butterworth.
35. Elsley FWH: Paper presented to the Pig Commission, EAAP, Vienna, 1973.
36. Elsley FWH and others: Anim Prod 11:225, 1969.
37. Elsley FWH and others: Anim Prod 13:257, 1971.
38. Endres B and others: Can J Anim Sci 68:225, 1988.
39. England DC: J Anim Sci 63:1297, 1986.
40. English PR: Anim Prod 12:375, 1970.
41. English PR, Morrison V: Pig News and Information 5:369, 1984.
42. English PR, Robb CM, and Dias MFM: Anim Prod 30:496, 1980.
43. English PR and Smith B: Vet Annual 15:95, 1975.
44. English PR, Smith WJ, and Maclean A: The sow—improving her efficiency, Suffolk, England, 1984, Farming Press Ltd.
45. English PR and Wilkinson V: In Cole DJA and Foxcroft GR, editors: Control of pig reproduction, London, 1982, Butterworth.
46. Esbenshade KL and others: J Anim Sci 62:309, 1986.
47. Etienne M, Camous S, and Cuvillier A: Reprod Nutr Dev 23:309, 1983.
48. Etienne M, Noblet J, and Desmoulin B: Reprod Nutr Dev 25:341, 1985.
49. Fahmy MH and Bernard C: Can J Anim Sci 51:351, 1971.
50. Fahmy MH and Dufour JJ: Anim Prod 23:103, 1976.
51. Fowler SH and Robertson EL: J Anim Sci 13:949, 1954.
52. Fowler V and others: Anim Prod 44:488, 1987.
53. Fowler VR and others: Anim Prod 28:439, 1979 (abstract).

54. Friend DW: J Anim Sci 37:701, 1973.
55. Gatel F, Castaing J, and Lucbert J: Livestock Prod Sci 17:247, 1987.
56. Greenhalgh JFD and others: Anim Prod 24:307, 1977.
57. Greenhalgh JFD and others: Anim Prod 30:395, 1980.
58. Gueblez R, Fleha JY, and Boulard J: Journés de la Recherches Porcine en France 17:29, 1985.
59. Hall DD and others: J Anim Sci (suppl)1:388, 1984.
60. Hays VW and Muir WM: Can J Anim Sci 59:447, 1979.
61. Henry Y and Etienne M: Journés de la Recherches Porcine en France 10:19, 1978.
62. Henry Y and Etienne M: L'Elevage (edition porcine) 75:25, 1978.
63. Hillyer GM and Philips P: Anim Prod 30:469, 1980 (abstract).
64. Hughes PE and Varley MA: Reproduction in the pig, London, 1980, Butterworth.
65. Jones RD and Maxwell CV: J Anim Sci 39:1067, 1974.
66. King RH: Pig News and Information 8:15, 1987.
67. King RH and Dunkin AC: Anim Prod 42:319, 1986.
68. King RH and Williams IH: Anim Prod 38:241, 1984.
69. King RH and Williams IH: Anim Prod 38:249, 1984.
70. King RH and Williams IH: Theriogenology 21:677, 1984.
71. Kirkwood RN and others: J Anim Sci 60:1518, 1985.
72. Kirkwood RN and Hughes PE: Pig News and Information 3:389, 1981.
73. Kirkwood RN and Thacker PA: Pig News and Information 9:15, 1988.
74. Kirkwood RN and others: Can J Anim Sci 68:283, 1988.
75. Knudson BK and others: J Anim Sci 65(suppl 1):293, 1987.
76. Kornegay ET: Feedstuffs, p 24, November 21, 1977.
77. Leece JG: J Anim Sci 33:47, 1971.
78. Leece JG: J Anim Sci 45:659, 1975.
79. Leibbrandt VD, Kemp RA, and Crenshaw TD: J Anim Sci 65(suppl 1):129, 1987.
80. Lellis WA and Speer VC: J Anim Sci 56:1334, 1983.
81. Lewis AJ, Speer VC, and Haught DG: J Anim Sci 47:634, 1978.
82. Libal GW and Wahlstrom RC: J Anim Sci 45:286, 1977.
83. Lightfoot AL, Miller BG, and Spechter HH: Anim Prod 44:490, 1987 (abstract).
84. Lima J and Cline TR: J Anim Sci 65(suppl 1):319, 1987.
85. Lodge GA: Anim Prod 11:133, 1969.
86. Lodge GA and Hardy B: J Reprod Fertil 15:329, 1968.
87. Lucas IAM and Lodge GA: British animal nutrition technical communication 22, 1961.
88. Lynch PB: Ir J Agric Res 16:123, 1977.
89. Lynch PB: Proceedings of the British Society of Animal Products, Leeds, 1988.
90. Maclean CW: Vet Rec 85:675, 1969.
91. Mahan DC and Mangan TL: J Nutr 105:1291, 1975.
92. McGlone JJ, Stansbury WF, and Tribble LF: J Anim Sci 66:885, 1988.
93. Miller BG and others: Am J Vet Res 45:1730, 1984.
94. Miller BG and others: Pig Res Vet Sci 36:187, 1984.
95. Miller BG and others: Anim Prod 46:493, 1988 (abstract).
96. Moody NW, Speer VC, and Hays VW: J Anim Sci 25:1250, 1966 (abstract).
97. Moser BD and Lewis AJ: Feedstuffs 59(9):36, 1980.
98. Moser RC: Proceedings of the forty-fifth Minnesota nutrition conference, 1984.
99. Moser RL and others: Livestock Prod Sci 16:91, 1987.
100. Mullan BP: Doctoral thesis, 1987, Department of Animal Science, University of Western Australia.
101. National Research Council: Predicting feed intake of food producing animals, Washington, DC, 1986, National Academy of Science.
102. National Research Council: Nutrient requirements of domestic animals, No 2: Nutrient requirements of swine, ed 9, Washington, DC, 1988, National Academy of Science.
103. NCR-42, Committee on Swine Nutrition: J Anim Sci 57(suppl 1):263, 1983.
104. Noblet J and Etienne M: J Anim Sci 63:1888, 1986.
105. O'Grady JF and others: Anim Prod 17:65, 1973.
106. Okai DB, Aherne FX, and Hardin RT: Can J Anim Sci 56:573, 1976.
107. Partridge GG and Findlay M: Anim Prod 46:519, 1988 (abstract).
108. Paterson AM and Lindsay DR: Anim Prod 31:291, 1980.
109. Perrin DR: J Dairy Res 22:103, 1955.
110. Pettigrew JE: J Anim Sci 53:107, 1981.
111. Pettigrew JE and others: J Anim Sci 62:601, 1986.
112. Pike IH and Boaz TG: Anim Prod 15:147, 1972.
113. Reese DE and others: J Anim Sci 55:867, 1982.
114. Reese DE and others: J Anim Sci 55:590, 1982.
115. Reese DE, Peo ER Jr, and Lewis AJ: J Anim Sci 58:1236, 1984.
116. Seerley RW: In Wiseman J, editor: Fats in animal nutrition, London, 1984, Butterworth.
117. Shields RG Jr, Mahan DC, and Maxson PF: J Anim Sci 60:179, 1985.
118. Shurson MG and others: J Anim Sci 62:672, 1986.
119. Speer VC: Feedstuffs p 25, March 8, 1976.
120. Speer VC: In Woods W, editor: Symposium on management of food producing animals, vol 2, West Lafayette, Ind, 1982, Purdue University.
121. Stahley TS, Cromwell GL, and Simpson WS: J Anim Sci 65:1507, 1979.
122. Stansbury WF, McGlone JJ, and Tribble LF: J Anim Sci 65:1507, 1987.
123. Stokes CR and others: Vet Immunol Immunopathol 17:413, 1987.
124. Svendsen J, Bengtsson ACH, and Svendsen LS: Pig News and Information 7(2):159, 1986.
125. Taverner MR, Reale TA, and Campbell RG: In Farrell DJ, editor: Recent advances in animal nutrition, Hanover, NH, 1987, University Press of New England.

126. Taylor JA, Low AG, and Partridge IG: Anim Prod 46:523, 1988 (abstract).
127. Toplis P, Ginesi MFJ, and Wrathall AE: Anim Prod 37:45, 1983.
128. Tribble LF and Orr DE: J Anim Sci 55:608, 1982.
129. Van Kempen GJM and others: Neth J Agric Sci 33:23, 1985.
130. Verstegen MWA and others: J Anim Sci 60:731, 1985.
131. Verstegen MWA, Verhagen JMF, and den Hartog LA: Livestock Prod Sci 16:75, 1987.
132. Wahlstrom RC and Libal GW: J Anim Sci 45:94, 1979.
133. Whittemore CT and others: Anim Prod 46:494, 1988.
134. Whittemore CT, Smith WC, and Phillips P: Anim Prod 47:123, 1988.
135. Wilson MR and others: J Anim Sci 62:576, 1986.
136. Winship HL and others: J Anim Sci 55(suppl 1):204, 1982.
137. Young LG and King GJ: Ontario Swine Res Rev, Publ No 0388, 1987.
138. Zimmerman DR: J Anim Sci 62(suppl 3):6, 1986.
139. Zimmerman DR, Peo ER, and Hudman DB: J Anim Sci 26:514, 1967.

Feeding the Weaned Pig for Optimal Productivity

P. A. THACKER

The period after weaning is an extremely critical time that may greatly influence the overall performance and profitability of the market hog. Healthy, vigorous pigs that are eating well adjust rapidly to weaning. Slower-doing, unthrifty pigs find weaning an extreme shock and may exhibit a prolonged postweaning slump. Studies have shown that as much as 30% of the variation in age at market weight can be attributed to the time it takes pigs to reach a weight of 23 kg (50 lb).[14]

Weaning at any age is stressful, but the stress of weaning is greater the earlier the pig is weaned. It is important to minimize all possible stresses at weaning. One method is to ensure that pigs are properly fed.

COMPOSITION OF STARTER DIETS

There are many factors to be considered in the formulation of practical weaner pig diets. Factors such as overall nutritional objectives (maximum growth, minimum feed cost per pig, minimum feed cost per kg gain), market conditions, feed mixing ability, and cost all come into play.

In determining which feedstuffs to include in a starter diet, consideration must be given to the age and weight of the pig for which the diet is formulated. A high nutrient-density diet is a must for pigs under 10 kg (25 lb).[13] These diets should contain a large proportion of milk-based products such as whey, dried skim milk, and casein, which are compatible with the underdeveloped digestive system of the recently weaned pig.

Once the pig has reached a body weight of approximately 10 kg (25 lb), a more economical starter diet can be provided with lower energy concentration and less than half the quantity of milk products. High-energy grains, such as corn or wheat, as well as readily digested protein supplements, such as soybean meal, should form the basis of such a diet. In addition, consideration should be given to ingredients such as oat groats, dried whey, sugar, and animal fat to increase the palatability of the diet.[2,8,11,17] Table 43-1 shows the composition of two relatively inexpensive starter diets that have been quite successful in commercial use.

NUTRIENT REQUIREMENTS OF THE WEANED PIG

The nutrient requirements of the weaned pig depend upon the age and weight of the pig at weaning.[5,12] For pigs weaned at 3 weeks of age or younger, a diet containing at least 21% protein (as-fed basis) is recommended (Table 43-2). Once pigs have reached a body weight of 10 kg (25 lb), the protein content of the starter diet can be reduced to about 18%.[12]

Some producers are of the opinion that feeding high-protein diets to weaner pigs can increase the incidence of scours. However, this contention is not supported by experiments utilizing weaner diets with a high protein content. In fact, protein levels as high as 30% have been fed without inducing scours. If scouring occurs, producers are advised to improve their hygiene and alter management practices rather than looking to the feed composition as the source of their problem.

It should be pointed out that pigs do not require protein per se but rather individual amino acids. Cereal grains, which form the basis for most swine starter diets, contain inadequate levels of several essential amino acids required for the 5 to 20 kg (11 to 45 lb) pig. They must be supplemented in order to obtain a balanced diet.

Lysine generally is considered to be the first limiting amino acid in swine diets because it is usually present in the lowest amount relative to requirement. Care must be taken to ensure that diets formulated for starter pigs contain sufficient amounts of lysine.[10] Current estimates of the lysine requirement of the 5 to 10 kg (11 to 45 lb) pig indicate the potential to increase the performance of starter pigs by providing up to 1.25% dietary lysine (Table 43-3). This requirement can often be met more economically by using synthetic lysine than by using a protein concentrate. However, care should be taken to ensure that lysine is not supplemented at too high a level, as this practice causes other amino acids in the diet to become limiting; instead of improving performance, growth rates actually decline.

The energy level of the diet is a very important factor affecting the performance of starter pigs. Most estimates of requirement suggest that starter pig diets should provide a minimum of 3400 kcal DE per kg if maximum growth rates are to be obtained.[12] In order to obtain these levels, starter diets must be formulated using high-energy feedstuffs such as corn, wheat, or tallow. High-fiber feeds such as oats, alfalfa, or

Table 43-1 *Formulation of starter diets*

	Pig weight	
	5-10 kg	10-20 kg
Ingredients		
Wheat	30.0%	23.0%
Barley	—	15.0%
Oat groats	10.0%	25.0%
Sugar	5.0%	—
Tallow	5.0%	2.0%
Dried whey	20.0%	20.0%
Skim milk	10.0%	—
Fish meal	10.0%	—
Soybean meal	7.0%	12.0%
Premix	3.0%	3.0%
Antibiotic	+	+
Lysine	+	+
Composition (as-fed basis)		
Digestible energy (kcal/kg)	3500	3400
Crude protein (%)	21.0	18.0
Lysine (%)	1.25	1.10
Calcium (%)	.90	.80
Phosphorus (%)	.70	.70

Table 43-2 *Effect of dietary protein levels in starter feeds on postweaning performance*

	% Protein in starter			
	10	14	18	22
Performance 5-20 kg (11-45 lb)				
Average daily gain (g)	167	304	367	411
Average daily feed (g)	880	857	807	743
Feed conversion	5.27	2.82	2.20	1.81
Days to 20 kg	107	68	60	56
Performance 20-90 kg (45-200 lb)				
Average daily gain (kg)	0.67	0.65	0.63	0.65
Days to 90 kg	211	176	171	163

Modified from Nielsen and others, 1976, unpublished data.

Table 43-3 *Performance of weaner pigs fed various lysine levels*

	Lysine level (%, as-fed basis)					
	0.95	1.05	1.15	1.25	1.35	1.45
Average daily gain (g)	287	349	367	399	358	358
Average daily feed (g)	608	644	680	712	653	648
Feed conversion	1.96	1.86	1.83	1.79	1.86	1.82

Lewis AJ and others: J Anim Sci 51:361, 1980.

canola meal are to be avoided. Specific recommendations regarding other nutrients required in starter pig diets are provided in Table 43-4.

EFFECT OF ANTIBIOTICS IN STARTER DIETS

Antibiotics have played a major role in the growth and development of the swine industry for more than three decades. Their efficiency in increasing growth rate, improving feed utilization, and reducing mortality from clinical disease is well documented.[6] As a consequence, most starter diets formulated for swine are medicated.

The effect of utilizing various antibiotics in starter diets is shown in Table 43-5. It can be seen that over a wide range of test conditions the improvement in growth and feed efficiency resulting from the addition of antibacterial compounds to starter diets is very impressive. On average, a 12% to 15% improvement in daily gain and a 5% to 6% improvement in feed efficiency can be expected as a result of medicating swine starter diets.

Medication of starter diets is beneficial even in facilities with a high health status (Table 43-6). Considerable improvement in starter pig performance has been observed as a result of medication even in minimal-disease herds.[15] It therefore is apparent that producers who feed an unmedicated starter diet are missing an opportunity to improve the overall efficiency of their herd.

EFFECT OF PROCESSING ON WEANER PIG PERFORMANCE

In order to maximize growth rate and feed efficiency, weaner pigs should be fed a pelleted feed as opposed to a meal type of feed. On average, about a 5% improvement in daily gain and a 10% improvement in feed efficiency by providing a pelleted diet to starter pigs can be expected. The reason for the improved performance of starter pigs as a result of pelleting is subject to debate. Some feel the improvements are due to a reduction in feed wastage, whereas others suggest that the improvements result from an increase in feed consumption or from an increase in nutrient digestibility. In addition to its effects on performance, pelleting is beneficial because it improves the handling of the feed. Other benefits are reduced dustiness, less wind loss, and less settling of particles during transportation. Improved environment also may account to some degree for the improvement in

Table 43-4 *Nutrient requirements of starter pigs*

	Live weight (kg)	
Requirements	5-10	10-20
Digestible energy (kcal per kg)	3500	3370
Crude protein (%)	20	18
Lysine (%)	1.15	0.95
Threonine (%)	0.68	0.56
Methionine (%)	0.58	0.48
Tryptophan (%)	0.17	0.14
Calcium (%)	0.80	0.70
Phosphorus (%)	0.65	0.60

Values are on a 90% dry matter basis (air dry).
From National Research Council: Nutrient requirements of swine, ed 9, Washington, DC, 1988, National Academy of Sciences, National Academy Press.

Table 43-5 *Use of antibiotics in starter diets*

Antibiotic	Number of trials	Improvement (%) Daily gain	Gain per unit feed
Tetracycline-sulfamethazine-penicillin	333	22.5	8.5
Mecadox	292	18.6	8.6
Tylosin-sulfamethazine	76	17.6	6.8
Penicillin-streptomycin	95	14.8	7.4
Tylosin	124	14.8	6.0
Lincomycin	8	11.1	7.6
Virginiamycin	90	11.0	5.0
Tetracycline	234	10.8	6.2
Bacitracin	54	9.7	3.3
Penicillin	14	9.4	8.7
Nitrofurans	66	8.0	2.3

Hays VW and Muir WM: Can J Anim Sci 59:447, 1979.

Table 43-6 *Effect of medication on the performance of weaner pigs in a minimal-disease swine herd*

	Control	Medicated
Average daily gain (kg)	0.37	0.47
Average daily feed (kg)	0.58	0.70
Feed conversion	1.56	1.48

Patience JF and Christison GI: Prairie Swine Centre Report, 1988, University of Saskatchewan.

piglet performance as a result of pelleting.

Pellet size is important in determination of the response obtained from weaner pigs fed a pelleted diet. Most research indicates that the best performance can be obtained by using a 2.5-mm (0.1 inch) pellet for young pigs 5 weeks old and younger. For older animals, pellet size does not appear to be so important. It is possible to further improve performance by supplying the diet in the form of a crumble. Producers should obtain starter feeds in a crumble form whenever possible.

Various other feed processing techniques have been promoted by feed manufacturers. Recent trials evaluated the performance of pigs fed extruded, steam-flaked, or micronized cereal grain. These trials indicate little benefit from these additional processing techniques. Because feeds processed by these techniques are higher in cost, it is not cost-effective to utilize them in swine diets.

FEEDING METHODS TO REDUCE POSTWEANING SCOURS

The development of postweaning scours in pigs is commonly associated with increased numbers of certain strains of *E. coli* in the small intestine. One factor associated with the increase of *E. coli* and the subsequent development of scours is the sudden change of environment and diet at weaning. The weaned pig may go off feed and then, once it gets very hungry, overeat. The digestive system is not accustomed to a large intake of any meal, and the piglet scours.

This problem can be overcome by restricting the feed intake of pigs for the first few days following weaning.[3] Although the growth rate may be slower for the first week as a result of the restriction in feed intake, several experiments have demonstrated that the incidence and severity of diarrhea is much lower, and overall performance to market weight is improved.

The composition and complexity of the diet are frequently discussed in relation to postweaning pig performance and the occurrence of postweaning diarrhea. Increasing the level of fiber in starter pig diets has been suggested as a way to reduce weaning stress and to avoid the lag period in performance normally associated with weaning.[16] Because this method of restricting energy intake involves the formulation of an additional diet, however, it is not as simple to implement as restricting access to the self-feeder.

Concerns about scours should not be overemphasized, however. Some looseness can be accepted, providing it does not develop into diarrhea. The stool of pigs eating high nutrient density diets is always loose because of the nature of the diet.

EFFECT OF THE ENVIRONMENT ON THE POSTWEANING PERFORMANCE OF WEANER PIGS

The recently weaned pig is very sensitive to cold. A baby pig has a large surface area compared to its body size and tends to lose heat very rapidly. In addition, the baby pig has few body fat reserves, resulting in a low level of thermal insulation.

At weaning, pigs generally are moved from a warm creep area with a heat lamp to a pen in which the temperature is at least 5° C (10° F) lower. This constitutes a severe shock for the young pig and can result in poor performance.[4] The effects of a low environmental temperature on piglet performance are clearly demonstrated by the results of the experiment presented in Table 43-7. At an environmental temperature below 20° C (70° F), growth rate and feed efficiency are significantly reduced. In addition, mortality can be significantly higher at low environmental temperatures. Therefore, to minimize cold stress, the weaning environment should be kept at a temperature of 27° C to 28° C (82° F), and variations of more than 2° C (5° F) should be avoided.

Although air temperature is important, a number of other factors can alter the effective temperature to which the pig is subjected.[9] Air movement should be minimized, as a relatively minor draft can reduce the skin temperature of the baby pig by as much as 10° C (20° F). Damp conditions also lower the effective temperature. In contrast, the provision of bedding can increase the effective temperature.

Overcrowding weaner pigs is a serious mistake that occurs all too often in commercial practice. When pigs are crowded, feed intake is decreased and performance is impaired. A floor space allowance of 0.38 m² (4 sq ft) per pig is adequate for animals weighing up to 25 kg (55 lb).[7] With slatted floors and lighter pigs, this area can be reduced. Table 43-8 provides some guidelines as to the space requirements of weaner pigs.

There is little information regarding the optimal group size for weaner pigs. As group size increases, there is an increase in the aggressive behavior of the pigs. In addition, the variation in pig size within the pen also tends to increase in large groups. This may reduce overall pig performance. It is recommended that group sizes be less than 20 pigs per pen.

The amount of feeder space available to the pig also affects performance. If feeder space is insufficient, aggressive behavior increases, and some timid pigs may not be able to obtain adequate amounts of feed. Most research trials have indicated that weaner pens should be equipped to provide a minimum of 6 cm (2.3 inch) of feeder space per pig.[1]

Table 43-7 *Effect of air temperature on the performance of piglets weaned at 5 weeks*

	Temperature (°C)		
	13	20	26
Daily gain (g)	281	298	331
Feed intake (g)	1520	1470	1410
Mortality (%)	10	0	0

Modified from Volozchik and Morozov: Svinovodstvo 12:30, 1972.

Table 43-8 *Space requirements for weaner pigs*

Weight	Space
Up to 14 kg	0.15 m²
Up to 18 kg	0.20 m²
Up to 20 kg	0.25 m²

Table 43-9 *Performance targets for weanling pigs from 7 to 20 kg body weight*

	Good	Better	Best
Average daily gain (g)	400	475	550
Average daily feed (g)	640	715	770
Feed conversion	1.6	1.5	1.4
Mortality (%)	2.5	1.5	0.5

From Patience JF and Thacker PA: Swine nutrition guide, Prairie Swine Center, Saskatoon, 1989.

PERFORMANCE TARGETS

The nutritional program provided for the weaner pig greatly affects its performance all the way to market. Every effort should be made to ensure that the diet being fed meets the pig's requirements. Producers should keep records of the performance of their starter pigs in order to ensure that their pigs are growing to their genetic potential. Reasonable performance targets for starter pigs from 7 to 20 kg (15 to 45 lb) include a growth rate of at least 475 g (1 lb)/day, a feed intake of 715 g (1.5 lb)/day, and a feed conversion of 1.5 (Table 43-9). If pigs are not achieving this level of performance, then steps should be taken to improve productivity. Improvements in diet quality are one method of obtaining better performance. However, disease status, environmental factors, and genetic quality also should be evaluated.

REFERENCES

1. Aherne FX: Proceedings of the thirty-first annual swine short course, Lubbock, 1983, Texas Tech University.
2. Aherne FX, Danielsen V, and Nielsen HE: Acta Agricul Scand 32:151, 1982.
3. Ball RO and Aherne FX: Can J Anim Sci 67:1105, 1987.
4. Feenstra A: Pig News and Information 6:295, 1985.
5. Fowler VR: Pig News and Information 1:11, 1980.
6. Hays VW, Muir WW: Can J Anim Sci 59:447, 1979.
7. Kornegay ET and Notter DR: Pig News and Information 5:23, 1984.
8. Kornegay ET, Thomas HR, and Kramer CY: J Anim Sci 39:527, 1974.
9. Le Dividich J and Aumaitre A: Livestock Production Science 5:71, 1978.
10. Lewis AJ and others: J Anim Sci 51:361, 1980.
11. Lewis CJ and others: J Anim Sci 14:1103, 1955.
12. National Research Council: Nutrient requirements of domestic animals, no 2: Nutrient requirements of swine, ed 9, Washington, DC, 1988, National Academy of Science.
13. Nelssen JL: Forty-seventh Minnesota nutrition conference, 1986.
14. Patience JF: Proceedings of the Alberta pork seminar, 1989.
15. Patience JF and Christison GI: Prairie Swine Centre report, Saskatoon, 1988, University of Saskatchewan.
16. Rivera ER and others: J Anim Sci 46:1685, 1978.
17. Wahlstrom RC, Reiner LJ, and Libal GW: J Anim Sci 45:948, 1977.

44

Feeding the Growing-Finishing Pig for Optimal Production

P. A. THACKER

Pigs traditionally have been considered "opportunity feeders," and as such they have been fed all manner of feedstuffs. Many swine producers believe that pigs can be raised on almost anything. However, in today's profit-oriented world, pig feeding has become highly sophisticated, and a good working knowledge of nutritional principles is a must if maximum productivity and profit are to be obtained.

The growing-finishing period accounts for 60% to 70% of the cost of raising a pig to market weight. Therefore, any improvements in this area go a long way toward improving the overall profitability of a swine operation. Because feed represents the largest single expense incurred during this phase of the production cycle, it is imperative that every effort be made to meet the nutritional requirements of the growing-finishing pig at the lowest possible cost.

NUTRITIONAL REQUIREMENTS

The quality of the diet fed during the growing-finishing period can have a tremendous influence on growth rate, feed conversion efficiency, and carcass quality. There are no magical ingredients. Good performance results from feeding diets formulated to meet the nutritional needs of the pig and then mixed correctly using high-quality feedstuffs. Each nutrient has a specific function to perform in the body; if these nutrients are not supplied or if they are supplied in insufficient amounts, pig performance is reduced.

The best available estimates of the nutritional requirements of the pig are those provided by the National Research Council (NRC).[57] However, the environmental conditions of commercial swine operations may differ greatly from the experimental conditions under which the estimates of requirement were obtained. In addition, the NRC requirements make no allowance for variability in the nutrient content of feedstuffs, nutrient availability, or nutrient interactions. They also make no allowance for stressful conditions and do not take into account the higher genetic potential of some pigs. Therefore, NRC recommendations should be regarded as minimum values, and a margin of safety should be added to ensure that optimum performance is obtained. For most nutrients, it is recommended that at least 15% to 20% be added to the NRC requirements when formulating swine diets.

Energy requirements

Energy is usually the most costly nutrient in the diet of the pig. Other nutrients may cost more per unit but are required in much smaller amounts. Most swine diets are formulated on the basis of digestible energy (DE).[40] In theory, a more accurate estimate of the amount of energy that is in a form that can actually be utilized by the pig can be obtained by measuring metabolizable energy. However, the amount of energy lost through urine is relatively minor, and most estimates of metabolizable energy equal about 95% of digestible energy.[19] Therefore, metabolizable energy usually is not measured in swine.

The amount of energy required by the growing-finishing pig is dependent on a number of factors. The size of the animal is important because energy needs for maintenance are directly related to body size.[9] In addition, the environment in which the pig is housed affects energy requirements. In colder environments or in wet, drafty conditions, the amount of energy required to maintain body temperature rises.[33] If the pigs can huddle, then their energy requirements are reduced.

Growing-finishing diets should be formulated to supply about 3400 kcal/kg air dry (90% DM) diet.[57] However, pigs tend to eat to satisfy their energy requirements. This means that they eat more of a diet containing a low energy concentration than of a diet containing a higher energy level.[62] As a consequence, it is not as critical to maintain dietary energy as it is to maintain the levels of other nutrients.

Of greater importance than the actual energy content of the diet is the ratio of energy to other nutrients. The ratio of protein to energy is particularly important, and diets should be formulated to provide protein-calorie ratios of 47.3, 41.4, and 38.1 g of protein per Mcal DE (11.3, 9.9, and 9.1 g/MJ), for pigs of 20 to 35 kg (44 to 77 lb), 35 to 60 kg (77 to 132 lb) and 60 to 100 kg (132 to 220 lb), respectively.[47]

The digestible energy content of feeds is affected by many factors. High fiber levels reduce the energy content of the feed and adversely affect the quality of the ration.[42] Pigs do not digest fiber readily, which may interfere with the utilization of other nutrients.[48] As the level of crude fiber in the diet increases, daily gain and feed utilization decrease.[41] Therefore, high-fiber feeds, such as oats or alfalfa, should be used sparingly, if at all, in the diet of growing-finishing pigs.

The bushel weight (weight of a known volume of grain) of the cereal grain utilized also affects the energy content of the ration.[13] As the bushel weight of the grain increases, its digestible energy content increases. Therefore, producers should be encouraged to obtain high-bushel-weight cereal grains for use in pig rations. Recommended minimum values are 0.59 kg/L (46 lb/bushel) for barley, 0.68 kg/L (53 lb/bushel) for corn, and 0.73 kg/L (57 lb/bushel) for wheat.

The fat content of the diet also affects the concentration of digestible energy. Fats can be used to increase the caloric density of diets fed to growing-finishing swine.[45] This increase in diet density usually is associated with a decrease in feed intake without any adverse effect on daily gain. As a result, feed efficiency is improved by supplementing the diet with fat.[45]

Unfortunately, fat usually is quite expensive, and it often is difficult to justify the incorporation of fat into growing-finishing diets. However, pigs tend to reduce feed intake during periods of hot weather. Therefore, fat supplementation can be an effective method of maintaining energy intake in hot climates.[75] Fat also is useful in reducing dust. A recent study indicated that the addition of 2% fat reduced dust generation at feeder filling time by 17.6%.[58] The improved working conditions for the swine producer and the beneficial effects on the health of the herd usually are sufficient to justify the increase in cost associated with the incorporation of low levels of fat (1% to 2%) into the diet of the growing-finishing pig.

Amino acid requirements

Protein supplements normally are one of the most expensive ingredients in a hog ration. As a result, many producers try to reduce the amount of protein supplement going into the diet in the hope of reducing feed costs. This is usually a false economy, as pig performance is greatly reduced if adequate levels of amino acids are not provided.

If the diet contains adequate levels of lysine, then the remaining amino acids are usually present at high enough levels to meet requirements. Care must be taken to ensure that diets formulated for growing-finishing pigs contain sufficient amounts of lysine. Current recommendations suggest that at least 0.8% lysine be provided during the growing period and 0.7% lysine during the finishing period.[57] Recommended levels of other amino acids are shown in Table 44-1.

Not all of the amino acids present in a feed-

Table 44-1 *Recommended levels of amino acids for growing-finishing swine*

Amino acid (%)	Growing	Finishing
Arginine	0.25	0.10
Histidine	0.22	0.18
Isoleucine	0.46	0.38
Leucine	0.60	0.50
Lysine	0.75	0.60
Methionine + cysteine	0.41	0.34
Phenylalanine	0.66	0.55
Threonine	0.48	0.40
Tryptophan	0.12	0.10
Valine	0.48	0.40

Values are on an air dry (90% dry matter) basis.
From National Research Council: Nutrient requirements of swine, ed 9, Washington, DC, 1988, National Academy of Sciences, National Academy Press.

stuff are biologically available to the pig. The availability of amino acids has been shown to be reduced by incomplete digestion and absorption, by the presence of digestive enzyme inhibitors, and by heat damage.[17] Therefore, a considerable amount of research has been conducted to determine the availability of the amino acids in various feedstuffs.[38,73,86] Experience has demonstrated that adjusting feed formulas for amino acid availability improves the performance of swine.

Unfortunately, amino acid analysis is relatively expensive, and rarely does a producer have data available for the specific feedstuffs used. When an amino acid analysis is not available, it is preferable to formulate diets using the results of an actual total (crude) protein analysis rather than using book values for amino acids. When formulating on a crude protein basis, most experiments indicate that a minimum of 16% protein should be fed during the growing (20 to 50 kg or 45 to 110 lb) period and a minimum of 14% protein during the finishing (50 to 100 kg or 110 to 220 lb) period.[11]

Mineral requirements

Pigs require at least 14 inorganic mineral elements[57]: calcium, chlorine, cobalt, copper, iodine, iron, magnesium, manganese, phosphorus, potassium, selenium, sodium, sulfur, and zinc. Minerals are commonly provided in the form of a vitamin-mineral premix or hog concentrate incorporated into the ration. Feed companies usually formulate premixes so that if sufficient premix is used to meet calcium and phosphorus requirements then sufficient trace minerals also are present. Therefore, the remaining discussion relates only to calcium and phosphorus.

Inadequate calcium and phosphorus levels can reduce growth rate and feed efficiency as well as impair bone mineralization.[28,29,64] Diets fed to growing-finishing pigs should contain at least 0.9% calcium and 0.7% phosphorus. The ratio of calcium to phosphorus also is important, and best results are obtained by maintaining a ratio of approximately 1.3:1.[57] It should also be pointed out that the levels of calcium and phosphorus that result in maximum growth rate may not be adequate for maximum bone mineralization. The calcium and phosphorus requirements of gilts also appear to be slightly higher than those of barrows.[87] Therefore, it may be wise to formulate a special diet for replacement gilts that contains greater amounts of premix than that in diets fed to market hogs.

Excessive levels of a nutrient can be just as detrimental to animal performance as deficiencies, and care should be taken not to overfeed calcium and phosphorus. For example, a high level of calcium to market pigs may result in a secondary zinc deficiency and the development of parakeratosis.[50] Excess calcium in the diet can also result in poor growth and feed efficiency.[26]

Vitamin requirements

Some vitamins are present in the diet naturally and need not be supplemented. However, at least nine different vitamins are not present in adequate amounts in most cereal-based diets and require supplementation. Vitamins requiring supplementation in the diet include the fat-soluble vitamins A, D, E, and K, as well as the water-soluble vitamins biotin, niacin, pantothenic acid, riboflavin, and B_{12}. Other vitamins may be supplemented as a safety factor.

Like the trace minerals, vitamins usually are

supplemented in the diet relative to the levels of calcium and phosphorus. Therefore, if the level of these two minerals is adequate, then an individual vitamin deficiency is unlikely in fresh feeds. However, vitamin supplements deteriorate with time; factors such as moisture, light, heat, air, and the presence of various oxidizing and reducing agents all speed vitamin degradation. The presence of trace minerals also tends to increase the breakdown of some vitamins. Because vitamins are relatively unstable, premix management should be closely scrutinized. Premixes should be stored in cool, dry, dark places to improve their shelf life. In addition, supplies should be rotated routinely, and the maximum storage time should not exceed 90 days.

SELECTION OF DIETARY INGREDIENTS
Factors to consider in selecting ingredients

Once a pig reaches 20-kg (45-lb) body weight, the selection of dietary ingredients is not as critical as during the weaning period. A wide variety of ingredients can be successfully utilized. It is most important that the ingredients selected meet the nutritional requirements of the pig at as low a cost as possible. However, consideration also should be given to the nutrient composition of the feedstuff, whether it contains any toxins, its cost relative to other ingredients, how readily available it is, and whether the feedstuff is palatable.

Although the selection of dietary ingredients is not as critical during the growing-finishing period, the choice of ingredients can greatly affect the overall performance of the pig. There is little point in providing all the necessary nutrients in the correct amounts and proportions if the pigs will not consume the diet. The use of unpalatable feedstuffs should be avoided. Of particular concern is the use of cereal grains contaminated with weed seeds such as mustard or stinkweed. These weed seeds should be removed or serious palatability problems result.

Toxic substances also must be avoided in compounding diets. This means that producers should be aware of the various injurious substances that feed ingredients may contain. Glucosinolates in rapeseed, gossypol in cottonseed meal, hemagglutinins in faba beans, and ergot in rye and triticale are examples of some of these substances. These feedstuffs can be used in swine diets, but there are limitations to the amounts that can be incorporated.

Mycotoxins also must be avoided. Mycotoxins are a diverse class of chemical compounds that occur in grain as a result of mold growth. Aflatoxin is perhaps the best known of all mycotoxins, and its high degree of toxicity is well established.[79] Other important toxins include zearalenone, vomitoxin, and ochratoxin. Toxins can cause feed refusal, impaired growth rate, and a reduction in feed efficiency,[76] and at high levels an increase in mortality has been reported. Susceptibility to infection is increased. However, not all moldy feed contains toxins. Moreover, not all feeds containing mycotoxins are moldy. If there is doubt about the quality of a feed, it should be tested.

Recommended inclusion levels for commonly used feedstuffs

In order to maximize growth rate and feed efficiency, it is best to feed a combination of ingredients to growing or finishing pigs. Most diets formulated for growing or finishing pigs contain an energy source (cereal grain), a protein source, and a supplementary source of vitamins and minerals (premix or concentrate). It is beyond the scope of this chapter to provide a detailed analysis of the nutritional value of each and every feedstuff available for use in swine production. However, some guidelines for those most commonly utilized are given in Table 44-2. For more information, a text on swine feeding should be consulted.[63,71]

Sample diets

The diets shown in Table 44-3 are typical of those that might be used in areas where barley and wheat predominate, whereas those shown in Table 44-4 are more typical for areas where corn is the predominant grain. These formulations are presented to provide examples only of the types of diets that may be fed to pigs. However, they should not be used as recipes. A feed test must be conducted on all ingredients before the diet is formulated; variations in nutrient content cause the diet formulation to change.

Table 44-2 *Recommended inclusion levels for commonly used feedstuffs*

Feedstuffs	Maximum level	Potential problems
Energy sources		
Corn	All of cereal portion	Molds and mycotoxins
Wheat	All of cereal portion	Wide lysine-energy ratio
Barley	All of cereal portion	Low energy
Rye	25% of diet	Ergot, pentosans
Triticale	50% of diet	Ergot, pentosans
Oats	25% of diet	Low energy
Protein sources		
Soybean meal	All of supplementary protein	Sulfur-containing amino acids
Canola meal	50% of supplementary protein	Glucosinolates, high fiber
Field peas	All of supplementary protein	Trypsin inhibitor
Faba beans	20% of diet	Trypsin inhibitor, tannins
Lentils	All of supplementary protein	
Meat meal	5% of diet	Poor amino acid balance
Sunflower seed	10% of diet	Unsaturated fat, low lysine
Sunflower meal	50% of supplementary protein	Low lysine
Blood meal	5% of diet	Poor amino acid balance
Feather meal	3% of diet	Poor amino acid availability

DIET PREPARATION
Importance of feed testing

Results from various feed-testing laboratories indicate that a fairly large proportion of home-mixed swine diets do not meet the nutrient requirements of the pigs for which they were formulated.[80] Fewer than 25% of swine producers use some form of feed-testing service.[80] Uncertainty as to the nutrient content of the ingredients used in formulating swine diets is a major factor leading to many of the deficiencies observed in home-mixed swine diets.

The nutrient content of cereal grains is dependent on a number of factors including:
- The variety of the cereal
- The stage of maturity at harvest
- The climatic conditions
- The fertilization practices
- The location where grown

There is a tremendous amount of variation in the nutrient content of cereal grains. For example, locally grown barley may vary in protein content from 9% to 15%.[54] This amount of variation has a significant impact on the nutrient content of diets.

The problem of nutrient variation is not restricted to cereal grains but also occurs in various protein supplements. Soybean meal guaranteed at 44% protein has been shown to range from 40% to 48% protein.

Unless the nutrient composition of a feed ingredient is known, formulation can be done only by guesswork. Producers therefore should be advised to feed test. Use of a feed-testing laboratory helps to maximize the use of feed resources and minimize feed costs.

Importance of proper mixing

Many producers do not weigh feed ingredients into the mixer. They know the volume that the mixer holds and add supplement by the bag. This practice can lead to nutritional deficiencies as the bushel weight of cereal grains can vary tremendously and alter the level of nutrients going into the mix. Therefore, all ingredients should be weighed into the feed mixer or the bushel weight guaranteed before inclusion. If feed is measured by volume, then the mill should be calibrated regularly.

Mixing time is important in obtaining a good

Table 44-3 Diets for market hogs (20 to 100 kg or 45 to 220 lb) based on barley and wheat

	Diets									
	1	2	3	4	5	6	7	8	9	10
Ingredients										
Barley	830†	835	735	415	555	678	380	415	415	678
Wheat	—	—	—	400	225	—	400	—	—	—
Peas	—	—	—	—	—	200	—	—	—	—
Lentils	—	—	—	—	—	—	—	—	—	200
Screenings (#1 feed)	—	—	—	—	—	—	—	—	415	—
Screenings (#1 wheat)	—	—	—	—	—	—	—	400	—	—
Soybean meal (44%)	—	130	—	—	—	—	—	—	—	—
Soybean meal (47%)	135	—	—	150	85	12	160	150	135	12
Canola meal	—	—	230	—	100	75	—	—	—	75
Vegetable fat	—	—	—	—	—	—	25	—	—	—
Premix	35	35	35	35	35	35	35	35	35	35
Nutrients (as-fed basis)										
DE (kcal/kg)	3085	3030	2960	3200	3100	3050	3330	3200	3085	3050
Crude protein (%)	15.0	14.5	16.4	16.7	16.6	15.0	16.8	16.7	15.0	15.0
Lysine (%)	0.74	0.74	0.73	0.78	0.80	0.79	0.80	0.78	0.74	0.79
TSAA (%)*	0.51	0.54	0.54	0.59	0.56	0.45	0.59	0.59	0.51	0.45
Tryptophan (%)	0.22	0.20	0.21	0.23	0.22	0.19	0.23	0.23	0.22	0.19
Threonine (%)	0.57	0.57	0.66	0.59	0.62	0.58	0.60	0.59	0.57	0.58
Calcium (%)	0.90	0.90	0.90	0.90	0.90	0.90	0.90	0.90	0.90	0.90
Phosphorus (%)	0.75	0.75	0.75	0.75	0.75	0.75	0.75	0.75	0.75	0.75

*Total sulfur amino acids.
†If values are used as kg makes 1 metric tonne of feed, if used as lb makes 1000 lb of feed.
From Patience JP and Thacker PA: Swine nutrition guide, 1989.

mix. Overmixing can result in a poor-quality ration. Mixing feed for too long a period causes denser materials, such as minerals, vitamins, and antibiotics, to shift to the center of the mixer. This may result in an unbalanced ration when taken out of the mixer. If a vertical mixer is being used, 15 to 30 minutes of mixing after the addition of the last ingredient should be sufficient. With a horizontal mixer, 3 to 5 minutes should be allowed.

The following mixing procedure will help to obtain the most uniform distribution of the feed ingredients:

Add one half of the cereal grain.
Add the protein supplement.
Add the vitamin-mineral premix.
Add the remainder of the grain.

The commonly used vertical and horizontal mixers both have nonmix zones located at their top and bottom. When adding premixes to the mixer, the nonmix zone should be avoided by adding in the middle of the mixer. Minimum size of premix is also an area that is continually being debated. The smaller the premix package, the greater the difficulty in acquiring adequate ingredient dispersion. However, with adequate mixing time, a 10 lb (4.5 kg) premix can be fairly well dispersed.

Table 44-4 *Diets for market hogs (20 to 100 kg or 45 to 220 lb) based on corn*

	Diets				
	1	2	3	4	5
Ingredients					
Corn	760†	765	450	665	740
Wheat shorts	—	—	400	—	—
Soybean meal (47%)	205	—	115	—	125
Soybean meal (44%)	—	200	—	—	—
Soybeans (full fat)	—	—	—	300	—
Canola meal	—	—	—	—	100
Vegetable fat	—	—	—	—	—
Premix	35	35	35	35	35
Nutrients (as-fed basis)					
DE (kcal/kg)	3435	3300	3235	3560	3300
Crude protein (%)	15.90	16.20	15.70	16.6	16.7
Lysine (%)	0.81	0.78	0.78	0.84	0.78
TSAA (%)*	0.61	0.57	0.62	0.57	0.66
Tryptophan (%)	0.17	0.17	0.19	0.19	0.15
Threonine (%)	0.68	0.59	0.63	0.66	0.63
Calcium (%)	0.90	0.90	0.90	0.90	0.90
Phosphorus (%)	0.75	0.75	0.75	0.75	0.75

*Total sulfur amino acids.
†If values are used as kg makes 1 metric tonne of feed, used as lb makes 1000 lb of feed.
From Patience JP and Thacker PA: Swine nutrition guide, 1989.

PROCESSING FEED

The efficiency with which feedstuffs are utilized can be greatly improved by the application of various feed-processing techniques.[81] Many potentially useful sources of nutrients for swine would remain unused if appropriate processing were not possible. Raw soybean contains a trypsin inhibitor that drastically reduces protein digestibility.[71] However, heating raw soybean destroys the trypsin inhibitor and produces a product of high nutritional value.

Selection of the method of processing feedstuffs for growing-finishing pigs can be as critical as the selection of the ingredients themselves. The digestibility of the nutrients within each feedstuff may be improved if the surface area is increased to allow greater exposure to digestive enzymes. However, the altered feed surface may also reduce diet acceptance. Processing may also lower the availability of various nutrients or result in the destruction of vitamins. The cost and benefits of the various processing techniques should be evaluated to determine their suitability.

Grinding

Grinding is the most common method of processing cereal grains for swine. Grinding allows easier and more uniform mixing of minerals, vitamins, proteins, and grains. It minimizes sorting of the ingredients during mixing or by the pigs when feeding. In addition, grinding increases the surface area of the grain and improves its digestibility.

Numerous experiments have been conducted to determine the optimum level of grinding for many of the common cereal grains.[39,46,66] The results of these experiments indicate little difference in feed utilization as a result of grinding wheat. However, finely ground wheat has the consistency of flour and is relatively unpalatable. In addition, an increase in the incidence of gastric ulcers has been observed when wheat is ground too fine.[65] Finely ground wheat has a tendency to go pasty and bridge (clog) in the feeder. Therefore, it is recommended that wheat should be coarsely ground when included in feeder pig diets.

Oats contain a high amount of fiber that inhibits digestion. Grinding partially breaks down the fiber in the hull and allows the digestive enzymes to come in contact with more of the soluble cellular contents. A fine grind is recommended for oats.[46]

A response intermediate to that obtained with wheat and oats is obtained when barley is ground.[46] Because a fine grind tends to result in greater wear on the blades of the grinder and mill output is increased when using a coarse grind, it would appear best to utilize a medium screen when grinding barley. Suggested screen sizes for grinding cereal grains are as follows:

Oats	2 or 3 mm (0.08 to 0.11 inch)
Barley and Corn	4 or 5 mm (0.16 to 0.2 inch)
Wheat	6 mm (0.24 inch)

If numerous whole kernels can be seen in the processed ration, then the condition of the screen and hammers should be checked or the screen size reduced.

Pelleting and crumbling

Pelleting and crumbling are commonly used feed-processing techniques for swine.[12,28,35] Feed wastage is reduced by pelleting or crumbling; dust is not as great a problem, and pigs cannot sort out feed ingredients in the feeder. Settling out during storage does not occur, and less space is needed for storage. A survey of 30 experiments that compared meal (ground grain) to pellets showed an improvement of 6.2% in daily gain and 4.9% in feed conversion efficiency with pelleted diets.[52] (See Chapter 43, p. 477)

The major disadvantage of pelleting is the increased cost. The pelleting process requires energy, and therefore is relatively expensive. The potential benefits of pelleting may be offset by the additional cost. In economic terms, the feeding value of pellets and meal may be equal. Pelleting should be considered when feed prices are high and even a low level of feed wastage is undesirable.

FEEDING MANAGEMENT OF NEWLY PURCHASED FEEDER PIGS

Newly purchased feeder pigs are subject to the stresses of fatigue, hunger, thirst, temperature changes, ration changes, different surroundings, and social antagonism. The aim of management should be to reduce these stresses and assist the pig to recover as quickly as possible. Pens should be cleaned and disinfected prior to arrival. This practice allows the pigs to gradually become accustomed to the bacteria in the new environment. A dirty pen is a serious hazard when the pigs are already stressed from shipping. It is also important to keep the pigs as clean as possible. Feeding and watering devices should be easily accessible. One water nipple should be provided for every 15 to 20 pigs and one feeder hole for every four pigs.

A warm welcome should be provided with the environmental temperature maintained between 21° C to 24° C (70° F to 75° F) for the first few days following arrival. Pigs should be provided with at least 0.4 m² (4 square feet) of floor space per head and sorted by size with no more than 20 pigs in a pen. This reduces stress and makes observation of the pigs easier.

Restricted feeding is recommended for the first few days after arrival as the pigs may have been off feed for several hours prior to arrival. Overconsumption at this time may lead to scours. Floor feeding encourages proper dunging habits. Water medication is recommended as stressed pigs may go off feed but are more likely to consume water.

FEEDING MANAGEMENT FOR OPTIMAL PRODUCTIVITY

Benefits of separate growing and finishing diets

There are large differences in nutrient requirements between growing and finishing pigs.[57] Growing pigs are depositing less fat and have a lower gut capacity in comparison with older pigs. Therefore, a diet with a narrow ratio between energy and the essential amino acids is required. Finished pigs are depositing higher levels of body fat as they approach maturity and have a greater gut capacity. Therefore, a wider ratio between energy and amino acids can be utilized.

When only one diet is fed during the entire growing-finishing period, it inevitably means that the producer is either underfeeding or overfeeding during some stage of the production cycle and causing either a reduction in performance or an increase in costs. It is recommended that separate diets be fed during the growing and finishing periods. The changeover in diet should occur when the pigs reach a weight of approximately 50 kg (110 lb). Table 44-5 provides recommendations regarding nutrient levels required during each period.

Feeding methods

Two methods of feeding are commonly employed in commercial swine production: *Full feeding*, in which a self-feeder is used, and *restricted* or *limited feeding*, in which feed is fed once or several times a day by hand or automated machinery. Most experiments indicate that floor-fed pigs grow more slowly and with a poorer feed efficiency than pigs fed ad libitum.

The poorer performance of the floor-fed pigs generally is attributed to some degree of feed restriction. It is impossible to full-feed pigs on the floor without a great deal of the feed going down the slats and being wasted. When feed intake is restricted, slower growth rate and longer days to market result. However, some of the reduction in performance may be compensated for by a reduction in carcass backfat that may improve grades.

Several manufacturers of drop-feeding systems have claimed that dropping feed 9 to 12 times daily, in small amounts, reduces the feed wastage that inevitably occurs from dropping larger amounts of feed two or three times daily. This reduction is based on the "appearance of little waste." Although 10% of a large amount dropped two or three times may be easily visible, 10% waste of a small amount dropped 9 to 12 times may be difficult to observe. However, 10% is still wasted.

When choosing self-feeders, great care should be taken to choose a model that minimizes waste. A U.S. study compared six makes of self-feeders and showed great variation in wastage from model to model. Waste for the test period ranged from 1% to 20% of the feed offered. Generally feeders with adjustable openings and mechanical agitators were superior in preventing wastage. Lips over self-feeder chambers or cups also helped to reduce feed wastage by forcing pigs to keep their heads in the chamber so that any feed that falls from their mouths falls back into the trough. Recently, a wet-dry feeder has been developed that appears to be useful in preventing wastage.

Table 44-5 *Nutrient requirements of growing-finishing pigs*

Nutrient (90% DM basis)	Growing	Finishing
DE (kcal/kg)	3400	3400
Crude protein (%)	16.0	14.0
Calcium (%)	0.90	0.80
Phosphorus (%)	0.70	0.60
Lysine (%)	0.80	0.70

Feeding barrows and gilts separately

It is well documented that barrows grow faster than gilts but have poorer carcass quality because of their tendency to fatten earlier.[49] As a consequence, it has been suggested that there may be benefits from feeding barrows and gilts separately. Housing barrows and gilts in separate pens allows a producer the opportunity to restrict the feed intake of barrows to overcome problems associated with increased carcass fat. The system should be implemented only in situations where carcass grades are routinely poor, and the degree of restriction should never exceed 85% of ad libitum.

There is evidence that gilts respond to higher levels of lysine than castrates.[5] Barrows have been shown to achieve their maximum growth rate on about 10% less lysine than gilts. Thus, feeding the same diet to barrows and gilts either wastes protein on the castrates or robs the gilts of needed nutrients. Unfortunately, very few barns have the capability to feed separate diets during both the growing and finishing periods, and most producers are not in a position to take advantage of this phenomenon.

Use of medication

Antibiotics have been widely used as feed additives in swine rations for the past 30 years. Their ability to increase growth rate, improve feed utilization, and reduce mortality from clinical disease is well documented.[7,30,94] The widespread use of antibiotics is one of the major reasons for the successful development of highly intensive confinement systems.

The major disadvantage of using antibiotics is cost, and it must be justified by improvements in growth or health. In most cases, changes in management or changes in quality of stock are a less expensive means to achieve the same end result. Antibiotics should be used only when an economic response can be obtained and other methods of disease control are inappropriate.

As the pig becomes older, it develops greater immunological protection and is better able to cope with disease organisms. As a result, older pigs do not respond to antibiotics to the same extent as younger animals. Most experiments on the effects of antibiotic inclusion during the growing-finishing period have observed little or no response. If barn conditions are particularly dirty or if disease levels are elevated, then it may be cost-effective to add antibiotics to the ration. In most instances, however, it would be better to improve the health status of the herd and institute more hygienic management practices than to rely on antibiotics during the growing-finishing period.

Reducing feed wastage

Reliable authorities have estimated that in the majority of swine operations somewhere between 5% and 20% of the feed is wasted. Assuming that it takes 337 kg (750 lb) of feed to get a pig to market weight, then even at the conservative estimate this wastage is adding approximately $3 to the cost of producing a market hog. Most of this waste could be avoided by good management.

Self-feeder adjustment can help to reduce feed wastage. With properly adjusted feeders, feed wastage can be as low as 2%. It is important to keep the level of feed in the feeding chamber low. Pigs should have to work feed down into the chamber. However, feeders that are too tight result in limited feeding, which increases the time required to get pigs to market weight. As the pigs grow, their daily feed requirements increase and the rate of flow must increase to meet this need. Feeder plate openings between 1.8 and 3.1 cm (0.75 and 1.25 inches) have been shown to reduce wastage and improve feed efficiency.[90]

Feed wastage also has been reduced by allowing pigs access to feed for only a portion of the day. This can be accomplished by removing the feeder or by putting a lid on the self-feeder and allowing access to the feed only during certain portions of the day. The improvement in feed efficiency seems to arise from the fact that pigs consume feed that is spilled during the remainder of the day.

Unpalatable feed ingredients can cause increased feed wastage. Pigs simply sort out and eat what they like and push the unappetizing ingredient out of the feeder. Unpalatable feedstuffs should be avoided.

Worn screens in grinders also can lead to wastage, as whole kernels of grain are poorly utilized by the pig. Feces should be inspected to see if traces of whole grain are being passed. If whole kernels of grain are found, then grinder screens should be inspected for holes and replaced.

The amount of feed consumed by mice and other rodents constitutes only a limited economic loss to producers. However, rodents contaminate considerable amounts of feed with their urine and feces. The amount of contaminated feed consumed by swine has been shown to be significantly lower than when fed uncontaminated feed.[89] Because rodents represent a vector of disease transmission, every effort should be made to reduce their numbers.

Feed accounts for a high percentage of the total cost of producing a market hog. It is worthwhile to take time to go through any operation and identify where feed may be wasted and take steps to prevent this wastage.

THE PROPER ENVIRONMENT IN THE FEEDER BARN

Effect of temperature

Pigs are very poorly equipped to deal with extremes of temperature. They possess few sweat glands for cooling in hot weather and have little hair for protection against the cold. Therefore, swine barns must be well maintained, with good insulation and control over the ventilation system. This is necessary to maintain the zone of thermoneutrality, within which pigs are comfortable. Above or below this temperature zone, a pig is stressed and feed efficiency declines.

Kansas State University workers[61] studied the effects of environmental temperature on weight gain and feed utilization during the growing-finishing period. The results of their experiments would indicate that 20° C (68° F) is the optimal temperature for growing-finishing pigs. As temperature declines, the pig is forced to eat more, and a larger proportion of the feed energy is diverted to heat production in place of meat production. If the temperature goes too high, the pig reduces its intake.

Whereas air temperature is the prime factor affecting the rate of heat loss, other components of the environment and characteristics of the pig itself make a significant contribution to the thermal comfort of the pig. Heat loss can be increased by air movement, damp or dirty floors, lack of insulation, and consumption of large volumes of cold liquid. Heat loss can be decreased by provision of bedding, increased animal density, increased subcutaneous fat, radiant heating, and solid partitions to modify or impede air movement. It is important that pigs be monitored to determine their reaction to a given temperature in order to ensure that they are comfortable.

Effect of crowding

Overcrowding of growing pigs can impair performance significantly. Depending on the severity, crowded conditions can decrease feed intake, rate of gain, and feed efficiency of growing pigs.[34-36] Optimal floor space allowances have been established for various weights of growing-finishing pigs housed in confinement (Table 44-6), and these allowances should be followed if optimal feed utilization is to be obtained.

Effect of group size

Group size also is important if feed efficiency is to be maximized.[35] This work compared the performance of pigs housed 16 or 32 to a pen during the growing stage and 8 or 16 to a pen during

Table 44-6 *Optimal space allowance for growing-finishing pigs*

Pig weight	Floor space
23 kg (50 lb)	0.18 m² (2 sq ft)
45 kg (100 lb)	0.37 m² (4 sq ft)
68 kg (150 lb)	0.55 m² (6 sq ft)
90 kg (200 lb)	0.72 m² (8 sq ft)

the finishing stage. Space allowance was similar under both systems of rearing. Performance was improved significantly in smaller groups. Tail biting and fighting also were reduced with pigs housed in the smaller groups.

NEWER CONCEPTS IN FEEDER PIG NUTRITION
Use of synthetic amino acids

Ideally, each amino acid should be present at the exact level it is needed in the pig's body. If an amino acid is deficient, it limits the utilization of other amino acids in the diet, and growth is impaired. If an amino acid surplus exists, the body has to get rid of it. This process uses energy and reduces feed efficiency.

When a diet is formulated to meet the pig's requirement for lysine, using protein supplements, a surplus of the other essential amino acids in the diet is inevitable. This occurs because dietary proteins do not contain the essential amino acids in the exact proportions required by the pig. As the cost of protein supplements increases, it becomes more economical to add the specific amino acids necessary to meet the pig's requirements, rather than feeding surplus protein.

A barley-soybean meal diet, formulated to contain 16% crude protein, contains all the essential amino acids required by the growing pig. When the protein content of this diet is reduced to 14%, all of the essential amino acids, with the exception of lysine, still are present in adequate amounts. However, less protein concentrate is used in the formulation of the diet. The requirement for lysine can be met by fortifying the diet with synthetic lysine.

A considerable amount of research has been done on the performance of pigs fed diets containing synthetic lysine versus complete proteins.[2,5,7,8] The results of these experiments indicate that the protein level in the diet can be lowered by approximately two percentage points without adversely affecting performance if synthetic lysine (0.15%) is added to the diet. Lowering the protein level beyond 2% causes other amino acids in the diet to become limiting and impairs growth. Although very small amounts of lysine usually are required in the formulation of balanced diets, synthetic lysine is very expensive. Whether the addition of synthetic lysine is an economical practice depends on the prevailing prices of protein supplements and synthetic lysine.

Caution should be exercised when adding synthetic lysine to swine diets. If too much lysine is added, it could create an amino acid imbalance that impairs the utilization of other amino acids in the diet. In addition, most producers cannot accurately mix the small quantities of lysine required to balance a ration. It is not recommended that producers formulate their own rations using synthetic lysine. However, many feed manufacturers are producing a high-lysine premix or protein supplement that can be added in larger amounts to allow producers to take advantage of the benefits of synthetic lysine. Synthetic tryptophan and threonine have recently become available at competitive prices.

Probiotics

Probiotics are a relatively new concept in the regulation of intestinal bacteria that have been widely touted as an alternative to the use of antibiotics in swine ration.[25,67,93] They are supposed to have the opposite effect to antibiotics on the intestinal microorganisms in the digestive tract. Whereas antibiotics control the microbial population in the intestine by inhibiting or destroying microorganisms, probiotics actually introduce live bacteria into the intestinal tract.[68]

Both beneficial and potentially harmful bacteria normally can be found in the digestive tract of swine. Examples of harmful bacteria are certain serotypes of *Salmonella, Escherichia coli,* and *Clostridium perfringens.* Not only can these bacteria produce specific diseases known to be detrimental to the host but also, through competition for essential nutrients, may decrease animal performance. In contrast to the effects of these disease-causing microorganisms, bacteria such as *Lactobacillus* and vitamin B complex–producing species theoretically could be beneficial to the host. By encouraging the proliferation of these bacteria in the intestinal tract, it might be possible to improve animal performance.

The ideal situation would be to have specific numbers of beneficial bacteria constantly present in the intestinal tract. However, physiological and environmental stress can create an imbalance in the intestinal flora of the intestinal tract and allow pathogenic bacteria to multiply. When this occurs, disease and poor performance may result. Probiotics increase the numbers of desirable microflora in the gut and thereby swing the balance toward a more favorable microflora.

The mode of action of probiotics has not been clearly defined. It has been suggested that probiotics increase the synthesis of lactic acid in the gastrointestinal tract of the pig.[88,91] This increased production of lactic acid is postulated to lower the pH in the intestine and thereby prevent the proliferation of harmful bacteria such as *E. coli*.[55] The decrease in the number of *E. coli* might also reduce the amount of toxic amines and ammonia produced in the gastrointestinal tract.[32] In addition, reports suggest that probiotics might produce an antibiotic-like substance[75] and also stimulate the early development of the immune system of the pig.

Research on the value of probiotics in swine diets, however, has not shown obvious benefits.[69,70] The value of adding probiotics to growing-finishing rations would appear to be questionable; see the data from Fralick and Cline[22] in Table 44-7.

Some researchers have speculated that probiotics actually might have some negative effects on pig performance during the growing-finishing phase by competing for nutrients with indigenous organisms of the digestive tract, decreasing carbohydrate utilization and increasing the intestinal transit rate of digesta.[67] Therefore, although the theoretical concept of probiotics appears promising, the evidence suggests that the search must continue for a workable alternative to antibiotics.

Table 44-7 *Performance of growing pigs fed diets containing a probiotic*

	Control	Probiotic
Average daily gain (kg)	0.71	0.70
Average daily feed (kg)	2.33	2.37
Feed conversion	3.28	3.38

From Fralick C and Cline TR: Purdue University Swine Day, 1982.

Feed flavors

The use of flavors in animal feeds has increased considerably in the past decade as more attention is being paid to ingredient and diet palatability. This increase in feed flavor usage has been paralleled by a dramatic increase in the number of commercially available products ranging from simple spices and tonics to aroma modifiers, sweeteners, flavor intensifiers, and artificial flavors.[6]

Unfortunately, feed flavors tend to be incorporated into swine feeds because of marketing appeal and consumer preference rather than proven efficacy. At present, very little is known about which specific flavors pigs find attractive, and too often flavors are chosen for inclusion in swine feeds because they are attractive to the human palate rather than to that of the pig. Even if a particular flavor has been shown to be preferred by swine in free-choice or stimulus tests, this preference does not necessarily result in improved performance.

Very few studies have demonstrated a consistent improvement in feed intake or growth rate as a result of the inclusion of feed flavors in the diet.[44] For example, one researcher compared 129 different feed flavors to determine which specific flavors were preferred by pigs.[51] Five of the flavors that were shown to be most preferred by pigs then were used in a feeding trial. However, none of the flavors significantly increased feed intake or growth rate. Because of the cost associated with the inclusion of flavors in the diet, it would seem wise to avoid them until further research is conducted and a more consistent response is obtained.

Organic acids

Organic acids are widely utilized to inhibit mold activity in stored feedstuffs as well as finished feed. By treating with organic acids, it is possible to harvest and store grains at a higher moisture content without spoiling. Propionic and acetic acids are the most commonly used acids for this purpose.

Most of the research conducted with organic acids has focused on the performance of starter pigs fed rations treated with fumaric, citric, or propionic acid.[16] Supplementation with organic acids at levels between 0.5% and 3% of the total diet has been consistently shown to improve feed efficiency without affecting growth rate.

The mechanism by which the beneficial effects of organic acid supplementation are achieved has not been determined. The improvements in performance of pigs fed acid-treated grains can not be attributed solely to the antifungal properties of these acids.[18,23,31,43] It has been suggested that the reduction in dietary pH may increase the activity of pepsin in the stomach. The reduction in dietary pH also may reduce gastric pH, resulting in greater bactericidal activity in the stomach that reduces nutrient-robbing bacterial loads in the intestinal tract. Organic acids also may act as chelating agents that increase the absorption of minerals in the intestine of the pig. Finally, it has been suggested that the reduction in pH may slow gastric emptying and allow greater time for proteolysis (digestion of protein) to occur in the stomach.

As pigs age, their ability to produce gastric acid is increased. There is little benefit, in terms of growth rate or feed efficiency, from supplementing the diets of growing pigs with organic acids.[16]

However, it may be possible to improve carcass quality through organic acid supplementation of growing pigs. Recent evidence has suggested that methylmalonyl-CoA, a breakdown product of propionic acid metabolism, inhibits some of the enzymes involved in fat synthesis.[82,83] Pigs fed high levels (3% to 9%) of propionic acid have been shown to have significantly lower levels of backfat than control pigs. If the current consumer demand for reduced carcass backfat continues, propionic acid supplementation of diets fed to market hogs may increase in the future.

Repartitioning agents

The recent discovery of synthetic agents that have a repartitioning effect on nutrient utilization in adipose tissue and skeletal muscle has caused a considerable amount of interest in the livestock industry.[3] These repartitioning agents, known as β-agonists, have the ability to stimulate the production of lean muscle while limiting the synthesis and deposition of subcutaneous and internal fat.[15,37]

The β-agonists derive their name from the way they act on individual cells in the body. Cells have receptors on their outer surfaces that bind blood-borne messengers.[77] The receptors are divided into α and β types and are very specific as to which messengers they accept. When the right one is present, however, its action on the receptor causes the whole cell to alter its metabolism.

A β-agonist is a chemical messenger that activates β-adrenergic receptors. β-Agonists affect muscle contractility, stimulate the breakdown of fat in the cell, and increase the rate at which released fatty acids are oxidized.[53] Under normal conditions, a large part of the energy obtained from the oxidation of fatty acids would be lost as heat.[78] Under the influence of β-agonists, however, more of the energy obtained from fatty acid oxidation is made available to the body for protein synthesis.[4] The end result is a decrease in the amount of fat and an increase in the amount of protein in the body.

The β-agonists that have been most widely studied are clenbuterol, ractopamine, and cimaterol. Experimental results indicate that pigs fed diets containing cimaterol gain at approximately the same rate as control pigs but consume less feed. As a consequence, pigs treated with cimaterol exhibit a trend toward improved feed conversion efficiency. In addition, there is a reduction in carcass fat and an increase in loin eye area.[14,27,56]

Unfortunately, the withdrawal of cimaterol from the diet for as short a period as 7 days has been shown to result in compensatory accumulation of fat in subcutaneous and internal depots.[37] Therefore, if a withdrawal period of 7 days or longer is legislated, it is unlikely that there would be any benefit from including β-agonists in pig diets. In addition, several experiments have indicated a greater incidence of hoof lesions when β-agonists were included in the diet.[14,37] The etiology of these hoof lesions has not been determined. More research is needed before repartitioning agents can be recommended for routine inclusion in commercial swine diets. However, recent research with ractopamine shows promise.

Enzyme supplementation

Enzymes have been discovered that can inactivate deleterious compounds such as β-glucans in barley and soluble pentosans in rye. This has rekindled interest in the use of enzyme supplementation as a means of growth promotion.

The β-glucans are water-soluble polysaccharides found in the aleuron layer and endosperm of barley kernels.[72] They consist of glucose units linked together by β-1,4 and β-1,3 linkages.[20] The β-1,3 linkages confer upon the molecule a steplike structure that interferes with hydrogen bonding between adjacent chains and results in increased water solubility of the glucan. Solutions of glucan are very viscous, which may impede enzyme attack or affect the rate at which released nutrients approach the mucosal surface for absorption. It has been suggested that β-glucans also may allow microbial populations to assimilate a greater proportion of the nutrients contained in the feed into their own system and thereby reduce the availability of these nutrients to the host.

The level of β-glucan in a barley sample can vary from 1.5% to 8%, depending on the cultivar of barley and the environmental conditions under which it was grown.[24,92] Barley grown in areas with low rainfall has higher levels of β-glucan than that grown under conditions of adequate moisture.[1] In addition, hulless varieties of barley contain higher levels of β-glucans than the hulled varieties;[21] malting varieties of barley may be lower in β-glucans than the so-called feed varieties. These variations in β-glucan content may explain some of the difference in feeding value often seen among barley cultivars.

Table 44-8 Goals for the feeder barn

	Good	Better	Best
Age at 105 kg (days)	175	160	145
Days in feeder barn	110	100	90
Average growth rate (g)	775	850	950
Feed conversion	3.3	3.0	2.7
Mortality (%)	2	0.5	0.1
Canadian carcass index	105	107	110

From Patience JF and Thacker PA: Swine nutrition guide, Prairie Swine Center, Saskatoon, 1989.

Enzymes capable of breaking down β-glucans are termed β-glucanases. Treatment of barley with these enzymes improves the nutritive value of barley for poultry.[10,91] However, there has been little research conducted to determine the effects of β-glucans on the performance of pigs. The few reports indicate that the supplementation of barley-based pig diets with β-glucanase may be beneficial.[59,60,84]

Enzyme supplementation may also have potential for improving the nutritive value of rye. At the present time, rye is not widely used in swine diets. The classic explanation for its absence relates to its potentially high ergot content. However, research with poultry indicates that high levels of soluble pentosans in rye may pose an even greater problem than ergot. These pentosans are solubilized during digestion and result in a highly viscous intestinal fluid that interferes with digestion in a manner similar to the β-glucans found in barley. One experiment indicated that the performance of pigs fed rye-based diets supplemented with an enzyme capable of breaking down soluble pentosans can equal that of pigs fed barley-based diets.[85]

CONCLUSION

Pig production continues to be an industry concerned with the profitable and efficient conversion of feed into meat. There are many factors that reduce the efficiency with which feed is used for productive purposes, including unbalanced rations, poor genetics, a high disease or parasite level, moldy grain, poor-quality water, poor environmental conditions, and poor management. Producers who keep accurate records can rapidly detect the presence of these adverse factors and take steps to overcome them.

Swine producers should set goals for their operation. The goals listed in Table 44-8 are obtainable in properly fed pigs and can be used as targets of excellence toward which to strive.

REFERENCES

1. Aastrup S: Carsberg Res Commun 44:381, 1979.
2. Acker DC, Catron DV, and Hays VW: J Anim Sci 18:1053, 1959.
3. Asato G and others: Agricul Biochem Chem 48:2883, 1984.
4. Baker PK and others: The 4th international congress on obesity, New York, 1983.
5. Batterham ES, Giles LR, and Dettman EB: Anim Prod 40:331, 1985.
6. Bradley B: Feed Management 34:52, 1983.
7. Braude RH, Wallace D, and Cunha TJ: Antibiot Chemother 3:271, 1953.
8. Braude RH and others: Br J Nutr 27:169, 1972.
9. Brody S: Bioenergetics and growth, New York, 1945, Reinhold Publishing Corporation.
10. Burnett GS: Br Poult Sci 7:55, 1966.
11. Combs GE and Coffey MT: University of Florida 27th Annual Swine Field Day, 1982.
12. Conrad JH: Feedstuffs 30(35):38, 1958.
13. Crampton EW: Sci Agricul 22:326, 1942.
14. Cromwell GL, Kemp JD, and Stahly TS: Feed Management 38:8, 1987.
15. Dalrymple RH and Ingle DL: Proceedings of the 47th annual Minnesota nutrition conference, 1986.
16. Easter RA: In Haresign W and Cole DJ: Recent advances in animal nutrition, London, 1988, Butterworth.
17. Eggum BO: Doctoral thesis, Copenhagen, Denmark, 1973, National Institute of Animal Science.
18. Falkowski JF and Aherne FX: J Anim Sci 58:935, 1984.
19. Farrell DJ: J Aust Inst Agricul Sci 34:21, 1979.
20. Fleming M and Kawakami K: Carbohydr Res 57:15, 1977.
21. Fox GL: Doctoral dissertation, Missoula, 1981, Montana State University.
22. Fralick C and Cline TR: Purdue University Swine Day, 1982.
23. Geisting DW and Easter RA: J Anim Sci 60:1288, 1985.
24. Gohl G and Thomke S: Poult Sci 55:2369, 1976.
25. Hale OM and Newton GL: J Anim Sci 48:770, 1979.
26. Hall DD, Cromwell GL, and Stahly IS: J Anim Sci 61(Suppl):319, 1985.
27. Hanrahan JP and others: In Haresign W and Cole DJ, editors: Recent advances in animal nutrition, London, 1986, Butterworth.
28. Hays VW: Proceedings of the twenty-second annual Texas nutrition conference, College Station, Texas, 1967.

29. Hays VW: Phosphorus in swine nutrition, West Des Moines, Iowa, 1976, National Feed Ingredients Association.
30. Hays VW and Muir WM: Can J Anim 59:447, 1979.
31. Henry RW, Pickard DW, and Hughes: Anim Prod 40:505, 1985.
32. Hill IR, Kenworthy R, and Porter P: Res Vet Sci 11:320, 1970.
33. Holmes CW and Close WH: In Haresign W, Swan H, and Lewis D, editors: Nutrition and the climatic environment, London, 1977, Butterworth.
34. Jensen AH: Feedstuffs 38(31):24, 1966.
35. Jensen AH: Hog Farm Management :10, 1976.
36. Reference deleted in proof.
37. Jones RW and others: J Anim Sci 61:905, 1985.
38. Jorgensen H, Sauer WC, and Thacker PA: J Anim Sci 58:926, 1984.
39. Just A: Proceedings of the 29th annual meeting of the European Society of Animal Production, 1978.
40. Just A: Pig News Information 2:401, 1981.
41. Kennelly JJ and Aherne FX: Can J Anim Sci 60:385, 1980.
42. King RH and Tavverner MR: Anim Prod 21:275, 1975.
43. Kirchgessner M and Roth FX: Pig News and Information 3:259, 1982.
44. Kornegay ET, Tinsley SE, and Bryant KL: J Anim Sci 48:99, 1979.
45. Kuryvial MS, Bowland JP, and Berg RT: Can J Anim Sci 42:23, 1962.
46. Lawrence TL: Anim Feed Sci Tech 3:179, 1978.
47. Lewis AJ: Haresign W and Cole DJ, editors: Recent advances in animal nutrition, London, 1984, Butterworth.
48. Livingstone RM: Anim Prod 32:396, 1981.
49. Lucas IM and Calder AF: J Agricul Sci (Camb) 47:287, 1956.
50. Luecke RW and others: J Anim Sci 16:3, 1957.
51. McLaughlin CL and others: J Anim Sci 56:1287, 1983.
52. Meade RJ: Hog Farm Management 3:14, 1965.
53. Mersmann HL: Proceedings of the reciprocal meat conference 32:93, 1979.
54. Mitchall KG: Feed analysis for swine, Saskatchewan agriculture swine facts, 1979.
55. Mitchell I and Kenworthy R: J Appl Bacteriology 4:163, 1976.
56. Moser RL and others: J Anim Sci 62:21, 1986.
57. National Research Council: Nutrient requirements of domestic animals, No 2: Nutrient requirements of swine, ed 9, Washington, DC, 1988, National Academy of Sciences.
58. Nelssen JL: Kansas State University swine day, 1985.
59. Newman CW and others: Nutr Rep Inter 22:833, 1980.
60. Newman CW, Eslick RF, and El-Negoumy AM: Nutr Rep Inter 28:139, 1983.
61. Nichols DA, Ames DR, and Hines RH: Kansas State University Swine Day, 1980.
62. Owen JB and Ridgeman WJ: Anim Prod 9:107, 1967.
63. Patience JP and Thacker PA: Swine nutrition guide, 1989.
64. Peo ER: Calcium in swine nutrition. West Des Moines, Iowa, 1976, National Feed Ingredients Association.
65. Perry TW: Effect of processing on the nutritive value of feeds, Washington, DC, 1973, National Academy of Science.
66. Pickett RA and others: J Anim Sci 28:37, 1969.
67. Pollman DS: Guelph pork symposium, 1985.
68. Pollman DS: Haresign W and Cold DJ, editors: Recent advances in animal nutrition, London, 1986, Butterworth.
69. Pollmann DS: Haresign W and Cole DJ, editors: Recent advances in animal nutrition, London, 1987, Butterworth.
70. Pollman DS, Danielson DM, and Peo ER: J Anim Sci 51:577, 1980.
71. Pond WG and Maner JH: Swine production nutrition, Westport, 1984, AVI Publishing Co.
72. Prentice N and Faber S: Cereal Chem 58:77, 1981.
73. Sauer WC and Thacker PA: Anim Feed Sci Tech 14:183, 1986.
74. Seerley RW, McDaniel MC, and McCambell HC: J Anim Sci 47:427, 1978.
75. Shahani KM, Uakil JR, and Kilara A: J Cultured Dairy Prod 11:14, 1976.
76. Southern LL and Clawson AJ: J Anim Sci 49:1006, 1979.
77. Stiles GL, Caron MG, and Lefkowitz RJ: Physiol Rev 64:661, 1984.
78. Stock MJ and Rothwell NJ: Biochem Soc Trans 9:525, 1981.
79. Swick RA: J Anim Sci 58:1017, 1984.
80. Thacker PA: Western Hog Journal 5(3):18, 1983.
81. Thacker PA: Western Hog Journal 5(4):22, 1984.
82. Thacker PA and Bowland JP: Can J Anim Sci 60:971, 1980.
83. Thacker PA and Bowland JP: Can J Anim Sci 61:775, 1981.
84. Thacker PA, Campbell GL, and Grootwassink JW: Nutr Rep Inter 38:91, 1988.
85. Thacker PA, Campbell GL, and Grootwassink JW: Department of animal science research reports, Saskatoon, 1988, University of Saskatchewan.
86. Thacker PA, Jorgensen H, and Sauer WC: J Anim Sci 59:409, 1984.
87. Thomas HR and Kornegay ET: J Anim Sci 52:1041, 1981.
88. Thomlinson JR: Vet Rec 109:120, 1981.
89. Timm RM and Moser BD: Nebraska swine report, 1980.
90. Tribble LG and Stansbury WF: Texas Tech University swine report, 1985.
91. White WB and others: Poult Sci 62:853, 1983.
92. Willingham HE and others: J Poult Sci 39:103, 1960.
93. Wren WB: Large Animal Veterinarian 28, 1987.
94. Zimmerman DR: J Anim Sci 62(suppl 3):6, 1986.

Appendix

Energy Requirements

Table A-1 *Predictive equations for net energy requirements in beef cattle*

Maintenance energy requirements for steers, heifers, bulls, and cows

NEm, Mcal/day = $0.077 \times$ Body Weight$^{0.75}$

Outside the thermoneutral zone (15° C to 25° C but sometimes as low as $-35°$ C) the coefficient (0.077) should be increased by 0.0007 for every 1° C change from 20° C.

Net energy requirements for fattening mature thin cows

NEg, Mcal = $6.2 \times$ Live Weight Gain, kg

Net energy requirements for fattening in beef calves

NEg, Mcal = (Coefficient $\times$ Body Weight, kg$^{0.75}$) (Live Weight Gain, kg$^{1.097}$)

Sex	Animal type	Coefficient
Steer	Medium frame	0.0557
Steer	Large frame/ Compensating medium frame	0.0493
Steer	Compensating large frame	0.0437
Bull	Medium frame	0.0493
Bull	Large frame	0.0437
Heifer	Medium frame	0.0686
Heifer	Large frame/ Compensating medium frame	0.0608

Energy required for pregnancy

	NEm requirements for pregnancy (Mcal/day)		
Month	Small calf, 30 kg	Medium calf, 40 kg	Large calf, 50 kg
7	1.24	1.55	1.94
8	2.05	2.56	3.2
9	2.76	3.45	4.31

Energy required for lactation

NEm, Mcal/kg milk = $(0.1 \times$ Fat, %) + 0.35

Note that live weight and live weight gain are "shrunk" values (overnight without feed and water); they are approximately 96% of weight measured in the early morning.

Compiled from data in National Research Council: Nutrient requirements of beef cattle, ed 6, Washington, DC, 1984, National Academy of Sciences, National Academy Press.

Table A-2 *Predictive equations for dry matter intake in beef cattle*

For breeding females

DMI (kg dry matter) = $W^{0.75}$ (0.1462 NEm $-$ 0.0517 NEm^2 $-$ 0.0074)

For growing and finishing cattle

DMI (kg dry matter) = $W^{0.75}$ (0.1493 NEm $-$ 0.0460 NEm^2 $-$ 0.0196)

Adjustments for frame size

a. No adjustments are necessary for medium-frame steer calves, large-frame heifers, and medium-frame bulls.
b. Add 10% for large-frame steer calves and medium-frame yearling steers.
c. Add 5% for large-frame bulls.
d. Reduce 10% for medium-frame heifers.

DMI, Dry matter intake; W, live weight, kg; NEm, net energy for maintenance, Mcal/kg diet dry matter.
Compiled from data in National Research Council: Nutrient requirements of beef cattle, ed 6, Washington, DC, 1984, National Academy of Sciences, National Academy Press.

Table A-3 *Suggested energy requirements and dry matter intake for dairy cows*

Maintenance energy requirement

NEl, Mcal/day = $0.080 \times W^{0.75}$

(Add 20% if heifer in first lactation or pregnancy, 10% if heifer in second lactation or pregnancy.)

Lactation

NEl, Mcal/kg Milk = $0.3512 + (0.0962 \times \%$ Fat in Milk)

Pregnancy (210 days +)

NEl, Mcal/day = $0.024 \times W^{0.75}$

Dry matter intake

Production (kg of 4% Fat Milk)	Body weight, Kg			
	500	600	700	800
10	2.4*	2.2	2.0	1.9
20	3.2	2.9	2.6	2.4
30	3.9	3.5	3.2	2.9
40	4.6	4.0	3.6	3.3
50	5.4	4.7	4.1	3.7

*Values are % of body weight.
W, Live weight, kg.
Compiled from data in National Research Council: Nutrient requirements of dairy cattle, ed 6, Washington, DC, 1989, National Academy of Sciences, National Academy Press.

Table A-4 *Energy requirements for goats*

Maintenance

ME, Mcal/day = $0.1014 \times$ Live Weight$^{0.75}$

Multiply by 1.25, 1.5, or 1.75 for light grazing activity, semiarid rangelands, or mountainous pasture, respectively.

Pregnancy

ME, Mcal/day = $0.080 \times$ Live Weight

(Add to maintenance value.)

Growth

ME, Mcal = $7.25 \times$ Live Weight Gain, kg

Lactation

ME, Mcal = $1.246 \times 4\%$ Fat Corrected Milk, kg

All weights are in kg.
Compiled from data in National Research Council: Nutrient requirements of goats, 1981, National Academy of Sciences, National Academy Press.

Table A-5 *Energy requirements for sheep*

Maintenance

NEm, Mcal/day = $0.056 \times$ Live Weight$^{0.75}$

Growth

For breeds having medium (115-kg) mature ram weights:

NEg, Mcal = $0.276 \times$ Live Weight Gain $\times$ Live Weight$^{0.75}$

Increase this amount by 7.6% for every 10 kg decrease in expected ram weight. Decrease by 7.6% for every 10 kg increase in expected ram weight.

Rams and lambs tend to be leaner than ewes and require about 82% as much energy for gain than ewes. Castrated males are also more efficient than ewes.

Pregnancy

During the last 6 weeks of pregnancy the requirements for ewes bearing single lambs increase by $0.5 \times$ maintenance and are double maintenance for ewes bearing twins.

Ewes require an additional 125% to 190% of the maintenance DE requirement during the first 6 to 8 weeks of lactation, depending on whether suckling single or twin lambs and on ewe size (smaller ewes require more additional energy).

All weights are in kg.
Compiled from data in National Research Council: Nutrient requirements of sheep, ed 6, Washington, DC, 1985, National Academy of Sciences, National Academy Press.

Table A-6 Energy requirements for swine

Maintenance for growing pigs and gestating sows

DE, Mcal/day = $0.110 \times W^{0.75}$ (The coefficient can vary from 0.096 to 0.167)

NE, Mcal/day = $0.075 \times W^{0.75}$ (The coefficient can vary from 0.071 to 0.078)

Maintenance energy requirements may be 5% to 10% higher in lactating sows.

For cold thermogenesis

Growing pigs

DE kcal/day = $(0.326 \times W) + 24 (T_c - T)$

Sows

Add 4% to the DE requirement for maintenance for each 1° C below the lower critical temperature (20° C).

Growth

DE, Mcal = $12.5 \times$ kg of fat or protein dry matter deposited

Muscle tissue is 20% to 25% protein; adipose tissue is about 85% fat.

Gestation

Net weight gain of sow (assumed to be 25% fat, 15% protein)

DE, Mcal = $5 \times$ Weight gain, kg

e.g., 1.1 Mcal/day for a 25-kg gain over a 114-day gestation plus 0.19 Mcal/day for growth of conceptus.

Lactation

DE, Mcal = $2 \times$ milk yield, kg

An additional 200% to 300% of maintenance DE may be required to support lactation.

Dry matter intake

DE intake declines by 0.017% for 1° C above the upper critical temperature in growing pigs.

T, Ambient temperature, °C; Tc, critical temperature, °C (usually about 20° C).
Compiled from data in National Research Council: Nutrient requirements of swine, ed 9, Washington, DC, 1988, National Academy of Sciences, National Academy Press.

Table A-7 Energy requirements for horses

Maintenance (200 to 600 kg horses)

DE, Mcal/day = $1.4 + (0.03 \times W)$

Breeding stallions

Additional $0.25 \times$ maintenance DE requirement

Pregnancy (9 to 11 months of term)

Additional 0.1 to 0.2 of maintenance DE requirement

Lactation (400 to 900 kg mares)
0 to 3 Months post foaling

Additional $0.03 \times W \times 0.792$

3 Months post foaling to weaning

Additional $0.02 \times W \times 0.792$

Work

Additional 0.25 to 1.0 $\times$ maintenance DE requirement

W, Body weight, kg.
Compiled from data in National Research Council: Nutrient requirements of horses, ed 5, Washington, DC, 1989, National Academy of Sciences, National Academy Press.

Beef Requirements

Table A-8 *Nutrient requirements of breeding cattle (metric)*

Weight[a] (kg)	Daily Gain[b] (kg)	Daily DM[c] (kg)	ME (Mcal)	TDN (kg)	NEm (Mcal)	NEg (Mcal)	ME (Mcal/kg)	TDN (%)	NEm (Mcal/kg)	NEg (Mcal/kg)	Daily (g)	In diet DM (%)	Daily (g)	In diet DM (%)	Daily (g)	In diet DM (%)	Vitamin A[d] Daily (1000s IU)
Pregnant yearling heifers—last third of pregnancy																	
325	0.4	7.1	14.2	3.9	8.04	NA[e]	2.00	55.2	1.15	NA[e]	591	8.4	19	0.27	14	0.20	20
325	0.6	7.3	15.7	4.3	8.04	0.77	2.15	59.3	1.29	0.72	649	8.9	23	0.32	15	0.21	20
325	0.8	7.3	17.2	4.8	8.04	1.67	2.35	64.9	1.47	0.88	697	9.5	27	0.37	16	0.22	20
350	0.4	7.5	14.8	4.1	8.38	NA	1.99	55.0	1.14	NA	616	8.3	20	0.27	15	0.21	21
350	0.6	7.7	16.5	4.6	8.38	0.81	2.14	59.1	1.28	0.71	674	8.8	24	0.32	16	0.21	22
350	0.8	7.8	18.1	5.0	8.38	1.76	2.34	64.6	1.46	0.88	720	9.3	27	0.35	17	0.22	22
375	0.4	7.8	15.5	4.3	8.71	NA	1.98	54.7	1.13	NA	641	8.2	21	0.27	15	0.19	22
375	0.6	8.1	17.2	4.8	8.71	0.86	2.13	58.8	1.27	0.70	697	8.6	25	0.31	17	0.21	23
375	0.8	8.2	19.0	5.2	8.71	1.86	2.32	64.1	1.45	0.86	743	9.1	27	0.33	18	0.22	23
400	0.4	8.2	16.1	4.5	9.04	NA	1.97	54.4	1.12	NA	664	8.1	22	0.27	16	0.20	23
400	0.6	8.5	18.0	5.0	9.04	0.90	2.12	58.6	1.26	0.69	721	8.5	25	0.30	18	0.21	24
400	0.8	8.6	19.8	5.5	9.04	1.95	2.31	63.8	1.44	0.85	764	8.9	28	0.33	18	0.20	24
425	0.4	8.6	16.8	4.6	9.36	NA	1.96	54.1	1.11	NA	687	8.0	23	0.27	17	0.20	24
425	0.6	8.9	18.7	5.2	9.36	0.94	2.11	58.3	1.25	0.69	743	8.4	26	0.30	18	0.21	25
425	0.8	9.0	20.7	5.7	9.36	2.04	2.30	63.5	1.43	0.84	786	8.8	28	0.31	19	0.21	25
450	0.4	8.9	17.3	4.8	9.67	NA	1.95	53.9	1.10	NA	710	8.0	23	0.26	18	0.20	25
450	0.6	9.2	19.4	5.4	9.67	0.98	2.10	58.0	1.25	0.68	765	8.3	26	0.29	19	0.21	26
450	0.8	9.4	21.5	5.9	9.67	2.13	2.29	63.3	1.42	0.84	807	8.6	28	0.30	20	0.21	26
Dry pregnant mature cows—middle third of pregnancy																	
350	0.0	6.8	11.9	3.3	6.23	NA	1.76	48.6	0.92	NA	478	7.1	12	0.16	12	0.18	19
400	0.0	7.5	13.1	3.6	6.89	NA	1.76	48.6	0.92	NA	525	7.0	13	0.17	13	0.17	21
450	0.0	8.2	14.3	4.0	7.52	NA	1.76	48.6	0.92	NA	570	7.0	15	0.17	15	0.18	23
500	0.0	8.8	15.5	4.3	8.14	NA	1.76	48.6	0.92	NA	614	7.0	17	0.19	17	0.19	25
550	0.0	9.5	16.7	4.6	8.75	NA	1.76	48.6	0.92	NA	657	6.9	18	0.19	18	0.19	27
600	0.0	10.1	17.8	4.9	9.33	NA	1.76	48.6	0.92	NA	698	6.9	20	0.20	20	0.20	28
650	0.0	10.7	18.9	5.2	9.91	NA	1.76	48.6	0.92	NA	739	6.9	22	0.21	22	0.21	30
Dry pregnant mature cows—last third of pregnancy																	
350	0.4	7.4	14.7	4.1	8.38	NA	1.98	54.7	1.13	NA	609	8.2	20	0.27	15	0.20	21
400	0.4	8.2	16.0	4.4	9.04	NA	1.96	54.1	1.11	NA	657	8.0	22	0.27	16	0.20	23
450	0.4	8.9	17.2	4.8	9.67	NA	1.94	53.6	1.10	NA	703	7.9	23	0.26	18	0.21	24
500	0.4	9.5	18.3	5.1	10.29	NA	1.92	53.1	1.08	NA	746	7.8	25	0.26	20	0.21	27
550	0.4	10.2	19.5	5.4	10.90	NA	1.91	52.8	1.07	NA	790	7.8	26	0.25	21	0.21	29
600	0.4	10.8	20.6	5.7	11.48	NA	1.90	52.5	1.06	NA	832	7.7	28	0.26	23	0.21	30
650	0.4	11.5	21.7	6.0	12.06	NA	1.89	52.2	1.05	NA	872	7.6	30	0.26	25	0.22	32

Beef Requirements

Two-year-old heifers nursing calves—first 3-4 months postpartum—5.0 kg milk per day

Weight												
300	0.2	6.9	16.6	4.6	9.30[f]	0.72	2.41	66.6	1.53	814[g]	11.8	26	0.38	17	0.25	27
325	0.2	7.3	17.4	4.8	9.64[f]	0.77	2.37	65.5	1.49	841[g]	11.5	27	0.37	18	0.25	28
350	0.2	7.8	18.1	5.0	9.98[f]	0.81	2.34	64.6	1.46	866[g]	11.2	27	0.35	19	0.24	30
375	0.2	8.2	18.9	5.2	10.31[f]	0.86	2.31	63.8	1.44	892[g]	10.9	28	0.34	19	0.23	32
400	0.2	8.6	19.7	5.4	10.64[f]	0.90	2.29	63.3	1.42	916[g]	10.7	28	0.33	20	0.23	34
425	0.2	9.0	20.4	5.6	10.96[f]	0.94	2.27	62.7	1.40	939[g]	10.5	29	0.32	21	0.23	35
450	0.2	9.4	21.1	5.8	11.27[f]	0.98	2.25	62.2	1.38	963[g]	10.3	29	0.31	22	0.23	37

Cows nursing calves—average milking ability—first 3-4 months postpartum—5.0 kg milk per day

350	0.0	7.7	16.6	4.6	9.98[f]	NA	2.15	59.4	1.29	814[g]	10.6	23	0.30	18	0.23	30
400	0.0	8.5	17.9	4.9	10.64[f]	NA	2.11	58.3	1.25	864[g]	10.2	25	0.29	19	0.22	33
450	0.0	9.2	19.1	5.3	11.27[f]	NA	2.08	57.5	1.23	911[g]	9.9	26	0.28	21	0.23	36
500	0.0	9.9	20.3	5.6	11.89[f]	NA	2.05	56.6	1.20	957[g]	9.7	28	0.28	22	0.22	39
550	0.0	10.6	21.5	5.9	12.50[f]	NA	2.03	56.1	1.18	1001[g]	9.5	29	0.27	24	0.23	41
600	0.0	11.2	22.6	6.2	13.08[f]	NA	2.01	55.5	1.16	1044[g]	9.3	31	0.28	26	0.23	44
650	0.0	11.9	23.9	6.6	13.66[f]	NA	2.00	55.3	1.15	1086[g]	9.1	33	0.28	27	0.23	46

Cows nursing calves—superior milking ability—first 3-4 months postpartum—10.0 kg milk per day

350	0.0	6.2	18.5	5.1	13.73[f]	NA	3.00	82.9	2.03	1009[g]	16.4	36	0.58	24	0.39	24
400	0.0	7.6	21.4	5.9	14.39[f]	NA	2.80	77.4	1.86	1099[g]	14.4	37	0.49	25	0.33	30
450	0.0	9.1	23.2	6.4	15.02[f]	NA	2.56	70.7	1.66	1186[g]	13.1	39	0.43	26	0.29	35
500	0.0	10.0	24.6	6.8	15.64[f]	NA	2.45	67.7	1.56	1246[g]	12.4	40	0.40	28	0.28	39
550	0.0	10.9	25.8	7.1	16.25[f]	NA	2.38	65.8	1.50	1299[g]	12.0	42	0.39	30	0.27	42
600	0.0	11.6	27.0	7.5	16.83[f]	NA	2.32	64.1	1.45	1348[g]	11.6	43	0.37	31	0.27	45
650	0.0	12.4	28.2	7.8	17.41[f]	NA	2.28	63.0	1.41	1394[g]	11.3	45	0.36	33	0.26	48

[a] Average weight for a feeding period.
[b] Approximately 0.4 ± 0.1 kg of weight gain per day over the last third of pregnancy is accounted for by the products of conception. Daily 2.15 Mcal of NEm and 55 g of protein are provided for this requirement for a calf with a birth weight of 36 kg.
[c] Dry matter consumption should vary, depending on the energy concentration of the diet and environmental conditions. These intakes are based on the energy concentration shown in the table and assuming a thermoneutral environment without snow or mud conditions. If the energy concentrations of the diet to be fed exceed the tabular value, limit feeding may be required.
[d] Vitamin A requirements per kilogram of diet are 2800 IU for pregnant heifers and cows and 3900 IU for lactating cows and breeding bulls.
[e] Not applicable.
[f] Includes 0.75 Mcal NEm per kg of milk produced.
[g] Includes 33.5 g protein per kg of milk produced.

Adapted from National Research Council: Nutrient requirements of beef cattle, ed 6, Washington, DC, 1984, National Academy of Sciences, National Academy Press.

Continued.

Table A-8 Nutrient requirements of breeding cattle (metric)—cont'd

Weight[a] (kg)	Daily Gain[b] (kg)	Daily DM[c] (kg)	Energy Daily				Energy In diet DM				Total protein		Calcium		Phosphorus		Vitamin A[d]
			ME (Mcal)	TDN (kg)	NEm (Mcal)	NEg (Mcal)	ME (Mcal/kg)	TDN (%)	NEm (Mcal/kg)	NEg (Mcal/kg)	Daily (g)	In diet DM (%)	Daily (g)	In diet DM (%)	Daily (g)	In diet DM (%)	Daily (1000s IU)

Bulls, maintenance and regaining body condition

650	0.4	12.3	24.3	6.7	9.91	2.06	1.98	54.8	1.13	0.57	904	7.4	25	0.20	23	0.19	48
650	0.6	12.6	26.7	7.4	9.91	3.21	2.11	58.4	1.25	0.69	957	7.6	27	0.21	24	0.19	49
650	0.8	12.8	28.7	7.9	9.91	4.40	2.24	62.0	1.37	0.79	998	7.8	29	0.23	25	0.20	50
700	0.4	13.0	25.7	7.1	10.48	2.18	1.98	54.8	1.13	0.57	942	7.3	26	0.20	25	0.20	51
700	0.6	13.4	28.2	7.8	10.48	3.40	2.11	58.4	1.25	0.69	994	7.4	29	0.22	26	0.20	52
700	0.8	13.5	30.3	8.4	10.48	4.66	2.24	62.0	1.37	0.79	1032	7.6	30	0.22	26	0.19	53
800	0.0	12.9	22.6	6.3	11.58	NA	1.75	48.4	0.91	NA	882	6.8	27	0.21	27	0.21	50
800	0.2	13.7	25.5	7.1	11.58	1.12	1.86	51.5	1.02	0.47	956	7.0	27	0.20	27	0.20	53
900	0.0	14.1	24.7	6.8	12.65	NA	1.75	48.4	0.91	NA	958	6.8	30	0.21	30	0.21	55
900	0.2	15.0	27.9	7.7	12.65	1.23	1.86	51.5	1.02	0.47	1031	6.9	31	0.21	31	0.21	58
1000	0.0	15.3	26.8	7.4	13.69	NA	1.75	48.4	0.91	NA	1032	6.8	33	0.22	33	0.22	60

[a]Average weight for a feeding period.
[b]Approximately 0.4 ± 0.1 kg of weight gain per day over the last third of pregnancy is accounted for by the products of conception. Daily 2.15 Mcal of NEm and 55 g of protein are provided for this requirement for a calf with a birth weight of 36 kg.
[c]Dry matter consumption should vary, depending on the energy concentration of the diet and environmental conditions. These intakes are based on the energy concentration shown in the table and assuming a thermoneutral environment without snow or mud conditions. If the energy concentrations of the diet to be fed exceed the tabular value, limit feeding may be required.
[d]Vitamin A requirements per kilogram of diet are 2800 IU for pregnant heifers and cows and 3900 IU for lactating cows and breeding bulls.
[e]Not applicable.
[f]Includes 0.75 Mcal NEm per kg of milk produced.
[g]Includes 33.5 g protein per kg of milk produced.
Adapted from National Research Council: Nutrient requirements of beef cattle, ed 6, Washington, DC, 1984, National Academy of Sciences, National Academy Press.

Table A-9 Net energy requirements of growing and finishing beef cattle (Mcal/day)*

Body weight, kg: NEm required:	150 3.30	200 4.10	250 4.84	300 5.55	350 6.24	400 6.89	450 7.52	500 8.14	550 8.75	600 9.33
Daily gain, kg					NEg required					
Medium-frame steer calves										
0.2	0.41	0.50	0.60	0.69	0.77	0.85	0.93	1.01	1.08	
0.4	0.87	1.08	1.28	1.47	1.65	1.82	1.99	2.16	2.32	
0.6	1.36	1.69	2.00	2.29	2.57	2.84	3.11	3.36	3.61	
0.8	1.87	2.32	2.74	3.14	3.53	3.90	4.26	4.61	4.95	
1.0	2.39	2.96	3.50	4.02	4.51	4.98	5.44	5.89	6.23	
1.2	2.91	3.62	4.28	4.90	5.50	6.69	6.65	7.19	7.73	
Large-frame steers, compensating medium-frame yearling steers and medium-frame bulls										
0.2	0.36	0.45	0.53	0.61	0.68	0.75	0.82	0.89	0.96	1.02
0.4	0.77	0.96	1.13	1.30	1.46	1.61	1.76	1.91	2.05	2.19
0.6	1.21	1.50	1.77	2.03	2.28	2.52	2.75	2.98	3.20	3.41
0.8	1.65	2.06	2.43	2.78	3.12	3.45	3.77	4.08	4.38	4.68
1.0	2.11	2.62	3.10	3.55	3.99	4.41	4.81	5.21	5.60	5.98
1.2	2.58	3.20	3.78	4.34	4.87	5.38	5.88	6.37	6.84	7.30
1.4	3.06	3.79	4.48	5.14	5.77	6.38	6.97	7.54	8.10	8.64
1.6	3.53	4.39	5.19	5.95	6.68	7.38	8.07	8.73	9.38	10.01
Large-frame bull calves and compensating large-frame yearling steers										
0.2	0.32	0.40	0.47	0.54	0.60	0.67	0.73	0.79	0.85	0.91
0.4	0.69	0.85	1.01	1.15	1.29	1.43	1.56	1.69	1.82	1.94
0.6	1.07	1.33	1.57	1.80	2.02	2.23	2.44	2.64	2.83	3.02
0.8	1.47	1.82	2.15	2.47	2.77	3.06	3.34	3.62	3.88	4.15
1.0	1.87	2.32	2.75	3.15	3.54	3.91	4.27	4.62	4.96	5.30
1.2	2.29	2.84	3.36	3.85	4.32	4.77	5.21	5.64	6.06	6.47
1.4	2.71	3.36	3.97	4.56	5.11	5.65	6.18	6.68	7.18	7.66
1.6	3.14	3.89	4.60	5.28	5.92	6.55	7.15	7.74	8.31	8.87
1.8	3.56	4.43	5.23	6.00	6.74	7.45	8.13	8.80	9.46	10.10
Medium-frame heifer calves										
0.2	0.49	0.60	0.71	0.82	0.92	1.01	1.11	1.20	1.29	
0.4	1.05	1.31	1.55	1.77	1.99	2.20	2.40	2.60	2.79	
0.6	1.66	2.06	2.44	2.79	3.13	3.46	3.78	4.10	4.40	
0.8	2.29	2.84	3.36	3.85	4.32	4.78	5.22	5.65	6.07	
1.0	2.94	3.65	4.31	4.94	5.55	6.14	6.70	7.25	7.79	
Large-frame heifer calves and compensating medium-frame yearling heifers										
0.2	0.43	0.53	0.63	0.72	0.81	0.90	0.98	1.06	1.14	1.21
0.4	0.93	1.16	1.37	1.57	1.76	1.95	2.13	2.31	2.47	2.64
0.6	1.47	1.83	2.16	2.47	2.78	3.07	3.35	3.63	3.90	4.16
0.8	2.03	2.62	2.98	3.41	3.83	4.24	4.63	5.01	5.38	5.74
1.0	2.61	3.23	3.82	4.38	4.92	5.44	5.94	6.43	6.91	7.37
1.2	3.19	3.97	4.69	5.37	5.03	6.67	7.28	7.88	8.47	9.03

*Shrunk live weight basis.
Reproduced with permission from National Research Council: Nutrient requirements of beef cattle, ed 6, Washington, DC, 1984, National Academy of Sciences, National Academy Press.

Table A-10 *Protein requirements of growing and finishing cattle (g/day)**

Daily gain, kg	Body weight, kg									
	150	200	250	300	350	400	450	500	550	600
Medium-frame steer calves										
0.2	343	399	450	499	545	590	633	675	715	
0.4	428	482	532	580	625	668	710	751	790	
0.6	503	554	601	646	688	728	767	805	842	
0.8	575	621	664	704	743	780	815	849	883	
1.0	642	682	720	755	789	821	852	882	911	
1.2	702	735	766	794	822	848	873	897	921	
Large-frame steer calves and compensating medium-frame yearling steers										
0.2	361	421	476	529	579	627	673	719	762	805
0.4	441	499	552	603	651	697	742	785	827	867
0.6	522	576	628	676	722	766	809	850	890	930
0.8	598	650	698	743	786	828	867	906	944	980
1.0	671	718	762	804	843	881	918	953	988	1021
1.2	740	782	822	859	895	929	961	993	1023	1053
1.4	806	842	877	908	938	967	995	1022	1048	1073
1.6	863	892	919	943	967	989	1011	1031	1052	1071
Medium-frame bulls										
0.2	345	401	454	503	550	595	638	680	721	761
0.4	430	485	536	584	629	673	716	757	797	835
0.6	509	561	609	655	698	740	780	819	856	893
0.8	583	632	677	719	759	798	835	871	906	940
1.0	655	698	739	777	813	849	881	914	945	976
1.2	722	760	795	828	860	890	919	947	974	1001
1.4	782	813	841	868	893	917	941	963	985	1006
Large-frame bull calves and compensating large-frame yearling steers										
0.2	355	414	468	519	568	615	661	705	747	789
0.4	438	494	547	597	644	689	733	776	817	857
0.6	519	574	624	672	718	761	803	844	884	923
0.8	597	649	697	741	795	826	866	905	942	979
1.0	673	721	765	807	847	885	922	958	994	1027
1.2	745	789	830	868	904	939	973	1005	1037	1067
1.4	815	854	890	924	956	986	1016	1045	1072	1099
1.6	880	912	943	971	998	1024	1048	1072	1095	1117
1.8	922	942	962	980	997	1013	1028	1043	1057	1071
Medium-frame heifer calves										
0.2	323	374	421	465	508	549	588	626	662	
0.4	409	459	505	549	591	630	669	706	742	
0.6	477	522	563	602	638	674	708	741	773	
0.8	537	574	608	640	670	700	728	755	781	
1.0	562	583	603	621	638	654	670	685	700	

*Shrunk live weight basis.

Table A-10 *Protein requirements of growing and finishing cattle (g/day)—cont'd**

Daily gain, kg	Body weight, kg									
	150	200	250	300	350	400	450	500	550	600
Large-frame heifer calves and compensating medium-frame yearling heifers										
0.2	342	397	449	497	543	588	631	672	712	751
0.4	426	480	530	577	622	665	707	747	787	825
0.6	500	549	596	639	681	721	759	796	832	867
0.8	568	613	654	693	730	765	799	833	865	896
1.0	630	668	703	735	767	797	826	854	881	907
1.2	680	708	734	758	781	803	824	844	864	883

*Shrunk live weight basis.
Reproduced with permission from National Research Council: Nutrient requirements of beef cattle, ed 6, Washington, DC, 1984, National Academy of Sciences, National Academy Press.

Table A-11 *Calcium and phosphorus requirements of growing and finishing cattle (g/day)**

Daily gain, kg	Mineral	Body weight, kg									
		150	200	250	300	350	400	450	500	550	600
Medium-frame steer calves											
0.2	Ca	11	12	13	14	15	16	17	19	20	
	P	7	9	10	12	13	15	16	18	19	
0.4	Ca	16	17	17	18	19	19	20	21	22	
	P	9	10	12	13	14	16	17	18	20	
0.6	Ca	21	21	21	22	22	22	22	23	23	
	P	11	12	13	14	15	17	18	19	20	
0.8	Ca	27	26	25	25	25	25	24	24	24	
	P	12	13	14	15	16	17	19	20	21	
1.0	Ca	32	31	29	29	28	27	26	26	25	
	P	14	15	16	16	17	18	19	20	21	
1.2	Ca	37	35	33	32	31	29	28	27	26	
	P	16	16	17	17	18	19	20	21	21	
1.4	Ca	42	39	37	35	33	32	30	29	27	
	P	17	18	18	19	19	20	20	21	22	
Large-frame steer calves, compensating medium-frame yearling steers, and medium-frame bulls											
0.2	Ca	11	12	13	14	16	17	18	19	20	22
	P	7	9	10	12	13	15	16	18	20	21
0.4	Ca	17	17	18	19	19	20	21	22	23	24
	P	9	10	12	13	15	16	17	19	20	22
0.6	Ca	22	22	23	23	23	24	24	24	25	25
	P	11	12	13	15	16	17	18	20	21	22
0.8	Ca	28	27	27	27	27	27	27	27	27	27
	P	13	14	15	16	17	18	19	20	22	23
1.0	Ca	33	32	31	31	30	30	29	29	29	28
	P	14	15	16	17	18	19	20	21	22	23
1.2	Ca	38	37	36	35	34	33	32	31	30	30
	P	16	17	18	18	19	20	21	22	23	24
1.4	Ca	44	42	40	38	37	36	34	33	32	31
	P	18	18	19	20	20	21	22	22	23	24
1.6	Ca	49	47	44	42	40	38	37	35	34	32
	P	20	20	20	21	21	22	22	23	24	24
Large-frame bull calves and compensating large-frame yearling steers											
0.2	Ca	11	12	13	15	16	17	18	20	21	22
	P	7	9	10	12	13	15	17	18	20	21
0.4	Ca	17	18	19	19	20	21	22	23	24	25
	P	9	11	12	13	15	16	18	19	21	22
0.6	Ca	23	23	23	24	24	25	25	26	27	27
	P	11	12	14	15	16	18	19	20	22	23
0.8	Ca	28	28	28	28	28	29	29	29	29	30
	P	13	14	15	16	18	19	20	21	22	24

*Shrunk live weight basis.

Table A-11 *Calcium and phosphorus requirements of growing and finishing cattle (g/day)—cont'd*

Daily gain, kg	Mineral	Body weight, kg									
		150	200	250	300	350	400	450	500	550	600
Large-frame bull calves and compensating large-frame yearling steers—cont'd											
1.0	Ca	34	34	33	33	32	32	32	32	32	32
	P	15	16	17	18	19	20	21	22	23	24
1.2	Ca	40	39	38	37	36	36	35	35	34	34
	P	17	17	18	19	20	21	22	23	24	25
1.4	Ca	45	44	42	41	40	39	38	37	36	36
	P	18	19	20	20	21	22	23	24	25	26
1.6	Ca	51	49	47	45	44	42	41	40	39	38
	P	20	21	21	22	23	23	24	25	25	26
1.8	Ca	56	54	51	49	47	45	44	42	41	39
	P	22	22	22	23	23	24	25	25	26	26
Medium-frame heifer calves											
0.2	Ca	10	11	12	13	14	16	17	18	19	
	P	7	9	10	11	13	14	16	17	19	
0.4	Ca	15	16	16	16	17	17	18	19	19	
	P	9	10	11	12	14	15	16	18	19	
0.6	Ca	20	20	19	19	19	19	19	19	19	
	P	10	11	12	13	14	16	17	18	19	
0.8	Ca	25	23	23	22	21	20	20	19	19	
	P	12	12	13	14	15	16	17	18	19	
1.0	Ca	29	27	26	24	23	22	20	19	19	
	P	13	14	14	15	16	16	17	18	19	
Large-frame heifer calves and compensating medium-frame yearling heifers											
0.2	Ca	11	12	13	14	15	16	17	18	20	21
	P	7	9	10	12	13	15	16	18	19	21
0.4	Ca	16	16	17	17	18	19	19	20	21	22
	P	9	10	11	13	14	15	17	18	20	21
0.6	Ca	21	21	21	21	21	21	21	21	22	22
	P	10	12	13	14	15	16	17	19	20	21
0.8	Ca	26	25	24	24	23	23	23	22	22	22
	P	12	13	14	15	16	17	18	19	20	21
1.0	Ca	31	29	28	27	26	25	24	23	23	22
	P	14	14	15	16	17	18	18	19	20	21
1.2	Ca	35	33	31	30	28	27	25	24	23	22
	P	15	16	16	17	17	18	19	20	20	21

*Shrunk live weight basis.
Reproduced with permission from National Research Council: Nutrient requirements of beef cattle, ed 6, Washington, DC, 1984, National Academy of Sciences, National Academy Press.

Dairy Cattle Requirements

Table A-12 Recommended nutrient content of diets for dairy cattle

Cow wt (kg)	Fat (%)	Wt gain (kg/d)	Lactating cow diets Milk yield (kg/d)					Early lactation (weeks 0-3)	Dry, pregnant cows	Calf milk replacer	Calf starter mix	Growing heifers and bulls[a] 3-6 mo	6-12 mo	>12 mo	Mature bulls	Maximum tolerable levels[b,c]
400	5.0	0.220	7	13	20	26	33									
500	4.5	0.275	8	17	25	33	41									
600	4.0	0.330	10	20	30	40	50									
700	3.5	0.385	12	24	36	48	60									
800	3.5	0.440	13	27	40	53	67									
Energy																
NEl, Mcal/kg			1.42	1.52	1.62	1.72	1.72	1.67	1.25	—	—	—	—	—	—	—
NEm, Mcal/kg			—	—	—	—	—	—	—	2.40	1.90	1.70	1.58	1.40	—	1.15
NEg, Mcal/kg			—	—	—	—	—	—	—	1.55	1.20	1.08	0.98	0.82	—	—
ME, Mcal/kg			2.35	2.53	2.71	2.89	2.89	2.80	2.04	3.78	3.11	2.60	2.47	2.27	2.00	—
DE, Mcal/kg			2.77	2.95	3.13	3.31	3.31	3.22	2.47	4.19	3.53	3.02	2.89	2.69	2.43	—
Protein equivalent																
Crude protein, %			12	15	16	17	18	19	12	22	18	16	12	12	10	—
UIP, %			4.4	5.2	5.7	5.9	6.2	7.0	—	—	—	8.2	4.4	2.1	—	—
DIP, %			7.8	8.7	9.6	10.3	10.4	9.7	—	—	—	4.6	6.4	7.2	—	—
Fiber content (min.)[d]																
Crude fiber, %			17	17	17	15	15	17	22	—	—	13	15	15	15	—
Acid detergent fiber, %			21	21	21	19	19	21	27	—	—	16	19	19	19	—
Neutral detergent fiber, %			28	28	28	25	25	28	35	—	—	23	25	25	25	—
Ether extract (min.), %			3	3	3	3	3	3	3	10	3	3	3	3	3	—

Dairy Cattle Requirements

Minerals														Maximum tolerable
Calcium, %	0.43	0.51	0.58	0.64	0.66	0.77	0.39[c]	0.70	0.60	0.52	0.41	0.29	0.30	2.00
Phosphorus, %	0.28	0.33	0.37	0.41	0.41	0.48	0.24	0.60	0.40	0.31	0.30	0.23	0.19	1.00
Magnesium, %[f]	0.20	0.20	0.20	0.25	0.25	0.25	0.16	0.07	0.10	0.16	0.16	0.16	0.16	0.50
Potassium, %[g]	0.90	0.90	0.90	1.00	1.00	1.00	0.65	0.65	0.65	0.65	0.65	0.65	0.65	3.00
Sodium, %	0.18	0.18	0.18	0.18	0.18	0.18	0.10	0.10	0.10	0.10	0.10	0.10	0.10	—
Chlorine, %	0.25	0.25	0.25	0.25	0.25	0.25	0.20	0.20	0.20	0.20	0.20	0.20	0.20	—
Sulfur, %	0.20	0.20	0.20	0.20	0.20	0.25	0.16	0.29	0.20	0.20	0.20	0.16	0.16	0.40
Iron, ppm	50	50	50	50	50	50	50	100	50	50	50	50	50	1000
Iodine, ppm[h]	0.60	0.60	0.60	0.60	0.60	0.60	0.25	0.25	0.25	0.25	0.25	0.25	0.25	50.00[i]
Vitamins[j]														
A, IU/kg	3200	3200	3200	3200	3200	4000	4000	3800	2200	2200	2200	2200	3200	66000
D, IU/kg	1000	1000	1000	1000	1000	1000	1200	600	300	300	300	300	300	10000
E, IU/kg	15	15	15	15	15	15	15	40	25	25	25	25	15	2000

The values presented in this table are intended as guidelines for the use of professionals in diet formulation. Because of the many factors affecting such values, they are not intended and should not be used as a legal or regulatory base.

[a] Cobalt requirements are 0.1 ppm, copper 10 ppm (modified by molybdenum and sulfate in ration), manganese 40 ppm, and selenium 0.30 ppm. The approximate weight for growing heifers and bulls at 3 to 6 mos is 150 kg; at 6 to 12 mos, it is 250 kg; and at more than 12 mos, it is 400 kg. The approximate average daily gain is 700 g per day.

[b] The maximum safe levels for many of the mineral elements are not well defined and may be substantially affected by specific feeding conditions. Additional information is available in *Mineral Tolerance of Domestic Animals* (NRC, 1980).

[c] Vitamin tolerances are discussed in detail in *Vitamin Tolerance of Animals* (NRC, 1987).

[d] It is recommended that 75% of the NDF in lactating cow diets be provided as forage. If this recommendation is not followed, a depression in milk fat may occur.

[e] The value for calcium assumes that the cow is in calcium balance at the beginning of the dry period. If the cow is not in balance, then the dietary calcium requirement should be increased by 25% to 33%.

[f] Under conditions conducive to grass tetany, magnesium should be increased to 0.25% or 0.30%.

[g] Under conditions of heat stress, potassium should be increased to 1.2%.

[h] If the diet contains as much as 25% strongly goitrogenic feed on a dry basis, the iodine provided should be increased two times or more.

[i] Although cattle can tolerate this level of iodine, lower levels may be desirable to reduce the iodine content of milk.

[j] The following minimum quantities of B-complex vitamins are suggested per unit of milk replacer: niacin, 2.6 ppm; pantothenic acid, 13 ppm; riboflavin, 6.5 ppm; pyridoxine, 6.5 ppm; folic acid, 0.5 ppm; biotin, 0.1 ppm; vitamin B_{12}, 0.07 ppm; thiamine, 6.5 ppm; and choline, 0.26 percent. It appears that adequate amounts of these vitamins are furnished when calves have functional rumens (usually at 6 weeks of age) by a combination of rumen synthesis and natural feedstuffs.

Adapted from National Research Council: Nutrient requirements of dairy cattle, Washington, DC, 1989, National Academy of Sciences, National Academy Press.

Table A-13 *Daily nutrient requirements of lactating and pregnant cows*

Live weight (kg)	Energy			Total crude protein (g)	Minerals		Vitamin 1000 IU	
	NEl (Mcal)	ME (Mcal)	DE (Mcal)		Ca (g)	P (g)	A	D
Maintenance of mature lactating cows*								
400	7.16	12.01	13.80	318	16	11	30	12
450	7.82	13.12	15.08	341	18	13	34	14
500	8.46	14.20	16.32	364	20	14	38	15
550	9.09	15.25	17.53	386	22	16	42	17
600	9.70	16.28	18.71	406	24	17	46	18
650	10.30	17.29	19.86	428	26	19	49	20
700	10.89	18.28	21.00	449	28	20	53	21
750	11.47	19.25	22.12	468	30	21	57	23
800	12.03	20.20	23.21	486	32	23	61	24
Maintenance plus last 2 months of gestation of mature dry cows†								
400	9.30	15.26	18.23	890	26	16	30	12
450	10.16	16.66	19.91	973	30	18	34	14
500	11.00	18.04	21.55	1,053	33	20	38	15
550	11.81	19.37	23.14	1,131	36	22	42	17
600	12.61	20.68	24.71	1,207	39	24	46	18
650	13.39	21.96	26.23	1,281	43	26	49	20
700	14.15	23.21	27.73	1,355	46	28	53	21
750	14.90	24.44	29.21	1,427	49	30	57	23
800	15.64	25.66	30.65	1,497	53	32	61	24
Milk production—nutrients per kg of milk of different fat percentages								
(Fat %)								
3.0	0.64	1.07	1.23	78	2.73	1.68	—	—
3.5	0.69	1.15	1.33	84	2.97	1.83	—	—
4.0	0.74	1.24	1.42	90	3.21	1.98	—	—
4.5	0.78	1.32	1.51	96	3.45	2.13	—	—
5.0	0.83	1.40	1.61	101	3.69	2.28	—	—
5.5	0.88	1.48	1.70	107	3.93	2.43	—	—
Live weight change during lactation—nutrients per kg of weight change‡								
Weight loss	−4.92	−8.25	−9.55	−320	—	—	—	—
Weight gain	5.12	8.55	9.96	320	—	—	—	—

NEl, Net energy for lactation; ME, metabolizable energy; DE, digestible energy; TDN, total digestible nutrients.

*To allow for growth of young lactating cows, increase the maintenance allowances for all nutrients except vitamins A and D by 20% during the first lactation and 10% during the second lactation.

†Values for calcium assume that the cow is in calcium balance at the beginning of the last 2 months of gestation. If the cow is not in balance, then the calcium requirement can be increased from 25% to 33%.

‡No allowance is made for mobilized calcium and phosphorus associated with live weight loss or with live weight gain. The maximum daily nitrogen available from weight loss is assumed to be 30 g or 234 g of crude protein.

Adapted from National Research Council: Nutrient requirements of dairy cattle, Washington, DC, 1989, National Academy of Sciences, National Academy Press.

Table A-14 Daily nutrient requirements of lactating cows using absorbable protein

Live weight (kg)	Fat (%)	Milk (kg)	Live weight change (kg)	Dry matter intake (kg)	NEIDM (Mcal/kg)	Energy NEI (Mcal)	TDN (kg)	Protein UIP (g)	DIP (g)	Minerals Ca (g)	P (g)
\multicolumn{12}{l}{Intake at 100% of the requirement for maintenance, lactation, and weight gain}											
400	4.5	8.0	0.220	10.14	1.43	14.55	6.44	511	753	44	28
400	4.5	14.0	0.220	12.66	1.52	19.26	8.48	710	1052	65	41
400	4.5	20.0	0.220	14.91	1.61	23.96	10.51	880	1355	85	54
400	4.5	26.0	0.220	16.94	1.69	28.67	12.54	1026	1662	106	67
400	4.5	32.0	0.220	19.41	1.72	33.37	14.58	1220	1962	127	80
400	5.0	8.0	0.220	10.36	1.44	14.94	6.60	525	778	46	30
400	5.0	14.0	0.220	13.00	1.53	19.93	8.77	730	1096	68	43
400	5.0	20.0	0.220	15.35	1.62	24.93	10.93	902	1419	90	57
400	5.0	26.0	0.220	17.44	1.72	29.92	13.07	1048	1745	112	71
400	5.0	32.0	0.220	20.30	1.72	34.91	15.25	1277	2061	134	84
400	5.5	8.0	0.220	10.57	1.45	15.32	6.77	538	803	48	31
400	5.5	14.0	0.220	13.33	1.55	20.61	9.07	748	1140	71	45
400	5.5	20.0	0.220	15.77	1.64	25.89	11.34	923	1483	95	60
400	5.5	26.0	0.220	18.13	1.72	31.17	13.62	1091	1826	118	75
400	5.5	32.0	0.220	21.20	1.72	36.45	15.92	1334	2160	142	89
500	4.0	9.0	0.275	11.59	1.42	16.49	7.30	540	883	49	32
500	4.0	17.0	0.275	14.78	1.51	22.38	9.86	797	1257	75	48
500	4.0	25.0	0.275	17.62	1.61	28.27	12.40	1015	1635	101	64
500	4.0	33.0	0.275	20.14	1.70	34.15	14.93	1201	2018	126	80
500	4.0	41.0	0.275	23.29	1.72	40.04	17.49	1453	2392	152	95
500	4.5	9.0	0.275	11.84	1.43	16.92	7.49	556	911	51	33
500	4.5	17.0	0.275	15.20	1.53	23.20	10.21	821	1310	79	50
500	4.5	25.0	0.275	18.16	1.62	29.47	12.92	1043	1715	107	68
500	4.5	33.0	0.275	20.79	1.72	35.74	15.61	1230	2124	134	85
500	4.5	41.0	0.275	24.44	1.72	42.02	18.35	1526	2519	162	102
500	5.0	9.0	0.275	12.08	1.44	17.36	7.68	571	939	53	35
500	5.0	17.0	0.275	15.60	1.54	24.01	10.57	844	1364	83	53
500	5.0	25.0	0.275	18.68	1.64	30.67	13.44	1069	1795	113	71
500	5.0	33.0	0.275	21.71	1.72	37.33	16.31	1289	2226	142	89
500	5.0	41.0	0.275	25.58	1.72	43.99	19.21	1599	2646	172	108
600	3.0	10.0	0.330	12.52	1.42	17.79	7.87	533	974	52	34
600	3.0	20.0	0.330	16.20	1.49	24.18	10.67	845	1375	79	51
600	3.0	30.0	0.330	19.37	1.58	30.58	13.43	1102	1784	106	68

NEIDM, Net energy for lactation per kg of dry matter; NEI, net energy for lactation; TDN, total digestible nutrients; UIP, undegraded intake protein; DIP, degraded intake protein.

Continued.

Table A-14 Daily nutrient requirements of lactating cows using absorbable protein—cont'd

Live weight (kg)	Fat (%)	Milk (kg)	Live weight change (kg)	Dry matter intake (kg)	Energy			Protein			Minerals	
					NEIDM (Mcal/kg)	NEl (Mcal)	TDN (kg)	UIP (g)	DIP (g)		Ca (g)	P (g)

Intake at 100% of the requirement for maintenance, lactation, and weight gain—cont'd

600	3.0	40.0	0.330	22.21	1.67	36.98	16.19	1323	2198		133	84
600	3.0	50.0	0.330	25.23	1.72	43.38	18.95	1565	2608		161	101
600	3.5	10.0	0.330	12.86	1.42	18.27	8.08	557	1004		54	35
600	3.5	20.0	0.330	16.70	1.51	25.15	11.08	874	1438		84	54
600	3.5	30.0	0.330	20.04	1.60	32.03	14.06	1137	1879		113	72
600	3.5	40.0	0.330	23.00	1.69	38.90	17.01	1360	2326		143	90
600	3.5	50.0	0.330	26.63	1.72	45.78	20.00	1654	2763		173	109
600	4.0	10.0	0.330	13.20	1.42	18.75	8.30	581	1034		56	37
600	4.0	20.0	0.330	17.19	1.52	26.11	11.50	902	1501		89	57
600	4.0	30.0	0.330	20.69	1.62	33.47	14.68	1170	1975		121	77
600	4.0	40.0	0.330	23.78	1.72	40.83	17.84	1395	2454		153	96
600	4.0	50.0	0.330	28.03	1.72	48.19	21.05	1744	2918		185	116
700	3.0	12.0	0.385	14.46	1.42	20.54	9.09	607	1154		61	40
700	3.0	24.0	0.385	18.75	1.50	28.21	12.44	968	1638		94	60
700	3.0	36.0	0.385	22.48	1.60	35.89	15.76	1269	2129		127	81
700	3.0	48.0	0.385	25.80	1.69	43.57	19.05	1525	2627		159	101
700	3.0	60.0	0.385	29.81	1.72	51.25	22.39	1857	3114		192	121
700	3.5	12.0	0.385	14.86	1.42	21.11	9.34	636	1190		64	42
700	3.5	24.0	0.385	19.34	1.52	29.37	12.94	1002	1713		100	64
700	3.5	36.0	0.385	23.26	1.62	37.62	16.50	1309	2244		135	86
700	3.5	48.0	0.385	26.72	1.72	45.88	20.04	1567	2781		171	108
700	3.5	60.0	0.385	31.48	1.72	54.13	23.65	1964	3300		207	130
700	4.0	12.0	0.385	15.20	1.43	21.69	9.60	658	1227		67	44
700	4.0	24.0	0.385	19.92	1.53	30.52	13.44	1035	1789		105	68
700	4.0	36.0	0.385	24.02	1.64	39.35	17.25	1347	2359		144	91
700	4.0	48.0	0.385	28.03	1.72	48.19	21.05	1648	2930		182	115
700	4.0	60.0	0.385	33.16	1.72	57.02	24.91	2071	3485		221	139
800	3.0	14.0	0.440	16.36	1.42	23.24	10.29	682	1331		71	46
800	3.0	27.0	0.440	20.93	1.51	31.56	13.91	1064	1857		106	68
800	3.0	40.0	0.440	24.95	1.60	39.88	17.50	1388	2390		142	90
800	3.0	53.0	0.440	28.54	1.69	48.20	21.08	1665	2928		177	112
800	3.0	66.0	0.440	32.87	1.72	56.51	24.69	2022	3457		213	134
800	3.5	14.0	0.440	16.78	1.42	23.92	10.58	710	1374		74	49
800	3.5	27.0	0.440	21.59	1.52	32.86	14.47	1102	1942		113	72
800	3.5	40.0	0.440	25.82	1.62	41.80	18.33	1432	2517		151	96
800	3.5	53.0	0.440	29.57	1.72	50.75	22.17	1711	3099		190	120

Dairy Cattle Requirements

800	3.5	66.0	0.440	34.72	1.72	59.69	26.07	2140	3661	228	144	
800	4.0	14.0	0.440	17.17	1.43	24.59	10.88	734	1418	77	51	
800	4.0	27.0	0.440	22.24	1.54	34.16	15.03	1139	2027	119	76	
800	4.0	40.0	0.440	26.66	1.64	43.73	19.16	1474	2644	161	102	
800	4.0	53.0	0.440	31.00	1.72	53.29	23.28	1800	3263	203	128	
800	4.0	66.0	0.440	36.56	1.72	62.86	27.46	2259	3865	244	154	

Intake at 85% of the requirement for maintenance and lactation

400	4.5	20.0	−0.696	11.62	1.67	19.41	8.49	687	1066	85	54	
400	4.5	26.0	−0.840	14.02	1.67	23.41	10.24	931	1310	106	67	
400	4.5	32.0	−0.983	16.41	1.67	27.41	11.99	1187	1554	127	80	
400	5.0	20.0	−0.726	12.11	1.67	20.23	8.85	720	1118	90	57	
400	5.0	26.0	−0.878	14.65	1.67	24.47	10.71	987	1377	112	71	
400	5.0	32.0	−1.030	17.20	1.67	28.72	12.56	1255	1635	134	84	
400	5.5	20.0	−0.755	12.60	1.67	21.05	9.21	761	1169	95	60	
400	5.5	26.0	−0.916	15.29	1.67	25.54	11.17	1042	1443	118	75	
400	5.5	32.0	−1.077	17.98	1.67	30.03	13.14	1323	1717	142	89	
500	4.0	25.0	−0.819	13.67	1.67	22.83	9.99	810	1286	101	64	
500	4.0	33.0	−0.998	16.67	1.67	27.83	12.18	1134	1590	126	80	
500	4.0	41.0	−1.178	19.66	1.67	32.84	14.37	1458	1894	152	95	
500	4.5	25.0	−0.856	14.28	1.67	23.85	10.44	864	1350	107	68	
500	4.5	33.0	−1.047	17.48	1.67	29.18	12.77	1205	1674	134	85	
500	4.5	41.0	−1.238	20.67	1.67	34.52	15.10	1546	1998	162	102	
500	5.0	25.0	−0.892	14.89	1.67	24.87	10.88	917	1414	113	71	
500	5.0	33.0	−1.095	18.28	1.67	30.53	13.36	1275	1758	142	89	
500	5.0	41.0	−1.298	21.67	1.67	36.19	15.83	1633	2103	172	108	
600	3.0	30.0	−0.881	14.71	1.67	24.56	10.74	860	1399	106	68	
600	3.0	40.0	−1.076	17.96	1.67	30.00	13.12	1223	1728	133	84	
600	3.0	50.0	−1.271	21.22	1.67	35.44	15.50	1585	2057	161	101	
600	3.5	30.0	−0.925	15.44	1.67	25.79	11.28	924	1476	113	72	
600	3.5	40.0	−1.135	18.94	1.67	31.63	13.84	1308	1830	143	90	
600	3.5	50.0	−1.344	22.44	1.67	37.48	16.40	1692	2184	173	109	
600	4.0	30.0	−0.969	16.17	1.67	27.01	11.82	988	1552	121	77	
600	4.0	40.0	−1.193	19.92	1.67	33.27	14.55	1393	1932	153	96	
600	4.0	50.0	−1.418	23.67	1.67	39.52	17.29	1798	2311	185	116	
700	3.0	36.0	−1.034	17.26	1.67	28.83	12.61	1054	1669	127	81	
700	3.0	48.0	−1.268	21.17	1.67	35.36	15.47	1489	2064	159	101	
700	3.0	60.0	−1.502	25.08	1.67	41.88	18.32	1924	2458	192	121	
700	3.5	36.0	−1.087	18.15	1.67	30.30	13.26	1131	1761	135	86	
700	3.5	48.0	−1.339	22.35	1.67	37.32	16.33	1591	2186	171	108	
700	3.5	60.0	−1.590	26.55	1.67	44.34	19.40	2052	2611	207	130	

NEIDM, Net energy for lactation per kg of dry matter; NEl, net energy for lactation; TDN, total digestible nutrients; UIP, undegraded intake protein; DIP, degraded intake protein.

Continued.

Table A-14 *Daily nutrient requirements of lactating cows using absorbable protein—cont'd*

Live weight (kg)	Fat (%)	Milk (kg)	Live weight change (kg)	Dry matter intake (kg)	NEIDM (Mcal/kg)	Energy NEI (Mcal)	TDN (kg)	Protein UIP (g)	DIP (g)	Minerals Ca (g)	P (g)
700	4.0	36.0	−1.140	19.03	1.67	31.78	13.90	1208	1853	144	91
700	4.0	48.0	−1.409	23.52	1.67	39.28	17.19	1694	2308	182	115
700	4.0	60.0	−1.678	28.02	1.67	46.79	20.47	2180	2764	221	139
800	3.0	40.0	−1.147	19.15	1.67	31.98	13.99	1176	1871	142	90
800	3.0	50.0	−1.342	22.41	1.67	37.42	16.37	1538	2200	169	107
800	3.0	60.0	−1.537	25.66	1.67	42.86	18.75	1900	2529	196	124
800	3.5	40.0	−1.206	20.13	1.67	33.62	14.71	1261	1973	151	96
800	3.5	50.0	−1.416	23.63	1.67	39.46	17.27	1645	2327	181	114
800	3.5	60.0	−1.625	27.13	1.67	45.31	19.82	2028	2682	211	133
800	4.0	40.0	−1.264	21.11	1.67	35.25	15.42	1346	2075	161	102
800	4.0	50.0	−1.489	24.86	1.67	41.51	18.16	1751	2455	193	122
800	4.0	60.0	−1.713	28.60	1.67	47.76	20.90	2156	2835	225	142

NEIDM, Net energy for lactation per kg of dry matter; NEI, net energy for lactation; TDN, total digestible nutrients; UIP, undegraded intake protein; DIP, degraded intake protein.

Table A-15 Daily nutrient requirements of growing dairy cattle and mature bulls

Live weight (kg)	Gain (g)	Dry matter intake* (kg)	Energy NEm (Mcal)	Energy NEg (Mcal)	Energy ME (Mcal)	Energy DE (Mcal)	Protein UIP (g)	Protein DIP (g)	Protein CP (g)	Minerals Ca (g)	Minerals P (g)	Vitamins A (1000 IU)	Vitamins D
Growing large-breed calves fed only milk or milk replacer
40	200	0.48	1.37	0.41	2.54	2.73	—	—	105	7	4	1.70	0.26
45	300	0.54	1.49	0.56	2.86	3.07	—	—	120	8	5	1.94	0.30
Growing large-breed calves fed milk plus starter mix
50	500	1.30	1.62	0.72	5.90	6.42	—	—	290	9	6	2.10	0.33
75	800	1.98	2.19	1.30	8.98	9.78	—	—	435	16	8	3.20	0.50
Growing small-breed calves fed only milk or milk replacer
25	200	0.38	0.96	0.37	2.01	2.16	—	—	84	6	4	1.10	0.16
30	300	0.51	1.10	0.52	2.70	2.90	—	—	112	7	4	1.30	0.20
Growing small-breed calves fed milk plus starter mix
50	500	1.43	1.62	0.72	6.49	7.06	—	—	315	10	6	2.10	0.33
75	600	1.76	2.19	0.96	7.98	8.69	—	—	387	14	8	3.20	0.50
Growing veal calves fed only milk or milk replacer
40	200	0.45	1.37	0.55	1.89	2.07	—	—	100	7	4	1.70	0.26
50	400	0.57	1.62	0.57	2.39	2.63	—	—	125	9	5	2.10	0.33
60	540	0.80	1.85	0.81	2.84	3.17	—	—	176	13	8	2.60	0.40
75	900	1.36	2.19	1.47	4.82	5.39	—	—	300	16	9	3.20	0.50
100	1250	2.00	2.72	2.26	6.22	7.06	—	—	440	20	11	4.20	0.66
125	1250	2.38	3.21	2.44	7.40	8.40	—	—	524	22	13	5.30	0.82
150	1100	2.72	3.69	2.29	8.46	9.60	—	—	598	24	15	6.40	0.99

NEm, Net energy for maintenance; NEg, net energy for gain; ME, metabolizable energy; DE, digestible energy; TDN, total digestible nutrients; UIP, undegraded intake protein; DIP, degraded intake protein; CP, crude protein.

*The data for DMI are not requirements per se, unlike the requirements for net energy maintenance, net energy gain, and absorbed protein. They are not intended to be estimates of voluntary intake but are consistent with the specified dietary energy concentrations. The use of diets with decreased energy concentrations will increase dry matter intake needs; metabolizable energy, digestible energy, and total digestible nutrient needs; and crude protein needs. The use of diets with increased energy concentrations will have opposite effects on these needs.

Adapted from National Research Council: Nutrient requirements of dairy cattle, Washington, DC, 1989, National Academy of Sciences, National Academy Press.

Continued.

Table A-15 *Daily nutrient requirements of growing dairy cattle and mature bulls—cont'd*

Live weight (kg)	Gain (g)	Dry matter intake* (kg)	Energy NEm (Mcal)	NEg (Mcal)	ME (Mcal)	DE (Mcal)	Protein UIP (g)	DIP (g)	CP (g)	Minerals Ca (g)	P (g)	Vitamins A (1000 IU)	D
Large-breed growing females													
100	600	2.63	2.72	1.22	7.03	8.13	317	57	421	17	9	4.24	0.66
100	700	2.82	2.72	1.44	7.54	8.72	346	75	452	18	9	4.24	0.66
100	800	3.02	2.72	1.66	8.06	9.32	374	92	483	18	10	4.24	0.66
150	600	3.51	3.69	1.45	9.14	10.61	283	150	562	19	11	6.36	0.99
150	700	3.75	3.69	1.71	9.76	11.33	307	173	600	19	12	6.36	0.99
150	800	3.99	3.69	1.97	10.39	12.07	331	196	639	20	12	6.36	0.99
200	600	4.39	4.57	1.65	11.14	12.99	254	239	631	20	14	8.48	1.32
200	700	4.68	4.57	1.95	11.87	13.84	274	267	686	21	14	8.48	1.32
200	800	4.97	4.57	2.25	12.62	14.71	294	295	741	22	15	8.48	1.32
250	600	5.31	5.41	1.84	13.10	15.33	229	326	637	22	16	10.60	1.65
250	700	5.65	5.41	2.18	13.94	16.32	246	359	678	23	17	10.60	1.65
250	800	5.99	5.41	2.51	14.79	17.32	263	393	726	24	17	10.60	1.65
300	600	6.26	6.20	2.02	15.05	17.69	209	413	752	23	17	12.72	1.98
300	700	6.66	6.20	2.39	16.00	18.81	223	452	799	24	18	12.72	1.98
300	800	7.06	6.20	2.77	16.97	19.95	236	490	848	25	19	12.72	1.98
350	600	7.29	6.96	2.20	17.01	20.09	193	501	874	24	18	14.84	2.31
350	700	7.75	6.96	2.60	18.09	21.36	204	545	930	25	19	14.84	2.31
350	800	8.21	6.96	3.01	19.18	22.64	214	590	985	26	20	14.84	2.31
400	600	8.39	7.69	2.37	19.03	22.58	182	592	1007	25	19	16.96	2.64
400	700	8.92	7.69	2.80	20.23	24.00	190	641	1070	26	20	16.96	2.64
400	800	9.46	7.69	3.24	21.44	25.44	198	692	1135	26	21	16.96	2.64
450	600	9.59	8.40	2.53	21.12	25.18	176	686	1151	28	19	19.08	2.97
450	700	10.20	8.40	2.99	22.46	26.78	182	742	1224	28	20	19.08	2.97
450	800	10.82	8.40	3.46	23.81	28.40	187	799	1298	29	21	19.08	2.97
500	600	10.93	9.09	2.69	23.32	27.96	175	785	1311	28	20	21.20	3.30
500	700	11.63	9.09	3.18	24.81	29.74	179	848	1395	28	21	21.20	3.30
500	800	12.33	9.09	3.68	26.32	31.55	182	913	1480	29	21	21.20	3.30
550	600	12.42	9.77	2.84	25.67	30.95	180	891	1490	28	20	23.32	3.63
550	700	13.22	9.77	3.37	27.33	32.95	183	963	1587	28	20	23.32	3.63
550	800	14.04	9.77	3.90	29.02	34.99	185	1035	1685	29	21	23.32	3.63
600	600	14.11	10.43	3.00	28.23	34.24	193	1007	1694	28	20	25.44	3.96
600	700	15.05	10.43	3.55	30.09	36.50	194	1088	1805	28	21	25.44	3.96
600	800	15.99	10.43	4.11	31.98	38.79	195	1170	1919	29	21	25.44	3.96

Dairy Cattle Requirements

Small-breed growing females

BW	ADG												
100	400	2.41	2.72	0.91	6.34	7.35	249	38	386	15	8	4.24	0.66
100	500	2.64	2.72	1.16	6.92	8.03	275	59	422	16	8	4.24	0.66
100	600	2.86	2.72	1.40	7.51	8.71	300	80	458	17	9	4.24	0.66
150	400	3.31	3.69	1.09	8.39	9.78	222	129	512	17	10	6.36	0.99
150	500	3.60	3.69	1.39	9.12	10.63	243	156	567	18	11	6.36	0.99
150	600	3.89	3.69	1.69	9.86	11.50	263	185	622	19	11	6.36	0.99
200	400	4.24	4.57	1.26	10.38	12.16	201	217	513	19	13	8.48	1.32
200	500	4.60	4.57	1.60	11.25	13.19	217	251	562	20	13	8.48	1.32
200	600	4.96	4.57	1.95	12.14	14.23	232	286	611	20	14	8.48	1.32
250	400	5.24	5.41	1.41	12.36	14.57	185	305	629	21	15	10.60	1.65
250	500	5.68	5.41	1.80	13.38	15.78	197	346	681	21	16	10.60	1.65
250	600	6.12	5.41	2.20	14.43	17.01	209	389	735	22	16	10.60	1.65
300	400	6.34	6.20	1.56	14.38	17.06	176	395	761	22	16	12.72	1.98
300	500	6.87	6.20	1.99	15.57	18.48	184	445	824	23	17	12.72	1.98
300	600	7.40	6.20	2.43	16.79	19.92	192	495	888	23	17	12.72	1.98
350	400	7.57	6.96	1.71	16.50	19.71	173	490	909	23	17	14.84	2.31
350	500	8.20	6.96	2.18	17.87	21.35	178	548	985	23	18	14.84	2.31
350	600	8.85	6.96	2.66	19.28	23.03	183	608	1062	24	18	14.84	2.31
400	400	8.98	7.69	1.84	18.77	22.58	177	592	1078	24	18	16.96	2.64
400	500	9.74	7.69	2.35	20.36	24.50	181	661	1169	24	19	16.96	2.64
400	600	10.52	7.69	2.87	21.98	26.45	183	730	1263	25	19	16.96	2.64
450	400	10.64	8.40	1.98	21.27	25.80	191	706	1276	27	18	19.08	2.97
450	500	11.56	8.40	2.52	23.12	28.04	193	786	1387	28	19	19.08	2.97
450	600	12.50	8.40	3.08	25.01	30.33	194	867	1500	28	19	19.08	2.97

Large-breed growing males

BW	ADG												
100	800	2.80	2.72	1.42	7.48	8.66	401	65	448	18	10	4.24	0.66
100	900	2.97	2.72	1.60	7.92	9.16	433	79	475	19	10	4.24	0.66
100	1000	3.13	2.72	1.79	8.36	9.67	465	93	501	20	11	4.24	0.66
150	800	3.60	3.69	1.64	9.52	11.03	364	155	576	20	12	6.36	0.99
150	900	3.80	3.69	1.85	10.03	11.63	393	172	607	21	13	6.36	0.99
150	1000	3.99	3.69	2.07	10.55	12.22	422	190	639	22	13	6.36	0.99
200	800	4.43	4.57	1.84	11.48	13.34	333	241	709	22	15	8.48	1.32
200	900	4.66	4.57	2.08	12.06	14.02	359	262	745	23	15	8.48	1.32
200	1000	4.89	4.57	2.33	12.66	14.71	385	284	782	24	16	8.48	1.32

NEm, Net energy for maintenance; NEg, net energy for gain; ME, metabolizable energy; DE, digestible energy; TDN, total digestible nutrients; UIP, undegraded intake protein; DIP, degraded intake protein; CP, crude protein.

*The data for DMI are not requirements per se, unlike the requirements for net energy maintenance, net energy gain, and absorbed protein. They are not intended to be estimates of voluntary intake but are consistent with the specified dietary energy concentrations. The use of diets with decreased energy concentrations will increase dry matter intake needs; metabolizable energy, digestible energy, and total digestible nutrient needs; and crude protein needs. The use of diets with increased energy concentrations will have opposite effects on these needs.

Adapted from National Research Council: Nutrient requirements of dairy cattle, Washington, DC, 1989, National Academy of Sciences, National Academy Press.

Continued.

Table A-15 Daily nutrient requirements of growing dairy cattle and mature bulls—cont'd

Live weight (kg)	Gain (g)	Dry matter intake* (kg)	Energy NEm (Mcal)	Energy NEg (Mcal)	Energy ME (Mcal)	Energy DE (Mcal)	Protein UIP (g)	Protein DIP (g)	Protein CP (g)	Minerals Ca (g)	Minerals P (g)	Vitamins A (1000 IU)	Vitamins D
Large-breed growing males—cont'd														
250	800	5.27	5.41	2.03	13.37	15.58	305	325	778	24	17	10.60	1.65
250	900	5.53	5.41	2.30	14.03	16.35	329	350	837	25	18	10.60	1.65
250	1000	5.80	5.41	2.57	14.70	17.13	352	375	897	26	18	10.60	1.65
300	800	6.13	6.20	2.21	15.22	17.80	281	408	771	25	19	12.72	1.98
300	900	6.43	6.20	2.51	15.96	18.66	302	436	827	25	19	12.72	1.98
300	1000	6.73	6.20	2.80	16.70	19.53	323	464	884	26	20	12.72	1.98
350	800	7.02	6.96	2.38	17.06	20.02	261	490	843	26	20	14.84	2.31
350	900	7.36	6.96	2.70	17.88	20.98	280	522	883	26	20	14.84	2.31
350	1000	7.70	6.96	3.02	18.70	21.94	298	554	924	27	21	14.84	2.31
400	800	7.96	7.69	2.55	18.91	22.27	244	572	955	26	21	16.96	2.64
400	900	8.34	7.69	2.89	19.80	23.32	260	608	1001	27	21	16.96	2.64
400	1000	8.72	7.69	3.24	20.71	24.39	277	644	1046	28	22	16.96	2.64
450	800	8.95	8.40	2.71	20.78	24.56	230	656	1074	29	21	19.08	2.97
450	900	9.37	8.40	3.08	21.76	25.72	245	696	1125	29	22	19.08	2.97
450	1000	9.80	8.40	3.44	22.75	26.89	259	736	1176	29	23	19.08	2.97
500	800	10.00	9.09	2.87	22.69	26.92	220	742	1201	29	22	21.20	3.30
500	900	10.48	9.09	3.25	23.76	28.19	233	786	1257	29	23	21.20	3.30
500	1000	10.95	9.09	3.64	24.84	29.47	246	830	1314	29	21	21.20	3.30
550	800	11.14	9.77	3.02	24.66	29.38	213	831	1336	29	22	23.32	3.63
550	900	11.66	9.77	3.43	25.82	30.76	225	879	1399	29	23	23.32	3.63
550	1000	12.19	9.77	3.84	27.00	32.16	236	927	1463	30	21	23.32	3.63
600	800	12.36	10.43	3.17	26.71	31.95	211	923	1483	29	22	25.44	3.96
600	900	12.95	10.43	3.60	27.97	33.47	221	976	1554	29	23	25.44	3.96
600	1000	13.54	10.43	4.03	29.25	34.99	231	1029	1624	30	21	25.44	3.96
650	800	13.54	11.07	3.32	28.86	34.67	212	1020	1643	29	22	27.56	4.29
650	900	13.69	11.07	3.77	30.24	36.33	222	1078	1722	29	23	27.56	4.29
650	1000	14.35	11.07	4.22	31.63	38.00	230	1137	1801	30	23	27.56	4.29
700	800	15.01	11.70	3.46	31.14	37.59	219	1124	1820	29	22	29.68	4.62
700	900	15.16	11.70	3.93	32.64	39.40	227	1187	1907	29	22	29.68	4.62
700	1000	15.90	11.70	4.40	34.16	41.23	235	1252	1996	30	23	29.68	4.62
750	800	16.63	12.33	3.60	33.59	40.73	232	1235	2015	29	22	31.80	4.95
750	900	16.79	12.33	4.09	35.23	42.73	239	1305	2114	29	23	31.80	4.95
750	1000	17.62	12.33	4.58	36.89	44.74	246	1376	2213	30	23	31.80	4.95
800	800	18.45	12.94	3.74	35.12	42.59	216	1303	2107	29	22	33.92	5.28
800	900	17.56	12.94	4.25	36.83	44.67	221	1377	2210	29	23	33.92	5.28
800	1000	18.41	12.94	4.76	38.55	46.76	227	1451	2313	30	23	33.92	5.28
		19.28											

Dairy Cattle Requirements

Small-breed growing males

		NEm		NEg		ME	DE	TDN	CP	DIP	UIP
100	500	2.45	2.72	1.02	6.54	7.56	287	41	392	16	8	4.24	0.66
100	600	2.64	2.72	1.23	7.04	8.15	316	58	422	17	9	4.24	0.66
100	700	2.83	2.72	1.45	7.55	8.74	345	75	453	18	9	4.24	0.66
150	500	3.28	3.69	1.20	8.55	9.92	257	129	525	18	11	6.36	0.99
150	600	3.52	3.69	1.46	9.16	10.64	282	151	563	19	11	6.36	0.99
150	700	3.76	3.69	1.71	9.78	11.36	306	174	601	19	12	6.36	0.99
200	500	4.12	4.57	1.37	10.45	12.18	232	213	573	20	13	8.48	1.32
200	600	4.40	4.57	1.66	11.17	13.02	252	241	629	20	14	8.48	1.32
200	700	4.69	4.57	1.96	11.90	13.87	273	268	684	21	14	8.48	1.32
250	500	4.99	5.41	1.53	12.31	14.41	210	296	598	21	16	10.60	1.65
250	600	5.32	5.41	1.86	13.14	15.38	228	328	638	22	16	10.60	1.65
250	700	5.66	5.41	2.19	13.97	16.35	245	361	679	23	17	10.60	1.65
300	500	5.89	6.20	1.68	14.15	16.64	193	378	707	23	17	12.72	1.98
300	600	6.28	6.20	2.04	15.09	17.74	207	415	754	23	17	12.72	1.98
300	700	6.68	6.20	2.41	16.04	18.85	221	453	801	24	18	12.72	1.98
350	500	6.86	6.96	1.82	16.01	18.91	180	461	823	23	18	14.84	2.31
350	600	7.31	6.96	2.22	17.06	20.15	191	503	877	24	18	14.84	2.31
350	700	7.76	6.96	2.62	18.13	21.41	203	547	932	25	19	14.84	2.31
400	500	7.90	7.69	1.96	17.91	21.25	171	545	947	24	19	16.96	2.64
400	600	8.41	7.69	2.39	19.08	22.64	180	594	1010	25	19	16.96	2.64
400	700	8.94	7.69	2.82	20.27	24.06	189	644	1073	26	20	16.96	2.64
450	500	9.03	8.40	2.10	19.87	23.70	166	634	1083	28	19	19.08	2.97
450	600	9.62	8.40	2.55	21.18	25.26	174	689	1155	28	19	19.08	2.97
450	700	10.23	8.40	3.01	22.51	26.84	180	744	1227	28	20	19.08	2.97
500	500	10.28	9.09	2.23	21.93	26.29	167	726	1233	28	19	21.20	3.30
500	600	10.96	9.09	2.71	23.39	28.04	173	788	1315	28	20	21.20	3.30
500	700	11.65	9.09	3.20	24.87	29.81	177	851	1398	28	20	21.20	3.30
550	500	11.67	9.77	2.36	24.12	29.08	174	825	1400	28	19	23.32	3.63
550	600	12.46	9.77	2.87	25.75	31.05	178	895	1495	28	20	23.32	3.63
550	700	13.26	9.77	3.39	27.40	33.03	181	966	1591	28	20	23.32	3.63
600	500	13.25	10.43	2.48	26.50	32.14	187	933	1590	28	19	25.44	3.96
600	600	14.16	10.43	3.02	28.32	34.35	190	1012	1699	28	20	25.44	3.96
600	700	15.08	10.43	3.57	30.17	36.59	192	1091	1810	28	21	25.44	3.96

NEm, Net energy for maintenance; NEg, net energy for gain; ME, metabolizable energy; DE, digestible energy; TDN, total digestible nutrients; UIP, undegraded intake protein; DIP, degraded intake protein; CP, crude protein.

*The data for DMI are not requirements per se, unlike the requirements for net energy maintenance, net energy gain, and absorbed protein. They are not intended to be estimates of voluntary intake but are consistent with the specified dietary energy concentrations. The use of diets with decreased energy concentrations will increase dry matter intake needs; metabolizable energy, digestible energy, and total digestible nutrient needs; and crude protein needs. The use of diets with increased energy concentrations will have opposite effects on these needs.

Adapted from National Research Council: Nutrient requirements of dairy cattle, Washington, DC, 1989, National Academy of Sciences, National Academy Press.

Continued.

Table A-15 Daily nutrient requirements of growing dairy cattle and mature bulls—cont'd

Live weight (kg)	Gain (g)	Dry matter intake* (kg)	Energy				Protein			Minerals		Vitamins	
			NEm (Mcal)	NEg (Mcal)	ME (Mcal)	DE (Mcal)	UIP (g)	DIP (g)	CP (g)	Ca (g)	P (g)	A (1000 IU)	D

Maintenance of mature breeding bulls

500	—	7.89	9.09	—	15.79	19.15	161	472	789	20	12	21.20	3.30
600	—	9.05	10.43	—	18.10	21.95	155	573	905	24	15	25.44	3.96
700	—	10.16	11.70	—	20.32	24.64	148	670	1016	28	18	29.68	4.62
800	—	11.23	12.94	—	22.46	27.24	142	764	1123	32	20	33.92	5.28
900	—	12.27	14.13	—	24.53	29.76	135	854	1227	36	22	38.16	5.94
1000	—	13.28	15.29	—	26.55	32.20	129	943	1328	41	25	42.40	6.60
1100	—	14.26	16.43	—	28.52	34.59	122	1029	1426	45	28	46.64	7.26
1200	—	15.22	17.53	—	30.44	36.92	115	1113	1522	49	30	50.88	7.92
1300	—	16.16	18.62	—	32.32	39.21	108	1196	1616	53	32	55.12	8.58
1400	—	17.09	19.68	—	34.17	41.45	102	1277	1709	57	35	59.36	9.24

NEm, Net energy for maintenance; NEg, net energy for gain; ME, metabolizable energy; DE, digestible energy; TDN, total digestible nutrients; UIP, undegraded intake protein; DIP, degraded intake protein; CP, crude protein.
*The data for DMI are not requirements per se, unlike the requirements for net energy maintenance, net energy gain, and absorbed protein. They are not intended to be estimates of voluntary intake but are consistent with the specified dietary energy concentrations. The use of diets with decreased energy concentrations will increase dry matter intake needs; metabolizable energy, digestible energy, and total digestible nutrient needs; and crude protein needs. The use of diets with increased energy concentrations will have opposite effects on these needs.
Adapted from National Research Council: Nutrient requirements of dairy cattle, Washington, DC, 1989, National Academy of Sciences, National Academy Press.

Goat Requirements

Table A-16 Daily nutrient requirements of goats

Body weight (kg)	Feed energy			Crude protein			Ca (g)	P (g)	Vitamin A (1000 IU)	Vitamin D IU	Dry matter per animal			
	DE (Mcal)	ME (Mcal)	NE (Mcal)	TP (g)	DP (g)						1 kg = 2.0 Mcal ME		1 kg = 2.4 Mcal ME	
											Total (kg)	% of kg BW	Total (kg)	% of kg BW

Maintenance only (includes stable feeding conditions, minimal activity, and early pregnancy)

10	0.70	0.57	0.32	22	15	1	0.7	0.4	84	0.28	2.8	0.24	2.4
20	1.18	0.96	0.54	38	26	1	0.7	0.7	144	0.48	2.4	0.40	2.0
30	1.59	1.30	0.73	51	35	2	1.4	0.9	195	0.65	2.2	0.54	1.8
40	1.98	1.61	0.91	63	43	2	1.4	1.2	243	0.81	2.0	0.67	1.7
50	2.34	1.91	1.08	75	51	3	2.1	1.4	285	0.95	1.9	0.79	1.6
60	2.68	2.19	1.23	86	59	3	2.1	1.6	327	1.09	1.8	0.91	1.5
70	3.01	2.45	1.38	96	66	4	2.8	1.8	369	1.23	1.8	1.02	1.5
80	3.32	2.71	1.53	106	73	4	2.8	2.0	408	1.36	1.7	1.13	1.4
90	3.63	2.96	1.67	116	80	4	2.8	2.2	444	1.48	1.6	1.23	1.4
100	3.93	3.21	1.81	126	86	5	3.5	2.4	480	1.60	1.6	1.34	1.3

Maintenance plus low activity (= 25% increment, intensive management, tropical range, and early pregnancy)

10	0.87	0.71	0.40	27	19	1	0.7	0.5	108	0.36	3.6	0.30	3.0
20	1.47	1.20	0.68	46	32	2	1.4	0.9	180	0.60	3.0	0.50	2.5
30	1.99	1.62	0.92	62	43	2	1.4	1.2	243	0.81	2.7	0.67	2.2
40	2.47	2.02	1.14	77	54	3	2.1	1.5	303	1.01	2.5	0.84	2.1
50	2.92	2.38	1.34	91	63	4	2.8	1.8	357	1.19	2.4	0.99	2.0
60	3.35	2.73	1.54	105	73	4	2.8	2.0	408	1.36	2.3	1.14	1.9
70	3.76	3.07	1.73	118	82	5	3.5	2.3	462	1.54	2.2	1.28	1.8
80	4.16	3.39	1.91	130	90	5	3.5	2.6	510	1.70	2.1	1.41	1.8
90	4.54	3.70	2.09	142	99	6	4.2	2.8	555	1.85	2.1	1.54	1.7
100	4.91	4.01	2.26	153	107	6	4.2	3.0	600	2.00	2.0	1.67	1.7

Goat Requirements

Maintenance plus medium activity (= 50% increment, semiarid rangeland, slightly hilly pastures, and early pregnancy)

Weight													
10	1.05	0.86	0.48	33	23	1	0.7	0.6	129	0.43	4.3	0.36	3.6
20	1.77	1.44	0.81	55	38	2	1.4	1.1	216	0.72	3.6	0.60	3.0
30	2.38	1.95	1.10	74	52	3	2.1	1.5	294	0.98	3.3	0.81	2.7
40	2.97	2.42	1.36	93	64	4	2.8	1.8	363	1.21	3.0	1.01	2.5
50	3.51	2.86	1.62	110	76	4	2.8	2.1	429	1.43	2.9	1.19	2.4
60	4.02	3.28	1.84	126	87	5	3.5	2.5	492	1.64	2.7	1.37	2.3
70	4.52	3.68	2.07	141	98	6	4.2	2.8	552	1.84	2.6	1.53	2.2
80	4.98	4.06	2.30	156	108	6	4.2	3.0	609	2.03	2.5	1.69	2.1
90	5.44	4.44	2.50	170	118	7	4.9	3.3	666	2.22	2.5	1.85	2.0
100	5.90	4.82	2.72	184	128	7	4.9	3.6	723	2.41	2.4	2.01	2.0

Maintenance plus high activity (= 75% increment, arid rangeland, sparse vegetation, mountainous pastures, and early pregnancy)

Weight													
10	1.22	1.00	0.56	38	26	2	1.4	0.8	150	0.50	5.0	0.42	4.2
20	2.06	1.68	0.94	64	45	2	1.4	1.3	252	0.84	4.2	0.70	3.5
30	2.78	2.28	1.28	87	60	3	2.1	1.7	342	1.14	3.8	0.95	3.2
40	3.46	2.82	1.59	108	75	4	2.8	2.1	423	1.41	3.5	1.18	3.0
50	4.10	3.34	1.89	128	89	5	3.5	2.5	501	1.67	3.3	1.39	2.7
60	4.69	3.83	2.15	146	102	6	4.2	2.9	576	1.92	3.2	1.60	2.7
70	5.27	4.29	2.42	165	114	6	4.2	3.2	642	2.14	3.0	1.79	2.6
80	5.81	4.74	2.68	182	126	7	4.9	3.6	711	2.37	3.0	1.98	2.5
90	6.35	5.18	2.92	198	138	8	5.6	3.9	777	2.59	2.9	2.16	2.4
100	6.88	5.62	3.17	215	150	8	5.6	4.2	843	2.81	2.8	2.34	2.3

Additional requirements for late pregnancy (for all goat sizes)

1.74	1.42	0.80	82	57	2	1.4	1.1	213	0.71		0.59	

Additional requirements for growth—weight gain at 50 g per day (for all goat sizes)

0.44	0.36	0.20	14	10	1	0.7	0.3	54	0.18		0.15	

Additional requirements for growth—weight gain at 100 g per day (for all goat sizes)

0.88	0.72	0.40	28	20	1	0.7	0.5	108	0.36		0.30	

Additional requirements for growth—weight gain at 150 g per day (for all goat sizes)

1.32	1.08	0.60	42	30	2	1.4	0.8	162	0.54		0.45	

Continued.

Table A-16 *Daily nutrient requirements of goats—cont'd*

Body weight (kg)	Feed energy			Crude protein		Ca (g)	P (g)	Vitamin A (1000 IU)	Vitamin D IU	Dry matter per animal			
										1 kg = 2.0 Mcal ME		1 kg = 2.4 Mcal ME	
	DE (Mcal)	ME (Mcal)	NE (Mcal)	TP (g)	DP (g)					Total (kg)	% of kg BW	Total (kg)	% of kg BW

Additional requirements for milk production per kg at different fat percentages (including requirements for nursing single, twin, or triplet kids at the respective milk production level) (% Fat)

2.5	1.47	1.20	0.68	59	42	2	1.4	3.8	760				
3.0	1.49	1.21	0.68	64	45	2	1.4	3.8	760				
3.5	1.51	1.23	0.69	68	48	2	1.4	3.8	760				
4.0	1.53	1.25	0.70	72	51	3	2.1	3.8	760				
4.5	1.55	1.26	0.71	77	54	3	2.1	3.8	760				
5.0	1.57	1.28	0.72	82	57	3	2.1	3.8	760				
5.5	1.59	1.29	0.73	86	60	3	2.1	3.8	760				
6.0	1.61	1.31	0.74	90	63	3	2.1	3.8	760				

Additional requirements for mohair production by Angora at different production levels
Annual fleece yield (kg)

2	0.07	0.06	0.03	9	6								
4	0.15	0.12	0.07	17	12								
6	0.22	0.18	0.10	26	18								
8	0.29	0.24	0.14	34	24								

Adapted from National Research Council: Nutrient requirements of goats, Washington, DC, 1981, National Academy of Sciences, National Academy Press.

Sheep Requirements

Table A-17 Daily nutrient requirements of sheep

Body weight (kg)	Weight change per day (g)	Dry matter per animal[a] (kg)	Dry matter per animal[a] (% body weight)	Energy[b] TDN (kg)	Energy[b] DE (Mcal)	Energy[b] ME (Mcal)	Crude protein (g)	Ca (g)	P (g)	Vitamin A activity (IU)	Vitamin E activity (IU)
Ewes[c]											
Maintenance											
50	10	1.0	2.0	0.55	2.4	2.0	95	2.0	1.8	2350	15
60	10	1.1	1.8	0.61	2.7	2.2	104	2.3	2.1	2820	16
70	10	1.2	1.7	0.66	2.9	2.4	113	2.5	2.4	3290	18
80	10	1.3	1.6	0.72	3.2	2.6	122	2.7	2.8	3760	20
90	10	1.4	1.5	0.78	3.4	2.8	131	2.9	3.1	4230	21
Flushing—2 weeks prebreeding and first 3 weeks of breeding											
50	100	1.6	3.2	0.94	4.1	3.4	150	5.3	2.6	2350	24
60	100	1.7	2.8	1.00	4.4	3.6	157	5.5	2.9	2820	26
70	100	1.8	2.6	1.06	4.7	3.8	164	5.7	3.2	3290	27
80	100	1.9	2.4	1.12	4.9	4.0	171	5.9	3.6	3760	28
90	100	2.0	2.2	1.18	5.1	4.2	177	6.1	3.9	4230	30
Nonlactating—first 15 weeks gestation											
50	30	1.2	2.4	0.67	3.0	2.4	112	2.9	2.1	2350	18
60	30	1.3	2.2	0.72	3.2	2.6	121	3.2	2.5	2820	20
70	30	1.4	2.0	0.77	3.4	2.8	130	3.5	2.9	3290	21
80	30	1.5	1.9	0.82	3.6	3.0	139	3.8	3.3	3760	22
90	30	1.6	1.8	0.87	3.8	3.2	148	4.1	3.6	4230	24
Last 4 weeks gestation (130% to 150% lambing rate expected) or last 4-6 weeks lactation of suckling singles[d]											
50	180 (45)	1.6	3.2	0.94	4.1	3.4	175	5.9	4.8	4250	24
60	180 (45)	1.7	2.8	1.00	4.4	3.6	184	6.0	5.2	5100	26
70	180 (45)	1.8	2.6	1.06	4.7	3.8	193	6.2	5.6	5950	27
80	180 (45)	1.9	2.4	1.12	4.9	4.0	202	6.3	6.1	6800	28
90	180 (45)	2.0	2.2	1.18	5.1	4.2	212	6.4	6.5	7650	30
Last 4 weeks gestation (180% to 225% lambing rate expected)											
50	225	1.7	3.4	1.10	4.8	4.0	196	6.2	3.4	4250	26
60	225	1.8	3.0	1.17	5.1	4.2	205	6.9	4.0	5100	27
70	225	1.9	2.7	1.24	5.4	4.4	214	7.6	4.5	5950	28
80	225	2.0	2.5	1.30	5.7	4.7	223	8.3	5.1	6800	30
90	225	2.1	2.3	1.37	6.0	5.0	232	8.9	5.7	7650	32

Sheep Requirements

Body weight (kg)	Weight change/day (g)	Dry matter per day (kg)	Dry matter per day (% body wt)	TDN (kg)	DE (Mcal)	ME (Mcal)	Crude protein (g)	Ca (g)	P (g)	Vit. A (IU)	Vit. E (IU)
First 6 to 8 weeks lactation of suckling singles or last 4 to 6 weeks lactation of suckling twins[a]											
50	−25 (90)	2.1	4.2	1.36	6.0	4.9	304	8.9	6.1	4250	32
60	−25 (90)	2.3	3.8	1.50	6.6	5.4	319	9.1	6.6	5100	34
70	−25 (90)	2.5	3.6	1.63	7.2	5.9	334	9.3	7.0	5950	38
80	−25 (90)	2.6	3.2	1.69	7.4	6.1	344	9.5	7.4	6800	39
90	−25 (90)	2.7	3.0	1.75	7.6	6.3	353	9.6	7.8	7650	40
First 6 to 8 weeks lactation of suckling twins											
50	−60	2.4	4.8	1.56	6.9	5.6	389	10.5	7.3	5000	36
60	−60	2.6	4.3	1.69	7.4	6.1	405	10.7	7.7	6000	39
70	−60	2.8	4.0	1.82	8.0	6.6	420	11.0	8.1	7000	42
80	−60	3.0	3.8	1.95	8.6	7.0	435	11.2	8.6	8000	45
90	−60	3.2	3.6	2.08	9.2	7.5	450	11.4	9.0	9000	48
Ewe lambs											
Nonlactating—first 15 weeks gestation											
40	160	1.4	3.5	0.83	3.6	3.0	156	5.5	3.0	1880	21
50	135	1.5	3.0	0.88	3.9	3.2	159	5.2	3.1	2350	22
60	135	1.6	2.7	0.94	4.1	3.4	161	5.5	3.4	2820	24
70	125	1.7	2.4	1.00	4.4	3.6	164	5.5	3.7	3290	26
Last 4 weeks gestation (100% to 120% lambing rate expected)											
40	180	1.5	3.8	0.94	4.1	3.4	187	6.4	3.1	3400	22
50	160	1.6	3.2	1.00	4.4	3.6	189	6.3	3.4	4250	24
60	160	1.7	2.8	1.07	4.7	3.9	192	6.6	3.8	5100	26
70	150	1.8	2.6	1.14	5.0	4.1	194	6.8	4.2	5950	27
Last 4 weeks gestation (130% to 175% lambing rate expected)											
40	225	1.5	3.8	0.99	4.4	3.6	202	7.4	3.5	3400	22
50	225	1.6	3.2	1.06	4.7	3.8	204	7.8	3.9	4250	24
60	225	1.7	2.8	1.12	4.9	4.0	207	8.1	4.3	5100	26
70	215	1.8	2.6	1.14	5.0	4.1	210	8.2	4.7	5950	27

[a] To convert dry matter to an as-fed basis, divide dry matter values by the percentage of dry matter in the particular feed.
[b] One kg TDN (total digestible nutrients) = 4.4 Mcal DE (digestible energy); ME (metabolizable energy) = 82% of DE.
[c] Values are applicable for ewes in moderate condition. Fat ewes should be fed according to the next lower weight category and thin ewes at the next higher weight category. Once desired or moderate weight condition is attained, use that weight category through all production stages.
[d] Values in parentheses are for ewes suckling lambs the last 4-6 weeks of lactation.
[e] Lambs intended for breeding; thus, maximum weight gains and finish are of secondary importance.
[f] Maximum weight gains expected.

Reproduced with permission from National Research Council: Nutrient requirements of sheep, ed 6, Washington, DC, 1985, National Academy of Sciences, National Academy Press.

Continued.

Table A-17 Daily nutrient requirements of sheep—cont'd

Body weight (kg)	Weight change per day (g)	Dry matter per animal[a] (kg)	Dry matter per animal[a] (% body weight)	Energy[b] TDN (kg)	Energy[b] DE (Mcal)	Energy[b] ME (Mcal)	Crude protein (g)	Ca (g)	P (g)	Vitamin A activity (IU)	Vitamin E activity (IU)
First 6 to 8 weeks lactation of suckling singles (wean by 8 weeks)											
40	−50	1.7	4.2	1.12	4.9	4.0	257	6.0	4.3	3400	26
50	−50	2.1	4.2	1.39	6.1	5.0	282	6.5	4.7	4250	32
60	−50	2.3	3.8	1.52	6.7	5.5	295	6.8	5.1	5100	34
70	−50	2.5	3.6	1.65	7.3	6.0	301	7.1	5.6	5450	38
First 6 to 8 weeks lactation of suckling twins (wean by 8 weeks)											
40	−100	2.1	5.2	1.45	6.4	5.2	306	8.4	5.6	4000	32
50	−100	2.3	4.6	1.59	7.0	5.7	321	8.7	6.0	5000	34
60	−100	2.5	4.2	1.72	7.6	6.2	336	9.0	6.4	6000	38
70	−100	2.7	3.9	1.85	8.1	6.6	351	9.3	6.9	7000	40
30	227	1.2	4.0	0.78	3.4	2.8	185	6.4	2.6	1410	18
40	182	1.4	3.5	0.91	4.0	3.3	176	5.9	2.6	1880	21
50	120	1.5	3.0	0.88	3.9	3.2	136	4.8	2.4	2350	22
60	100	1.5	2.5	0.88	3.9	3.2	134	4.5	2.5	2820	22
70	100	1.5	2.1	0.88	3.9	3.2	132	4.6	2.8	3290	22
Replacement ram lambs[e]											
40	330	1.8	4.5	1.1	5.0	4.1	243	7.8	3.7	1880	24
60	320	2.4	4.0	1.5	6.7	5.5	263	8.4	4.2	2820	26
80	290	2.8	3.5	1.8	7.8	6.4	268	8.5	4.6	3760	28
100	250	3.0	3.0	1.9	8.4	6.9	264	8.2	4.8	4700	30

Sheep Requirements

Lambs finishing—4 to 7 months old[f]

Weight										
30	295	1.3	4.3	0.94	4.1	3.4	191	6.6	3.2	1410	20
40	275	1.6	4.0	1.22	5.4	4.4	185	6.6	3.3	1880	24
50	205	1.6	3.2	1.23	5.4	4.4	160	5.6	3.0	2350	24

Early weaned lambs—moderate growth potential[f]

10	200	0.5	5.0	0.40	1.8	1.4	127	4.0	1.9	470	10
20	250	1.0	5.0	0.80	3.5	2.9	167	5.4	2.5	940	20
30	300	1.3	4.3	1.00	4.4	3.6	191	6.7	3.2	1410	20
40	345	1.5	3.8	1.16	5.1	4.2	202	7.7	3.9	1880	22
50	300	1.5	3.0	1.16	5.1	4.2	181	7.0	3.8	2350	22

Early weaned lambs—rapid growth potential[f]

10	250	0.6	6.0	0.48	2.1	1.7	157	4.9	2.2	470	12
20	300	1.2	6.0	0.92	4.0	3.3	205	6.5	2.9	940	24
30	325	1.4	4.7	1.10	4.8	4.0	216	7.2	3.4	1410	21
40	400	1.5	3.8	1.14	5.0	4.1	234	8.6	4.3	1880	22
50	425	1.7	3.4	1.29	5.7	4.7	240	9.4	4.8	2350	25
60	350	1.7	2.8	1.29	5.7	4.7	240	8.2	4.5	2820	25

[a] To convert dry matter to an as-fed basis, divide dry matter values by the percentage of dry matter in the particular feed.
[b] One kg TDN (total digestible nutrients) = 4.4 Mcal DE (digestible energy); ME (metabolizable energy) = 82% of DE.
[c] Values are applicable for ewes in moderate condition. Fat ewes should be fed according to the next lower weight category and thin ewes at the next higher weight category. Once desired or moderate weight condition is attained, use that weight category through all production stages.
[d] Values in parentheses are for ewes suckling lambs the last 4-6 weeks of lactation.
[e] Lambs intended for breeding; thus, maximum weight gains and finish are of secondary importance.
[f] Maximum weight gains expected.

Reproduced with permission from National Research Council: Nutrient requirements of sheep, ed 6, Washington, DC, 1985, National Academy of Sciences, National Academy Press.

Table A-18 Nutrient content of diets for sheep (nutrient concentration on 100% dry feed)

Body weight (kg)	Daily gain or loss (g)	Daily dry matter		Energy			Crude protein (%)	Calcium (%)	Phosphorus (%)	Vitamin A (IU/kg)	Vitamin E (IU/kg)
		Per animal (kg)	% live weight	TDN (%)	DE (Mcal/kg)	ME (Mcal/kg)					

Ewes
Maintenance*

50	10	1.0	2.0	55	2.4	2.0	9.5	0.2	0.2	2354	15
60	10	1.1	1.8	55	2.4	2.0	9.6	0.2	0.2	2590	15
70	10	1.2	1.7	55	2.4	2.0	9.6	0.2	0.2	2789	15
80	10	1.3	1.6	55	2.4	2.0	9.3	0.2	0.2	2859	15
90	10	1.4	1.6	55	2.4	2.0	9.4	0.2	0.2	3009	15

Flushing, 2 weeks prebreeding and first 3 weeks of breeding*

50	100	1.6	3.2	60	2.6	2.2	9.4	0.3	0.2	1479	15
60	100	1.7	2.8	60	2.6	2.2	9.2	0.3	0.2	1680	15
70	100	1.8	2.6	60	2.6	2.2	9.0	0.3	0.2	1814	15
80	100	1.9	2.4	60	2.6	2.2	9.0	0.3	0.2	1973	15
90	100	2.0	2.2	60	2.6	2.2	8.9	0.3	0.2	2118	15

Nonlactating, first 15 weeks gestation*

50	32	1.2	2.4	55	2.4	2.0	9.6	0.2	0.2	1993	15
60	32	1.3	2.2	55	2.4	2.0	9.3	0.2	0.2	2143	15
70	32	1.4	2.0	55	2.4	2.0	9.4	0.2	0.2	2339	15
80	32	1.5	1.9	55	2.4	2.0	9.4	0.3	0.2	2511	15
90	32	1.6	1.8	55	2.4	2.0	9.4	0.3	0.2	2665	15

Last 4 weeks gestation (130% to 150% lambing rate expected) or last 4 to 6 weeks lactation of suckling singles*†

50	181 (45)	1.6	3.2	60	2.6	2.2	10.9	0.4	0.3	2676	15
60	181 (45)	1.7	2.8	60	2.6	2.2	10.8	0.4	0.3	3038	15
70	181 (45)	1.8	2.6	60	2.6	2.2	10.5	0.3	0.3	3280	15
80	181 (45)	1.9	2.4	60	2.6	2.2	10.5	0.3	0.3	3569	15
90	181 (45)	2.0	2.2	60	2.6	2.2	10.7	0.3	0.3	3834	15

Last 4 weeks gestation (180% to 225% lambing rate expected)*

50	227	1.7	3.4	65	2.9	2.4	11.6	0.4	0.2	2533	15
60	227	1.8	3.0	65	2.9	2.4	11.3	0.4	0.2	2811	15
70	227	1.9	2.7	65	2.9	2.2	11.2	0.4	0.2	3124	15
80	227	2.0	2.5	65	2.9	2.4	11.1	0.4	0.3	3406	15
90	227	2.1	2.3	65	2.9	2.4	11.1	0.4	0.3	3666	15

Sheep Requirements

First 6-8 weeks lactation of suckling singles or last 4 to 6 weeks lactation of suckling twins*†

50	−27 (91)	2.1	4.2	65	2.9	2.4	14.6	0.4	0.3	2037	15
60	−27 (91)	2.3	3.9	65	2.9	2.4	13.7	0.4	0.3	2205	15
70	−27 (91)	2.5	3.6	65	2.9	2.4	13.3	0.4	0.3	2385	15
80	−27 (91)	2.6	3.2	65	2.9	2.4	13.3	0.4	0.3	2630	15
90	−27 (91)	2.7	3.0	65	2.9	2.4	13.2	0.4	0.3	2859	15

First 6 to 8 weeks lactation of suckling twins*

50	−59	2.4	4.8	65	2.9	2.4	16.2	0.4	0.3	2079	15
60	−59	2.6	4.3	65	2.9	2.4	15.6	0.4	0.3	2321	15
70	−59	2.8	4.0	65	2.9	2.4	14.8	0.4	0.3	2489	15
80	−59	3.0	3.7	65	2.9	2.4	14.5	0.4	0.3	2672	15
90	−59	3.2	3.5	65	2.9	2.4	14.1	0.4	0.3	2835	15

Ewe lambs
Nonlactating—first 15 weeks gestation

40	159	1.4	3.5	59	2.6	2.2	11.0	0.4	0.2	1336	15
50	136	1.5	3.0	59	2.6	2.2	10.6	0.3	0.2	1570	15
60	136	1.6	2.7	59	2.6	2.2	10.0	0.3	0.2	1777	15
70	127	1.7	2.4	59	2.6	2.2	9.7	0.3	0.2	1960	15

Last 4 weeks gestation (100% to 120% lambing rate expected)

40	181	1.5	3.7	63	2.8	2.3	12.4	0.4	0.2	2271	15
50	159	1.6	3.2	63	2.8	2.3	12.0	0.4	0.2	2676	15
60	159	1.7	2.8	63	2.8	2.3	11.4	0.4	0.3	3038	15
70	150	1.8	2.6	63	2.8	2.3	10.8	0.4	0.3	3280	15

Last 4 weeks gestation (130% to 175% lambing rate expected)

40	227	1.5	3.7	66	2.9	2.4	13.3	0.5	0.2	2271	15
50	227	1.6	3.2	66	2.9	2.4	12.9	0.5	0.2	2676	15
60	227	1.7	2.8	66	2.9	2.4	12.4	0.5	0.3	3038	15
70	213	1.8	2.6	66	2.9	2.4	11.5	0.5	0.3	3280	15

First 6 to 8 weeks lactation of suckling singles (wean by 8 weeks)

40	−50	1.7	4.2	66	2.9	2.4	15.1	0.4	0.3	2026	15
50	−50	2.1	4.2	66	2.9	2.4	13.5	0.3	0.2	2037	15
60	−50	2.3	3.9	66	2.9	2.4	12.7	0.3	0.2	2205	15
70	−50	2.5	3.6	66	2.9	2.4	12.4	0.3	0.2	2185	15

*Values are applicable for ewes in moderate condition. Fat ewes should be fed according to the next lower weight category and thin ewes at the next higher weight category.
†Values in parentheses are for ewes suckling lambs the last 4 to 6 weeks of lactation.

Table A-18 Nutrient content of diets for sheep (nutrient concentration on 100% dry feed)—cont'd

Body weight (kg)	Daily gain or loss (g)	Daily dry matter Per animal (kg)	Daily dry matter % live weight	TDN (%)	Energy DE (Mcal/kg)	Energy ME (Mcal/kg)	Crude protein (%)	Calcium (%)	Phosphorus (%)	Vitamin A (IU/kg)	Vitamin E (IU/kg)
First 6-8 weeks lactation of suckling twins (wean by 8 weeks)											
40	−100	2.1	5.2	69	3.1	2.5	14.6	0.4	0.3	1918	15
50	−100	2.3	4.6	69	3.1	2.5	13.9	0.4	0.3	2160	15
60	−100	2.5	4.2	69	3.1	2.5	13.5	0.4	0.3	2405	15
70	−100	2.7	3.9	69	3.1	2.5	12.8	0.3	0.3	2573	15
Replacement ewe lambs											
30	227	1.2	3.9	65	2.9	2.4	15.8	0.5	0.2	1195	15
40	181	1.4	3.5	65	2.9	2.4	12.6	0.4	0.2	1336	15
50	118	1.5	3.0	58	2.6	2.1	9.1	0.3	0.2	1570	15
60	100	1.5	2.5	58	2.6	2.1	9.1	0.3	0.2	1885	15
70	100	1.5	2.1	58	2.6	2.1	8.8	0.3	0.2	2198	15
Replacement ram lambs											
40	330	1.8	4.5	63	2.8	2.3	13.5	0.4	0.2	1036	15
60	320	2.4	4.0	63	2.8	2.3	10.9	0.3	0.2	1173	15
80	290	2.8	3.5	63	2.8	2.3	9.5	0.3	0.2	1336	15
100	250	3.0	3.0	63	2.8	2.3	8.8	0.3	0.2	1570	15
Growing lambs finishing—4 to 7 months old											
30	295	1.3	4.4	72	3.2	2.5	14.5	0.5	0.2	1071	15
40	272	1.6	4.0	77	3.3	2.7	11.7	0.4	0.2	1184	15
50	204	1.6	3.2	77	3.4	2.8	10.0	0.4	0.2	1479	15
Early weaned lambs—moderate growth potential											
10	200	0.5	5.0	80	3.5	2.7	25.5	0.8	0.4	941	20
20	250	1.0	5.0	80	3.5	2.7	16.8	0.5	0.3	941	20
30	300	1.3	4.4	77	3.3	2.6	14.7	0.5	0.2	1071	15
40	345	1.5	3.7	77	3.5	2.6	13.5	0.5	0.3	1257	15
50	300	1.5	3.0	77	3.3	2.6	12.1	0.5	0.3	1570	15
Early weaned lambs—rapid growth potential											
10	250	0.6	5.9	80	3.5	2.9	26.9	0.8	0.4	798	20
20	300	1.2	5.9	78	3.4	2.8	17.3	0.6	0.2	798	20
30	326	1.4	4.7	77	3.3	2.7	15.5	0.5	0.2	1003	15
40	399	1.5	3.7	76	3.3	2.7	15.5	0.6	0.3	1257	15
50	426	1.7	3.4	76	3.3	2.7	14.3	0.6	0.3	1400	15
60	349	1.7	2.8	76	3.3	2.7	14.3	0.5	0.3	1680	15

Reproduced with permission from National Research Council: Nutrient requirements of sheep, ed 6, Washington, DC, 1985, National Academy of Sciences, National Academy Press.

Horse Requirements

Table A-19 *Daily nutrient requirements of ponies (200 kg mature weight)*

Animal	Weight (kg)	Daily gain (kg)	DE (Mcal)	Crude protein (g)	Lysine (g)	Calcium (g)	Phosphorus (g)	Magnesium (g)	Potassium (g)	Vitamin A (1000 IU)
Mature ponies										
Maintenance	200		7.4	296	10	8	6	3.0	10.0	6
Stallions (breeding season)	200		9.3	370	13	11	8	4.3	14.1	9
Pregnant mares	200									
9 months			8.2	361	13	16	12	3.9	13.1	12
10 months			8.4	368	13	16	12	4.0	13.4	12
11 months			8.9	391	14	17	13	4.3	14.2	12
Lactating mares										
Foaling to 3 months	200		13.7	688	24	27	18	4.8	21.2	12
3 months to weaning	200		12.2	528	18	18	11	3.7	14.8	12
Working ponies										
Light work	200		9.3	370	13	11	8	4.3	14.1	9
Moderate work	200		11.1	444	16	14	10	5.1	16.9	9
Intense work	200		14.8	592	21	18	13	6.8	22.5	9
Growing ponies										
Weanling, 4 months	75	0.40	7.3	365	15	16	9	1.6	5.0	3
Weanling, 6 months										
Moderate growth	95	0.30	7.6	378	16	13	7	1.8	5.7	4
Rapid growth	95	0.40	8.7	433	18	17	9	1.9	6.0	4
Yearling, 12 months										
Moderate growth	140	0.20	8.7	392	17	12	7	2.4	7.6	6
Rapid growth	140	0.30	10.3	462	19	15	8	2.5	7.9	6
Long yearling, 18 months										
Not in training	170	0.10	8.3	375	16	10	6	2.7	8.8	8
In training	170	0.10	11.6	522	22	14	8	3.7	12.2	8
Two-year-old, 24 months										
Not in training	185	0.05	7.9	337	13	9	5	2.8	9.4	8
In training	185	0.05	11.4	485	19	13	7	4.1	13.5	8

Reproduced with permission from National Research Council: Nutrient requirements of horses, ed 5, Washington, DC, 1989, National Academy of Sciences, National Academy Press.

Horse Requirements

Table A-20 *Daily nutrient requirements of horses (500 kg mature weight)*

Animal	Weight (kg)	Daily gain (kg)	DE (Mcal)	Crude protein (g)	Lysine (g)	Calcium (g)	Phosphorus (g)	Magnesium (g)	Potassium (g)	Vitamin A (1000 IU)
Mature horses										
Maintenance	500		16.4	656	23	20	14	7.5	25.0	15
Stallions (breeding season)	500		20.5	820	29	25	18	9.4	31.2	22
Pregnant mares	500									
9 months			18.2	801	28	35	26	8.7	29.1	30
10 months			18.5	815	29	35	27	8.9	29.7	30
11 months			19.7	866	30	37	28	9.4	31.5	30
Lactating mares										
Foaling to 3 months	500		28.3	1427	50	56	36	10.9	46.0	30
3 months to weaning	500		24.3	1048	37	36	22	8.6	33.0	30
Working horses										
Light work	500		20.5	820	29	25	18	9.4	31.2	22
Moderate work	500		24.6	984	34	30	21	11.3	37.4	22
Intense work	500		32.8	1312	46	40	29	15.1	49.9	22
Growing horses										
Weanling, 4 months	175	0.85	14.4	720	30	34	19	3.7	11.3	8
Weanling, 6 months										
Moderate growth	215	0.65	15.0	750	32	29	16	4.0	12.7	10
Rapid growth	215	0.85	17.2	860	36	36	20	4.3	13.3	10
Yearling, 12 months										
Moderate growth	325	0.50	18.9	851	36	29	16	5.5	17.8	15
Rapid growth	325	0.65	21.3	956	40	34	19	5.7	18.2	15
Long yearling, 18 months										
Not in training	400	0.35	19.8	893	38	27	15	6.4	21.1	18
In training	400	0.35	26.5	1195	50	36	20	8.6	28.2	18
Two-year-old, 24 months										
Not in training	450	0.20	18.8	800	32	24	13	7.0	23.1	20
In training	450	0.20	26.3	1117	45	34	19	9.8	32.2	20

Reproduced with permission from National Research Council: Nutrient requirements of horses, ed 5, Washington, DC, 1989, National Academy of Sciences, National Academy Press.

Swine Requirements

Table A-21 *Nutrient requirements of breeding swine*

Intake levels	Bred gilts, sows, and adult boars	Lactating gilts and sows
Digestible energy (kcal/kg diet)	3340	3340
Metabolizable energy (kcal/kg diet)	3210	3210
Crude protein (%)	12	13
Requirement (% or amount per kg diet)*		
Nutrient		
Indispensable amino acids (%)		
Arginine	0.00	0.40
Histidine	0.15	0.25
Isoleucine	0.30	0.39
Leucine	0.30	0.48
Lysine	0.43	0.60
Methionine + cystine	0.23	0.36
Phenylalanine + tyrosine	0.45	0.70
Threonine	0.30	0.43
Tryptophan	0.09	0.12
Valine	0.32	0.60
Linoleic acid (%)	0.1	0.1
Mineral elements		
Calcium (%)	0.75	0.75
Phosphorus, total (%)	0.60	0.60
Phosphorus, available (%)	0.35	0.35
Sodium (%)	0.15	0.20
Chlorine (%)	0.12	0.16
Magnesium (%)	0.04	0.04
Potassium (%)	0.20	0.20
Copper (mg)	5.00	5.00
Iodine (mg)	0.14	0.14
Iron (mg)	80.00	80.00
Manganese (mg)	10.00	10.00
Selenium (mg)	0.15	0.15
Zinc (mg)	50.00	50.00

Table A-21 *Nutrient requirements of breeding swine—cont'd*

Intake levels	Bred gilts, sows, and adult boars	Lactating gilts and sows
Vitamins		
Vitamin A (IU)	4000	2000
Vitamin D (IU)	200	200
Vitamin E (IU)	22	22
Vitamin K (menadione) (mg)	0.50	0.50
Biotin (mg)	0.20	0.20
Choline (g)	1.25	1.00
Folacin (mg)	0.30	0.30
Niacin, available (mg)	10.00	10.00
Pantothenic acid (mg)	12.00	12.00
Riboflavin (mg)	3.75	3.75
Thiamin (mg)	1.00	1.00
Vitamin B_6 (mg)	1.00	1.00
Vitamin B_{12} (μg)	15.00	15.00

Knowledge of nutritional constraints and limitations is important for the proper use of this table.

*These requirements are based upon corn–soybean meal diets, feed intakes, and performance levels listed in Tables A-22, A-24, and A-25. In the corn–soybean meal diets, the corn contains 8.5% protein; the soybean meal contains 44%.

Reproduced with permission from National Research Council: Nutrient requirements of swine, ed 9, Washington, DC, 1988, National Academy of Sciences, National Academy Press.

Table A-22 *Daily nutrient intakes and requirements of intermediate-weight breeding animals*

	Mean gestation or farrowing weight (kg) of:	
	Bred gilts, sows, and adult boars	Lactating gilts and sows
Intake and performance levels	162.5	165.0
Daily feed intake (kg)	1.9	5.3
Digestible energy (Mcal/day)	6.3	17.7
Metabolizable energy (Mcal/day)	6.1	17.0
Crude protein	228	689
Requirement (amount per day)		
Nutrients		
Indispensable amino acids (g)		
Arginine	0.0	21.2
Histidine	2.8	13.2
Isoleucine	5.7	20.7
Leucine	5.7	25.4
Lysine	8.2	31.8
Methionine + cystine	4.4	19.1
Phenylalanine + tyrosine	8.6	37.1
Threonine	5.7	22.8
Tryptophan	1.7	6.4
Valine	6.1	31.8
Linoleic acid (g)	1.9	5.3
Mineral elements		
Calcium (g)	14.2	39.8
Phosphorus, total (g)	11.4	31.8
Phosphorus, available (g)	6.6	18.6
Sodium (g)	2.8	10.6
Chlorine (g)	2.3	8.5
Magnesium (g)	0.8	2.1
Potassium (g)	3.8	10.6
Copper (mg)	9.5	26.5
Iodine (mg)	0.3	0.7
Iron (mg)	152	424
Manganese (mg)	19	53
Selenium (mg)	0.3	0.8
Zinc (mg)	95	265
Vitamins		
Vitamin A (IU)	7600	10,600
Vitamin D (IU)	380	1060
Vitamin E (IU)	42	117
Vitamin K (menadione) (mg)	1.0	2.6
Biotin (mg)	0.4	1.1

Table A-22 *Daily nutrient intakes and requirements of intermediate-weight breeding animals—cont'd*

	Mean gestation or farrowing weight (kg) of:	
	Bred gilts, sows, and adult boars	Lactating gilts and sows
Intake and performance levels	162.5	165.0
Choline (g)	2.4	5.3
Folacin (mg)	0.6	1.6
Niacin, available (mg)	19.0	53.0
Pantothenic acid (mg)	22.8	63.6
Riboflavin (mg)	7.1	19.9
Thiamin (mg)	1.9	5.3
Vitamin B_6 (mg)	1.9	5.3
Vitamin B_{12} (μg)	28.5	79.5

Reproduced with permission from National Research Council: Nutrient requirements of swine, ed 9, Washington, DC, 1988, National Academy of Sciences, National Academy Press.

Table A-23 *Requirements for several nutrients of breeding herd replacements allowed feed ad libitum*

	Weight (kg) of			
	Developing gilts		Developing boars	
Intake levels	20-50	50-110	20-50	50-110
Energy concentration (kcal ME/kg diet)	3255	3260	3240	3255
Crude protein (%)	16	15	18	16
Nutrient				
Lysine (%)	0.80	0.70	0.90	0.75
Calcium (%)	0.65	0.55	0.70	0.60
Phosphorus, total (%)	0.55	0.45	0.60	0.50
Phosphorus, available (%)	0.28	0.20	0.33	0.25

Reproduced with permission from National Research Council: Nutrient requirements of swine, ed 9, Washington, DC, 1988, National Academy of Sciences, National Academy Press.

Table A-24 *Daily energy and feed requirements of pregnant gilts and sows*

Intake and performance levels	Weight (kg) of bred gilts and sows at mating*		
	120	140	160
Mean gestation weight (kg)†	142.5	162.5	182.5
Energy required (Mcal DE/day)			
Maintenance‡	4.53	5.00	5.47
Gestation weight gain§	1.29	1.29	1.29
Total	5.82	6.29	6.76
Feed required/day (kg)‖	1.8	1.9	2.0

*Requirements are based on a 25 kg maternal weight gain plus 20 kg increase in weight due to the products of conception; the total weight gain is 45 kg.
†Mean gestation weight is weight at mating + (total weight gain/2).
‡The animal's daily maintenance requirement is 110 kcal of DE per $kg^{0.75}$.
§The gestation weight gain is 1.10 Mcal of DE per day for maternal weight gain plus 0.19 Mcal of DE per day for conceptus gain.
‖The feed required per day is based on a corn–soybean meal diet containing 3.34 Mcal of DE per kg.
Reproduced with permission from National Research Council: Nutrient requirements of swine, ed 9, Washington, DC, 1988, National Academy of Sciences, National Academy Press.

Table A-25 *Daily energy and feed requirements of lactating gilts and sows*

Intake and performance levels	Weight (kg) of lactating gilts and sows at postfarrowing		
	145	165	185
Milk yield (kg)	5.0	6.25	7.5
Energy required (Mcal DE/day)			
Maintenance*	4.5	5.0	5.5
Milk production†	10.0	12.5	15.0
Total	14.5	17.5	20.5
Feed required per day (kg)‡	4.4	5.3	6.1

*The animal's daily maintenance requirement is 110 kcal of DE per $kg^{0.75}$.
†Milk production requires 2.0 Mcal of DE per kg of milk.
‡The feed required per day is based on a corn–soybean meal diet containing 3.34 Mcal of DE per kg.
Reproduced with permission from National Research Council: Nutrient requirements of swine, ed 9, Washington, DC, 1988, National Academy of Sciences, National Academy Press.

Table A-26 *Daily nutrient intakes and requirements of swine allowed feed ad libitum*

Intake and performance levels	Swine live weight (kg)				
	1-5	5-10	10-20	20-50	50-110
Expected weight gain (g/day)	200	250	450	700	820
Expected feed intake (g/day)	250	460	950	1900	3110
Expected efficiency (gain/feed)	0.800	0.543	0.474	0.368	0.264
Expected efficiency (feed/gain)	1.25	1.84	2.11	2.71	3.79
Digestible energy intake (kcal/day)	850	1560	3230	6460	10,570
Metabolizable energy intake (kcal/day)	805	1490	3090	6200	10,185
Energy concentration (kcal ME/kg diet)	3220	3240	3250	3260	3275
Protein (g/day)	60	92	171	285	404
Requirement (amount/day)					
Nutrient					
Indispensable amino acids (g)					
Arginine	1.5	2.3	3.8	4.8	3.1
Histidine	0.9	1.4	2.4	4.2	5.6
Isoleucine	1.9	3.0	5.0	8.7	11.8
Leucine	2.5	3.9	6.6	11.4	15.6
Lysine	3.5	5.3	9.0	14.3	18.7
Methionine + cystine	1.7	2.7	4.6	7.8	10.6
Phenylalanine + tyrosine	2.8	4.3	7.3	12.5	17.1
Threonine	2.0	3.1	5.3	9.1	12.4
Tryptophan	0.5	0.8	1.3	2.3	3.1
Valine	2.0	3.1	5.3	9.1	12.4
Linoleic acid (g)	0.3	0.5	1.0	1.9	3.1
Mineral elements					
Calcium (g)	2.2	3.7	6.6	11.4	15.6
Phosphorus, total (g)	1.8	3.0	5.7	9.5	12.4
Phosphorus, available (g)	1.4	1.8	3.0	4.4	4.7
Sodium (g)	0.2	0.5	1.0	1.9	3.1
Chlorine (g)	0.2	0.4	0.8	1.5	2.5
Magnesium (g)	0.1	0.2	0.4	0.8	1.2
Potassium (g)	0.8	1.3	2.5	4.4	5.3
Copper (mg)	1.50	2.76	4.75	7.60	9.33
Iodine (mg)	0.04	0.06	0.13	0.27	0.44
Iron (mg)	25	46	76	114	124
Manganese (mg)	1.00	1.84	2.85	3.80	6.22
Selenium (mg)	0.08	0.14	0.24	0.28	0.31
Zinc (mg)	25	46	76	114	155
Vitamins					
Vitamin A (IU)	550	1012	1662	2470	4043
Vitamin D (IU)	55	101	190	285	466
Vitamin E (IU)	4	7	10	21	34
Vitamin K (menadione) (mg)	0.02	0.02	0.05	0.10	0.16
Biotin (mg)	0.02	0.02	0.05	0.10	0.16
Choline (g)	0.15	0.23	0.38	0.57	0.93
Folacin (mg)	0.08	0.14	0.28	0.57	0.93
Niacin, available (mg)	5.00	6.90	11.88	19.00	21.77
Pantothenic acid (mg)	3.00	4.60	8.55	15.20	21.77
Riboflavin (mg)	1.00	1.61	2.85	4.75	6.22
Thiamin (mg)	0.38	0.46	0.95	1.90	3.11
Vitamin B_6 (mg)	0.50	0.69	1.42	1.90	3.11
Vitamin B_{12} (μg)	5.00	8.05	14.25	19.00	15.55

Reproduced with permission from National Research Council: Nutrient requirements of swine, ed 9, Washington, DC, 1988, National Academy of Sciences, National Academy Press.

Trace Minerals and Vitamins

Table A-27 *Approximate dietary trace mineral requirements (mg/kg or ppm) of diet dry matter) of farm animals*

Nutrient	Animal					
	Pig	Horse	Sheep	Beef	Dairy	Calf
Iodine*	0.14	0.1	0.1-0.8†	0.5	0.25-0.5†	0.25
Iron	40-100‡	40-50†,‡	40	50	50	100
Copper§	3-6‡	10-11‡	7‖	10	10	10
Molybdenum		0.1	0.5			
Cobalt		0.1	0.1	0.1	0.1	0.1
Manganese	2-10†	40	30	40	40	40
Zinc	50-100‡	40-50‡	30	30	40	40
Selenium	0.1-0.3‡	0.1	0.1-0.2	0.2	0.1	0.1
Sulfur				0.1	0.2-0.1†	0.2-0.3‡

*Higher levels required if goitrogens are present in feed.
†Higher levels are for pregnancy and lactation.
‡Higher levels are for young, growing animals.
§Requirement is variable, depending on the levels of molybdenum, sulfur, zinc, calcium, iron, and other antagonistic factors in the diet.
‖Requirements are highly variable and depend on sheep breed and levels of antagonistic factors in the diet. Some breeds are adapted to low-copper diets and may be poisoned by "normal" intakes of copper. This is particularly likely to happen if the dietary molybdenum concentration is low.
Adapted from National Research Council: Nutrient requirements of swine, 1988; Nutrient requirements of beef cattle, 1984; Nutrient requirements of dairy cattle, 1978; and Nutrient requirements of horses, 1989.

Table A-28 *Approximate vitamin concentrations required in the diet to ensure adequacy and approximate toxic concentrations*

Vitamin	Cattle		Equine		Porcine
	Adequate	Toxic	Adequate	Toxic	Adequate
A (IU/kg)	4000	66,000	2000-3000*	16,000	1300-4000*
D (IU/kg)	300-1200*	10,000	300-800†	2200	220
E (IU/kg)	15-60	2000	80	1000	22

*Higher levels are required for pregnancy or lactation.
†Higher levels are required for growth.
Adapted from data in National Research Council publications.

Feed

Table A-29 *Descriptive terms for blooming plants*

Term	Definition
Germinated	Sprouted seed
Early vegetative	Fresh growth, prior to stem elongation
Late vegetative	Stems beginning to elongate, just prior to blooming; first bud to first flower
Early bloom	Beginning to bloom, less than 10% of plants in bloom
Midbloom	Between 10% and 70% of plants in bloom
Full bloom	More than 70% of plants in bloom
Late bloom	Blossoms falling, seeds beginning to form
Milk stage	Seeds are well formed but soft and immature
Dough stage	Seeds have doughlike consistency
Mature (dent)	Plant can be harvested for seed (ripe)
Post ripe	Seeds are ripe and plants are cast or weathered
Stem cured	Dormant, seeds lost, plants weathered
Regrowth, early vegetative	Plant regrows due to increased moisture, or end of hot season, or following harvesting; no flowering activity or stem elongation
Regrowth, late vegetative	Regrowth period from stem beginning to elongate to just before blooming

Adapted from National Research Council: Nutrient requirements of sheep, ed 6, Washington, DC, 1985, National Academy of Sciences, National Academy Press.

Table A-30 Approximate nutritional value of selected feeds

	DM %	DE[a] Mcal/kg	DE[b] Mcal/kg	DE[c] Mcal/kg	ME[d] Mcal/kg	NEm[e] Mcal/kg	NEg[b] Mcal/kg	NEl[d] Mcal/kg	Crude protein %	Lysine %	Methionine %	Crude fiber %	NDF %	Ca %	P %	Mg %
Alfalfa																
Fresh, late veg	23		2.94	2.56	2.27	1.41	0.83	1.38	22.2	1.24		24	31	1.71	0.30	0.36
Hay, mid bloom	90		2.28	2.47	2.13	1.24	0.68	1.25	18.7			28	47	1.37	0.24	0.35
Meal, dehydrated	92	2.04	2.36		2.31	1.34	0.77	1.40	18.9	0.92	0.29	26	45	1.51	0.25	0.32
Barley																
Grain	89	3.51	3.68	3.79	3.29	2.06	1.40	1.94	13.2	0.45	0.18	6	19	0.05	0.38	0.15
Hay	88		2.01		2.04	1.18	0.61	1.25	8.8			27		0.24	0.28	0.16
Straw	91		1.62	2.12	1.73	0.60	0.08	1.08	4.4			42	73	0.30	0.07	0.23
Canola																
Meal Seed, solv. ext.	91	3.19	3.11	3.26	2.62	1.60	1.00	1.57	40.9	2.29	0.75	12	24	0.69	1.30	0.61
Carrot																
Roots, fresh	12		3.78		3.29	2.06	1.40	1.94	10.0			10	12	0.40	0.35	0.20
Citrus																
Pomace, dehydrated, no fines	91		2.81	3.70	2.98	2.00	1.35	1.77	6.7	0.22		13	23	1.88	0.13	0.17
Clover																
Alsike hay	88		1.95	2.56	2.09	1.24	0.68	1.28	14.2			30		1.30	0.25	0.45
Clover																
Red, fresh, early bloom	20		2.53	3.00	2.58	1.60	1.00	1.55	20.8			23	40	2.26	0.38	0.51
Fresh regrowth	24		3.19			1.57	0.97		22.3				27	1.71	0.26	0.48
Hay	88		2.22	2.78	2.45	1.14	0.58	1.47	15.0			31	47	1.38	0.24	0.38

Feed

Feed	DM%															
Corn																
Cobs ground	90		1.36	2.25	1.78	0.97	0.42	1.11	2.8			35	87	0.12	0.04	0.07
Corn and cob-meal (ground ears)	86	3.64	3.29	3.66	3.25	2.03	1.37	1.91	9.0	0.20	0.16	9	28	0.07	0.27	0.15
Fodder, with ears, sun-cured	81		2.06	2.78	2.44	1.47	0.88	1.47	8.9			25	48	0.50	0.25	0.29
Grain	88	4.011	3.84	3.84	3.11	2.24	1.55	1.84	10.4	0.28	0.21	3	11	0.05	0.31	0.12
Silage	30		2.68	3.09	2.85	1.63	1.03	1.69	7.9	0.43		24	49	0.31	0.23	0.20
Stover/straw, no ears or husks	87		1.62	2.60	1.78	0.97	0.42	1.11	5.6			35	79	0.52	0.10	0.31
Cotton																
Hulls	90		1.89	2.16	1.55	0.68	0.15	0.98	4.2			48	88	0.15	0.09	0.14
Seeds, solv. ext, ground	91	2.93	3.01	3.13	2.93	1.82	1.19	1.74	45.4	1.85	0.54	13	28	0.18	1.22	0.59
Fats																
Animal, hydrolyzed	99		8.00		7.30	4.75	3.51	5.84								
Oil, vegetable	100		9.00		7.30	4.75	3.51	5.84								
Flax																
Meal seeds, solv. ext	90		3.04	3.48	3.02	1.88	1.24	1.79	38.4	1.28		10	25	0.43	0.89	0.66
Meadow plants																
Intermountain, hay	95		1.69	2.56	1.24	0.68			8.7			33		0.60	0.18	0.17

DE, Digestible energy; NEm, net energy maintenance; NEg, net energy gain; NEl, net energy lactation; NDF, neutral detergent fiber; Ca, calcium; P, phosphorus; Mg, magnesium.
[a] Pigs.
[b] Horses and ponies.
[c] Sheep.
[d] Dairy cattle.
[e] Beef cattle.

Adapted from National Research Council: Nutrient requirements of swine, 1988; Nutrient requirements of beef cattle, 1984; Nutrient requirements of dairy cattle, 1978; Nutrient requirements of horses, 1989; Nutrient requirements of sheep, 1985; and Nutrient requirements of dairy cattle, 1989.

Continued.

Table A-30 *Approximate nutritional value of selected feeds—cont'd*

	DM %	DE[a] Mcal/kg	DE[b] Mcal/kg	DE[c] Mcal/kg	ME[d] Mcal/kg	NEm[e] Mcal/kg	NEg[b] Mcal/kg	NEl[d] Mcal/kg	Crude protein %	Lysine %	Methionine %	Crude fiber %	NDF %	Ca %	P %	Mg %
Milk																
Bovine, skimmed, dehydrated	94		4.05		3.34	2.07	1.37	1.96	35.5	2.70		0	0	1.36	1.09	0.13
Millet, foxtail																
Hay	87		1.53		2.18	1.28	0.71	1.33	8.4			37				
Molasses																
Sugar beet	78	3.22	3.40	3.40	2.89	1.91	1.27	1.72	8.5			0	0	0.15	0.03	0.29
Oats																
Grain	89	3.10	3.20	3.40	2.98	1.85	1.22	1.77	13.3	0.44	0.20	12	27	0.09	0.38	0.16
Hay	91		1.92	2.34	2.73	1.14	0.58	1.35	9.5			32	63	0.32	0.25	0.29
Straw	92		1.62	2.07	1.78	0.79	0.25	1.11	4.4			40	74	0.23	0.06	0.17
Orchard grass																
Hay, late bloom	91		1.90	2.16	1.96	1.11	0.55	1.20	8.4			37	65	0.26	0.30	0.11
Pea																
Seeds	89		3.45		3.42	2.15	1.48	2.01	26.3	1.72	0.30	6	20	0.14	0.46	0.14
Peanut																
Hulls	91		1.25	0.88					8.1	0.40		63	72	0.26	0.07	0.15
Meal seeds, solv. ext	92	3.41	3.25	3.40	2.98	1.85	1.22	1.77	53.0	1.57	0.48	8		0.32	0.66	0.17
Prairie plants, midwest																
Hay	91		1.62	2.12		1.00	0.45		6.4			34		0.35	0.14	0.26

Feed															
Sorghum															
Grain (milo)	90	3.79	3.56	3.88	3.12	2.06	1.40	1.84	12.7	0.29	3	23	0.04	0.36	0.17
Soybean															
Meal, solv. ext	90	3.88	3.73	3.88	3.29	2.06	1.40	1.94	54.0	3.44	4	8	0.29	0.71	0.33
Timothy															
Fresh, mid bloom	29		2.00	2.73	2.31	1.41	0.83	1.40	9.1		34		0.38	0.30	0.14
Hay, late bloom	88		1.80	2.43	1.95	1.18	0.61	1.20	7.8		36	69	0.38	0.15	0.09
Urea	99								28.0						
Wheat															
Bran	89	2.66	3.30	3.13	2.67	1.63	1.03	1.60	17.4	0.63	11	43	0.14	1.27	0.63
Fresh, early veg	22		2.88	3.31	2.80	1.73	1.11	1.67	27.0		17	46	0.42	0.40	0.21
Grain, red, hard	89	3.82	3.86	3.97	3.47	2.18	1.50	2.04	14.6	0.45	3	12	0.05	0.42	0.14
Hay	89		1.90	2.29	2.13	1.24	0.68	1.30	8.7		29	68	0.15	0.20	0.12
Straw	91		1.62	1.81	1.51	0.64	0.11	0.96	3.5		42	79	0.17	0.05	0.12
Wheatgrass, crested															
Fresh, early veg.	29		2.54	3.31		1.79	1.16		21.0		22		0.44	0.33	0.28
Whey															
Dehydrated	93	3.46	4.06		3.16	1.97	1.32	1.87	14.0	1.00	0	0	0.92	0.81	0.14

DE, Digestible energy; NEm, net energy maintenance; NEg, net energy gain; NEl, net energy lactation; NDF, neutral detergent fiber; Ca, calcium; P, phosphorus; Mg, magnesium.
[a] Pigs.
[b] Horses and ponies.
[c] Sheep.
[d] Dairy cattle.
[e] Beef cattle.

Adapted from National Research Council: Nutrient requirements of swine, 1988; Nutrient requirements of beef cattle, 1984; Nutrient requirements of dairy cattle, 1978; Nutrient requirements of horses, 1989; Nutrient requirements of sheep, 1985; and Nutrient requirements of dairy cattle, 1989.

Continued.

Table A-30 *Approximate nutritional value of selected feeds—cont'd*

					Dry matter basis									
	DM %	DE[a] Mcal/kg	DE[b] Mcal/kg	DE[c] Mcal/kg	ME[d] Mcal/kg	NEm[e] Mcal/kg	NEg[b] Mcal/kg	NEl[d] Mcal/kg	Crude protein %	Lysine %	Methionine %	Crude fiber %	NDF %	Ca %	P %	Mg %

Yeast

Brewer's, dehydrated	93	3.54	3.30		3.07	1.91	1.27	1.82	46.6	3.47	0.71	3.50		0.15	1.47	0.26

DE, Digestible energy; NEm, net energy maintenance; NEg, net energy gain; NEl, net energy lactation; NDF, neutral detergent fiber; Ca, calcium; P, phosphorus; Mg, magnesium.
[a] Pigs.
[b] Horses and ponies.
[c] Sheep.
[d] Dairy cattle.
[e] Beef cattle.
Adapted from National Research Council: Nutrient requirements of swine, 1988; Nutrient requirements of beef cattle, 1984; Nutrient requirements of dairy cattle, 1978; Nutrient requirements of horses, 1989; Nutrient requirements of sheep, 1985; and Nutrient requirements of dairy cattle, 1989.

Table A-31 Ruminal undegradability of protein in selected feeds

Feed	Number of determinations	Undegradability		
		Mean	SD	CV
Alfalfa, dehydrated	8	0.59	0.17	29
Alfalfa hay	12	0.28	0.07	25
Alfalfa silage	6	0.23	0.08	36
Alfalfa-bromegrass	1	0.21		
Barley	16	0.27	0.10	37
Barley, flaked	1	0.67		
Barley, micronized	1	0.47		
Barley silage	1	0.27		
Bean meal, field	1	0.46		
Beans	2	0.16	0.02	14
Beet pulp	4	0.45	0.14	30
Beet pulp molasses	2	0.35	0.03	8
Beets	3	0.20	0.03	16
Blood meal	2	0.82	0.01	1
Brewer's dried grains	9	0.49	0.13	27
Bromegrass	1	0.44		
Casein	3	0.19	0.06	32
Casein, HCHO	2	0.72	0.08	11
Clover, red	3	0.31	0.04	12
Clover, red, silage	1	0.38		
Clover, white	1	0.33		
Clover-grass	2	0.54	0.11	21
Clover-grass silage	7	0.28	0.06	22
Coconut	1	0.57		
Coconut meal	5	0.63	0.07	11
Corn	11	0.52	0.18	34
Corn, 0% cottonseed hulls	1	0.46		
Corn, 7% cottonseed hulls	1	0.43		
Corn, 14% cottonseed hulls	1	0.59		
Corn, 21% cottonseed hulls	1	0.48		
Corn, 10.5% protein, 0% $NaHCO_3$	1	0.36		
Corn, 10.5% protein, 3.5% $NaHCO_3$	1	0.30		
Corn, 12% protein, 0% $NaHCO_3$	1	0.29		
Corn, 12% protein, 3.5% $NaHCO_3$	1	0.24		
Corn, dry rolled	6	0.60	0.07	12
Corn, dry rolled, 0% roughage	1	0.54		
Corn, dry rolled, 21% roughage	1	0.49		
Corn, flaked	1	0.58		
Corn, flakes	1	0.65		
Corn, high-moisture acid	1	0.56		
Corn, high-moisture ground	1	0.80		
Corn, micronized	1	0.29		
Corn, steam flaked	1	0.68		
Corn, steam flaked, 0% roughage	1	0.51		
Corn, steam flaked, 21% roughage	1	0.47		
Corn gluten feed	1	0.25		
Corn gluten feed dry	2	0.22	0.11	51
Corn gluten feed wet	1	0.26		
Corn gluten meal	3	0.55	0.08	14

SD, Standard deviation; SV, coefficient of variation; HCHO, formaldehyde treatment.
Reproduced with permission from National Research Council: Nutrient requirements of dairy cattle, ed 6, Washington, DC, 1989, National Academy of Sciences, National Academy Press.

Continued.

Table A-31 *Ruminal undegradability of protein in selected feeds—cont'd*

Feed	Number of determinations	Undegradability		
		Mean	SD	CV
Corn silage	3	0.31	0.06	20
Cottonseed meal	21	0.43	0.11	25
Cottonseed meal, HCHO	2	0.64	0.15	23
Cottonseed meal, prepressed	2	0.36	0.02	6
Cottonseed meal, screwpressed	2	0.50	0.10	20
Cottonseed meal, solvent	6	0.41	0.13	32
Distiller's dried grain with solubles	4	0.47	0.18	39
Distiller's dried grains	1	0.54		
Distiller's wet grains	1	0.47		
Feather meal, hydrolyzed	1	0.71		
Fish meal	26	0.60	0.16	26
Fish meal, stale	1	0.48		
Fish meal, well-preserved	1	0.78		
Grapeseed meal	1	0.45		
Grass	4	0.40	0.10	26
Grass pellets	2	0.46	0.05	11
Grass silage	20	0.29	0.06	20
Guar meal	1	0.34		
Linseed	1	0.18		
Linseed meal	5	0.35	0.10	27
Lupin meal	1	0.35		
Manioc meal	1	0.36		
Meat and bonemeal	5	0.49	0.18	37
Meat meal	1	0.76		
Oats	4	0.17	0.03	15
Palm cakes	6	0.66	0.06	9
Peanut meal	8	0.25	0.11	45
Peas	4	0.22	0.03	15
Rapeseed meal	10	0.28	0.09	31
Rapeseed meal, protected	1	0.70		
Rye	1	0.19		
Ryegrass, dehydrated	4	0.22	0.14	66
Ryegrass, dried artificially	1	0.71		
Ryegrass, dried artificially, chopped	1	0.30		
Ryegrass, dried artificially, ground	1	0.73		
Ryegrass, dried artificially, pelleted	1	0.54		
Ryegrass, fresh	1	0.48		
Ryegrass, fresh or frozen	3	0.41	0.18	44
Ryegrass, frozen	1	0.52		
Ryegrass silage, HCHO	1	0.93		
Ryegrass silage, HCHO dried	1	0.83		
Ryegrass silage, unwilted	1	0.22		
Sanfoin	1	0.81		

SD, Standard deviation; SV, coefficient of variation; HCHO, formaldehyde treatment.
Reproduced with permission from National Research Council: Nutrient requirements of dairy cattle, ed 6, Washington, DC, 1989, National Academy of Sciences, National Academy Press.

Continued.

Table A-31 *Ruminal undegradability of protein in selected feeds—cont'd*

Feed	Number of determinations	Undegradability		
		Mean	SD	CV
Sorghum grain	2	0.54	0.02	4
Sorghum grain, dry ground	1	0.49		
Sorghum grain, dry rolled	2	0.64	0.08	12
Sorghum grain, micronized	1	0.64		
Sorghum grain, reconstituted	2	0.42	0.32	75
Sorghum grain, steam flaked	2	0.47	0.07	15
Soybean meal	39	0.35	0.12	33
Soybean meal, dried 120° C	1	0.59		
Soybean meal, dried 130° C	1	0.71		
Soybean meal, dried 140° C	1	0.82		
Soybean meal, 35% concentrate	1	0.18		
Soybean meal, 65% concentrate	1	0.46		
Soybean meal, HCHO	3	0.80	0.11	14
Soybean meal, unheated	1	0.14		
Soybean-rapeseed meal, HCHO	2	0.78	0.02	3
Soybeans	2	0.26	0.11	40
Subterranean clover	2	0.40	0.18	45
Sunflower meal	9	0.26	0.05	20
Timothy, dried artificially, chopped	1	0.32		
Timothy, dried artificially, pelleted	1	0.53		
Wheat	4	0.22	0.06	27
Wheat bran	4	0.29	0.10	34
Wheat gluten	1	0.17		
Wheat middlings	3	0.21	0.02	11
Yeast	1	0.42		
Zein	1	0.60		

SD, Standard deviation; CV, coefficient of variation; HCHO, formaldehyde treatment.
Reproduced with permission from National Research Council: Nutrient requirements of dairy cattle, ed 6, Washington, DC, 1989, National Academy of Sciences, National Academy Press.

Table A-32 *Average mineral composition of some commonly used macromineral supplements*

Feed	Na	K	Mg	Ca	P	Cl	S
Bonemeal, steamed	5.5	0.2	0.3	29.8	12.5		
Calcium carbonate, $CaCO_3$				38.0			
Calcium phosphate, dibasic, defluorinated			0.6	21.3	18.7		
Calcium phosphate, monobasic, defluorinated				22.0	23.0		
Magnesium carbonate, $MgCO_3Mg(OH)_2$			30.2				
Magnesium oxide, MgO			54.9	3.0			
Magnesium sulfate, heptahydrate, $MgSO_4 \cdot H_2O$			9.6				12.8
Potassium chloride, KCl	1.0	50.0				47.3	
Sodium chloride, NaCl	39.3					60.7	
Sodium phosphate, monobasic, monohydrate, $NaH_2PO_4 \cdot H_2O$	16.2				21.8		

Adapted from National Research Council: Nutrient requirements of swine, Washington, DC, 1988, National Academy of Sciences, National Academy Press.

Table A-33 *Average mineral composition of some commonly used trace mineral supplements*

Supplement	Fe	Cu	Zn	Se	S	I
Copper chloride, dihydrate $CuCl_2 \cdot 2H_2O$		36.9				
Copper oxide, CuO		79.1				
Copper sulfate, pentahydrate, $CuSO_4 \cdot 5H_2O$		25.4			12.8	
Ferrous carbonate, $FeCO_3$	39.6					
Ferrous oxide, FeO	75.4					
Ferrous sulfate, heptahydrate, $FeSO_4 \cdot 7H_2O$	21.4				12.1	
Ferrous sulfate, monohydrate, $FeSO_4 \cdot H_2O$	32.3				18.0	
Potassium iodide, KI						68.2
Sodium selenite, Na_2SeO_3				44.7		
Zinc carbonate, $ZnCO_3$			51.6			
Zinc oxide, ZnO			78.0			
Zinc sulfate, heptahydrate, $ZnSO_4 \cdot 7H_2O$			22.3		10.9	
Zinc sulfate, monohydrate, $ZnSO_4 \cdot H_2O$			36.0		17.5	

Adapted from National Research Council: Nutrient requirements of swine, Washington, DC, 1988, National Academy of Sciences, National Academy Press.

Index

A

Abomasal digestion, 111
Abomasal impaction, 215, 216
Abrubt weaning of pigs, 467, 468
Acid(s), 8; *see also* Amino acid(s)
    organic, 492
Acid detergent fiber, 9
Acid detergent fraction, 369
Acidity of water, 93
Acremonium coenophialum, 186
Acylesters, 102
Ad libitum access to feed, 201, 202
Adaptation, digestive system and, 111, 112
Additives
    feed, 381, 382
    horses and, 428-431
    for lactating cows, 296-301
    replacement heifer feeding and, 271
Aerobic work, 417, 418t
Age
    of cattle, 159, 160
    of puberty in gilts, 456
Agent, repartitioning, 493
β-Agonists, 493
Alfalfa (*Medicago sativa*), 121, 122
    horses and, 410, 428
Algae, toxic blue green, 91, 92
Alimentation
    of clinically ill ruminants, 393-404
    for ill horses, 442-444
    of ill neonatal ruminant, 397-404
    modes of, 141, 142
Alkalinity of water, 93
Allocation, feed, 201, 202
Almond hulls, 199
Alsike (*Trifolium hybridum*), 122
Alterations in nutritional needs, 139-141
Amino acid(s), 22-31, 300, 301, 441
    essential, 22, 23
    importance of, for ruminants, 30
    lactating sows and, 465
    pigs and, 475, 481, 482
    synthetic, 490, 491
Ammonia toxicity, 33, 34

t indicates table

Ammonium chloride, sheep and, 382
Ammonium sulfate, sheep and, 382
Amylase, 101
Anabolic steroids, 443
Anaerobic work, 417, 418t
Analysis
    of hair for trace minerals, 442
    laboratories for, 133t, 134t
    methods of, 133-137
    ration, principles of, 131-137
Anasarca, 71
Aneurine; *see* Thiamin
Angora (hair) goats, 341, 342
    fiber production and, 349, 350
Animal production, disease and, 138, 139
Animal proteins, livestock feed and, 130
Animal welfare, 154, 155
Anorexia, 117-119
    ill animals and, 141, 142
Antibiotic(s), 189, 381, 382
    beef industry and, 153, 154
    effect of, in starter diets, 476, 477
    horses and, 431
    oral, biotin (vitamin H) and, 83
    swine rations and, 489
Antioxidants, 75
Appendix, 497-558
Appetite
    control of, 116-118
    sows and, 463
    specific, 114, 115
    stimulation of, 118
Arachidonic acid, 8
Arginine, 22, 23
    food intake and, 116
Artificial feeding, 245, 246
Artificial rearing
    of lambs, 374-376
    of pigs, 470, 471
Ascorbic acid (vitamin C), 86
Aspartate serum transferase (AST), 76, 77
Aspergillus oryzae, 295, 298
AST; *see* Aspartate serum transferase
Atony, stomach, llamas and, 363, 364
Average daily gain (ADG), 173
Aversions, learned, nutrition and, 115
Azoturia, 425, 426

559

B

B vitamins, 79-83
  appetite and, 118
Backgrounding, 152
Bahia grass *(Paspalpum notatum)*, 124
Bakery waste, 199, 200
Barley *(Hordeum vulgare)*, 126, 195, 196
  horses and, 410
Barn(s)
  feeder, 489, 490, 494t
  temperature of, sows and, 464
Barrows, feeding of, 488
Basal metabolism, energy required for, 6
Basic concepts of bioenergetics, 5-14
Basic language program(s), 210-213
  daily feed allotments, 203t, 204t
  for estimating daily gain, 209t
  for estimating diet NEm and NEg, 211t
Basic principles of nutrition, 2-143
BCAA; *see* Branched-chain amino acids, 441
BCS; *see* Body condition scoring
Bedding, 439
Beef, human health and, 153, 154
Beef cattle, 147-229
  body condition scoring in, 169-178
  equations for dry matter intake in, 499t
  forage-associated diseases in, 183-189
  net energy requirements of, 499t, 507t
Beef cow, feeding of for optimal production, 157-168
Beef industry, 147-156
  animal welfare and, 154, 155
  efficiency in, 155
  human health and, 153, 154
  structure of, 147-153
Beef requirements, 503-511
Beet pulp, 200
  livestock feed and, 128
Beet tops, 228
Behavior, feeding, 114-116
Bermuda grass, 124
Bighead, 46, 426
Biliary secretions, 101
Bioenergetics, 3-21
  basic concepts of, 5-14
  history of, 3-5
  ration formulation and, 15-20
Biological value (BV), 25
Biotin (vitamin H), 83, 84
Bird's-foot trefoil *(Lotus corniculatus)*, 122
Birth weight, pigs and, 470
Blackstrap, 199
Bladder, stones in, 227
Bleeding, vitamin K and, 78, 79
Blended fats, 198
Blindness
  in ruminants, thiamin deficiency and, 81
  sodium deficiency and, 39
  vitamin A deficiency and, 70, 71
Blister beetles, 428
Bloat, 184, 196, 227, 228; *see also* Frothy bloat

Blood, trace mineral measurement and, 64
Blood clotting, vitamin K and, 78
Bluegrass, 124
Body composition and reproduction in swine, 457, 458
Body condition
  changing, 173-175
  dairy cows and, 277, 278
  determining, 169
  economic impact of, 174, 175
  growth rate and milk production and, 171-173
  predicting reproductive performance using, 175-177
Body condition scoring (BCS)
  for beef cattle, 169-178
  for dairy cattle, 316-322
  for ewes, 372, 373
  ill horses and, 442
  systems for, 169, 170t, 318-322
Body scoring system for horses, 420t
Body volume, 4, 5
Body weight, 4, 5, 191-193
  in farm animals, 12, 13
  illness and, 442
  regulation of, 114, 115t
Bomb calorimeter, 9, 10
Bones, calf, magnesium content of, 51t
Bound energy, 4
Bovine bonkers syndrome, 34
Bovine somatotropin (BST), 289-293
Bowel, resection of, 437, 438
Brain, food intake and, 117
Bran disease, 46, 426
Branched-chain amino acids (BCAA), 441
Brassica spp., 187-189
Breeding
  of goats, nutrition and, 346, 347
  of replacement heifer, 163
Breeding cattle, nutrient requirements of, 504-506
Breeds of horses, registration figures for, 408t
Brewer's grain, 128
Brewer's yeast, 79
Bromegrass, 124
Bromus inermis; see Bromegrass
Brood mare, feeding of, 412, 413
Browning, 121
BST; *see* Bovine somatotropin
Bucket feeding, 246
Buckwheat *(Fagopyrum esculentum)*, 126
Buffers, 199
  rumen, 298, 299
Bull(s)
  fertility of, 150, 151
  nutrient requirements of, 521t-526t
Bullets, trace mineral, 65
Burns, nutritional support for horses and, 432-436
Butyrivibrio fibrinosolvens, 305
BV; *see* Biological value, 25
By-product
  livestock feed and, 128
  metabolic, 418t
By-product feeds, feedlot cattle and, 199-201
Bypass protein, 15, 287

Index

C

Calcitonin, 41
Calcitriol, 41, 72, 73
Calcium, 40-47
  calves and, 272
  dairy cows and, 288, 289
  dairy goats and, 352
  dry cows and, 327
  feedlot cattle and, 194
  foals and, 415
  growing-finishing cattle and, 510t, 511t
  growing-finishing pigs and, 482
  horses and, 426, 427
  intravenous, 44
  mares and, 413
  sows and, 465
Calcium-phosphorus ratio, 49
Calcium supplements, analyses of, 130t
Calculation(s)
  of energy requirements of ill animals, 141
  of ration composition, 17t
  for ration formulations, 18, 19
  of wet matter and protein intakes, 19t
Calf; *see* Calves
California Net Energy System, 7
Caloric needs of sick animals, 141
Calorimeter, bomb, 9, 10
Calves, 8
  copper deficiency in, 216
  dairy, colostrum and feeding management of, 242-247
  diarrheic, 395-396
  diet of, 204, 205
  feedlot, feeding of for optimal production, 190-213
  hand reared, dietary management of, 248-260
  important nutrients in feeds for, 271-273
  liquid feeds for, 250-252
  nutritional problems and, 214-222
  physiology of, 268
  sick, nutritional support for, 393-396
  starter, 269, 270
  weaning of, 266-268
Calving season, sweet clover and, 79
Carbohydrates, 8, 9
Carboxylic ionophores, 431
Carcass grading in swine industry, 453, 454
Cardiac disease in horses, 441, 442
Carotene, 69-72, 300
Carotenoids, 69-72
Carriage horses, diet for, 421, 422
Carrots, 200
Casein (milk protein), 27, 28
Cattle
  background of, feedlot feeding and, 191-193
  beef, 147-229
  breeding, nutrient requirements of, 504-506
  dairy, 231-338
  feeder, feeding of, 190-213
  growing and finishing, 508t-511t
  at pasture, calcium deficiency and, 216-219
  purebred, 148
  sick, nutritional support for, 393-396
CCC; *see* Commodity Credit Corporation
CCN; *see* Cerebral cortical necrosis
Central nervous system cues to feeding activity, 117
Cereal forages, 125, 126
Cereal grains, grinding of, 486, 487
Cerebral cortical necrosis (CCN), 81
Cestrum diurnum, 427
CGE; *see* Combustible gas energy, 6
Chelates, 65
Chemical analysis of feeds, 135, 136
Chemical composition of feeds, 8, 9
Chlorine, 39, 40
Choke, esophageal, 364
Cholecalciferol, 72
Cholesterol, beef and, 153
Choline, 86
Chromium (Cr), 56t
Chronic obstructive pulmonary disease (COPD), horses and, 439
Citrus pulp, 200
Clinical nutrition, 139
Clover, 122-123
  sweet, 78, 79
Co; *see* Cobalt
Cobalamins; *see* Vitamin B_{12}
Cobalt (Co), 56t, 58t, 63t, 85, 86, 221, 222
Cold stress, 11, 12
Cold weather, calves and, 269
Colic in horses, 423, 424
Coliform counts, 91
Colitis X, 425
Colon, resection of, 437
Colostral immunoglobulin, 242-244
Colostrum, 242-247
  preserved, 251, 252
Combustible gas energy (CGE), 6
Commodity Credit Corporation (CCC), 237
Common clover, 123
Compensatory gain, 14
Composition
  chemical, of feeds, 8, 9
  milk, dietary fat and, 310
  mineral, of mature bovine carcass, 35t
Computer program(s); *see also* Basic language program(s)
  documentation and user friendliness of, 337, 338
  for ration balancing, 332-338
Concentrates, 10
  goats and, 344
  horses and, 423
  sampling of, 132
Conception rate of gilts, 456, 457
Condition; *see also* Body condition
  nutrition and, 159-161
Conditioners, 198, 199
Constipation, 362, 363
Consumption of meat, 450t
Consumption trends in dairy industry, 231, 232
Contaminants, 91-94
  horses and, 428-431
  mineral, 92-94

Contaminated feed, 79
Control of feed intake, 114-119
Convulsion(s), sodium deficiency and, 39
Copper (Cu), 56t, 58t, 61t, 63t
 deficiency of, 216-219
  foals and, 415
  goats and, 353, 354
  horses and, 413
  limits of, for livestock, 65t
  sheep and, 389-391
Copper glycinate, 219
Copper toxicity, sheep and, 390, 391
Corn, 126, 127, 164, 196
 horses and, 410
Corn gluten, livestock feed and, 128
Corn protein (Zein), 28
Corn stalks, 166
Coronilla varia; see Crown vetch
Cost(s)
 management practices to control, 163-167
 of protein supplements, 165t
 unit, of feed, 15, 16
Cotton gin trash, 200
Cottonseed hulls, 200
Cottonseed products, calves and, 272, 273
Cow(s); see also Beef cow; Dairy cow(s)
 annual cycle of, 149-151
 nutrient requirements of, 516t-520t
Cow-calf enterprise, 148-151
Cow-calf practice, nutritional problems and, 214-222
Cow worksheet, 176t, 177t
CPK; see Creatine phosphokinase
Cr; see Chromium
Cracked hooves, 440
Cramping, muscular, 426
Creatine phosphokinase (CPK), 76, 77
Creep feed, foals and, 414
Creep feeding, 161
 of lambs, 377-379
 of pigs, 467-469
Crimson clover (Trifolium incarnatum), 122, 123
Crop residue fields, 165, 166
Crown vetch (Coronilla varia), 123
Crude protein, 23, 24
Crumbling of feed, 487
Cu; see Copper
Cues to eating behavior, 116, 117
Custom feeding, 148
Cyanobacteria, 91, 92
Cycle, annual cow, 149-151
Cynodon dactylon; see Bermuda grass

D

Dactylis glomerata; see Orchard grass
Dairy calf, colostrum and feeding management of, 242-247
Dairy cattle, nutrient requirements of, 513-526
Dairy cow(s)
 body condition and, 277, 278
 dietary fat and, 304-315
 dry, nutritional problems and, 325-329
 energy needs of, 274-280
 energy requirements of, 500t
 feeding of for optimal production, 274-303
 grouping of, 278, 279
 heat stress and, 293-296
 human nutrition and, 222, 223
 lactation and, 329, 330
 mineral and vitamin needs and, 288, 289
 nitrogen metabolism and, 280-288
 nutritional requirements for, 301, 302
 protein needs of, 280-288
 recent advances and feeding implications and, 289-301
Dairy Fat Prills, 307, 308
Dairy goats, 339-341
Dairy Herd Improvement Association (DHIA), 311
Dairy industry
 body condition scoring (BCS) and, 316-322
 consumption and production trends in, 231, 232
 management systems in, 233-237
 in North America, 231-238
 nutritional problems encountered in, 323-331
 price support systems for, 237, 238
 replacement heifers and, 261-273
 veterinarian and, 239-244
DE; see Digestible energy
Death, sudden, 76
Decoquinate, 271
Deficiency
 biotin (vitamin H), 84
 calcium, 42-47, 352
 choline, 86
 copper, 216-219
  sheep and, 389, 390
 dietary, of goats, 351-354
 of energy and protein in sheep, 384, 385
 folacin (vitamin B_C), 84
 iodine, 353
 iron, 352, 353
 magnesium, 51, 216, 352
 micronutrient, in llamas, 363, 365
 niacin (vitamin PP), 83
 pantothenic acid (vitamin B_3), 82
 phosphorus, 48, 49, 216
 potassium, 36, 37
 protein, appetite and, 118
 pyridoxine (vitamin B_6), 83
 riboflavin (vitamin B_2), 82
 selenium, 219-222, 352
 signs of, 44t
 sodium, 37, 38, 40
 sulfur, 352
 thiamin, 80-82, 227
 trace mineral, 60-62, 216-222
 vitamin, 68
 vitamin A, 69-72, 226, 227
 vitamin B_{12}, 85
 vitamin D, 73, 74
 vitamin E, 76, 77, 219-222
 vitamin K, 78, 79
Deficiency syndromes, 426-428

Index

Deformities of limbs, 363
Degradable intake protein (DIP), 283
Delivery of feed, 202-204, 240, 241
Depraved appetite; see Pica
Detergent system analysis, 135, 136
Developmental orthopedic disease in horses, 414-416
Deworming, 139
DHIA; see Dairy Herd Improvement Association
Diagnosis of copper deficiency, 217-219
Diarrhea
   in horses, 436
   nutritional, 362
Diarrheic calf, 395-396
Diary cattle, 231-338
Diazepam, ill horses and, 443
Dicoumarol, 78, 79
Diet
   diagnosis of trace mineral deficiency and, 62
   fat content of, 481
   growing-finishing, 205, 206
   for growing-finishing lambs, 379-382
   for growing-finishing pigs, 487, 488
   receiving, 204
   sample
      for performance horses, 419-422
      for pigs, 483
   starter, for weaned pigs, 474-477
   transition, 205
Diet formulation
   ingredient considerations in, 195-201
   nutrient considerations in, 193-195
Dietary changes, 435
Dietary fat, 481
   dairy cows and, 304-315
   ration formulation and, 313, 314
   supplements for, 307-313
Dietary fiber, goats and, 351, 352
Dietary ingredients, protein values for, 29t
Dietary lipids, 304, 305
Dietary recommendations for clinically ill horses, 433t, 434t
Digestibility
   effect of protein on, 30, 31
   feed and, 104, 105
Digestible energy (DE), 6
   clinically ill horses and, 434, 435
   pigs and, 480, 481
Digestible energy content of feeds for horses, 411t
Digestion; see also Digestive system(s)
   gastric, 99-101, 105, 106
   horses and, 108, 109
   pigs and, 107, 108
   ruminants and, 109-111
   small intestine, 101-103
Digestive system(s), 97-113
   adaptation and, 111, 112
   fermentation related problems and, 112
   maturation of, 106, 107
   neonates and 105-107
   overview of, 99-104
DIP; see Degradable intake protein

Disease(s)
   developmental orthopedic, in horses, 414-416
   forage-associated, in beef cattle, 182-189
   in horses, 423-431, 436-442
   inflammatory, trace minerals and, 59
   nutritional, in sheep, 384-392
   nutritional needs and, 138-143
Disorders, metabolic, in horses, 423-426
Distiller's grain, 128
Disturbances, metabolic, in sheep, 385-391
DM; see Dry matter
Documentation of computer programs, 337, 338
Draft horses, diet for, 421, 422
Dressage, diet for horses in, 421
Drop feeding, 488
Dry cows, 325-329
Dry grass, 166
Dry matter (DM), 18, 274
   equations for intake of, 499
Dry matter intake, 14, 499
   effect of protein on, 30, 31
Duodenum, 101
Dutch clover, 123
Dysfunction, gastrointestinal, in horses, 436-438
Dystocia, 325-327

E

Eating behavior, grazing and, 115, 116
Ecology and plant physiology, 179-181
Economic evaluation
   computer programs for, 335
   of feed, 240
Economic impact of body condition in cows, 174, 175
Economic implications of calf crop percentage, 150t
Economic value of feed, 15, 16
EDTA, 219
Efficiencies of utilization (k values), 8
Efficiency
   of beef production, 155
   of food production, 13, 14
   reproductive, 157
Electrolyte(s)
   performance horses and, 418, 419
   sick cattle and, 393-395
Electrolyte fluxes, 103
Electrolyte mix, 419
Elements, maximum tolerable dietary levels of, 300t
Embryo survival, 457
Emphysema, pulmonary, *Brassica* spp. and, 188
Endurance races, diet for horses in, 421
Energy, 3-21
   dairy cow needs for, 274-280
   gilts and, 465
   goats and, 351
   ill neonatal ruminants and, 397, 398
   llamas and, 357
   mares and, 412
   measurement of, 5-8
   performance horses and, 417-419

Energy—cont'd
  pigs and, 480, 481
  and protein, 214, 215
  required for basal metabolism (Hb), 6
  required for normal activity (Ha), 6
  sheep and, 368, 384, 385
  sources of, in livestock feed, 128
  supplementation of, in ill animal, 141
Energy Booster, 307, 308
Energy content of feed, 9, 10
Energy deficiency, signs of, 10, 11
Energy deposited in growth and fat storage (NEg), 7
Energy deposited in milk (NEl), 7
Energy deposited in products of gestation (NEr), 7
Energy interconversions, 7
Energy requirements, 499-501
  of sick animals, 141
Energy used to perform work (NEw), 7
Enteral solutions, 398-400
Enteral support for ill neonatal ruminants, 398
Enteric organisms, 91
Enterolithiasis in horses, 438
Enterotoxemia, 362
Entropy, 4
Environmental factors
  dietary fat supplementation and, 312, 313
  eating activity and, 115
  nutritional requirements and, 160, 161
  trace minerals and, 59
Environmental temperature
  newly purchased pigs and, 487
  piglets and, 470
Enzyme supplementation, 493, 494
Equisetum spp., 80
Ergocalciferol, 72
Ergot, 228
Esophageal choke, llamas and, 364
Esophageal transport, 99
Esophageal tube feeding, 246
Esophagostomy, 443, 444
Essential amino acids, 22, 23
Estrus; *see* Postpartum estrus
Evaluation
  of forages, 121, 122
  nutritional, of ill horses, 442
Events, three day, diet for horses in, 421
Ewe(s)
  appropriate rations for, 375t
  nutrition of, 371-374
  parturient paresis in, 387, 388
Examination, diagnosis of trace mineral deficiency and, 62
Exertional myopathy, 77
Exophthalmos, 71

F

Fagopyrum esculentum; *see* Buckwheat
Farm animals, efficiency of growth and, 12-14
Farm size, 452
Farm visit by veterinarian, 239, 240

Fat(s), 8; *see also* Dietary fat
  digestibility of types of, 257t
  disease and, 140
  feedlot cattle and, 193, 197, 198
  food intake and, 116
  ill neonatal ruminants and, 397, 398
  livestock feed and, 128
  rumen bypass, 307, 308
  sow diets and, 462
  supplementation of, 465
  synthesis of, 306
Fat cow syndrome, 277
Fat-soluble vitamins, 68-79
Fatty acids, 8, 304-315
Fe; *see* Iron
FE; *see* Fecal energy
Fecal energy (FE), 6
Feed(s), 97-143, 549-558
  analysis of, 23, 24, 133-137
  by-product, feedlot cattle and, 199-201
  calf, important nutrients in, 271-273
  carotene content of, 70t
  chemical composition of, 8, 9
  computer reference library and, 332, 333
  contaminated, 79
  creep
    composition of, 468, 469
    foals and, 414
  crumbling and pelleting of, 487
  delivery of, 202-204, 240, 241
  digestibility and, 104, 105
  disease in horses and, 423-431
  economic evaluation of, 240
  economic value of, 15, 16
  energy content of, 9, 10
  evaluation of, 252-259
  flavors of, 492
  high concentrate, 225, 226
  high energy, 126-129
  for horses, 409-412
  liquid, for calves, 250-252
  for livestock, 120-130
  mixing of, 201, 484, 485
  moderate energy, 120-126
  molds and fungi associated with, 429t
  nutritional value of, 550-554
  odor, taste, and texture of, 116
  physical quality of, 240
  preserved, 120
  problematic or toxic, 429t, 430t
  processing of, 225, 476, 477, 486, 487
  ruminal undegradability of protein in, 555t-557t
  sampling of, 132, 133
  selection of, 15, 16
  testing of, 484
  vitamin D content of, 73t
  wastage of, 489
Feed additives, 381, 382
Feed allocation, 201, 202
Feed bunk space, 202-204
Feed conversion ratios, 138, 139

Index

Feed intake
  control of, 114-119
  pattern of, in sows, 461, 462
  sick cattle and, 393-396
  supplemental fat and, 309
Feed toxins, 228
Feeder barn, 489, 490
  goals for, 494t
Feeder cattle, feeding of, 190-213
Feeder enterprise, 151-153
Feeder pigs, newly purchased, 487
Feeding, 97-143; see also Feedlot feeding
  of beef cow for optimal production, 157-168
  of calves, 245, 246
  creep, 161
    of pigs, 467-469
  custom, 148
  of dairy cow for optimal production, 274-303
  of gilt and sow, 455-473
  of goats, 346-350
  of horses, 407-409
  of llamas, 359-365
  of sheep, 366-382
  of sick horses, 432-445
  tube, 394
  of weaned pig, 474-479
Feeding activity, substances that influence, 118t
Feeding behavior, 114-116
Feeding hay, 166, 167
Feeding problems for horses, 411t
Feeding programs
  for growing-finishing lambs, 379-382
  for horses, 412-416
Feeding standards for ruminants, 34
Feeding systems, 249, 250
  for llamas, 356
  for sows, 464, 465
Feedlot(s)
  nutritional problems and, 224-229
  nutritional requirements and, 226, 227
  problems in practices in, 227, 228
Feedlot feeding, 190-213
  ingredient considerations in diet formulations and, 195-201
  management of, 201-204
  monitoring feeding program and, 206-213
  nutrient considerations in diet formulations and, 193
  strategies for, 204-206
  targeting growth and performance in, 190-193
Feedstuffs
  goats and, 343-346
  for growing-finishing pigs, 483, 484
  llamas and, 357, 358
  mineral composition of, 36t
Fermentation
  digestion and, 112
  rumen, 305
Fertility, bull, 150, 151
Fescue, 124, 186
Fescue toxicity, 186, 187
Festuca arundinacea, 124, 186

Fetal membranes, retained, 327, 328
Fever; see also Milk fever
  shipping, 152
Fiber, 9
  dietary, goats and, 351, 352
  llamas and, 357
Fiber production, goats and, 349, 350
Fibrous osteodystrophy, 426
Fields, crop residue, 165, 166
Finishing pig; see Growing-finishing pig
First law of thermodynamics, 3
Flavors of feed, 492
Fluid, 103
Fluid evaluation, diagnosis of trace mineral imbalance and, 62-64
Fluid fluxes, 103
Fluoride, dairy cows and, 289
Fluxes, electrolyte and fluid, 103
Foal, feeding of, 414-416
Folacin (vitamin B_C), 84
Folate, 84
Food production, efficiency of, 13, 14
Forage(s), 120-126
  diseases in beef cattle associated with, 182-189
  dry or ensiled, 132
  feedlot cattle and, 196, 197
  goats and, 343, 344
  horses and, 410, 411
  for lactating cows, 279, 280
  llamas and, 358
  low quality, use of, 163, 164
  moisture of, 136, 137
Foraging, energy and, 12
Foraging behavior, 115
Free-choice salt, animal consumption of, 39t
Free energy, 4
Frequency of feeding, 202
Frothy bloat, 184, 185
Fungal additives, 298
Fungal culture, 295
Fungi, feeds and, 429t

G

Gain, compensatory, 14
Gastric digestion, 99-101, 105, 106
Gastrointestinal cues to feed intake, 116
Gastrointestinal dysfunction in horses, 436-438
Gastrointestinal infections, 139
Gastrointestinal resection, 436-438
Gastrointestinal tract of llamas, 357
Gastrointestinal ulceration, 363
GE; see Gross energy, 6
Geriatric horses, 438, 439
Gestation
  dairy goats and, 347
  sows and, 458-461
Giardia, 91
Gilt(s)
  energy and feed requirements of, 545t

Gilt(s)—cont'd
  feeding of, 455-473, 488
  nutrition and, 456-458
Glucose, 432
  disease and, 139, 140
Glycine, 22, 23
Goal(s)
  for feeder barn, 494t
  of heifer replacement program, 262-264
  of swine industry, 453, 454
Goat(s), 339-355
  angora (hair), 341, 342
  dairy, 339-341
  dietary deficiencies and, 351-354
  energy requirements of, 500t
  feeding of for optimal production, 346-350
  feedstuffs and, 343-346
  nutrient requirements of, 527-530
  nutritional problems and, 351-355
  sick, feeding of, 354, 355
  Spanish (meat), 342, 343
Goat industry, 339-343
Gossypol, 272, 273
Grading, carcass, in swine industry, 453, 454
Grain(s), 126-128, 195, 196, 228
  high-concentrate, goats and, 354
  horses and, 409, 410
  llamas and, 358
  sampling of, 132
Grain overload, 112, 225
Grain production in United States (1987), 127t
Grape pomace, 200
Grass staggers, 51
Grasses, 120, 121, 124, 125
Grazing
  eating behavior and, 115, 116
  systems for, 181, 182
Grazing distribution, 163
Grease, 198
Grinding of cereal grains, 486, 487
Gross energy (GE), 6
Grouping of cows for maximum milk production, 278, 279
Growing-finishing diet, 205, 206
Growing-finishing lambs, feeding programs and diets for, 379-382
Growing-finishing pig(s)
  feeding of, 480-495
  selection of dietary ingredients for, 483
Growth
  dairy goats and, 348, 349
  dairy heifers and, 262-264
  energy and, 12-14
  feedlot, targeting of, 190-193
  nutrition and, 157-167
Growth rate
  body condition and, 171-173
  disease and, 138, 139
Gymkhana, 419

H

Ha; *see* Energy required for normal activity
Haemonchus placei, 187
Hair goats, 341, 342
Hair trace mineral analysis, 442
Hand rearing, 248-260
  evaluation of feeds and, 251-259
Hardness of water, 93
Harness racing, diet of horses in, 419-421
Hay, 121, 122
  feeding, 166, 167
  horses and, 410
  sampling of, 132
Hb; *see* Energy required for basal metabolism
Health, human, beef and, 153, 154
Heat, energy and, 4
Heat increment (Hi), 6
Heat stress, 11, 12, 293-296
Heaves, 439
Heifer(s); *see also* Replacement heifer(s)
  first calf, nutritional requirements of, 164t
Heliotropium spp., 354
Hemoglobinuria, postparturient, 48
Hepatic failure, horses and, 440, 441
Hepatosis dietetica, 77
Heptamine, 401
Herbicides, 94
High-concentrate feeds, 225, 226
High-concentrate grain, 354
High energy feeds, 126-129
Histidine, 22, 23
Holstein calves, diet of, 204, 205
Home-made electrolyte mix, 419
Hominy, 200
Hooves
  cracked, 440
  crumbly, misshapen, soft, of horses, 84
Hordeum vulgare; *see* Barley
Horse(s), 407-445; *see also* Performance horse(s)
  additives, contaminants, and toxic feed and, 428-431
  alimentation and, 442-444
  B vitamin synthesis in, 80t
  body scoring system for, 420t
  cardiac disease in, 441, 442
  carriage, 421, 422
  chronic obstructive pulmonary disease and, 439
  cracked hooves and, 440
  deficiency and toxicity syndromes in, 426-428
  diet for, 419-422
  digestive systems of, 97-113
  diseases in, 423-431, 436-442
  draft, 421, 422
  energy requirements of, 501t
  feed for, 409-412
  feeding of for optimal performance and growth, 407-409
  feeding problems for, 411t
  feeding programs for, 412-416
  gastrointestinal dysfunction and, 436-438
  geriatric, 438, 439
  hepatic failure in, 440, 441

Index

nutrient requirements of, 539-541
quarter, 439, 440
registration figures for breeds of, 408t
renal disease and, 441
sick, feeding of, 432-445
Horse industry, 407
Housing
nutrition and, 266-268
in swine industry, 448-450
Human health, beef and, 153, 154
Human nutrition, role of dairy cow in, 222, 223
Hyperalimentation, 400
Hyperesthesia, 72
Hyperkalemic periodic paresis, quarter horses and, 439, 440
Hyperlipemia in horses, 425
Hyperlipidemia in horses, 425
Hyperparathyroidism, 46, 426, 427
Hypoalbuminemia, 442
Hypocalcemia, 42, 43, 46
ewes and, 387, 388
Hypocalcemic disease, 42
Hypomagnesemia, 50, 51
Hypomagnesemic tetany, 183, 184
Hypothalamus, 117

I

I; *see* Iodine
IgG1; *see* Colostral immunoglobulin
Ill animals, 141
Impaction
abomasal, 215, 216
of large intestine, 423
Implants, 381, 382
in beef industry, 154
suckling calves and, 161
Industry; *see also* Beef industry; Dairy industry
goat, 339-343
horse, 407
pork, 447, 448
swine, 447-454
Infection, gastrointestinal, 139
Inflammation, trace minerals and, 59
Ingestion, 99
Ingredient considerations in diet formulation, 195-201
Ingredients, dietary
for pigs, 483
protein values for, 29t
Inositol, 86
Intake
dry matter, equations for, 499
feed
control of, 114-119
dietary fat and, 309
sick cattle and, 393-396
of water, expected mean, 90t
Interconversions, energy, 7
Internal parasitism, 187
Intestinal absorption, ruminants and, 305, 306

Intestinal digestion, ruminants and, 305, 306
Intestine, impaction of, 423
Intoxication, salt, 38, 39
Intragastric supplementation, 142
Intravenous alimentation, 142
of horses, 444
Intravenous calcium, 44
Iodine (I), 56t, 58t, 61t, 63t
goats and, 353
limits of for livestock, 65t
mares and, 413
Iron (Fe), 56t, 58t, 63t
clinically ill horses and, 436
goats and, 352, 353
horses and, 428
limits of for livestock, 65t
supplementation with, 140
Isoacids, 296-298
Isobutyric acid, 296-298
Isoleucine, 22, 23
Isomaltase, 101
Isovaleric acid, 296-298

J

Johne's disease, 139

K

K values, 8
K_3; *see* Menadione
Kale, 126
kcal; *see* Kilocalories
Ketosis, 19
ovine, 385-387
Kidney, stones in, 227
Kids, 8
feeding of, 354, 355
iron and, 352, 353
Kilocalories (kcal), 5
Kinetic energy, 3
Kitchen grease, 198

L

Laboratories for forage analysis, 133t, 134t
Lacerations, rectal, 438
Lactase, 101
Lactating animals, food intake and, 14
Lactating cows, nutrient requirements of, 516t-520t
Lactating mares, 427
Lactation
additives and, 296-301
calcium and, 42, 43
dairy goats and, 347, 348
dietary fat and, 306
energy and, 19, 274-277
forages and, 279, 280

Lactation—cont'd
  horses and, 412, 413
  llamas and, 359-360
  magnesium and, 50
  mares and, 427
  nonprotein nitrogen and, 285, 286
  nutritional problems and, 329, 330
  sows and, 463-465
  swine industry and, 450
Ladino clover, 123
Lamb(s), 8, 371
  artificial rearing of, 374-376
  creep feeding of, 377-379
  energy requirements for, 369t
  growing-finishing, feeding and diet for, 379-382
  weaning of, 376, 377
Laminitis, 424, 425
Large colon resection, 437
Large intestine, 103
Lasalocid, 271, 431
Law(s)
  surface, 4
  of thermodynamics, 3, 4
Laxatives, sows and, 462, 463
Learned aversions, 115
Legumes, 122-124
  when to harvest, 121t
Leptospira, 91
Lespedeza, 124
Leucine, 22, 23
Library, computer, for feed reference, 332, 333
Limb, deformities of, 363
Limestone ($CaCO_3$), 199
Linoleic acid, 8
α-Linolenic acid, 8
Lipid(s), 8
  beef and, 153
  dietary, 304, 305
Liquid feeds for calves, 250-252
Liquid supplements, 285, 286
Liver
  digestion and, 103, 104
  vitamin A concentrations in, 71t
Liver disease, 440, 441
Livestock
  feeds for, 120-130
  limits of mineral elements for, 65t
Llama(s), 356-365
  anatomy of gastrointestinal tract of, 357
  energy, fiber, and protein requirements of, 357
  feeding requirements of, 359-361
  juvenile, nutritional problems and, 362, 363
  mature, nutritional problems and, 363-365
  nutritional problems of, 362-365
Lolium spp.; *see* Ryegrass
Lotus corniculatus; *see* Bird's foot-trefoil
Lysine, 22, 23, 25, 30t
  pigs and, 481, 482
  swine diets and, 475, 476
  synthetic, 490, 491

M

Macromineral(s), 35-54
  cattle at pasture and, 216
  sheep and, 367t
Macromineral supplements, 558t
Magnesium, 49-52, 216
  calves and, 272
  content of in calf bones, 51t
  goats and, 352
  imbalance of, foraging and, 183, 184
  supplemental, 183, 184
Magnesium oxide (MgO), 199
Maillard reaction, 121
Maintenance energy, 11
Maize, 127
Major minerals, 35-54
Malabsorption, 437, 438
Malnutrition, protein-energy, 214
Maltase, 101
Management
  of dry cows, 328, 329
  of feedlot feeding, 201-204
  pasture, 179-182
  in rearing of calves, 248-250
Management practices to control maintenance costs, 163-167
Management systems in dairy industry, 233-237
Manganese (Mn), 56t, 58t, 61t, 63t
  limits of for livestock, 65t
Mare(s); *see also* Brood mare
  lactating, 427
Masonex, 199
Mastication, 99
Mastitis, 325-327
Mastitis milk, 269
Mcal; *see* Megacalories
ME; *see* Metabolizable energy
Measurement
  of forms of vitamin E, 75
  systems of, 5-8
Measures of vitamin A activity, 69t
Meat consumption, per capita, 450t
Meat goats, 342, 343
Meat products, nutrient composition of, 454t
Medicago sativa; *see* Alfalfa
Medication, swine rations and, 489
Megacalories (Mcal), 5
Megalac, 307-310
Melilotus alba; *see* Sweet clover
Melilotus officinalis; *see* Sweet clover
Membranes, fetal, retained, 327, 328
Menadione (K_3), 78, 427
  clinically ill horses and, 435
Metabolic by-products, 418t
Metabolic cues to feeding activity, 116, 117
Metabolic disorders in horses, 423-426
Metabolic disturbances in sheep, 385-391
Metabolic fecal nitrogen, 24
Metabolism, digestion and, 104
Metabolizable energy (ME), 6

Index

Metabolizable protein, 27, 28t
Metalloenzymes of trace minerals, 56t, 57t
Methione hydroxy analogue (MHA), 299, 300
Methionine, 22, 23, 25, 30t, 300, 301
Methods
    of analysis, 133-137
    of feeding for pigs, 477, 478
2-Methyl butyric acid, 296-298
Metritis, 325-327
MHA; see Methione hydroxy analogue
Microangiopathy, 76
Microbial contaminants, 91-94
Microminerals, sheep and, 367t
Micronutrient deficiencies in llamas, 363-365
Microphthalmia, 72
Milk
    as calf feed, 268, 269
    composition of, 105
    digestion and, 105
    neonatal ruminants and, 398
    quantity of for calves, 269
    of sows, 466, 467
Milk composition, dietary fat and, 310
Milk fever, 42-44, 325-327
    goats and, 352
    vitamin D administration and, 74
Milk lameness; see Osteomalacia
Milk production, 159, 160t
    body condition and, 171-173
Milk protein; see Casein
Milk replacers, 252-259
    calves and, 269
    neonatal ruminants and, 398
Milk substitutes, 268-270
Milk yield, 309, 310
Miller's disease, 46, 426
Millet, 127
Milo, 410
Milo stover, 164
Mineral(s), 226; see also Trace mineral(s)
    calves and, 272
    cow-calf nutrition and, 166, 167
    daily requirements of in drinking water, 93t
    dairy cattle and, 288, 289
    disease and, 140
    feeding of, 289
    llamas and, 358
    major, 35-54
    mares and, 413
    performance horses and, 419
    pigs and, 482
    solubilization of in water, 92
    supplementation of, 130, 140, 165
    toxic concentrations of in milk replacers, 258t
    toxic levels of in water, 94
Mineral chemistries, 136
Mineral composition
    of foodstuffs, 36t
    of mature bovine carcass, 35t
Mineral contaminants, water and, 92-94
Mineral deficiency that can lead to pica, 36

Mix, electrolyte, home made, 419
Mixing of feed, 201, 484, 485
Mn; see Manganese
Mo; see Molybdenum
Moderate energy feeds, 120-126
Mohair, 341, 342
Moisture, forage, 136, 137
Moisture content of feed, 193
Molasses, 198, 199
    livestock feed and, 128
Mold, 228, 429t
Molybdenum (Mo), 56t, 58t, 63t
    goats and, 353
Monday morning disease, 426
Monensin, 271, 431
Monitoring
    feeding program, 206-213
    replacement heifer program, 264-266
Monogastrics, 8
Mortality of piglets, 469-471
Mulberry heart disease, 77
Muscular cramping, 426
Muscular dystrophy, nutritional, 76, 77
Muscular spasm, 426
Mycotoxins, 228, 483
Myodegeneration (muscle degeneration), 76, 77
Myopathy, exertional, 77

N

Na-KATPase, 35, 36
Nasogastric feeding, 443, 444
Naracin, 431
National Research Council (NRC)
    dietary fat and, 304
    energy calculations and, 7
    energy requirements and, 19
    feeds and, 131
    nutritional requirements for goats and, 346
    phosphorus requirements for beef cows of, 48
Natural sucking, 244, 245
NDF; see Neutral detergent fiber
NE; see Net energy
Near infrared reflectance spectroscopy (NIRS), 121, 134, 135
Nebraska system for body condition scoring, 169, 170t
NEg; see Energy deposited in growth and fat storage
NEl; see Energy deposited in milk
Neonatal ruminant(s)
    alimentation of ill, 397-404
    energy, fat, and protein requirements of, 397, 398
    enteral support for, 398
Neonates, digestion and, 105-107
NEp; see Net energy for production
Nephramine, 401
NEr; see Energy deposited in products of gestation
Net energy (NE), 6, 7, 20
    feedlot cattle and, 193
    for maintenance (NEm), 6
    for production (NEp), 7

Net energy requirements of beef cattle, 507t
Neutral detergent fiber (NDF), 9
NEw; *see* Energy used to perform work
Newborn calves, 266-270
Newly purchased feeder pigs, 487
Ni; *see* Nickel
Niacin (vitamin PP), 82, 83, 299
Nickel (Ni), 57t
Night blindness, 70
Nipple bottle feeding, 246
NIRS; *see* Near infrared reflectance spectroscopy
Nitrate poisoning, 185, 186
Nitrate toxicity, 33
Nitrogen; *see also* Nonprotein nitrogen
   metabolism of, 280-288
   metabolic fecal, 24
Nonprotein nitrogen (NPN), 22-34, 283-288
   utilization of by ruminants, 31-34
Nonruminants, 405-495
   absorption and digestion of proteins in, 24, 25
North America, dairy industry in, 231-238
NPN; *see* Nonprotein nitrogen
NRC; *see* National Research Council
NSH; *see* Nutritional secondary hyperparathyroidism
Nutrient(s), 3-95
   for performance horses, 417-419
   sources of, 418t
   total digestible, 10
Nutrient composition of meat products, 454t
Nutrient considerations in diet formulation, 193-195
Nutrient evaluation of feed, 240
Nutrient requirements
   of breeding cattle, 504-506
   of breeding swine, 459t, 460t
   computer programs and, 332
   of dairy cattle, 513-526
   of goats, 527-530
   of horses, 539-541
   of swine, 543-546
   of weaned pig, 474-476
Nutrition, 190
   basic principles of, 2-143
   calf growth and reproduction and, 157-167
   calves and, 241
   clinical, 139
   of dairy replacement heifers, 261-273
   disease in horses and, 423-431
   during first 3 weeks of life, 268-270
   feeder pig, newer concepts in, 490-494
   gilts and, 456-458
   housing and management and, 266-268
   human, role of dairy cow in, 222, 223
   parenteral, 400-404
   for performance horses, 417-422
   postpartum, 158
   prepartum, 157, 158
   protein, 22-34
   of replacement heifer from 3 weeks to calving, 270
   sow, 458-466
Nutritional diarrhea, 362
Nutritional disease in sheep, 384-392
Nutritional evaluation of clinically ill horses, 442
Nutritional muscular dystrophy, 76, 77
Nutritional needs
   alterations in, 139-141
   disease and, 138-143
Nutritional problems
   in cow-calf practice, 214-222
   in dairy practice, 323-331
   feedlot and, 224-229
   of llamas, 362-365
Nutritional program for weaner pigs, 478, 479
Nutritional requirements
   for dairy cows, 301, 302
   factors influencing, 158-161
   feedlot and, 226, 227
   of growing-finishing pig, 480-483
Nutritional secondary hyperparathyroidism (NSH), 46, 426, 427
Nutritional support, 139-142
   for sick calves and cattle, 393-396
Nutritional value of selected feeds, 550-554
Nutritional wisdom, 114, 115
Nutritive value of water, 92, 93

O

Obesity, 11
   in horses, 425
   llamas and, 364
Odor of feeds, eating behavior and, 116
Oil, 8
   livestock feed and, 128
Opisthotonos, 39, 81
Oral alimentation of horses, 442, 443
Oral antibiotics, 83
Oral supplementation, 141, 142
Oral supplements for trace minerals, 64, 65
Orchard grass *(Dactylis glomerata)*, 124, 125
Organic acids, 492
Organisms, enteric, 91
Oropharyngeal cues to eating behavior, 116
Orphaned llamas, 359
Orthopedic disease, developmental, 414-416
Oryza sativa; *see* Rice
Osteomalacia (osteoporosis), 44, 46
Osteopetrosis, 46
Osteoporosis; *see* Osteomalacia
Ostertagia ostertagi, 187
Outputs of computer programs, 335-337
Overcrowding of growing pigs, 490
Overfeeding of horses, 414, 415
Overload, grain, 112, 225
Ovine ketosis, 385-387
Ovulation rate of swine, 456
Oxytetracycline, 41

P

Palatability, urea and, 32

Index

Pancreatic secretions, 101
Pantothenic acid (vitamin B_3), 82
Papilledema, 70
Parasitism, 138, 139
   internal, 187
Parathormone, 41
Parenteral nutrition, 400-404
   complications of, 403, 404
Parenteral supplements, 65, 66
Paresis, 42-44, 439, 440
   parturient, in ewes, 387, 388
Parity, dietary fat and, 311, 312
Parturient complications, 325-329
Parturient paresis; see also Milk fever
   in ewes, 387, 388
Paspalpum notatum; see Bahia grass
Pasture(s)
   copper deficiency in cattle at, 216-219
   management of, 179-182
   use of, 163
   sampling of, 133
Pathophysiology
   of magnesium, 50
   of phosphorus, 47
   of sodium, 37, 40, 41
   of potassium, 35, 36
Pattern of feed intake in sows, 461, 462
Peanut hulls, 200
Pearson square, 18, 19
Pellagra, 83
Pelleting of feed, 487
Performance, feedlot, targeting of, 190-193
Performance horse, 417-422
   nutrients and, 417-419
   sample diets for, 419-422
Performance targets for weaner pig nutritional program, 478, 479
Periparturient disorders, 325-327
Pesticides, 94
Phalaris arundenacea; see Reed Canary grass
Phenylalanine, 22, 23
Phleum pratense; see Timothy
Phosphorus; 47-49, 216
   cow-calf nutrition and, 167
   dairy cows and, 288, 289
   feedlot cattle and, 194
   goats and, 353, 354
   growing and finishing cattle and, 510t, 511t
   growing-finishing pigs and, 482
   horses and, 426, 427
   mares and, 413
   sows and, 465
Phosphorus-calcium ratio, 49
Phosphorus supplements, analyses of, 130t
Physiological factors, trace minerals and, 64
Physiology
   of calves, 268
   plant, ecology and, 179-181
Physitis, 415
Pica, mineral deficiencies that can lead to, 36
Pig(s); see also Gilt(s); Sow(s); Swine
   creep feeding of, 467-469
   digestion and, 107, 108
   feeder
      newly purchased, 487
      nutrition and, 490-494
   finishing; see Growing-finishing pig
   growing; see Growing-finishing pig
   mortality of, 469-471
   weaned; see Weaned pig
Piglet, 469-471
Plant(s)
   descriptive terms for, 549t
   problematic or toxic, 429t, 430t
   trace minerals and, 57
Plant physiology and ecology, 179-181
Plasma, trace mineral measurement and, 64
Poa pratensis; see Bluegrass
Poisoning
   nitrate, 185, 186
   salt, 38, 39
Polioencephalomalacia, 81, 188, 189
Polo, diet of horses in, 419-421
Poloxalene, 185
Polydipsia, 37, 38
Polyuria, 37, 38
Pork industry, 447, 448
Postoperative period, 436-438
Postpartum estrus, body condition and, 169-171
Postpartum nutrition, 158
Postparturient hemoglobinuria, 48
Postweaning of sows, 465, 466
Postweaning scours, 477, 478
Potassium, 35-37
   feedlot cattle and, 194
   need for, in dairy cows during hot weather, 295
   and sick animals, 36
Potatoes, 228
Potential energy, 3
Preconditioning, 152, 153
Predicting reproductive performance using body condition, 175-177
Pregnancy
   body condition and, 169-171
   calcium and, 42, 43
   energy and, 19
   horses and, 412, 413
   llamas and, 359-360
   magnesium and, 50
   nutrition and, 158, 159
   protein-energy malnutrition and, 214
Pregnancy rate, 456, 457
Pregnancy testing, economic benefits of, 151t
Pregnancy toxemia, 385-387
Pregnant animals, food intake and, 14
Pregnant cows, nutrient requirements of, 516t
Pregnant gilts, energy and feed requirements of, 461t
Pregnant sows, energy and feed requirements of, 461t
Preparation of diet for swine, 484-486
Prepared feed, 344, 345
Prepared mixes, llamas and, 358
Prepartum nutrition, 157, 158

Preserved colostrum, 251, 252
Preserved feeds, 120
Preweaning, 161, 266, 267
Price support systems for dairy industry, 237, 238
Principles
   basic, of nutrition, 2-143
   of ration analysis, 131-137
Probiotics, 491, 492
Problem(s)
   nutritional
      in dairy practice, 223-331
      goats and, 351-355
      water, 91-94
Processing of feeds, 225, 476, 477, 486, 487
Production systems in swine industry, 448-451
Production trends in dairy industry, 231, 232
Program(s); *see also* Basic language program(s)
   computer, for ration balancing, 332-338
   feeding, for horses, 412-416
Prolificacy of gilts and sows, 456
Proteases, 101
Protein(s), 8, 226
   absorption and digestion of, 24-31
   analysis of feeds for, 23, 24
   animal, livestock feed and, 130
   bypass, 15
   dairy cows need for, 280-288
   disease and metabolism of, 140
   effect of on dry matter intake and digestibility, 30, 31
   and energy, 214, 215
   feedlot cattle and, 193
   ill neonatal ruminants and, 397, 398
   lactating sows and, 465
   llamas and, 357
   mares and, 412, 413
   performance horses and, 418
   ruminal undegradability of, 555t-557t
   sheep and, 368-371, 384, 385
   sows and, 461
   vegetable, 129, 130
Protein deficiency, 118
Protein-energy malnutrition, 214, 215
Protein nutrition, 22-34
Protein requirements of growing and finishing cattle, 508t, 509t
Protein supplements, 129, 130
   feedlot cattle and, 197
   use of, 165
Protein values for dietary ingredients, 29t
Proximate (Weende) analysis, 135
Pteridium aquilinum, 80
Puberty, age of, of gilts, 456
Pulmonary emphysema, 188
Purebred cattle, 148
Pyridoxine (vitamin B_6), 83

Q

QPM; *see* Quality protein maize
Quality of feed, 240

Quality protein maize (QPM), 127
Quantity of milk for calves, 269
Quarter horses, paresis and, 439, 440

R

Racehorses, 419
Races, endurance, diet for horses in, 421
Ralgro, 382
Range, use of, 163
Rape, 126
Rape blindness, 188
Ration(s)
   economic and nutrient evaluation of, 240, 241
   for ewes, 375t
   for horses, 416
   sample, 312t, 313, 314
Ration analysis, 131-137
Ration composition, calculation of, 17t
Ration evaluation, 16, 17, 240, 241
   computer programs for, 333, 334
Ration formulation, 15-20, 241
   calculations for, 18, 19
   computer programs for, 334, 335
   dietary fat and, 313, 314
Ration-balancing, computer programs for, 332-338
Ratios; *see* Feed conversion ratios
Rearing; *see also* Hand rearing
   artificial
      of lambs, 374-376
      of pigs, 470, 471
   systems of, 248, 249
Receiving diet, 204
Recommendations, dietary, for clinically ill horses, 433t, 434t
Rectal lacerations, 438
Red clover (*Trifolium pratense*), 123
Reed Canary grass (*Phalaris arundenacea*), 125
Registration figures for breeds of horses, 408t
Renal disease, 441
Repartitioning agents, 493
Replacement heifers
   additives used in feeding of, 271
   dairy, nutrition of, 261-273
   goals of replacement program for, 262-264
   housing and management and, 266-268
   monitoring program for, 264-266
   nutrition and, 150t, 270
   nutritional development of, 161-163
Replacers, milk, 398
Reproduction, 149-151
   dietary fat and, 310, 311
   ewes and, 373, 374
   nutrition and, 157-167
   swine and, 457, 458
Reproductive efficiency of gilts, 455
Reproductive performance, 149-151
   nonprotein nitrogen and, 32, 33, 286, 287
   predicting, 175-177
Requirement(s)

Index

beef, 503-511
calcium, 41, 42
energy, 499-501, 507t
magnesium, 50, 51
of minerals in drinking water, 93t
nutritional, factors influencing, 158-161
phosphorus, 47, 48
potassium, 36
sodium, 37, 40
water, 90, 91
Resection, gastrointestinal, 436-438
Restaurant grease, 198
Restricted feeding, 202
Retained fetal membranes, 327, 328
Retinol, 68-72
Return milk, 269
Rhabdomyolysis, 425, 426
Riboflavin (vitamin B_2), 82
Rice *(Oryza sativa)*, 127
Rice hulls, 200
Rickets, 44-46, 72
Roughage(s), 7, 10, 120-126
Rumen
  digestion in, 109-111
  protein and energy degradation and, 283
Rumen buffers, 298, 299
Rumen bypass fats, 307, 308
Rumen fermentation, 305
Ruminal tympany, 184, 185
Ruminant(s), 8, 145-404; *see also* Neonatal ruminant(s)
  absorption and digestion of proteins in, 25-31
  alimentation of clinically ill, 393
  amino acids and, 30
  digestion and, 109-111
  digestive systems of, 97-113
  feeding standards for, 34
  intestinal absorption and digestion of, 305, 306
  protein and, 23
  thiamin deficiency and, 81
  utilization of nonprotein nitrogen by, 31-34
  vitamins and, 68, 84, 85
Rust, 228
Rye *(Secale cereale)*, 127, 228
Ryegrass *(Lolium* spp.), 125

S

Saccharomyces cerevisiae, 298
Salmonella
  beef and, 153, 154
  water and, 91
Salt, 199
  calves and, 272
  as carrier for trace mineral supplementation, 65
  free-choice, animal consumption of, 39t
Salt intoxication, 38, 39
Salt poisoning, 38, 39
Sample diet(s)
  for performance horses, 419-422
  for pigs, 483

Sample ration(s), 312t, 313, 314
  for horses, 416
Samplers, suppliers of, 132t
Sampling of feeds, 132, 133
Sand colic, 423, 424
Scottish system for body condition scoring, 169, 170t
Scours, postweaning, 477, 478
Se; *see* Selenium
Season, sheep and, 12
Secale cereale; see Rye
Second law of thermodynamics, 3, 4
Secretions, biliary and pancreatic, 101
Segmental contractions, 101
Selenium (Se), 56t, 58t, 61t, 63t, 75-77
  calves and, 272
  dairy cows and, 289
  deficiency of, 219-222
  foals and, 416
  goats and, 352
  limits of for livestock, 65t
  llamas and, 364
  mares and, 413
  syndromes associated with deficiency of, 427, 428
Self-feeders, 488
Sepsis, nutritional support for horses and, 432-436
Serum
  trace mineral measurement and, 64
  vitamin A concentrations in, 71t
Serum tocopherol, 428
Sex, growth and performance and, 190, 191
Sheep, 366-392
  maintenance of, 373
  metabolic disturbances in, 385-391
  nutritional diseases in, 384-392
  reproduction and, 373, 374
  seasons and, 12
  water and, 366-368
Shipping fever, 152
Show animals, 361
Show jumping, diet of horses in, 419-421
Sick animals, potassium and, 36
Sick calves, nutritional support for, 393-396
Sick cattle, nutritional support for, 393-396
Sick goat, feeding of, 354, 355
Sick horses, feeding of, 432-445
Silage, 120
  horses and, 411
  sampling of, 132
Size
  cow, 159, 160t
  of farm, 452
Sling, 443
Slobbering disease, 410
Small bowel, resection of, 437, 438
Small intestine, digestion and, 101-103
Smut, 228
Sodium, 37-39
  need for in dairy cows during hot weather, 295
Sodium bicarbonate, 271
Soil, trace minerals and, 57-59
Solanum, 427

Solids, total dissolved, in water, 93, 94
Solubilization of minerals in water, 92
Solutions, enteral, 398-400
Sorghum, 196
Sorghum grain *(Sorghum vulgare)*, 127, 410
Sources
    of ascorbic acid (vitamin C), 86
    of biotin (vitamin H), 83
    of choline, 86
    of folacin (vitamin B_C), 84
    of niacin (vitamin PP), 83
    of nutrients, 418t
    of pantothenic acid (vitamin B_3), 82
    of riboflavin (vitamin B_2), 82
    of thiamin (vitamin B_1), 80
    of vitamin A, 69
    of vitamin B_{12}, 85
    of vitamin D, 73
    of vitamin E, 75, 76
Sow(s)
    energy and feed requirements of, 545t
    feeding of, 455-473
    milk of, 466, 467
    nutrition and, 458-466
    postweaning and, 465, 466
Space allowances for growing-finishing pigs, 490
Spanish (meat) goats, 342, 343
Spasm, muscular, 426
Specific appetites, 114, 115
Specific dynamic activity (SDA), 6
Spelt, 127
Stage of production, nutrition and, 158, 159
Starter calf, 269, 270
Starter diets for weaned pigs, 474-477
Steatitis (yellow fat disease), 76
Steeplechase, diet of horses in, 419-421
Steroids, 443
Stillbirths of pigs, 469
Stimulation of appetite, 118
Stocking rate(s), 163, 181, 182
Stomach atony, 363, 364
Stones in bladder or kidney, 227
Straw, 120, 163, 164, 200, 201
Stress
    cold and heat, 11, 12, 293-296
    in newly purchased feeder pigs, 487
    of transportation, 224, 225
Substances that influence feeding activity, 118t
Sucking, natural, 244, 245
Sudan grass, 125
Sudden death, 76, 77
Sulfate, water and, 94
Sulfur
    dairy cows and, 289
    feedlot cattle and, 194, 195
    goats and, 352
    polioencephalomalacia, and, 188, 189
Supplement(s)
    analyses of calcium and phosphorous, 130t
    dietary fat, 307-313
    goats and, 345, 346, 349t, 350t
    for horses, 434t
    liquid, 285, 286
    llamas and, 358
    macromineral, 558t
    mineral, 130, 165
    parenteral, 65, 66
    for performance horses, 421t
    protein, 129, 130
    trace mineral, 558t
Supplementation
    with amino acids, 31
    ascorbic acid (vitamin C), 86
    biotin (vitamin H), 84
    calcium, 47
    choline, 86
    enzyme, 493, 494
    fat, 465
    of fat and carbohydrates in late gestation, 462
    folacin (vitamin B_C), 84
    intragastric, 142
    magnesium, 52, 183, 184
    mineral, 140
    niacin (vitamin PP), 83
    oral, 141, 142
    pantothenic acid (vitamin B_3), 82
    phosphorus, 49
    potassium, 37
    pyridoxine (vitamin B_6), 83
    rations and, 18
    riboflavin (vitamin B_2), 82
    selenium, 77
    sodium, 39, 40
    thiamin (vitamin B_1), 82
    trace mineral, 64-66
    vitamin A, 72
    vitamin B_{12}, 85, 86
    vitamin E, 77
    vitamin K, 79
Suppliers of samplers, 132t
Support, nutritional, 139-142
Surface law, 4
Surgery, vaginal, 438
Surgical procedures, sweet clover and, 79
Survival of embryo, 457
Sweet clover, 78, 79, 123
Swine, 447-495
    B vitamins intake and, 80t
    breeding, nutrient requirements of, 459t, 460t
    diet preparation for, 484-486
    digestive systems of, 97-113
    energy requirements of, 501t
    nutrient requirements of, 543-546
    ovulation rate of, 456
Swine industry, 447-454
    goals in, 453, 454
    history of, 447, 448
    production systems in, 448-451
    structure of, 451-453
Swine rations, energy and, 20
Syndrome(s)
    bovine bonkers, 34

Index

deficiency and toxicity, 426-428
fat cow, 277
Synthesis of fat, 306
Synthetic amino acids, 490, 491
Synthetic lysine, 490, 491
System(s)
  body scoring, for horses, 420t
  of energy measurement, 5-8
  of feeding, 249, 250

T

Tapioca, 201
Targets, 478,
TDN; see Total digestible nutrients
Temperature
  of barn, sows and, 464
  environmental
    energy and, 11, 12
    pigs and, 470, 489, 490
Testers, forage moisture, 137t
Testing of feed, 484
Tetany
  ewes and, 388, 389
  hypomagnesemic, 183, 184
Tetracyclines, 271
Texture of feeds, eating behavior and, 116
Thermodynamics, laws of, 3, 4
Thiamin (vitamin B_1), 80-82
  deficiency of, 188, 189, 227
Thiamine, 353
Three-day events, 421
Threonine, 22, 23
Time of feeding, 202
Timothy (Phleum pratense), 125
Tissue evaluation, 62-64
TMR; see Total mixed ration
Tocopherol, serum, 428
Tocotrienols, 75
Tomato pomace, 201
Total digestible nutrients (TDN), 10
  of hay, 121
  sheep and, 368-371
Total dissolved solids in water, 93, 94
Total mixed ration (TMR), 235
Toxemia, 385-387
Toxic blue green algae, 91, 92
Toxic feeds and plants, 428-431
Toxic substances, limits of concentration of in water, 94t
Toxicity
  ammonia, 33, 34
  calcium, 47
  copper, 390, 391
  fescue, 186, 187
  goats and, 353, 354
  magnesium, 51, 52
  nitrate, 33
  phosphorus, 49
  potassium, 37
  signs of, 10, 11
  sodium, 38-40
  thiamin (vitamin B_1), 81, 82
  vitamin A, 72
  vitamin D, 74, 75, 427
  vitamin E, 77
  vitamin K, 79, 427
Toxicity syndromes, 426-428
Toxins, 483
  feed, 228
Trace mineral(s), 55-67, 547t
  analysis of hair for, 442
  clinical problems associated with, 60-64
  deficient state of, 55-57
  diagnosing deficiency of, 60-64
  factors that influence status of, 57-59
  feedlot cattle and, 195
  normal function of, 55, 56t, 57t
  nutritional problems and, 216
  supplementation of, 64-66, 558t
Trail rides, competitive, diet for horses in, 421
Transfaunation, 394
Transition diet, 205
Transphylloquinone, 78
Transport, esophageal, 99
Transportation, stress of, 224, 225
Trauma, nutritional support for horses and, 432-436
Trends in dairy industry in North America, 231, 232
Trifolium hybridum: see Alsike
Trifolium incarnatum; see Crimson clover
Trifolium pratense; see Red clover
Trifolium repens; see White clover
Triticale, 127
Triticum dicoccum; see Spelt
Triticum spelta; see Spelt
Tryptophan, 22, 23
Tube feeding, 394
  of horses, 443, 444
Tumors, 141
Turnips, 126
Twin lamb disease, 385-387
Tympany, ruminal, 184, 185
Type, cattle, growth and performance and, 191

U

UE; *see* Urinary energy
UIP; *see* Undegradable intake protein
Ulceration, gastrointestinal, 363
Undegradable intake protein (UIP), 283
Underfeeding, 202
Unit costs of feed, 15, 16
Urea, 31-34, 226, 287t, 370, 371
Urinary calculi, 227, 353, 354
Urinary energy (UE), 6
Urolithiasis, 39
U.S.-Canadian Tables of Feed Composition, 131
User friendliness of computer programs, 337, 338

V

Vaginal surgery, 438
Valeric acid, 296-298
Valine, 22, 23
Values, protein, for dietary ingredients, 29t
Vegans (vegetarians), 85
Vegetable proteins, 129, 130
Vegetarians, 85
Veterinarian, role of, 239-244
Viruses, water and, 91
Vitamin(s), 68-89, 547t; *see also* specific vitamins
    calves and, 271-273
    dairy cattle and, 288, 289
    disease and, 140, 141
    fat-soluble, 68-79
    feedlot cattle and, 195
    hepatic damage and, 441
    mares and, 413
    performance horses and, 419
    pigs and, 482, 483
    water-soluble, 79-86
Vitamin A, 68-72, 226, 227
    clinically ill horses and, 436
    cow-calf nutrition and, 167
Vitamin assays, 136
Vitamin B_1, 80-82; *see also* Thiamin
Vitamin B_2, 82; *see also* Riboflavin
Vitamin B_3, 82; *see also* Pantothenic acid
Vitamin B_6, 83; *see also* Pyridoxine
Vitamin B_{12} (cobalamins), 84-86
Vitamin B_C, 84
Vitamin C, 68, 86
    clinically ill horses and, 435
Vitamin D, 41, 72-75
    clinically ill horses and, 436
    dairy cows and, 288, 289
    milk fever prevention and, 326, 327
    toxicity of, in horses, 427
Vitamin deficiency, 68, 226, 227
Vitamin E, 75-78
    clinically ill horses and, 435, 436
    dairy cows and, 289
    deficiency of, 219-222
    llamas and, 364
    potencies of forms of, 75
    syndromes associated with deficiency of, 427, 428
Vitamin F; *see* Thiamin
Vitamin G; *see* Riboflavin
Vitamin H, 83; *see also* Biotin
Vitamin K, 78, 79
    clinically ill horses and, 435
    toxicity of, in horses, 427
Vitamin M; *see* Folacin
Vitamin P complex, 86
Vitamin PP, 82, 83; *see also* Niacin

W

Wastage of feed, 489

Water
    expected mean intake of, 90t
    feeding of sheep and, 366-368
    herbicides and, 94
    nutritive value of, 92, 93
    performance horses and, 418, 419
    pesticides and, 94
    problems with, 91-94
    requirements for, 90, 91
    sick cattle and, 393-395
    suckling pigs and, 469
Water quality, 93, 94
Water-soluble vitamins, 79-86
Weaned pig(s)
    feeding of, 474-479
    environment and, 478
Weaning, 267, 268
    to breeding, 161-163
    of horses, 414-416
    of lambs, 376, 377
Weather, cold, calves and, 269
Weight, birth, of piglets, 470
Welfare, animal, 154, 155
Wet chemical analyses, 135, 136
Wheat, 127, 128, 196, 410
Wheat bran, 128, 129
Wheat germ, 128, 129
Wheat middlings, 201
Wheat mill run, 201
Wheat pasture poisoning, 51
Wheat red dog, 129
Wheat shorts, 128, 129
Wheat straw, 163, 164
White clover (*Trifolium repens*), 123
Worksheet, cow, 176t, 177t

X

Xerophthalmia, 71

Y

Yellow grease, 198
Yield, milk, supplemental fat and, 309, 310

Z

Zea mays; *see* Corn
Zein (corn protein), 28
Zeranol, 382
Zinc (Zn), 57t, 59t, 61t, 63t, 226
    clinically ill horses and, 436
    foals and, 413
    goats and, 353
    limits of for livestock, 65t
Zinc methionine, 301
Zn; *see* Zinc